PRINCIPLES OF LITHOGENESIS

VOLUME 3

PRINCIPLES OF LITHOGENESIS

Academician

N. M. STRAKHOV

Institute of Geology, Academy of Sciences of the U.S.S.R.

Volume 3

Translated by

J. PAUL FITZSIMMONS, PH.D.
Professor of Geology, University of New Mexico

Edited by

The Late S. I. TOMKEIEFF, D.SC., F.G.S., F.R.S.E.
Emeritus Professor, University of Newcastle upon Tyne

and

J. E. HEMINGWAY, PH.D., F.G.S.
Professor of Geology, University of Newcastle upon Tyne

PLENUM PUBLISHING CORPORATION

NEW YORK

OLIVER & BOYD

EDINBURGH

PLENUM PUBLISHING CORPORATION
227 West 17th Street
New York 11

OLIVER AND BOYD

Tweeddale Court Edinburgh EH1 1YL
A Division of Longman Group Ltd.

This book is a translation of ОСНОВЫ ТЕОРИИ
ЛИТОГЕНЕЗА by N. M. STRAKHOV, first published
by ИЗДАТЕЛЬСТВО АКАДЕМИИ НАУК СССР
Moscow 1962

FIRST ENGLISH EDITION · 1970
For this joint publication the
spelling follows current U.S.
conventions

CONTENTS OF VOLUME 3

PART ONE
THE INITIAL STAGE OF ARID LITHOGENESIS AND ITS CHARACTERISTICS IN MODERN AND ANCIENT EPOCHS

Chapter 1
TERRIGENOUS SEDIMENTATION IN ARID ZONES AND ITS FEATURES

Chapter 2
ACCUMULATIONS OF Cu-Pb-Zn; THEIR ORIGIN AND DISTRIBUTION IN ARID REGIONS

Chapter 3
P-CaCO$_3$-MgCO$_3$-SiO$_2$ DEPOSITION IN WEAKLY MINERALIZED BASINS OF ARID ZONES

Chapter 4
ACCUMULATION OF ORGANIC MATTER IN ARID ZONES

PART TWO
BASIC FEATURES OF PRESENT-DAY HALOGENESIS

Chapter 1
TYPES OF PRESENT-DAY SALINE BASINS AND THEIR SEDIMENTS

Chapter 2
THE PHYSICOCHEMICAL MECHANISM OF PRESENT-DAY HALOGENIC SEDIMENTATION

Chapter 3
GEOLOGICAL CONDITIONS CONTROLLING THE FORMATION OF LAKES OF DIFFERENT HYDROCHEMICAL TYPES AND THEIR GENETIC INTERRELATIONSHIPS

PART THREE
HALOGENESIS IN PHANEROZOIC TIME

Chapter 1
FACIES CORRELATIVES OF RECENT DEPOSITS AMONG ANCIENT SALINE FORMATIONS

Chapter 2
HALOGENIC FORMATIONS IN THE MARGINAL ZONES OF OPEN CONTINENTAL SEAS

Chapter 3
FORMATIONS OF INTRACONTINENTAL MARINE SALINE BASINS

Chapter 4

THE MORPHOLOGY, HYDROLOGY AND DISTRIBUTION OF SEDIMENTS IN ANCIENT MARINE HALOGENIC BASINS, AND THE INFLUENCE OF THESE FACTORS ON THE CONSTITUTION OF SALINE FORMATIONS

Chapter 5

THE COMPOSITION AND ORIGIN OF THE ROCKS OF THE HALOGENIC FORMATIONS

Chapter 6

THE DISTRIBUTION OF SALINE FORMATIONS IN ARID ZONES. DEVELOPMENT OF HALOGENESIS THROUGHOUT EARTH HISTORY

PREFACE

Four peculiarities distinguish lithogenesis of arid type from lithogenesis of humid type: (1) separation into two subtypes—autochthonous, effected by the water resources and material of the arid zone itself, and allochthonous, effected by supply from neighboring (horizontal and vertical) humid zones; (2) completeness of precipitation of sedimentary material and participation in the sedimentation not only of fragmental and components of low solubility but also of readily dissolved salts; (3) combination of sharply developed stages in the development of arid weathering zones; and (4) very weak, and progressively weaker, effect of living organisms on the precipitation of substances from solution, and the replacement of biogenic processes by chemical processes in proportion to increasing salinity of the basin. Arid lithogenesis at average and high salinities is a classic example of pure chemical processes controlled by physico-chemical laws.

All these features not only make arid lithogenesis a distinctive, markedly individual type of sedimentary process when compared with humid and glacial types, but they give it a greater complexity than the other two. In arid lithogenesis we are concerned with a higher development of the lithogenic process in the series of climatic types:

$$\text{Glacial} \rightarrow \text{humid} \rightarrow \text{arid} \Big\langle {{\nearrow \text{autochthonous}} \atop {\searrow \text{allochthonous.}}}$$

The distinctive and progressive character of arid lithogenesis, however, is not equally well defined in all its various aspects. It differs most feebly in the process of mechanical sedimentation, but is clearly expressed in chemical-biogenic and purely chemical lithogenesis. This circumstance determines the content of the present volume of this monograph. We will touch only slightly on terrigenous rocks and will have to do almost exclusively with authigenic rocks.

The principal task here is to point out the evolution of biochemical and chemical lithogenesis with progressively increasing salinity of the basin, and to define the pattern of composition and distribution of *arid rocks and formations* within the arid zone. The treatment of these problems has been facilitated by advances in Soviet geology, particularly in the investigation of salt. But no attempt to *create a whole image of arid lithogenesis as a special type of sedimentary lithogenesis* has yet been undertaken, because the very idea of types of lithogenesis in general has been lacking. The introduction and discussion of this idea in the first volume of this monograph allowed us for the first time to present a view of arid lithogenesis as a *single entire natural phenomenon.*

The basic method of investigation has been, as before, comparative lithology. Its application to deposits in arid zones has enabled us both to discover many new features of lithogenesis previously overlooked, and *to point out new possibilities of the method itself, not even yet used to their full effectiveness.*

The author wishes to express his sincere appreciation to M. G. Valyashko, A. A. Ivanov, and M. P. Fiveg, who read this volume and made a number of valuable suggestions. He is also deeply grateful to G. I. Buchinskii, who took on himself the entire labor of editing all three volumes of the monograph.

PART ONE

THE INITIAL STAGE OF ARID LITHOGENESIS AND ITS CHARACTERISTICS IN MODERN AND ANCIENT EPOCHS

CHAPTER 1

TERRIGENOUS SEDIMENTATION IN ARID ZONES AND ITS FEATURES

The concept of "deposits of arid zones" is generally associated with the idea of saline rocks, forming from more or less mineralized waters. This view is valid only in part. In regions of arid steppes, semi-deserts, and deserts, sediments that formed in essentially fresh or in but slightly mineralized water (from parts of a percent to 4–4·5%) are widespread at the present time. These include deluvium, proluvium, alluvium, and the sediments of many lakes and seas (Caspian, Aral, Balkhash, Issyk-Kul, and others). Fossil analogues of these are known from many stratigraphic horizons and regions (Upper Permian red beds of the Russian platform, the Middle and Upper Devonian rocks of the principal Devonian area, and others).

Before the introduction of the concept of lithogenetic types, these deposits attracted no special attention, because no general ideas were associated with them. In the light of studies on arid and humid types of lithogenesis, however, deposits in the arid zone associated with slightly mineralized waters immediately acquire exceptional significance. They are important, primarily as the *initial stage in the evolution of arid lithogenesis, from which, during further increase in salinity, the stage we have called halogenesis developed.* Clearly it is impossible to understand the specific features of arid lithogenesis as a whole without knowing this incipient stage. The deposition of arid-zone sediments from waters of low mineralization is important also because it took place at a salinity closely similar, and at times identical, to that of sedimentation in basins of humid zones. *The specific features of arid lithogenesis can therefore be compared with humid lithogenesis from their initial stages.*

It is clearly established that, in arid lithogenesis, the processes of mechanical sedimentation of particles transported to the site of deposition in the solid phase must be distinguished from processes of chemico-biogenic precipitation of solid phases from solution. The former group of phenomena will be examined first.

1. General Characteristics of Terrigenous Arid Deposits

The facies types of aqueous terrigenous deposits in arid regions are essentially the same as those in humid belts: deluvium, proluvium, alluvium, and lacustrine and marine deposits. The physical mechanism by which these deposits were formed was fundamentally the same in both environments also, especially in regard to basin sediments. This means that any detailed description of the above types of terrigenous sediments need not be considered here;

only a short review of some details that are specific for arid deposits is therefore necessary.

One such specific feature is the *greater development of deluvial and proluvial deposits as compared with humid zones of similar relief*. Deluvial cover in arid regions of little dissection is thick, but alluvial fans and proluvial belts of slopewash form around the mountains. This is caused by the decrease of plant cover, which facilitates the removal of finely-divided clastic material.

At the same time, new features appear in the composition of the arid deluvial-proluvial deposits. Episodic rains, by impact and lubrication, in combination with the disintegration of exposed rocks, initiate mud flows in dissected regions of deserts and semi-deserts, which, in turn, gives a coarse-grained structure to the alluvial fans and piedmont deposits. After such rains about the proluvial margins, transient lakes appear, quickly converted to playa flats, with their characteristic silt-clay sediments and the development of polygonal mud cracks. The clays in these flats at times dry into shard-like, curved plates, forming a frequent member in the distal part of alluvial fans.

The alluvium of arid zones acquires distinctive features. Only the largest rivers, flowing from mountains and from neighboring plains of humid belts, carry sufficient water to reach the terminal discharge basin. The alluvium in such rivers is closely similar in composition and structure to the alluvium of humid zones, though with some additional specific features. One of these is the variability and frequent lateral migration of the channel (Syr-Darya, Amu-Darya), with consequent complexity in the structure of the alluvium, especially in its lateral variation. Medium-sized rivers and very small streams, because of their low volume and high rate of evaporation, fail to reach the terminal discharge basin consistently. They tend to spread out on the plains, losing their water, with the formation of sub-aerial fans (Fig. 1). In essence, these deposits are modified alluvial or slope-wash fans, with long ribbons of alluvium forming on the upper surface. These delta-like fans are distinguished from proluvium proper by their fine grain-size.

The clastic sediments of lakes and seas of arid regions are closely similar to humid-climate deposits in general type, composition, and structure; which is natural, since the hydrodynamics of slightly mineralized arid-climate basins are similar in most details to those of basins in humid regions.

A distinguishing feature of all these types of arid-climate terrigenous deposits is that they form in water oversaturated in $CaCO_3$. This generally leads to more or less abundant precipitation of carbonates, mixed with the clastic components. Hence, *terrigenous deluvial-proluvial-alluvial and basin sediments of arid zones always contain some carbonate, commonly to such an extent that they can no longer be referred to the group of clastic sediments proper ($CaCO_3 < 5\%$), but rather to clastic-authigenic rocks ($CaCO_3$ 10–20%)*. Another specific feature of clastic-authigenic (and clastic proper) rocks of arid regions is the variety of coloring. Sedimentary material brought into arid zones from humid zones always includes iron oxides. The organic material necessary for their reduction, however, is not always present in arid regions in sufficient quantity. This material is found most commonly and most abundantly in basin deposits, where the reducing process is active, and the basin sediments are therefore chiefly grey, greenish grey, or green. But even in these basin sediments, organic material has at times been insufficient to reduce the iron completely and to eliminate the red colors. Red bed facies then develop, generally along the margin of the basin, but at times extending into the basin among the grey deposits. It is noteworthy that *clays or marls prove to be red most commonly, whereas silts and, especially, sands are grey.* The cause of this lies in the relative proportions of Fe_2O_3 and C_{org}. Although

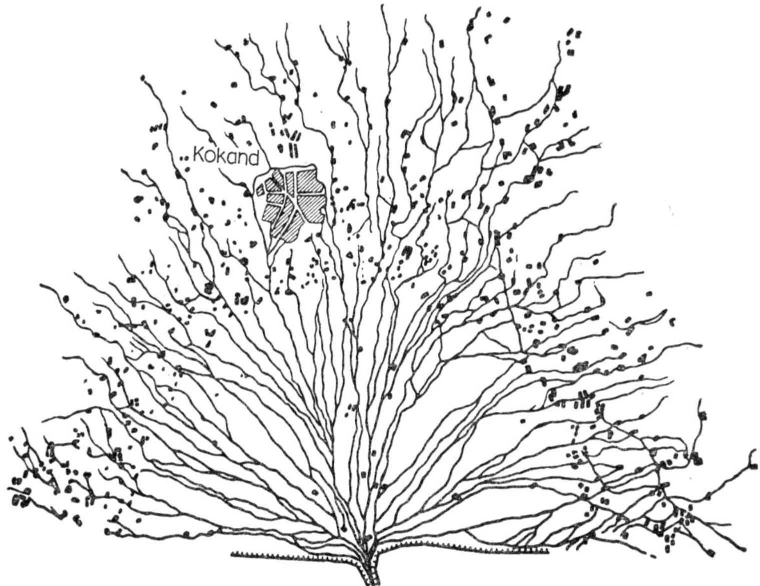

FIG. 1. Continental alluvial fan (after V.N. Veber). Black spots represent populated areas.

the amount of C_{org} in sandstones is small, it is adequate to reduce all the available Fe^{3+} and so to eliminate the red color. The content of C_{org} is somewhat greater in clays than in sandstones, but the Fe_2O_3 content is many times greater; the available organic material is insufficient for complete reduction of the iron, and the clays therefore remain red. In sub-aerial deposits (deluvial-proluvial, in part alluvial and eolian), because of the scarcity of plant cover, the accumulation of organic matter is very low. Reduction processes here are markedly less effective. When the sediments contain much iron they remain red or brownish. When the iron content is low, the rocks assume a color determined by that of the mineral components, such as rose-colored or yellowish. Dehydration of Fe and Mn oxides, because of strong insolation of the sub-aerial deposits and the low humidity of the air, favor brightness and variety in the rock colors. Thus, *red and variegated rocks, with various and*

B

frequent changes in color within short distances, are characteristic of arid regions and commonly serve as a diagnostic feature of such regions. On the other hand, it is necessary to keep in mind that this feature in itself is not decisive, because red deposits are also found in humid belts. But in arid regions the red and variegated appearance of all types of rocks is associated with carbonate content, in places even with increased salinity. Such rock associations are excluded from humid regions by the very essence of the environment and, particularly, by the very low mineralization of streams and other continental waters.

Thus, *although the facies types of terrigenous rocks and the mechanism of their formation in arid regions are the same as in humid belts, the invariable presence of carbonate rocks in the former, sometimes in considerable quantities, the wide distribution of red and variegated colors, because of the weak development of reduction processes, and, finally, the sporadic, sometimes notable, salt content give the terrigenous deposits of arid regions a well-defined individual character.* Further investigation serves to define a number of other distinguishing features. To discuss them, the terrigenous components in clastic-carbonate deposits of arid zones, in particular their grain-size distribution and mineralogy, will be considered.

2. Grain-Size Distribution and Mineral Composition of Terrigenous Rocks of the Arid Class

The Fig. 2*A* shows the grain-sizes of different facies types of arid rocks. A comparison with the corresponding diagram for humid rocks (Vol. 2, Fig. 170) reveals that grain-size types of the same name are almost identical. In both humid and arid zones poor sorting characterises deluvial, proluvial, and terrestrial deposits, but much better sorting is found in basin sediments.

The principal cause of the similarity is the identical hydrodynamic conditions of deposition. In addition, part of the proluvial and alluvial deposits and all the basin terrigenous rocks are deposited from material introduced to arid zones from humid. *Essentially, these are the same constituents as those that make up humid-climate terrigenous rocks. It is not surprising, therefore, that the grain-size characteristics of arid deposits are the same as those of humid rocks.* Here, for the first time, we meet the phenomenon of *inheritance* by arid rocks of compositional features established in the primary material which originated in an entirely different climatic zone. This inheritance factor applies also to the mineral composition of the sand-silt rocks.

The work of V. N. Gridnev (1955) on the mineralogy of the Cenozoic arid-climate sandstone of Central Asia established that the majority of such rocks are typically polymictic, in particular arkosic, and they fall in the arkose field on the diagram (Fig. 2*B*), though some varieties may be with the mesomictic sands. The same is true of sandstones and siltstones in the productive sequence of Azerbaizhan and Kabristan which are typically polymictic for the most part and fall partly in the field of lithites, partly in the graywacke field (Fig. 3). Individual horizons of the sequence, however, include not only mesomictic sandstones with a quartz content of 60–35% but even oligomictic sandstones, in which the quartz content reaches 75–90%. "Oligomictic sandy and silty rocks are found chiefly on the Apsheron Peninsula, mostly in the Balakhany series, rarely in the Infra-Kirmakinskoe (Podkirmakinskoe) or Supra-Kirmakinskoe (Nadkirmakinskoe) series. . . . Quartz is present in well-rounded grains" (Aliev 1949, p. 194). Similar variations in mineral

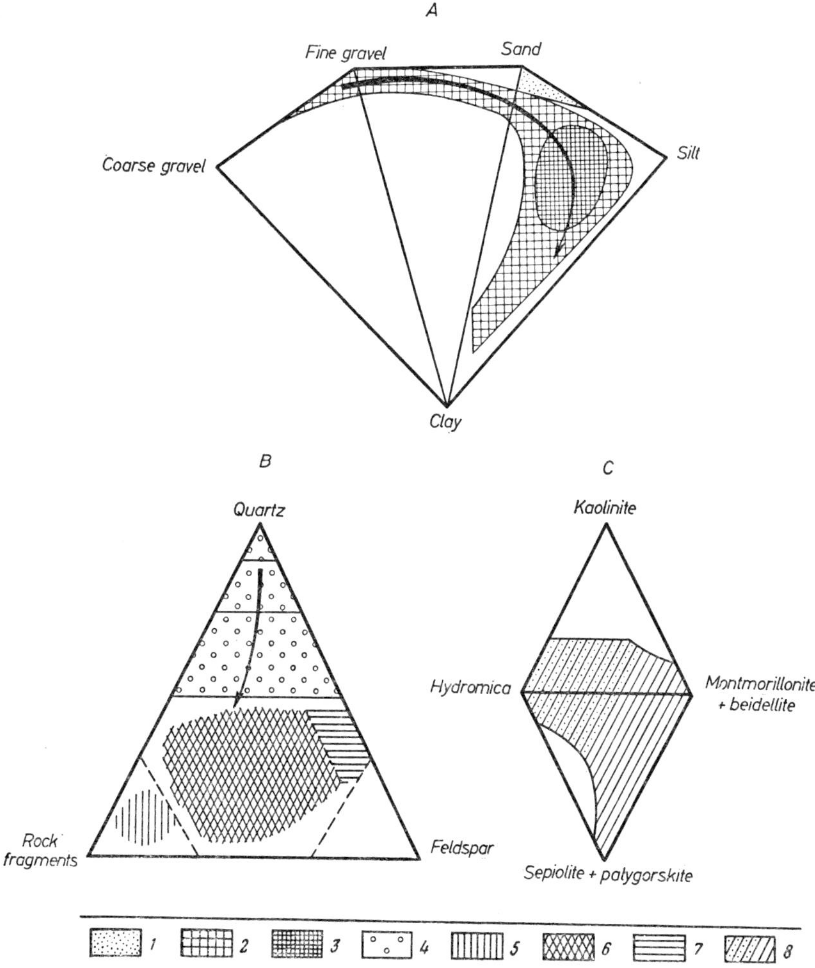

FIG. 2. Grain-size and mineral composition of terrigenous rocks of the arid class. *A.* Grain-size characteristics of fragmental rocks; *B.* Mineral composition of fragmental rocks; *C.* Composition of clay components in mixed allochthonous rocks. 1. Eolian sands; 2. rocks of alluvial fans; 3. most common compositions; 4. composition of sandy material initially absent in autochthonous terrigenous rocks and introduced from eroded deposits of humid zones; 5–7 polymictic sands: 5. lithite; 6. graywacke; 7. arkoses; 8. enrichment of hydromicas in potassium and sodium. Arrows indicate the direction of compositional evolution of the fragmental rocks.

composition of sandstones may be found in any rather thick and areally widely developed arid sandy sequence. Superficially the composition of fragmental rocks in arid zones is the same as that in humid belts, but, in detail, the relationship proves to be much more complex, and some differences are apparent. For example, *mesomictic, and especially oligomictic, varieties of arid clastic rocks are much less extensively and less abundantly developed than in humid belts; they are found chiefly in restricted, local bodies.* Qualitative differences are found in the different origin of beds rich in quartz in the two zones. In humid zones most quartzose rocks are primary, and arise through intense

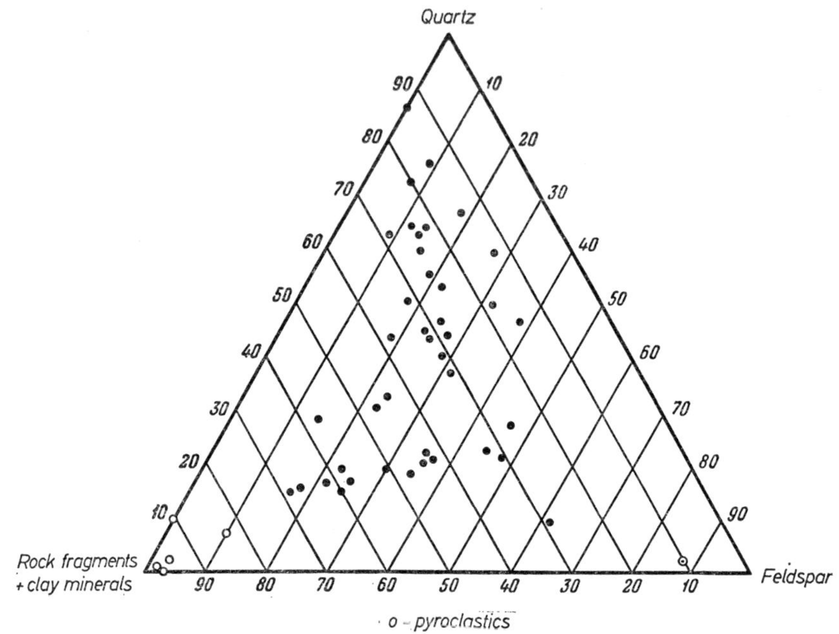

FIG. 3. Mineral composition of sand-silt rocks of the productive series of Azerbaizhan (from A. G. Aliev).

chemical weathering during times of weak tectonic activity. Secondary redeposition of quartz and silt fragments is of little importance here. In arid regions the situation is reversed. Because of marked reduction in chemical weathering in those regions and because of the prevalence of solely physical disintegration of bed rock, sand-silt sediments derived from intrusive, volcanic, and metamorphic rock-types must inevitably be polymictic, of arkosic, graywacke, or lithite types, and even when formed in tectonically passive regions. Deposits of increased quantities of quartz may be explained only by inheritance, by the presence of considerable volumes of quartz-rich sedimentary rock in the drainage areas. This secondary character of mesomictic and oligomictic fragmental rocks in arid zones is in agreement with the form of the quartz grains themselves, being commonly well rounded as a result of the reworking of an older sequence that formed under humid conditions.

Thus, *the hereditary high concentration of quartz in terrigenous rocks of the arid zone and the allochthonous origin of mesomictic and oligomictic rocks cannot be doubted.* Therefore, the sand-silt rocks with a quartz content of more than 50% fall in the field of inherited composition, whereas those with a quartz content less than 50% may have initially developed by destruction of magmatic and metamorphic parent rocks. It may be that the introduction of derived material from the destruction of sedimentary rocks begins to appear even with a quartz content greater than 35%. This is shown by the arrow, which indicates the direction of influx of derived minerals (Fig. 2*B*).

The relative composition of clayey rocks in arid and humid zones is distinguished by other features. Clays of arid zones include the same minerals as are found in deposits of humid regions: hydromica, montmorillonite, ferromontmorillonite, beidellite, chlorite, halloysite, kaolinite, sepiolite, and palygorskite in various quantitative combinations (M. A. Rateev 1956, 1958 and A. G. Kossovskaya 1954). Pure kaolinitic, or even monothermitic, deposits are, however, absent in clayey rocks of the arid zone. Kaolinite and halloysite occur only as minor constituents with other clay minerals. At the same time, the magnesian silicates and aluminosilicates sepiolite and palygorskite, forming at times almost monomineralic deposits, are to be found among such clays. Thus, though the general assemblage of clay minerals is the same, the composition of clay rocks is noticeably different in humid and in arid regions (Fig. 2*C*). It is indeed clear that the composition of arid clay deposits, although in part overlapping the field of humid clays (Vol. 2, Fig. 170), is almost a mirror image of it.

Genetically, the clay minerals of arid clayey rocks differ among themselves. Keeping in mind that kaolinite and halloysite require an acid environment with the extensive removal of alkalies and silica for their formation, and that the climatic conditions in arid regions retard chemical weathering, stopping it at the alkaline stage, we may be sure that kaolinite and halloysite in clayey deposits of arid zones must be derived from humid regions. *These are typical allochthonous minerals.* Hydromica, chlorite, and montmorillonite may also be allochthonous, as well as autochthonous, forming in the arid zone itself. By contrast, in basin sediments, in which all sedimentary material is allochthonous, these minerals are undoubtedly derived. As for sepiolite and palygorskite, their common, and in places extensive, accumulation in sediments of arid zones demonstrates their undoubted formation within the zone of arid conditions. These are autochthonous, and commonly authigenic, minerals. It is this circumstance that lends an individual character to the clay rocks of arid zones as compared with those of humid zones.

3. EOLIAN DEPOSITS OF ARID ZONES AND THEIR CHARACTERISTICS

In deserts and semi-deserts, the fundamental factor in the formation of terrigenous deposits is the wind. Since eolian deposits arise from the winnowing action of older sandy or sandy-silt rocks, it follows that their initial grain-size and mineral composition should be inherited essentially from the parent rocks; they are, however, more or less extensively altered during eolian re-working.

The reworking differs for different circumstances, and, according to A. V. Sidorenko (1949, 1956), it depends on whether the sands remain in contact with the parent rocks, i.e. whether the sand rests either upon the parent rocks and is but slightly removed from them, or whether the connection is lost, by transport over great distances. In sands that have not been transported far the fine fraction (>0.05 mm, $0.05–0.10$ mm, $0–0.10–0.15$ mm) is small, and the

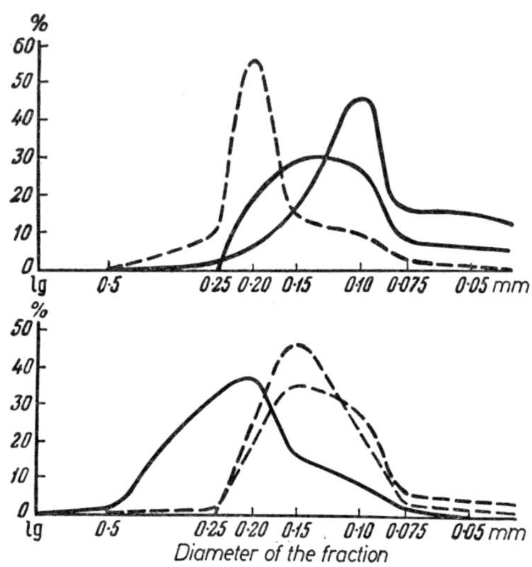

FIG. 4. Change in grain-size of two types of eolian sands (after A. V. Sidorenko). The solid line indicates original sands, the dashed line barkhan sands; the upper diagram represents sands not greatly displaced, the lower transported sands.

coarse fraction large. The grain-size of untransported sands essentially reflects the grain-size of the parent rocks, from which the wind has removed only the finest fraction. At the same time, the average grain-size increases and the sorting of the sands is improved. In transported sands the content of silty and fine-grained particles (less than 0.1 mm) does not diminish during winnowing, but may even increase. In this process the proportion of coarse-grained and medium-grained particles decreases somewhat. The over-all change in composition of transported sands is toward relative increase in the mass of silty and fine-grained material, and sorting also improves (Fig. 4).

During the winnowing of sands the mineral composition also changes, but differently for each group. In undisplaced eolian sands the content of heavy minerals of the relatively primary bedrock sands increases (Fig. 5) by as much as one and a half times, rarely as much as twofold. In transported sands, however, the proportion of the heavy fraction decreases noticeably in comparison with that of the original rocks, commonly to as little as 20% of the original value. As for the composition of the clastic minerals, "during winnowing of either type of sand, the proportion of easily abraded minerals decreases. The amounts of calcite, gypsum, hornblende, pyroxene, feldspar, and epidote decrease because of the relative increase in the proportion of minerals resistant to mechanical abrasion: quartz, garnet, zircon, kyanite, sillimanite, and magnetite" (Sidorenko 1956). Eolian sands are generally free of mica; when present it is found in small individual flakes, concentrated in wind shadows. Over much of the highly micaceous alluvial sands of the Kara Kum series, for example, material winnowed by the wind is almost completely mica-free. In transported eolian sands the number of mica flakes remains almost the same as in the original sands, or, during the first stages of winnowing, even increases somewhat because of splitting along the cleavage. In both types of eolian sands the mica is characterised by rounded forms.

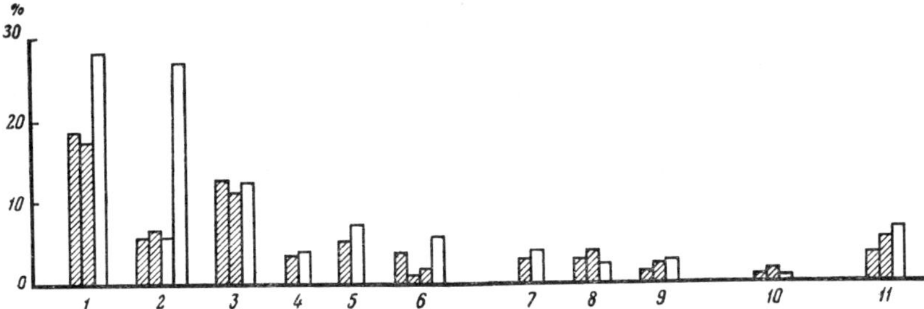

FIG. 5–6. Heavy minerals in original and eolian sands in the fraction 0·05–0·25 mm in correlative samples (from A. V. Sidorenko). 1–6. Kara Kum series; 7–9. Obruchev series; 10. Upper Karabil series; 11. Murgap series. *Shading* signifies original sands; *White* eolian sands.

The longer the sand is winnowed, the smaller the proportion of mica. The eolian sands of the Kara Kum series in Central Kara Kum, having been formed much earlier than the eolian sands of the Amu Darya barchan belt, contain almost no mica, whereas the eolian sands in the Amu Darya belt, the age of which is recent, almost invariably contain flakes of muscovite and biotite. Biotite is destroyed more rapidly than muscovite. The absence of mica, or its presence only in small rounded flakes, is one of the characteristic features of eolian sands. Chlorite undergoes similar changes, but its plates are better rounded and less cleaved (Sidorenko 1956, pp. 10-12).

On the whole, *the total number of minerals in eolian sands is much smaller than in the original sands*, and consequently, eolian winnowing changes a polymictic mineral assemblage toward mesomictic, and mesomictic toward oligomictic. It is impossible, however, to state with confidence how far this process can go. In the rocks studied by A. V. Sidorenko, the changes in composition were quantitatively of only secondary significance, and did not alter the mineral type of sand; polymictic original sands, for example, gave rise to polymictic eolian sands. Polymictic eolian sands have also been described from other modern deserts, such as the Arabian and the Sahara. These facts are grounds for considering the intensity of eolian reworking the mineral composition of sandy deposits to be weak. In other words, *the mineral assemblage of the original parent deposits, inherited by the eolian sands, is the principal characteristic of the mineralogy of these sands; modification of the inherited material is restricted.*

4. Secondary Salinization of Terrigenous Deposits

One of the characteristic features of modern arid regions is the widespread occurrence of ground water at different depths. Hydrological investigations in Kara Kum have shown the presence of a single water table in that region. Intake of water takes place chiefly (86·4%) by infiltration from the channels of the Amu Darya, Murgab, and Tedzhen Rivers and by subsurface flow from the Kopet Dag. Infiltration of direct rainfall is insignificant, amounting to only 13·6%. Up to 3×10^6 tons of various salts enter Kara Kum annually in solution (Sidorenko 1956; Kumin 1947, 1959). Similar relations are observed in Kyzyl Kum and, in general, in all deserts traversed by major streams and surrounded by mountain chains (the Great Basin of the U.S.A., and others). By contrast, deserts without bordering mountains and without traversing streams are poor in ground water (Western Sahara). Similar distributions have obtained, in all probability, in arid regions of the geologic past.

When ground water occurs in clastic rocks far below the surface, it has no specific effect on the composition of the enclosing rocks, but the nearer the water table is to the surface, the greater is the effect the water has on the rocks. A cycle of specific processes involves a secondary salinization of the clastic sediments—processes that characterize only arid climates and are therefore of exceptional significance from the comparative lithologic point of view.

Two types of salinization may be distinguished. The first occurs in regions where the water table approaches the surface but does not reach it, being separated by only 0·5–1·5 m of dry rock. Because of insolation and high temperature, water from the surface of the water table constantly evaporates, but is replaced by capillarity which keeps the lower part of the unsaturated zone moist. Prolongation of this process leads to increasing salinity of the ground water and then to precipitation of solid salt phases in the intergranular pores in the clastic rocks. *Salinization occurs in the surface zone, generally in the capillary fringe of ground water, i.e. at depths from 0·5 to 2·5–3 m. This phase is sometimes deposited in such abundance that a distinctive, more or less compact horizon forms (caliche, hardpan), following topography at shallow depth.* These horizons consist most commonly of calcite and gypsum (A. V. Sidorenko 1956, 1958), rarely of dolomite, and very rarely, apparently, of silica.

Salinization by calcium carbonate occurs in three distinctive forms: calcareous crusts, nodular horizons, and pseudomorphous growths after plant roots. Calcareous crusts (caliche, hardpan, and the like) are best developed in the deserts of Mexico and Egypt, in lesser degree in Kara Kum and Kyzyl Kum. In Mexico, calcareous crusts occur only in the northern Mexican Highland, in large gently sloping depressions or bolsons, underlain by proluvial gravels. They occur at shallow depths (down to 0·5 m), though locally, after erosion, they lie directly on the surface. The thickness of these calcareous crusts may reach 2·5–3·0 m. The upper part (1–1·5 m) forms a moderately dense limestone, the structure of which depends on the texture of the primary rock. When gravels have become calcified, the upper part of the caliche zone forms a dense conglomerate, cemented by calcite. Around the pebbles and between individual beds may occur isolated zones of $CaCO_3$, crudely concentric zones of calcite, which replaces the original sand-clay matrix. Should such rocks be encountered in the fossil state, they might be erroneously referred to typical conglomerates with calcareous cement. The metasomatic replacement of the fine-grained fraction in sandy loams has led to the formation of dense limestones with only an occasional, individual

FIG. 7. Modes of occurrence of calcareous horizons (black) among conglomerates
(after A. V. Sidorenko).

granule or pebble. Rocks of this kind in the fossil state might well be considered typical limestones.

The density of the calcareous horizon and the degree of recrystallization gradually decrease downward. Below 1·5–2·0 m friable, unrecrystallized zones appear, grading into a continuous, powdery, white, friable mass. The thickness of this powdery horizon is 1–2 m, and the content of $CaCO_3$ in it may reach 95–96%.

The Mexican caliche marks ancient surfaces of sediment accumulation, and for this reason Recent and near-Recent sedimentation surfaces cut across caliche horizons (Fig. 7).

Calcareous crusts are not forming today in Kara Kum and Kyzyl Kum. These crusts are confined exclusively to rocks of Pliocene age (the Lower Karabil and Zaunguze series). In contrast to the Mexican crusts, these consist of calcareous nodules, which are compacted in the upper part to a considerable extent, forming a continuous horizon; below, the nodules are individual and the zone grades into a friable calcareous mass. The content of $CaCO_3$ in the different calcareous nodular zones ranges from 21·4 to 78·38%. Locally, in the lower part of the calcareous horizon, palygorskite occurs in films and selvages.

A variety of calcareous crust is the so-called akkyrshi, which represents

$CaCO_3$ metasomatic replacement of plant roots. These are of Recent origin, being found only in semi-deserts. In the U.S.S.R. they are distributed chiefly in the vicinity of the Aral Sea, and are not found in the deserts of Kara Kum and Kyzyl Kum.

In the Etosha Pan in the Kalahari Desert carbonate deposits occur that are apparently calcareous crusts but, unlike others, contain marked propor- tions of $MgCO_3$, probably in the form of dolomite, although this has not yet been established.

All the various types of carbonate deposition within clastic deposits are characteristically localized in regions of arid steppes and semi-deserts and are associated with bicarbonate ground water. With the small content of $MgCO_3$ in these waters and the absence of Na_2CO_3, the carbonate crusts are calcareous. With higher contents of $MgCO_3$, dolomite appears in the crusts. Calcretes in ancient rocks, occurring in modern deserts, are regarded as indicative of a noticeably moister climate in the past, probably that of a semi-desert at the time the crusts were formed (A. V. Sidorenko 1958).

In contrast to calcretes, gypsum crusts are characteristic features of desert regions themselves, where they develop widely. Several varieties may be recog- nized. The so-called Repetek gypsum in southeastern Kara Kum forms a continuous horizon to a depth of 1·5–2·0 m below the surface; more rarely it will form an horizon of sands rich in gypsum crystals. These crystals are

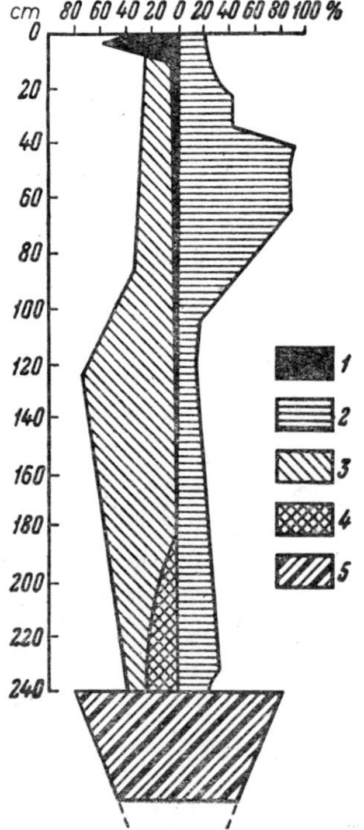

FIG. 8. Profile of soil salinization in deserts (from V. A. Kovda). 1. Readily soluble salts; 2. CaSO$_4$ 2H$_2$O; 3. CaCO$_2$; 4. R$_2$O$_3$; 5. salts in ground water.

commonly very large, with poikilitic structure, including many sand grains. In loamy or clayey sediment, the gypsum crust is essentially an horizon of fine-grained gypsum alone. It is clear that in such rocks the pores were insufficiently large for the free growth and crystallization of $CaSO_4.2H_2O$. In deserts, where the water table is near the surface, gypsum replacement of plant roots occurs, similar to the calcareous forms—akkyrshi. These pseudomorphous forms are usually composed of porous gypsum, the coarse parallel fibrous structure apparently replacing the wood-fibers.

Gypsum crusts of all varieties develop where the ground water contains an abundance of calcium sulfate. Gypsum crusts are therefore localized in the central parts of arid zones, i.e. in deserts proper, whereas calcretes are formed along the margins of these zones, i.e. in semi-deserts and on arid steppes.

TABLE 1

Characteristics of Accumulation of Soil Salts in Various Parts of Arid Zones

Segment of the arid zone	Highest concentration of readily soluble salts in upper horizon of salt bottom (solonchak), %	Composition of characteristic salts
Arid steppe	1–3 (5–10)	$Na_2SO_4·10H_2O$; NaCl; **$Na_2CO_3·10H_2O$**; $CaCO_3$
Semi-desert	5–8 (5–10)	NaCl; **$Na_2SO_4·10H_2O$**; $MgSO_4·7H_2O$; **$CaSO_4·2H_2O$**; $CaCO_3$
Desert	15–25 (10–15)	**NaCl**; $NaNO_3$; $MgCl_2·6H_2O$; $MgSO_4·7H_2O$; $CaSO_4·2H_2O$; $CaCO_3$

Parentheses indicate $CaCO_3$ content. Bold-faced type designates dominant salt

Another type of salinization arises when the water table reaches the surface in depressed or basin areas, where swamps develop, comparable to swampy basins in humid belts. However, the difference in climatic conditions impresses very different features on the arid paludal process from those in moist zones. In the latter, peat bogs develop, which are later converted to coal. In arid zones, the development of swamps leads to surface accumulation of salt and to subsequent development of solonchak. This process is similar to that for the first type of salinization. A distinguishing feature of solonchaks is the abundance of salt precipitation in them, which reaches a maximum in the uppermost (25–50 cm) part of the succession (Fig. 8).

Such salt horizons are developed in intracontinental basins, on coastal lowlands (such as the lowlands along the Sivash and the Caspian Sea), on fans (or continental deltas) and marine deltas of rivers (such as those of the Kura, Araks, Amu Darya, and Volga), on flood plains, alluvial depositional plains, and elsewhere. The intensity of salinization and the composition of the salts vary systematically from the margins of the arid region to the center (Table 1, after V. A. Kovda 1946).

Thus, salinization is insignificant on the margins of arid zones (from 1–1·5 to 3·0%), and calcium carbonate is the dominant salt. In semi-deserts, the salt content rises to 5–8% and gypsum is dominant. In deserts, the salt content reaches 15–25% and halite becomes the principal component (Fig. 9). "The uppermost horizons, which are those with the greatest salt content, in salt floors of deserts, reaching a thickness of 10–20 cm, commonly consist of 60–70% of various salts: 15–25% of readily soluble salts, 20–40% gypsum, and 10–15% calcium carbonate" (Kovda 1946, Vol. 1, p. 75).

Since the clastic minerals in the rocks of the solonchak have long been impregnated with salt water, the possibility of change in the composition of the clay minerals must be considered. This has not yet been adequately studied, but two facts are noteworthy. I. D. Sedletskii has described a specific hydromica from soda solonchaks, gedroitzite, which is rich in sodium (in place of potassium), and M. A. Rateev has noted the wide development of sepiolite in solonchak clays. This demonstrates that progressive salinization of the solonchak horizon is apparently accompanied, at least in part, by

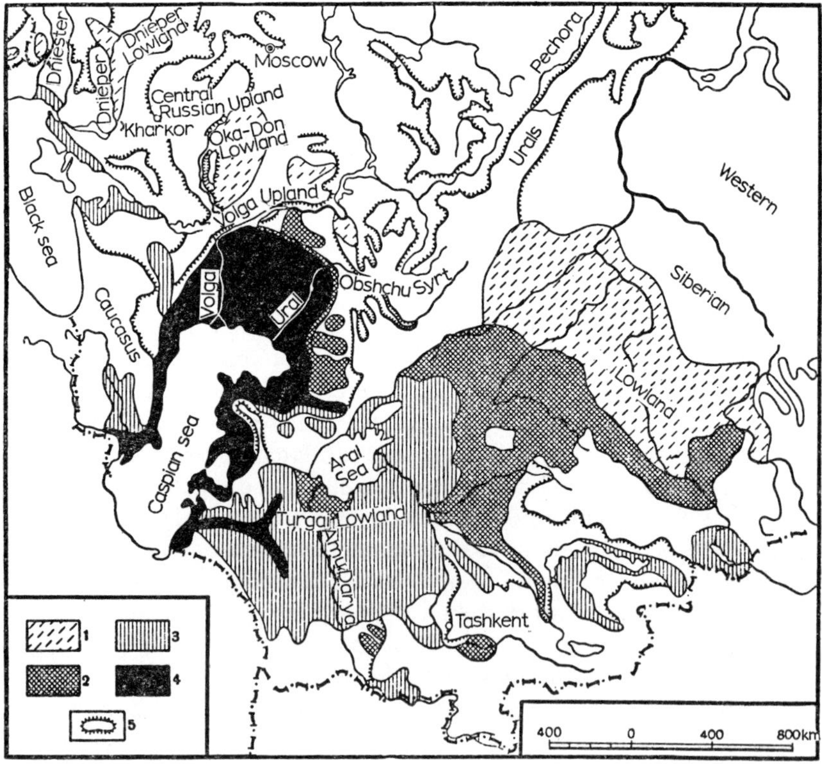

FIG. 9. Recent salinization of soil horizons (from V. A. Kovda). 1. Sodium-sulfate salinization; 2. chloride-sulfate salinization; 3. sulfate-chloride salinization; 4. chloride salinization; 5. outline of uplands.

the alteration of the original hydromica, by a replacement of the potassium by sodium, and in part by the authigenic growth of magnesium silicates. The extent of these diagenetic changes is not yet known.

Since salinization is affected chiefly by very soluble salts, salt floors are stable only in rather dry climates, where the solid salt phases may be preserved for long periods of time. When the climate becomes moist, the salt deposits begin to be leached and to become desalinized. A solonchak gives way to solonets and then to solod'.* The salinized clastic rocks, being leached of their salts, apparently return to their original state. Soil scientists, however, have shown that this process is accompanied by a substantial change in the colloidal fraction of the rocks and of the clay minerals. On the whole, therefore, salinization and subsequent desalinization change the original sediments irreversibly, and this change is the greater the richer the rocks are in clay minerals and colloidal components.

5. FACIES TYPES OF TERRIGENOUS FORMATIONS

As in humid regions, the sediments of the initial stage of lithogenesis in arid regions form natural association; these are the arid formations. The essential conditions for their development are: (1) a single type of tectonic conditions lasting over a prolonged period and extending over a considerable segment of the arid zone, and (2) a single type of topographic and climatic conditions over a prolonged period in the region of sedimentation.

Arid formations can be discussed in all their aspects only after chemical-biogenic and chemical sedimentation has been assessed (see below). Only the clastic rocks that are representative of arid formations will therefore be discussed here. A general classification with nomenclature is given in Table 2.

This list of arid clastic formations agrees completely with the classifications of humid formations given in Vol. 1 of this monograph (p. 98). *No new specific features of arid formations arise during the initial stage of arid lithogenesis, at least among clastic rock associations.* This is not surprising because during the initial stage of lithogenesis all the topographic, tectonic, and physicochemical elements are still essentially similar to those of clastic deposition in humid zones. Nevertheless, in detail some differences between clastic arid and clastic humid formations may be detected. These will now be considered.

Formations of arid plains are complexes normally of red or variegated sand-silt-clay rocks, containing some carbonate commonly interbedded with sulfates (salinization), sometimes with veinlets of halite, and more often with halite casts and moulds. Sandstones and siltstones are generally polymictic or mesomictic. Eolian or torrential cross-bedding is characteristic. Mud cracks, clay balls, and clay galls (traces of playas) are common in clay rocks. Organic remains are lacking, or are very rare. The thickness of such formations is measured in tens of meters, sometimes as much as 200–700 m. Such

---

* *Translator's note:* These Russian terms have no good equivalents in English. They represent progressively lower salt contents. One more step would yield a *podzol.*

are the early Middle Devonian and middle Upper Devonian red beds in the principal Devonian region, the red Tatarian series of the Russian platform, among others. All these sediments of arid plains are partly lacustrine and partly alluvial, but they are reworked by the wind to a great extent. The internal structure of the formations is variable, lacking systematic changes in any direction. The formations of arid plains are correlatives of the formations of humid plains, but the different climatic conditions give rise to completely

TABLE 2

Facies Types of Clastic Formations in the Initial Stage of Arid Lithogenesis

Formational group	Type of formation	Series of formations			
		Platform	Geosynclinal, normal stage	Geosynclinal, closing stage	
				Foredeep	Intermontane basin
I. Intra-continental	1. Formations of arid plains (epeirogenic)	+++	−	−	−
	2. Formations of intermontane basins (continental molasse, orogenic)	++	+?	−	+++
II. Paralic, on plains inclined toward the sea	3. Epeirogenic, far from developing mountain chains	+	+?	−	−
	4. Orogenic, near developing mountain chains	+	−	+++	++++
III. Marine	*a.* Epeirogenic 5. Marine formations from moderately dissected drainage areas (terrigenous-carbonate)	++	++	−	−
	b. Orogenic 6. Flysch and flysch-like	−	+++	+++	+?
	7. Marine molasse formations	−	−	+++	+++

different facies, petrographic, and geochemical characters. These differences are found in the clastic mineral composition—polymictic, possibly mesomictic character of the sand-silt rocks instead of oligomictic (quartzose) composition in humid formations, in the absence of kaolinitic clays and in the complete wedging out of chemically weathered crusts and of Fe-Mn-Al ores, in the absence of organic deposits to form coal, and in other features.

Terrigenous formations of intermontane basins of the arid zone are thick (hundreds or thousands of meters), extremely variable fragmental rocks, from conglomerate to fine-grained clay. Colors are variegated, for the most part red, crimson, brown-red, but with grey, green and yellow beds. The fragmental rocks are typically polymictic; the clays are hydromica-montmorillonite, sometimes including magnesian silicates. Along the margins of the basins, the deposits are proluvial fans, commonly thick. In the middle part the sediments are fluvial, though with a series of small lacustrine deposits in the valley (as in the modern Fergana depression), or they are major lacustrine units (as in the modern Issyk Kul depression). The rocks are carbonate- and commonly sulfate-bearing, with small lenticular deposits of salts (the remains of ground salinization or of small lakes). This unusual circumstance may give rise to occasional thin seams of coal—traces of local peat bogs, where ground water emerged, confined to the margins of alluvial fans. In cross-sections through a formation, the structure is generally more or less lenticular, with coarser-grained rocks forming large lenses in the marginal zones, and finer sediments occurring toward the center (Fig. 10). Again we see, in essence, a repetition of

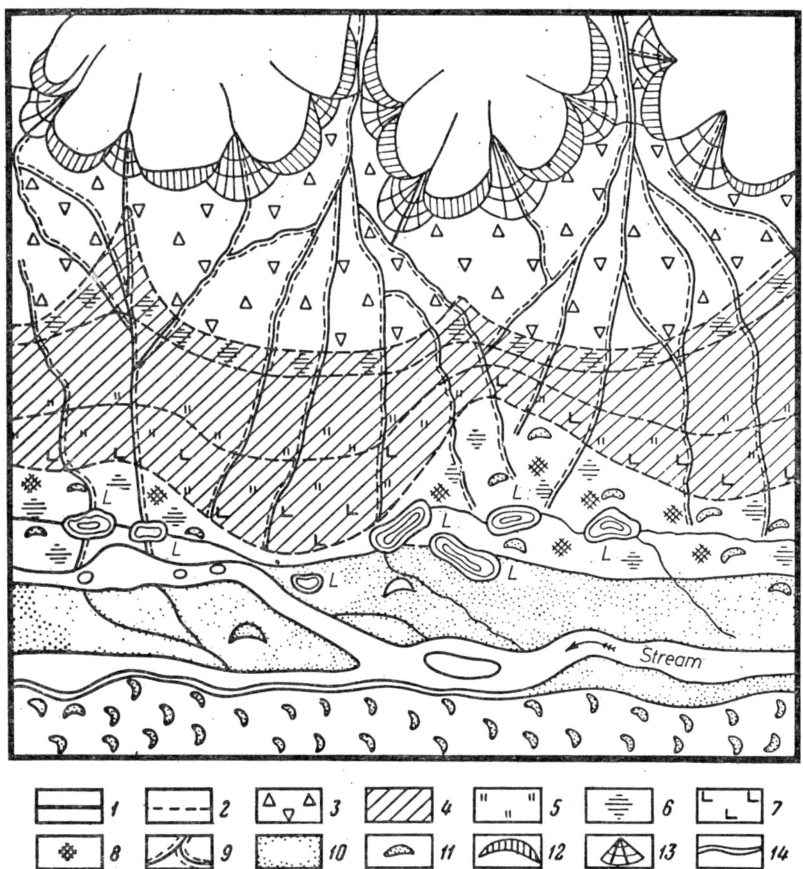

FIG. 10. Facies of modern sediments in intermontane basins of arid regions (from V. I. Popov 1950). 1. Boundary of belts; 2. boundary of zones and subzones; 3. rubbly and gravelly sediment; 4. loess and loessal rocks; 5. meadow; 6. swamp; 7. gypseous sediment; 8. salt-bearing sediment; 9. fan gravel, sand, silt; 10. plains and valley sand and silt; 11. eolian sand; 12. piedmont talus and deluvium; 13. mud flow; 14. torrential valley gravels, L = lake.

the composition and structure that characterize intermontane formations of humid regions. The similarity here is even closer, because the fragmental rocks in both regions are polymictic. The principal differences lie in the virtual absence of coals, so typical of intermontane formations of humid regions. Only rarely will there appear the very thinnest coaly streak. The arid-climate rocks are rich in carbonates, however, and may contain individual layers of salt. The suppression of reducing processes because of the low organic content commonly causes variegated and red colors to appear, either in the formations as a whole or in individual horizons. *The specific characteristics of arid conditions, while not changing the principal content of a formation, still lend it well-defined features, differing from those of formations in humid regions.*

Arid paralic formations are not widespread. Representatives of orogenic varieties are the red Ufa deposits of the Russian platform and their correlatives of Kazanian age. An epeirogenic variety is found in the ribbon of Ordovician red beds about the margin of the Irkutsk amphitheater.

In petrographic and facies character, these complexes differ little from their correlatives in humid zones: they have the same polymictic composition, the alternation of marine and continental sediments in the succession, the well-defined cyclicity, and lastly, the asymmetrical structure, with continental deposits occurring along the upper margin of the seaward slope and marine deposits along the lower margin. Differences are found only in compositional details: variegated coloration, widespread carbonate cement, and the absence of coal.

Marine terrigenous formations that are deposited during the initial stage of arid lithogenesis, are so similar to humid clastic formations that it is very difficult to distinguish them. Flysch deposits may serve as a classic example. As was shown by I. V. Khvorova (1956, 1961), although the Upper Carboniferous and Lower Permian flysch in the southern part of the Ural foredeep is known to have been deposited in the arid zone, its composition and structure show no clear differences from the composition and structure of terrigenous flysch of other localities characterised by a humid climate, such as the Eocene flysch of the Alps and the Lower Cretaceous flysch of the Carpathians.

From the above discussion it follows that *the distinctive features of arid terrigenous formations appear most clearly in the intracontinental group, more faintly in the paralic group, and most feebly in the marine group.* This is a natural consequence of the climatic differences during sedimentation which are most distinct under continental conditions and least clearly defined under marine conditions.

6. CHARACTERISTICS OF TERRIGENOUS PETROGENESIS DURING THE INITIAL STAGE OF ARID LITHOGENESIS

To summarize this discussion, the specific features of terrigenous petrogenesis during the initial stage of arid may be readily compared with those of humid lithogenesis.

(1) *Arid terrigenous rocks and formations are rich in $CaCO_3$, normally to such an extent ($> 5\%$ to 20%) that, in general, the rocks are transitional to the terrigenous-authigenic group.* Furthermore, because of the low content of organic matter, reducing processes are minimized; which leads to a dominance of red and variegated coloration. Individual horizons of a formation may contain salt or gypsum, partly from the action of ground water, partly from the sedimentation of halite and gypsum in small, short-term saline lakes.

(2) *The terrigenous minerals in sand-silt rocks are the same as those in the same grade of sediments of the humid zone, but the quantitative distribution of the various minerals differs.* Polymictic sands and silts are most widespread; mesomictic accumulations are rarer, and oligomictic types are extremely rare. Polymictic types are found not only in intermontane basins but also on plains. The origin of mesomictic and oligomictic sandstones and siltstones in humid belts also differs substantially from that in arid regions. Humid clastic rocks, rich in quartz (mesomictic and oligomictic), are almost entirely primary formations, characterizing regions of weak tectonic activity and of intense chemical weathering. The latter condition, properly speaking, gives rise to the direct production of oligomictic sediments. Rarely do oligomictic sands arise in humid zones by derivation from, and reworking of older primary quartzose rocks. *In arid regions the enrichment of quartz, up to the stage of making the sands and silts mesomictic, and increasingly to the stage of oligomictic composition, is always a hereditary phenomenon, resulting from reworking of ancient primary quartzose rocks of humid zones.* The difference in the process of enriching rocks in quartz in arid and in humid regions cannot be doubted.

(3) Pure accumulations of kaolinite and halloysite are absent from clay rocks in arid zones, where these minerals are found only as minor constituents. Almost pure deposits of magnesian clay minerals (sepiolite and palygorskite) are widespread, however. These latter minerals also are common as impurities in clays. On the whole, the compositional diagrams of humid and arid clays are complementary to each other (see Fig. 2*C*).

The genetic significance of kaolinite and halloysite enrichment in clays is fundamentally different in humid zones from that in arid regions. In humid regions the development of kaolinite and halloysite is a primary process, genetically associated with the formation of quartz accumulations. In arid regions kaolinite and halloysite are inherited, by reworking from older humid-climate sequences. The deposition of magnesian minerals in arid-climate clays, on the other hand, is a primary process, and a specific characteristic of arid regions.

C

CHAPTER 2

ACCUMULATIONS OF Cu-Pb-Zn; THEIR ORIGIN AND DISTRIBUTION IN ARID REGIONS

As in humid regions authigenic sedimentation in slightly mineralized waters of arid basins produces a number of *monoclimatic, specifically arid* deposits in addition to *biclimatic* deposits, types that may also form in moist climates. The first group includes the sedimentary ores of Cu-Pb-Zn with their accompanying elements: Cd, Ag, As, Sb, Au, Pt, and others. The second includes accumulations of $P\text{-}CaCO_3\text{-}MgCO_3\text{-}SiO_2$ and of organic matter in marine oil shales. Each group will be examined.

1. CRITERIA FOR THE RECOGNITION OF Cu-Pb-Zn SEDIMENTARY ORES, WITH EXAMPLES; THEIR RESTRICTION TO ARID REGIONS

Until recently all deposits of lead and zinc and very many copper deposits were considered to be hydrothermal. However, both foreign (Schneiderhöhn and others) and Soviet investigators, primarily M. M. Konstantinov (1951, 1952, 1954, 1960), V. M. Popov (1955a, 1955b, 1957, 1959, and others), and V. S. Domarev (1948, 1949, 1958), have been able to show that a number of deposits of these metals in their geologic setting can in no way be connected with magma, but must be considered genuinely sedimentary. The principal features of these deposits are as follows.

(1) *They are all strictly stratified* and are confined to a definite, sometimes very thin, horizon of sedimentary rock, which may persist over a considerable area, extending laterally for tens and even a few hundreds of kilometers and having an area of hundreds of square kilometers. Sometimes the ore content and persistence are confined to only a relatively thin part of the stratigraphic horizon, amounting to only 1 m or even 0·5 m. Horizons enclosing the mineralized zone are in chemical composition, structure, texture, and physical-mechanical properties (porosity, fracturing) indistinguishable from the ore-bearing horizon. In other words, *the stratigraphic control of mineralization cannot be explained by any special properties of the rocks containing the ore.*

(2) *The distribution of ore bodies, as a rule, does not depend on fold or fault structures.* The ore bodies are found in complexly deformed zones as well as in flat-lying strata, near faults as well as at great distances from them. It is clear that here *faulting is a post-ore phenomenon.*

(3) *No spatial connection between intrusive rocks and ore bodies is found when such intrusive rocks may occur in the vicinity of the deposit.* Ores may be found immediately next to the intrusion and at considerable distance from the intrusion, without change in character of the mineralization. Many intru-

sive bodies, for which some hypothetical connection with the mineralization might be ascribed, occur at distances of tens and hundreds of kilometers.

(4) *Ore-bearing horizons generally show no indication of contact metamorphism from the intrusion or from hydrothermal solutions.* When ore deposits include hydrothermal veins with mineralization somewhat similar to the stratified ores, detailed comparison of the two discloses fundamental geochemical differences, as was shown by E. L. Abramovich (1957) for the lead ores of Kalkan-ata in the Tashkent region and by A. M. Lur'e (1960) for the Sumsar lead deposit.

(5) *A. I. Tugarinov (1957) has shown by a study of isotopic composition of the lead that in such lead deposits of the U.S.S.R. as he has examined the absolute age of the lead conforms to that of the ore-containing rock and is considerably greater than the age of the postulated "parent" intrusions.*

These features of stratified Pb-Zn or Cu-Pb-Zn deposits, taken either separately or together, cannot be explained from the viewpoint of the hydrothermal hypothesis; on the other hand, they can be understood, and indeed be self-evident by interpreting the deposits as normal sedimentary accumulations. For this reason the hypothesis of a sedimentary origin for stratified Zn-Pb-Cu ores, advocated by M. M. Konstantinov and V. M. Popov, is gaining more and more adherents, including the author of the present monograph.

The following deposits appear to the author to be indisputably, or at least very probably sedimentary.

For copper: Lena (on the Siberian platform), Urals, Mansfeld, Kirgiz Range, Mangyshlak, Artemovsk, Dzhezkazgan, the Permian and Jurassic ores in Arizona, Idaho, Utah, Colorado, Bolivia, Durham, and others.

For lead and zinc: Kalkan-ata (in the Tashkent region), Sumsar (in the Chatkal Range), Dzhergalan (Terskei Alatau), Western Balkhash, Truskavets, Lower Silesia, Upper Silesia, Artemovsk (in the Donbas), and Bogdo.

By including all these deposits among sedimentary ores the problem of contemporary climatic control must be considered: are the ores of Cu-Pb-Zn biclimatic or monoclimatic, and if the latter, are they related to a humid or arid climate? In order to solve these questions the above deposits were plotted on paleoclimatic maps of the Ordovician, Middle Devonian, Lower Carboniferous, Permian, Triassic, Jurassic and Cretaceous, and Paleogene and Neogene (Figs. 11-17). *Almost without exception, the sedimentary deposits of Pb, Zn, and Cu lie strictly within arid zones.* Individual discrepancies from this generalization are due either to inaccuracy in drawing the boundary of the arid zone or to an appreciable stratigraphic difference among the ores relative to the horizon chosen for comparison on the paleoclimatic map. Since discrepancies are few and deposits that lie within arid zones are dominant, it may be assumed that *deposits of Pb-Zn-Cu are typically monoclimatic, particularly arid, formations.*

Two further points must be emphasized. *It is essential to distinguish carefully between ore accumulations of Pb-Zn-Cu and rocks that carry only low concentrations of these elements: sandy, silty, clayey, and carbonate rocks that*

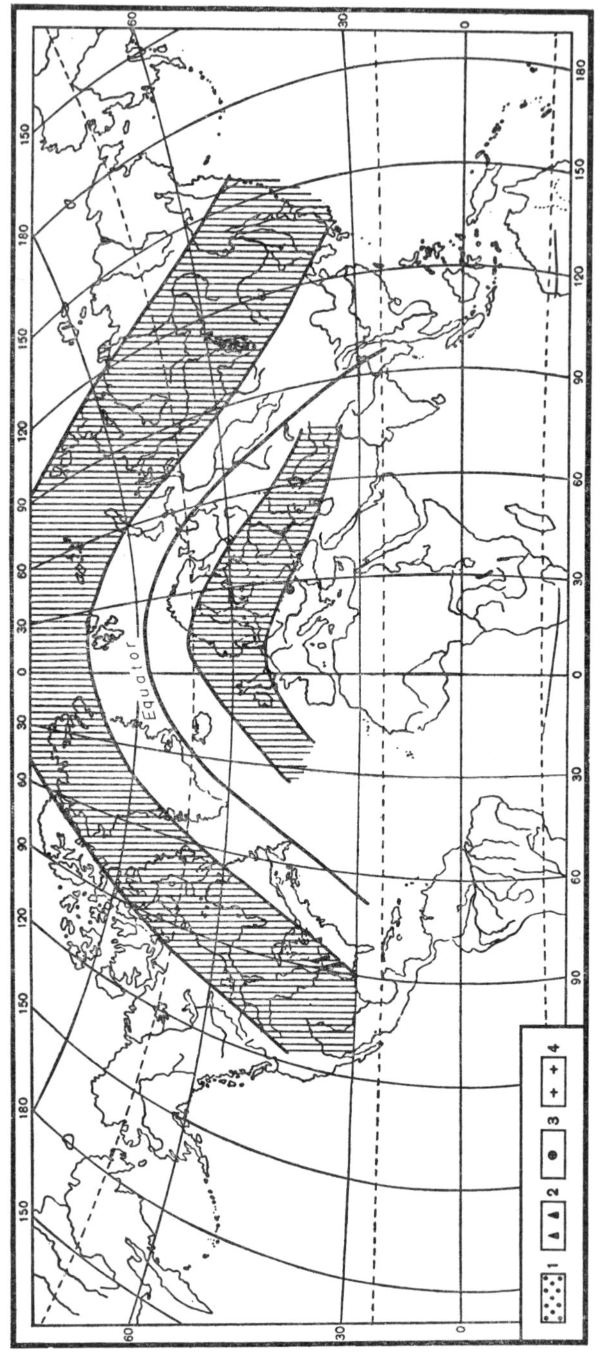

Fig. 11. Climatic distribution of Cu-Pb-Zn ores in the Caledonian stage. 1. Cu ores; 2. Pb-Zn ores; 3. Pb-Zn ores in the Cambrian; 4. disseminated Pb. In this and the following figures, up to and including Fig. 17, the hatching indicates arid regions.

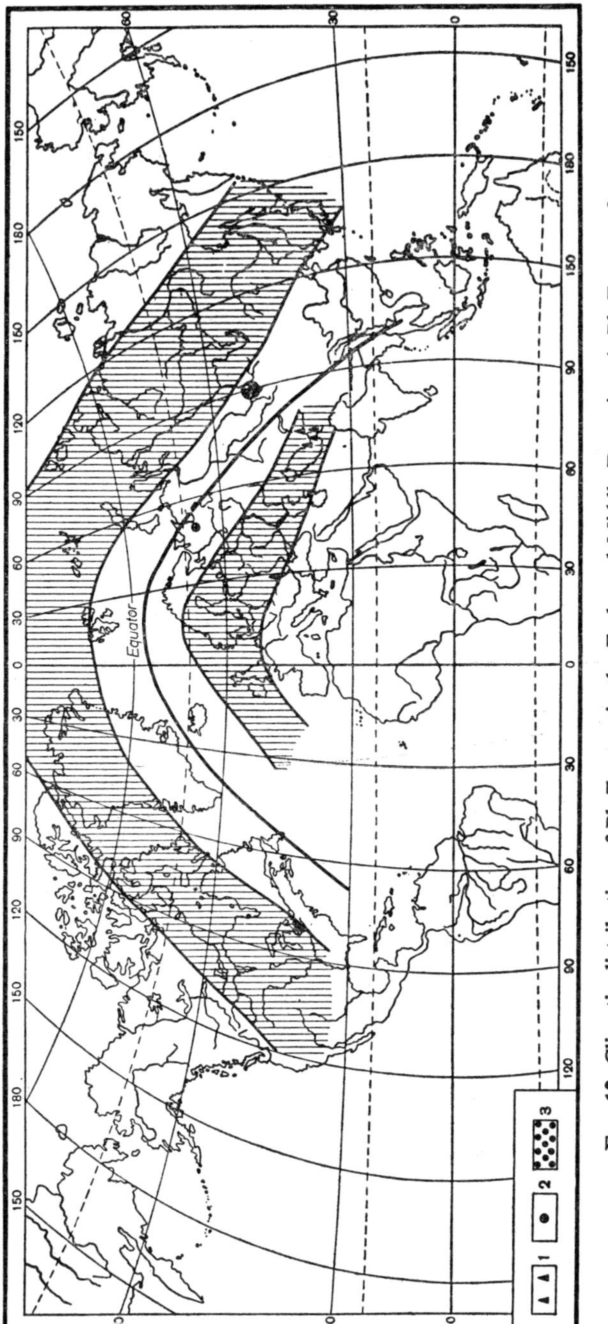

FIG. 12. Climatic distribution of Pb-Zn ores in the Early and Middle Devonian. 1. Pb-Zn ores; 2. disseminated Cu; 3. cupriferous sandstones.

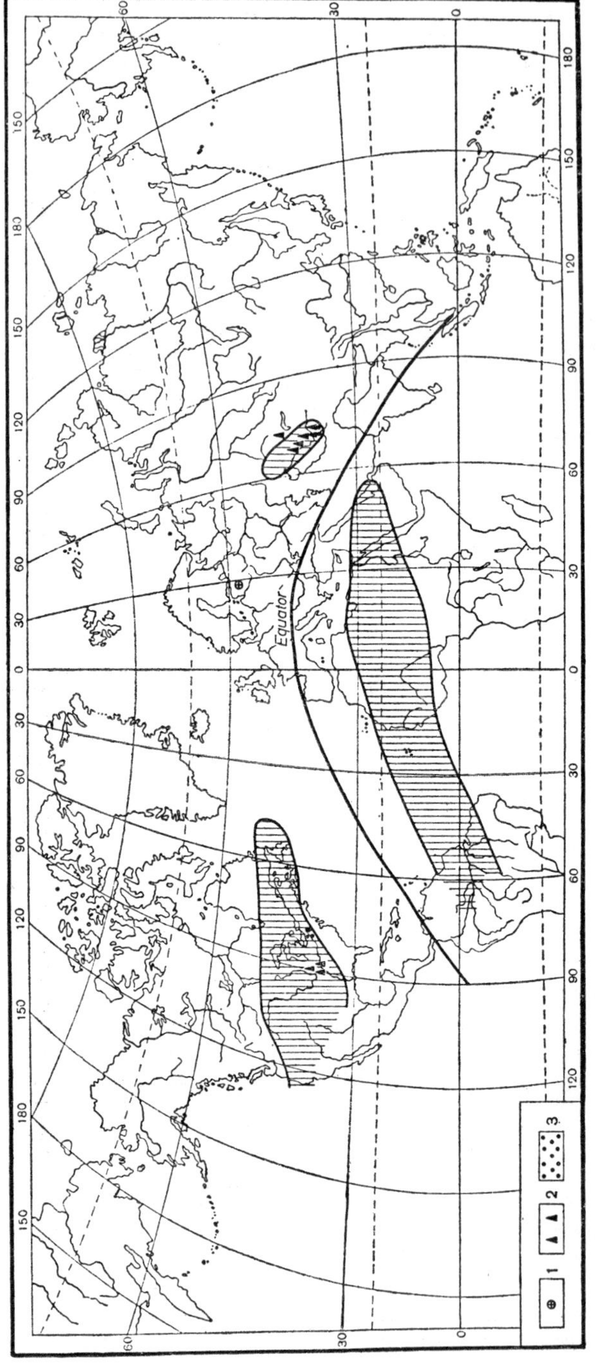

Fig. 13. Climatic distribution of Cu-Pb-Zn ores in the Early Carboniferous. 1. Cu-Pb-Zn contaminations; 2. Pb-Zn ores; 3. Cu ores.

26

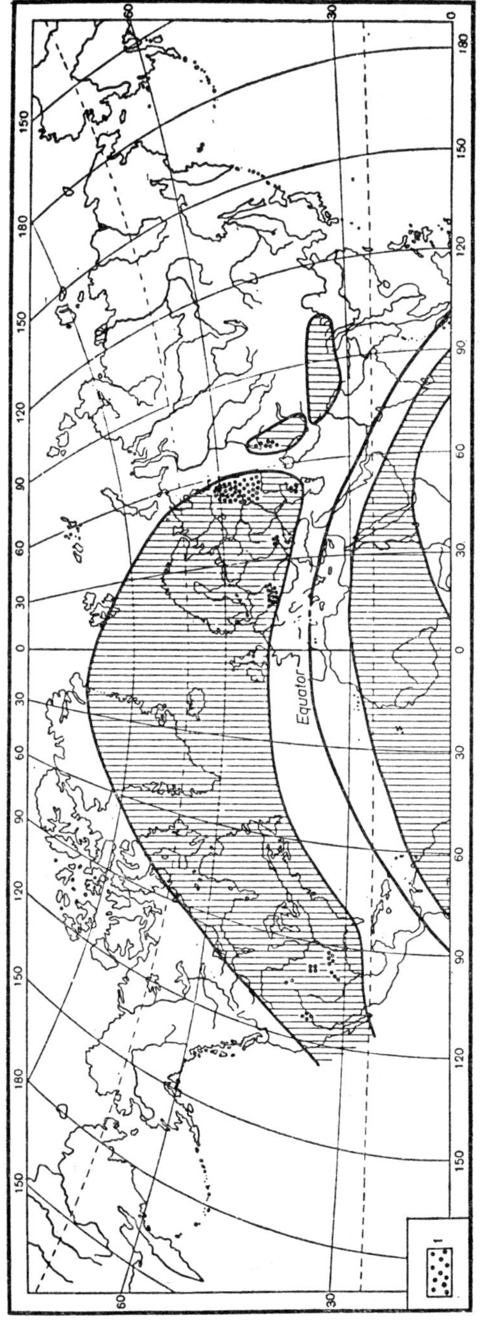

Fig. 14. Climatic distribution of Cu-Pb-Zn ores (black dots) in Late Carboniferous and Permian time.

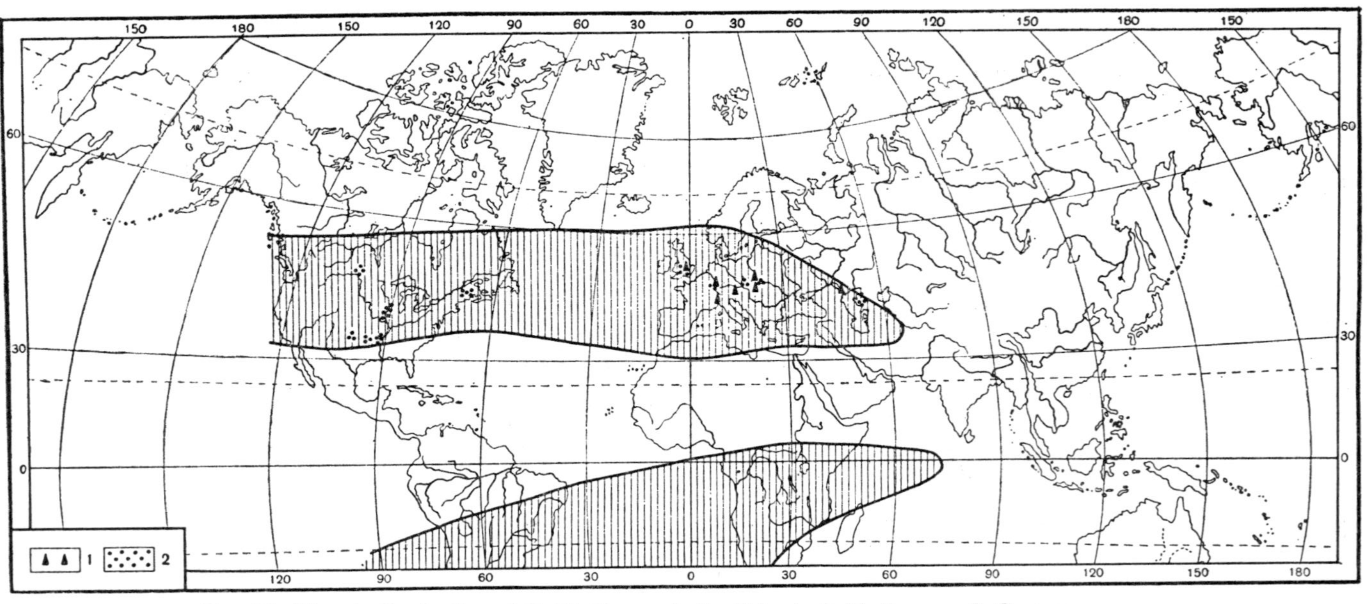

FIG. 15. Climatic distribution of Cu-Pb-Zn ores in the Triassic. 1. Pb-Zn ores; 2. Cu ores.

28

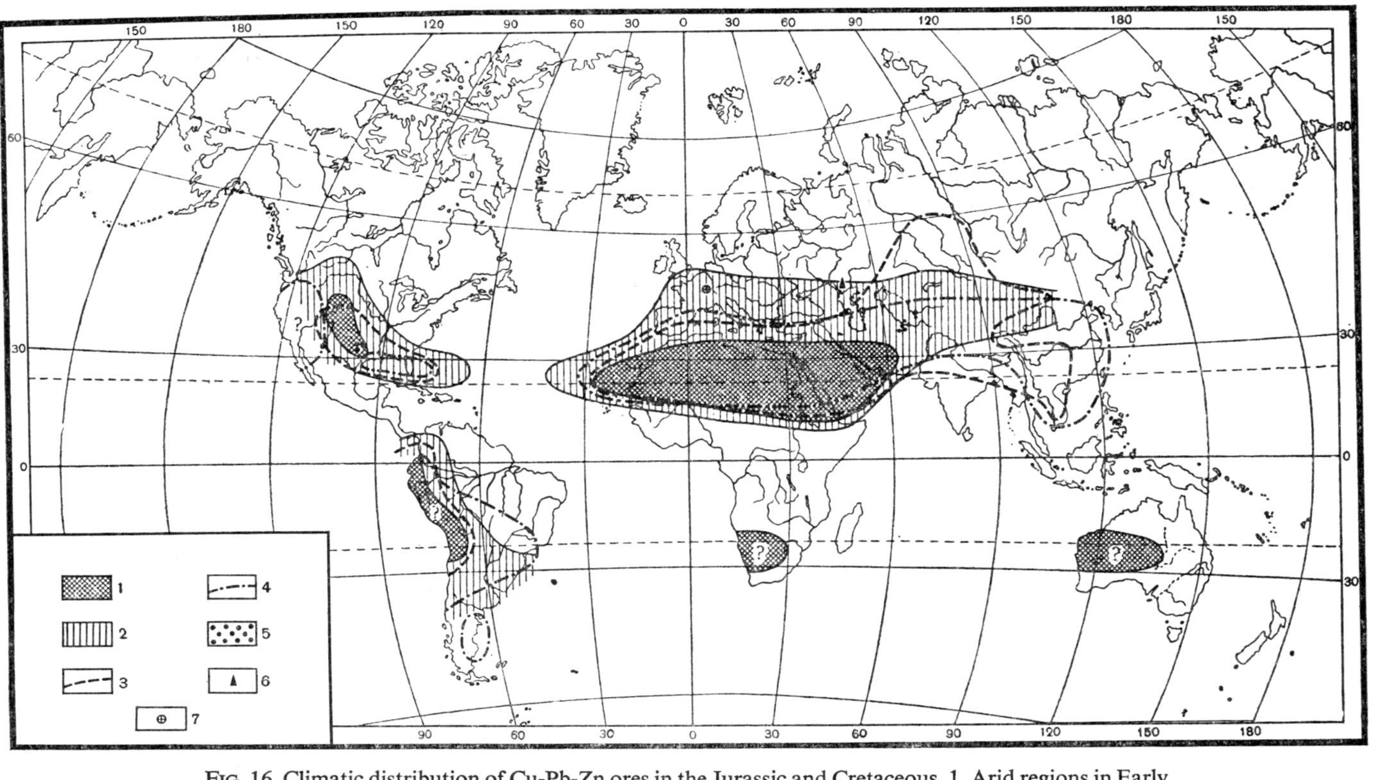

Fig. 16. Climatic distribution of Cu-Pb-Zn ores in the Jurassic and Cretaceous. 1. Arid regions in Early and Middle Jurassic time; 2. additional area of arid regions in Late Jurassic time; 3. boundary of arid regions in Lower Cretaceous time; 4. boundary of arid regions in Late Cretaceous time; 5. cupriferous sandstones; 6. Pb-Zn ores; 7. weak contamination of Pb + Zn.

29

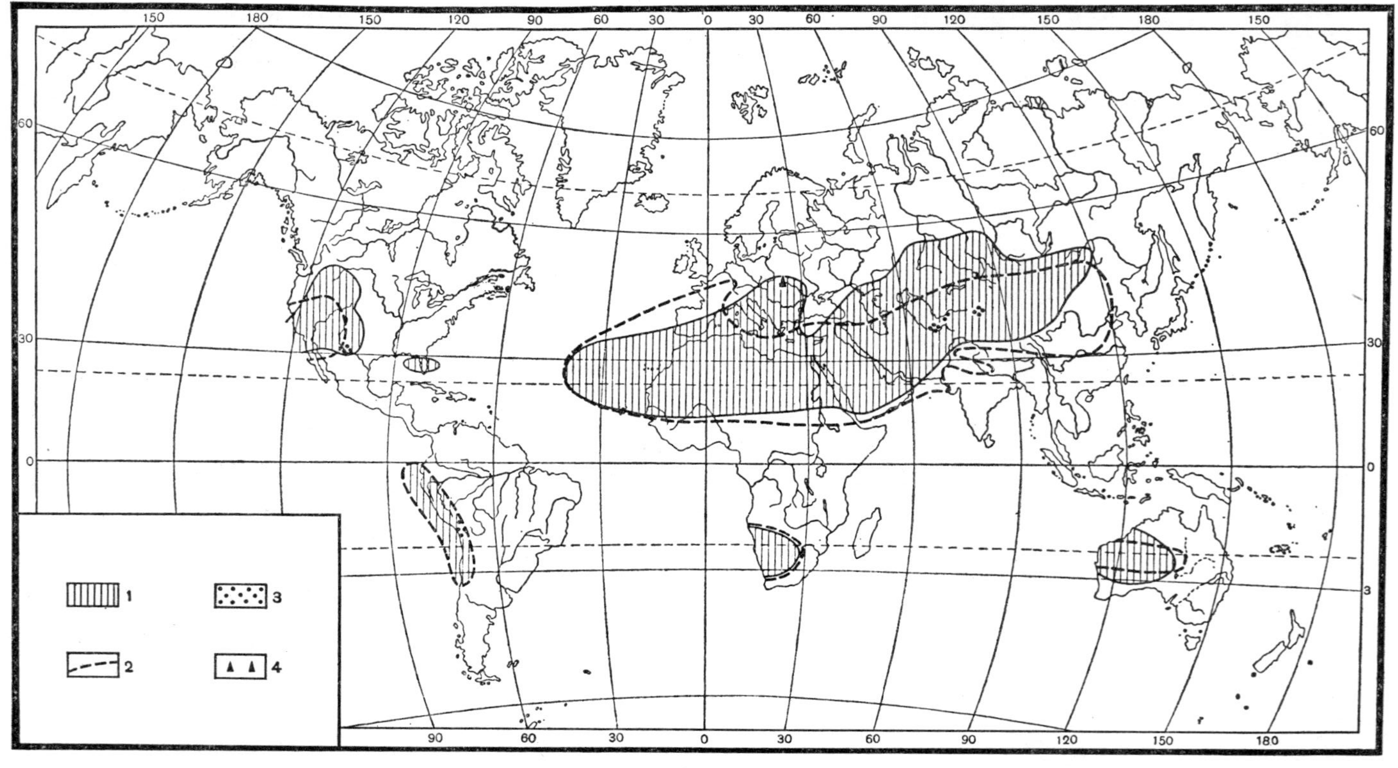

Fig. 17. Climatic distribution of Cu-Pb-Zn ores in the Cenozoic. 1. Arid regions of the Neogene; 2. Boundary of arid zones in the Paleogene; 3. Cu ores; 4. Pb-Zn ores.

may contain individual crystals of galena, sphalerite, or chalcopyrite. Such "contamination", as it is now made clear, is widespread and occurs in deposits of both arid and humid zones; i.e. it is typically biclimatic. Ore concentrations of these elements are climatically narrowly restricted and occur only in arid zones.

In the wide range of arid-climate rocks, however, Pb-Zn-Cu ores never occur among anhydrite or halite deposits, still less in rocks with potassium salts. They always occur in sand-silt-clay rocks or in dolomites. This means that *major deposits of these elements in arid basins form only during the initial stage of arid lithogenesis.*

It must be stated that this paragenesis was first recognized by M. M. Konstantinov in 1954. It was based on the petrography, or lithofacies, of the strata in which Cu-Pb-Zn ores occurred. "Concentrations of lead and zinc sulfides are most frequently restricted to the time of transition from a moist temperate climate to a hot arid climate, when features of the latter are beginning to be distinct. The range through which concentrated copper sulfides are deposited is somewhat broader, but is also restricted to the time of change in climatic conditions, of progressive salinity in the basins, and so forth" (Konstantinov 1954, p. 77).

2. Geological Controls on the Formation of Cupriferous Sequences; Their Petrographic Composition and Structure

Figure 18 shows the stratigraphic distribution of cupriferous sequences in the U.S.S.R. and foreign countries (from V. M. Popov 1956). It is clear that they have formed essentially without interruption throughout the history of the earth. Only in Late Cretaceous and Silurian are cupriferous strata unknown, and this may well be due to the incompleteness of the geological record. However, the distribution of cupriferous strata within the stratigraphic column is very irregular. Whereas in the Cambrian, Ordovician, Lower Devonian, Upper Carboniferous, Upper Triassic, Jurassic and Paleogene the cupriferous strata are individual deposits, in the Neogene, Lower Triassic, Permian, Middle and Lower Carboniferous, and Middle and Upper Devonian they are very extensive and are found in markedly different parts of the lithosphere. *The copper-accumulating epochs may therefore be contrasted with those in which the formation of sedimentary copper deposits was weak.* There were five copper-forming epochs: one embraced the end of the Cambrian and the beginning of the Ordovician; another the Middle and Upper Devonian and the Lower and Middle Carboniferous; the third the Permian and Lower Triassic; the fourth the Lower Cretaceous; and the fifth the Neogene. The intervening time intervals were either epochs of weak ore formation or, for parts of these intervals, of no ore formation at all. In considering the tectonic relationship of copper-bearing sequences, and also the nature of the tectonism of copper-accumulating regions during ore-formation, it may be concluded that *cupriferous sandstones are usually formed immediately after major orogenic epochs, within the fold zones, or in immediately adjacent parts of the*

Ng	1– Naukat (northern Fergana) 2– central Tien Shan 3– southern Tadzhik depression 4– Kashgaria (Sinkiang) 5– Mexico
Pg	1– Coro-Coro (Bolivia) 2– Aral region
K	1– Eastern Fergana 2– Altai Range. 3– southern Tadzhik depression
J	1– Colorado
Ҡ	1– South of Manchester 2– Lotharingia 3– Rhine district 4– Mangyshlak 5– China (Prov. of Kwangsi, northern Yunnan, Kweichow) 6– New Mexico, 7– Arizona, 8– Utah, 9– Colorado, 10– Connecticut, 11– New Jersey, 12– Australia (New South Wales), 13– Bavaria,
P	1– Donbass, 2– western Ural region, 3– Mangyshlak, 4– Inder region, 5– central Kazakhstan, 6– Mansfeld, 7– Durham (England), 8– Westphalia, 9– northeast Bohemia, 10– Lower Silesia, 11– China (Prov. Kwangsi, northern Yunnan), 12– New Mexico, 13– Arizona, Texas, Oklahoma, Colorado, Idaho, 14– Nova Scotia,
C	1– Ketmen Range, 2, 3, 4, Kirgiz Range, 5– Dzhezkazan, 9– Betpakdala, 10– Brunswick, 11– Arizona,
D	1– Podolia, 2– Pennsylvania, 3– Minusinsk basin, 4– Pechora region, 5– Kazakhstan (Atbasar, Tersakan, Argapatak, Ulutavka, Dzhezkazgan, upper Ishim, and upper Chider region)
S	
O	1– Ust-Kut series of the Lena region
Є	1– Usolye series in the Lena region, 2– Upper Lena series in the Lena region
	1– Rhodesia, 2– Katanga, 3– Uc'okan

platform (Sapozhnikov 1948). *In other words, cupriferous strata have formed against a background of marked tectonic activity.*

This circumstance has controlled both the great thicknesses of strata and the petrographic type of ore-bearing rock.

The petrographic composition of cupriferous strata is variable. Chiefly they are sandstones and siltstones, with subordinate clays or mudstones and also thin beds of detrital and pelitomorphic limestones. Intraformational conglomerates also occur. The fragmental rocks always belong to the *well-defined polymictic type,* and they normally contain a high proportion of volcanic rock fragments as well as feldspar grains. Clayey rocks are composed mainly of hydromica, with some montmorillonite. These rocks are characterised by their high carbonate content, a feature that is typical of formations of the arid zones. The carbonates are calcite and dolomite, which act as cementing agents. Silicification is very rare. The rocks are most commonly reddish (brick red, brownish red, sometimes violet), but greyish varieties may be present, in places markedly subordinate in quantity, elsewhere forming half the entire thickness of the sequence. Despite the intense red coloration of these rocks, the content of iron is generally found at the average (clarke) level. The iron is not reduced because of the negligible organic content (a few hundredths of a percent). The iron content in the gray layers is the same as in the red, but the iron has been reduced by the organic matter, which is present in notable quantities, as large fragments of leaves and branches of plants, as fine detritus, or as microscopic inclusions, spores, and pollen.

Arid formations that include cupriferous strata are variable in facies type. They include intracontinental types (such as the Middle Devonian of the Minusinsk basin), paralic types (the fore-Ural and others), and marine types (the lower Zechstein of Germany, deposits of Dzhezkazgan, Dzhergalan, and elsewhere). Some of these formations lie in intermontane basins, others in marginal depressions, and still others on the margins of platforms. *In all, however, the position of the cupriferous strata within the formation remains the same: the strata are confined to the extreme periphery of the formation,* i.e. *to the margins of the former sedimentary regions, to the areas near the drainage areas of these regions* (Figs. 19 and 20). Only rarely, and only in the

---

FIG. 18. Stratigraphic distribution of sedimentary copper deposits (after V. M. Popov 1956, with some modifications).

FIG. 19. Paleogeography of the western part of Central Kazakhstan during deposition of the Dzhazkazgan series (from V. M. Popov). 1. Land; 2. divide areas composed of Precambrian and Caledonian structures; 3. divide areas composed of Variscan structures; 4. epi-continental seas; 5. gulfs, lagoons, straits, lakes; 6–8. present position of cupriferous sandstone deposits: 6. in rocks of the red Lower Permian Kiima series; 7. in rocks of the Middle Carboniferous Dzhezkazgan (Vladimirovka) series; 8. in rocks of the red Upper Devonian sequence.

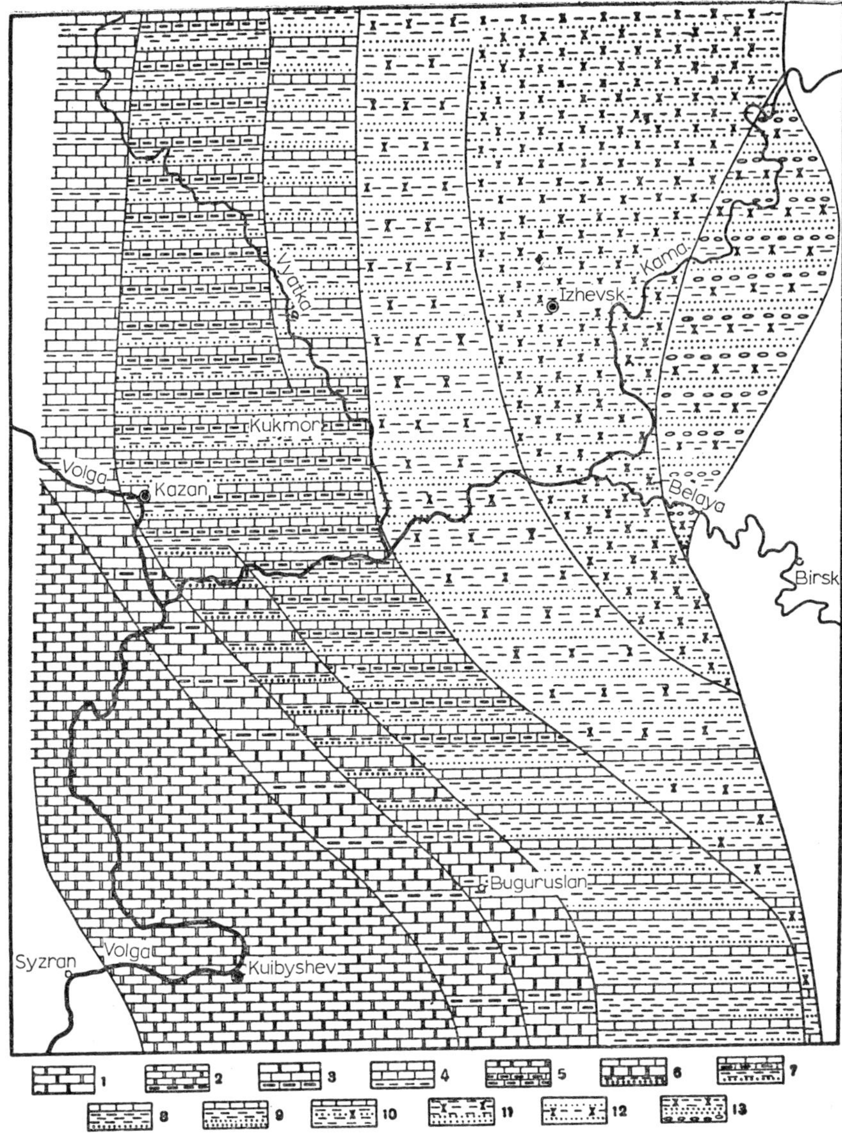

Fig. 20. Paleogeography of the Kazan basin during deposition of the cupriferous sandstones (from N. N. Forsh). *I.* Zone of marine carbonate deposits. *Subzone A_1*: 1. dolomite (constituting more than 90% of the total thickness of the subseries); *subzone A_2*: 2. inter-bedded dolomite and marl, with marked dominance of dolomite; 3. interbedded dolomite, limestone, and marl, with dominance of dolomite and limestone; 4. interbedded limestone and grey clay, with marked dominance of limestone.

II. Zone of marine carbonate and clastic deposits, *subzone B_1*: 5. interbedded marl limestone, and dolomite, with considerable content of marl; 6. interbedded grey clay, sandstone, marl, limestone, and dolomite; *subzone B_2*: 7. interbedded grey clay, sandstone, marl, and limestone, with considerable content of grey clay; 8. interbedded grey clay, sandstone, and limestone, with dominance of grey clay; 9. interbedded grey clay, sandstone, and limestone, with dominance of grey clay and sandstone; 10. interbedded grey clay, sandstone, and red clay, with dominance of grey clay.

III. Zone of littoral-marine lagoonal and continental deposits; 11. Interbedded grey and red clays and sandstone, with much red clay.

IV. Zone of continental deposits. *Subzone D_1*: 12. interbedded red clays and sandstone; *subzone D_2*: 13. interbedded red clay, sandstone, and conglomerate, with dominance of red clay and sandstone.

paralic group, do cupriferous rocks extend into the central parts of a formation. Such facies-tectonic localization of cupriferous strata does not prevent their attaining large dimensions, both laterally and in thickness. The sequence of Permian cupriferous sandstones extends along the western slope of the Urals for more than 1500 km. The copper-ore belt of Katanga-Northern Rhodesia may be traced for 500 km, and this by no means embraces the entire extent of its distribution. On Mangyshlak cupriferous deposits have been followed for about 350 km, but this represents only the exposed part. In Bolivia the copper-bearing strata of Coro-Coro have been traced for 750 km. In Central Kazakhstan, cupriferous beds extend for over 600 km, and on the Siberian platform for more than 350 km. Only in small intramontane basins do red cupriferous deposits form bodies of small areal extent (Domarev 1958).

The thickness of the cupriferous sequences also may be great. The ore series in the copper belt of Katanga-Northern Rhodesia is about 1000 m thick. In central Kazakhstan the thickness of the Upper Paleozoic copper-bearing rocks reaches 635–680 m. On Mangyshlak the exposed Permo-Triassic rocks with cupriferous horizons have a total thickness of 3520 m. In the Coro-Coro deposits of Bolivia copper occurs through more than 1000 m of the succession, and in the region of Kugitang, for some 129 m.

The constitution of the cupriferous sequences partly depends on the facies type of formation of which it is a peripheral part and partly on specific features of that peripheral position itself. Generally speaking, the structure of the cupriferous sequences is complex. Individual units occur as more or less large lenses, wedging out or truncating neighboring units. Laterally such units measure from a few tens of meters up to several kilometers. The thickness ranges from parts of a meter to several meters or even a few tens of meters. Cupriferous strata are thus characterized, in general, by small and moderate dimensions and, hence, by an irregularity of composition. Copper-bearing strata are especially irregular in this sense when they occur among continental arid formations of marginal downwarps (Urals) or in intermontane basins (Minusinsk basin). Paralic or marine formations with cupriferous beds, by contrast, show a uniform structure, and cyclicity is more or less clearly marked, as was demonstrated by V. M. Popov (1960) in the Dzhezkazgan and other deposits.

In the Dzhezkazgan succession, 36 colored beds have been recognized, forming 18 cycles. The base of each cycle, which begins with a grey unit, rests upon an eroded surface of the underlying red bed. The lower parts of a grey horizon therefore generally contain lenses of conglomerate, which pass upward to coarse- and medium-grained sandstones. This part of the cycle is generally well sorted; which indicates that the material was deposited in the shallow-water zone and that the finer fraction was transported to the deeper part of the basin. In the upper parts of the grey units, fine-grained sandstone gradually gives way to siltstone and mudstone, interfingering with them and forming finer rhythms. This part of the section is characterized by different types of bedding: parallel-horizontal, lenticular and cross-laminated, undulatory, etc. The upper part of the grey unit is characterized by wave-cut features, mudcracks, rain prints, and imprints of vertebrate tetrapods. Ripple marks are found only in the medium- and fine-grained sandstones. They commonly occur over areas of the order of several square kilometers. These ripple marks are asymmetric with wavelengths ranging from 3 to 30 cm, and amplitudes from 1 to 5 cm. Interference ripple systems also occur. Mud cracks and rain prints are generally confined to argillaceous sediments that occur above the ripple-marked deposits. The cracks themselves are usually filled with a sandy sediment.

The passage of the grey units in the cycle to the overlying red beds is gradual. The red beds in the Dzhezkazgan series are generally of finer grain-size than the grey rocks. Mudstones and siltstones predominate, and such sandstones as occur are fine-grained.

Another distinguishing feature of the red sandstones is poor sorting, with grains up to 0·5 mm in diameter set in dominant matrix of fine sand, silt, and clay. These sandstones commonly exhibit cross-bedding. This feature is seen in the thin clays within the main mass of sandy material, the lithic types being emphasized by color differences: cherry red in the clay layers, grey-red in the sand. From this evidence, it is concluded that the red beds in the Dzhezkazgan series are of continental origin, formed under subaerial conditions.

Thus, two basic facies types are found in the Dzhezkazgan cupriferous sequence: a near-shore, shallow-water basin facies, and a terrestrial facies. These combine to form a sedimentary cycle. The grey beds in each cycle formed in a gulf during subsidence of the region (Fig. 21), the red during regression, when the gulf was drained and the region became an arid plain. A distinctive feature of these cycles is the development of nearly identical thicknesses of the grey and red units, which interdigitate as long, narrow tongues and wedges. The general inclination of the series of sediments is toward the central part of the basin, i.e. toward the south-southwest.

According to V. M. Popov, a similar cyclic structure is typical of the Lower Permian cupriferous sandstones of the Donbas (8 cycles), the deposits of northern Kirgizia in the Kirgiz Range (3–6 cycles), deposits of Mangyshlak (up to 14 cycles), strata at Coro-Coro (up to 20 cycles), and other deposits. Apparently cyclicity in complex cupriferous sequences is widespread, although it is well developed only when the strata are part of a marine or paralic sequence.

3. FACIES TYPES AND ESSENTIAL COMPOSITION OF COPPER ORE ACCUMULATIONS

It must be first emphasized that sedimentary copper ores are by no means restricted to the minerals of a single metal (Cu). Chemical and spectroscopic analyses always reveal the presence of Pb, Zn, Ag, Mo, As, Sb, Co, Ni, Bi, U, and V, and commonly metals of the Pt group as well. These impurities may be quantitatively so considerable that the ores acquire a bi-, tri-, or polymetallic character. It is especially characteristic to find marked accumulations of Pb and Zn, Ag, more rarely Mo, Bi, Co, or U, with the copper. Thus, *copper ores*

D

in arid zones are geochemical bodies just as complex as, commonly even more complex than ore accumulations in humid zones; a fact that is significant in their genetic interpretation.

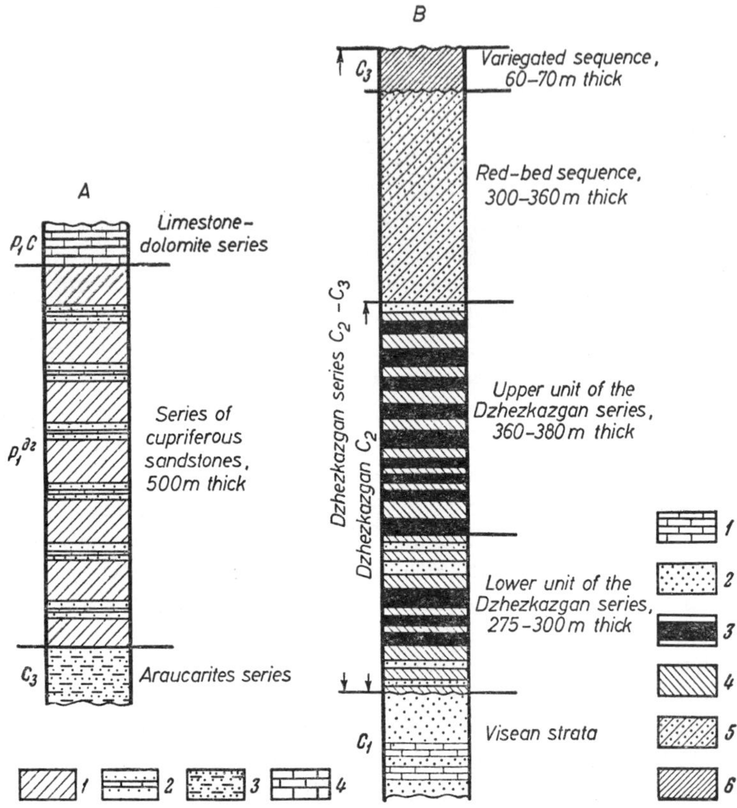

FIG. 21. Cyclicity in the cupriferous sequences of the Donbas and of Dzhezkazgan (from V. M. Popov). *A.* The Donbas, eastern limb of the Bakhmut basin in the vicinity of Yama: 1. red-brown sandstone and siltstone; 2. "grey zones"—grey sandstone and mudstone with thin limestones (5–10 cm) and copper mineralization; 3. coarse-grained sandstone; 4. limestone and dolomite. *B.* Dzhezkazgan, central ore district: 1. Viséan limestone; 2. grey Viséan limestone and the Dzhezkazgan series; 3. grey ore-bearing sandstone; 4. red mudstone, siltstone, and sandstone of the Dzhezkazgan series; 5. sandy clays of the red-bed sequence; 6. Variegated, gypseous shales, marl, dolomitized limestone, and gypsum.

It is characteristic that high copper values (1–5%), making a rock an ore, are related to the bulk color of the rock: *copper ores of primary deposition always occur in grey deposits and are markedly absent from red and variegated rocks.* When high Cu values occur in red beds, they always originate by supergene processes. Since repeated alternations of grey and red-variegated rocks make up the cupriferous successions, the copper ores in such deposits will occur in several stages. *At the same time, by no means all the grey layers in a deposit necessarily contain ore some indeed are barren.* In Dzhezkazgan the Cu content of the grey beds depends on the thickness ratio of grey to red beds. Where the ratio is near unity, the ore is richest in Cu. Where the ratio falls below unity, the Cu content also falls (Ivankov *et al.* 1957).

In their facies development, copper-bearing rocks include alluvium, as shown not only by their internal structures but also by their ribbon-like areal distribution (Figs. 22 and 23*A*, *B*), deltaic deposits, and basin-filling (lacustrine or near-shore marine) deposits. Alluvium as a copper-bearing medium is rare. The other two groups are commonly cupriferous, especially near-shore marine deposits. The facies profile of copper-bearing deposits is consequently narrow, including only those facies that form inshore.

Within facies that are favorable to Cu emplacement, the distribution of petrographic types is irregular. Ore usually occurs in sandstone or siltstone, rarely in clay, and only very rarely in limestone or in conglomerate at the base of a cycle. The grain size of cupriferous sandstones is variable. In most regions (according to V. S. Domarev 1958), the clastic grains range in diameter from a few hundredths to a few tenths of a millimeter. In the cupriferous Kazanian sandstones of the Ural region, nearly 50% of the rock consists of grains of 0·25 to 1 mm diameter. In the medium-grained sandstones of the Dzhezkazgan region, the clastic material averages 0·3–0·5 mm, and ranges from 0·2 to 1 mm. In siltstones of the same region, the grain size ranges from 0·02 to 0·1 mm, with an average value of 0·05–0·06 mm. The diameter of clastic grains in the cupriferous sandstones of Coro-Coro (Bolivia) is mostly 0·3 mm.

No definite pattern has been established for the distribution of the various copper minerals within cupriferous rocks, particularly where the copper-bearing grey rocks occur in small lenses. In cyclic cupriferous sequences with extensive and persistent grey horizons, however, a well-defined mineral zoning is usually present. This was first recognized in the Katanga copper belt deposits of Northern Rhodesia. At the Chambishi deposit, the chief minerals in the foot wall of the ore beds are bornite and chalcocite. Upwards these pass to bornite and chalcopyrite, then to chalcopyrite and pyrite, and finally to pyrite. In other words, *mineral associations with a marked predominance of copper over iron gradually give way upward to associations with an ever increasing iron content, until the latter becomes the only metal.* A similar change in minerals may be traced also within the ore deposit: copper sulfides also give way laterally to iron sulfides. This is accompanied by a progressive decrease in grain-size of the rock, together with some enrichment in organic content (Garlick 1953). A contrasting zonal distribution of sulfides, however, characterises the Roan Antelope copper deposits (G. R. Davis 1954). Pyrite is here present at the base of the copper horizon, with some chalcopyrite. The pyrite disappears upwards, and bornite appears with the chalcopyrite, which in its turn gives way to bornite-chalcocite, and finally to a chalcocite zone. The boundaries between zones lie at some angle to the bedding planes, and in consequence the sulfide distribution varies not only vertically but also horizontally. These changes are accompanied by some increase in grain size, and consequently by a change in organic content.

The significance of this mineral zoning is clear: in the inshore and coarsest-grained parts of a bed, with smaller quantities of organic material in the initial sediments, copper sulfides are dominant; where the grains are smaller and the sediment is richer in organic material, iron sulfides gradually replace the copper sulfides. When copper beds are deposited during transgression,

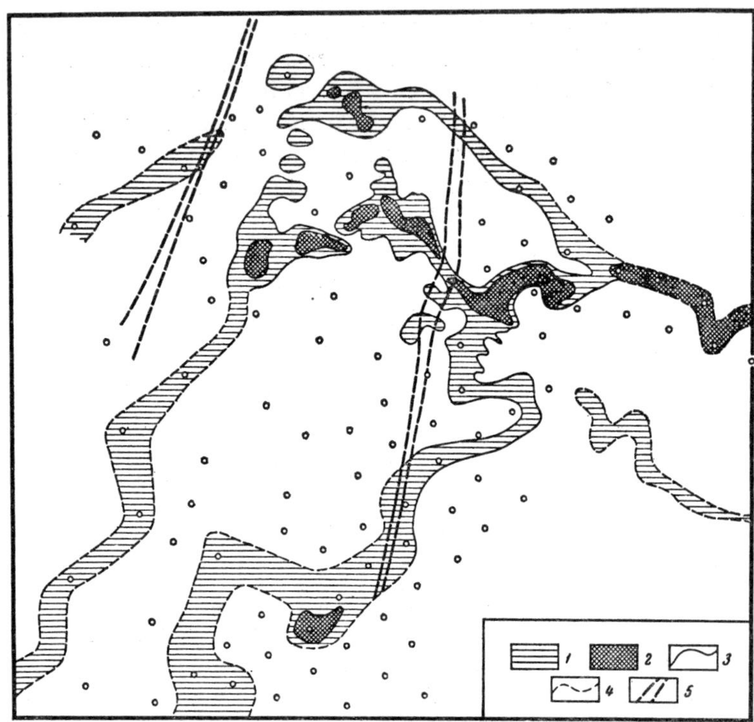

FIG. 22. Distribution of metal in deposits of the Pokro vicinity, southwest of the Dzhezkazgan deposit (from L. I. Ivankov). 1. Average ore; 2. very rich ore; 3. boundary established by stoping; 4. inferred boundary; 5. fold.

therefore, zoning of the Chambishi type develops, but regression gives rise to the reverse type of zoning, such as that at Roan Antelope (Fig. 24).

The same type of mineral zoning in copper-ore horizons occurs in the Permo-Triassic deposits of Mangyshlak (Domarev 1956), the Udokan deposit (Krendelev *et al.* 1958), and Dzhezkazgan (Ivankov *et al.* 1957). At one district in the eastern Karatau (Mangyshlak), the cupriferous unit is 0·5 m thick and varies in grain size from siltstone at the base, through mudstone, to fine-grained sandstone. The ore mineral at the base is chalcocite, occurring in amygdaloidal aggregates. Upwards, bornite appears, occurring in the central

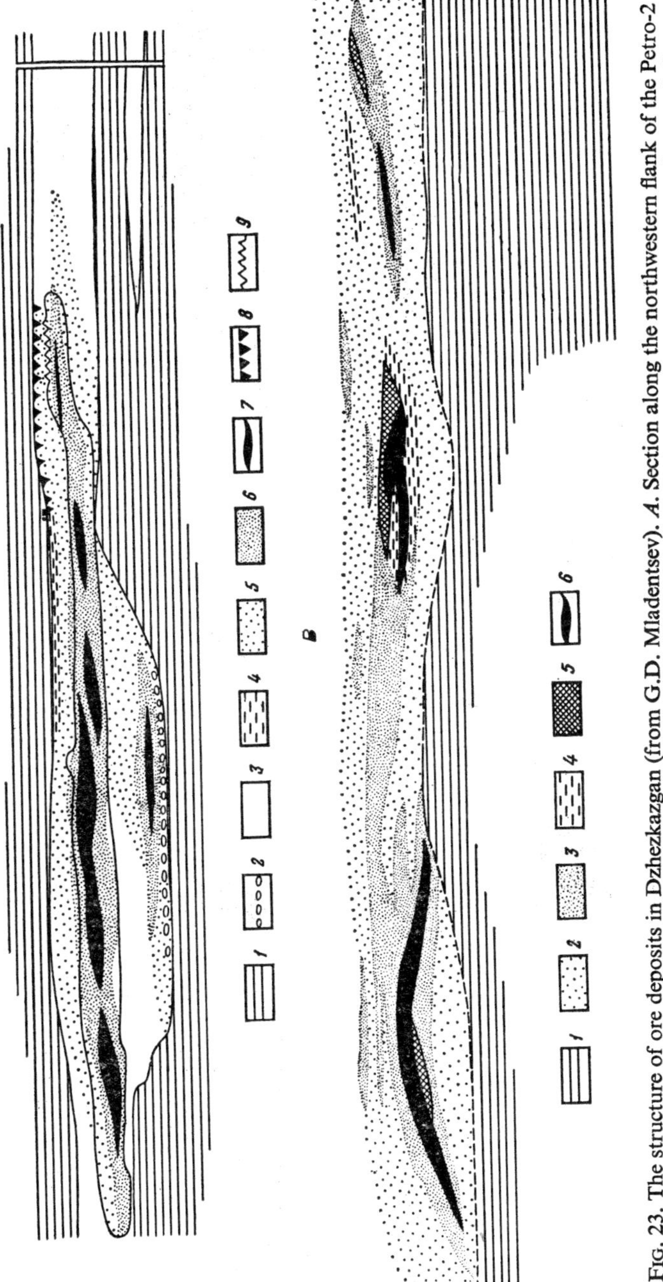

Fig. 23. The structure of ore deposits in Dzhezkazgan (from G.D. Mladentsev). *A*. Section along the northwestern flank of the Petro-2 deposit: (1) red siltstone and mudstone; 2. intraformational conglomerate; 3. barren grey sandstone; 4. grey mudstone; 5–7. ore-bearing grey sandstone: (5) with uneconomic mineralization; 6. with moderate mineralization; 7. with rich mineralization; 8. mud cracks; 9. bedding plane with ripple marks and with leg and tail impression of tetrapods. *B*: Section along chambers 19–18 IS in the Petro-Ts mine of the Petro-2 deposit: 1. red siltstone and mudstone; 2–6. ore-bearing grey sandstone; 2. with uneconomic mineralization; 3. poor; 4. moderate; 5. rich; 6. very rich mineralization.

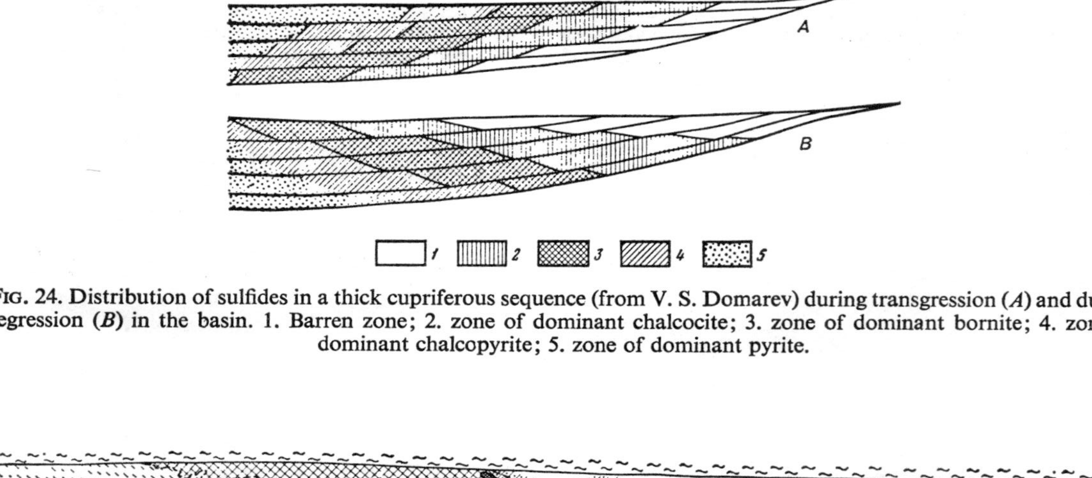

Fig. 24. Distribution of sulfides in a thick cupriferous sequence (from V. S. Domarev) during transgression (A) and during regression (B) in the basin. 1. Barren zone; 2. zone of dominant chalcocite; 3. zone of dominant bornite; 4. zone of dominant chalcopyrite; 5. zone of dominant pyrite.

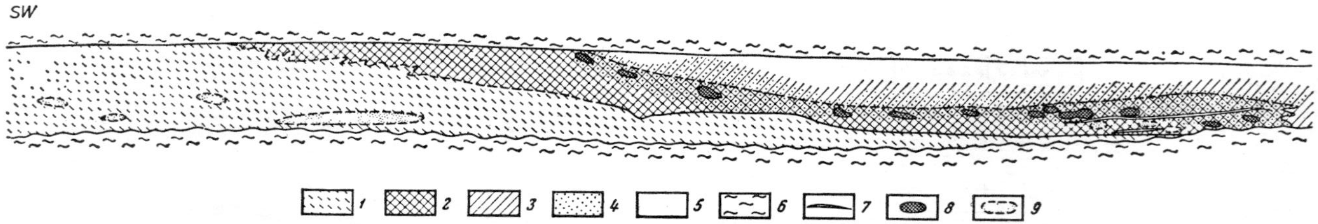

Fig. 25. Zonal distribution of ore minerals in the Pokro-8 ore body at the Dzhezkazgan deposit (from L. F. Narkelyun). 1. chalcocite; 2. bornite; 3. chalcopyrite; 4. galena; 5. barren grey sandstone; 6. red siltstone and mudstone; 7. intraformational calcite vein; 8. pockets of ore; 9. pockets in red-brown sandstone.

parts of the aggregates. Still higher, these aggregates consist of chalcopyrite and bornite or of chalcopyrite alone, and, near the top, pyrite is dominant, with insignificant amounts of chalcopyrite. This distribution of sulfides persists laterally throughout the copper-bearing horizon.

An additional feature of the Dzhezkazgan deposit is the sequence of minerals within the succession from bottom to top: chalcocite—bornite—chalcopyrite—pyrite. The pattern of vertical distribution in areas of lead-zinc mineralization is, from bottom to top: copper—lead—zinc. The same sequences are also found in plan. Bornite, rarely chalcocite, mineralization at any deposit forms a comparatively narrow copper-rich zone (the "enriched zone"). "Transitions from ore of one mineral type to that of another are gradual and are characterized by minerals of both zones. The zone with both minerals may be wide (20–100 m), but the pattern extends for many hundred meters along the deposit. On thinning out, the chalcopyrite-galena zone is normally confined to the lower part of the productive bed, but the chalcocite zone favors the upper part." (Fig. 25) "There is also an association between the mineral composition of ores and the color of the host sandstone. Ore-bearing sandstones of the chalcocite and, in part, the bornite zones generally have reddish tints, and patchy remnants of red color, and they locally include lenses and thin beds of fine-grained red rock. Sandstone of the chalcopyrite-pyrite zones, as a rule, have grey and greenish-grey tones. It is well known that pyrite and chalcopyrite form in a more strongly reducing environment, with chalcocite and bornite in a more weakly reducing environment. The color of the host rocks depends strictly on the valency of the iron, and is a sensitive indicator of the oxidation-reduction potential" (Ivankov *et al.* 1957, pp. 261-262).

The mineral species in which copper occurs, including their zonal distribution, are undoubtedly features that develop during diagenesis and survive in the rocks without subsequent change. Moreover, all deposits exhibit features of later origin—katagenetic or even metamorphic. These include veins both transcurrent and concurrent to the bedding of the ore beds. These are usually only a few millimeters in width. The number of veins increases in those cupriferous rocks that have undergone relatively strong secondary alteration, but even here the vein development of copper minerals is quantitatively very subordinate. The veinlets are composed either of single ore minerals or of gangue (quartz, carbonate, barite, etc.) with some sulfide ores. Ore minerals are present in veins only within the ore horizon or where some part of a vein cuts the ore horizon. The relation of veins to ores indicates that the veins were formed after the ore was emplaced and that they are the so-called veins of Alpine type.

4. FACIES TYPES AND ESSENTIAL COMPOSITION OF LEAD-ZINC ORES

In the main the stratigraphic distribution of sedimentary lead and zinc ores is the same as that of copper ores. *Epochs of lead-zinc mineralization are mostly very close to those of copper accumulation, or they may be identical,*

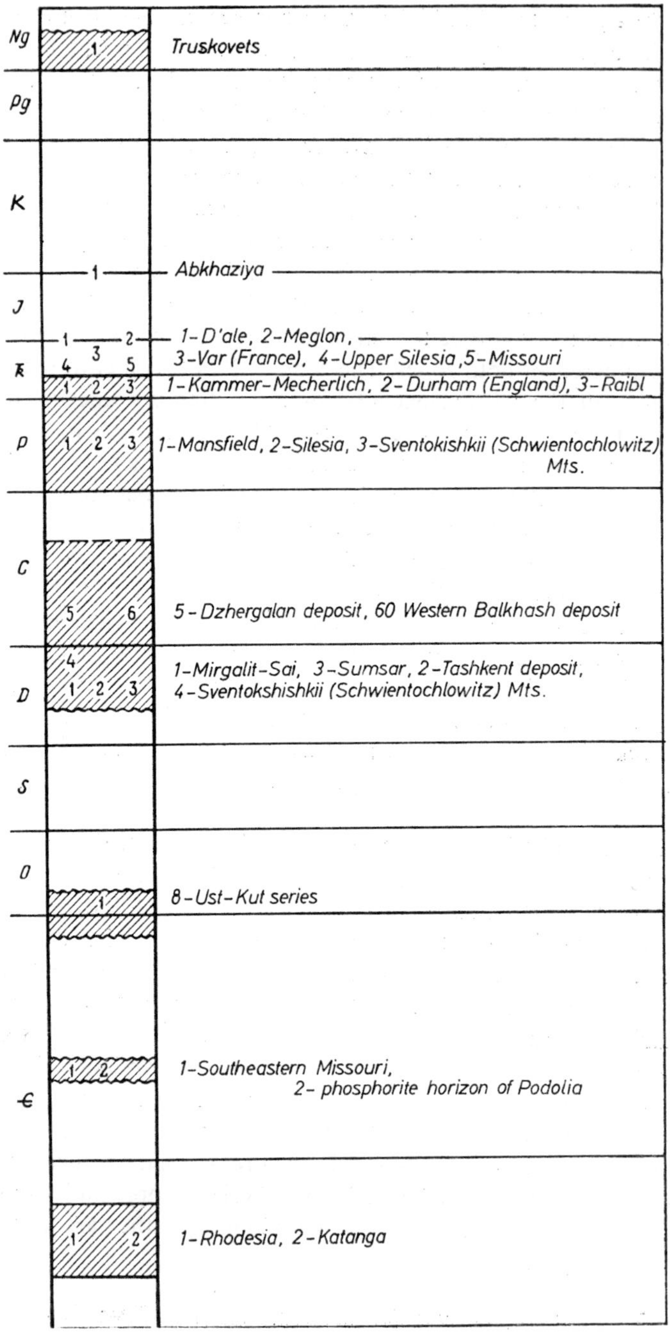

FIG. 26. Stratigraphic distribution of sedimentary lead and zinc ores (after M. M. Konstantinov 1959).

Any marked difference in the geochronology of Pb and Zn is rare, and may be considered exceptional. Lead-zinc sedimentary ores, consequently, also formed against a background of active tectonism.

Like copper ores, lead-zinc ores are always complex. Together with Pb and Zn, many other elements, including Cu, Cd, Ag, V, Ni, Bi, As, and P, are present. The assemblage of these elements varies considerably from one deposit to another, and the content of each is also variable. In general, however, accompanying elements are quantitatively markedly less than the ore components. The lead content in ores averages 2–5%, and sometimes reaches 7–10%. The zinc content is normally less than lead, but it ranges between the same limits. The associated accompanying elements in lead-zinc ores are generally present in quantities ranging from traces to hundredths, and even tenths, of a percent, but rarely more.

The facies range of Pb-Zn accumulations and accompanying elements is narrower than that for copper ores. Independent ores of these metals have not yet been recognised in any continental facies, particularly in deluvial, proluvial, alluvial, or deltaic sediments. Lead-zinc ores invariably accumulate in marine basins. Thus, *the facies profile from the central parts of drainage areas to the terminal discharge basins shows a distinct shift of lead and zinc deposits seawards as compared with copper accumulations.*

Within marine basins Pb-Zn ores are always confined to the peripheral shallow-water zone. In this they resemble copper ores, but by contrast to the latter, *lead and zinc accumulations generally occur in carbonates, chiefly in dolomites and calcareous dolomites, rarely in dolomitized limestones, and not in sandy or silty clay sediments.* Petrographically the carbonate rocks are variable: micritic and fine-grained to coarse-grained, commonly bioclastic, oolitic, or algal. Since carbonate deposits normally form far off-shore, as compared with sand-silt-clay deposits, *the localization of Pb-Zn ores in carbonate rocks is additional confirmation of the general pelagic shift of Pb-Zn deposits as compared with copper ores.* It is noteworthy that even where Pb-Zn mineralization is found in sand-clay rocks the metals are "more noticeably associated with carbonate concentrations that formed either as individual beds or as cement in sandstones, conglomerates, and similar rocks" (Konstantinov 1954, p. 75).

Mineralization is generally observed in thin units of carbonate rocks, 2 to 20 m thick, which may occasionally be traced for some tens of kilometers (Dzhergalan deposit), or for only several hundred meters (Lena region of the Siberian platform). These large areas, however, prove only low lead and zinc values. Such "contamination" is by no means always accompanied by specific lead and zinc minerals such as galena, sphalerite, or their oxidation products. Even where the metal content may reach 1% specific minerals of the metals may be lacking (Middle Triassic rocks of Upper Silesia). Clearly the Pb and Zn in such rocks occur isomorphously in the Ca and Mg carbonates, or are adsorbed by clays.

Within an ore horizon ore zones proper are lenticular bodies, some irregular in plan, which may be traced in places for 100–200 m, elsewhere 400–900 m, and in some localities for kilometers, with thicknesses of 1–3 m. In section such lenses may form one, two, or three layers, separated by beds with not more than a strong contamination of Pb and Zn. In small deposits, the ore-lenses are generally stratigraphically persistent, although even here local

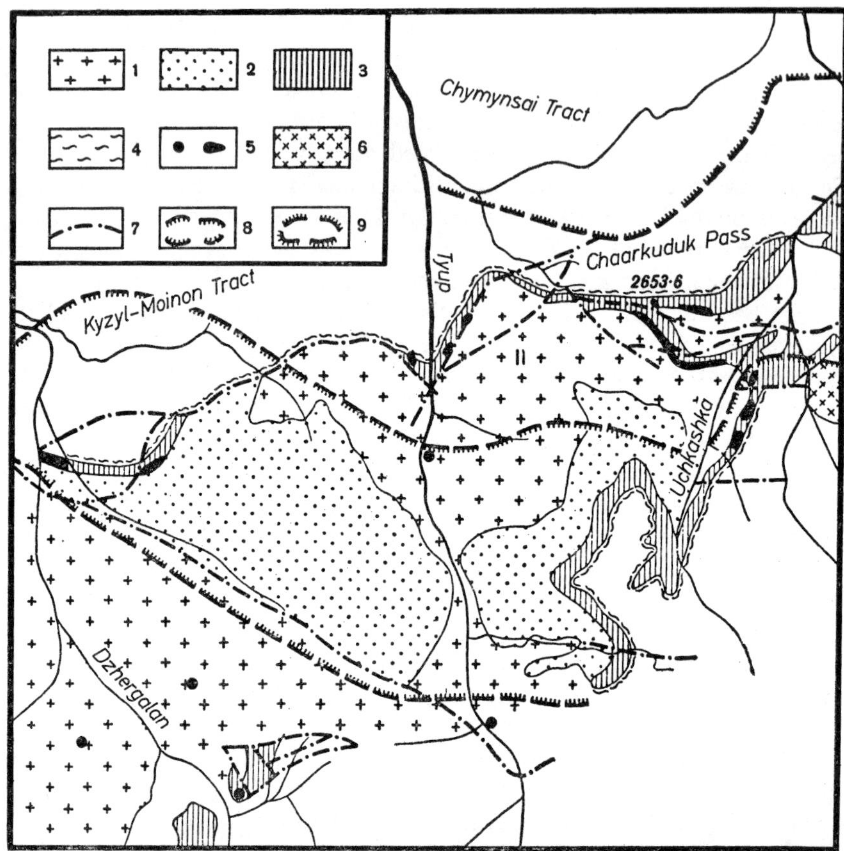

FIG. 27. Distribution of lead mineralization of the Dzhergalan group (from G. P. Bogomazov). 1. Ancient basement (granitoidal rocks, schist); 2. Upper Namurian series of lower conglomerates and sandstones; 3. Upper Namurian calcareous series; 4. Upper Namurian gypsiferous series; 5. lead mineralization; 6. Upper Paleozoic intrusions; 7. Alpine faults; 8. Upper Paleozoic depressions (*I*); 9. Upper Paleozoic uplifts (*II*).

supplementary ore lenses may occur, appearing abruptly above or below the principal horizon (Figs. 27, 28). In huge ore basins the distribution of ore is much more complex. In the Middle Triassic deposits of Upper Silesia, for example, the following are present in a shelly limestone: (a) *a lower coquinoid rock*, 90 m thick, the lower horizon of which consists of undulatory-bedded limestone, marl, and limestone conglomerate, and the upper horizon of ore-bearing dolomite; (b) *diplopore dolomite*, and (c) *alternating layers of*

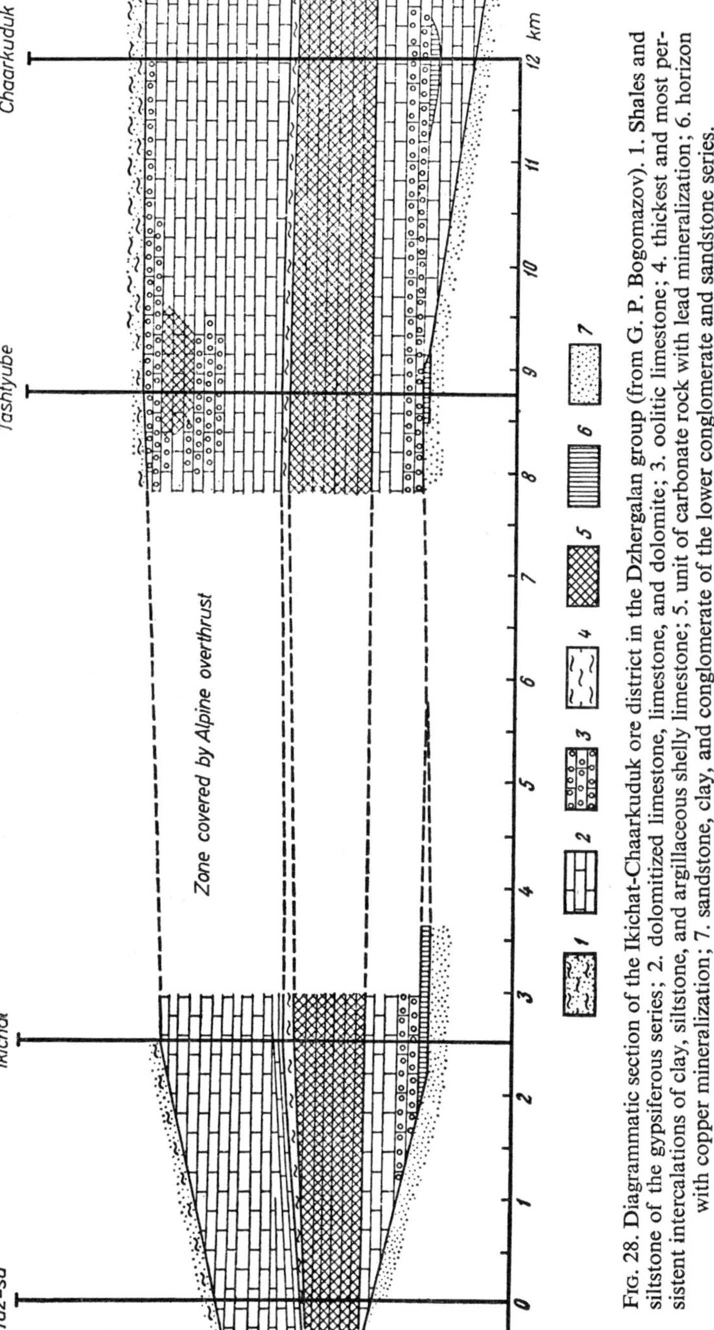

Fig. 28. Diagrammatic section of the Ikichat-Chaarkuduk ore district in the Dzhergalan group (from G. P. Bogomazov). 1. Shales and siltstone of the gypsiferous series; 2. dolomitized limestone, limestone, and dolomite; 3. oolitic limestone; 4. thickest and most persistent intercalations of clay, siltstone, and argillaceous shelly limestone; 5. unit of carbonate rock with lead mineralization; 6. horizon with copper mineralization; 7. sandstone, clay, and conglomerate of the lower conglomerate and sandstone series.

limestone, marl, and sandstone. The mineralization is concentrated in several horizons, irregularly developed in various parts of the ore district (Fig. 29). In the Bytom, Khshanuv, and Dlugoshina synclines, two ore beds occur in the ore dolomites: the lower, principal bed is 4 m thick, and may be traced with few interruptions over the entire area; the second bed is 12 m above the first, about 1 m thick, and is less persistent. In the Sevezh-Olkush zone the principal ore horizon lies in the higher horizons of the ore-bearing dolomites. Here an ore bed occurs in the overlying diplopore dolomite. Beside the chief ore beds, small deposits appear locally in other horizons.

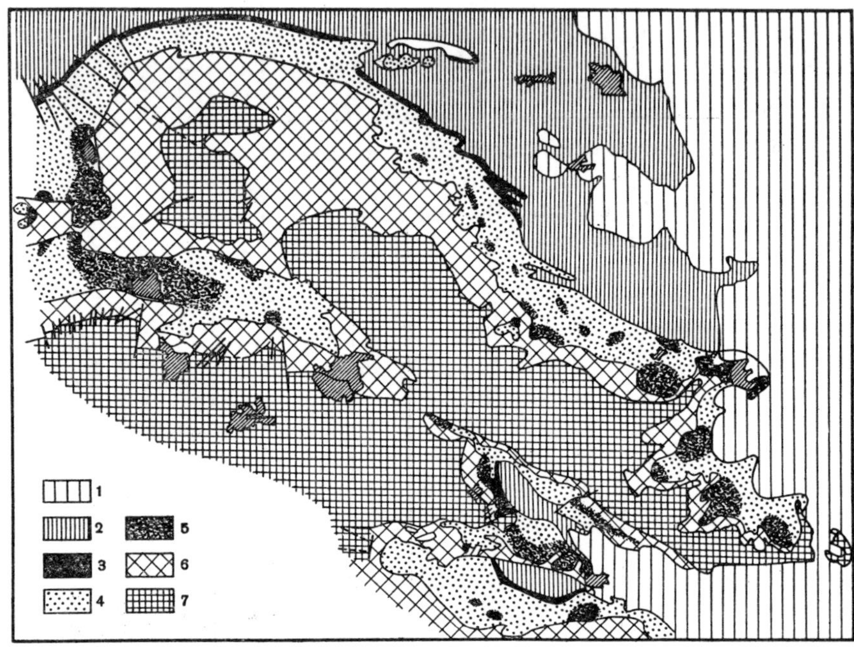

FIG. 29. Distribution of lead ores in Upper Silesia (from M. M. Konstantinov). 1. Jurassic; 2–6. Triassic: 2. Keuper; 3. Shelly limestone above the ore-bearing series; 4. Ore series; 5. Economic deposits of lead and zinc; 6. Rhaetic and mottled sandstone; 7. Paleozoic.

The distribution of ore lenses in the ore horizon varies markedly. Out of a total area of several hundred square kilometers, an ore zone proper may cover 10–20 square kilometers. Some ore zones extend for only 5 kilometers. In the Upper Silesian ore basin, where the combined extent of the ore belts is 100 km, the ore zones make up 33 km. Many of these zones are very large, extending for 20 km, whereas others are negligible (about 1 km). This emphasizes that an ore-bearing horizon, even where it has considerable extent and the ore zones cover a large area, may be still far from rich in ore; concentrations of ore minerals are indeed widely scattered.

The ore bodies themselves are always concordant with the host rocks, and the ore composition is generally simple. The dominant minerals are normally galena and sphalerite, with minor amounts of chalcopyrite, pyrite and, rarely, tetrahedrite. In the oxidation zone the assemblage is complicated by the development of Pb, Zn, Cu, and Fe oxides and carbonates. In some deposits (Lena, Artemovsk, and others) these ore minerals are accompanied by no specific gangue minerals, but elsewhere (the Central Asian Sumsar, Dzhergalan, and Western Balkhash deposits) the ore minerals are always associated with authigenic quartz and barite, in places with fluorite.

The textures and structures of the rocks vary, and these show a close connection with the geologic structure of the ore district. The simplest form occurs in platform areas or areas of little deformation, such as in the Lena district, at the Artemovsk deposit, where the ores are typically disseminated. Galena and sphalerite are found in grains 1·0–1·5 mm across, locally in oolitic carbonate rock, elsewhere in algal varieties. Some ooliths are formed from concentric zones of sphalerite and galena, and more rarely of pyrite. The centers of small ooliths are sometimes a dark brown, almost black, uncrystallized gel. Ooliths composed entirely of sphalerite occur rarely. Composite ooliths are found, consisting of 3–5 small ooliths covered with a common shell. Extremely small veinlets where present are parallel to the bedding. The simplicity of the rock texture corresponds to the simplicity of the formational history. By whatever processes the ore components were emplaced in the sediment, there is no doubt that they were uniformly and diffusely scattered at the stage of sedimentogenesis. Segregations developed only during diagenesis, as a result of redistribution of the ore minerals, and this pattern remained unchanged throughout the subsequent history of the rock.

The textures and structures are much more complex in ores of tectonically, disturbed zones, especially in deposits that have been affected by hydrothermal solutions, where the disseminated texture of the ores is supplemented by well-developed veinlets and fracturing. In places, mineralization is found associated with stylolitic seams. Redeposition of ores is observed in karst belts. The Central Asian stratified lead-zinc deposits in the Turkestan Range, in the Chatkal Range (Sumsar deposit), in the Terskei Alatau (Dzhergalan group), among others, show such structures.

The most complex type of ore structure is found in the Dzhergalan group of deposits (G. P. Bogomazov 1957 and Z. E. Burykhina 1957). Limestones, carrying lead mineralization, include micritic, medium- and coarse-grained dolomitized, oolitic, and nodular varieties. All these are stylolitic to some degree. Most varieties of the ore-bearing carbonate rocks are also brecciated to some extent.

In their textural characters, sulfide-carbonate ores are divided into disseminated and clustered veinlet varieties. In dimension, the disseminations range from dust-size to coarse (5–8 cm and more). The finest galena is irregularly distributed in the rock and is not associated with gangue minerals such as barite and calcite. The medium, and particularly the coarse, disseminations form bands parallel to the bedding of the rock, and are always accompanied by veins of barite and calcite.

In ores showing well-defined banding, one may almost always detect an asymmetric zonal distribution of ore and gangue minerals (Fig. 30).

At the very base of elongated, irregularly equidimensional and isolated pockets of greenish grey montmorillonitic clay 1–2, rarely 5–10 mm thick usually occur. Pyrite, marcasite and galena, ranging from dust size to 0·1–0·5 mm, occur here. Galena increases in grain size upwards, making up the lower part or even the entire deposit. Coxcomb barite and calcite grains occur above the galena, and the upper part of the deposit usually consists of calcite. Within a single pod of ore clay frequently underlies the pocket and continues beyond it, forming a stylolitic surface. Large galena grains or calcite aggregates with galena appear at intervals on these surfaces, and the clay beyond the pockets almost always carries small grains of galena, marcasite, and pyrite. Above the aggregates of galena and calcite are occasional aureoles of coarse, recrystallized limestone, grading downward into the calcite of the pocket

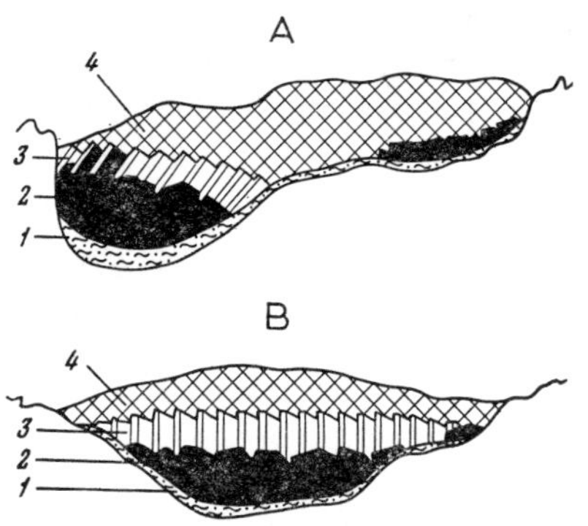

FIG. 30. Relations of ore and gangue minerals in pockets within ore-bearing limestone (from G. P. Bogomazov). *A*. In a vertical section along the dip of the bed. *B*. In a vertical section along the strike. 1. Aggregate of clay minerals, concretions of iron sulfide, and grains of galena; 2. coarse-grained aggregates of galena; 3. barite; 4. calcite.

and upward into fine-grained varieties. Beneath a pocket the rock texture is unchanged. This relationship among clay borders, ore minerals, and veins points to a close spatial and genetic connection among the stylolitic surfaces, the calcite-barite pockets, and ore impregnations.

Mineralized veinlets are extensively developed in the Dzhergalan group of deposits where three well-defined types may be recognized according to their field relationships.

Veinlets of the first group are closely associated with stylolites and are sinuous features, extending either below the stylolite for 2–10 cm, or above it. Some straight veinlets occur between adjacent stylolites, and only rarely touch them. The veinlets are 0·5–5·0 mm thick (Fig. 31) and are of clay minerals and calcite with some galena.

Veinlets of the second group are 0·5–2·5 m long, 5–30 mm wide, and may be traced through an entire ore body, sometimes along a single ore horizon (Figs. 32 and 33). They are straight and are genetically related to the Alpine fracture tectonics, particularly to a large Alpine overthrust, which passes through some of the southern deposits. Only a minority of the veinlets show

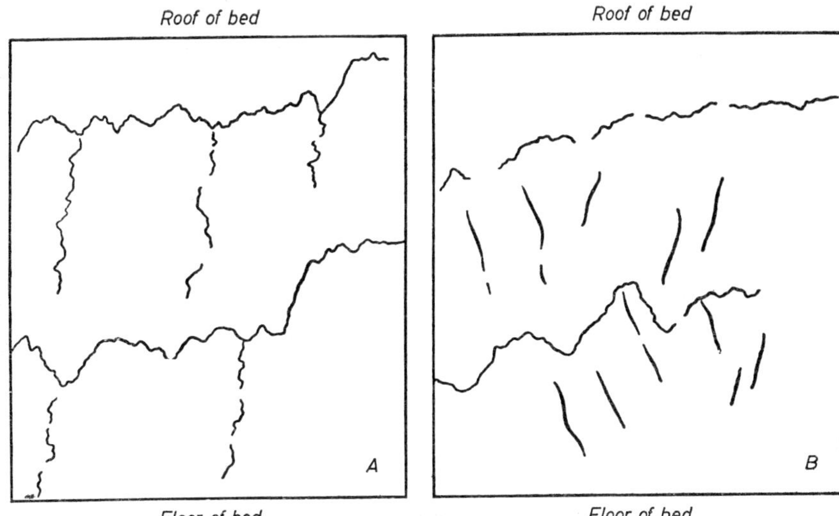

FIG. 31. Types of veinlets in ore horizon. *A*. Veinlets of calcite with galena branching from stylolites towards the base. *B*. Veinlets occurring between stylolites.

FIG. 32. The relationship of the cross-cutting veinlets to stylolites and to clay partings. 1. Limestone; 2. cross-cutting calcite veinlet; 3. clay parting; 4. disseminated galena; 5. stylolite surface.

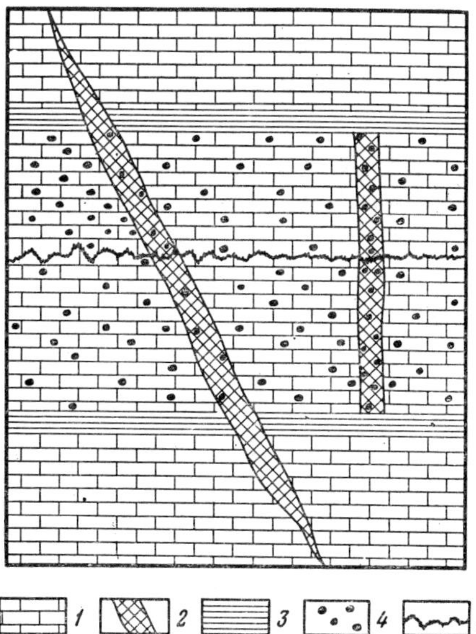

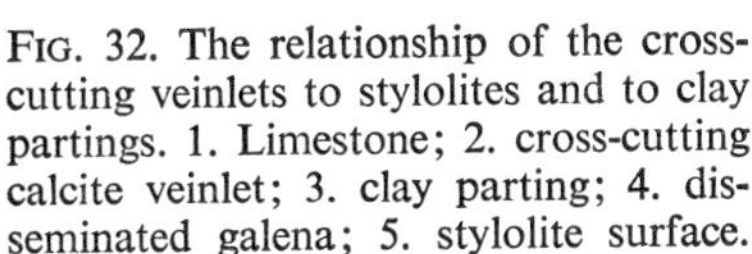

ore mineralization, and these are confined to fracture zones passing through the ore-bearing horizons. Elsewhere the veinlets are barren of ore.

Veinlets of the third group are found in wide faulted zones with strongly crushed rock, such as that 190 m. wide in the Chaarkuduk deposit. Within bedded ore bodies this zone is ore-bearing. As a rule it contains mineralized lenses and marked impregnations of galena, together with veinlets and pockets of calcite and barite.

In the barren limestones between ore deposits only occasional impregnations of galena occur. At distances exceeding 10–15 m from the bedded ore bodies no galena is found.

From comparison of these examples it is clear that the structures and textures of the Dzhergalan deposit point to a long history of ore growth. Dust-like disseminations of galena clearly represent the primary form of the

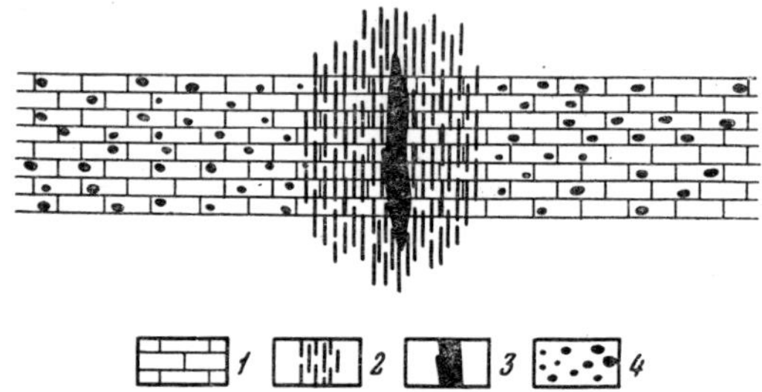

FIG. 33. The relationship of the disseminated lead in a bedded body and a galena lens in a cross-cutting zone. 1. Limestone; 2. cross-cutting fractures; 3. galena lens; 4. disseminated galena.

ore, only slightly redistributed during diagenesis. Restriction of impregnations and pockets to stylolite surfaces is regarded as the second stage of Pb and Zn redistribution, probably during katagenesis. The process occurred along stylolite surfaces, which had developed earlier (during diagenesis, according to G. P. Bogomazov). The veinlet mineralization has been associated with the Alpine epoch, i.e. it occurred during late katagenesis.

A characteristic feature of the Dzhergalan deposit, common to all the Central Asian deposits (Mirgalit-sai, Kalkan-ata, Western Balkhash), is the invariable presence of barite. *Since barite is by no means characteristic of marine calcareous deposits of arid zones, it may be concluded that hydrothermal waters may have contributed to the Central Asian deposits. These need not necessarily be associated with magmatic intrusions; they more probably influenced the redistribution of the ore masses, the recrystallization of limestone adjacent to the individual deposits, and the introduction of barite into these deposits.*

The source of the barite may have been effusive sedimentary sequences and, in particular, arkosic sandstones underlying the ore horizons. In some places, such as the Sumsar deposit, the ore has been displaced for considerable distances by hydrothermal solutions, and is concentrated in the anticlinal structures, leaving the synclines barren (Lur'e 1960).

In passing from simplest geologic conditions to the more complex, a parallel trend of growing complexity at ever-increasing intervals of time may be recognized in the formation of the ore bodies. The katagenic stage becomes increasingly important and in ore formation features may be recognized suggesting a growing hydrothermal influence, not only in the redistribution of ore but also in the emplacement of a new constituent, such as barite.

By comparing the mode of occurrence of the ore bodies and the morphology, structure, and texture of copper and lead-zinc deposits it is clear that the two have much in common. In addition, the primary sedimentary-diagenetic features in copper ores have been more fully and clearly preserved than in Pb-Zn ores, whereas the secondary features of redistribution resulting from tectonic influences (veins and veinlets) are more weakly expressed. In other words, copper ores are less susceptible to secondary influences than are Pb-Zn ores. It seems likely that this reflects to some extent the lower geochemical mobility of copper than that of lead and zinc. Further, since the secondary structures of sedimentary Cu-Pb-Zn ores are very similar to the structures of hydrothermal deposits of these metals, it is natural that Pb-Zn bedded ores have been more usually regarded as hydrothermal than have copper ores. Whereas cupriferous sandstones in many deposits, for example, have long been recognized as of primary sedimentary origin, only in the last ten years have bedded deposits of Pb-Zn been established as being of sedimentary origin. The leading role in this interpretation was played by Soviet geologists, including the late M. M. Konstantinov.

5. Interrelationship between Cu and Pb+Zn Concentrations in Time and Space

Since lead and zinc are ubiquitous in cupriferous sandstones, sometimes in considerable quantities, and copper is always present in lead and zinc ores, their interrelationship in space and time is an outstanding problem. Is there any systematic pattern in the distribution of these elements? This question will be studied first with regard to the Mansfeld Shales of Germany, which are of outstanding importance.

The Mansfeld Shales represent one of the basal facies of the Lower Zechstein basin of Germany, which formed a gulf of about 20,000 km² in the area now bordered by the Rhenish Schiefergebirge, the Thuringian Forest, the Harz Mountains, and the Flechtingen Mountains (Fig. 34). Throughout the area the Upper Rothliegende underlies the Zechstein. Its surface is irregular, with elevations up to 10–15 m in height, separated by depressions. In the Mansfeld trough the elevations have been considered as the relics of ancient sand dunes, spreading from northwest to southeast in the basin; the depressions are the inter-dune hollows (Gillitzer 1935; Kautsch 1942; Eisenhut and Kautsch 1954). The oldest Zechstein deposits resting on this irregular surface form the impersistent *Zechstein conglomerate*, (2–3 cm), a fine

E

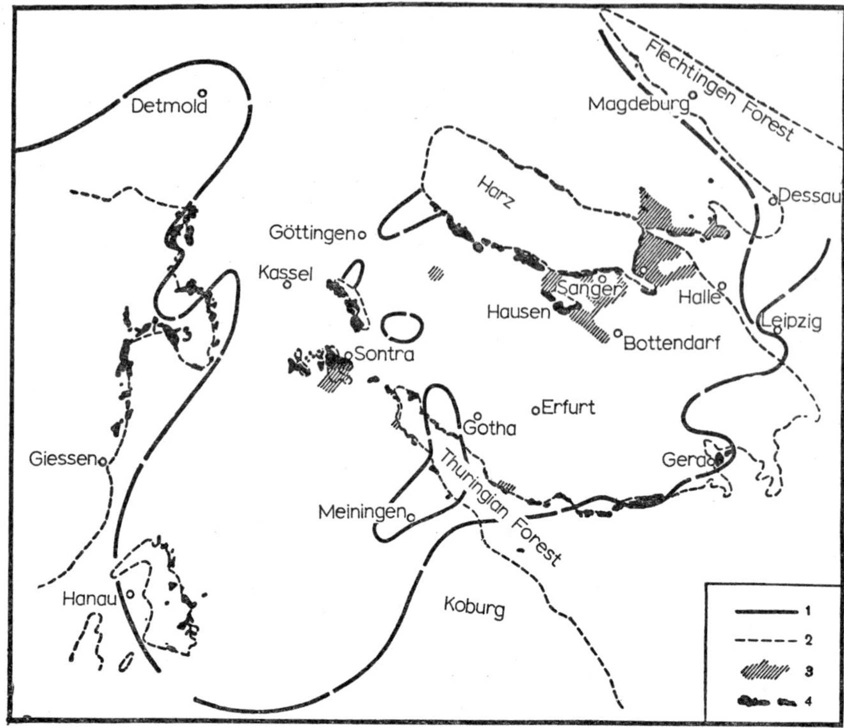

FIG. 34. Paleogeography of the basin (from A. Schüller). 1. Margin of sea; 2. boundary of Hercynian basement; 3. deposits encountered in boreholes; 4. exposures of the Zechstein.

gravel in a sandy matrix. Where this conglomerate is absent, the overlying beds rest on the bleached surface of the Rothliegende (the so-called Weissliegende).

The conglomerate is succeeded by the Mansfeld Kupferschiefer, black, bituminous, and rich in organic material, at most 32–48 cm thick. Despite its thinness, the unit varies sharply in petrographic character from bottom to top, and is divided into specifically named horizons, which vary in different parts of the basin (Table 3, after Eisenhut and Kautsch, 1954).

TABLE 3

Subdivisions of the Kupferschiefer (thicknesses in cm)

Mansfeld basin	Sangerhausen basin
Dachklotz, 20–30	
Schwarze Berge, 12–17	Noberge und Unterwald, 22–30
Köpfen, 0·8–1·2	
Grauer Kopf, 4–6·5	
Schwarzer Kopf, 2·5–4	Schieferkopf, 10
Kammschale, 2·6–4	Blattschiefer, 5–6
Grobe Lette, 4–6	Schramchiefer, 5–6
Feine Lette, 2–5	Erzschiefer, 5–6

Changes in composition parallel changes in the petrographic character of the rock.

The clastic component, which is abundant in the basal shale layer (up to 35% SiO_2), decreases upwards with increase in the carbonate (Fig. 35). C_{org} reaches 10% in the lower part, decreases upwards, in places to as little as 0·45%. In general the shale is markedly rich in organic material, being, in the lower part at least, sufficiently rich to form a bituminous shale.

The facies character of the shales is not completely clear. From the fineness of the clastic fraction (clay with a small amount of silt) and by the preservation of delicate bedding, it seems clear that the deposits formed in relatively deep parts of the basin, in any case near a zone of turbulence. Estimated depths range from 10–15 to 50 m, and only I. W. Pompecki (1921) considered the basin to be deep, such as the Black Sea. This view, however, has received little attention. Unfortunately, the inshore facies to these shales is rarely preserved. Only where rocks are increasingly sandy around the periphery of the shale area is there any indication that towards the shore the shale facies gave

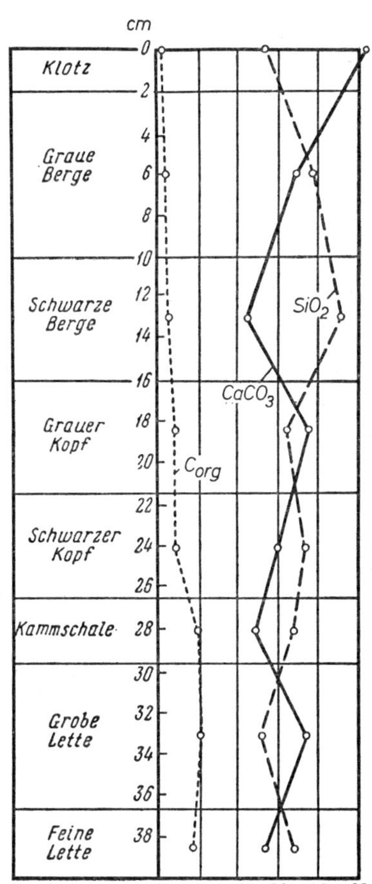

FIG. 35. Distribution of $CaCO_3$, SiO_2, and C_{org} in the Mansfeld Shales (from Kautsch).

way to normal silty, and then to sandy, deposits forming a belt of variable width. A more interesting facies transition of the shales is to red beds, the so-called Rote Fäule. This is a distinctive shale, and is differentiated from the normal type by its red color and its lower density. The transition is gradual: first red spots appear in the black shale, then increase in size and number, finally merge into a mass of red rock. The Rote Fäule clearly belongs to the margin of the marine basin, along which it forms broad, discontinuous belts, though sometimes of large size. Areas of Rote Fäule are sometimes found in the middle of the black shale deposits (Fig. 36). According to Gillitzer, the Rote Fäule facies represents areas where fresh river water flowed into the Zechstein marine basin.

After the appearance of I. W. Pompecki's paper (1921), in which the Kupferschiefer basin was compared to the Black Sea, the idea of a "cupriferous shale sea" as a basin contaminated with hydrogen sulfide on its floor was developed. Despite the spread of this view, it must be noted that it does not correspond to the facts. Among the organic remains found in the cupriferous shale, besides numerous fish, cephalopods, and terrestrial plant fragments, *shells of unquestionably bottom forms* occur: one echinoderm species, four species of bryozoans, and ten species of pelecypods. These forms could not have existed if the bottom water had been contaminated with hydrogen sulfide. The same conclusion is suggested by the occurrence of Rote Fäule in localities *far from shore, in the midst of the black rocks.*

It is important to note also that the Kupferschiefer was influenced by the irregular surface of the Weissliegende, upon which it was deposited. Thus, its thickness increases in hollows and decreases over elevations on this floor (Fig. 37). In consequence, some horizons in the succession, specifically the upper part (Schwarze Berge, Dachklotz, Fäule), are absent over the hills. It is clear that, when these shales were formed, the prominences were at the water-level in the zone of turbulence, which inhibited the accumulation of fine-grained sediments.

Since this study is concerned with sedimentation in a shelf sea in the "cupriferous shale basin", where the typical average yearly sedimentation rate is 0·02–0·03 mm, it follows that the time for accumulation of the Kupferschiefer was very brief, lasting only about 30,000 years. The duration of deposition of the individual members of the shale, on this reckoning, was extremely short: 1400 years for the Feine Lette, 2000 years for Grobe Lette, 1320 years for Kammschale, 3400 years for Schieferkopf, 400 years for Köpfen, 7000 years for Schwarzer Kopf, and 15,000 years for Dachklotz. In terms of geologic time, *the interval for accumulation of the Kupferschiefer represents only a geological instant.*

The scale of mineralization, though restricted to a unit negligible in thicknesses but spread over considerable area, is indeed impressive. "We may compute the copper content with some degree of accuracy for an area of about 500 km^2 if we use data from mining and drilling operations. This area contains about 5 million tons of metallic copper. If we now take as an average

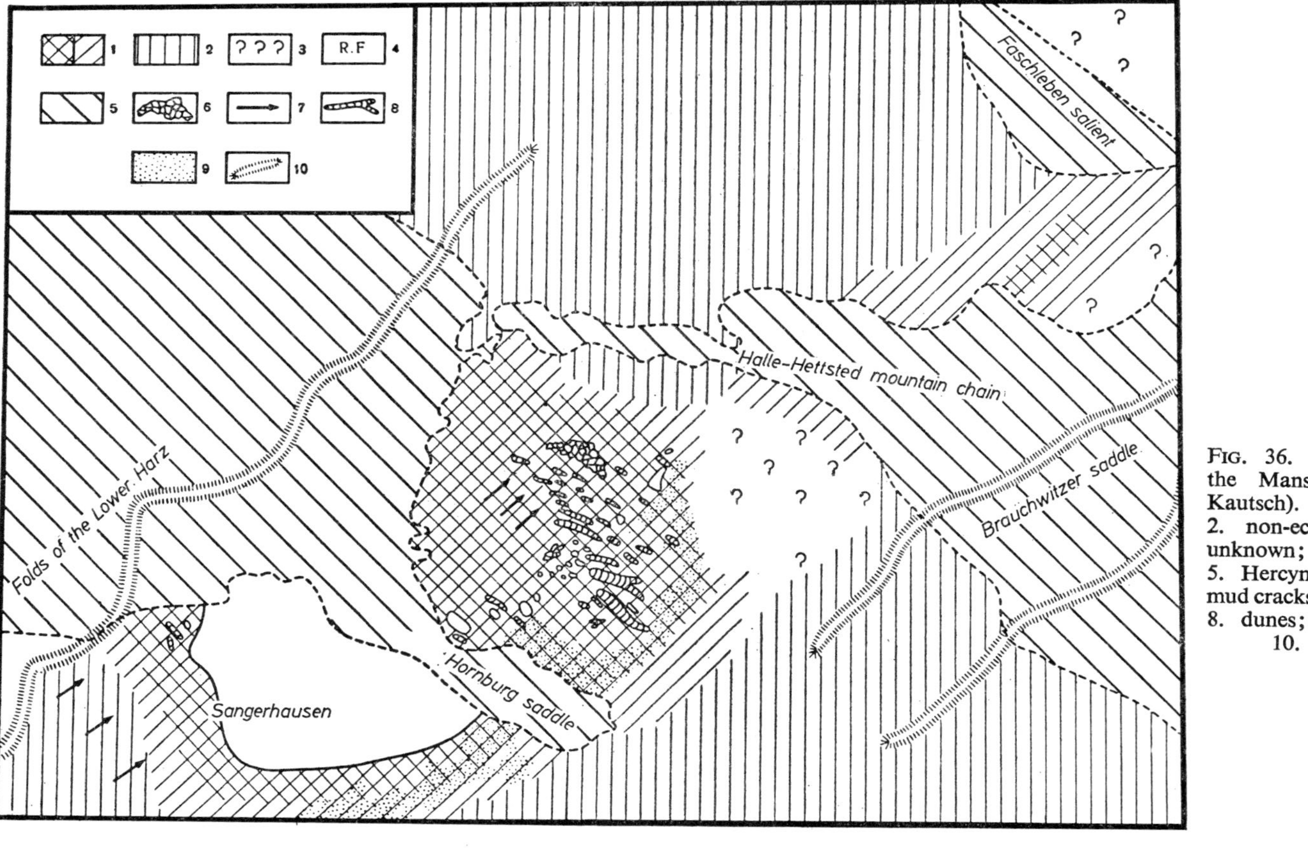

Fig. 36. Paleogeography of the Mansfeld basin (from Kautsch). 1. Economic beds; 2. non-economic beds; 3. unknown; 4. Rote Fäule; 5. Hercynian structures; 6. mud cracks; 7. wind direction; 8. dunes; 9. conglomerate; 10. hilly zone.

but one-fourth of this value, we obtain about 50 million tons of copper for the area between Upper Fulda and Flechtingen Mountain, an area of approximately 20,000 km². Correspondingly, the same area would contain about 200–250 million tons of zinc and about 100–150 million tons of lead" (Richter 1941, p. 53). These calculations are admittedly approximate, but they clearly show not only the vast size of the deposit but also its complex polymetallic nature, with the dominant development not of copper, but of lead and zinc. The polymetallic nature of the deposits makes it of special significance for studying the spatial relations in the emplacement of the various metals.

It is widely accepted that *the mineralization of the Kupferschiefer is undoubtedly primary, more precisely sedimentary-diagenetic.* This clearly appears from the forms in which the ore minerals are found. The cupriferous shales

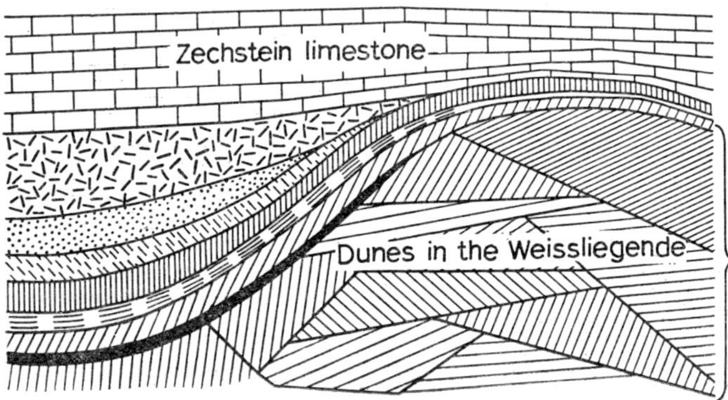

FIG. 37. Cross-section through the Mansfeld shales (after Kautsch).

are typical disseminated ores. Macroaccumulations are distinguished by disseminations and lineal ore: Erzlineale. The *disseminations* are rounded bodies and lenses of bornite and chalcocite, fractions of a millimeter thick and up to several millimeters long. In places sphalerite and galena occur. An Erzlineale is nothing but a thin accumulation of disseminations, elongated in the bedding plane. They may extend for several centimeters, and may be several millimeters thick.

Together with these relatively coarse forms of ore, a vast number of microsegregations of ore also occur, of which H. Schneiderhöhn (1921) has recognized four types, as follows:—(i) "Ore points," having a maximum diameter of 0.1μ and composed of chalcopyrite. These are scattered in large numbers through the clay-bituminous substance of the shales. (ii) Rounded or rod-shaped bodies 4–8μ in diameter and up to 20μ in length. These represent aggregates of numerous rounded bodies of the first type. They consist of chalcopyrite. (iii) Rounded, elliptical, and compressed bodies of the same dimensions as those in the second group, but composed of bornite or chalcocite and without any indication of their internal structure. This type is associated by gradual transition with disseminations and, through these, with Erzlineale. (iv) Similar to the third, but distinguished by somewhat larger size and by rims of very small chalcopyrite crystals.

The relation of ore minerals to carbonates is characteristic. Calcite, dolomite, and siderite are found in the rock in the same type of rounded and lenticular bodies as the ore. The ore minerals locally form thin crusts around carbonate segregations, or replace the segregations in varying degrees, sometimes completely. It is extremely important to note that *the adjacent rock appears to be moulded round the ore-bodies, adapting itself to their form.* This means that the *ore-bodies and the carbonate segregations formed very early, when the sediments were unlithified and plastic, and when they could adapt themselves to the solid fragments.*

H. Schneiderhöhn (1921) believes that his second type of ore microsegregations are mineralized bodies of sulfur bacteria, and the "ore points" may even be mineralized aggregates of sulfur, formed by sulfur bacteria, but extending beyond the bacterial cell. A. Schüller (1957) considers the carbonate bodies and lenses to be remains of calcareous algae. In a manner of speaking, a connection has been established between mineralization and organisms in the Zechstein; which emphasizes the syngenetic character of the mineralization in the host rocks. It should be stated, however, that the interpretations of Schneiderhöhn and Schüller are entirely speculative, with no reliable basis in fact. It would be more accurate, on analogy with similar segregations of diagenetic minerals in modern sediments, to consider the ore-bodies and carbonate lenses as inorganic, formed during early diagenesis, during microredistribution of authigenic minerals (see Vol. 2, Chapter 10). Whichever genetic interpretation of ore dissemination and microsegregation proves to be correct, one matter is certain: the ore accumulations are extremely early features, and it is possible that the *mineralization of the Kupferschiefer is of primary sedimentary origin, and not epigenetic.*

For the sake of completeness, it may be pointed out that west- and northwest-trending fractures developed in the Kupferschiefer in Late Cretaceous and Tertiary times. These were then filled with calcite and barite, and they include several ore associations: Bi-Ni-Co the earliest, U-Mo-Re later, and Cu-Pb-Zn last. The ore elements in the latest association were carried somewhat beyond the limits of the veins, which led to some slight enrichment in Cu, Ag, Zn, and Pb of the adjacent host rock. This was restricted to rock so near the veins that for practical purposes there was no re-distribution of primary mineralization. This is of primary significance.

The distribution of Cu, Pb, and Zn over the general region of the Kupferschiefer is shown on the geochemical maps (G. Richter and E. Kautsch 1941-42). High, and very high, copper values in the shales (>8 kg/ton) occur in four restricted localities: the Mansfeld basin (between the southeastern spurs of the Harz), in the Sangerhausen basin (south of the Harz), in the vicinity of Riechelsdorf (northwest of the Thuringian Forest ridge), and at Kassel and Grodizberg, along the northeast border of the same ridge (Fig. 38).

Outside these districts, i.e. throughout most of the region where the shales occur, the copper content is very low (<5 kg/ton).

In relation to the paleogeography of the Kupferschiefer basin and to the structure of its floor, it may be seen that *the regions of strong copper concentration lie in the near-shore zones of the basin, associated partly with the prominences on the Variscan basement (near the Harz and the Thuringian ridge), partly with the marginal zone of the lower-lying Central Germanic uplift*, which then appeared as relatively large islands and shoals. This distribution is important,

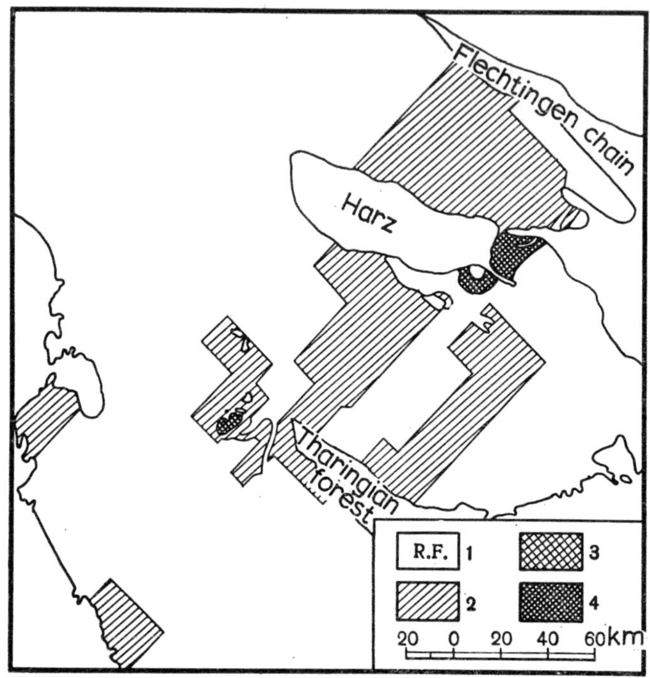

FIG. 38. Distribution of copper in the district of Mansfeld shales (from Kautsch). 1. Rote-Fäule; 2. 0–5 kg/m²; 3. 5–12 kg/m²; 4. 12 kg/m² and more.

because it points to *the rather limited geochemical mobility of copper in the physico-geographic environment of the Zechstein basin.*

A number of more distinctive patterns, controlling the high concentrations of copper, may be noted in the inshore zone. "It has been established," wrote E. Kautsch (1942), "that in zones nearest to the elevations* there are no copper ores in the Kupferschiefer of economic value. The barren inshore zone (taube) gives way, towards the interior of the basin, to a belt richer in copper and somewhat farther removed from the shore, and this again gives way, seawards to an area with rich copper mineralization."

Thus, *though the copper deposits of the Zechstein basin may generally be*

* Near the shore line—N. M. Strakhov.

considered inshore deposits, they are still somewhat removed from the shore line itself.

An essential fact is the well-defined relationship between basin-floor topography in the copper-bearing belt and the copper concentration in the sediments. Not only does the thickness of the Kupferschiefer decrease over the hilly features, and increase in the hollows, but the copper content also varies sympathetically. This is illustrated by G. Richter (1941) (see Fig. 39*A* and *B*). The first of these shows the floor of the Kupferschiefer. A narrow belt with highly irregular borders occurs at Kornburg, in which the base of the shales rests on Rothliegende conglomerates (Fig. 39*A*). The Lower Zechstein thickens markedly into the trough; which indicates a subsiding character of the belt at the beginning of Zechstein time. The maximum distribution of copper in the Kupferschiefer and the upper Sanderze horizon corresponds exactly to this zone. It may therefore be concluded that *in the inshore zone of potential copper deposition the concentrations of copper ore show clear preference for negative elements of basin floor relief, irrespective of their origin.*

An even more significant relationship in copper mineralization is found in the zones of Rote Fäule. Of themselves, these red facies prove only very low, uneconomic copper values, the only ore mineral being bornite (Hoffman 1924). However, the zones of black cupriferous shales immediately adjacent to the Rote Fäule usually show high copper values, forming rich ore, as in the Sangerhausen basin (Fig. 40). The inshore sandy zone adjacent to the Harz is barren of copper ore. The succeeding Rote Fäule zone also carries no ore, but the black shales immediately west of the Rote Fäule prove maximal copper values—more than 10 kg/ton. The same relationship occurs elsewhere. *The preference of copper-ore concentrations in contact zones of the red and black facies of the Kupferschiefer is highly characteristic.*

Thus, *the inshore position of the copper belt near Hercynian massifs, the preferred accumulation of ore in negative topographic elements along the belt, the reduced deposition on positive elements, and, lastly, the preference of ore concentrations for the zone where red and black facies join—these are the basic patterns of copper accumulation in the Lower Zechstein cupriferous deposits of Germany.* The same relationships are observed in the distribution of silver (Fig. 41*A* and *B*).

The distribution of lead and zinc is substantially different. *Lead accumulation clearly lies relatively seaward of copper, and zinc increasingly so* (Figs. 38 and 42). In other words, Cu-Pb-Zn *form an ore triad in the Zechstein basin, within which Cu shows the lowest geochemical mobility, lead higher, and zinc the highest.* Three ore facies within the Early Zechstein basin may therefore be recognized, each spatially defined: copper, lead, and zinc.

The vertical distribution of the three elements through the Kupferschiefer is also distinctive. G. Richter 1941 has shown that in all major areas where the cupriferous shales occur, in specific parts of the Riechelsdorf district, the Riechelsdorf district as a whole, the Mansfeld basin as a whole, the Sangerhausen district, and Kassel in Lower Silesia, the vertical distribution is

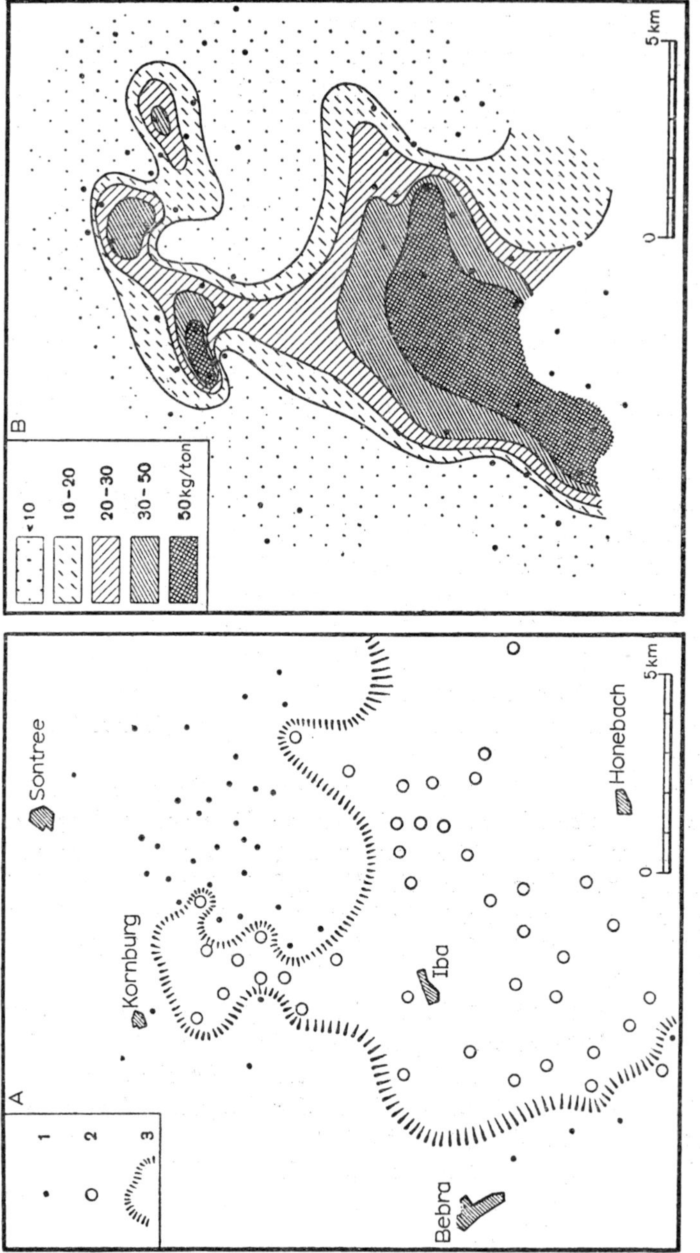

FIG. 39. Facies and copper distribution in the Riechelsdorf Mountains (from G. Richter). *A.* Facies: 1. Kornburg dune sand; 2. conglomerate; 3. boundary of basin. *B.* Copper content.

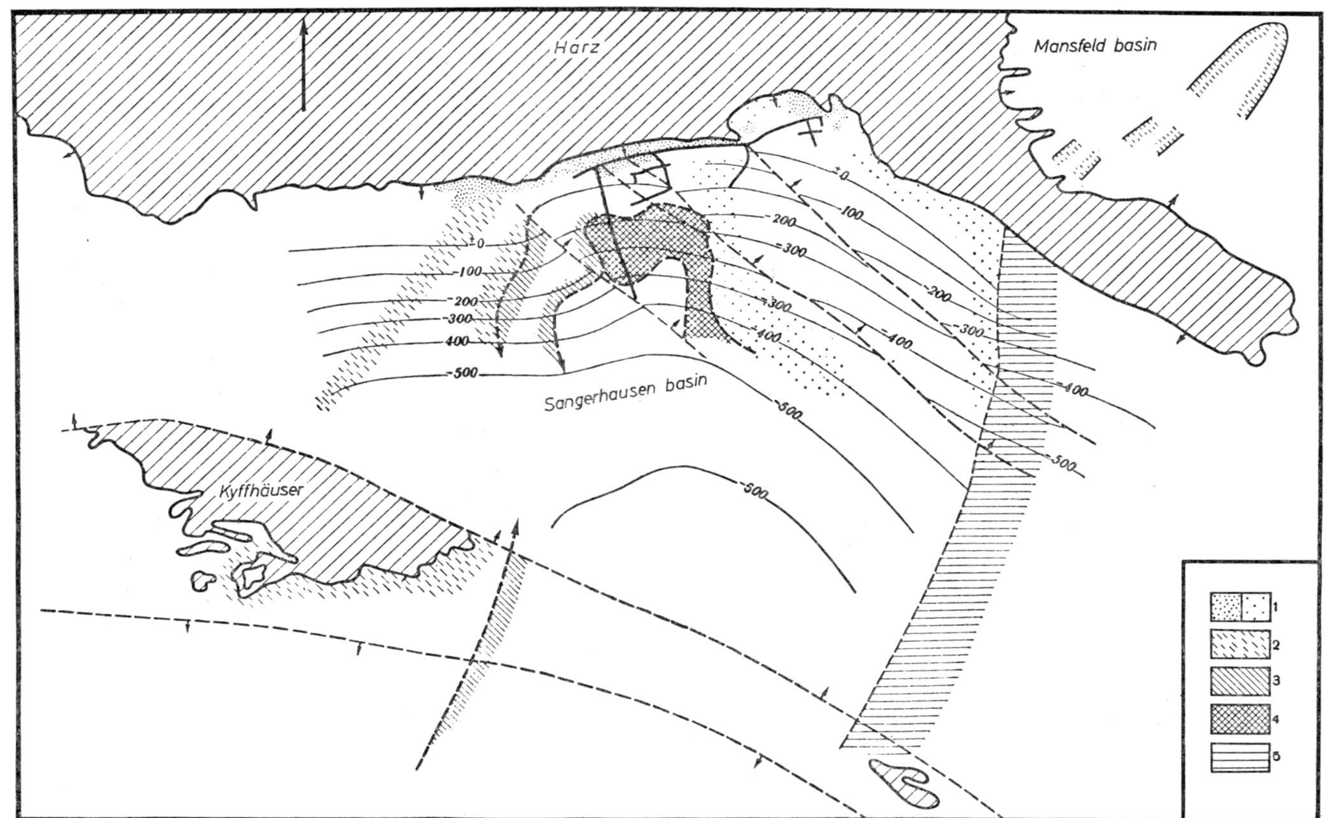

FIG. 40. Distribution of copper in the Sangerhausen district (from Hoffman). Copper content (kg/m² Cu): 1. from 0 to 8; 2. from 8 to 10; 3. over 10; 4. Rote Fäule; 5. open-sea facies.

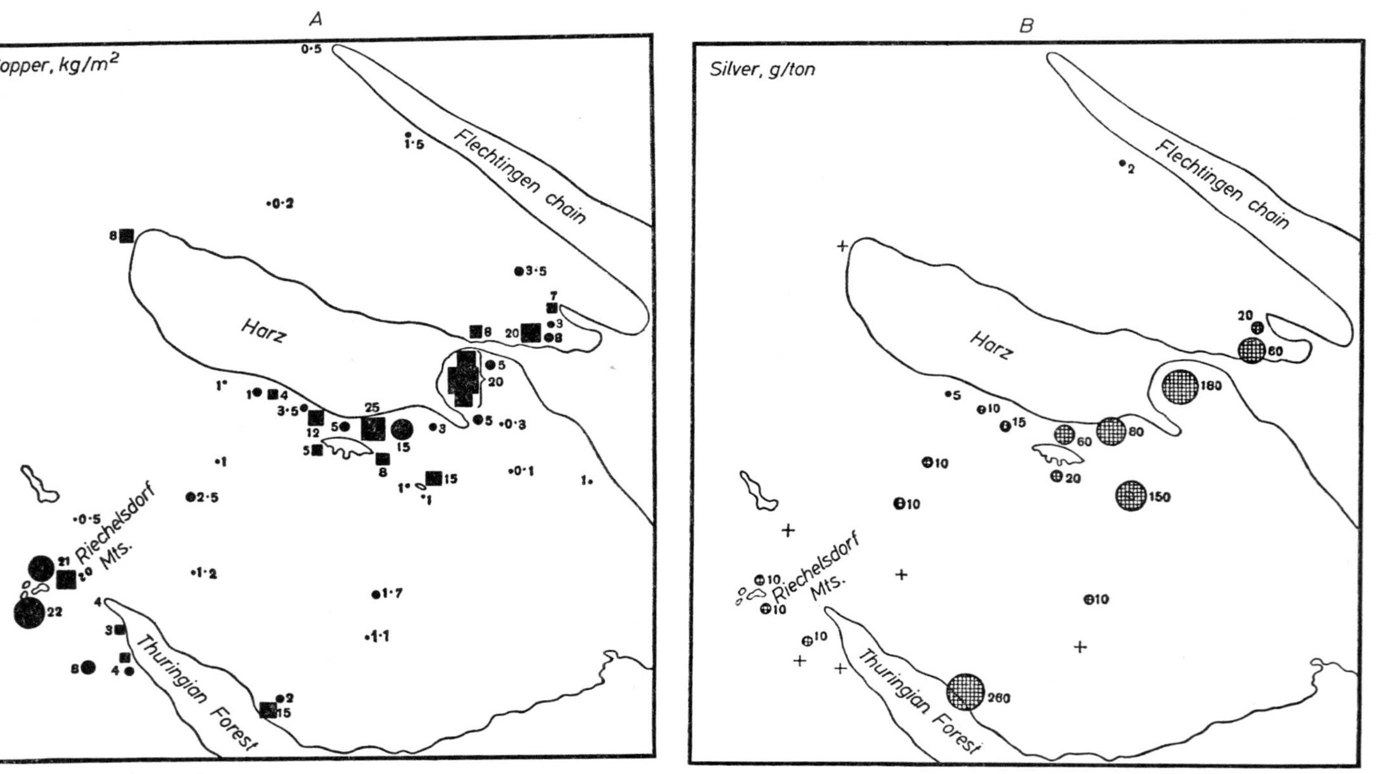

Fig. 41. Distribution of copper (*A*), and silver (*B*), in the Mansfeld shales (from Richter).

A

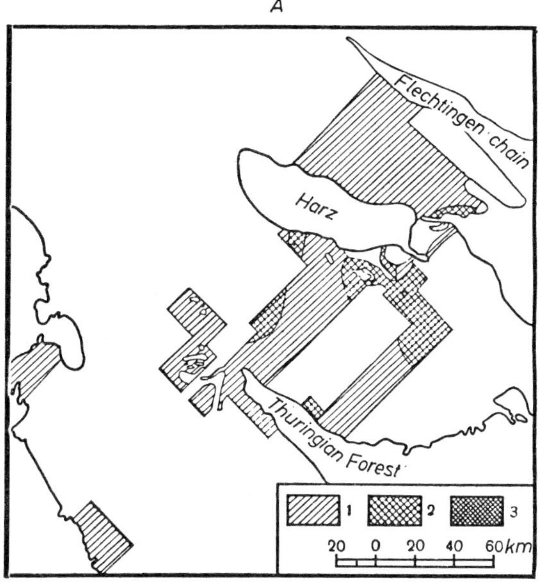

B

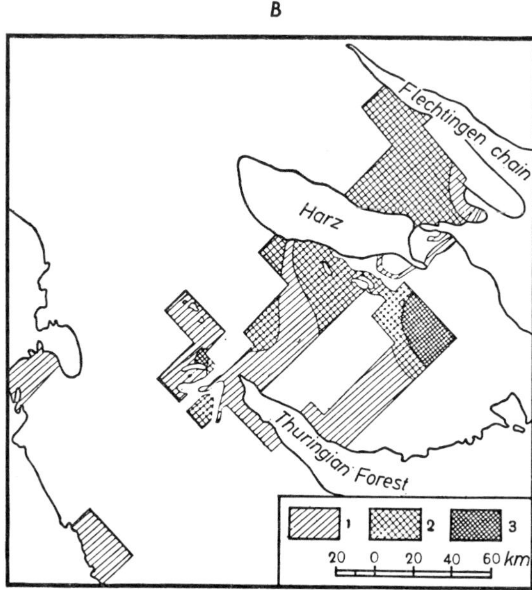

FIG. 42. Distribution of lead (*A*), and zinc (*B*), in the Mansfeld shales (from Kautsch). 1. 0–5 kg/m²; 2. 5–30 kg/m²; 3. 30 kg/kg/m² and more.

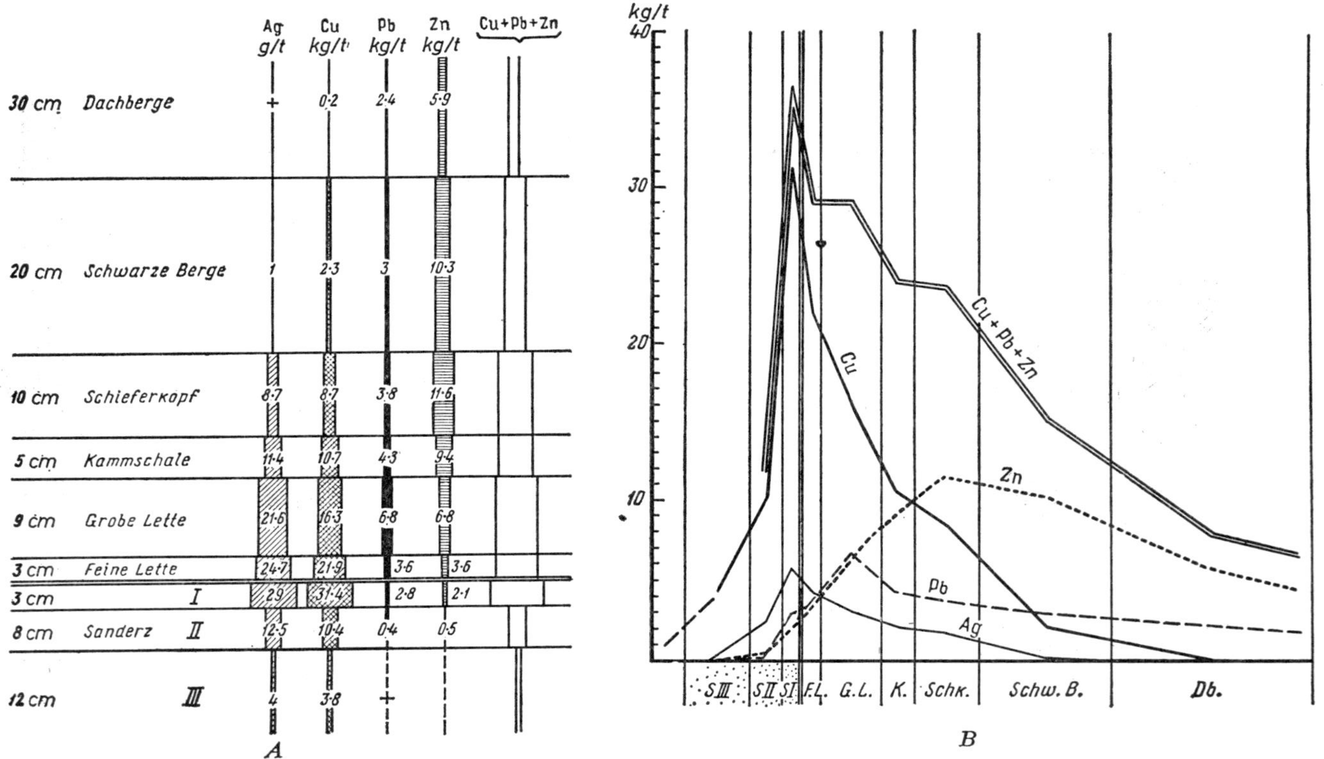

FIG. 43. *A.* Distribution of metals in a section at a deposit in the Riechelsdorf Mountains (from G. Richter). *B.* Distribution of metals from averaged data for all drilling in the Riechelsdorf Mountains (from G. Richter).

fundamentally uniform (Figs. 43 and 44): the copper reaches maximal concentration in the lowest shale horizon, the Feine Lette (in places even in the underlying Weissliegende); lead is most abundant in a higher horizon, the Grobe Lette; and zinc is concentrated chiefly in the upper part of the Schwarzer Kopf shale member. There *the more mobile an element is geochemically, the higher the shale member in which it is deposited.*

Since the absolute time interval for deposition of the Mansfeld shales was negligible, *the processes of Cu-Pb-Zn distribution provide a clear example of chemical differentiation within the basin. The vertical distribution of these elements also demonstrates chemical differentiation with time, conditioned by successive changes in composition of the aqueous solutions reaching the basin.*

In any attempt to apply the distributional pattern of elements in the

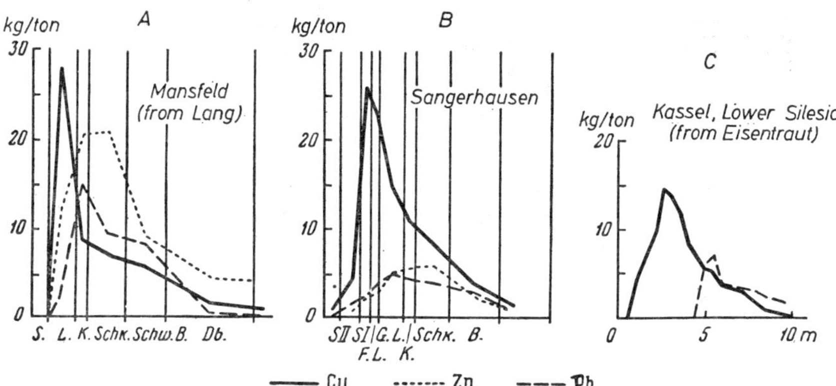

Fig. 44. Distribution of metals: *A.* in Mansfeld shales; *B.* in cupriferous shales of Sangerhausen; *C.* in cupriferous marls at Kassel, Lower Silesia (from Richter). (*S*) sandy ore; (*F.L*) Feine Lette; (*G.L.*) Grobe Lette; (*L.*) Lette in general; (*K.*) Kammschale, (*Schk.*) Schieferkopf; (*Schw. B.*) Schwarze Berge, *B.* Rock in general; (*Db.*) Roof beds.

Kupferschiefer of Germany to other regions, it is necessary to recognize two phenomena: those where Cu-Pb-Zn accumulations form bodies intimately related to each other, grading one into the other, and those deposits which are spatially and stratigraphically separated from each other.

The latter is much more widespread. It clearly indicates that the supply of ore constituents to each deposit was independent of the others, and differed sharply with time. This circumstance essentially solves the problem concerning the genetic relationships between Cu ores on one hand and Pb-Zn ores on the other. However, where the ores are intimately associated, a single common source for the ore constituents is undoubted, the relationships of the various ore elements in time and space are of first importance. Unfortunately no sedimentary polymetallic deposits have been studied in such detail as the Kupferschiefer of Germany. Some valuable observations have, however, been made elsewhere.

According to L. I. Ivankov and others (1957), in the Dzhezkazgan deposit "the vertical distribution in polymetallic mineralization is, from bottom to top: copper→lead→zinc". The same sequence is apparent in plan. The restricted position of silver in the series of base metals is noteworthy. Semi-quantitative data from spectrograph analysis and assays indicate that the

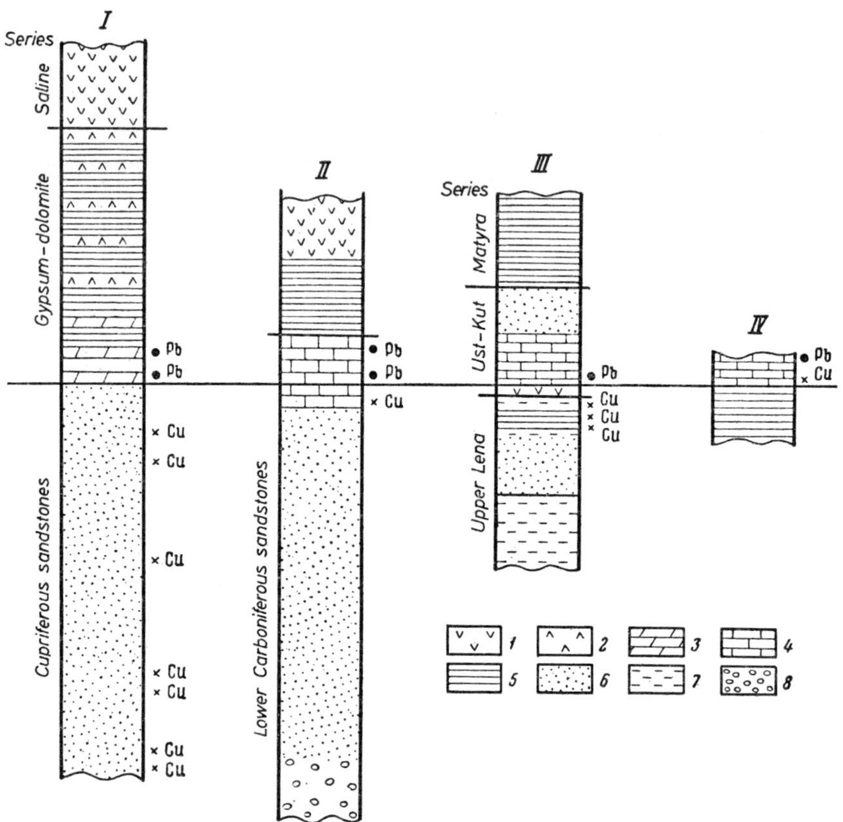

FIG. 45. Distribution of Cu and Pb. *I.* The Artemovsk basin; *II.* Dzhergalan region; *III.* Lena region; *IV.* Great Bogdo. 1. Saline deposits; 2. gypsum; 3. dolomite; 4. limestone; 5. clays and shales; 6. sandstone, 7. marl; 8. conglomerate.

silver concentration in chalcocite-bornite ores is 4 or 5 times that in chalcopyrite-pyrite-galena ores. Horn silver was found in a zone of purely chalcocite ore within quartz-calcite veins. It is clearly not associated with any supergene processes (Petro-8 deposit).

"Individual analyses on the silver of mineralogically pure chalcocite and galena show that the concentration of silver in chalcocite is more than 10–15 times that in galena. Thus, silver and lead are clearly separated from each other in the Dzhezkazgan ores" (Ivankov 1957, pp. 261-262).

All this closely resembles the relations observed in the Mansfeld deposit.

The distribution of copper, lead, and zinc in sections from the Artemovsk basin, the Lena region, the Dzhergalan region, and Bogdo Mountain follows the same sequence, with *copper restricted to the lower part of the ore-bearing sequence, lead and zinc to higher horizons*. The copper is concentrated chiefly in fragmental rocks, mainly sandy clays; the Pb + Zn deposits are in carbonates. The vertical distance between the different ore-bearing horizons also varies markedly.

Again, fundamentally the same relationship is repeated, but there a vertical shift in mineralization is also reflected in a lateral shift of the same ore facies.

The spatial relations between Pb and Zn in Pb–Zn deposits proper are generally complex. In the Upper Silesian Middle Triassic basin, according to M. M. Konstantinov (1959), "their separation from each other, both vertically and areally, is very clear. In the principal ore bed sphalerite is concentrated in the middle and lower part, and galena in the upper. This distribution may be explained in great measure by secondary redistribution. The trend of areal distribution is such that zinc is chiefly concentrated in the Bytom syncline (Fig. 29), and lead in the Tarnowice syncline. The Khshanuv, Dlugoshina, Charnov-Zhurad synclines occupy a middle position, but the Sevezh-Olkush zone carries abundant lead." In other deposits the Pb-Zn relationships are different: Pb is concentrated towards the bottom of the bed, and Zn towards the top. It must be emphasized that *in comparison with lead, zinc is generally much more widespread, both stratigraphically and areally*. This is apparently because of the greater geochemical mobility of zinc.

From the foregoing, it is clear that some degree of chemical differentiation in space and time takes place among elements which travel in multicomponent solutions. Polymetallic deposits of Cu-Pb-Zn represent graphic examples of the process.

6. The Origin of Cu-Pb-Zn Deposits in Formations of Arid Zones

The processes of distribution and deposition of Cu-Pb-Zn in arid regions will now be considered, and particularly the sources of the metals, the forms of their transfer and precipitation, the development of authigenic minerals, and relationships between these metals in space and time.

The sources of Cu-Pb-Zn and the accompanying elements may only rarely be determined, and then only in a general way. The cupriferous sandstones of the Ural region have long been ascribed to erosion of the copper-sulfide belt of the Urals. The cupriferous Mansfeld shales were long related in origin to the granites and accompanying ore veins of the Harz, but the discovery by Schüller (1957) of fragments of propylitized volcanic rocks with traces of ore minerals extends the source rocks to include the volcanic rocks of the barren Rothliegende. The range of copper sources for the Dzhezkazgan deposit is still wider. "Within the Dzhezkazgan-Ulutau region, rocks of the ancient crystalline basement are known to contain numerous endogene deposits of non-ferrous and ore metals associated with Caledonian granitoidal

F

rocks, and in Devonian red beds sedimentary copper deposits of cupriferous sandstones similar to the Dzhezkazgan ores are widespread. The erosional products of these deposits were carried into the Dzhezkazgan-Sarysu basin and its peripheral depressions, including the half-closed Dzhezkazgan gulf. Late Caledonian adamellites and granite porphyries of the Eskulinskii uplift, containing copper (up to 0·16%), barium, and sulfur, were undoubtedly sources of these elements in the Dzhezkazgan ores" (Popov, 1959, p. 203). Still other possible continental source rocks for copper ores are the contemporary volcanics, such as those in the basins of the Dzhaksy-kon and Syurtu-su Rivers. The source of the lead for the Devonian deposits of the Tashkent region and of Sumsar lay in older Devonian volcanics of the adjacent regions, according to E. L. Abramovich and A. M. Lur'e.

These examples scarcely exhaust the range of deposits that may represent sources of metal, however generally. Usually, however, it is impossible to supply even this general indication, and some original hydrothermal copper deposit or some magmatic complexes enriched in these elements but now eroded are often mentioned. The characteristic feature of these views is, however, the same: *the necessity of primary Cu-Pb-Zn enrichment of the source rocks.* This idea is advanced because the average metallic content of rocks is low, and there is no mechanism in the sedimentary process that would concentrate these negligibly small fractions of ore into ore accumulations. The possibility appears even less likely in view of the opposing tendency —a decrease in ore concentrations as a result of mixing with additional clastic material. As for all microelements, the formation of sedimentary deposits of Cu-Pb-Zn therefore requires a preliminary and extensive, enrichment of the parent rocks by the ore components.

The form of the ore components during transport—whether in solution or by mechanical suspension—for long remained an unsolved problem. It now appears to the present author that *the different geochemical mobilities in the triad Cu-Pb-Zn, as expressed by the preference of copper for fragmental rocks and of lead and zinc for carbonates, as well as the spatial differentiation of these elements within polymetallic deposits, such as the Mansfeld shales and the Dzhezkazgan district, could appear only if the ore-forming elements migrated in the dissolved state.* If the elements were also transported in suspension, the proportion must have been negligible.

It is well known, however, that in arid zones chemical weathering is sharply restricted and is replaced by mechanical processes. The migration of Cu-Pb-Zn in solution is here only possible because of two favorable circumstances. Ores of Cu-Pb-Zn develop either during or soon after folding, against a background of tectonic activity in the ore-forming region. It follows that the *topographic relief is considerable and there must be some degree of vertical climatic zoning;* the area of sedimentation undoubtedly belongs to the zone of arid climate, but the drainage area may be humid to some extent (Strakhov 1946). *Ore components pass into solution in this humid zone.* The formation of these solutions is greatly facilitated by the fact that Cu-Pb-Zn occurred initially as sulfides rather than silicates, which are easily oxidized to yield relatively soluble forms. Thus, thanks to a fortunate association of circumstances, in an environment characteristic of polymictic deposition and generally unfavorable to transport in solution, some elements may be transported chiefly in the dissolved state. This is a specific characteristic of Cu-Pb-Zn ore emplacement.

Since solutions of Cu-Pb-Zn form by oxidation of sulfides, it is clear that the *initial chemical forms of migration are the sulfates*—$CuSO_4$, $PbSO_4$, and $ZnSO_4$. Copper and lead sulfates are unstable, however, and even in neutral or slightly alkaline environments they quickly hydrolyze, and in the presence of $CaCO_3$ change into the slightly soluble basic carbonates, which are then precipitated. The stability of copper sulfate is especially low. It may exist only at a pH below 6·63; at higher values it hydrolyzes and is precipitated. The solubility of basic copper carbonate is about 0·1–0·2 mg/liter. $PbSO_4$ is somewhat more stable, its hydrolysis beginning at pH > 7. The solubility of $PbCO_3$ is about 1·1 mg/liter. $ZnSO_4$ is the most stable of this group. It hydrolyzes weakly and only at pH > 7·5 (?), and the solubility is 40 mg/liter. This variation in stability of the sulfates determines the fate of the elements: copper is precipitated first in sediments, followed by lead and finally zinc. *Though the original solution was multicomponent and complex, during its transport chemical differentiation of the ore constituents took place to a greater or lesser extent.* It must be recalled, however, that this degree of differentiation is always limited: some of the Pb and Zn has precipitated with the Cu, and, finally, some Cu with the lead and zinc. During precipitation of the main ore components, co-precipitation of other associated elements occurs, partly by isomorphous replacement, partly by adsorption. These are silver, cobalt, arsenic, antimony, and other elements that passed into solution with the main ore components during weathering of the source rocks.

The facies type of ore that forms depends to a considerable extent on the length of the streams along which the ore solutions migrate. This is especially true of copper. In long rivers copper is precipitated in the alluvium; where streams are short, copper precipitation is shifted to the deltaic zone; but where streams are very short, the solid phase of the basic copper carbonate is transported into the basin, but even here is deposited in the inshore zone, forming marine ore. In general, *copper ores, by the very mechanism of their formation, are associated with relatively coarse-grained deposits; cupriferous sandstones and siltstones represent the principal type of such ores.* With the more mobile Pb and Zn, deposition in clastic sediments of the delta or the inshore zone occurs only when the streams are very long. Where streams are shorter, the basic lead and zinc salts are carried farther seaward, to be deposited with the carbonate sediments. In the arid ore triad, consequently, the relationship formed in the humid ore triad Al-Fe-Mn is repeated, especially with regard to Fe and Mn (see Vol. 2, Chap. 4).

The high rate of precipitation of the solid phases from solutions of copper sulfate is demonstrated by observations made at 10 km intervals on streams draining from a mine in pyrite deposits, i.e. on streams that closely resemble those that have given rise to the ancient deposits of Cu-Pb-Zn (Table 4).

Copper almost disappeared from solution in the stream water within 10 km of the point where mine waters seeped into the stream. This does not mean that all of it was deposited in the sediments within this distance. Solid phases coming out of solution are transported for some time in suspension before they become part of the bottom sediment. Furthermore, the alluvial sediments are deposited and eroded repeatedly by the meandering stream. As a result, copper deposits move much farther downstream than is possible according to the migrational capacity of copper-sulfate solution. Each successive redeposition of copper, however, unavoidably causes impoverishment. Although mechanical transport of ore generally extends the zone of ore accumulation

TABLE 4

Content of Cu and Other Components in Stream Water (in mg/liter)
(from A. F. Efremov)

Components	Water Sample No.					
	1*	2	3	4	5	6
pH	6·5–6·7	3·3	3·3	4·1	6·3	6·9
Free H SO$_4$	None	142·5	58·8	None	—	None
Dense residue	164	928	470	246	142	124
Residue after roasting	—	580	280	176	—	66
Sulphates	7·2	5·28	183	101·0	—	20
Iron	9·0	133·5	0·87	9·0	4·5	1·4
Copper	None	68·15	19·20	9·04	1·04	None
Oxidizability	99·3	9·9	8·2	11·2	—	15·0

* Water sample from above entrance of mine waters; remainder below.

in a stream, the process is still of limited range and fails to separate the copper ores from the clastic sediment. The same factors, *mutatis mutandis*, control the distribution of lead and zinc.

These conclusions concerning the non-uniform migrational capacity of the Cu-Pb-Zn group have been fully confirmed by observations on the distribution of these elements in the waters of polymetallic deposits of Kazakhstan. The Cu-Pb-Zn content in streams is here controlled by the pH of the water. "As seen in the figure (Fig. 46 in the present volume), the lead and copper content in the water clearly depends on the concentration of hydrogen ions. The control is less well defined for zinc. The lead content in waters of the ore bodies, other conditions remaining unchanged, is sharply reduced in slightly alkaline solutions—at pH $\geqslant$ 7·2. In such solutions, even when in direct contact with the ore, the lead content does not exceed the regional background, and at pH $\geqslant$ 8 it is reduced to mere traces undetectable by ordinary spectrographic methods. The highest lead content is found in acid waters with pH $\leqslant$ 5·5. The copper content in water of ore veins, other factors remaining unchanged, drops to the regional hydrochemical background at pH $\sim$7. The maximal lead content is found in acid waters with pH $\leqslant$ 5·5 and the greatest zinc values in weakly acid waters (pH 5·5–6·5). In zones of alkaline waters the zinc content drops in proportion to the increase of alkalinity, but reaches the background value only at pH $\geqslant$ 7·5. Thus, these elements . . . may accumulate in waters corresponding to the limits of their stability in the aqueous environment, as determined by the pH: copper in acid waters with pH < 7, lead in acid, neutral, and slightly alkaline waters with pH < 7·2, and zinc in acid, neutral, and alkaline waters with pH < 7·5" (Belyakova 1961, pp. 102-103). In other words, whatever the migrational form of Cu-Pb-Zn in stream waters, copper proves to be least stable and precipitates at pH > 7·0, lead at pH > 7·2, and zinc at pH > 7·5.

It has already been noted that the solubility of basic copper carbonate in stream waters is about 0·1–0·2 mg/liter (Yakovleva 1952); the solubility of

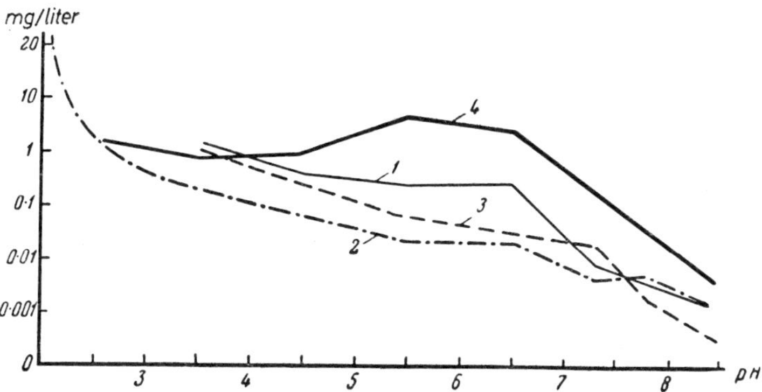

FIG. 46. Dependence of Cu, Zn, and Pb content, and of their total (Zn + Cu + Pb), on the concentration of hydrogen ions (pH) in water. 1. zinc; 2. copper; 3. lead; 4. total metals.

lead carbonate is about 1·1 mg/liter, and of zinc carbonate 40 mg/liter. Consequently, the Cu, Pb, and Zn content in stream waters at the time the above salts are precipitated must be greater than these values. The concentrations of these elements are many times those normally observed in streams. The possibility, therefore, must be borne in mind that such high concentrations of Cu, Pb, and Zn in streams could be obtained if the adsorption effect of clay colloids and organic material were considered. It may be asked whether such concentrations do occur in streams and sea water, considering that copper and lead have a toxic effect on organisms and also that organic remains are always found in copper and lead-bearing sediments.

For determining the role of adsorption as a potential limiting factor in the content of Cu-Pb-Zn ions, Freundlich's general equation of adsorption is critical:

$$X_0 = KC^{1/n},$$

where X_0 is the total quantity of adsorbed material, C is the equilibrium concentration of the solution, and n and K are constants.

The value of $1/n$ depends on the concentration of material C and so varies that, as the solution is diluted, $1/n$ tends toward unity, and when C increases, $1/n$ tends towards zero. Hence, it follows that in greatly diluted solutions the general adsorption equation may be expressed as $X_0 = KC$. In this case, consequently, the amount of adsorbed material increases proportionally with increase in concentration of the adsorbable ion. In strongly concentrated solutions the adsorption equation takes the form $X_0 = K$; that is to say, the amount of adsorbed component no longer depends on the concentration of the ion in the solution. In the transitional zone between greatly diluted and highly concentrated solutions, the increase in mass of the adsorbed ion is slower, and the rate becomes slower and slower as the concentration of the dissolved ion increases (Fig. 47). *It permits the solutions of minor elements to become gradually concentrated in natural waters despite the adsorbing activity of colloidal clay minerals.*

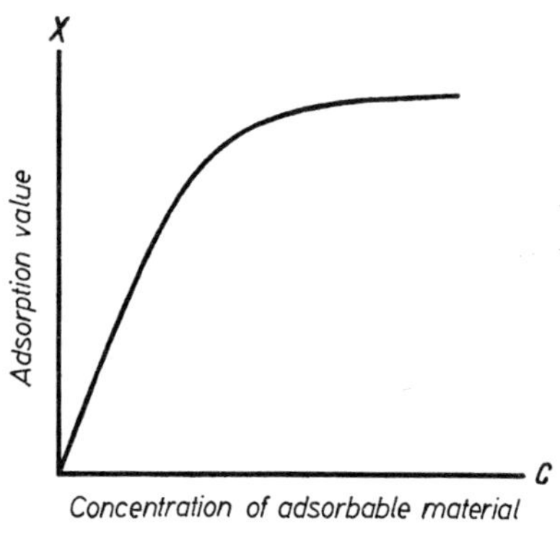

FIG. 47. Curve of the adsorption equation.

This has been clearly shown for copper (M. N. Yakovleva 1952). When the copper concentration in water is 50·0 mg/liter, 0·208 mg/equiv. is absorbed by loam, and when it is 174·8 mg-liter, 0·285 mg-equiv. is adsorbed. This by no means exhausts the adsorption capacity of 1200 mg of loam, which is equal to 0·480 mg-equiv.

Organic matter proves to have even less effect on the concentrations of Cu-Pb-Zn in the natural waters of arid zones. In a region of tectonic activity it occurs mostly in mechanical suspensions, i.e. in a physicochemically inactive state. Besides, as M. N. Yakovleva (1952) has shown, even dissolved organic material adsorbs Cu to a very small degree.

Thus, *neither adsorption by mineral colloids nor adsorption by dissolved organic material prevents the development of high, even very high, concentrations of Cu-Pb-Zn in natural stream and basin waters when their sulfates derive from a weathering crust on source rocks rich in the sulfides of these elements.*

It has been shown experimentally that Cu-Pb solutions appear to be toxic only at relatively high concentrations. With *Bacterium coli*, toxic effects of $CuSO_4$ appeared only at concentrations above 16 mg/liter. At lower concentrations the copper seemed to foster bacterial life rather than reduce it

(Schneiderhöhn 1921). Thus, toxic effects of copper appear at concentrations many times greater than that at which copper carbonate is precipitated in the sediment. In other words, *the geochemical processes of copper concentration and the formation of copper deposits take place without any suppression of the normal activity of living organisms; the processes may even stimulate such activity.* This is true also of Pb and Zn.

Thus, the weathering of Cu-Pb-Zn sulfides and of the sulfides of the associated metals in parent rocks that are to some extent enriched in these elements gives rise to an increased concentration of the elements in natural waters. The low stability of Cu and Pb sulfates and, in part, of Zn sulfates leads to hydrolysis and precipitation as basic carbonates. This process takes place first with copper sulfate, later with lead sulfate, and still later with zinc sulfate. This has resulted in spatial differentiation of the elements in basins and stream networks.

The basic Cu-Pb-Zn carbonates, buried in the process of sedimentation, are subsequently subjected to complex chemico-mineralogical change and redistribution, processes of great significance in forming deposits of the elements.

These chemico-mineralogical transformations involve the conversion of all oxides of Cu-Pb-Zn to sulfides, when chalcocite, bornite, chalcopyrite, covellite, galena, sphalerite, tetrahedrite, and related minerals form. *This universal and complete replacement of the oxides of heavy metals by sulfides is a characteristic diagenetic transformation.* Individual exceptions may be found in impurities of native copper and native silver, but their occurrence in the sulfide paragenesis is associated simply with the sulphide decomposition.

This remarkable tendency for the ores of heavy metals to pass into the sulfide, and only the sulfide form, with rare developments of oxide, is combined in this case with suppression of the independent sulfide form of iron, an element that generally shows an affinity toward these metals. Manganese is even more completely suppressed. Among the authigenic minerals of some copper deposits, such as the Mansfeld Shales, pyrite may be entirely absent, or found in negligible quantities. On the other hand, siderite, the chief form of authigenic iron in deposits of Cu-Pb-Zn, may be extensively developed. Alabandite is never found.

These authigenic mineral associations in ore accumulations are founded in the physicochemical characteristics of heavy metal sulfides, particularly in their very low solubility (Table 5).

The solubility values as determined by the various investigators diverge appreciably, probably because of the low solubilities and experimental difficulties. Nevertheless, the sequence of the various sulfides remains the same, and the results may therefore be used for comparison.

The solubilities of the sulfides of Cu, Pb, Zn, and other heavy metals are but hundredths or thousandths that of the basic carbonates of these elements —the form in which they enter the sediments. *This leads to the result that, as soon as H_2S or, more likely S^{2-}, begins to form in the waters of the bottom muds,* *as a result of sulfate reduction, the heavy metal sulfides begin to be precipitated and the more soluble carbonates are converted to these less soluble forms.** Since competitors for S^{2-} in the bottom mud water are numerous, we may descriptively speak of the "struggle for the sulfide anion". The "winners" *in this competition are those cations which yield the least soluble sulfides, and at a higher* Eh. Since Cu, Pb, Zn, and related cations yield with S^{2-} the least soluble solid phases and at not very low values of Eh, it follows that these cations attract practically the entire volume of H_2S generated. Thus, there is no surplus sulfide ion for the formation of the more soluble FeS, much less for MnS, and these sulfides are consequently entirely absent in Cu-Pb-Zn deposits, or sulfides such as pyrite may be present in minute amounts. Alabandite is entirely absent. Iron is found in such copper-iron sulfides as

TABLE 5

Solubility of Heavy Metal Sulfides in Distilled Water

Sulfide	From Bruner and Zawadzki 1931, mol/liter	From Krauskopf 1955, mol/liter	From F. V. Chikhrov, g/liter
MnS	$3{\cdot}8 \times 10^{-8}$	—	$3{\cdot}3 \times 10^{-6}$
FeS	$3{\cdot}9 \times 10^{-10}$	4×10^{-9}	$3{\cdot}4 \times 10^{-6}$
ZnS	$3{\cdot}5 \times 10^{-12}$	$4{\cdot}5 \times 10^{-14}$	$3{\cdot}3 \times 10^{-10}$
Cu_2S	—	3×10^{-20}	$8{\cdot}0 \times 10^{-10}$
CoS	—	$1{\cdot}0 \times 10^{-18}$	—
NiS	$1{\cdot}2 \times 10^{-12}$	3×10^{-11}	$7{\cdot}0 \times 10^{-11}$
CdS	$6{\cdot}0 \times 10^{-15}$	1×10^{-18}	$8{\cdot}6 \times 10^{-13}$
PbS	$2{\cdot}0 \times 10^{-14}$	7×10^{-19}	$4{\cdot}9 \times 10^{-12}$
CoS	$9{\cdot}2 \times 10^{-23}$	8×10^{-27}	$8{\cdot}8 \times 10^{-12}$
Ag_2S	$3{\cdot}4 \times 10^{-17}$	7×10^{-21}	$3{\cdot}3 \times 10^{-15}$
HgS	$6{\cdot}3 \times 10^{-27}$	1×10^{-44}	$1{\cdot}5 \times 10^{-24}$

bornite and chalcopyrite, but only because these minerals are much less soluble than other forms of FeS.

Thus, *"the struggle for the sulfide ion" results in all the heavy metals passing into the sulfide form during diagenesis, whereas authigenic iron minerals are almost entirely carbonates or silicates, and manganese minerals are carbonates. A distinctive association is formed, superficially suggesting hydrothermal associations, but genetically having nothing in common with such processes.* These new aspects of mineral diagenetic growth are additional to those touched on in the analysis of diagenesis in humid zones (Vol. 2, Chap. 8).

The initial distribution of Cu-Pb-Zn in sediments was probably irregular, a characteristic feature of sedimentation in general. *During diagenesis, parallel with conversion of the carbonate forms of heavy metals to sulfides, the metals were extensively redistributed, and numerous points of concentration appeared, leading finally to the deposition of ore minerals and to diffuse impregnations,*

* The mechanism of this transition was described in detail in Vol. 2, Chap. 8, and will not be repeated.

lenticular, linear, or tabular in form. The factor controlling such distribution and localization of ore bodies was undoubtedly organic material.

It is known that even in microscopic form—the finest fragments, colloidal clots, and the like—organic matter is irregularly distributed in sediments. Macroscopic remains of organic material—leaf and wood débris—are even more erratically distributed. This irregularity has led to the migration of large amounts of sulfides during diagenesis. During later stages of this process, when sulfide nuclei have formed, redistribution is aided by the forces of selective crystallization. The range and character of redistribution varies markedly under different conditions. We may consider three types.

In the simplest case, redistribution takes place within a single "grey" bed— in cupriferous sandstones, limestones, or dolomites of lead-zinc deposits. Fragments of plant tissue and dead benthonic and planktonic bodies become foci of accelerated H_2S formation during early diagenesis, corresponding to a decline in value of Eh. *At such points sulfides are first precipitated in solid phases.* The decrease in their concentration in waters of bottom muds, because of this process, causes the migration of the ions of Pb, Cu, and Zn and other metals to these centers from zones poor in organic material. Impregnations of chalcocite, bornite, chalcopyrite, galena, sphalerite, and other minerals appear; and where these impregnations are locally concentrated, ores may form.

This process cannot be, and has not been, limited by these simple migrations. For example, the red facies in the Mansfeld Shales, the Rote Fäule, partly lying inshore of the black facies and, farther offshore, forming patchy zones within this facies, is undoubtedly the result of primary bleaching of the black rocks by organic material. It is barren of ore, with Cu-Pb-Zn content at most only very slightly higher than the average for the crust. In the adjacent black bituminous shales, however, copper and the other metals are concentrated, the richest ores being found at the contact with the Rote Fäule. In explanation, it is postulated that the Rote Fäule is deposited where rivers rich in oxygen flowed into the sea, carrying Cu-Pb-Zn salts in solution. The Zechstein sea itself, in the view of German geologists, was rich in H_2S, and sulfides of the heavy metals were precipitated in the contact zone, forming a belt of rich ores. Concentration in this manner is open to serious doubt, however. Firstly, the traditional view that the Early Zechstein sea was a basin with hydrogen-sulfide bottom water is incorrect, because benthos lived on the floor of this sea—though not in great abundance—and the basin itself was too shallow and too flat to allow any stable, prolonged contamination of the bottom waters by hydrogen sulfide. Furthermore, the Rote Fäule is found not only along the margin of the black shale facies but, in places, within that facies (Fig. 40). This cannot be explained by streams carrying fresh water. Lastly, precipitation of the basic Cu-Pb-Zn carbonates in oxygenated waters is not inhibited, as the German investigators state it to be but actually does take place (A. F. Efremov). In consequence, this leads to an entirely different interpretation of the Rote Fäule and the Cu-Pb-Zn ores about its margin.

The Rote Fäule represents zones of muddy sediments in the Early Zechstein basin in which, either because of hydrodynamic or some other conditions, only minimal quantities of organic matter collected, but in which at first the concentration of Cu-Pb-Zn was on average the same as at other localities. Reducing processes in the zones of Rote Fäule essentially failed to develop, and the red muds were preserved. The contents of Cu-Pb-Zn ions in the interstitial waters of the muds, according to the solubilities of their carbonates, were considerable. When reducing processes and sulfide precipitation began in the black muds rich in C_{org} adjoining the Rote Fäule, the Cu-Pb-Zn content of the interstitial waters in the muds in the immediate vicinity was reduced almost to zero. Migration of Cu-Pb-Zn was then initiated from the interstitial waters of the Rote Fäule into these black sediments containing water rich in H_2S, and they were deposited as sulfides, immediately next to the red facies. The areas of Rote Fäule were thus depleted of their metals which migrated to the black sediments, to create a marginal zone rich in sulfides, long recognized empirically.

It may be thought that transfer of ore components from red sediments to the black or grey layers is by no means limited to the Rote Fäule. *The very essence of the process suggests that it must inevitably have occurred wherever grey ore-producing beds have been adjacent, laterally or vertically, to red beds.* The latter have lost their ore constituents, having become entirely barren, whereas the former have become enriched by the transfer of the Cu-Pb-Zn. This means that *in cupriferous sandstones the juxtaposition of red barren horizons with grey ore horizons is not a primary feature, but was developed during diagenesis. The dominant factor in the removal of the ore component from the red to the grey horizon was the difference in solubilities of the basic carbonates and sulfides of Cu-Pb-Zn, giving rise to a great diffusion potential.* These diffusion processes are especially typical of cupriferous sandstones, in which the red and grey units are clearly differentiated and are very numerous. Such combinations are not found in carbonate sequences with lead-zinc mineralization. In such sequences the processes of element redistribution are less strongly developed, and the ore horizons are essentially features of primary sedimentation. Secondary redistribution of metals has been restricted almost exclusively to interstratal redistribution (the first type described above).

It further appears that authigenic zoning of ore minerals, briefly described above, is associated with diagenetic redistribution of sulfides in cupriferous sandstones. Unfortunately this phenomenon has not yet been adequately studied. Nevertheless, it has been established that *this zoning is in places distinctly developed, though absent elsewhere.* It is apparently suppressed where the distribution of organic matter in the sediments is irregular and where, accordingly, the foci of high and low Eh, of large and small concentrations of H_2S are also erratic. At points of high Eh and low H_2S concentrations, chalcocite is formed (with or without galena and sphalerite). Where the Eh is low and the H_2S concentration high, bornite appears, then chalcopyrite, and, lastly, pyrite. The erratic distribution of physicochemical conditions in the sediments gives rise to a similar distribution of the iron-copper sulfides. Zoning of the authigenic minerals does not develop. If the horizons primarily rich in copper (i.e. during sedimentation) have a distinctive zonal arrangement of organic matter, such as vertical and horizontal changes from siltstone, with low C_{org}, to clays with high C_{org}, a zonal distribution of minerals will result. In the silts with low C_{org}, and consequently with a low H_2S content, solely copper minerals are formed, since Ag, Pb, Zn, and other heavy metals form sulfides at a somewhat higher Eh. The sulfide ion in this environment does not form FeS_2 but rather chalcocite. With increase in amount of organic material and of H_2S generated by it, the ever increasing amount of S^{2-} does combine with iron: chalcocite is found to be mixed with bornite (Cu_5FeS_5), chalcopyrite ($CuFeS_2$), and finally with pyrite (FeS_2).

At the same time the elements in the beds begin to be actively redistributed. After the zones with high Eh have lost their copper, ions of copper migrate into them both vertically and horizontally. This impoverishes these neighboring zones in copper minerals and further increases the concentration in the chalcocite zone. At the same time, zones with low Eh and abundant hydrogen sulfide inevitably receive iron from the adjacent beds, where it cannot go into the sulfide form (because of insufficiently low Eh) and where a high concentration of Fe^{2+} is consequently maintained in the interstitial water, a concentration that is in equilibrium with the solid iron carbonates and leptochlorites. The vertical and horizontal authigenic mineral zoning indicated by this hypothesis is actually found in fact. The vertical disposition of zones (from chalcocite to pyrite in some sections, from pyrite to chalcocite in others) is entirely predetermined by the varying C_{org} content of the rock. The same applies to horizontal variations.

Yet another explanation of the origin of authigenic mineral zoning has been proposed (V. S. Domarev 1958, 1960). This author sees in zoning traces of chemical differentiation of the elements, taking place during sedimentation. It is not explained, however, why this differentiation appears in some places, but is absent elsewhere. It appears more nearly correct to me, therefore, to associate the authigenic mineral zoning in cupriferous sandstones not with sedimentation but with diagenesis, with sulfide generation and selective redistribution of these sulfides.

From this discussion it becomes clear that diagenesis in general and the redistribution of authigenic Cu-Pb-Zn sulfides in particular play a critical role in the formation of ore deposits of these elements. *During this stage the uniform, average, low content of ore elements gives way to a sharply defined irregularity. Copper, lead, and zinc migrate from many parts of the sediments and accumulate in excess in others, creating highly concentrated ores.* In other words diagenesis here is a typical, supplementary and very important process, and ores of Cu-Pb-Zn prove to be characteristic representatives of sedimentational-diagenetic ores, the second process being much more significant than the first.

The concentration of copper in the inshore zone, lead farther seaward, and zinc still farther into the pelagic zone, of which the Mansfeld Shales is a classic example, is a basic result of chemical differentiation during sedimentation. The lateral or vertical change from copper to lead deposits must also be associated with sedimentation. By contrast, authigenic mineral zoning of iron-copper minerals in cupriferous sandstones is undoubtedly a diagenetic phenomenon.

However important the diagenetic stage in the formation of ore accumulations of Cu-Pb-Zn, the history of the ore process does not end with it. During burial and deformation of the crust, the ore deposits are further altered to some extent. Two cycles of phenomena are known. *First,* the various fractures that develop are healed, partly by redeposition of ore minerals, partly by vein formation. By either method, the substance is derived from the existing ore accumulations or the host rocks. *Secondly,* the ore minerals are locally redistributed within the ore horizons, concentrating chiefly in anticlines and migrating from the synclines, as in the Sumsar deposit. This redistribution is controlled by the circulation of ground water through the ore horizons. At depths below 2 km the temperature of these waters is undoubtedly above 60°C; i.e. from this point of view it becomes hydrothermal. Genetically, however, it has nothing in common with hydrothermal solutions of volcanic origin. These hydrothermal solutions are vadose and their activity is consequently not generally accompanied by any introduction of new components to the ore horizon, but solely with redistribution of the pre-existing minerals.

The extent of this katagenetic redistribution varies strongly from one deposit to another. As a rule it is slight, and the sedimentational-diagenetic character of the deposit is unaltered. In places, however, this distribution becomes appreciable, and the deposit becomes multistaged: sedimentational-diagenetic-katagenetic. It should be stated that copper-ore deposits are less susceptible to katagenetic processes than lead-zinc ores, probably because copper has a lower geochemical mobility than lead or zinc.

The distribution of Cu-Pb-Zn ore provinces within arid zones and their absence from humid zones remains to be explained. The association between Cu-Pb-Zn ores and arid regions appears to originate in the following manner. The components begin their migration as solutions of sulfates, forming through chemical weathering of sulfides in parent rocks. The conversion of sulfates in the sediments is brought about by the pH and the presence of carbonates in natural waters. The higher the pH and the greater the carbonate concentration in natural waters the more quickly and completely the Cu-Pb-Zn sulfates are changed into solid phases of basic carbonates. The presence of large amounts of dissolved organic matter, lowering the pH of the waters, retards the precipitation of these basic carbonates. *The waters of arid regions, both in streams and in basins, are characterized by the abundance of dissolved carbonates* ($CaCO_3$, $MgCO_3$, *and, at times,* Na_2CO_3) *and by high pH. In addition, these waters contain practically no dissolved acid organic matter.* Such waters are natural percipitants of Cu-Pb-Zn from sulfate solutions. The waters of humid zones, by contrast, are generally poor in dissolved carbonates and rich in dissolved organic matter, giving them a neutral or even slightly acid reaction, especially in the tropical moist zone. These waters are hence very poor precipitants of Cu-Pb-Zn sulfates. Furthermore, the abundance of moisture, constantly percolating through the crust of weathering, greatly dilutes the solutions of copper, lead, and zinc salts. This also has strong negative effect on any depositional tendency. As a result, the Cu-Pb-Zn mobilized in the crust of weathering is for the most part not precipitated during further migration, but is disseminated in natural waters. Only under the best conditions, when the content rises slightly above the average in the crust, do rare and isolated segregations of chalcopyrite, galena, or sphalerite appear in humid formations, or perhaps microscopic impurities in pyrite and marcasite concretions. *The hydrochemical characters of natural waters, especially stream waters, differ so sharply between humid and arid regions, that ore accumulations of Cu-Pb-Zn form only in arid zones.*

The distribution of these ores within arid regions has been controlled chiefly by the petrographic character and tectonic regime of the drainage areas. For ores of Cu-Pb-Zn to develop, it is necessary that the rocks of the drainage areas should originally carry a high content of these metals in a form readily accessible to chemical weathering and readily mobilized under arid conditions, i.e. as sulfides. However, when tectonic activity is slight and desert and semi-desert peneplains develop, there is no water to effect mobilization and transfer of Cu-Pb-Zn compounds. When the tectonic regime is more active, and the topography is dissected, when the parent rocks containing the Cu, Pb, and Zn lie in a moister climatic zone, chemical weathering of sulfide metals and the transport of these products by streams is possible, and ores are formed.

The specific hydrochemical features of natural waters have thus restricted Cu-Pb-Zn ores to regions of arid climates; and the petrographic character and tectonic regime of various parts of arid regions have determined the precise location of ore deposition. Cu-Pb-Zn deposits are located partly in intermontane basins, partly in marginal depressions of developing fold zones and at times on the adjoining platform (Figs. 11-19).

7. HOMOLOGOUS FEATURES AND DIFFERENCES IN THE ORE TRIADS Al-Fe-Mn AND Cu-Pb-Zn

In comparing what has been said about Cu-Pb-Zn ores with what is known about the ore triad Al-Fe-Mn (Vol. 2, Chap. 4), one is readily convinced that there is much in common in the geology of both groups. Both form because of direct supply of the ore material from the continent. Both are localized to some extent in continental deposits, but chiefly in the near-shore marine zone. In both groups the first member is least mobile geochemically, the second more mobile, and the third most mobile. Because of this, facies profiles of ore accumulations, in going from the first member of a triad to the last, show an ever increasing displacement from continental facies to marine facies. In both groups the main bulk of the ore has been precipitated from solution, and the generation of solid phases has occurred purely by chemical means. In both groups the ore process during sedimentation took place with no detectable effect of organisms. These latter began to be effective only during diagenesis, through creation of oxidation-reduction conditions. Each group, lastly, is monoclimatic, although the two form in substantially different climates.

These two groups may therefore be regarded as homologous, the one replacing the other in transitions from one climatic regime to another—from humid to arid. The restriction of the triad Al-Fe-Mn to the humid zone is explained by the fact that only in such climate do all three elements acquire relative geochemical mobility, permitting mass migration and, hence, concentration. In arid climates these elements are immobile, and their concentration is consequently impossible. For the triad Cu-Pb-Zn, however, humid conditions cause *excessive mobility and furnish too diffuse conditions for mass precipitation.* Solutions of these elements are therefore disseminated. In arid climates, the hydrochemical conditions limit the mobilities of Cu, Pb, and Zn, make possible their mass precipitation and accumulation to the extent of producing ore bodies.

Together with the many similar features of the two homologous groups there are also distinct differences. The ores of the humid triad are chiefly sedimentational. Diagenesis as an ore-forming factor appears only in manganese ores, which are therefore referred to as sedimentational-diagenetic. But even here the relative value of secondary diagenetic accumulation is only of minor importance. The ores of the arid triad are all sedimentational-diagenetic, and the role of diagenesis as an ore-forming factor is always large and even dominant. In some cases katagenesis may be an important factor as a specific ore-forming process. Thus, the ore forms in several stages, spread over a considerable length of time. All these differences are based fundamentally in the much lower average content in the crust of Cu-Pb-Zn than Al-Fe-Mn and also in the greater geochemical mobility of members of the arid triad than of the humid triad. This means that the arid group is more sensitive to geological processes during the various stages of sedimentary lithogenesis.

CHAPTER 3

P-CaCO₃-MgCO₃-SiO₂ DEPOSITION IN WEAKLY MINERALIZED BASINS OF ARID ZONES

In the description of lithogenesis in humid zones (Vol. 2), brief reference was made to the specific features that distinguish the history of P-$CaCO_3$-$MgCO_3$-SiO_2 in basins of arid zones from that in humid zones. In this chapter these features will be discussed in greater detail, in that they allow a more precise evaluation of the distinctive aspects of arid lithogenesis at the incipient stage of its development.

1. INCREASE IN PHOSPHATE DEPOSITION IN BASINS OF ARID ZONES, THE PALEOGEOGRAPHY OF PHOSPHATIC BASINS, AND THE PETROGRAPHIC TYPES OF PHOSPHATES

The accumulation of phosphates in humid zones has been a process of low intensity. This is manifest both in the small total reserves of P_2O_5 forming in humid seas and in the petrography of phosphorites themselves. They are exclusively nodular and poorly sorted: both diagenetic processes and reworking have been important ore-forming factors. Phosphate deposits in basins of arid zones acquire sharply different features: *the process of formation is much more intense, leading to such vast accumulations of phosphorus as are found in the Karatau deposit of Central Asia (and its continuation in the Ulutau Range), the Permian Phosphoria formation in North America, the Tertiary phosphorites of North Africa, and others. At the same time these phosphorites are of high grade, as is clear from their petrographic character. Thus, the weakly mineralized marine basins of arid regions have proved to be the principal sites of phosphate accumulation and these accumulations form more than* 80% *of the total phosphate reserves.*

The paleogeography of arid phosphate accumulation has been variable, and therefore several of the largest phosphate deposits will be described.

The Lower Eocene phosphorite basin of North Africa (Fig. 48) extended from the Atlantic coast (between Tangier and Rabat) to Tunisia. On the west it undoubtedly connected with the Atlantic Ocean, but at a number of places on the northern coast, especially in Tunisia, it was linked with a geosynclinal basin occupying the present site of the Mediterranean Sea. Accumulations of phosphorites were confined to a gulf in the region of Algeria and Tunisia. Southwest of the North African basin, between the Moroccan Meseta and the High Atlas, lay a second basin, a large gulf of the Atlantic Ocean, in which phosphates were also deposited. In both deposits the bedded phosphates lie between grey marls and are characterized by abrupt changes in thickness

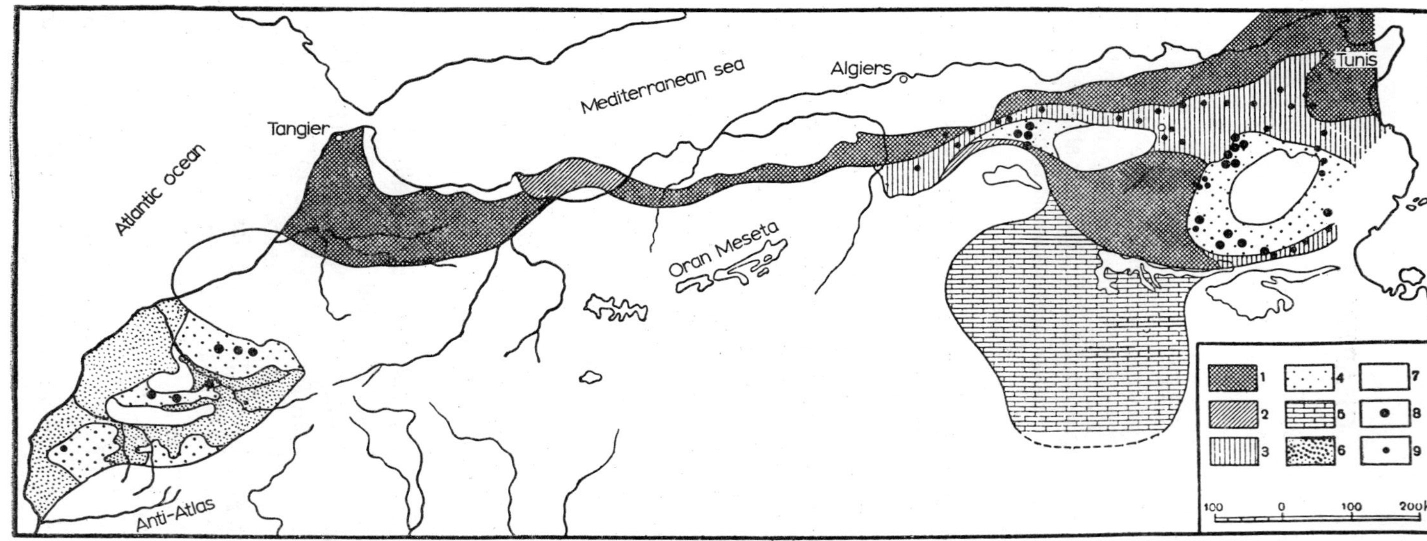

Fig. 48. Paleogeography of the Lower Eocene phosphorite basin of North Africa (from E. V. Orlova). 1. Deep-water sediments with *Globigerina*; 2. inferred continuation of basin; 3. shallow-water sediments with *Nummulites* and deposits of poor phosphorites; 4. shallow-water sediments with deposits of rich phosphorites; 4. shallow-water sediments with deposits of rich phosphorites; 5. lagoonal sediments; 6. zone of thick Recent alluvium; 7. land; 8. deposits of workable phosphorites; 9. deposits of uneconomic phosphorites.

(from 2 to 4 m), and lateral variation (Fig. 49*A* and *B*). The typical bedded phosphorites are hard, granular or oolitic, dark gray, grayish brown, or black rocks. They give off a bituminous odor when struck. Normally they include many foraminifera and radiolaria together with coprolites, teeth, and bone fragments of fish. The oolitic grains are spherical or oval, from 0·05 to 0·5 mm in diameter. The tri-basic phosphate content of these oolitic grains is about 32%. The phosphate sequence yields a rich fauna, both the phosphorite horizons themselves and the interbedded sediments. "The oyster *Ostrea multicostata* and other forms are very abundant. *Thersitea* is very characteristic. Radiolaria and diatoms appear abundantly in the phosphorite grains and in the cement. The radiolaria (*Spumellaria*) are shallow-water forms, requiring constantly moving water. The presence of diatoms as nuclei of phosphoritic ooliths suggests that a cold current was present, possibly introducing a northern flora into this subtropical basic. Sublittoral species of fish are abundant" (Orlova 1951, pp. 61-62). Thus, the phosphorite sequence is undoubtedly a shallow-water unit. In the Moroccan gulf of the Atlantic Ocean the phosphorite deposits line the marginal shore zone of the basin, but in the Algeria-Tunisian gulf they are confined to a shallow-water uplift forming two large islands. North and south of the uplift the water is relatively deep. The thickness of the phosphorite series in central Tunisia averages 20–25 m, locally reaching 40 m. In southern Tunisia it increases to 90 m, with an average thickness of about 46 m. This strikingly indicates the intensity of the phosphorite-forming process. The phosphorites wedge out towards the deep-water deposits, and stratified forms give way to nodular forms.

"The deposits of southern Tunisia are of a shallower-water character than those in central Tunisia, with a more clearly defined shore line. This was responsible for the development of gravels, coquina, and gypsiferous beds" (Orlova 1951, p. 60).

"*Red beds and lagoonal deposits are widely developed in the zones bordering the Algeria-Tunisian basin on the south; an Eocene gypsiferous series is developed to the south and east of the phosphorite region of the Gafsa district in Tunisia, and to the south of Negrin in the Jabel Oqq deposit in eastern Algeria.* The red beds cover the region of the Moroccan Meseta, and are also known in the Anti-Atlas and the High Atlas" (Orlova 1951, p. 49). Both phosphorite regions are characterized by insignificant fractions or even absence of clastic material. From this we may conclude that the surrounding land area and islands were of low relief.

Highly distinctive circumstances gave rise to the phosphorites in the Permian Phosphoria formation (Fig. 50). A large geosynclinal basin occupied the region of the North American Cordillera and the Rocky Mountains. In the northern third it was a single basin with its greatest subsidence in the central part; in the remainder it was divided by the Manhattan anticlinorium into western and eastern basins. The western basin was filled with thick sedimentary-volcanic sequences, the eastern with normal sedimentary strata.
    G

The volcanic centers (Schuchert 1957) were on the extreme west, at the present Pacific coast.

The deposition of phosphate took place along the eastern shore of the geosyncline, in the region extending from Banff to the Uinta Mountains, over

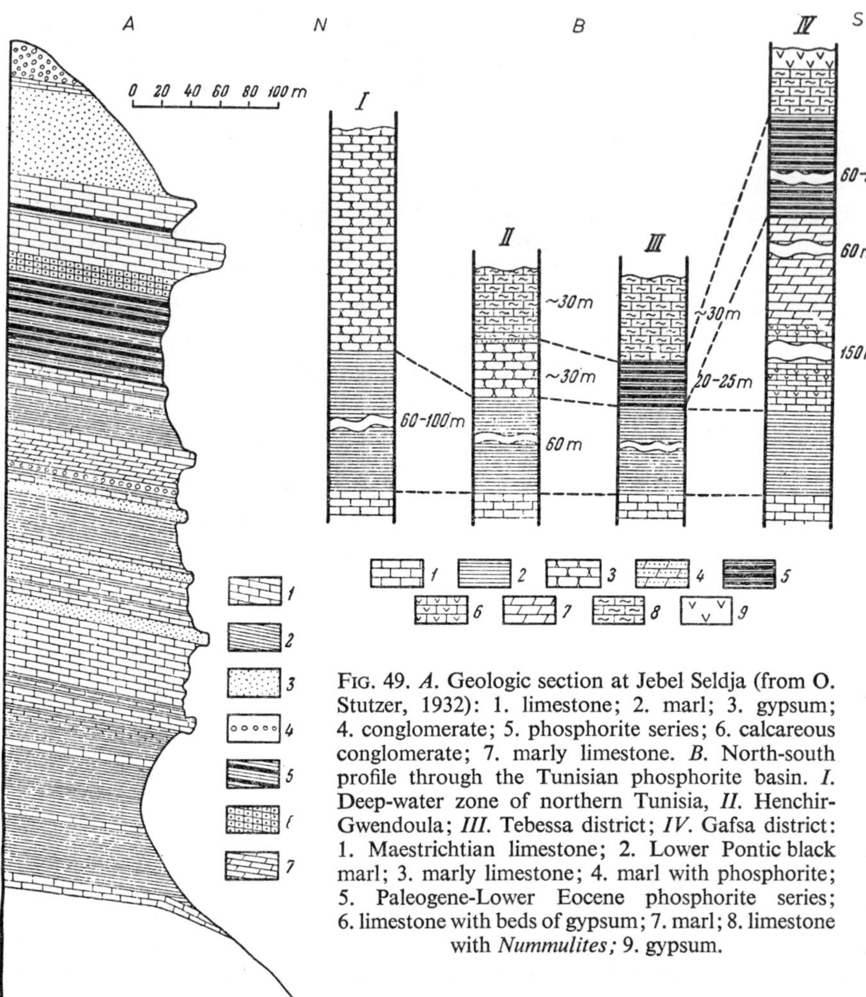

FIG. 49. *A*. Geologic section at Jebel Seldja (from O. Stutzer, 1932): 1. limestone; 2. marl; 3. gypsum; 4. conglomerate; 5. phosphorite series; 6. calcareous conglomerate; 7. marly limestone. *B*. North-south profile through the Tunisian phosphorite basin. *I*. Deep-water zone of northern Tunisia, *II*. Henchir-Gwendoula; *III*. Tebessa district; *IV*. Gafsa district: 1. Maestrichtian limestone; 2. Lower Pontic black marl; 3. marly limestone; 4. marl with phosphorite; 5. Paleogene-Lower Eocene phosphorite series; 6. limestone with beds of gypsum; 7. marl; 8. limestone with *Nummulites;* 9. gypsum.

a length of about 1700 km (Fig. 50). In Wyoming two gulfs—the Lander on the north and the Uinta on the south—branched off from the main marine basin.

Substantial changes in composition of the deposits occur from north to south along the main phosphorite belt. In Canada the phosphorite group is in part quartzitic. A phosphorite bed has an average thickness only of 0·3 m and consists chiefly of phosphatic shells and bone fragments. No oolites are found. Shales and to some extent the quartzites are also phosphatized. Southward, in Montana, the phosphorite formation is subdivided into two horizons: a lower shale horizon, which is the phosphorite horizon proper, and an upper cherty horizon. Their joint thickness increases to the south, up to a total of about 100 m in the vicinity of Portneuf-Randolf. The

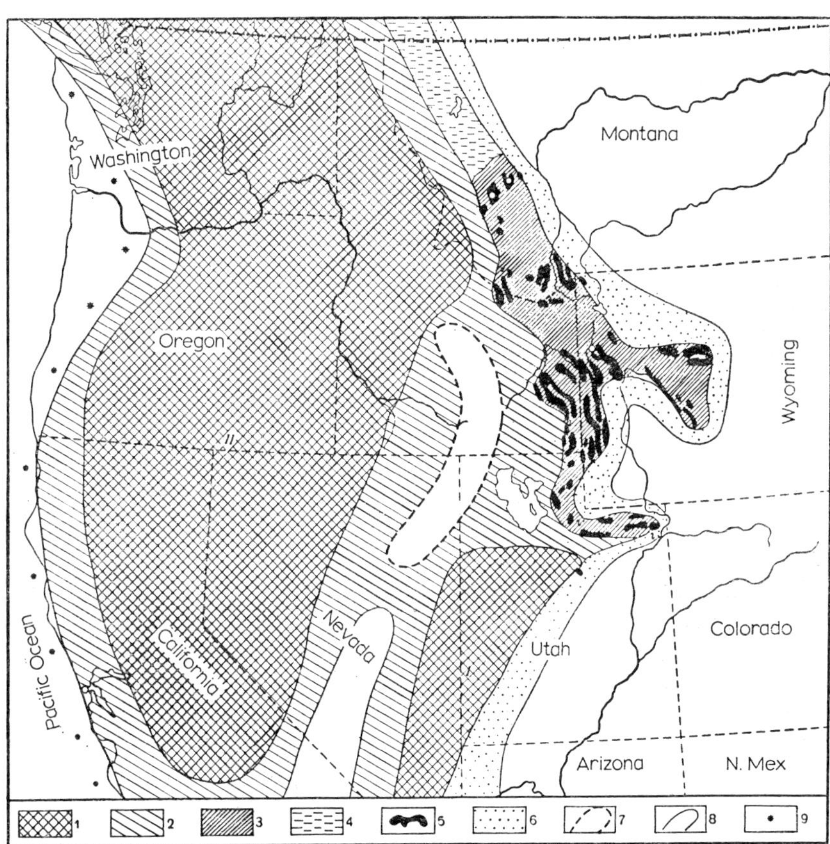

FIG. 50. Paleogeography of the Permian phosphorite basin of North America (from E. V. Orlova). 1. Region of downwarping and accumulations of sediments over 300 m thick: *I.* zone of normal marine sediments, *II.* zone of sedimentary-volcanic formations; 2. region of accumulations of sediments up to 300 m thick; 3. area where phosphate facies developed; 4. area of inferred distribution of phosphorite facies; 5. distribution of phosphorite formation; 6. shallow-water zone; 7. land; 8. under-water islands; 9. Paleozoic volcanic region.

thickness of the phosphatic beds here is highly variable, ranging up to 3·6 m. The phosphate concentration is also variable; beds of well-sorted phosphorite with P_2O_5 contents of 32–36% alternate with fine-bedded shales in which individual beds of phosphates contain 10–25% P_2O_5, and with fine-grained sandy clays, in which the P_2O_2 content is no more than 1–7%. The average P_2O_5 content for the entire shale sequence is 12%. This concentration of phosphorus consequently reaches tremendous proportions and is concentrated between the Manhattan anticline on the west and the margin of the platform on the east. The phosphorites are oolitic, ranging up to pisolitic (12 mm). The phosphorites contain almost no fauna, although locally, at the base of the series, abundant fragments of phosphatized shells occur. The beds enclosing the phosphorites contain *Lingula, Productus* (three species), *Lingulidiscina, Chonetes, Pustula, Pugnax, Leda, Aviculopecten,* and *Omphalotrochus.* The phosphorite series of this region is characterized by a high vanadium content (0.5–2.0% V_2O_5). South of the Portneuf-Georgetown region both the upper and lower horizons include increasing proportions of limestone. South of Randolf, approaching the deep-water Oquirrh basin, phosphatized rocks decrease sharply in importance.

The two gulfs extending into the interior of the continent are represented by a special type of phosphorite deposit. The host rocks here are limestones. The phosphorites lose their oolitic content and take on a coquinoid texture. The quality of ore declines and the P_2O_5 content decreases to 25–16%. Glauconite appears, and the phosphorites acquire a platform aspect.

In individual parts of the phosphorite belt, along the eastern shore of the sea, red beds including local gypsiferous varieties are found. At Maxville (Fig. 51) these beds are similar to the recent formations in the vicinity of Cutch in India, and they indicate that broad shallow-water lagoons lay to the east of the Permian sea, being periodically connected with the sea. The drainage areas were low, supplying minimal amounts of red clastic material. Apparently only in isolated zones did sandy material reach the sea.

The nearest correlative of the Phosphoria formation is the Karatau phosphorite series (Turkestan). Here also the phosphorite beds are thick and are associated with dolomites and siliceous rocks. The phosphorite itself is granular and oolitic. The paleogeography during their deposition has, however, not yet been determined.

The paleogeography of the Lower Miocene Hawthorn formation in Florida is highly characteristic. The sea then extended over the northern shore of the present Gulf of Mexico to the east of the Appalachian Hercynian structures, which had been greatly reduced in elevation by that time (Fig. 52). The basin as a whole was an open oceanic shelf, communicating freely with ocean water. The Hawthorn formation covers an area of about 17,500 km², is 180 m thick, and is composed of alternating beds of white and gray marls with intercalations of hard limestone and green clay.

Almost all the constituent rocks are rich in phosphorus, in the form of small brown grains of phosphate or phosphate pebbles. The P_2O_5 content is normally 4–14%, but may rise to 31–34%. Considering the great thickness and areal distribution of the Hawthorn formation, the vast concentration of phosphates in it, even though the grade is low, is apparent.

Although the Middle Ordovician phosphorites of the Siberian platform have little practical value, the paleogeography of their deposition is of importance in an understanding of the variations in environmental conditions for accumu-

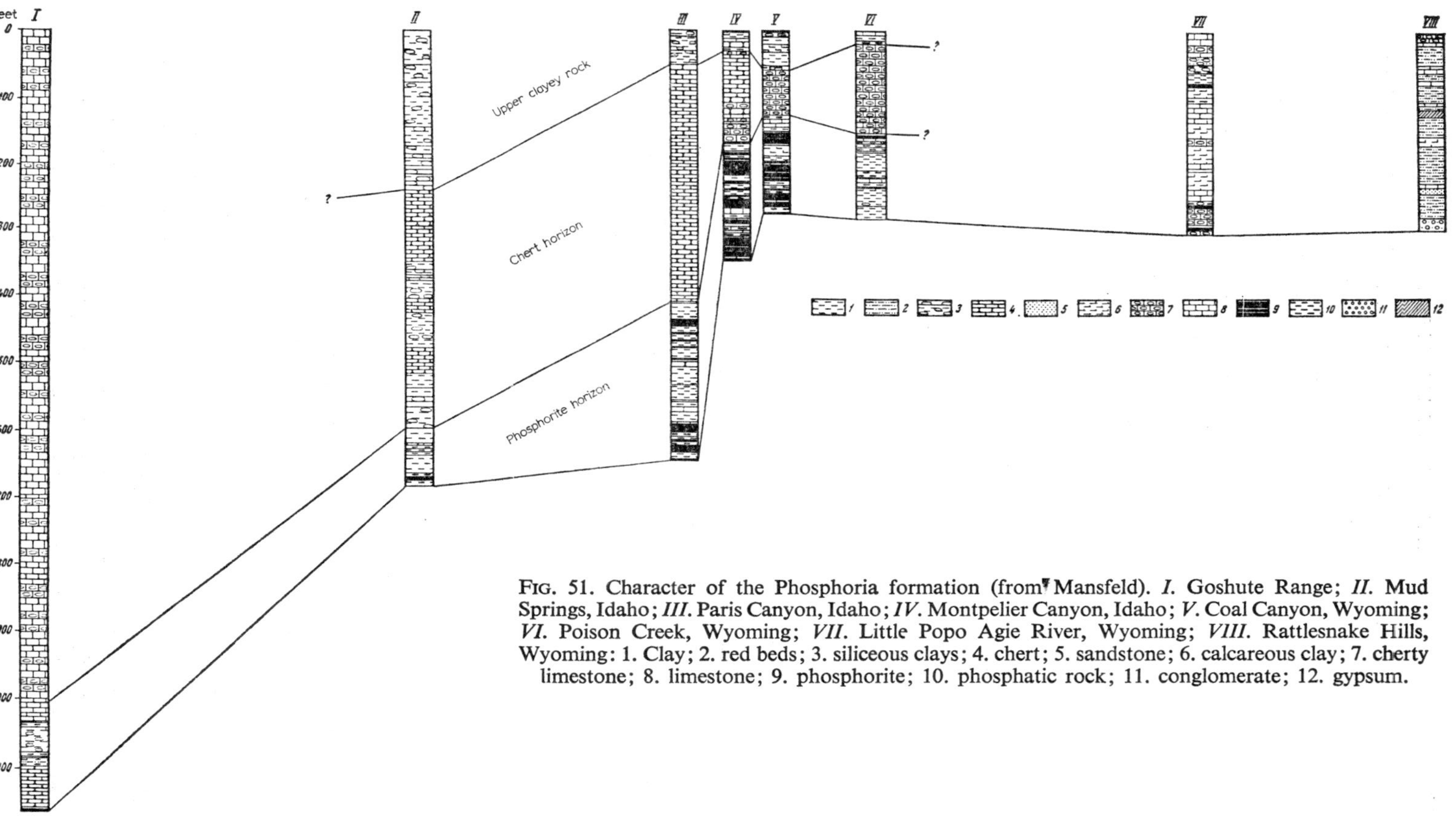

FIG. 51. Character of the Phosphoria formation (from Mansfeld). *I*. Goshute Range; *II*. Mud Springs, Idaho; *III*. Paris Canyon, Idaho; *IV*. Montpelier Canyon, Idaho; *V*. Coal Canyon, Wyoming; *VI*. Poison Creek, Wyoming; *VII*. Little Popo Agie River, Wyoming; *VIII*. Rattlesnake Hills, Wyoming: 1. Clay; 2. red beds; 3. siliceous clays; 4. chert; 5. sandstone; 6. calcareous clay; 7. cherty limestone; 8. limestone; 9. phosphorite; 10. phosphatic rock; 11. conglomerate; 12. gypsum.

lation of phosphates in arid regions (Nikiforova, 1955). During the Krivaya Luka epoch four well-defined facies were present within the huge marine basin (Fig. 53). Red sandstones, commonly coarse-grained, with red, locally green, siltstone and rarer mudstone, were deposited during Krivaya Luka time in the Irkutsk amphitheater and along the western margin of the platform to the mouth of the Nizhnyaya Tunguska. Mud cracks are common in the siltstones and clays, indicating a shallow-water environment for

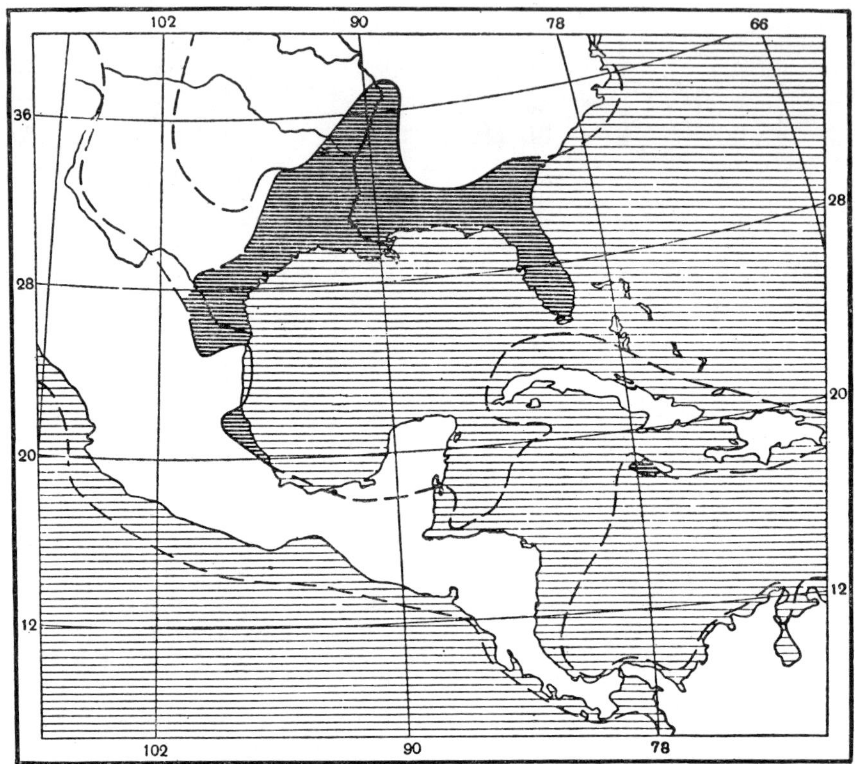

FIG. 52. Paleogeography of the Lower Eocene Hawthorn formation (from Schuchert). Lined area represents marine deposits.

deposition. Littoral cross-bedding and occasional current bedding are observed in the sandstones. These strata represent the near-shore part of the basin, the zone adjoining uplifts that bordered the basin to the south and west, and supplied the clastic material. In the more central parts of the platform finer-grained clastic-carbonate sediments accumulated, many of them red. These consist of alternating siltstones, sandstones, sandy dolomites, rare clayey marls and clays, and rarer limestones. The total thickness of the deposits does not exceed 50 m. The most open, and possibly the deepest, part of the basin lay on the north (Moiero River region), where a limestone sequence was deposited, thicker than elsewhere. These limestones yield the most varied and abundant fauna of the Krivaya Luka series, attesting to favorable conditions for animal life. Lastly, red sand-silt-dolomite deposits again appear on the northeastern part of the platform: these include lenses and beds of gypsum. They are found along the Markoki River and in the basin of the middle part of the Vilyui River. It is clear that this part of the sea was temporarily divided into a series of lagoons. A similar facies zoning was also

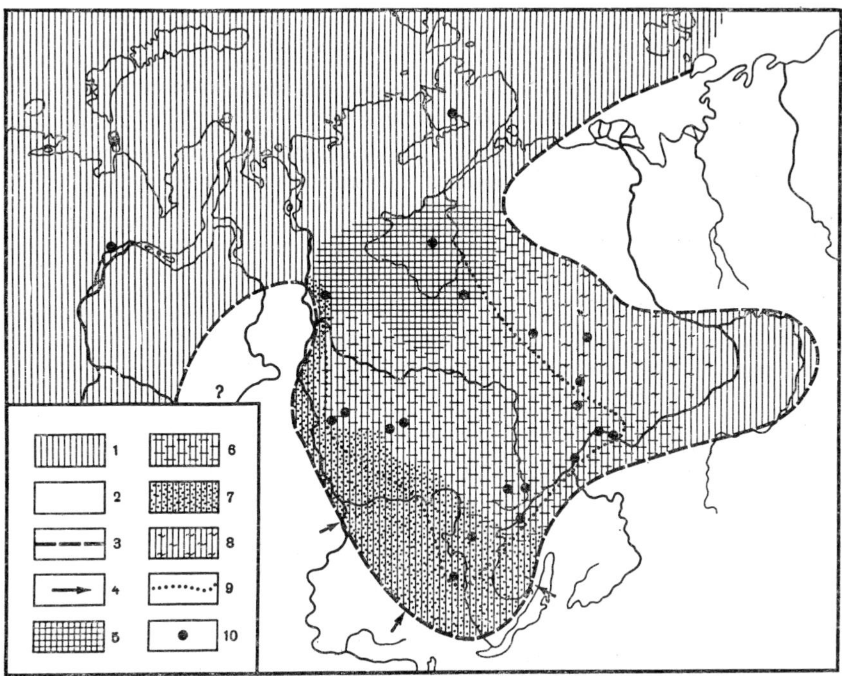

FIG. 53. Paleogeography of the Krivaya Luga epoch on the Siberian platform (from O. I. Nikiforova).
1. Sea; 2. Land; 3. tentative boundary of land; 4. direction of discharge; 5. platform sea with predominant accumulation of carbonate sediments; 6. platform sea with predominant accumulation of fine-grained clastic-carbonate sediments; 7. platform sea in immediate zone of discharge, with predominant accumulation of coarse-grained carbonate-clastic sediments; 8. platform sea with periodic development of lagoonal conditions; 9. tentative boundary of area of phosphate deposition; 10. localities of sections examined.

characteristic of the Mangazeya epoch, but with a somewhat different spatial distribution.

The phosphate of the Krivaya Luga and Mangazeya deposits is confined to some intermediate sedimentational zone in deposits of the Irkutsk amphi-theater (Uda, Oka, Ilim, and Lena Rivers), along the Podkamennaya Tunguska and Kureiki Rivers, on the Moeiro River, and, lastly, on the Markoki and Markhi Rivers. O. I. Nikiforova has recorded the inner boundaries of the phosphate zone (Fig. 53). The width of the zone is unknown, but it has been demonstrated that *phosphate deposition decreased toward the interior of the basin, in the deeper-water deposits; this is clearly seen in the Irkutsk amphitheater* (Dominikovskii and Librovich 1957).

The types of phosphate rock differ. Some are sandy phosphatic coquinas up to 0·6 m thick. Many lingulid shells and fragments lie within the cross-bedded structures, which are beautifully developed. No organic remains other than lingulids have been found. The cement is phosphatic and the P_2O_5 content ranges from 6·33 to 19·96%, values that are characteristic of coquinoid phosphorites (see Vol. 2). "Lingulid coquinas are shoal deposits near the mouths of streams that bring considerable sandy material into the sea. The greatest accumulation of present-day lingulids in phosphatic coquina is on sandy shoals that are exposed at low tide, as along the Kumano coast (Japan). The lingulids may appear where there is no competition from other marine fauna, in this case in the fresher waters" (Dominikovskii and Librovich 1957, p. 9).

Detrital phosphorites also occur: these are aggregates of rounded fragments of phosphatic lingulid shells; the fragments are 0·2–0·5 mm in diameter, mixed with smaller quartz grains. They also include fragments of silty phosphorite and, more rarely, phosphorite nodules with concentric structure. These phosphorites are characterized by a fairly high P_2O_5 content, from 13·80 to 19·38%. They are almost completely free of clayey material and are distinguished from the coquinoid type by the presence of unidirectional cross-bedding of current bedding type. "Thus," Dominikovskii and Librovich wrote, "these phosphorites are primarily deposits of currents and wave action in shallow water, concentrated by the sorting action of moving water."

The third type is represented by sandy brachiopod coquina and oolitic phosphatic-ferruginous rocks. The brachiopod coquina forms repeated lenticular units ranging in thickness from 15–30 to 3–5 cm within distances of 15 to 40 cm. The oolitic phosphatic-ferruginous rock generally rests on the brachiopod coquina. It consists essentially of iron ore somewhat enriched in P_2O_5 (0·49–2·92%). The presence of a rich marine fauna suggests that these rocks were formed in a shallow-water sector of the basin with normal salinity. Regionally they occur in those parts of the sea where no large streams empty into the basin.

Thus, in the Middle Ordovician sea of the Siberian platform, phosphates were essentially biogenetic in origin, and were deposited at very shallow depths. Within the Irkutsk amphitheater the greatest deposition of biogenic phosphates (coquinoid and detrital-grained types) took place in the fresher parts of the basin. In the northeast, however, near the zone of lagoons, these rocks might have formed in more salty water; a purely lingulid biocoenose may develop equally well in a somewhat fresher zone as in a zone of salty water, as is demonstrated by the present-day *Cardium edule* L.

In comparing the paleogeography of phosphate accumulation in seas of arid zones with that for humid belts, the fundamental similarity is clear. In both regions the phosphates accumulate in open marine shelves (the Hawthorn formation), in a broad marine gulf (Morocco), in less open gulfs

(Lander and Uinta), on the shelf of a geosynclinal sea (Phosphoria), in marginal depressions (Tunisia, Algeria), and, lastly, in shallow-water epicontinental marine basins (the Siberian platform). The critical differences have not been paleogeographic, but rather the intensity of the phosphate-forming processes, as shown primarily in the petrographic types of phosphorites.

The petrographic types of phosphorites in arid regions are varied: bedded, nodular, and coquinoid forms occur, though fundamentally they are the same types as in humid belts, but with a substantially varied distribution. In humid zones nodular types are dominant and stratified types are scarce. In the deposits of arid belts, on the contrary, *stratified high-grade deposits of oolitic, microgranular, or granular structure are dominant. Independent ferruginous phosphates are entirely absent.* Nodules are normally found only along the margins of the stratified, oolitic or granular deposits, where these stratified units begin to wedge out and the P_2O_5 content declines. Coquinoid phosphorites are negligible. It is important to note also that the reworking and redeposition of the phosphate, which is so extensive in humid zones and is undoubtedly important as an ore-forming factor, is normally absent in stratified deposits of arid zones. Where they occur, they are confined chiefly to peripheral deposits, particularly to a belt near the wedge out. In other words, *neither diagenetic redistribution of phosphates in the sediments nor reworking, with removal of the diluent clastic material, has been demonstrated to be of any real significance in arid zones as supplementary to phosphate accumulation. High concentrations of phosphate have arisen during initial sedimentation.* Furthermore, in contrast to humid phosphate deposits, the stratified phosphorites are commonly almost barren of fauna (Phosphoria, Karatau, Seleuk, and others). The granular varieties contain organic remains, some in abundance (North Africa), some in small quantities (Karatau). On the whole there is no doubt that phosphate accumulation in arid zones is purely a chemical process. The intensity of precipitation of phosphorus compounds from the bottom water has been repeatedly higher than that typical of basins in humid zones. Biogenic concentration of shells such as the Middle Ordovician deposits on the Siberian platform are rare and represent only an insignificant part of the total mass of phosphorites in arid zones.

During the deposition of arid phosphates, however, not only the petrographic aspect of the phosphates changes but also the paragenesis of the rocks with the associated phosphates.

The accumulation of phosphates is frequently associated with the formation of dolomite or with cherts. This was especially true in the Paleozoic. The clastic rocks containing the phosphates are commonly red or variegated. Gypsum appears immediately next to the phosphatic rocks in the section, and deposits of salty lagoons become well defined areally (Phosphoria, North Africa, and elsewhere). All these factors cause the petrographic association of phosphorites in arid zones to differ sharply from those typical of humid belts. Phosphate deposits of arid zones are also characterized by a tectonic pattern somewhat different from those in humid regions. Arid phosphorites form not only on platforms but chiefly in geosynclines, where great thicknesses of phosphatic rocks may develop.

On the whole, arid phosphorites differ from humid deposits by many striking features, to form a group that may be very clearly distinguished.

2. The Deposition of Phosphorite in Arid Regions

In interpreting the features of phosphate formation in seas of arid regions it must be borne in mind that some aspects of phosphate paragenesis are obvious from the arid conditions and need no special discussion. Such are the associations with red beds, dolomites, gypsum and other lagoonal deposits. But the great intensity of phosphate accumulation, the change from nodular to stratified type, and the appearance and extensive development of phosphates in geosynclines rather than on platforms—these are specific features of phosphate accumulation in arid regions that contrast with humid belts, the origin of which is by no means self-evident.

The theoretical basis here is the genetic conception of A. V. Kazakov. It is necessary to determine the processes by which phosphate accumulation is intensified in seas of arid zones. Two circumstances favor this. Whereas phosphates are only "very near saturation" in the surface waters of humid-climate seas (A. I. Smirnov 1958), the waters in arid zones are more highly mineralized, reaching saturation and even oversaturation as a consequence of evaporation, with the chemical precipitation of phosphates, as the final result. On the other hand, arid regions are characterized by the remarkable constancy of strong trade winds that control stable and strong currents, resulting in the upwelling of deep waters rich in phosphates. This has not only augmented the precipitation of phosphates and created high concentrations, but it has also extended precipitation over a longer interval of time, producing a thicker sequence of phosphatic rocks.

These specific features of arid-climate seas make the intensification of phosphate accumulation inevitable. Increased phosphate production has led to its localization not only on platforms, as is characteristic of humid zones, but also in geosynclines. Regions with deep basins in geosynclinal seas, below the levels of P_2O_5 enrichment, can only be positive indicators of phosphate accumulation. Increase in phosphate deposition leads to the displacement of nodular phosphorites, which correspond to a low intensity of phosphate formation, by stratified deposition, and to the displacement of the nodular type to the margins of the stratified deposits. Change in the petrographic type of phosphorite inevitably follows from acceleration of the phosphate process in arid zones, as surely as the connection with dolomites, red beds, and gypsum follows from the very nature of arid-climate phosphate accumulation.

3. The Increased Precipitation of $CaCO_3$ in Arid-Climate Basins and its Cause; Morphological Types of Calcite in Sediments

In common with phosphates, the history of $CaCO_3$ in weakly mineralized arid-climate basins differs from that in humid belts primarily by a sharp intensifi-

cation in precipitation. This is clearly shown in an analysis of Recent deposits of arid basins (Strakhov 1951): (Table 6). The precipitation of calcium carbonate in basins of arid zones is many times that in the ocean, and is even more pronounced than that in lakes and intracontinental seas of humid zones.

This sharp intensification of calcite deposition is due to several causes. Primarily, streams supplying the basins of arid zones normally carry a much higher concentration of dissolved carbonates than streams in humid belts. Thus, the water of the Volga at Astrakhan carries 2·15 mg-equiv $CaCO_3$; the water of the Ural has 2·6, the Terek 2·7, the Sulak 2·06, the Kura 2·7, the Don 4·7, the Dnieper 2·8, the Syr-Darya 2·63, the Amu-Darya 2·7, and the Ili 4·4 mg-equiv. These figures are generally 1·5–2 times those for the Amazon, Congo, Indus, Brahmaputra, and other rivers flowing into basins in humid regions. Further, the *water in streams flowing through arid regions is saturated*

TABLE 6

Average Intensity of $CaCO_3$ accumulation, in g/cm per year (Strakhov 1951)

Basin	B/α*	$CaCO_3$	Remarks
Atlantic ocean, equatorial belt .	0·3	0·5 ⎫	From map of distribution
Black Sea	3·6	7·4 ⎭	of $CaCO_3$
Caspian Sea	4·3	11·7 ⎫	
Aral Sea	7·0	19·4 ⎬	From balance of discharge
Lake Balkhash . . .	8·9	30·9 ⎭	(liquid and solid)

* B represents drainage area, α area of basin.

and oversaturated with $CaCO_3$. As a result carbonates begin to migrate in streams not only in solution but also in suspension. Direct analyses of stream suspensions from a number of rivers have shown that the carbonate content ranges from 2 to 25% of the dry weight (Strakhov 1951). Lastly, it must be borne in mind that the basins in arid zones generally have no outlets, whereas basins in humid zones are through-flowing, and consequently lose part of their load, including carbonates. These factors together increase the intensity of carcareous accumulation in arid-climate basins over that in the slightly mineralized basins of humid belts.

Besides the intensification of calcite deposition there occurs a marked change in the several morphological forms of the calcite. In arid deposits these are varied, and may be classified in five groups: (1) ooliths and oolitic bodies, (2) carbonate incrustations, (3) shell carbonate, (4) drewite, (5) pelitomorphic carbonate.

Ooliths are widespread and are found to some extent in all the basins examined, but best developed in the Aral and Caspian Seas. Ooliths are restricted to the northwestern part of the Aral Sea and occur with coquina along the western border of the central shoal, south of Kulandy Peninsula.

These ooliths are 0·1–0·4 mm in diameter, and developed as two to six concentric zones of aragonite round a sand grain or shell fragment. Several oolitic grains may be combined inside a common shell.

In the Caspian Sea ooliths are found at almost all the stations along the shelf of the western shore from the Apsheron Peninsula to the Kura River, on the Apsheron-Krasnovodsk shallow-water ridge, and in the eastern shelf zone from Ogurchinskii Island to the middle Caspian.

In the sediments of Balkhash and Issyk-Kul Lakes the oolitic deposits consist of individual grains, the ooliths themselves being only poorly developed, and consisting at most of one or two zones.

Oolitic deposits, together with shell fragments, are cemented into *dense hard masses* in several regions in the Caspian and Aral Seas. In some places, as in the Apsheron region, the upper layer of sediment itself is cemented, forming a *carbonate crust.*

Ooliths and the carbonate crust are undoubtedly of chemical origin, the crust apparently being associated with diagenetic migration of $CaCO_3$ and the ooliths chiefly by primary precipitation of $CaCO_3$ from water.

Shell carbonate in arid-climate basins is represented by entire or fragmental shells of molluscs, foraminifers, and ostracods. Some of this material is more or less densely covered by calcareous algae (Black Sea, Krasnovodsk Gulf of the Caspian Sea). There is a general decrease in shell carbonate from the shore, chiefly in silts, towards the deeper part of the basin.

Drewite is most widespread in the Black Sea. It consists of ellipsoidal masses of calcite from 0·15 to 1·5 mm long and from 0·03 to 0·44 mm thick. Each mass is a disordered aggregate of minute crystals. Such material is found in the *Modiola phaseolina* mud and in deep-water clay, but it is irregularly distributed. Where such aggregates are abundant, the clay has a microbrecciated aspect. In transitional, and especially in calcareous mud, drewite aggregates are very abundant. They occur in layers separated from each other by thin films of clayey sapropel. In these places the drewite lenses commonly touch each other or may join into larger seams. The carbonate layers as observed in a core are found to be closely packed in places, but rare in others, forming an indistinct banded texture. In the muds that are richest in $CaCO_3$ the calcite layers range from 17 to 27 per cm of dry core, averaging about 22 layers per cm. Drewite is also found in deep-water deposits of the Caspian Sea, but it is here very subordinate. It is also very rare in the Aral Sea.

Pelitomorphic carbonate is the variety most widely developed in sediments of arid-climate basins. It is found practically everywhere, in near-shore sands, silts, in coquina belts, in oolitic accumulations, and in deep-water muds. In near-shore sediments the amount of pelitomorphic carbonate is normally small, especially in coquina, but toward the deep-water zones, except for shell material and oolites, it becomes progressively more abundant. *Deep-water muds represent therefore the principal zone of pelitomorphic carbonate development.*

Pelitomorphic carbonate consists of minute grains, one or two microns in diameter, diffusely disseminated among aluminosilicate clays of the sediment. In the Aral argillaceous-calcareous sediments several varieties of fine-grained calcite have been recognized (Brodskaya 1949). The most widely developed types are formless grains and acicular crystalline individuals. The first range from 0·003 to 0·15 mm in maximum dimension, the second from 0·0028 to 0·008 mm. The formless grains constitute the great bulk of fine-grained carbonate in the peripheral zones of the sea (Fig. 54A); the acicular forms dominate the central parts.

In two small zones, furthermore, both forms are developed approximately equally. Very small but distinctly spherulitic grains are also found.

The acicular forms of calcite are found in abundance in the sediments of Lake Balkhash and the Caspian Sea.

Being a characteristic component of deep-water pelitic sediments, pelitomorphic carbonate commonly forms thin laminated beds, a feature most distinctly seen in the Black Sea. In the Caspian Sea these laminations are less pronounced, but may occasionally be found in deep-water deposits.

Genetically the pelitomorphic calcite (or aragonite) is chemically precipitated during evaporation as acicular grains and spherulitic aggregates or as a very fine clastic mud, transported into the basin as formless grains.

In marine basins of humid regions, by far the great bulk of $CaCO_3$ forms from biogenic deposits. In arid regions this is not so. A broad picture of the balance of calcite formation in these zones is seen in the distribution of shell and micritic carbonate in various types of sediment (Fig. 54B). Since the zone of sand and silt in intracontinental seas and lakes is generally narrow and repeatedly grades into a zone of calcareous muds, it is natural that micritic calcite should dominate the carbonate. In a sense, it forms the matrix through which individual shells or shell fragments are scattered. *This dominance of micrite over all other forms is a criterion for distinguishing sediments of arid-climate basins from those of moist regions, where the main bulk of carbonate is organogenic, with little or no micrite.*

In evaluating the role of biogenic calcite in the Black Sea we started from the following considerations. Shell material forms an almost pure coquina in the northwestern part and near the Kerch Peninsula, and elsewhere is included in shallow-water muds. The mass of $CaCO_3$ in these peripheral zones of coquina, accumulating at the present stage of Black Sea history, reaches a maximum of 16 g/cm². This figure was reached for calculating $CaCO_3$ of shell zones as a whole, although the negligible thickness of coquinas and their extraordinary friability (low bulk weight) lead us to believe that the actual mass of $CaCO_3$ in the shell zones must be smaller. There are few data on the shell content of the muds, though sieve analysis with 0·1-mm screens, with corrections for detritus in varieties rich in shells, has shown that shell material here makes up from 25 to 50% of the total mass of $CaCO_3$.

In order to avoid any prejudice against a dominant role of shell material, maximum figures for shell content in the muds were used in all calculations.

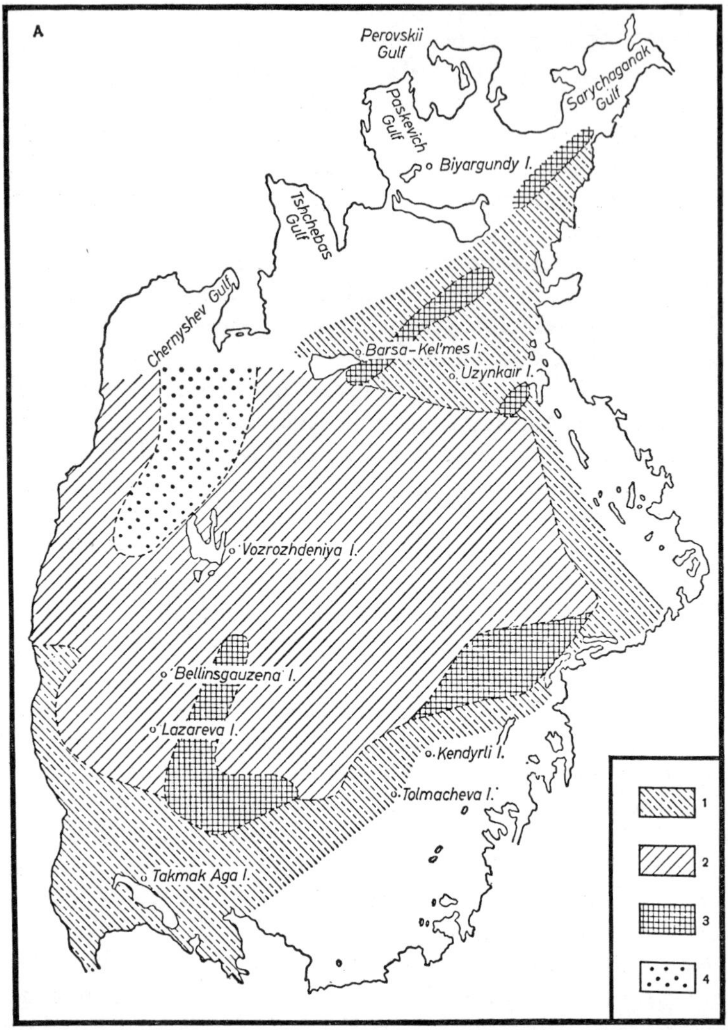

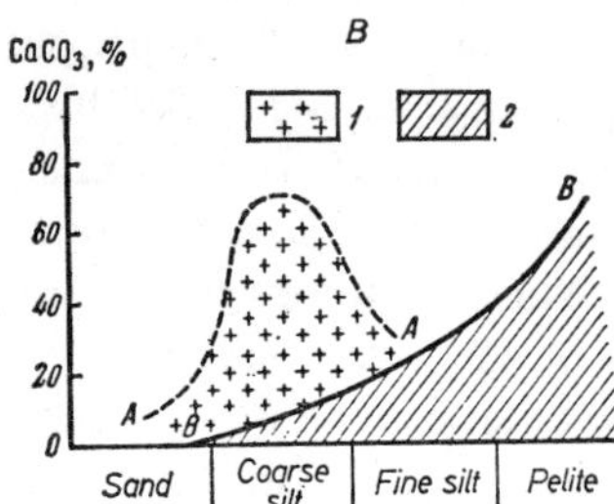

Fig. 54. Distribution of various types of $CaCO_3$ in the Aral and Caspian Seas. *A*. Map showing distribution of morphological types of carbonates: 1. Zone of dominant fragmental carbonate; 2. zone of dominant chemically formed carbonate; 3. zone of equivalent development of chemical and fragmental material; 4. zone of ooliths. *B*. Distribution of various genetic types of carbonate in sediments of the Caspian Sea: 1. shell carbonate; 2. pelitomorphic carbonate.

Using weighted values, it was found that shell material makes up about 70% of the total mass of $CaCO_3$ in shelf deposits. In the Black Sea, biogenic calcite may be considered to represent 17% of the total $CaCO_3$. Thus, *direct biological calcite in the Black Sea is definitely only a minor factor.*

Similar calculations for the Caspian Sea cannot yet be made because of lack of precise data, but judging from indirect data, the role of shell calcite can hardly be much different from that in deposits of the Black Sea. In sediments of the Aral Sea, shell calcite makes up at most 12% of the total carbonate. After correcting for loss during sieving, the figure may be increased by approximately half, and the role of organogenic calcite is seen to be similar to that observed for the Black Sea. In Lake Balkhash and Lake Issyk-Kul, the content of shell calcite is negligible. The amount is not easy to compute, and measurements would probably indicate only parts of a percent of the total carbonate, never exceeding 1–2%. Thus, *the sharp predominance of pelitomorphic carbonate over biogenic, up to complete suppression of the latter by the former in lakes, is fully demonstrated.*

The decreased role of organogenic $CaCO_3$ in continental basins of arid zones, however, does not result simply from a decrease in the intensity of the biogenic process. The tremendous populations of the few species living in these basins actually cause a greater biogenic precipitation of calcite than that in humid zones (Fig. 55). On the whole, however, the rate of increase is less than that of $CaCO_3$ supply to the basin and the farther the basin diverges from the normal marine type the greater is this discrepancy. *The organic community, becoming increasingly impoverished is increasingly unable to extract the annual increment of dissolved $CaCO_3$ to the basin, and the chemical precipitation of micrite results.*

Pelitomorphic calcite in the sediments of continental basins in arid zones might also be of fragmental origin, however; it might come from carbonate river muds. It is therefore essential to calculate, however approximately, the ratio of chemically precipitated form to clastic type in the pelitomorphic mass.

The amount of clastic calcite in each basin is clearly equal to that reaching the basin in the suspended state. In the Black Sea this amounts to less than 30% of the total annual increment of $CaCO_3$; in the Caspian Sea and in Lake Balkhash it is also 30%, though in the Aral Sea about 65%. Micritic calcite of chemical origin makes up the remainder. In the Black Sea it is known to be more than 50%, in the Caspian about 50%, in the Aral Sea about 20%, and in Lake Balkhash about 70%.

Bacterial precipitation of calcite may be determined for the Black Sea, where, because of hydrogen-sulfide content of the bottom water, the activity of desulfatizing bacteria producing $CaCO_3$ by reduction of $CaSO_4$, reaches its maximal development and may be measured. The amount of bacterial calcite must be proportional to the amount of H_2S produced. The amount of H_2S along two traverses across the Black Sea was plotted (Fig. 56), together with the $CaCO_3$ content in muds at the respective stations. The curves are

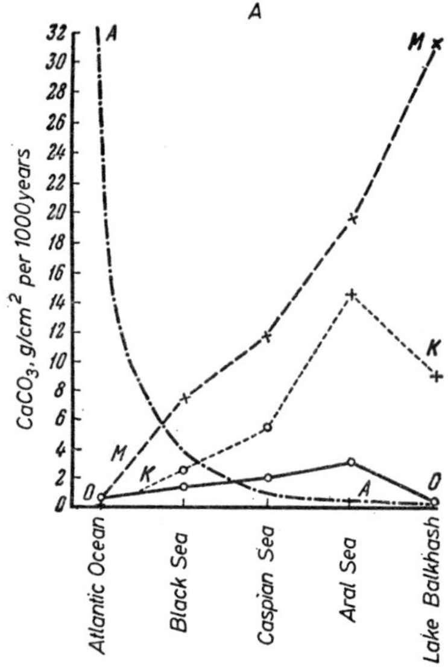

FIG. 55. The distribution of $CaCO_3$ in basins of arid zones. *A*. Rate of carbonate formation in different basins. *M—M* Rate of carbonate formation; the field between the abscissa axis and the *O—O* curve represents organogenic $CaCO_3$; the field between the *O—O* and *K—K* curves represents clastic $CaCO_2$ introduced by streams; the remainder represents $CaCO_3$ chemically precipitated in the basins; *A—A* lime-secreting organisms. *B*. Features of carbonate formation in basins of different physio-geographic type: 1. In basins with strong hydrogen-sulfide contamination of the water (Black Sea); 2. in individual basins (Aral Sea).

Type of $CaCO_3$	Basins of moist zones			Oceans	Basins of arid zones		
	Lake	Gulf	Intracontinental sea		Intracontinental sea	Gulf	Lake
Total intensity of carbonate formation							
Direct organogenic segregation						Chiefly in shallow-water near-shore zones; falling to zero in pelagic zone	
Chemical precipitation of $CaCO_3$	None	None	None	Chiefly in near-shore zones; falling to zero in pelagic zone			
Bacterial $CaCO_3$	None	None	None	Chiefly in hemipelagic clay sediments, rich in organic matter			
Introduced fragmental $CaCO_3$	None	None	None	None?, or accidental and very little			
Solution of $CaCO_3$				Only in deep parts of the sea (>1000 m)	None	None	None

interchangeable. This means that the bacterial precipitation of calcite has no noticeable effect on the distribution of $CaCO_3$ on the floor of the Black Sea.

A similar conclusion concerning the weakness of bacterial precipitation

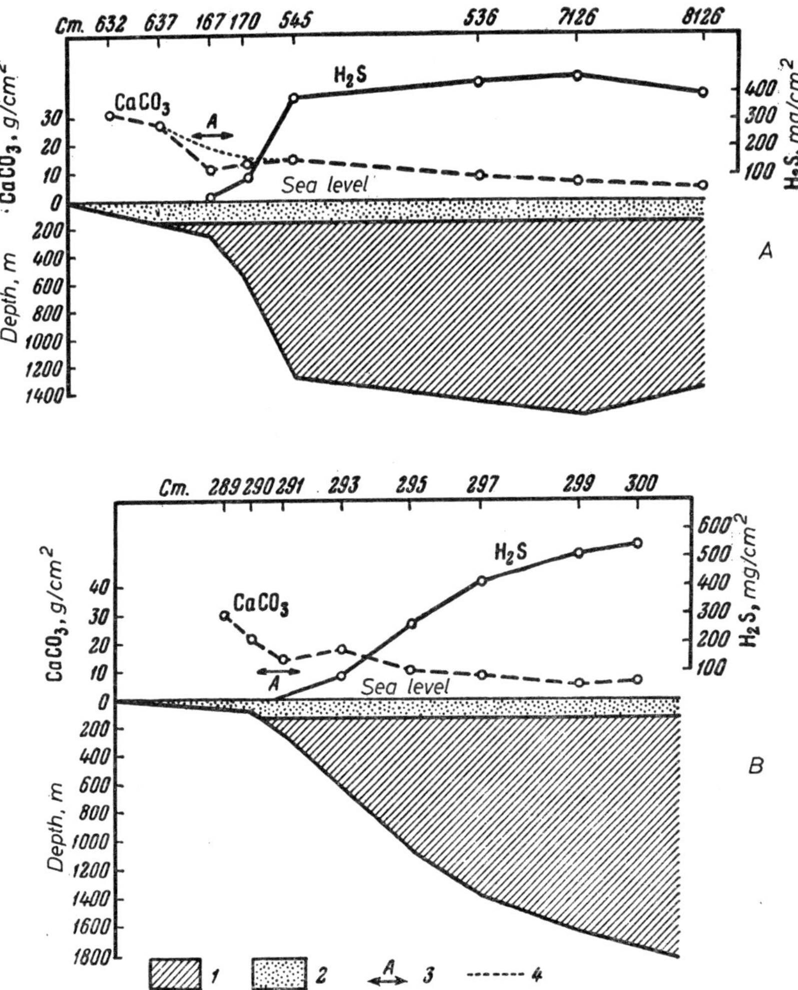

FIG. 56. Relationship between $CaCO_3$ in sediments and of H_2S in bottom water of the Black Sea. *A.* Hydrogen sulfide content in water of the Black Sea and $CaCO_3$ content in muds from Evpatoriya to Varna. *B.* Hydrogen sulfide content in Black Sea water, and $CaCO_3$ content in muds along the meridian of Tarkhankut Peninsula. 1. Hydrogen-sulfide zone; 2. oxygen zone; 3. zone of weak sedimentation of all components because of circular currents; 4. probable shape of $CaCO_3$ curve without effect of circular current.

was arrived at by analysis of the carbonate regime in the Black Sea, especially analysis of the calcite content and precipitation. If $CaCO_3$ formed by desulfatization settles to the floor, a decrease in total Ca concentration in the

H

hydrogen-sulfide zone should occur, corresponding to the decrease in SO_3^{4-} concentration. Actually, *the Ca content in water of the hydrogen-sulfide zone is generally higher, and only at the very bottom is there some slight decrease in concentration.* Accordingly, $CaCO_3$ produced by bacteria is held in solution in most of the hydrogen-sulfide zone with the simultaneously forming CO_2, and only on the floor is there some precipitation. However, the precipitation is only $0·265$ g/cm² for the entire time of hydrogen-sulfide contamination. This amounts to only $1·5\%$ of the average $CaCO_3$ in the deep-water sediments of the Black Sea.

These considerations lead to the conclusion that bacterial precipitation of calcite is negligible. The total activity of desulfatizing bacteria during the entire present stage of Black Sea history (2500 years) resulted in a maximal precipitation in the central, deepest parts of the basin of about $1·75$ g $CaCO_3$ per square centimeter (Strakhov, 1947, 1951). The effect of denitrifying bacteria has been still weaker—about one-fifth this figure. The total amount of calcite of bacterial origin may then reach $2·22$ g/cm². Since the average amount of calcite in the Black Sea is $17·2$ g/cm², the bacterial contribution constitutes about 13% as a maximum.

Another conclusion may be expressed. The clotted form of calcite in the deep-water deposits of the Black Sea is not of bacterial but of chemical origin. This consists of colloidal aggregates of $CaCO_3$, slowly settling out of oversaturated solutions in an exceptionally quiet environment. To regard this form of $CaCO_3$ as characteristic of bacterial calcite, as was assumed by A. D. Arkhangel'skii, is impossible. It may therefore be advisable to continue using the original name: drewite.

Thus, *a highly characteristic transformation of the calcite-forming process takes place in basins of arid zones. The total annual increment of $CaCO_3$ in the basins is sharply increased. In small basins (Lake Balkhash, Aral Sea) this is 30–40 times the increment of $CaCO_3$ in the equatorial part of the Atlantic Ocean. Biogenic extraction of $CaCO_3$, although increasing in absolute terms, plays a relatively reduced role. Chemical precipitation of $CaCO_3$ and the introduction of suspended material become dominant. Pelitomorphic calcite becomes the dominant type.*

4. The Distribution of $CaCO_3$ in Modern Basins of Arid Zones

The $CaCO_3$ distribution in the sediments of arid-climate basins acquires distinctive features, which differ substantially for the different morphological-genetic forms.

Skeletal $CaCO_3$, produced by the benthos, occurs in near-shore zones. Next to the shore the amount is minimal but in the zone of fine sands and silts it reaches a maximum, to fall off sharply toward the center of the basin, where it may entirely disappear. Fine-grained clastic calcite, introduced by streams, as Brodskaya (1952) pointed out for the Aral Sea, settles in deltas and in the silty zone, but its maximal deposition is in the peripheral zone of pelitic sediments. Toward the center of the sea, in the halistatic zones, the content decreases sharply. About the margins of the Black Sea the maximum clastic $CaCO_3$ lies off-shore from the mountainous areas, off well-drained parts of the coast that supply much $CaCO_3$ in suspension.

Precipitation of calcite takes place throughout the water area, but chiefly in the near-shore zones. Because of its insignificant amount, however, little of this material settles in the near-shore zone. It is carried to the more central parts of the basin, which form the principal deposition area. Since micritic easily transported $CaCO_3$ is the chief, and in some basins almost the only form of calcium carbonate, *it is natural that the chief factor in calcite distribution in the sediments is the hydrodynamic regime in the basin.*

The validity of this view is shown by: (*a*) the remarkable similarity in maps showing the distribution of $CaCO_3$ in Recent deposits of the Black and Aral Seas and the distribution there of noncarbonate clastic material (Figs. 57 and 58); and (*b*) the remarkable similarity, to the minutest detail, in the distribution in Lake Balkhash of the silicate pelitic fraction of sediments and of the carbonates (Fig. 59*A* and *B*). Such coincidence cannot be accidental. The carbonates in Lake Balkhash consist almost completely of easily transported micrite which is distributed in the basin in the same manner as the fine-grained clastic silicate. The complete dominance of the hydrodynamic factor in the carbonate distribution in Lake Balkhash is clearly seen.

Despite the fact that the $CaCO_3$ distribution in arid-climate basins follows the same laws as the distribution of quartz-silicate material, some difference may be seen on several parts of the basin floor. The mechanism of distribution may best be demonstrated in the Black Sea, where the high-carbonate zones correspond to two halistatic zones of the sea (Fig. 60). Between these zones lies a belt, corresponding to the current from the Asia Minor coast to the northeast, in which the muds are but weakly carbonate-bearing. A small area of maximal carbonate content is found also in the deep-water part of the sea south of the Kerch Strait, and this corresponds to a further minor halistatic zone. It is very likely that in the vicinity of Kerch Strait and in the north-western coquina zone the carbonate content of the sediments is high, approaching 90–95 and even 100%, but no observations have been made here.

To explain the mechanism by which the high concentrations of $CaCO_3$ developed along the two profiles, the $CaCO_3$ and the clastic contents in the Recent sediments proper were determined and compared. One of the profiles lay in the western part of the Black Sea (from Evpatoriya to Varna), and the other lay in the eastern half (at the meridian of Cape Chauda). In addition the $CaCO_3$ content in the muds was calculated as percentages of the dry weight (Fig. 61*A* and *B*).

Although the amounts of calcite and of clastic material both decrease considerably from the margins of the sea to the central pelagic zone, the decrease in calcite is less than that in clastic material. If the amount of $CaCO_3$ at the extreme near-shore station is taken as unity, the amount in the central parts of the sea along the profiles is 0·14–0·20–0·21. The distribution of clastic material, similarly expressed, is however only 0·02 in the center of the

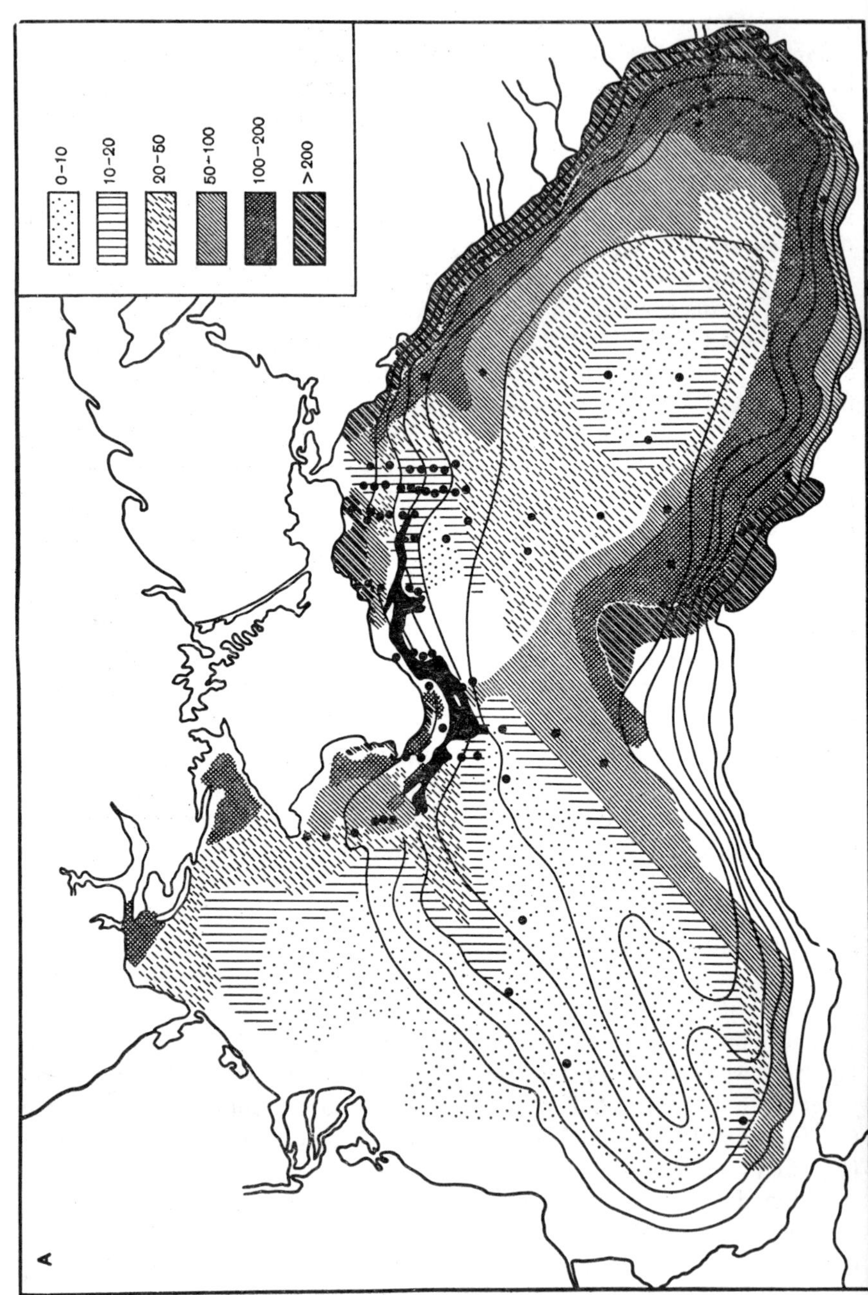

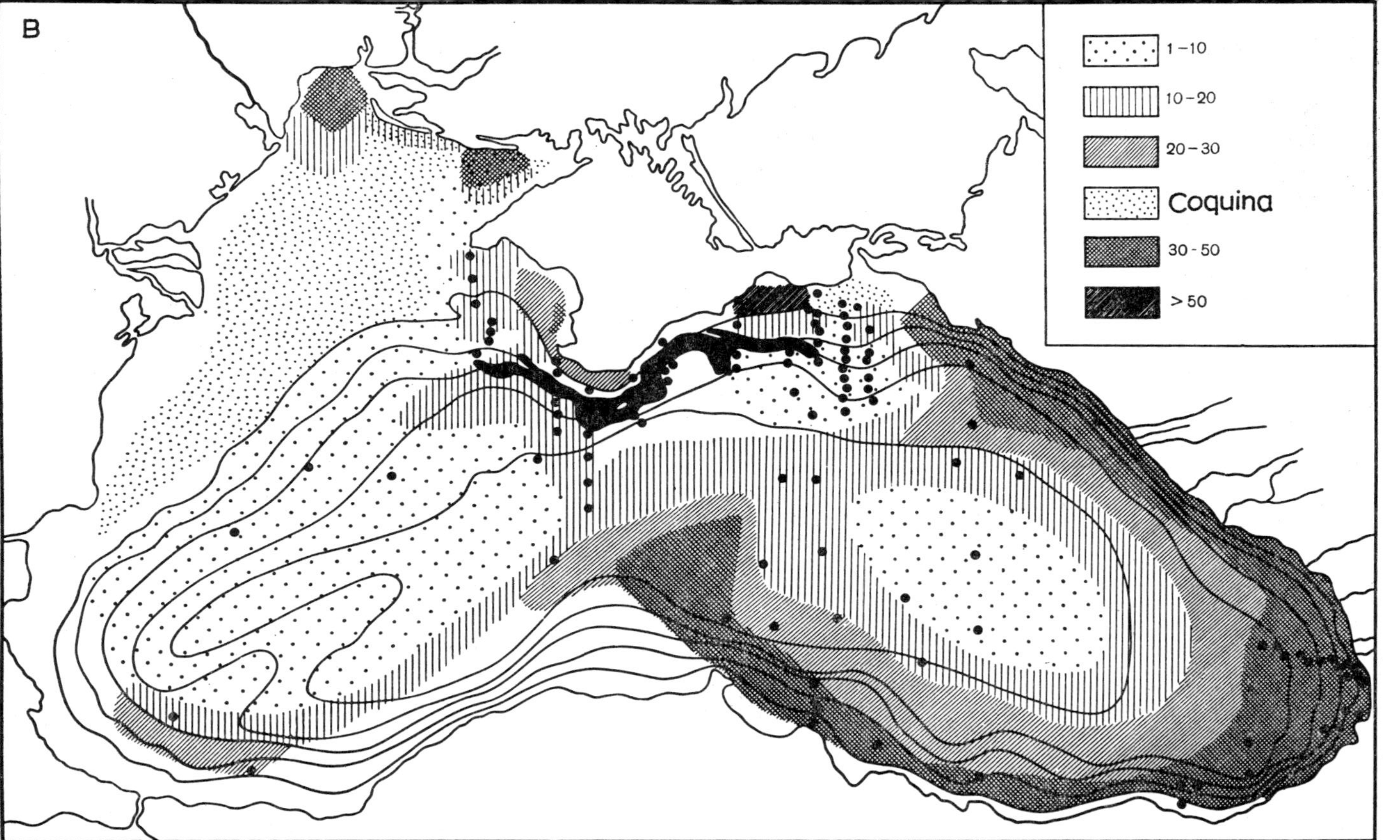

FIG. 57. Distribution of clastic material (*A*) and CaCO₃ (*B*) on the bed of the Black Sea (in g/cm²). The black indicates zones where Recent sediments are lacking. The large dots represent sample localities.

basin. In absolute values the difference is even more striking. The near-shore stations 632, 75, and 289, give 121, 142, and 130·8 g/cm² of clastic material as compared with 31·5, 30·5, and 29·3 g/cm² of $CaCO_3$. At the pelagic stations 8/27, 135/27, and 300, the clastic material is 2·3–3·12 g/cm², while the $CaCO_3$ is 4·7, 6·48, and 5·9 g/cm² respectively. In the pelagic muds $CaCO_3$ is markedly dominant over clastic material, although the absolute amounts of $CaCO_3$ (g/cm²) are much smaller than in the inshore zone. Thus the percentage curve of carbonate content should rise from near-shore to pelagic sediments. The cause of the pelagic shift is found in the difference in mobility of mechanical

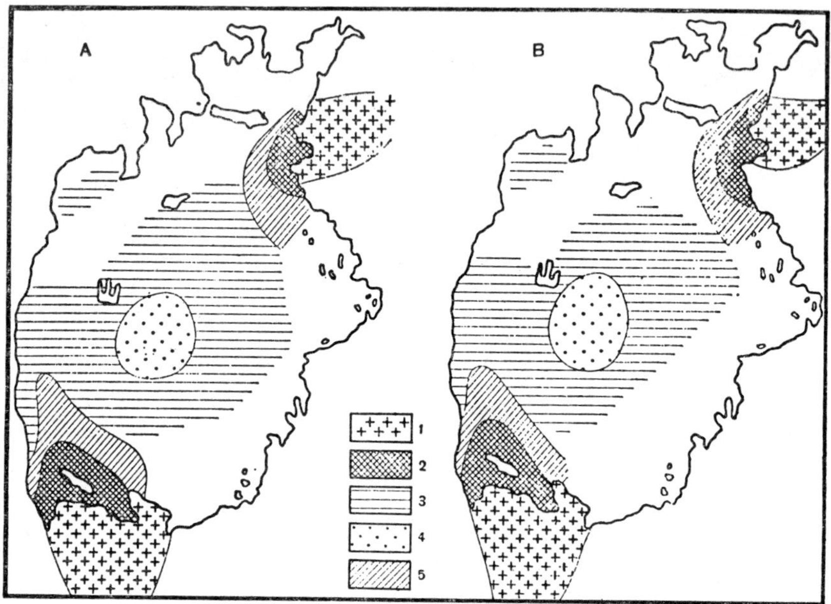

Fig. 58. Distribution of $CaCO_3$ *A.* and clastic material *B.* in sediments of the Aral Sea. 1. In deltaic zones; 2. maximal accumulation; 3 and 5. accumulations intermediate in amount; 4. minimal accumulations.

suspensions and dissolved substances. The first is mainly deposited in the near-shore zone and is but weakly deposited during halistasis. The second reaches the central parts of the sea in much greater measure, and, being deposited there, produces a higher percentage of $CaCO_3$.

In the zones of coquina the amount of clastic material per cm² decreases sharply, and may almost disappear. Clay-silt particles form only a film on the shells or in the cement, and in a very small measure they "dilute" the coquina. These coquina fields are essentially oozes, from which all clastic material is absent. *The high percentage of carbonate in coquinas therefore represents only an apparent and not a real intensification of biogenic carbonate fixation.* The distribution of coquinas in the Black Sea is indicative of their origin. As A. D. Arkhangel'skii and I originally noted (1938), the coquina zones are found in regions with flat, low coasts, undergoing only slight erosion. They are absent from mountainous coasts and parts of the continent that are being strongly eroded.

Thus, the *high concentrations* of $CaCO_3$ in Black Sea sediments do not arise by an active process; by contrast, they develop *passively, by decrease in detritus in certain parts of the sea and hence decrease in diluent clastic silicates.* This silicate material, being distributed, like the bulk of the $CaCO_3$, by water movement, is deposited chiefly nearest the shore, but $CaCO_3$ settles mainly in the pelagic zone.

The same mechanism controls the location of high $CaCO_3$ content in deposits of the Aral and Caspian Seas.

From the maps showing the distribution of clastic material it may be seen that more than 60 g/cm² accumulates near the delta of the Aral Sea, but only about 15 g/cm² in the middle of the sea. In the near-shore zone $CaCO_3$ is 30 g/cm², and about 15 g/cm² in the middle. *The clastic material decreases much more rapidly than the $CaCO_3$ from the near-shore to the pelagic zone of the basin.* The absolute amounts of calcium carbonate shift relative to the clastic material, and this again leads to a higher carbonate content in the middle of the sea.

In the Caspian Sea, the highest carbonate content is found along its eastern shore, and minimal along the western shore (Fig. 62). This is due to the asymmetrical supply of clastic material, which is abundantly supplied from the west, from the Caucasus. Very little clastic material is transported from the eastern desert shore and the carbonate percentage in consequence remains high. This mechanism of asymmetrical distribution of carbonate in the Caspian sediments was first pointed out by A. D. Arkhangel'skii in 1933. That the relationships are exactly the same in the deep-water part of the

TABLE 7

Insoluble Residues and $CaCO_3$ in the Southern Caspian, in g/cm²

Component	Western half of the basin	Eastern half of the basin	Ratio of components
Insoluble residue .	182·5	60·0	1 : 0·34
$CaCO_3$ . . .	45·0	34·0	1 : 0·75

southern Caspian may be seen from determinations made on changes in amounts of carbonates and clastic silicate material (Table 7). From west to east the amount of clastic material decreases very sharply. The amount of carbonate, though it also declines, falls at a much slower rate, resulting in the higher percentage of carbonate in muds toward the east.

Thus, the mechanism which controls the carbonates content in the Caspian muds is fundamentally the same as that in the Black and Aral Seas. The only difference is that in the Black and Aral basins clastic material is introduced more or less uniformly along the entire margin and is distributed parallel to

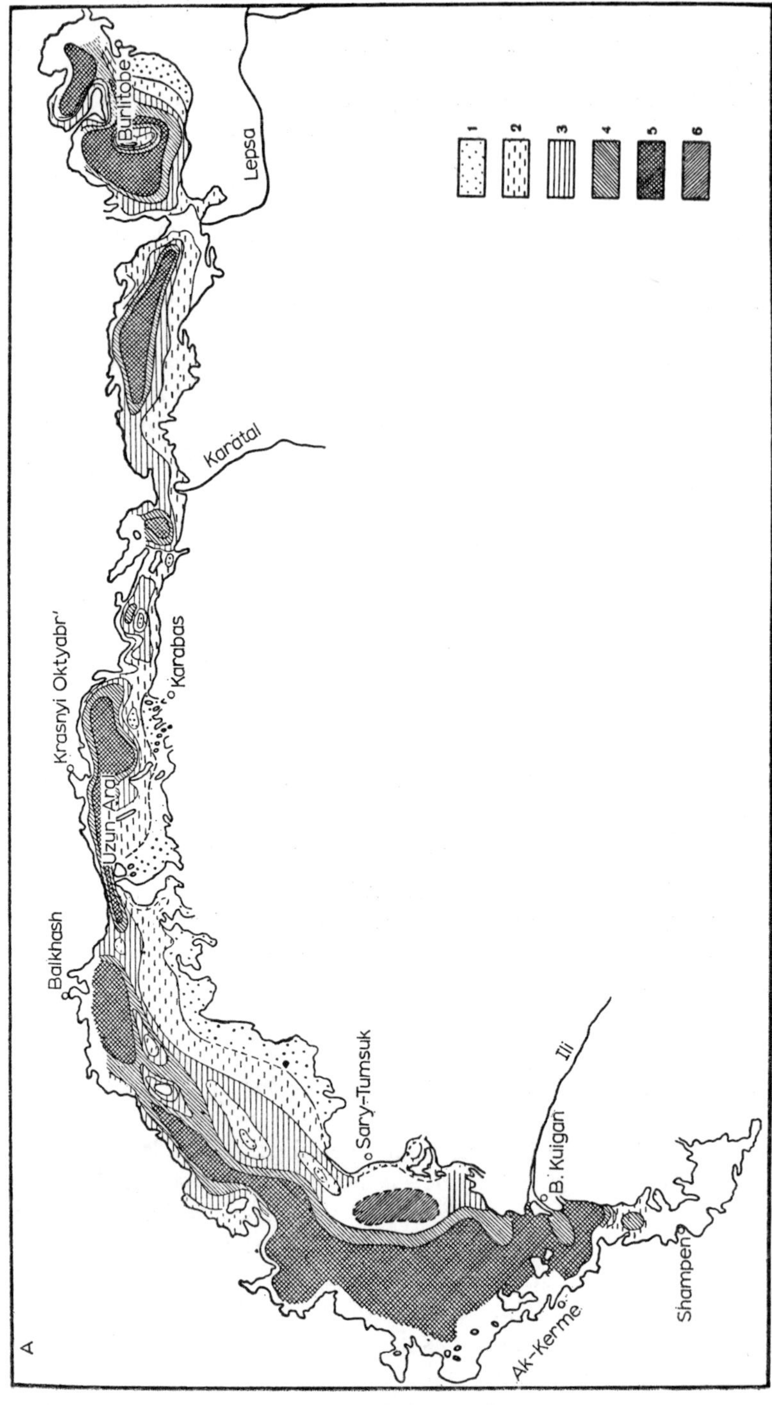
Burlitobe
Lepsa
Karatal
Krasnyi Oktyabr'
Karabas
Uzun-Aral
Balkhash
Sary-Tumsuk
Ili
B. Kuigan
Shampen?
Ak-Kerme
A
1
2
3
4
5
6

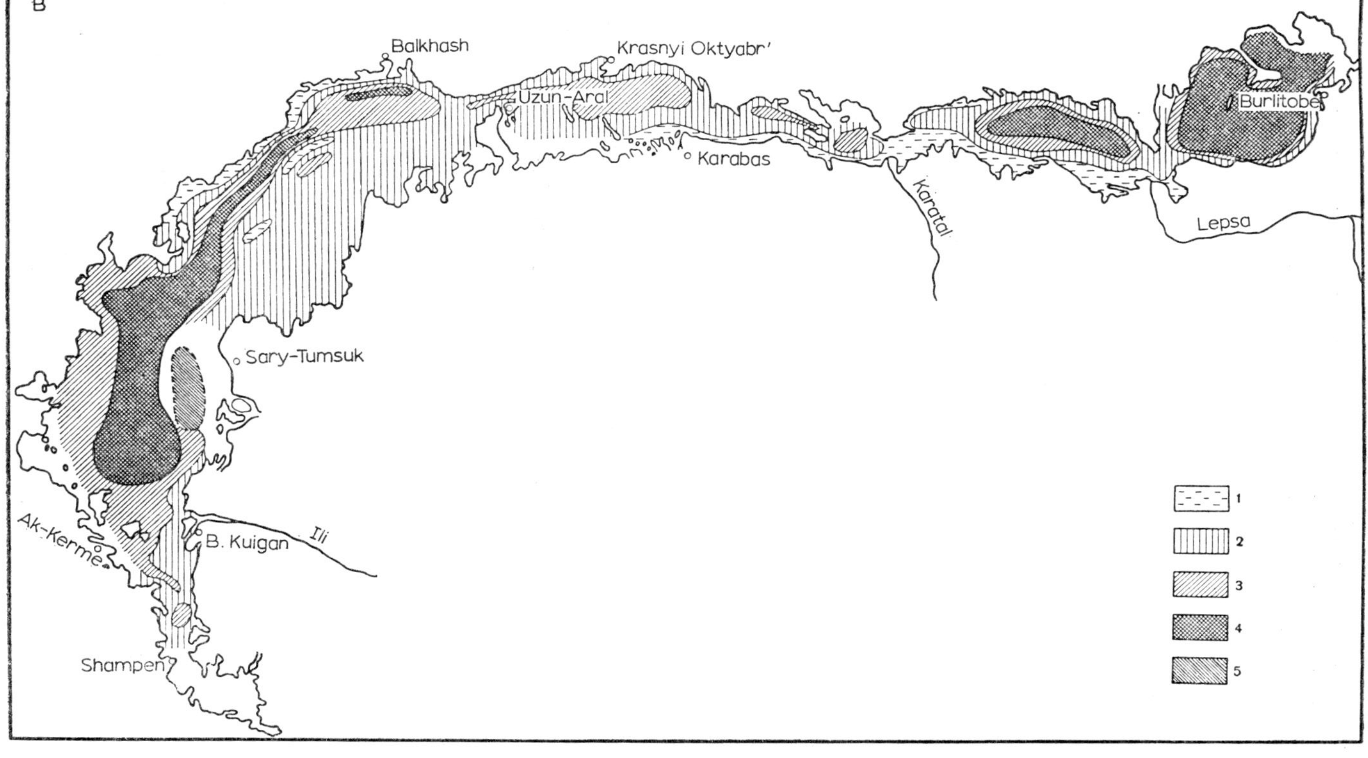

FIG. 59. Distribution of the clayey noncarbonate fraction and of $CaCO_3$ in Lale Balkhash sediments. *A.* Clay fraction: 1. less than 10%; 2. from 10 to 30%; 3. from 30 to 50%; 4. from 50 to 70%; 5. more than 70% (of insoluble residue); 6. sandy shoal built in recent years in the vicinity of the village of Sary-Tumsuk. *B.* $CaCO_3$; 1. less than 10%; 2. 10 to 30%; 3. 30 to 50%; 4. more than 50%; 5. sandy shoal built in recent years near Sary-Tumsuk.

the shore lines, whereas in the Caspian the clastic material is brought in from only the west side. Hence, a minimal percentage of carbonates is found in the muds on the west, and a maximal on the east.

The mechanism of $CaCO_3$ distribution in arid-climate basins may be

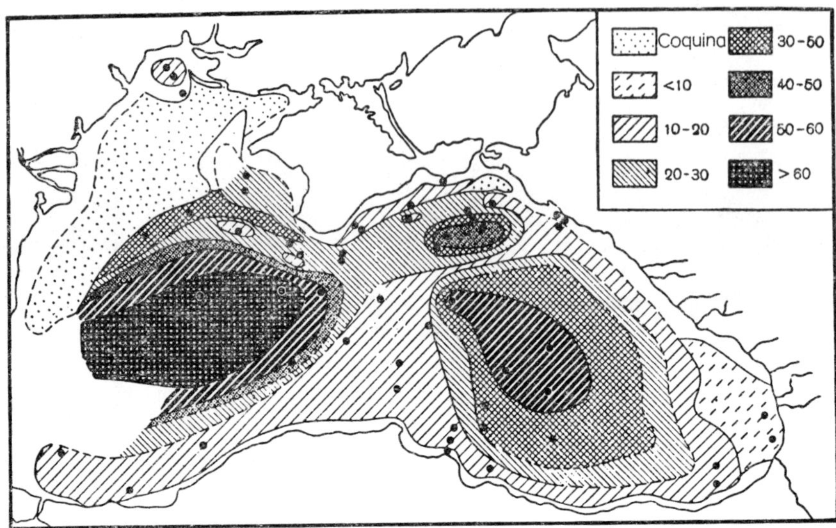

FIG. 60. Carbonate content in Black Sea deposits (in % $CaCO_3$ of dry sample).

compared with that in large seas and oceans of humid belts. There the distribution and accumulation of carbonates in sediments has been controlled (a) by climatic conditions in different parts of the sea, (b) by hydrodynamic condi-

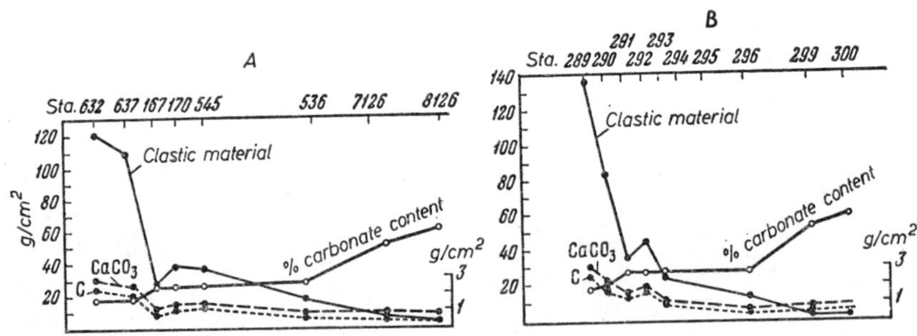

FIG. 61. Distribution of clastic material and $CaCO_3$ along profiles across the Black Sea. A. From Cape Chauda; B. from Tarkhankut Peninsula.

tions, (c) by relief in the adjacent land area, and (d) by depths of the basin. In arid basins the mechanism is considerably simpler. Since arid basins are found entirely within zones of warm climate, where calcite precipitation is favored, the climatic factor ceases to have any differentiating effect. At the

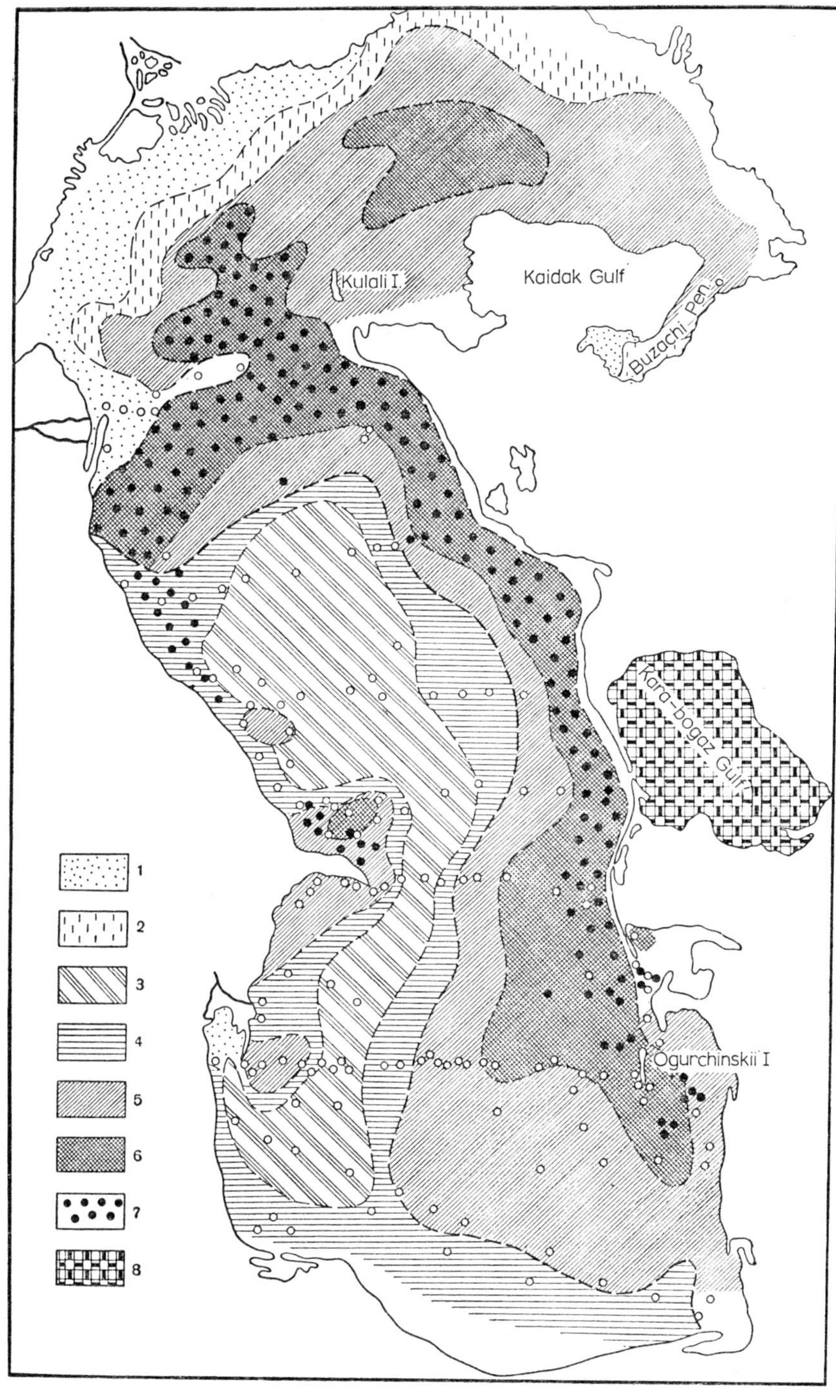

FIG. 62. Distribution of CaCO$_3$ in sediments of the Caspian Sea (in % of dry sample). 1. Less than 5%; 2. from 5 to 10%; 3. from 10 to 15%; 4. from 15 to 20%; 5. from 20 to 50%; 6. more than 50% (up to 95%); 7. coquina; 8. chemical sediments of Kara-bogaz Gulf.

same time, the depths of weakly mineralized arid basins are normally small, far less than those critical values at which depth influences the preservation $CaCO_3$ in the sediments. This factor, therefore, is also of no significance in arid zones. *There remains, consequently, only the hydrodynamic regime and the relief of the drainage areas, which are also effective in producing the distribution and accumulation of calcareous sediments in arid basins.*

Such are the specific features of calcite accumulation in present-day basins of arid zones. But, since the details of the process are closely related to the very essence of arid conditions and inevitably derived from them, it may be emphasized that *CaCO_3 deposition in ancient basins of arid regions essentially took place under the same controls that govern present-day accumulation.* The discovery of the specific features of the process, as recorded in ancient sediments, remains, however, incomparably more difficult than the recognition of present-day processes, where essentially any careful investigation may succeed.

5. ACCUMULATION OF DOLOMITE IN LAKES OF ARID REGIONS

Even more fundamental changes occur in the geochemistry of magnesium carbonate than of calcium carbonate.

In humid zones magnesium carbonate accumulates very sparsely and only biogenically. In this the geochemical cycle of magnesium is not associated with the geochemical cycle of silica. The accumulation of magnesium in dolomite is very strong in weakly mineralized basins of arid regions where biogenic processes are almost completely suppressed by the intensity of chemical precipitation. Furthermore, the geochemical cycle of magnesium crosses that of silica, and there is consequently initiated a massive accumulation of magnesium silicate (sepiolite, palygorskite, and other minerals). The facies profile of magnesium accumulations broadens sharply, including lacustrine deposits in addition to marine. This process, beginning with present-day lacustrine basins, will be first examined.

The waters in modern lakes may be classified in three geochemical groups according to the character of the carbonate: (*a*) soda lakes, (*b*) sulfate lakes with high magnesium carbonate content, tentatively called magnesium-carbonate lakes, and (*c*) sulfate lakes with water containing only $CaCO_3$, tentatively termed calcium-carbonate lakes.

The muds of soda lakes contain, according to optical, chemical, and X-ray determinations, calcite, dolomite, magnesium silicate such as sepiolite-cerolite, and, possibly, brucite.

The proportions of these minerals change substantially with change in salinity of the water. Table 8 shows the average mineral compositions of carbonate muds in lakes ranging from low salinity (almost fresh water) to 10–15% salinity.

In the initial stage of saline development, corresponding to a mineralization of 0·2–0·3% (Rublevo Lake), calcite is the dominant precipitate. Dolomite and magnesium silicate are much less abundant.

In Lake Tanatar IV, in which the salinity is about 1%, calcite is only secondary to dolomite and magnesium silicate, which are dominant. With further increase in salinity in soda lakes, calcite is reduced to negligible proportions in the mineral assemblage.

Significant relationships between authigenic minerals and the grain-size of sediments have been recorded in Lake Tanatar III (Table 9).

There is a marked enrichment in dolomite and sepiolite-cerolite with decrease in the median diameter of the grain.

Similar phenomena are observed in lakes with magnesium-carbonate water.

TABLE 8

Chemical Sedimentation in Soda Lakes, in %

Lake (in order of increasing mineralization)	Calcite (natural dry precipitate)	Dolomite (natural dry precipitate)	Dolomite proportion (of total carbonates)	Magnesium silicate (natural mud)	CaO/MgO in HCl extract
Rublevo .	34·47	22·31	38·5	12·02–17·31	1·72
Tanatar IV .	6·84	31·54	82·0	4·69–6·74	0·94
Tanatar V .	3·30	23·90	87·0	7·47–10·75	0·74
Tanatar III .	0·43	14·00	100·0	12·68–18·37	0·34

Optical, thermal, chromatographic, and chemical analyses have shown that the carbonate minerals here include calcite, dolomite and magnesite, with some hydromagnesite. The distribution of these varies systematically according to salinity (Table 10).

TABLE 9

Distribution of Authigenic Minerals in Different Types of Sediment in Lake Tanatar III, in %

Type of sediment	CaCO₃	Dolomite	Dolomite proportion of carbonates	Sepiolite-cerolite	Number of samples
Sand . .	0·73	1·24	63	2·65– 3·81	5
Muddy sand .	0·97	1·98	66	4·89– 7·04	4
Carbonate mud .	0·13	14·00	100	12·68–18·37	4

In the fresh-water stage of a lake, the carbonate is exclusively calcite, except for negligible proportions of dolomite (up to 1·5%) in a few instances. In some lakes a considerable part of this calcite may be organic, in the form of shell débris. With increase in mineralization and rise in alkali content and pH, the dolomite increases ultimately to become the dominant form, almost entirely displacing calcite. At high salinities, however, dolomite again disappears and the sediments consist of magnesite or hydromagnesite with admixtures of calcite. Unfortunately, the limited number of suitable lakes furnish only insufficient data to determine accurately the salinity at which dolomite ceases to accumulate in the sediment.

Furthermore, at least in the eastern arm of Lake Balkhash, where the accumulation of dolomite is concentrated, the muds are found to contain notable quantities of authigenic magnesium silicates, such as sepiolite or humite. Chemical analyses show this clearly (Zalmanzon 1951).

In calcium-carbonate lakes containing only $CaCO_3$ in solution, with 5–7% of dissolved salts, neither dolomite nor magnesium silicates are normally

TABLE 10

Paragenetic Associations of Carbonates in Magnesium-Carbonate Lakes of Different Salinities, in %

Lake	Salinity, in %	Calcite	Dolomite	Magnesite (hydro-magnesite)
Maloe Topol'noe	Fresh to 0·2	18·48	None	None
Khomutinoe		57·50	None	None
Paschanoe		25·68	None	None
Balkhash, western part		23·41	1·45	None
Bol'shoe Topol'noe		27·24	6·54	None
Balkhash, eastern part	0·5–4·5	28·43	37·75	None
The same		29·68	32·37	None
The same		21·52	33·88	None
Bol'shoe Kulundinskoe		3·85	18·25	None
The same		10·83	27·86	None
The same		3·01	20·06	None
Anjbulat, white ooze		7·03	None	30·95
The same		5·39	None	10·15
Lakes in Balkhash vicinity	Brine, 14	—		—
Insoluble residue of the-nardite		12·28	None	49·54
The same		21·29	None	44·17
The same		15·29	None	41·91

present in the muds. In some, however, such as the Aral Sea, magnesium silicates, specifically sepiolite, are still present (Rateev 1956). Sepiolite has been recognized as an appreciable impurity in suspensions carried into the lake by the Syr-Darya and Amu-Darya. In the Aral Sea deposits, needles of sepiolite are most numerous in the near-delta zone of the Syr-Darya and in stringers, corresponding to currents running from the river mouths to the center of the basin. A well-defined maximum is found in the central part of the western, deepest third of the Aral Sea. All these features clearly point to an allogenic origin for sepiolite deposits in this basin.

In order to show the distribution of $MgCO_3$ more clearly in lakes of different hydrochemical types, it is convenient to use the $MgCO_3/CaCO_3$ ratio in the sediments (Fig. 63). In soda lakes the ratio is near unity at very low mineralization, and attains unity on increase in mineralization; i.e. $MgCO_3$ becomes dominant over $CaCO_3$. In magnesium-carbonate lakes the ratio remains below unity for a long time, i.e. up to high salinities, exceeding unity only when mineralization is very great. In calcium-carbonate lakes of the Azov-Black Sea type the ratio remains negligible throughout the entire range of salinities, i.e. up to extreme mineralization corresponding to brines. Magnesium deposition in these sediments is very small. In lakes of the Aral-Caspian type, magnesium carbonate deposition is clearer, but only at high salinities and generally to a much less degree than in salt lakes of the carbonate II class.

Lakes of the different hydrochemical types thus form a continuous series according to amount of authigenic magnesium carbonate in the sediment. In

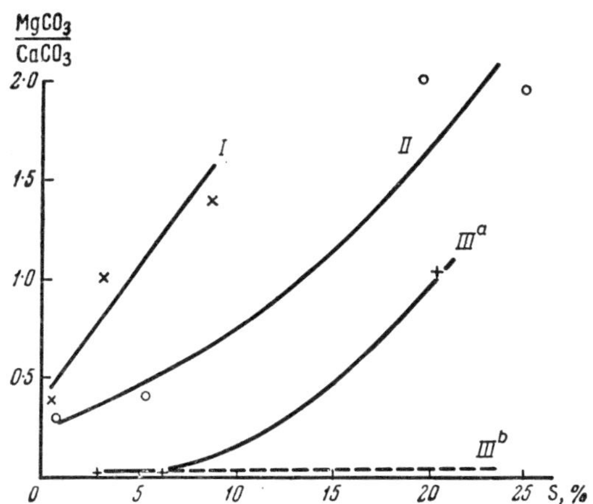

FIG. 63. Ratio of $MgCO_3/CaCO_3$ in sediments of different hydrochemical types of lakes. *I.* Soda lakes; *II.* magnesium-carbonate lakes; *IIIa.* Aral-Caspian group, *IIIb.* Azov-Black Sea group.

decreasing order the series is: soda lakes→magnesium-carbonate lakes→calcium-carbonate lakes. This corresponds exactly to the series of decreasing pH and alkali content. It graphically demonstrates the preference of magnesium carbonates for more alkaline types of basins.

Again it must be emphasized that *this pattern is by no means restricted to the present geological moment.* Soda and magnesium-carbonate lakes undoubtedly existed in arid zones of the past, as may be concluded from the occurrence of corresponding sediments among continental deposits of older periods.

In 1926-28, H. Klähn found, in the Upper Miocene continental deposits in Germany, undoubted fresh-water calcareous dolomite beds among clays and sands. The dolomite content of these beds is 60–70%. In their dolomite content these Miocene lacustrine muds strikingly resemble deposits in the eastern arm of Lake Balkhash, with which they should be compared. R. F.

Gekker (1948) described distinctive Upper Jurassic dolomitic rocks from the Rybnyi (Fish) sequence in the Turkestan Karatau. The rocks are thin-bedded with fish and insect impressions. In composition they correspond almost perfectly to the eastern Balkhash sediments, with 60–70% dolomite in very thin laminae interbedded with equally thin beds of calcite. Faunal and floral analyses by Gekker and M. F. Filippov indicate that the Upper Jurassic dolomites were deposited in a fresh-water lake in a distinctly arid climate. This circumstance and the molluscan fauna equally indicate that the Upper Jurassic carbonate rocks of the Karatau are undoubtedly close facies correlatives of the Lake Balkhash sediments, i.e. they formed in magnesium-carbonate waters. Such distribution of carbonate rock extends far back in the geologic record.

The first appearance of lacustrine calcareous-dolomitic deposits similar to the Balkhash sediments was undoubtedly in pre-Jurassic rocks. Magnesium-carbonate lake deposits are known from Upper Permian red beds of the Russian platform and from Middle and Upper Devonian rocks of the principal Devonian region. Carbonate rocks of this kind are probably as old as the platform structures that have formed the frameworks of large continental masses and on which continental lacustrine deposits accumulated.

Stratified deposits of magnesite of Tertiary age were found in California and Nevada at the beginning of the present century. In addition to magnesite, the carbonate rocks include dolomite, brucite, and magnesium silicates such as sepiolite (parasepiolite). This association very closely resembles that of the present-day soda lakes, and this appears to confirm the idea that magnesite deposits have formed in soda lakes (Gale 1914, Longwell 1928, and supported by Twenhofel). In 1944 soda-lake sediments were found by A. A. Aprodov in the Perm region of the Sub-Urals. A bed of soda was found in a Permian sequence at some depth, in a sequence of calcareous sandstones and marls, clearly representative of carbonate-bearing sediments in ancient soda lakes. The entire environment in which soda lakes now develop is encountered here: arid climate and a surrounding mountainous terrane undergoing weathering and erosion of exposed metamorphic and igneous rocks and supplying arkosic detritus and waters of soda type to the foothill region. The first appearance of soda lakes, together with the associated carbonate rocks belonging to this facies, must certainly have occurred much earlier.

Dolomite-depositing lake basins with soda and magnesium-carbonate waters are thus simply present-day representatives of very ancient types of lake basins, forming a "hydrologic relic," which aids in understanding the specific features of chemical sedimentation in weakly mineralized basins of arid regions.

6. DOLOMITE AND MAGNESIUM SILICATE DEPOSITION IN MARINE BASINS

Dolomite does not form in the seas of present-day arid regions, but such deposits were very characteristic of many Paleozoic marine basins, such as the Devonian, Carboniferous, and Permian seas of the Russian platform, the

Cambrian and Silurian seas of the Siberian, Chinese, and North American platforms, and others.

There are two petrographic types of marine dolomitic rocks: stratified and metasomatic.

Stratified dolomites are characterized by constancy of beds over great distances, measured not merely in kilometers but in tens and even a few hundreds of kilometers. Such are the Kashirian dolomites (Khvorova 1953) and the dolomitic bed with trace fossils at the boundary between the Steshevo and Protva beds in the northwestern part of the Moscow syneclise (Vishnyakov 1956). The rocks correspond in composition to normal dolomite, entirely free of calcite or with only a few percent of $CaCO_3$ as impurity. The texture is always pelitomorphic or microgranular with very faintly developed recrystallization. Organic fragments are either absent or very rare and poorly preserved. They are chiefly ostracods, rarely brachiopods, but of a distinctive type. The fauna as a whole forms a biocoenose that differs clearly from that of the underlying and overlying calcareous rocks. The dolomitic beds commonly exhibit traces of tabular gypsum crystals, sometimes merely as moulds. Fluorite is also met occasionally (Kashirian dolomite). The stratified dolomites commonly include some fine, argillaceous material that may be sufficiently abundant to form a dolomitic marl and may control the delicate bedded character of the rock. Montmorillonite is the common clay mineral in dolomites, but palygorskite and sepiolite are also present. *The most important diagnostic feature of stratified dolomites is the absence of any evidence of metasomatic replacement of calcium carbonate by dolomite.* Only where a dolomitic bed includes shell material consisting entirely or partly of dolomite may one justifiably speak of metasomatic replacement of calcite by dolomite. The scale of this phenomenon, however, is understandably negligible because of the small number of shells.

Metasomatic dolomites exhibit a great variety of forms (Figs. 64 and 65). Some show complex outlines but are essentially lensiform, extending up to a few tens of kilometers. Others are elongated or lenticular, extending up to a few hundred meters, commonly splitting into individual flats or branches. Still others are stock-like, irregular in outline, a few meters in length and thickness, or are in patches tens of centimeters long, or veinlets. In all these, however, three characteristics serve to differentiate this type of dolomitic rocks from stratified dolomites. These are:

(*a*) the striking variability in dolomite content within a lens, a stock, or even a patch, from 90–95% to 2–3% and even less. The distribution of high-dolomitic and low-dolomitic zones is entirely random.

(*b*) the clear evidence of metasomatic replacement of calcite by dolomite in the dolomite. The fine-grained cement is replaced first, forming regular rhombohedrons or aggregates of rhombohedrons, or alternatively dolomitic grains of irregular form. Organic remains are then replaced: shells of foraminifers, brachiopods, pelecypods, and similar forms, up to crinoid shells, which are the last to be dolomitized. Generally the coarser the grain size of the

I

calcite, whether of inorganic or biogenic origin, the less readily is it replaced by dolomite, and such material represents the final stage of dolomite metasomatism.

(c) the absence of any distinctive biocoenose in metasomatic dolomitic rocks. Dolomites of metasomatic type are commonly porous, cavernous, and friable, showing traces of ferruginous alteration.

In recent years two additional features of Paleozoic marine dolomites have been recognized: *their frequent association with deposits of Pb and Zn and*

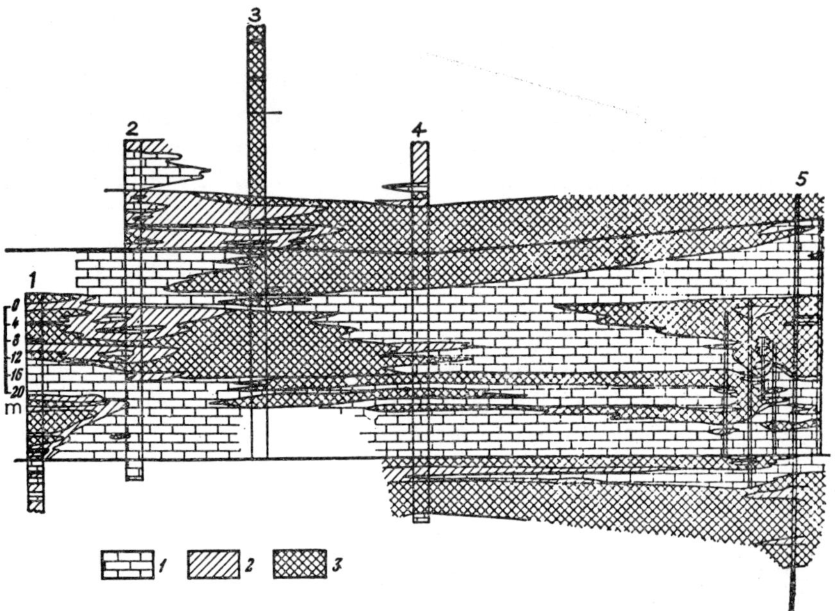

FIG. 64. Distribution of dolomitic rocks in the lower part of the Upper Carboniferous (C_3^{1c}) at Samarskaya Luka. 1. Limestone (containing less than 20% dolomite); 2. dolomitized limestone and calcareous dolomite (dolomite content from 20 to 80%); 3, dolomite (dolomite content greater than 80%).
  1. Dev'ya Mt.; 2. Yablonovyi Ravine; 3. Mogutova Mt.; 4. Bakhilova Glade; 5. Lipovaya Glade-Krestovyy Ravine.

with magnesium silicates. The first has already been discussed. As for magnesium minerals, I. I. Trofimov (1940) discovered palygorskite in many Middle Carboniferous dolomites in the Kashirian stage of the Moscow region. More recently M. A. Rateev has established the fact that sepiolite is even more characteristic of dolomites, and that associations with both palygorskite and sepiolite are frequent. These minerals, singly or together, are recorded through the uppermost Lower Carboniferous and the Middle and Upper Carboniferous strata: Lyskovo, Poretskaya, Krasnaya Polyana, Mt. Shatske, Mr. Vorotynsk, and elsewhere. Sepiolite and palygorskite are abundant in the Paleogene of Fergana, which represents a typically arid belt.

The distribution of large deposits of dolomite within Paleozoic seas varied:

some were concentrated in marginal areas, others in the central part; in the latter case the marginal zones have no, or little dolomite but are essentially calcareous. This dual distribution of dolomite is clearly seen on the maps of A. B. Ronov (1956) of the Upper Paleozoic of the Russian platform (Figs. 66 and 67) and of K. K. Zelenov (1957) for the Lower Cambrian of the Siberian platform (Fig. 68). Despite the differences in facies, all the waters of the dolomite-depositing basins were characterized by high salinity.

Increased mineralization in the zones of marginal dolomite deposition is demonstrated by the presence of fluorite in the dolomite sequences, in places

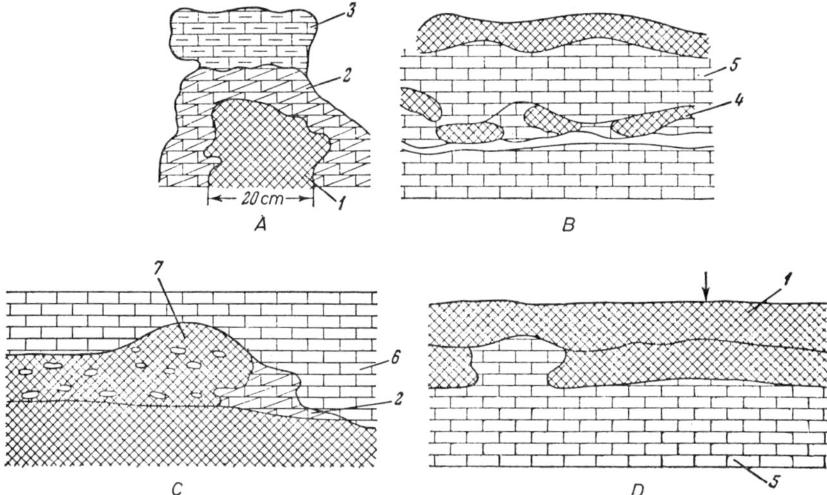

FIG. 65. Forms of metasomatic dolomite bodies in limestone beds in the lower part of the Upper Carboniferous (C_1) at Samarskaya Luka (from N. G. Brodskaya). *A.* Koz'ya Rozhka Ravine; *B* and *C* Koz'ya Rozhka Quarry; *D.* Krestovyi Ravine Quarry; 1. dolomite; 2. calcareous dolomite; 3. fusulinid limestone; 4. nodular dolomite; 5. limestone; 6. dense limestone; 7. dolomite without fusulinids.

by beds of gypsum or anhydrite, and, lastly, by the progressive faunal impoverishment as the dolomite content increases. The same features characterize dolomitic rocks deposited in the central parts of Upper Devonian and Carboniferous seas. By contrast, the fauna is abundant and varied in the calcareous deposits of the margins of the Carboniferous seas. It is indeed upon this fauna that our current views of Carboniferous fauna are based. In the dolomitic rocks of the central parts of the basin organic remains are much scarcer and the fauna more uniform. This faunal impoverishment, at least in a number of places, is not a secondary feature due to metasomatic dolomitization, but is primary. A classic example is the Upper Carboniferous of Samarskaya Luka, for which M. E. Noinskii pointed out fifty years ago that the upward faunal impoverishment, varying inversely with the dolomite content of the rocks, is a primary feature indicating progressive salinity of the

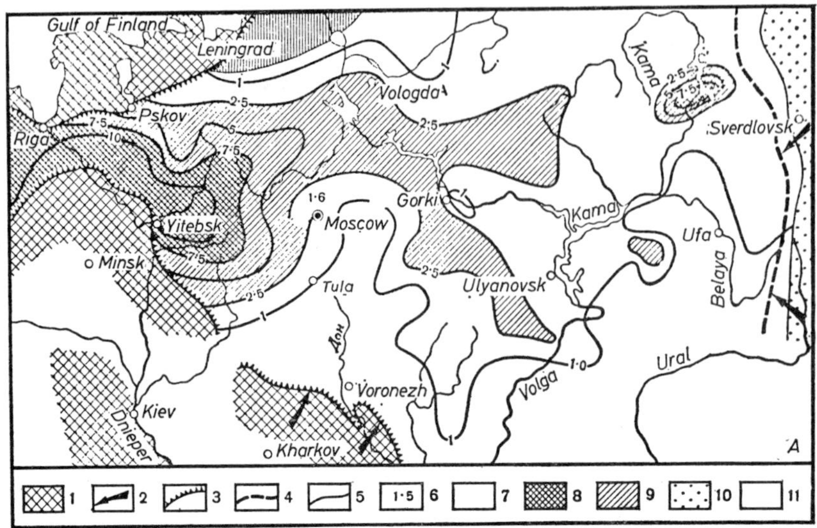

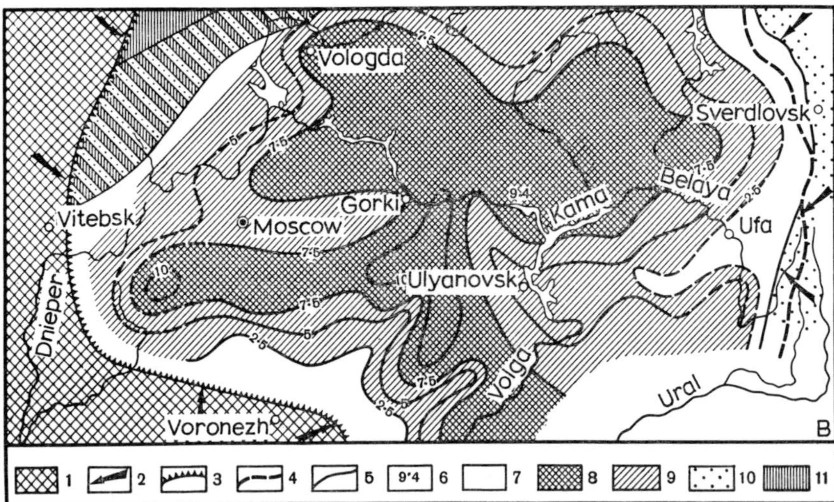

Fig. 66. Distribution of dolomite in Frasnian *A.* and Famennian *B.* rocks of the Russian platform (simplified from A. B. Ronov). 1. Region of erosion; 2. direction of detrital transport and probable course of fresh-water discharge; 3. boundary of erosional region; 4. boundary of the Russian platform and the Uralian geosyncline; 5. isopleth of magnesium content (in %); 6. average magnesium content in the section (in %); 7. zone of dominant limestone; 8. zone of dominant dolomite; 9. zone of carbonate rocks transitional between limestone and dolomite; 10. near-shore clastic marine sediments; 11. continental sediments.

Fig. 67. Distribution of dolomite in Viséan (*A*) and Middle Carboniferous (*B*) rocks of the Russian platform (simplified from A. B. Ronov). 1. Region of erosion; 2. direction of detrital transport and probable course of fresh-water discharge; 3. boundary of erosional region; 4. boundary between the Russian platform and the Uralian geosyncline; 6. average magnesium content in the section (in %); 7. zone of dominant limestone; 8. zone of dominant dolomite; 9. zone of carbonate rocks transitional between limestone and dolomite; 10. inshore clastic sediments.

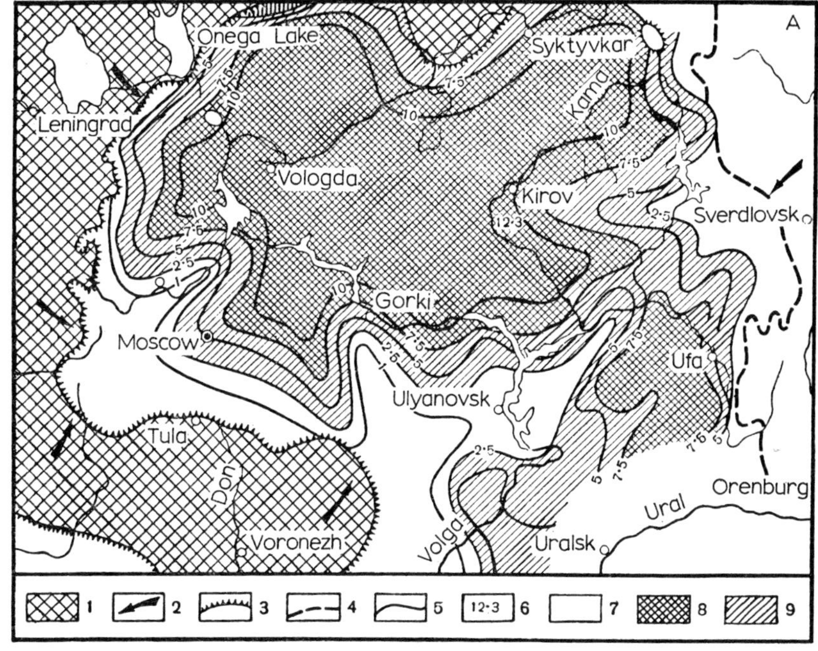

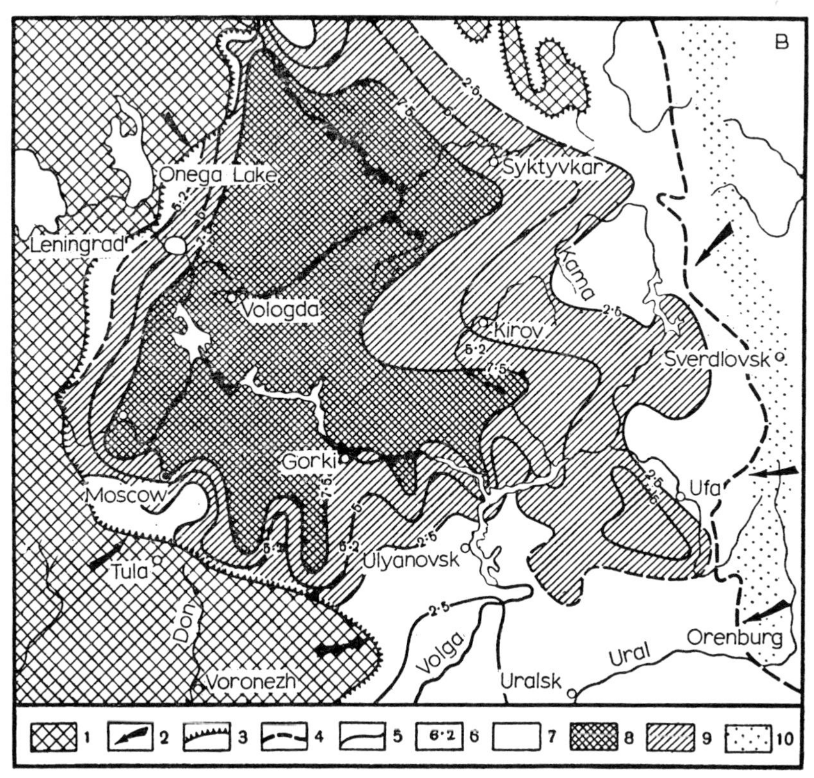

[Legend at foot of previous page

sea. A similar distribution of fauna and dolomite occurs in the Upper Famennian (see this volume, Part III, Chap. 2), in the upper part of the Lower Carboniferous (Viséan) and the Middle Carboniferous of the central parts of the Russian platform (A. B. Ronov 1956). Scattered crystals of fluorite and occasional local beds of gypsum in stratified dolomite afford supporting proof of an increased salinity.

It may therefore be reliably stated that *both the marginal and central parts of Upper Paleozoic basins on the Russian platform were characterized by*

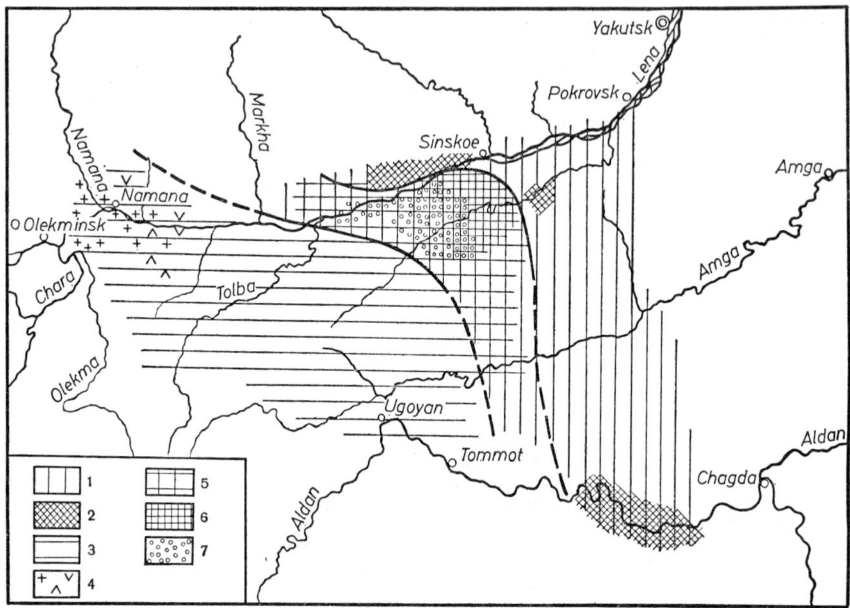

FIG. 68. Distribution of facies of the Zhura Substage. 1. Facies zone of normal open sea; 2. archaeocyathic-algal bioherms; 3. facies zone of lagoon with increasing salinity; 4. sulfate-dolomitic and saliniferous sediments of lagoon with increasing salinity; 5. transitional zone between lagoon and open sea; 6. patchy dolomitic limestones and calcareous dolomites in transitional zone between lagoon and open sea; 7. oolitic limestones and dolomites in transitional zone between lagoon and open sea.

somewhat increased salinity at the time of dolomite deposition. It is not possible to state the actual salinity, but it must be kept in mind that increase in salinity up to 38–42‰ above the average in the Red and Mediterranean Seas, i.e. an increase of 20% of the norm, brings about no noticeable restriction of the marine fauna. The fauna, however, was suppressed in ancient seas where dolomite accumulated. Accordingly, the salinity in the central parts of Upper Paleozoic basins on the Russian platform, at least at certain times, exceeded the marginal (normal) value by 8–15%, perhaps more, reaching one and one-half to twice the marginal value.

The presence of somewhat more saline waters in marginal dolomite-depositing parts of the Paleozoic seas in arid regions is not surprising. Such increase in salinity readily develops when water exchange with the open part of the basin becomes difficult and when evaporation from the surface is marked. But an increase in salinity in the central parts of a basin unaccompanied by any increase in the marginal part is also explained by increased evaporation of water from the surface in an arid climate. In the near-shore zone, however, this loss is compensated by inflow of fresh water from those adjacent continents which enjoy a humid climate. This influx has no effect on the central parts of the basin, because the water evaporates before it reaches the pelagic zone of the sea.

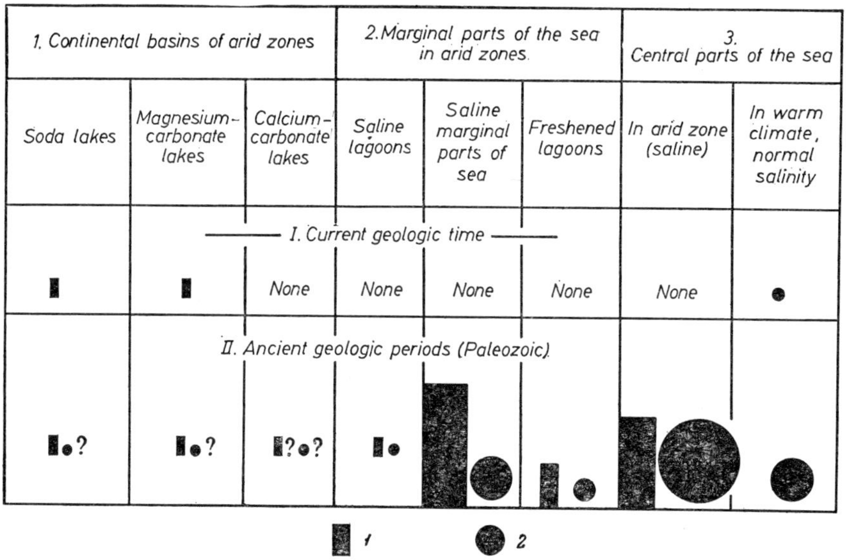

1. Continental basins of arid zones			2. Marginal parts of the sea in arid zones.			3. Central parts of the sea	
Soda lakes	Magnesium-carbonate lakes	Calcium-carbonate lakes	Saline lagoons	Saline marginal parts of sea	Freshened lagoons	In arid zone (saline)	In warm climate, normal salinity
I. Current geologic time							
▮	▮	None	None	None	None	None	●
II. Ancient geologic periods (Paleozoic)							
▮●?	▮●?	▮?●?	▮●	(bar + circle)	(bar + circle)	(bar + circle)	(circle)

FIG. 69. Facies types of dolomites. 1. Stratified dolomites (sedimentary); 2. patchy-lenticular metasomatic (sedimentary-diagenetic).

The irreversible evolution of dolomite throughout earth history is now generally accepted. This involves the elimination of a number of dolomite-forming facies, particularly those associated with seas in arid regions: lagoonal, marginal marine, pelagic marine, all widespread in the Paleozoic. It also involves a decline in the intensity of dolomite formation throughout geological time (Fig. 69). This was first established by Daly (1909), who compared a limited number of analyses of Paleozoic and Mesozoic-Cenozoic rocks of the U.S.A. It was later demonstrated by A. P. Vinogradov, A. B. Ronov, and V. M. Ratynskii (1952) for similar rocks of the Russian platform (Fig. 70*A* and *B*), on the basis of extensive numerical data. The same relation was further demonstrated by Chillinger (1956) for rocks in the U.S.A. Similar determinations have also been made for the Chinese platform.

In considering the significance of these data, I asked the question: might the sharp decrease in dolomite production during the Mesozoic be a consequence of the fact that all three platforms—North American, Russian, and Chinese— shifted from the arid zone, where they lay during Paleozoic time, to the humid

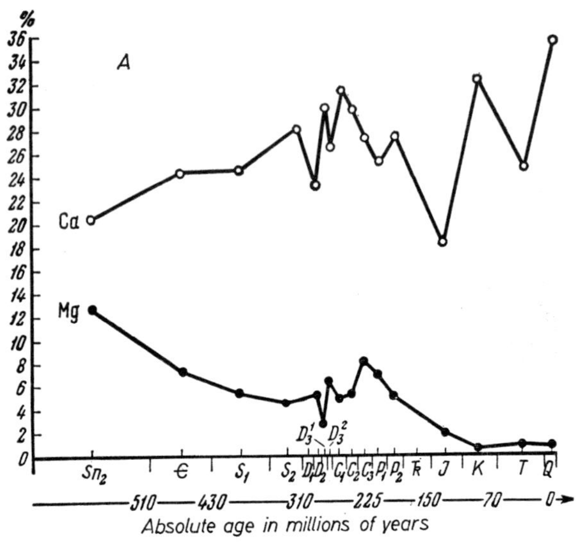

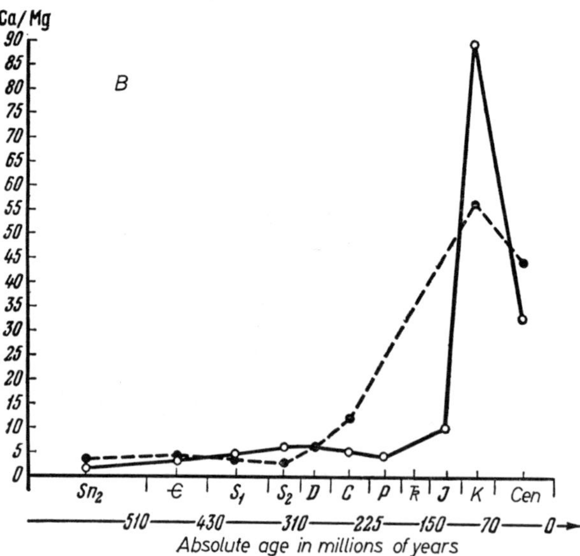

Fig. 70. Evolution of dolomite formation throughout earth history. *A.* Russian platform (after Vinogradov and Ronov); *B.* North American platform (from Daly).

zone in the Mesozoic and Cenozoic? Data on Cretaceous and Paleogene carbonate rocks of North Africa, where an arid zone existed during this time, were therefore compiled. Although it was impossible to gather extensive data, it is clear from the general description that no appreciable deposition of dolomite took place in the Mesozoic-Cenozoic seas and gulfs of the arid regions. Pure limestones, sometimes accompanied by gypsum, continued to dominate these regions. Therefore the sharp drop in dolomite formation reflects an elimination of marine dolomitic rocks in the past and their replacement by calcitic rocks. This process began in very ancient geologic epochs; by the end of the Paleozoic it was sharply defined and widespread.

7. The Principal Modes of Dolomite Formation

There is no single hypothesis concerning the origin of dolomitic rocks. Three essentially different points of view have been advanced.

According to the first, most clearly expressed by M. É. Noinskii (1913) and S. G. Vishnyakov (1956), dolomites are subdivided into two genetic groups. Stratified dolomites are admitted to be sedimentational, having been precipitated directly from water. The patchy metasomatic forms are later—epigenetic—having developed through the action of ground water on limestones.

According to the second, and much more widely held, concept, which is most completely discussed by G. I. Teodorovich (1950), stratified dolomites are also considered to be primary and sedimentational, but patchy metasomatic forms, which are markedly predominant, are diagenetic. They are formed in early diagenesis by replacement of some of the Ca^{2+} in the primarily calcareous muds by Mg^{2+}, supplied by the bottom water. Great emphasis is placed on the Haidinger reaction in this process, on the convergence of $CaCO_3$ and $MgCO_3$ solubilities during increase in pCO_2 in the sediments, and, lastly, on the effect of warm bottom currents on the sediments.

The third view was formulated in 1951-52, after I had worked on the Upper Carboniferous rocks of Samarskaya Luka (Strakhov, 1956). According to this hypothesis, there is no fundamental genetic difference between stratified and patchy (metasomatic) dolomites; in both, the dolomite is precipitated from the bottom water during sedimentogenesis. During the formation of stratified dolomites, however, $CaCO_3$ impurities are absent or negligible. During the formation of patchy metasomatic dolomites, by contrast, $CaCO_3$ is common or even abundant, the dolomitic material and the $CaCO_3$ being at first very uniformly distributed through the sediment. During diagenesis, the dolomite component is redistributed, forming lenses, patches, stocks, and other forms. In places of secondary concentration, dolomite replaces calcite, giving rise to a clear demonstration of dolomite metasomatism after primary carbonate sediment.

It is clearly necessary to assess each of these interpretations of dolomite origin in the light of known data on dolomitic rocks.

The hypothesis that the patchy metasomatic dolomites are epigenetic has three fundamental flaws. If these dolomites were truly epigenetic, they could not have features indicating formation in a limited physico-geographical environment. But they do have these features: by far most metasomatic dolomites formed in basins of arid zones, particularly in basins bearing indications of high salinity. Furthermore, it is impossible to understand how epigenetic dolomite formation could be so widespread during the Paleozoic and so sharply curtailed during the following Mesozoic. These two cardinal facts in the history of dolomite formation are not explained by the epigenetic hypothesis; they are simply ignored. Lastly, the epigenetic view cannot furnish a satisfactory interpretation of the purely quantitative aspect of the process. It must be borne in mind that the average dolomite content of horizons containing patchy dolomites is commonly very high—as much as 40–70%, as is observed in the Upper Carboniferous of Samarskaya Luka, and is noted on the maps of A. B. Ronov for other horizons. This requires the introduction of huge amounts of magnesium during epigenesis.

What is the source? The source of such magnesium is not satisfactorily explained by the epigenetic hypothesis. Where calculations have been made, as at Samarskaya Luka, no adequate source of magnesium for epigenetic replacement could be identified (Strakhov 1956). The vast development of patchy dolomites during epigenesis is therefore inexplicable, though the small patches and pore fillings in carbonate rocks, of negligible total volume, may be so formed.

The diagenetic, or the sedimentational hypothesis remains. According to the diagenetic view, calcareous muds are deposited first on the sea floor, and magnesium is introduced into the sediment during diagenesis by being "sucked" into the mud from the bottom water. Again, this interpretation furnishes no reasonable explanation either of the restriction of dolomite formation to arid zones or of the general evolution of dolomite formation throughout earth history.

Teodorovich attempts to explain the formation of dolomites in highly saline basins by saying that during increase in salinity of sea water the solubilities of $CaCO_3$ and $MgCO_3$ approach each other, and that this favors dolomite precipitation. However, as I pointed out in 1956, this idea is based on a misconception, and is indeed disproved experimentally. The Haidinger reaction is not effective, as this reaction takes place only under conditions of $CaSO_4$ saturation. In sea water, however, even at salinities up to 5–7%, $CaSO_4$ is known to be far from saturation point (it reaches saturation at a total salinity of 15%); and this is even more strikingly true of the interstitial water in the muds, since the sulfate ion in it is to some degree, perhaps entirely, lost because of desulfatizing bacteria (see this Vol., Chap. 8). The Haidinger reaction is thus excluded as a possible means of diagenetic dolomite formation. The restriction of metasomatic dolomites to arid zones and to the stage of slight increase in salinity of sea water—up to saturation of sea water by $CaSO_4$ and to precipitation of gypsum—remains unexplained by the view of Teodorovich. The explanation of the evolution of dolomite formation in earth history from the viewpoint of Teodorovich's hypothesis has, indeed, not been attempted.

Warm bottom currents that might enrich the muds in magnesium are equally useless for explaining the phenomenon. For such enrichment it is necessary to introduce some mechanism for "sucking" the magnesium into the mud, but Teodorovich points to no mechanism; and indeed none is known. Furthermore, shallow-water inshore zones of the sea are generally characterized by higher water temperature. Therefore, according to Rivière, considerable dolomitization of the near-shore carbonate muds would be expected; this is not found in present-day muds, nor did it occur in the past. The most graphic refutation of this view, however, is found in the maps showing carbonate content of carbonate rocks (Ronov 1956). These show that high dolomite content has developed, not in the nearshore zones, but in the great central areas of the basins. It is difficult to conceive of these regions being continuously enveloped by warm bottom currents. It is perfectly clear that the view of Rivière concerning the dolomitizing role of warm currents is an exaggeration and does not correspond to the true process of dolomite formation.

Thus, neither diagenetic nor epigenetic processes can explain the origin of patchy metasomatic dolomites in strata with any noticeable average dolomite content. There appears to be only one solution to the problem of how the patchy metasomatic dolomites originated, namely the third view indicated above; there is no fundamental difference between stratified and patchy metasomatic dolomites in manner of their origin. Both accumulate by precipitation of magnesium-bearing substances directly from bottom water, and any differences in origin between the two types involve details of the process, not its essence. The actual mechanism by which magnesium compounds are precipitated from water requires special examination.

8. The Mechanism of Dolomite Formation

Since dolomite is a carbonate, it is natural to infer that its precipitation must be directly connected with carbonate equilibrium in the system CaO-MgO-CO_2-H_2O and, particularly, with pCO_2 conditions. This system is not completely understood, but the following basic facts are known, partly from observations on natural processes and partly from laboratory experiments.

1. During increase in salinity of basins in arid zones, the formation of dolomites is restricted and systematic: *it always takes place after the deposition of calcite, with increase in the mineralization of the water*. This has been observed both in ancient marine deposits of all periods and in present-day lakes, *independent of the hydrochemical type*. The invariability of this sequence indicates very clearly that *the solubility of dolomite in natural waters is greater than the solubility of calcite, whatever the mechanism of dolomite precipitation*.

2. The formation of dolomite at the present time takes place not only where the salinity of a basin is high but where the alkali content is sharply increased, $MgCO_3$ being the principal constituent. In Lake Balkhash,

dolomite accumulates abundantly only where the alkali content reaches 12–13 mg-equiv/liter. The same is true of the Burla lakes and of soda lakes. Such conditions develop because of the accumulation of large amounts of $MgCO_3$ in the water, and of Na_2CO_3 also in soda lakes.

3. *A number of experiments have succeeded in synthesizing dolomite only in solutions with high alkali content and with abundant CO_2.* A.V. Kazakov (1957b) obtained dolomite at 150°C from an equilibrium solution of 28 mg/liter CaO, 86 mg/liter MgO, and 412 mg/liter total CO_2. The alkali content was 130 mg/liter, the pH 6·66, and the ratio of total CO_2 to alkali 3·07. It is understood that at ordinary temperatures the alkali content would have to be higher. G. V. Chillinger recently reported a method for obtaining dolomite: "A liter of sea water was placed in a container equipped with an electrical agitator. After adding 21·3 g $MgCO_3$ and 1·97 g $CaCO_3$ to this water (corresponding to the maximal solubility of $MgCO_3$ and $CaCO_3$ in sea water at a CO_2 pressure of 4 atmospheres), the container was placed in an autoclave with a CO_2 pressure of 4 atmospheres. The Ca:Mg ratio in the precipitate that had formed by the end of the second week was 1·8. Although the precipitated particles were too small to allow their optical properties to be determined, the index of refraction proved to be that of dolomite (1·7). X-ray analysis also indicated the presence of dolomite" (Chillinger 1955, pp. 2261-2268). G. Baron also obtained dolomite, and he too pointed out the necessity of high alkali content and abundant CO_2 in the solution (Baron, 1960).

Thus, all successful experimental synthesis of dolomite uniformly and un-equivocally points to the necessity of markedly abundant CO_2 and, conse-quently, of high alkali content for the formation of dolomite, as compared with calcite. Chillinger wrote that "preceding attempts of geologists and chemists to precipitate dolomite from sea water were unsuccessful because high CO_2 pressures were not employed".

4. More recently, Baron (1960) succeeded in showing experimentally that the formation of dolomite is a complex process developing in stages. When solid phases first come into existence there are no double bonds between $CaCO_3$ and $MgCO_3$; the two form a solid solution. But within a day the solid solution is recrystallized to a considerable extent and double bonds form, although the dolomite structure is still far from ordered. This incompletely ordered structure is termed "protodolomite." After two days the structure becomes completely ordered, and protodolomite changes to normal dolomite.

Thus, from observation in nature and by experiments, it is firmly established that *dolomite has formed (and is now forming) where somewhat increased salinity combines with a marked increase in alkali content and in pCO_2.* The role of salinity is undoubtedly of secondary importance here. It attests only to evaporative concentration in the water, producing in the solution the high alkali content necessary for the saturation point of dolomite.

On the basis of these data the mechanism of dolomite accumulation during the Paleozoic and later periods of earth history may be presented as follows. It is well known that during the Paleozoic the atmosphere was rich in CO_2, and, correspondingly, the alkali content in the sea was considerably higher than at present. We may therefore assume that dolomite was near saturation in seas of Paleozoic time. Now, since segments of the sea lose free connection with the main body of water only in arid climates, and thus become somewhat more salty, the water in these segments soon becomes saturated with dolomitic constituents and the dolomite is precipitated on the sea bed as a very fine-grained primary sediment.

Saturation is reached more easily when the solubility of dolomite is affected not only by $CaSO_4$ but also by $MgSO_4$ and $MgCl_2$, three salts with similar ions that are present in sea water. Because of this, the solubility of dolomite should decrease much more rapidly than that of $CaCO_3$, even where the salinity of the sea water is not great. Where the decline in $CaCO_3$ solubility amounts to 5% of the initial value (according to Trask) during an increase of 1% in salinity of the water, the corresponding figure for dolomite must be at least twice, if not three times, that value (Trask 1937).

Along with precipitation of dolomite, increase in salinity leads either to complete extinction or to a sharp reduction in the fauna, and to the appearance of a specific biocoenose. Stratified dolomites of normal composition formed during the Paleozoic, and these were merely compacted during diagenesis, with no appreciable redistribution or metasomatism. These dolomites are sedimentational in the fullest sense of the word.

The origin of patchy metasomatic dolomites is complex. Such dolomites have formed at a somewhat lower salinity of the water than the normal stratified dolomites. As a consequence, the intensity of primary dolomite precipitation from the water has been considerably lower and, consequently, the dolomite content in fresh sediment has been limited. Here dolomite has associated with large amounts of biogenic or chemically precipitated calcite. Great variation develops in the physicochemical environment, in Eh, pH, and concentration of individual components in the interstitial water of the muds during the course of earliest diagenesis of the sediments, by the action of bacteria (see Vol. 2, Chap. 8). As a result, extensive migration of the different constituents is initiated. This in turn leads to removal, concentration, or metasomatic replacement. The initially uniform distribution of dolomite in the sediment is also modified by diagenetic redistribution, as a result of which metasomatic concentrations of dolomite appear in the form of lenses, patches, and stocks in some parts of the sediments; whereas dolomite is removed from other parts.

Thus, *the formation of metasomatic dolomites during the Paleozoic occurred in two stages: in the sedimentational stage dolomite was formed as a mineral precipitated from the water of the basin; in the diagenetic stage it accumulated in many varied forms, and limestone-dolomitic rocks acquired their patchy texture.* In view of this complex course of metasomatism of Paleozoic dolomites, it is suggested that they be called "sedimentational-diagenetic".

The formation of patchy metasomatic dolomites in seas with somewhat lower salinity than that for stratified dolomites predetermines their location.

They accumulate chiefly in the marginal saline zones of the sea, where more mineralized zones grade into zones of normal sea water. In lagoons they lie in near-shore zones, along the margins of the middle, more saline part, where such a concentration is present. They also form at the opening of a lagoon, where less salty sea water enters. But metasomatic dolomites formed most extensively in the central parts of Carboniferous seas, slightly above-normal salinity on the Russian platform. It is probable that the low-degree and variability of salinity excess in these parts of the sea made dolomite precipitation possible only in some years, and even then only during the warmest and driest seasons. As a consequence, the sediments in the central part of the sea generally represented a mixture of various amounts of $CaCO_3$ and dolomite, and it was under such conditions that dolomite was redistributed and replaced calcite. The result was the formation of patchy metasomatic dolomites. Only where lagoon-like areas within the shoal belt in the central parts of Carboniferous seas became increasingly salty was $CaCO_3$ formation inhibited and dolomite precipitated during the yearly cycle, and for many years in succession. In these "lagoonal basins" in the sea, normal stratified sedimentational dolomite was deposited. Quantitatively, however, this type was subordinate to the patchy, metasomatic type because these "lagoons" occupied only limited areas where the salinity was only slightly higher than normal, as in the pelagic zones of the Carboniferous seas of the Russian platform.

When salinity increases strongly, the intense chemical precipitation of dolomite almost completely inhibits precipitation of calcite; stratified dolomites accumulate, with no evidence of metasomatic replacement after calcite. When the increase in salinity in a basin or in a part of it is weaker, primary precipitation of dolomite is less pronounced. $CaCO_3$ precipitates with the dolomite in some seasons of the year. A mixture of these minerals is deposited and furnishes an environment for diagenetic redistribution and for the development of patchy metasomatic dolomite.

Dolomitic constituents were much farther from saturation during the Mesozoic, but especially in the Cenozoic, than in the Late Paleozoic seas, because of a progressive decrease in pCO_2 in the atmosphere. Precipitation of dolomite from sea water that was becoming less and less mineralized in this component therefore became increasingly difficult, even when salinity increased markedly. The formation of dolomite in lagoons and in seas decreased sharply, even in arid regions, and continued to decline through succeeding geological epochs.

Conditions have arisen locally, chiefly in lagoons but also in the seas of arid regions, under which dolomite formation could be maintained at a high level. These have developed in parts of the sea where streams (or ground water) have supplied much $MgCO_3$ or $MgCO_3 + Na_2CO_3$. The alkali content and the pH of the water were therefore raised and dolomite reached saturation, with consequent precipitation. If no $MgCO_3$ is introduced into the basin, the precipitation of dolomite is retarded or stopped altogether. This is proved by the simultaneous development of basins with and without dolomitic stages. In the Upper Triassic (Tithonian), dolomite is found in considerable quantities along with gypsum in a red bed sequence in the Northern Caucasus. When the Tithonian lagoons in the region of Tadzhikistan grew salty, practically no dolomite was precipitated; limestones are now observed giving way directly to gypsum. The cause of this difference must be ascribed to difference in composition of the waters that fed the basins in the Caucasus and in Tadzhikistan. During the Paleogene, however, dolomite again formed in quantity in the lagoonal sediments of Fergana and Tadzhikistan. A. I. Osipova (1956) has shown convincing proof that waters rich in $MgCO_3$ were being introduced into the lagoon.

Thus, from an essential stage during increase in marine salinity, the formation of dolomite during the Mesozoic and Cenozoic entered an optional stage, taking place only when $MgCO_3$ or Na_2CO_3 was supplied to the basin. This has resulted in a sharp decrease in dolomite deposition during the Mesozoic, and especially the Cenozoic.

These facts help not only to interpret the history of dolomite formation in basins of arid regions, but also to identify the factors controlling the accumulation, however rare, of dolomitic rock in the seas of humid zones, chiefly in reef facies. Here (see Vol. 2, Chap. 6) the dolomite-precipitating factor is not so much salinity as other physicochemical characteristics of the environment. Dolomite is precipitated where the water is markedly warmed and the solubility of the salt is reduced, and where the water may thus become saturated with it. It is also precipitated where photosynthesis of phytoplankton or phytobenthos is intense. This raises the pH and also leads to saturation of the water with dolomitic material.

Both these causes of dolomite precipitation from water function in basins of increased salinity, or in parts of such basins. But they are only secondary factors in the precipitation of dolomite, which, without them, precipitated because of increased salinity in the basin. In humid regions, however, these factors are decisive in dolomite deposition. It is understood that temporary and local elevations of temperature and pH in sea water may give rise to only moderate dolomite content in the precipitating sediments. In humid regions this has led only to metasomatic patches and lenses of dolomite, neither large nor numerous. Only when some part of the platform basin, coming into an arid zone, is increasingly salty and passes into a stage of anomalous salinity does primary dolomite begin to form abundantly. Here dolomite lenses are larger and more numerous. This process may be traced in classical simplicity in the Carboniferous rocks of Samarskaya Luka.

9. THE ORIGIN OF MAGNESIUM SILICATES AND THEIR ASSOCIATION WITH
     DOLOMITE

The frequent association of stratified dolomites with montmorillonite and the magnesium minerals sepiolite and palygorskite is, in our view, due to the fact that the SiO_2 content (as also the CO_2 content) was much higher in Paleozoic seas than in modern seas. Because of this, magnesium silicates, such as $MgSiO_3$ or Mg_2SiO_4, closely approached saturation, but without reaching it. Then, when the salinity of the sea water began to increase either in the margins of the basin or in its central part, dolomite was precipitated together with magnesium silicates, such as sepiolite and cerolite. When clay minerals were absent from the sediments or were scarce, these magnesium minerals were preserved without diagenetic alteration. When clay minerals were present, they reacted with the $MgSiO_3$ (or Mg_2SiO_4) to form palygorskite. This mineral is thus apparently a function of the quantitative relations among the initial components.

Since the SiO_2 concentration in ancient seas, as in modern seas, *varied considerably, much more markedly than the carbonate content*, it is natural that magnesium silicates should in places accompany dolomite. They form only in those parts of the sea where the increasing salinity is characterized by excessive SiO_2 content, and where the water is saturated in magnesium silicates. It is characteristic that the patchy metasomatic sedimentational-diagenetic dolomites do not include these minerals. It is obvious that sepiolite and cerolite were precipitated from the water later than dolomite.

A noticeable decrease in SiO_2 apparently took place at the same time as the decrease in carbonate concentration in Late Paleozoic time (as a result of decreased chemical weathering?). This led to a reduction in the chemical precipitation of magnesium silicates; and consequently the association of these silicates with dolomite was rare.

10. Carbonate and Clastic-Carbonate Formations in Arid Regions

Like clastic rocks, rocks of the tetrad $P\text{-}CaCO_3\text{-}MgCO_3\text{-}SiO_2$ form large assemblages that may be called, after the principal formation-forming rocks, "arid carbonate" and "clastic-carbonate" formations. Their distribution is very irregular in the various structural units of the earth's crust (Table 11).

It is clear that the favorite areas for development of carbonate formations are platform regions, and much more rarely geosynclines, during their normal development. Marginal depressions (foredeeps) and intermontane basins are excluded from the zones of deposition of arid-climate carbonate formations. As in the deposition of carbonates in humid regions, a decisive role in the accumulation of arid-climate carbonates is played by tectonism. The more active the region, the more intense is the transport of diluent clastic material to the sea and the less the likelihood of pure carbonate concentration.

Platforms, typical regions of passive tectonic regimen, therefore, prove to be sites of the most extensive carbonate formations. A passive tectonic regimen is rare in geosynclines and is restricted in area. Carbonate formations are therefore rare in such regions.

Arid clastic-carbonate formations, like similar formations in humid regions, form during tectonic activity. Such rocks are less abundant on platforms, therefore, and are more widespread in geosynclines, but are absent in the very active tectonism of marginal depressions and intermontane basins of geosynclinal zones that are being folded and otherwise deformed.

The composition and structure of arid-climate carbonate formations are as yet little studied, and that only on platform formations. Only the platform type will therefore be described.

Characteristic representatives are the Middle and Upper Carboniferous deposits of the Russian platform, where the area occupied by Middle Carboniferous rocks is 2,782,700 km^2, at an average thickness of 179 m (Ronov 1956). The Upper Carboniferous rocks cover 2,204,300 km^2 and average 131 m in thickness. Clastic rocks are present in both sequences together with the carbonates. In the Middle Carboniferous, carbonate rocks total 462,400 km^3 in volume, clastic rocks 29,200 km^3. Corresponding figures for the Upper Carboniferous are 279,200 and 10,000 km^3. This means that the average

TABLE 11

Development of Arid Carbonate and Clastic-Carbonate Formations

Group of formations	Type of formation	Series of formation			
		Platform	Geo-synclinal, at stage of normal develop-ment	Geosynclinal, at terminal stage	
				Marginal depression	Inter-montane basin
Marine	Epeirogenic 1. Formations of seas with flat and low drainage areas (carbonates)	$++++$	$+$	$-$	$-$
	2. Formations of seas with moderately dissected drainage areas (clastic-carbonates)	$++$	$++$	$-$	$-$

accumulation rate of carbonate rocks was 2·75 g/1000 liters in the Middle Carboniferous and 2·00 g/1000 liters in the Upper Carboniferous.

The clastic sand-silt-clay rocks are generally red, always more or less carbonate-bearing. Regionally they are confined to the western margin of the formation, i.e. to the edge of the Moscow syneclise. Here they accumulated most extensively in Vereyan time. The zone occupied by these rocks in this region is narrow, but in individual horizons red clays and marls penetrate rather far into the formation. Genetically these are partly continental, partly deltaic and near-shore marine accumulations, marking a narrow shore-line belt. It is conceivable, however, that these clastic stringers might be absent, especially on gently sloping shores, and that carbonate sedimentation might appear directly at the edge of the shore. In some places lenses of gypsum accumulated in lagoons, and are found among near-shore deposits, such as in the Kashirian stage in the Kalinin region.

In the carbonate deposits proper in the Moscow syneclise, two essentially

K

different facies have been recognized (I. V. Khvorova 1953, and Fig. 71*A* and *B*). The western inshore zone was in the sphere of wave action, and it more or less corresponded to the zone of turbulence. Coarse cross-bedded calcareous sands and gravels formed at the upper edge of this zone. Erosion of these sediments occurred occasionally, and islands and shoals apparently existed, along the margins of which eolian limestones formed by the winnowing action of wind on the calcareous material. Great numbers of calcareous algae lived in this zone (oncolites) and there existed a poor fauna, clearly adapted to strong wave action, of *Meekella*, solitary and colonial corals, attaching foraminifers, and boring molluscs. Farther off-shore the reduced wave action allowed the fauna to become abundant and varied: brachiopods, bryozoans, crinoids, echinoids, some trilobites, bivalves, and others. Through movement of the water and the work of predators, shells were normally crushed and converted more or less to fine detritus, the most susceptible shells being ground to the finest powder. Locally, however, biohermal lenses of limestone were formed. The finest-grained silty and pelitic detrital carbonate material was brought into suspension and transported into deeper parts of the sea. Fine chemically precipitated calcite that formed in the zone of turbulence suffered the same fate. The marginal zones of carbonate formations were thus characterized by an extensive development of organic detrital and biohermal limestone and by subordinate pelitomorphic limestone. The average dolomite content of the rocks in this peripheral zone of the Middle Carboniferous sequence decreased, and dolomite itself occurred in metasomatic form, although units of normal pelitomorphic dolomite with fluorite (ratofkite) formed occasionally, as during Kashirian time in the Kalinin region.

Farther east, in the interior part of the formation, within the Moscow syneclise, micritic varieties formed among the carbonate rocks. Organisms were rare and were adapted to muddy bottoms. A characteristic feature of the inner part of the syneclise is the abundant development of dolomite, some of it normal, but most commonly metasomatic. Layers rich in magnesium sili-

---

FIG. 71. Facies zoning in calcareous-dolomitic formations of the Middle Carboniferous on the Russian platform (from I. V. Khvorova). *A.* At the time of stable sea level. 1. Land; 2. outer part of the marginal zone with sandy material; 3. outer part of the marginal zone with sand-silt sediment; 4. outer part of marginal zone with clayey sediment; 5. inner part of the marginal zone with sand, clay, and carbonate; 6. inner part of the marginal zone with clayey, marly, and organo-fragmental sediment; 7. inner zone with muddy and fine-foraminiferal material; 8. inner zone with chemically precipitated carbonate. *B.* At the time of strong shoaling. 1. Continental land and islands; inferred zone of clastic material; 3. littoral zone, covered with carbonate sediments; 4. bars, composed of calcitic sands; 5. lagoons with dolomitic muds; 6. quiet shallow-water zones, with accumulations of fine calcareous muds rich in faunal remains; 7. shoals of open sea, zone of coarse organofragmental sediments; 8. upper subzone of the inner edge of the shallow water zone, rich in fauna (corals, fusilinids, etc.); 9. lower subzone of the inner edge of the shallow-water zone, chiefly with fine-foraminiferal and fine-detrital sediment; 10. inner zone with fine carbonate (muddy) and fine-foraminiferal sediment.

cates are found in the dolomites, and also lenticular masses of gypsum of small extent. East of the Moscow syneclise the common type of carbonate rock is the same as in the syneclise itself, but the dolomite content decreases (according to Ronov's map), with approach to the Uralian geosyncline. It is

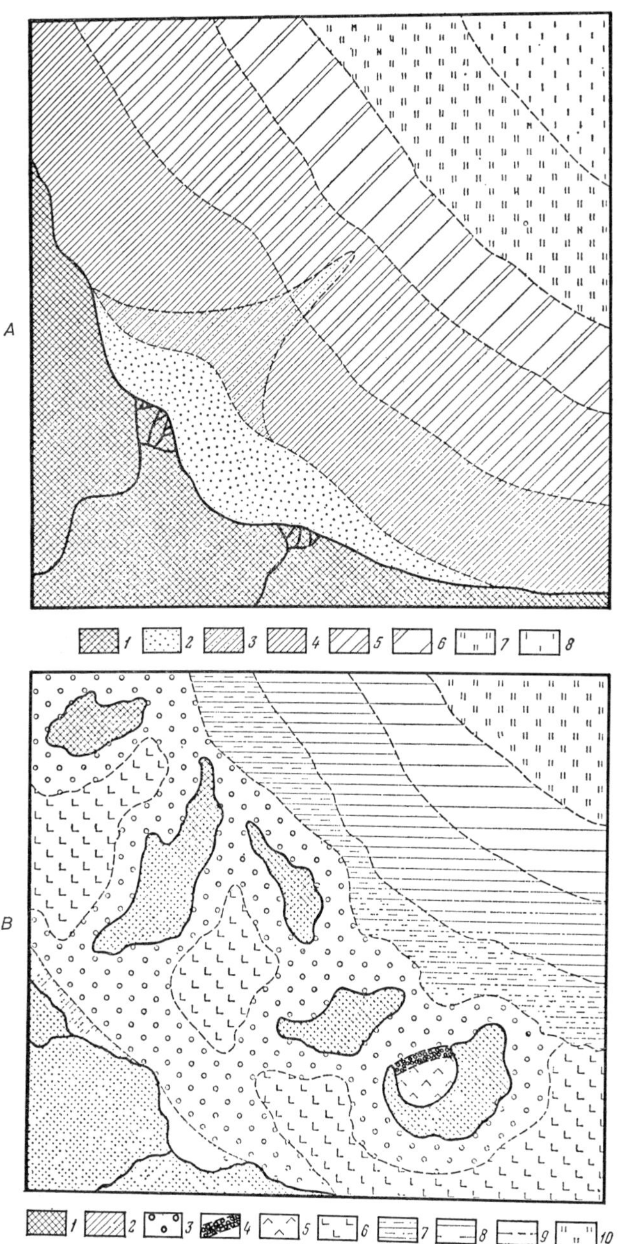

[Legend at foot of previous page

obvious that the salinity of the sea gradually decreased and the dolomite-forming process was suppressed.

A fundamental feature of this Middle Carboniferous limestone-dolomite formation is the *marked appearance of cyclothems*. The floor of the basin was frequently raised, with a temporary drying of the vast western part of the basin; after re-submergence carbonate sedimentation began anew. Such cycles, varying little over great distances, frequently covered almost the entire area of the Moscow syneclise.

Carbonate formations similar to those described for the Middle Carboniferous are widespread on other platforms, having formed at quite different times. These include the *Schwagerina* bed and the *Spirifer* beds of Upper Permian age on the Russian platform, the Lower Cambrian sequence of the Siberian platform (excluding the halite formation of the Irkutsk amphitheater and the Berezovka downwarp), the Ordovician, Silurian, and Devonian sequence of the Chinese platform and of North America (excluding again the zones of halogen formations), and others. Everywhere these are shallow-water deposits, commonly micritic (especially in the Cambrian), only locally organic. Everywhere dolomitization is extensive and in places sulfatization also. Traces of earth movements are ubiquitous, some minor, some more widespread, and everywhere the rate of carbonate accumulation has been slow, amounting to 1·45–4·12 g/1000 liters.

Together with these general features, some formations have their own individual peculiarities. These include a greater or lesser width and thickness of the marginal, red, sand-silt-clay rocks, and greater or lesser dolomite content and sulfatization. Even greater individuality among the formations is acquired by the distribution of dolomite and of associated minerals: fluorite, celestite, gypsum. In the above Middle Carboniferous carbonate formation, dolomite is located in the inner parts. This relationship also applies to other formations, such as the Viséan and Upper Carboniferous of the Russian platform. But other dolomitic rocks are confined to one edge or to the marginal zone in general, such as Givetian, Lower Permian, and Kazanian formations.

By comparing the characteristics of arid and humid platform carbonate formations, the two may readily be differentiated. The distinguishing features of arid formations are primarily the association with red sand-silt-clay rocks, the very sharp increase in the average dolomite content of the carbonate rocks, the general widespread distribution of fluorite and gypsum, the accumulation of magnesium silicates (palygorskite, sepiolite), and the sharply defined shallow-water character of the sediments, even in the central parts of the basin. Calcareous-dolomitic formations of arid climates have little in common with relatively deep-water chalks of humid zones.

In the clastic-carbonate rocks, and also in the geosynclinal carbonates, these distinguishing features are less clear and less complete. The characteristics of the formation gradually converge and become similar. This is natural because the growth of tectonic activity, as pointed out in Volume 2, reduces the effects of climatic differences, although it nowhere destroys them completely.

CHAPTER 4

THE ACCUMULATION OF ORGANIC MATTER IN ARID ZONES

1. GENERAL REMARKS

In humid regions the deposition of organic matter is widespread, both on the continents as coal, and in marine basins as oil shales. By contrast, in arid regions forests give way to steppes, and farther within the arid region even the continuous grass cover disappears, leaving only a poor, thin flora, or even in large areas no flora at all. (see Vol. 2, Fig. 134). *Under such conditions no coal will accumulate, since it forms only in places of forested growth.* Along the margins of deserts only small oases survive locally, under favorable hydrogeological conditions, and these may give rise to insignificant coal beds, a few centimeters thick, and negligible in area. In seas the situation is different. Since increased generation of organic matter, enriching the sediments and converting them into marine sapropels, has been associated not with the supply of biogenic elements from the continent but only with more active vertical mixing of the water, which has raised these elements from deep levels to the zone of photosynthesis, there is a real possibility of oil shales forming in the seas of arid regions. Large bituminous and oil shale sequences are actually formed within arid zones.

Such deposits are known from very different regions and stratigraphic horizons. The Sinyaya series on the Aldan Massif (Zelenov 1955) belongs to the Lower Cambrian of the Siberian platform, and similar deposits of the same age are found about the Anabar massif. Graptolitic shales were formed in the Silurian in many parts of the northwestern Siberian platform (Miroshnichenko). In the Devonian, the Domanik facies was deposited equally in humid and arid regions. It makes up a widespread bituminous horizon between two halogen series in the Pripyat downwarp, and apparently extends from there into the Dnieper-Donets basin. Domanik deposits have also been recognised on islands of the Arctic Ocean, in the then northern arid zone. Huge deposits of marine bituminous shales (black shales) are known from the Devonian of the U.S.A., from regions that lay in the arid zone. In the Permian the cupriferous Zechstein shales of Germany were deposited and the Abdulino and Yurezan' bituminous shales of the Soviet Union (Strakhov 1934, 1935). A small bituminous shale horizon in the Northern Caucasus (I. G. Kuznetsov 1928), formed in Kimmeridgian times. Numerous Tertiary bituminous horizons are known in the Maikopian and Sarmatian sequences on the northern slope of the Caucasus, in the Carpathian region (the Menilite series), and elsewhere. These examples are by no means all the known deposits of organic matter in seas of arid regions, but they are sufficient to demonstrate that such deposits have been extensive in arid regions in the geologic past.

On the whole, the history of organic matter in arid regions, like its history in the above biclimatic group, is distinctively and sharply changed: *continental sedimentation disappears in this history, but marine deposits remain, and these are as impressive as bituminous shales in the seas of humid zones.*

As for facies-genetic types of arid-climate bituminous shales, it is possible to distinguish planktonic and benthonic varieties, as in humid belts, but they are fundamentally different in their development. Whereas planktonic bituminous shales in arid seas are numerous and petrographically variable, benthonic types are rare, and those known are of negligible size and are restricted to specific environments. This is demonstrated in the following examples.

2. Benthonic Bituminous Shales in Arid Regions

The bituminous shale horizon at the village of Abdulino in the lower reaches of the Yurezan' River is apparently benthonic, and, of a more convincing example, will now be described.

The bituminous shale horizon of Abdulino rests on a coralline reef limestone of Lower Permian age (*Ps. urdalensis* zone) and reaches a maximum thickness of 4·5 m. It is restricted to about 3 km upstream and downstream

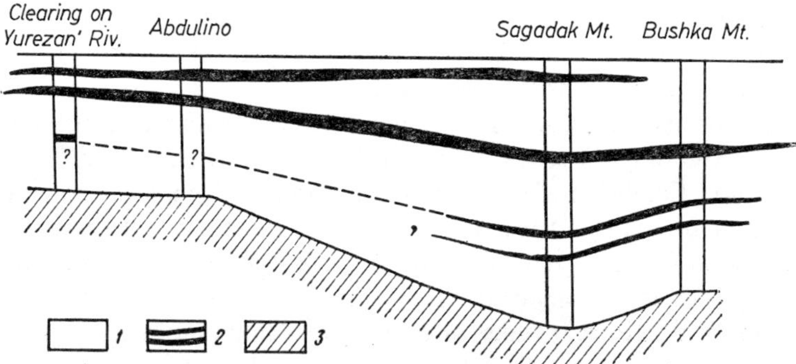

FIG. 72. Structure of the Abdulino oil shales. 1. Limestone; 2. bituminous shales; 3. coralline limestone substratum.

from Abdulino along the Yurezan' River. Four beds rich in organic matter occur in the shale. The lowest is known only from Mt. Sagadak and Mt. Bushka. The next two are more widespread, especially the third (Fig. 72). The beds are essentially rather large lenses, in thickness ranging from 10 to 30 cm, and occurring in a sequence of light-colored bioclastic limestone.

The bituminous shale lenses vary considerably. At the Mt. Sagadak section, in the central part of the third lens, for example, the shale is brownish-black, very dense, and very light and friable, with a resinous luster on fresh fracture.

It is very homogeneous and coal-like, and contains 26–29% insoluble mineral residues and 9–11% carbonate; the remainder is organic matter. Most of the $CaCO_3$ is in micritic calcite, which forms clearly defined individual masses in some places, and more or less large lenses in others. The organic matter occurs in films up to 5 mm thick, dark reddish brown, and structureless. In places the films are small and disconnected, but elsewhere they form distinctive layers. When analyzed, the organic material reveals 2·56% bitumen (bitumen A, 1·78%; bitumen C, 0·78%) and 35·78% humic acid. N. A. Ivanchina-Pisareva (Maslov 1950) demonstrated traces of the blue-green alga *Pila* and remains of limbate pores, thus demonstrating the contribution of higher plants to the whole. The only other organic remains identified are of ostracods. Towards the margins of the shale lens, the character of the rock changes rapidly. Downstream from Abdulino the bituminous shales are black, very dense and heavy, homogeneous, strongly bituminous and marly, and readily fissile. They contain about 30% clastic material, 45% carbonate, and 25% organic matter. It is clear that this part shows a marked enrichment in carbonate and a great reduction in organic matter. The large number of organic remains is characteristic. In the middle parts of the bed brachiopod fragments, chiefly productids, occur. The upper and lower surfaces of the bed contain so much shell débris that the bed becomes a distinctive coquina, abounding in organic material. Below Abdulino the bituminous shale lens grades into dark gray bioclastic limestone with little C_{org}. Here the terrigenous material amounts to 11–19%, carbonate to 66–81%, and C_{org} to 1·39–6·51%. The bulk of the rock consists of brachiopod fragments, crinoid stalks, foraminiferal shells (*Fusulina, Cribrostommum, Bigenerina, Rotalidae,* and others, bryzoan remains, and, lastly, the alga *Siphonales.* Upstream from Mt. Sagadak the lithological changes are generally the same, but they occur more rapidly.

Lithologically the bituminous shales differ markedly from the normal bioclastic limestones, but the two rock types pass gradually the one into the other in vertical sequence. The principal general difference between the bituminous shale and the enclosing rocks is the abundance of organic matter and, partly, of clastic material. In the shales the latter constitutes about 25%, in the enclosing rocks 10–15%.

The environmental conditions of formation of the Abdulino bituminous shales were distinctive. Immediately preceding their deposition the region was a site of coralline reef growth. On the resulting irregular reef surface a number of more or less extensive closed or half-closed basins probably formed, variable in outline, with individual high points forming islands. The Abdulino bituminous shales formed in such small quiet basins, which explains both the wedging out on the margins and the variability of the shales. The source of the organic matter in the shales may be humic material formed by decomposition of higher plants, and may be either autochthonous (in place) or allochthonous, with subsequent transport and redeposition of the organic matter. For the Abdulino shales there is indication of a continent with peat bogs either to the west or the east of the outcrop. The sea extended on every side for hundreds of kilometers, and it would be absurd to assume that suspended and dissolved organic substances—humic acids—were transported across such distances to be deposited in diminutive lagoons on a coral-floored shoal. The origin of the humic acids can only be explained on the assumption that higher plants grew within the basins where the shales were deposited. It is therefore concluded that the basins and quiet pools around the reef shoals were teeming not only with bottom-dwelling animals and with plankton, but also with bottom-dwelling plants, partly algae, partly higher plants, these latter being aqueous forms similar in their ecological characteristics to the modern *Zostera*. *From the abundance of humic compounds, it is concluded that the bottom-dwelling flora of higher plants was the principal source of organic matter in the bituminous shales*, although in individual basins the actual ratio of benthos to plankton might vary. The dominant role of bottom-dwelling plants leads us to consider the Abdulino oil shales benthonic.

3. PLANKTONIC BITUMINOUS SHALES OF ARID REGIONS

A characteristic representative of this type is the so-called Yurezan' shales (Strakhov and Osipov 1945), which are slightly younger than the Abdulino shales. They belong to the marine sequence of bituminous marls in the *Ps.*

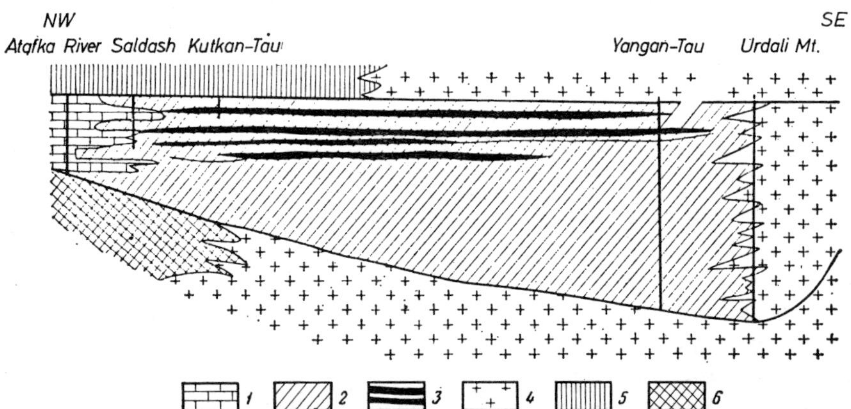

FIG. 73. Constitution of the Yurezan' bituminous shales. 1. Dark organic limestone; 2. bituminous marls with layers of clay and limestone; 3. bituminous shales; 4. sand-clay flysch, 5. marl of the Taimeevo series; 6. recrystallized limestone with *Productus cora*.

lutugini zone, which along the Atafka River on the west give way to organic, partly bituminous limestone. On the east, beginning with Urdali Mountain, these pass into sandy clay flysch (Fig. 73). The width of the lens from Mt. Saldash to Urdali Mountain is about 40 km. The thickness in the central part is unknown, but it exceeds 100 m, and farther east it reaches 250 m. The extent of the lens along the major axis has not yet been established, although

it is known that bituminous marls are found to the south in the Karatau
Range. To the northeast they have been traced to Mt. Krasnoufimsk. The
extent of the lens is thus at least 200–250 km.

The *bituminous marls*, constituting the main part of the marl sequence,
are almost black, dense rocks, generally with no macroscopic bedding, giving
a strong bituminous odor on being struck. The insoluble mineral residues
range in quantity from 15·03 to 50%, averaging 30·8% from 33 determinations.
The total carbonate ranges from 44·29 to 75·50%, averaging 67·89%, and
C_{org} ranges from 0·50 to 4·75%, with an average of 1·31%. Most of the carbonate has a clotted structure, but it is partly micritic, with grain diameters
of about 1μ. Some of it has recrystallized to grains ranging in size from 0·01
to 0·09 mm. The organic remains are uniform, consisting chiefly of sponge
spicules (uniaxial or tetraxial), generally replaced by $CaCO_3$, all stages of
replacement being present. In addition, individual, finely divided fragments
of brachiopod shells of somewhat rare occurrence, and, even more so, fragments of bryozoans and crinoids. Somewhat more frequently foraminiferal
shells such as *Nodosaria*, *Bigenerina*, and *Spirillina* occur and also "spheres"
—rounded bodies composed of aggregate calcite grains or of a single crystalline calcite grain. It is characteristic that entire shells of brachiopods (*Productus
irginae* most frequently) and pelecypods (*Edmondia*) are generally found
individually. Only rarely do they form large coquina-like beds. These forms
were found in all three successions, but most frequently at Mt. Saldash. The
amount of organic detritus is also greater in the western sections. Plant remains
are found in addition to the animal debris and are possibly more abundant.
These are generally *Calamites* impressions, rarely of leaf fragments. The fragmental material consists of a fine clay with rare grains of coarse silt and fine
sand, which increase toward the east. The organic matter is partly diffused
throughout the rock, staining it a light brown, and in part it forms reddish
brown translucent films 0·2 to 0·65 mm thick. Occasionally fusinized black
opaque shreds of plant tissue 0·1 to 0·3 mm across occur. Pyrite is an abundant
authigenic mineral, and is at times associated with chalcopyrite.

Locally the marl may grade into clayey limestones or calcareous shales,
normally accompanied by a decrease of C_{org} in the rock. These limestones
and shales are but rarely encountered, however, and their presence does not
alter the general monotony of the section.

Bituminous shales occur in the upper third of the marly sequence, and they
consist essentially of the same marl but with a considerable increase in organic
matter which ranges from 5·21 to 14·74%. This means that the organic
content of the shales is 5–8 times that of the marls proper, giving rise to a
denser brown color in thin section and to an increase in reddish brown films
of organic matter. Undoubted spores and broken cuticle occur in the bituminous shale, and are much more abundant than in the ordinary marl. The form
of the carbonates and their relationship with the clastic material in the bituminous shales are indistinguishable from the same features in the marl.

Limestones replace the marly sequence to the west, particularly along the

Atafka River. They are dark dense rocks with irregular fracture and are commonly packed with organisms. The commonest are foraminifers (*Fusulina, Nodosaria, Rotalia*), followed by crinoids, and then by brachiopods (*Marginifera, Productus, Spirifer*, and others), bryozoans, sponge spicules, ostracods, corals, and algae (*Mizzia*).

According to the development and association of various groups of organisms, several limestone types may be distinguished: *fusulinid-crinoidal* limestone, the commonest, *Marginifera bank* limestone, occurring as beds in the fusulinid-crinoidal type, and *bryozoan* limestone. The cement of the shelly detritus is mostly micritic carbonate, considerably contaminated with clay. Calcified sponge spicules may be abundant; silty and fine-sandy particles also occur. Recognizable organic matter is rarely observed. It is clear that the cement of the detrital limestones closely resembles the bituminous marl. In the western parts, the bituminous marl lens passes to detrital limestone by a sudden addition of organic fragments. When the increment of organic detritus was temporarily reduced, thin beds of marl were deposited in the zone along the Atafka River, in composition very similar to the marly sequence deposited farther east. The thickness of such beds along the Atafka River is small, but at Mt. Saldash marls dominate the section.

The Atafka facies is also characterized by abundant and variable forms of silicification. In this it differs sharply from the bituminous marl facies occurring farther east. There, abundant traces of SiO_2 removal and of its replacement by calcite are apparent, with subsequent calcification of sponge spicules. In the Atafka facies, however, SiO_2 extensively replaces calcite. It may be concluded that here are two genetically linked processes, that the silica from the bituminous marls migrates laterally to silicify the Atafka deposits. The sand-silt deposits and clays replacing the marly sequence on the eastern border have the polymictic composition normal for foredeeps.

The environmental conditions which control the formation of this sequence are revealed by the following facts. The organic assemblage inhabiting the basin as a whole was characteristically abundant and varied. Very many bryozoans, crinoids, foraminifers, sponges, corals, ostracods, pelecypods, and gastropods lived in the western part of the basin, where the Atafka facies developed, in addition to the numerous genera of brachiopods. Even trilobites and cephalods were fairly common. In its abundance, the organic assemblage of the *Ps. lutugini* stage differed in no way from a normal marine environment. To the east, in the area where the marly bituminous deposits were formed, organisms were much scarcer, apparently due to the greater depth of the basin and partly to the nature of the bottom mud. Nevertheless, brachiopods and cephalopods lived even here. It may be supposed, therefore, that *the chief factor determining the organic forms—salinity—was essentially normal throughout the basin.*

It is characteristic that, without exception, the rocks contain bottom forms (benthos). Even in the middle of the sequence where the organic remains are poorest, sponges, brachiopods, and foraminifers are present, and locally the brachiopods even form small coquinoid beds. This demonstrates that oxygen reached the floor in all parts of the basin; i.e. the gas regimen in the basin was also normal. If the oxygen content at the floor of the sea was somewhat low, it was still not extremely so.

A very important feature of the basin is demonstrated by the fauna and flora in the various rock types of the bituminous sequence. The richest and most variable fauna lived in the area where the bioclastic Atafka facies developed together with calcareous algae (*Mizzia*). This indicates that the western part of the sea was shallow. In the central part of the profile, where the bituminous marls accumulated, the organic assemblage was much poorer. Here there lived essentially only a sponge biocoenose, together with a few brachiopods and small foraminifers. There were no algae and no fusulinids, those typical representatives of late Paleozoic shallow-water forms. These relationships leave no doubt that the region of the bituminous marl deposits was a well-defined depression relative to the more western Atafka zone. On the east the depression included the zone of flysch deposition, where it probably was deepest. The eastern border of the depression undoubtedly lay somewhere east of the present exposures. Thus, the bituminous marl sediments accumulated in the central deeper part of the Uralian foredeep, a long trough in front of the growing Uralian fold system, a depression that was initiated long before the deposition of the bituminous marls.

At the time of their deposition, however, three features of environment and sedimentation were fundamental. *Firstly*, at the time the bituminous marls were deposited, the flysch facies formed much farther east than during the preceding and succeeding episodes of sedimentation. *Secondly*, the western bioclastic facies was also shifted somewhat to the west, and this permitted marls to extend somewhat farther west than previously. *Thirdly*, during accumulation of the bituminous marl, terrigenous material was brought into the central part of the depression in much smaller quantity than before. The insoluble mineral residues in the rocks underlying the bituminous marl constitute 60–80% of the rock, in the bituminous marl 30–32%, and in the succeeding rocks 37–58%.

These facts lead to the conclusion that the Uralian foredeep was temporarily expanded during deposition of the bituminous marl sequence, and that this expansion occurred both from the east and the west. This caused some advance of the shore-line eastward, which led to a decrease in the introduction of fragmental material to the central part of the basin.

In any consideration of the accumulation of organic matter in the bituminous marl, two questions must be answered: what determined the slight general increase in C_{org} (in %) that characterizes the bituminous marl; and what caused the development of bituminous shales in the upper part of the marl series?

For an answer to the first question, the variation in organic content may be compared with the changes in amount of clastic material (Fig. 74). From this it may be seen that a small enrichment of the bituminous marl in organic material generally coincides with a reduced influx of fragmental material.

This is not a coincidence. The negligible enrichment of the marls in C_{org} is due chiefly not to an active increase in the deposition of organic matter in the sediments but rather to a sharp decrease of diluent fragmental increment. The mechanism by which the oil shales were formed was different. These rocks show a considerable enrichment in C_{org}, from 5·2 to 14·4%. A characteristic feature is that petrographically they are identical to normal bituminous marls, containing merely a sharply increased number of films of organic matter and, in parts, of spores also. It is important to note that the amount of clastic material in the oil shales is the same as in normal marls. These facts prove that *in contrast to the marl, the organic enrichment of the oil shale is not controlled by an increase of derived clastic material, but by an increased increment of organic material to the sediment.*

This increased supply might be (1) an increase in episodic growth of plankton, or (2) a temporary increase in the introduction of organic material from

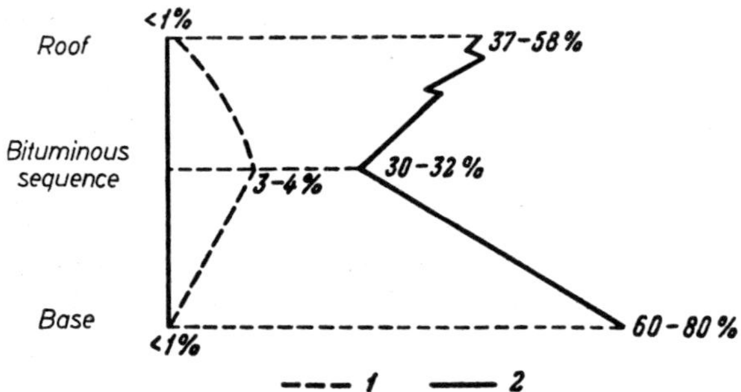

FIG. 74. Relation of C_{org} to clastic material in the bituminous marls of the Yurezan River. 1. Organic matter; 2. clastic material.

shore. The second possibility is not very probable, however, since the region of oil-shale accumulation was bordered on the west by a great water expanse, and on the east the Uralian zone of fold mountains was undergoing active erosion, a circumstance unfavorable to the deposition of organic matter. There were no coals that might have increased the supply of organic matter to the basin. It must be assumed, therefore, that the source of organic matter was within the basin itself, namely in the periodic "flowering of the plankton", a normal phenomenon in marine basins. The fact that the source was plankton and not phytobenthos follows from the notable depths in the central part of the basin and, in particular, from the absence of any indication of algal benthos.

Thus, the mechanism by which the organic matter accumulated in the *Ps. lutugini* zone of the bituminous marls was complex. *The slight enrichment of the normal marls in C_{org} was chiefly a negative phenomenon, caused by reduced influx of diluent clastic material into the basin of that time. The development of bituminous shales was the consequence of the active increase of organic contribution from the planktonic film, which underwent episodic periods of phytoplanktonic flowering.*

In the examples of bituminous shales—undoubted representatives of an arid climate—the constituents accumulated in basins unassociated with halogen facies, and they bear no trace of any excess mineralization of the water. Nevertheless, bituminous shales are found where such an association is present. Facies transitions from marine to saline basin are very clearly seen in the Lower Cambrian Sinyaya horizon on the Siberian platform (Zelenov 1955).

This horizon is exposed along the northern edge of the Aldan massif for a distance exceeding 300 km. Lithologically it is rather varied: shaly, silty marl, black, cherty, shaly limestone, and, lastly, bituminous dolomite. Three north-trending zones may be recognized (Figs. 75 and 76). The eastern zone

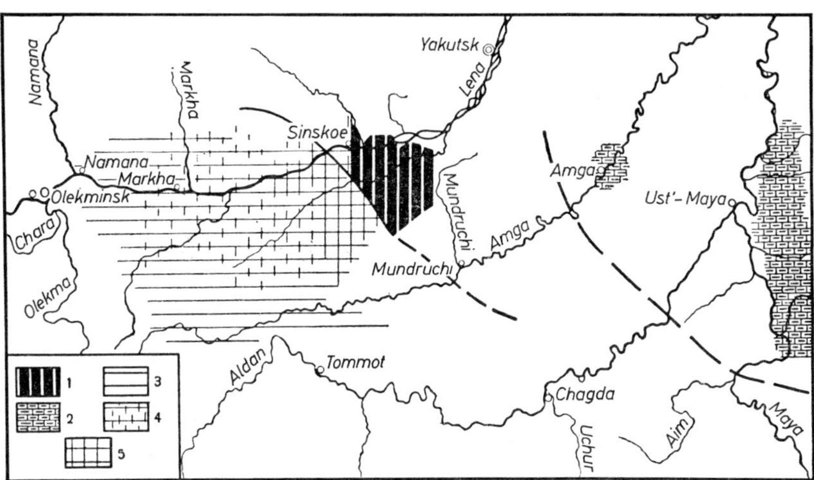

FIG. 75. Facies zones of the Sinyaya horizon (from K. K. Zelenov). 1. Shaly shelf limestones of the normal open sea, in organic matter; 2. shaly marl and cherty limestone of the normal open sea, rich in organic matter; 3. facies zone of a saline lagoon; 4. interbedded limestone and dolomite in the marginal part of the lagoon; 5. transitional zone between lagoon and the open sea.

includes the section on the Maya and Inikan Rivers (tributaries of the Aldan) and the bore-hole section on the Amga River. Bituminous rocks are developed here, containing up to 60% pelitic and silty clastic material. Some of the rocks are silicified. Many trilobites, brachiopods, pteropods, sponges, and foraminifers are found. To the west, in the central zone, the clastic fraction sharply declines, to occur almost exclusively in the middle of the section. The Sinyaya series is mostly limestone at this locality, normally black, crystalline or shaly, interbedded with bituminous aphanitic units. Silicification is markedly less. The organic content here reaches a maximum, as do the faunal remains. The western zone embraces the western sections, on the Lena, Amga, and Botoma Rivers, together with drill-hole successions at the mouth of Namana River and Russian Creek. The rocks here are fine-grained bituminous dolomites. In the Lena section limestones form about one-third of the succession. At Botoma the proportion of limestone is still less, and at the mouth of the Namana and on the Amga, limestone is absent. Organic material is much less abundant in the dolomitic unit than in the limestones of the Sinyaya series, but it is still more abundant than in the overlying or underlying rocks. Faunal remains are very rare.

This east-west change indicates that the drainage area of the basin supplying the clastic material was clearly east of the present exposures of the Sinyaya series. The bituminous limestone zone lay at some distance from the shore, in the pelagic zone of the basin, where little fragmental material reached, and where also little authigenic silica was emplaced. The fauna of sponges, trilobites, brachiopods, and pteropods leads to the conclusion that the salinity of the basin was normal in the eastern and central parts. The appearance of

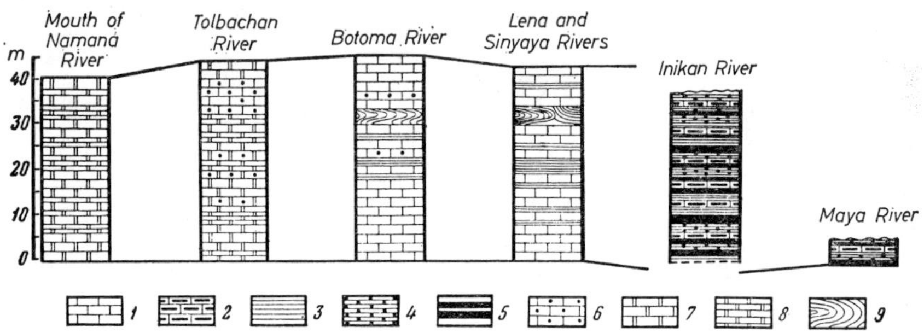

FIG. 76. Variations in the Sinyaya horizon (from K. K. Zelenov). 1. Aphanitic limestone; 2. coarse-crystalline limestone, 3. black shaly limestone; 4. shaly silty marl, 5. silicified shaly silty marl; 6. clotted limestone; 7. coarse-crystalline dolomite; 8. pelitomorphic dolomite; 9. shaly limestone, with underwater slumping.

normal dolomites toward the west means that the water was somewhat more salty, even if very slightly. Still farther west the salinity increased sharply, and gypsum was deposited. It is characteristic that *the accumulation of organic matter in sediments of the saline zone fell immediately to a low level, and the formation of oil shales ceased.*

4. BITUMINOUS SHALE DEPOSITION AND THE SALINITY OF ARID-CLIMATE MARINE BASINS

From the above examples it is clear that, *although these rocks are known to form in the arid zones, the salinity of basins in which the organic matter accumulated nowhere differs appreciably from normal sea water.* This is supported by other examples. The Domanik beds, which were also deposited in arid basins, are indistinguishable from beds forming in humid belts. The Zechstein and Abdulino bituminous shales show no indication of high salinity. The Maikopian and Neogene bituminous shales of the Caucasus are characterized not by high salinity of the basins, but, on the contrary, by freshened basins.

It must also be emphasized that *normal sedimentational dolomites generally have a negligible C_{org} content and nowhere include bituminous shales.* Any marked C_{org} content is also lacking in gypseous rocks, and even more so in halogen units.

These facts lead to the conclusion that *the accumulation of organic matter in marine sediments to the extent of bituminous shale production proves to be very sensitive to the mineralization of sea water and is attained only in those basins of arid regions where the salinity is essentially the same as that of normal sea water.* When mineralization increases, bituminous shales quickly cease to form in arid regions. This is characteristically reflected in the fact that *bituminous shales are paragenetically related only to limestones or calcareous clays, i.e. to rocks in which dolomitization is absent or very slight.* The composition of the carbonate component may here serve as a standard to measure the deviation of mineralization of the water from the norm. The cause of this sharp restriction of bituminous shales by salinity is not yet clear.

Marine accumulations of organic matter do not form independent formations, but in both humid and arid regions they contribute to clastic, clastic-carbonate, and carbonate formations. Planktonic types generally occupy the middle parts of formations, somewhat displaced to one side or the other; benthonic accumulations are, by contrast, scattered at random in the formations.

5. The Principal Features of Arid-Climate Lithogenesis in the First Stage of its Development

In the preceding part of the book arid-climate lithogenesis at the very beginning of its development has been analyzed, when this took place in basins of low mineralization. We have attempted to identify the specific characteristics of such lithogenesis as compared with humid-climate lithogenesis under similar salinity. These observations and conclusions will now be summarized.

1. *Grain-size sorting takes place in arid regions exactly as in humid regions,* but some characteristic differences are found in the composition of clastic rocks. *The accumulation of quartz in sand-silt sediments and of kaolinite in clay sediments does take place not by generation of the components within the arid zones themselves, but by the transport from humid zones or the redistribution from older oligomictic and kaolinitic rocks.* In other words, mesomictic and oligomictic clastic rocks and rich kaolinitic clays that are deposited in arid-climate basins at the stage of low mineralization are inherited features. *The wide development of inherited features in the essential composition of clastic deposits of arid zones, features inherited from a different (humid) lithogenesis, is a distinctive criterion of arid lithogenesis during its initial stage.*

2. Dryness of climate leads to the loss of mobility of Al-Fe-Mn in the zone of weathering, and ore facies of these elements do not develop. *They are replaced by deposits of Cu-Pb-Zn, which are thus the arid homologues of the humid ore triad.* The formation of these ores in arid regions does not, however,

take place everywhere, but only where a *more or less substantial primary concentration of Cu-Pb-Zn in the substratum, in the form of sulfides, combined with active tectonic movement, gives rise to vertical climatic zoning in the drainage areas.* Weathering of the sulfides in the parent rocks took place on moist uplifted areas, and the metals migrated in sulfate solutions. In dry depressed basins with rich carbonate water and with high pH, the sulfates changed to basic carbonate salts and were deposited in the sediment. During diagenesis these forms were replaced by sulfides and were actively redistributed, giving rise to sedimentary ores of Cu-Pb-Zn. These ores, being monoclimatic formations, are thus specific features of arid lithogenesis during its initial stage of development.

3. Accumulation of members of the biclimatic group P-$CaCO_3$-$MgCO_3$-SiO_2 in arid regions also changes characteristically. The precipitation of all members, especially of the first three members of the group, is sharply intensified. Biogenic processes in the accumulation of $CaCO_3$ and $MgCO_3$, on the other hand, are markedly reduced, being displaced by purely chemical precipitation. The geochemical cycle of magnesium cuts across the geochemical cycle of silica, and as a result magnesium silicates and aluminosilicates become widespread in the sediments.

4. Because of the disappearance of the forest cover on continental parts of arid regions, the formation of coal is curtailed. But concentrations of organic matter in the sediments of marine basins continues unweakened as compared with the concentration that takes place in humid regions. Bituminous shales accumulate as before, both benthonic and planktonic varieties. It is characteristic that bituminous shales in arid-climate basins are associated only with calcareous carbonates, and nowhere with dolomites. This indicates that bituminous shales are distinctive stenohaline facies, forming only in seas of normal salinity, not enduring even small deviations in mineralization from normal sea water.

5. The meagerness of plant life in arid regions leads to great poverty of C_{org} in the sediments, especially subaerial sediments, but commonly marine sediments also. Oxidized iron in such sediments is unreduced and the sediments acquire a red color. At the same time the sharp and widespread development of carbonate deposits leads to a noticeable deposition of the calcareous component in red beds. Such red carbonate-bearing sequences are very widespread in arid regions, and they differ sharply from the rarer carbonate-free red beds of humid regions.

6. Diagenetic processes in weakly mineralized basins of arid regions are generally similar to those in humid regions, but some differences appear. During the accumulation of Cu-Pb-Zn in the sediments, to the extent of ore formation, hydrogen sulfide is required by all, or almost all, these heavy metals. Independent sulfides of iron do not form or only in negligible quantities. Redistribution of authigenic minerals in Cu-Pb-Zn ores is very extensive during diagenesis; it becomes an important ore-forming factor, without which ore bodies proper would not form. In other words, the *ore triad Cu-Pb-Zn*

belongs entirely to the sedimentational-diagenetic type of accumulation, whereas Al-Fe-Mn in humid regions form distinctly sedimentational ores. In the biclimatic tetrad $P-CaCO_3-MgCO_3-SiO_2$ the secondary redistribution of dolomite and magnesium silicates may also be extensive, leading to the formation of vast lenses, beds, and stock-like masses of metasomatic dolomites with complex structures and to the accumulation of sepiolite horizons. One gets the general impression that *diagenetic redistribution of authigenic minerals in deposits of weakly mineralized arid-climate basins is much more intense than in basins of humid regions.* These redistributing components are, however, characterized at the same time by much greater geochemical mobility.

7. Formations that accumulate during the initial stage of arid-climate lithogenesis essentially parallel the groups, types, and even individual representatives of their counterparts in humid regions. Still, the presence of qualitative and quantitative features during the initial stage of arid-climate lithogenesis gives arid formations well-defined individual characters as compared with humid formations. Consideration of these characters is extremely important for a proper understanding of the relations between arid and humid types of lithogenesis in general. We may recall that arid lithogenesis takes place with sedimentary material brought into the arid-climate basin chiefly from humid regions that are either laterally or vertically adjacent. This material is naturally the same as that carried into basins of humid regions, and it may be thought, therefore, that the initial moments of arid lithogenesis will more or less duplicate the aspects of humid lithogenesis. From here it is not far to the idea that arid lithogenesis in general is perhaps little more than an extension of humid lithogenesis, a simple overgrowth of new features and processes. This is not so. Even at the beginning of arid lithogenesis, when the salinity in basins of both climatic zones is the same, or almost so, arid lithogenesis differs clearly from humid lithogenesis. *Here we see a new example of how the history of a single material essentially sedimentary may be changed. These changes and new features cause us to consider arid lithogenesis not simply an extension and expansion of humid lithogenesis but rather a special, new type, separate from humid lithogenesis even at the earliest stage of its existence.*

In the further history of increasing salinity in arid basins, the specific characteristics of lithogenesis in arid regions stand out ever more clearly. These specific features are readily detected even during the initial development of arid basins, by comparative lithology and particularly by making comparisons between sedimentogenesis and diagenesis in basins of arid and humid types.

L

PART TWO

BASIC FEATURES OF PRESENT-DAY HALOGENESIS

CHAPTER 1

TYPES OF PRESENT-DAY SALINE BASINS AND THEIR SEDIMENTS

With the progressive increase in salinity, the formation of sediments in basins of arid zones acquires distinctive features, showing an increasingly well-defined *halogen* aspect, i.e. an increasing proportion of soluble salts in the sediment.

In the study of halogenesis (the formation of salt deposits), one characteristic may immediately be noted: *the facies conditions for salt accumulation in present and ancient basins differ sharply; to a certain extent they are even antipathetic.* At present, halogen deposits accumulate most commonly in lakes and only rarely in saline gulfs and lagoons separated in part from the sea. In the past, such deposits were associated almost exclusively with marine basins of various types. From the facies point of view, modern and ancient halogeneses are more or less complementary. At first glance, this difference in environment excludes the application of the comparative-lithogenic method of analyzing salt deposits. In point of fact, halogenesis is a process that may very clearly demonstrate the fruitfulness of the comparative-lithologic method and the efficacy of this method in revealing new aspects of the complex natural phenomena. It is essential, however, that the principles of the method should not be applied mechanically but with due consideration of the distinctive features of each deposit.

The differences in facies conditions for ancient and modern halogenesis first necessitate a study of the course and pattern of this process in present-day saline basins, followed by an analysis of the specific characteristics of the process in basins of the geologic past.

1. Morphological Types of Continental Lakes

When they were forming from glacial melt waters at the end of the glacial epoch, lakes of modern arid belts were large, deep basins, such as Lake Lahontan in the U.S.A., the ancestor of several modern lakes (Fig. 77). Subsequently, as the arid climate persists, these initially large basins are progressively filled with sediments despite the inflow of water from neighboring humid regions. They evaporate to some extent and also become increasingly salty, eventually reaching their present condition.

In the areal reduction and increasing salinity of lakes, four clearly defined stages may be distinguished (Fig. 78). The first is represented by lakes in which *ordinary clastic-carbonate sediments are deposited, with no solid phases of soluble salts.* During this stage lakes possess both their greatest areal extent and depth. In regions of considerable tectonic activity, the consequent shrinking is marked by the appearance of small terraces (Bol'shoe Solenoe and Baskunchak Lakes) on the basin flanks. In lakes of exogenetic origin this shrinking is represented by a zone of salt bottoms and marshes between the shore and the present water level.

In the second stage the lakes become sufficiently salty for sulfates and chlorides to be precipitated, but they are still characterized by surface brines throughout the entire year. These are the so-called *briny lakes*. The areal dimensions of the lakes are sharply reduced at this stage, and the depths are negligible (0·5–0·4 m).

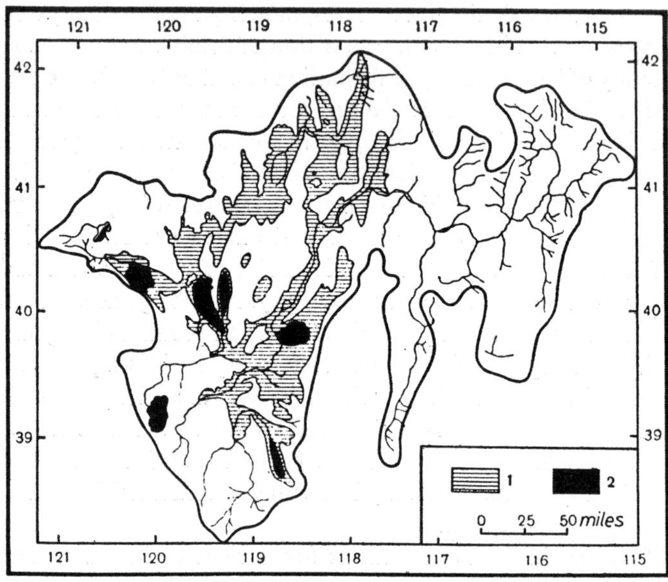

FIG. 77. Shrinking of post-glacial Lake Lahontan (from Lotze). 1. Initial area of lakes; 2. present-day relicts of the earlier basin (Pyramid, Walker, and other Lakes). The outer heavy line outlines the drainage area of Lake Lahontan.

In the third stage the lakes are characterized by a surface layer of brine only during the moist period of the year (autumn-winter-spring). During the summer months the surface of the bottom deposits (saline deposits) may be moist, but commonly they are entirely dry. Brine is found only in pores of a salt bed and is called *bottom* or *intercrystalline* brine. In the summer months, the brine level frequently lies below the surface of the bottom deposits. Lakes in this stage are called *dry*.

In the fourth and last stage, lakes are characterized by a complete absence of surface brines throughout the year. The level of intercrystalline brine lies much lower than the surface of solid lacustrine deposits. Where saline deposits are present in the lake, the upper part, above the level of intercrystalline brine, is modified by solution phenomena. A variable layer of detritus is generally deposited upon the saline deposits. This stage represents the transition from basin to salt deposit, and it is therefore called the stage of *infrasand lakes*.

This sequence of conversion of initially large, deep, fresh lakes to small infrasand salt deposits is essentially an idealized account that is never fully

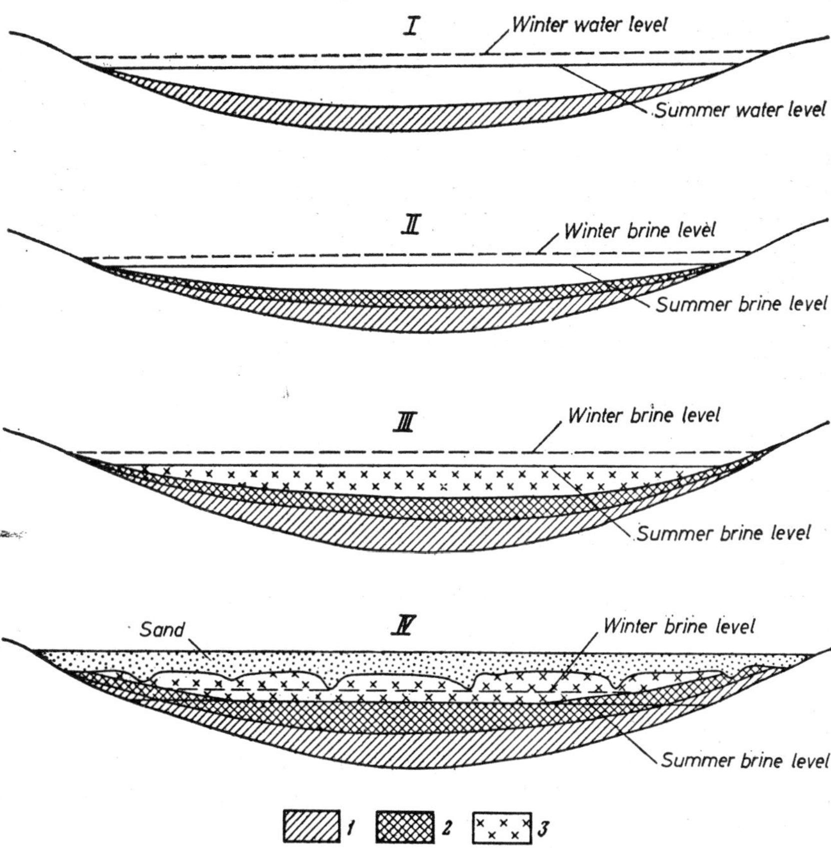

FIG. 78. Developmental stages of a lacustrine basin in an arid zone (modified from M. G. Valyashko). Stage *I*, low-salt lake; Stage *II*, briny lake; Stage *III*, dry lake; Stage *IV*, infrasand lake. 1. Clastic-carbonate sediments; 2. sulfate sediments (mirabilite, thenardite, gypsum, and other minerals); 3. halite.

realized. It must be borne in mind also that different lakes pass through their course of development at different rates; large lakes, in particular, develop much more slowly than small lakes. Hence, at any given moment various stages in the general evolution of salt lakes may be observed in different basins. This is one of the causes of the great morphological variation in present-day continental lakes of the arid zone.

The largest of the present-day lakes in an arid region is the brackish Lake Chad in north Africa. Depending on the water level, its area ranges from 22,000 to 11,000 km² with a corresponding variation in depth from 7 to 4 m. Although it is huge, this is an extremely shallow-water basin.

Brackish Lake Balkhash, second in size among arid-climate lakes, has an area of 17,200 km². Its length is approximately 600 km, and its maximal depth is 26 m. The lake has a number of segments connected by narrow straits. The depth of the western segment is only a few meters: the eastern segment, the Burlyu-Tobe, reaches a depth of 26 m.

The Great Salt Lake, with its highly mineralized water, has an area ranging from 5500 to 2900 km², depending on level. The average depth is 3–5 m, and the maximum depth is about 15 m. Lakes of about the same size are the Chany (2600 km², depth of 10 m), Tengiz (1200 km², depth of 6·8 m), and Kuku-Nor (4800 km², depth of 38 m). Despite their large size, all these are planar and shallow. The Dead Sea is exceptional, with a depth of 399 m and an area of 1000 km², and is a unique example of a trough-like saline basin.

Among the numerous lakes in arid zones, however, these large basins are rare. The typical salt-lake basin is 0·5–3 km long, 1–1·5 km wide, and 0·5–2 m deep.

The morphology and dimensions of ancient lakes in arid regions is not precisely known; but they are unlikely to differ appreciably from those of the present day.

2. Morphology of Present-Day Salt-Depositing Basins derived from the Sea

The morphology of marine-derived basins that become increasingly salty is very different. Among present-day basins the most important and extensive are *marine gulfs and lagoons* that connect with the sea through straits, and *maritime lakes* which develop from lagoons and are cut off from the sea by barriers of variable effectiveness.

Marine gulfs are segments of the sea that cut deep into the continent, and communicate with the principal water mass of the sea either freely or through narrow openings cut through bedrock. *Each marine gulf represents a more or less well-defined tectonic depression open toward the sea or, if closed, connected by a strait*. The Kaidak and Kara-bogaz are such gulfs, which are connected with the sea through channels and have become saline.

In the early thirties, before modern shrinking of the Caspian Sea, the Kaidak was an extension of Komsomolets Gulf (Mertvyi Kultuk) and had the form of a narrow elongate basin about 140 km long and 10–30 km wide. Its greatest depth, in the latitude of Kyzyl-Tash, was 4·4 m. The eastern shore of the gulf formed the Ust'-yurt escarpment; on the west the gulf was bounded by Buzachi Peninsula. The gulf was separated from the sea by submarine bars, which were almost exposed above water when winds from the gulf were strong. The direction of water movement was generally from sea to gulf, but when the wind was from the south a considerable amount of water was blown from the gulf to the Caspian.

The Gulf of Kara-bogaz is at present the largest saline gulf in the world (Fig. 79). It has an irregularly circular form and as late as 1930, before the recent shrinking, extended for 200 km from north to south and for 150 km from east to west. Its area was 18,000 km² and its maximum depth 13 m, with

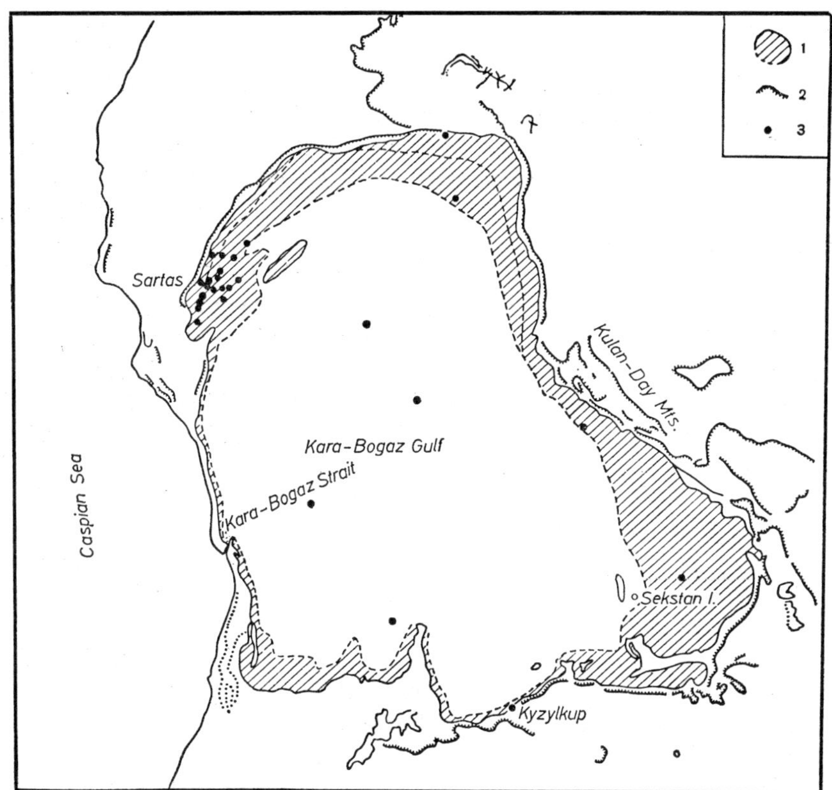

FIG. 79. Map of Gulf of Kara-bogaz (from A. I. Dzens-Litovskii). 1. Belt of the Kara-bogaz floor exposed from 1936 to 1953 (according to data of the Kara-bogaz Mining Industry), 2. escarpments, 3. drill holes.

an average of 9·19 m. In 1954, according to A. I. Dzens-Litovskii (1956), the length and width of the gulf had shrunk to 50–60 km, the area to 7000–8000 km², and the greatest depth was only 3·1 m (Fig. 79). Its western and southern shores are low, bordered by a belt of salt marshes, which have been considerably widened by the reduction in area of the gulf. The northern and eastern shores are cliffed. The Gulf of Kara-bogaz is separated from the sea by a bay bar founded on bedrock. This has a channel through the middle, connecting the gulf with the sea. The bar rises above sea-level to 5–6 m on the average, and ranging from 2 to 20 m. When the level of the Caspian fluctuates markedly, supplementary flow undoubtedly occurs across the low parts, causing some freshening of the gulf. A long narrow gulf extending parallel to the shore for 70–80 km existed in the recent past in the southwestern corner of the Gulf of Kara-bogaz. The salt lakes of Kuuli represent traces of this feature.

The Gulf of Kara-bogaz has no permanent stream discharge but is supplied by water from the Caspian through a strait, the length of which was 5·0 km before shrinking of the gulf; the width was 100–500 m. Now the strait has lengthened to 10·5 km. The current through the strait has a velocity of 25 to 44 m per minute and flows entirely lagoonward. Bottom counterflow, which normally exists in such places, is absent. In earlier times, however, before shrinking of the gulf, when the wind blew for long periods from the north, a considerable inflow of brines into the strait occurred, and a tongue of Kara-bogaz water in the northern part of the strait reached a thickness of 3 m, gradually wedging to the south, having a thickness of only a few centimeters at the opening into the Caspian sea.

The supply to the gulf has been extremely variable. For the period 1879-1932 it averaged 21·18 km³ a year. At present, because of shrinking of the Caspian it is only 9·5 km³. Despite the tremendous inflow of water, the level of the Gulf of Kara-bogaz remains below the level of the Caspian throughout the year because of intense evaporation. Before shrinking of the gulf, this difference in level was about 10 cm in winter and about 50 cm in summer. The difference in level is now 3 m.

Lagoons are generally segments of the sea, penetrating far into the land and separated from the sea by a sand bar. The bar may be a spit, formed by longshore drift, or it may be a barrier beach or an offshore bar formed by wave action from deeper parts of the sea. Lagoons are connected with the sea by narrow straits through the bars.

Lagoons most frequently develop along gently sloping shores. Normally, where the shore profile steepens sharply, at depths that are approximately twice wave height, sandy shoals are formed. Ultimately these are converted into sandy ridges and bars. The maximum depth of a lagoon in such places also is twice the wave height. Along oceanic coasts the wave height reaches 12–13 m; in marginal seas the height is only 6–7 m, and in intracontinental seas only 2–4 m. Thus, the depth of lagoons along oceanic coasts does not exceed 25–30 m; along marginal seas it is no more than 12–14 m, and in intracontinental seas but 4–8 m. Lagoons are most frequently formed on submergent coasts, but they also form on stable or emergent coasts. Bars generally form very quickly (tens of years), but they are also rapidly destroyed; lagoons are therefore only temporary geological phenomena. They are rarely found singly, but usually in a series forming what are lagoonal shores.

A classic example of a lagoon with complex morphology is the Sivash, or Putrid Sea. Its total area ranges from 2400 to 2542 km². Morphologically it is divided into the Eastern and Western Sivash (Fig. 80). The Eastern Sivash is separated from the Sea of Azov by the long Arabatskii Spit (107 km), and only at its northern end, near Genichesk, is it connected with the sea by a narrow channel, the Genichesk or Narrow Strait. On the west the Sivash is bounded by the Chongar Peninsulas and is connected with the Western Sivash through the Chongar Strait. The total area of the Eastern Sivash is 1430 km². The depth reaches 4 m, but is mostly about 2·0–2·5 m. The deepest point is in the northern part of the gulf. The shores are complex. The bays commonly have only poor communication with the main waters mass. In summer they dry out, and are converted to salt marshes or "zasukhas", but their total area is small, not exceeding 100 km². The total area of the several

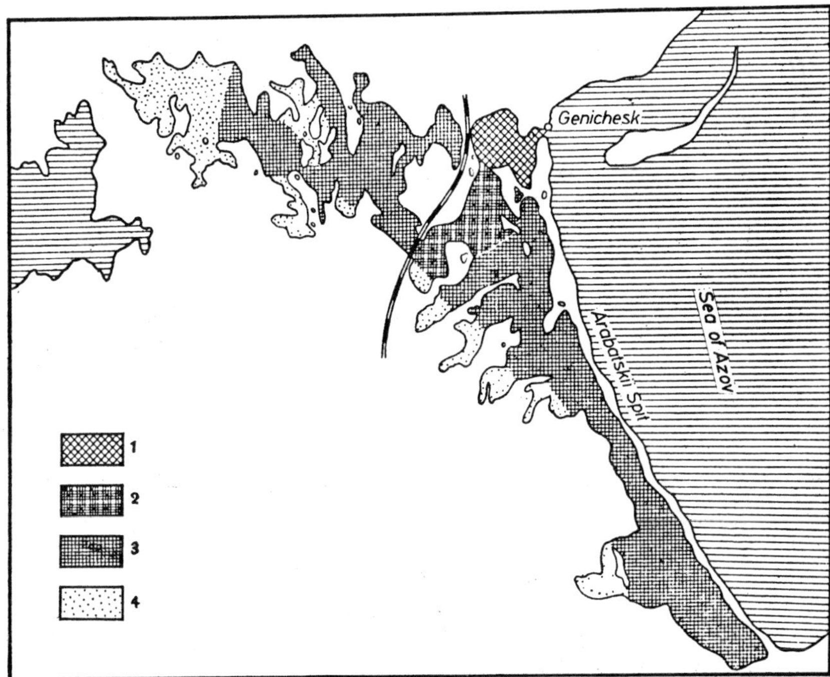

FIG. 80. The hydrochemical districts in the Sivash (from A. I. Proshkina-Lavrenko). 1. Moderately swampy Sivash; 2. very swampy Sivash; 3. extremely swampy Sivash; 4. shallow-water sand banks and salt marshes (zasukhas).

islands is also small (about 40 km²). Thus the water surface of the Eastern Sivash covers about 1210 km². The Western Sivash is substantially different. Its total area is approximately 1100 km², but much of it consists of islands and zasukhas, and the actual water surface covers only about 581 km², or about 52% of the total area. The extreme irregularity of the shore-line of the Western Sivash is due to the extreme shallowness of the basin (Fig. 80). South of Chongar Strait the depth reaches 0·67–1·00 m; beyond Karacha-Kitai it is no more than 0·67 m, and gradually it diminishes to the west, to only 0·2–0·25 m.

The Sivash is fed chiefly by water from the Sea of Azov, through Genichesk Strait. The direction of water movement is solely lagoonward, though, when

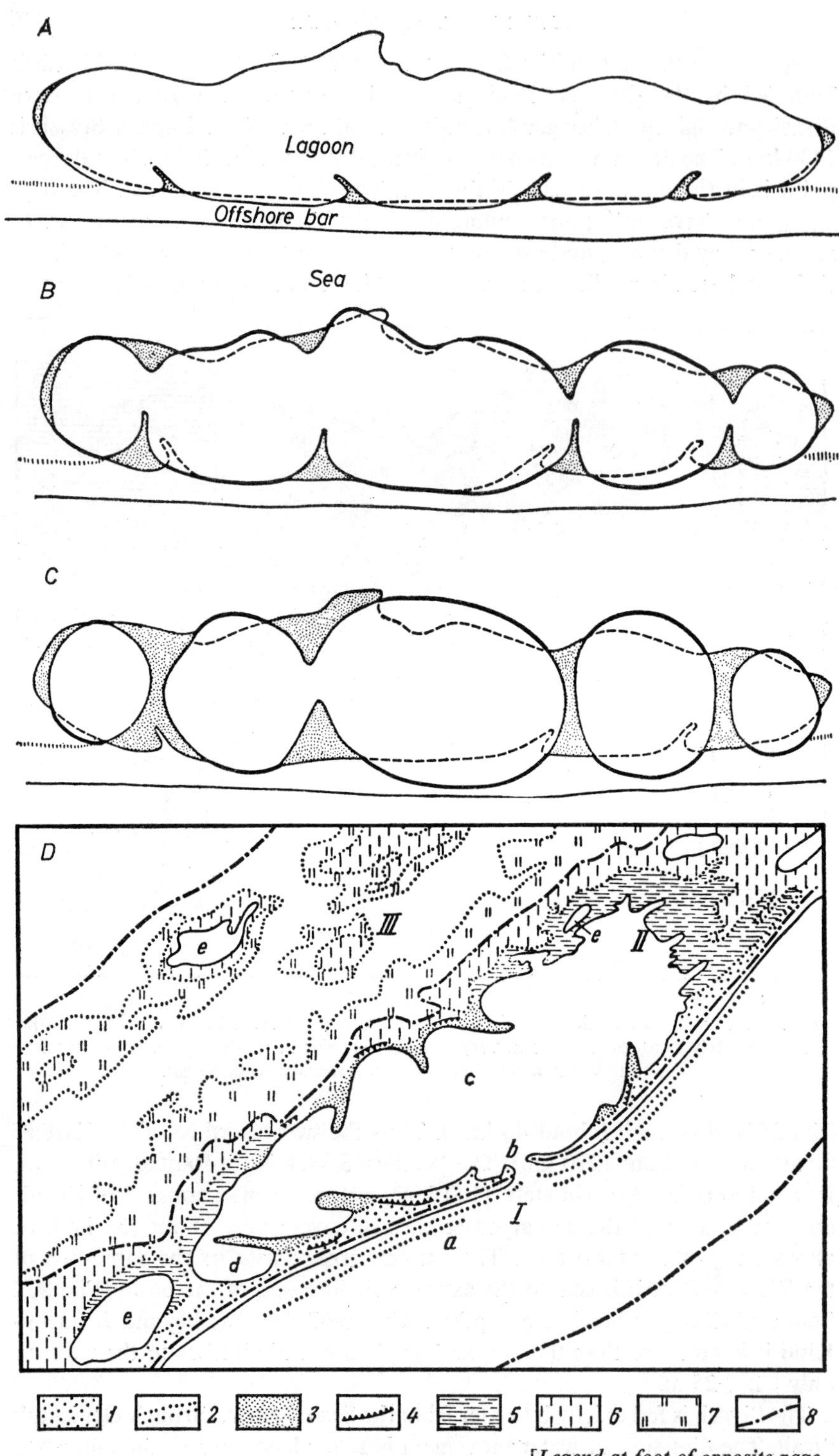

[Legend at foot of opposite page

strong winds blow from the west, saline water flows in considerable volume into the Sea of Azov, and penetrates far into the western part of this sea.

When a detrital bar, of whatever origin, completely separates a lagoon from the sea, it converts the lagoon to a maritime lake. This is how the maritime lakes of the Crimea formed: Bolshoe and Maloe Moinakskie, Sasyk-Sivash, Saki, Donuzlav, and others (Tables 12 and 13). Other such lakes include Kuuli, south of the Gulf of Kara-bogaz and Larnaca on Cyprus.

A maritime lake that is separated from the sea subsequently undergoes a complex morphological evolution, especially well demonstrated in elongate basins. This type of maritime basin is "commonly complicated by numerous depositional forms, spits and ridges which develop transverse to the length of the lagoon. These occur mainly on the inner side of the bar that separates the lagoon from the sea, but they are also found on the continental shore. These forms may occur at the same time on both shores of the lagoon. Such growing spits and bars may ultimately join, to divide the single lagoon into a number of increasingly isolated basins. These continue to evolve and, at times, to take on regular outlines. At the same time the bars that separate these small basins thicken" (Zenkovich 1952, p. 448). This evolution is illustrated in Fig. 81.

Maritime lakes vary markedly in size, ranging up to 20–30 km in length, up to several kilometers in width, and up to 40–70 km² in area (Sasyk-Sivash, Donuzlav). Most such lakes, however, are small, the depth generally ranging from 0·5 to 1·0 m. Only Donuzlav, with a depth of 25 m, is exceptional. The width of the sand bars commonly ranges up to a few hundreds of meters, and the depth up to a few meters. The level in a maritime lake is normally tens of centimeters, up to 1 meter below sea-level, and this together with the small width and the permeability of the bar, permits seepage of Black Sea water into the lake.

In the thirties an attempt was made to explain the hydrologic regime of the largest and most important Crimean lakes. It was shown (Table 13) that the principal source of water is surface and ground-water discharge from the drainage areas, together with precipitation. Seepage through the bars is negligible, only 2–11% of the total influx of water. Thus, after a lagoon is converted to a maritime lake, its subsequent existence is maintained almost

---

FIG. 81. Sequence of lagoonal development (*A, B* and *C* from V. P. Zen'kovich, *D* from O. M. Leont'ev). *A.* Initial growth of spits and formation of erosional bays on the inner side of the bar; simultaneous filling of both narrow ends of the lagoon by detritus. *B.* Spits on the inner side of the bar shifted and extended, with development of symmetrical form; depositional bars on the continental shore growing toward them (eroded segments of the shore indicated by solid line; initial shore line indicated by broken line). *C.* Three continuous barriers formed; lakes thus partitioned off acquire rounded outlines (initial shore line indicated by broken line). *D.* Diagram of a lagoonal coast: *a.* underwater slope of bar; *b.* strait; *c.* lagoon; *d.* second-order lagoon; *e,* lake. *I.* Outer shore zone; *II.* inner shore zone; *III.* paleolagoonal zone. 1. Bar; 2. underwater sand ridge; 3. spit of Azov type; 4. recent zones of erosion on lagoonal shore; 5. zasukha (dry lake); 6. crusted salt marsh; 7. saltwort meadows; 8. zonal boundary.

TABLE 12

Dimensions of Maritime Salt Lakes of the Crimea (from A. I. Dzens-Litovskii)

Lake	Length, km	Width (max.), km	Area, km²	Depth of brine (max.), m	Width of bar, km
Sasyk-Sivash	18	18	71	1·20	0·90–1·62
Kyzyl-Yar	5·65	2·24	6·85	0·03. At end of summer dries out and brine is driven by the wind from one shore to the other	2·24
B. Moinakskoe	2·39	0·98	1·76	0·85	—
Terekly Konratskoe	0·50	0·20	0·06	0·45. Normally dries up in fall	—
Galchasskoe	—	—	0·16	0·55	0·06
Adzhi-Baichi	1·50	0·70	1·20	0·65. Normally dry	0·50
Oiburskoe	4·00	1·50	5·00	?	0·30
Donuzlav	30·00	4·00	47·00	25·00	0·30–1·00
Karadzhinskoe	1·73	1·32	1·30	2·05	0·06–0·40
Mayakskoe	0·50	0·30	0·30	0·60	0·02
Ai-Mechetskoe	0·36	0·17	0·25	0·75. Spring area up to 3 km²	0·03
Sasyk	4·50	2·15	4·50	1·05	0·40–0·60
Dzharal-Arach	—	—	7·90	1·05	—
Karlov	—	—	1·60	0·75	—
Bakal	4·00	3·50	5·80	0·85	0·04

exclusively by the influx of fresh continental water. The salt regime is entirely different. Although the influx of sea water is small, its salinity, 18‰, is many times that of surface and ground water (0·3–0·5‰). It may readily be calculated that the bulk of salt arrives in the lake from the sea. *It is clear that, despite the dominant supply of river water, the Crimean maritime lakes steadily maintain the essential hydrochemical characters of the Black Sea* (see below).

TABLE 13

Supply of water to Crimean lakes (from A. I. Dzens-Litovskii)

Lake	Seepage of sea water through bar	Other influx of sea water	Ground water	Surface water	Precipitation on lake
Saki	47,410 m³	193,561 m³	1,859,280 m³	—	1,283,900 m³
Kyzyl-Yar	158,050 m³	—	4,027,300 m³		2,376,950 m³
B. Moinakskoe	31,610 m³	—	941,670 m³		610,720 m³
Saki	11·0%	—	54·0%	—	36·0%
Kyzyl-Yar	2·5%	—	61·9%		30·5%
B. Moinakskoe	2·0%	—	59·5%		38·5%

Maritime lakes of this type are widespread in arid zones of all continents, in all stages of development. On the shore of the Caspian Sea Lake Kuuli is of this type. It lies southwest of the Gulf of Kara-bogaz of which it was once a part. Its total length is about 55 km; the maximal width in the north is 2–2·5 km, widening to 3·5 km in the south. The lake is now dry. In winter it is covered with 10–30 cm of saline water, a consequence of the seasonal rainfall. In summer the brine above the lake floor dries out and the brine level falls to 5–6 cm below the surface of the salt. The lake is separated from the sea by

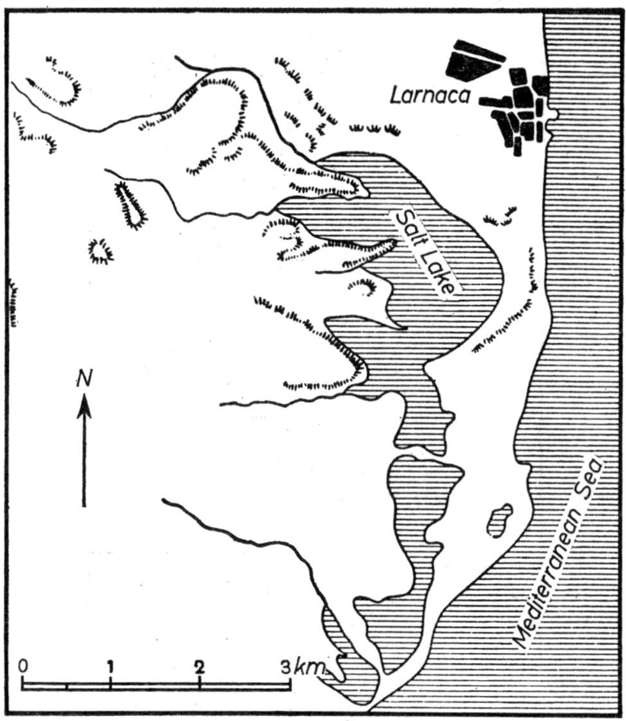

FIG. 82. Larnaca Lake, Cyprus (from Lotze).

a belt of barchan dunes 1–2 km wide. The dunes are as much as 10 m high, and most are fixed. They rest on a base of coarse-grained bioclastic limestone.

Two similar maritime lakes are found in the Mediterranean region: Larnaca in Cyprus (Fig. 82) and Torrevieja in Spain. Lake Larnaca occupies the lower part of a coastal depression. It has a total area of about 57 km² and is separated from the sea by a bar 1–1·5 km wide, overgrown with forest. This lake is dry. In summer the saline water completely disappears, leaving efflorescences of salts. In winter the basin holds up to 1 m. of brine. Even in winter, however, the lake level is 2 m below the level of the Mediterranean, and, as a result, Mediterranean water percolates through the bar into the lake.

The existence of this subsurface flow was demonstrated by drilling (Lotze 1938, p. 106).

The Salina de Torreviejo, east of Murcia, is a basin of 20 km², separated from the sea by a line of sand-hills 4–9 m high. Its surface is 3 m below sea-level, and this again leads to marine percolation. In summer, salts are extensively precipitated, and the lake almost dries up.

Similar salt lakes are known from the Dutch West Indies, the Cape Verde islands and elsewhere.

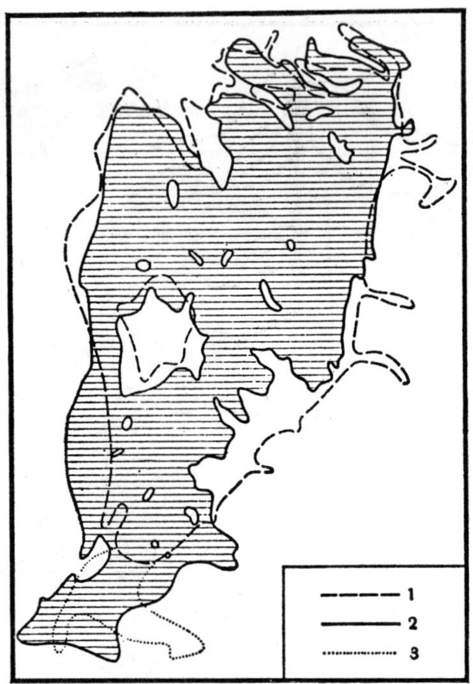

FIG. 83. Lake Dzhaksy-Klych in the Aral region (from G. S. Klebanov and others). 1. Outline in 1907; 2. outline in 1932; 3. outline of swamp in 1907.

Besides lakes in which the supply of sea water is by seepage through the barrier bar, other types may be recognized, *supplied periodically from the sea by breaching the obstructing bar or through normally dry channels*. According to the periodicity of this process we may distinguish: (*a*) a group of lakes with rarely occurring, but large, breaches (Dzhaksy-Klych, Bitter) and (*b*) lakes with an annual inflow of sea water (the Rann of Cutch in India, and others).

Lake Dzhaksy-Klych in the Aral region lies in a northeastern extension of the Sary-Cheganak Gulf, from which it has become separated. The lake basin (Fig. 83) now consists of two smaller basins separated by a barrier up to 3 km wide. Both sub-basins are elongated in a northeasterly direction.

The area of the southern basin is 54·7 km², that of the northern basin is 17·67 km². The southern basin abounds in islands, the total area of which amounts to 6·4 km². Like Lake Kuuli, Lake Dzhaksy-Klych (and a number of associated lakes also) is dry. In January and February saline waters stand 20–30 cm above the salt surface. In summer the level is 2–3 cm below this surface. Before the channel from the lake into Sary-Cheganak Gulf was sealed, the lake was fed from the Sea of Aral during high water. Since the channel has been closed, water has been supplied by rainfall and by discharge of surface water, but this has been insufficient to maintain the lake level (Klebanov, Korf, and Elovskaya 1937).

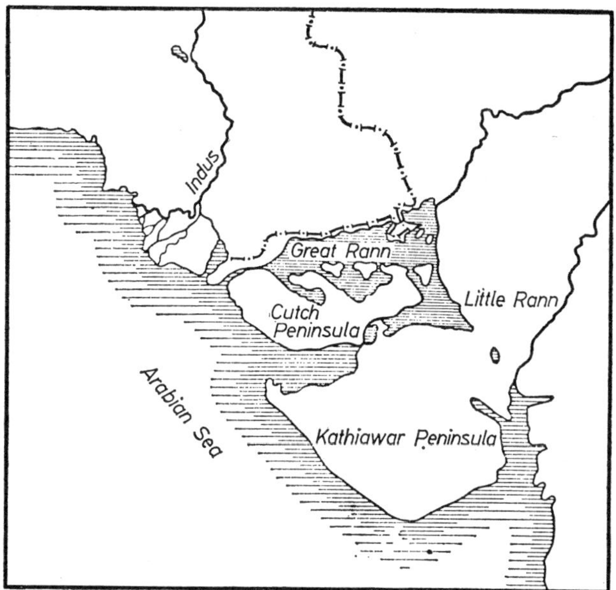

Fig. 84. The Rann of Cutch, India (from Grabau).

The Bitter Lakes, on the Isthmus of Suez, represent another characteristic example of a lake basin that fills periodically and for a considerable duration. In the sixties of last century, when the Suez Canal was being built, they filled a very saline basin with a thin layer of brine. In the central part of the basin there occurred a bed of salt nearly 13 km long, 6 km wide, and 8 m thick. It consisted of beds of halite, gypsum, and clay. The halite was 3–18 cm thick. There were two gypsum beds 7 and 11 cm thick. Individual clay beds were never more than a few millimeters thick, but nevertheless they frequently abounded in remains of the modern Red Sea fauna. It follows, therefore, that the Bitter Lakes represent a remnant of a gulf of the Red Sea, which at one time extended far to the north of the present sea. The separation occurred by the development of a broad shoal, cutting off the northern part of the gulf. Evaporation did not continue beyond the early part of the halite stage, which

M

was in all cases terminated by a major influx of the sea. This process was repeated many times. On cutting the Suez Canal, the salt bed was found to be covered by waters of increasing salinity (Grabau, 1920).

A typical example of supply to a lake through *annual breaching* is the Rann of Cutch in India (Fig. 84). It lies south of the delta of the Indus, where it occupies a flat plain up to 300 km long, 100 km wide, covering an area of about 18,000 km². It thus has the same area as the Kara-bogaz Gulf before its recent shrinking. The plain is underlain by sandy clay and the area is connected with the ocean through channels. In summer, during the southwest monsoons, sea-water floods the plain almost completely to a depth of 1 m. In

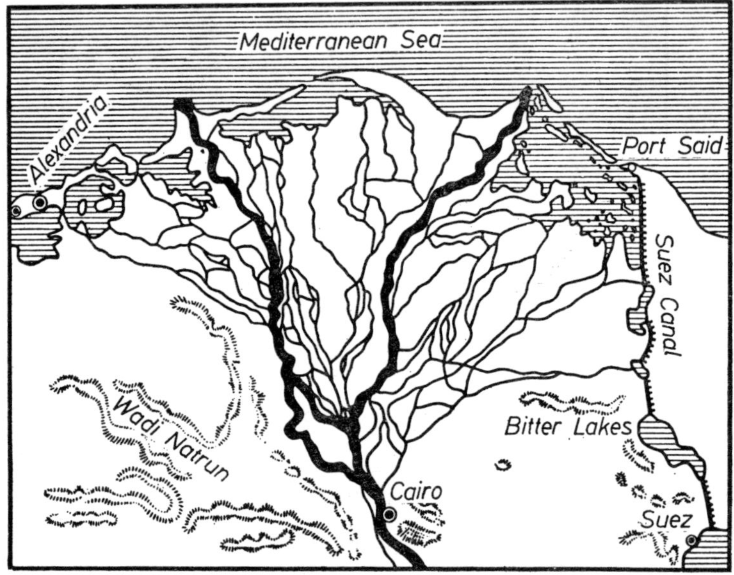

FIG. 85. Lakes on the Nile Delta (from Lotze).

winter, when dry northeasterly winds blow, the water is partly driven back into the sea, and partly evaporated. The low parts of the plain are then converted to salt pans, filled with newly precipitated halite to a depth of 10 cm, locally even reaching a meter. The higher parts between the depressions are also covered with a salt crust 2–3 cm thick. During the succeeding inundation all the salt is dissolved.

The Nile delta between Alexandria and Port Said also has flat-bottomed lakes that experience temporary flooding (Fig. 85). Mobile dunes frequently migrate into them, and the salt water absorbed by these sands commonly cements them. "Such small saline basins in the arid zone are observed on all coasts" (Lotze 1938, pp. 104-105).

From the foregoing it may therefore be seen how saline basins, variable in morphology, nature of water supply, and duration, are associated with the present shore zone of arid seas, and how varied and complex is the history of their development. Its significance in any analysis of the history of ancient basins and their salt deposits is apparent.

3. PERIODIC FLUCTUATIONS IN THE SIZE AND OUTLINE OF SALINE BASINS

One of the characteristic features of the geomorphology of present-day saline basins is the multiform fluctuation of water level and the corresponding variation in areal dimensions of the basins.

These fluctuations have different periods and different causes. The shortest fluctuations are the seasonal changes in level which are most clearly observed in lakes. The level rises in spring because of inflow of melt water. In summer, when the season of evaporation comes to an end, the water level is low, and the area decreases appreciably. In autumn water enters the basin again, and the level rises. Such seasonal fluctuations cause least areal variation in lakes that are large and contain considerable volumes of water, and that are in the early stage of development (Chany, Chad, Balkhash, and other lakes). In small lakes that have reached an advanced stage in degree of salinity, seasonal fluctuations in level may lead to very marked changes in area of the water surface; the depth of brine also will progressively increase. Such lakes are Elton and Baskunchak, which are full of water in the spring but dry up markedly by September, containing only brine in certain parts of their basins.

Similar annual fluctuations, though less obvious, are observed in present-day gulfs and lagoons. The level of the Gulf of Kara-bogaz has been 10 cm below the level of the Caspian in wintertime for thirty years, and 50 cm below in summertime. Where the floor of the basin is level, this fluctuation has caused variations in shore line amounting to several hundred meters. In the Sivash the average brine level in the gulf is always lower than the Sea of Azov, but it is especially low in August. These annual fluctuations in level of the Sivash are complicated by the effect of wind, which drives water from one part of the Sivash to another. Because the Sivash is very shallow, the effects are very marked.

In addition to seasonal fluctuations in lake level, others may last for longer periods. The fluctuations of 137 lakes in Kasakhstan and Western Siberia from A.D. 1700 to 1950 have been compiled (A. V. Shnitnikov 1949, 1950). It is apparent at once (Figs. 86 and 87) that in these two and a half centuries the lakes have passed through seven periodic fluctuations of level, ranging in duration from 21 to 47 years. The amount of variation in large lakes with inlets (such as Chany Lake) is about ± 3 m; for typical average lakes this value is about ± 2 m, and for saline, briny lakes, it is about $\pm 1\cdot2$–$1\cdot5$ m.

In size, salinity, and the entire regimen in general, lakes are affected differently over periods of many years. The largest of the lakes in Kazakhstan and Western Siberia are preserved as bodies of water even during a sharp fall in level, but the mineralization of the water varies considerably. Some of them may change from fresh to brackish and may lose their fish population. Moderately mineralized lakes may become strongly saline, even to saturation

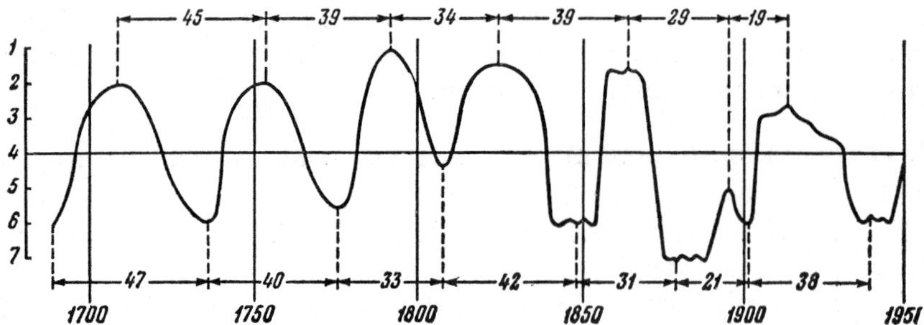

FIG. 86. Fluctuations of lake levels in Western Siberia and Northern Kazakhstan (from A. V. Shnitnikov). 1. Very high volume of water; 2. high volume of water; 3. average high-water stage; 4. the norm; 5. average low-water stage; 6. small volume of water; 7. very small volume of water. The arrows indicate the duration of cycles (in years).

of some salts. Lakes of average size may almost dry up when the level goes down, and small lakes may dry up entirely. Increased salinities and evaporate precipitation inevitably result.

The converse, however, also applies as in Lake Anzhbulat. In 1925 it held

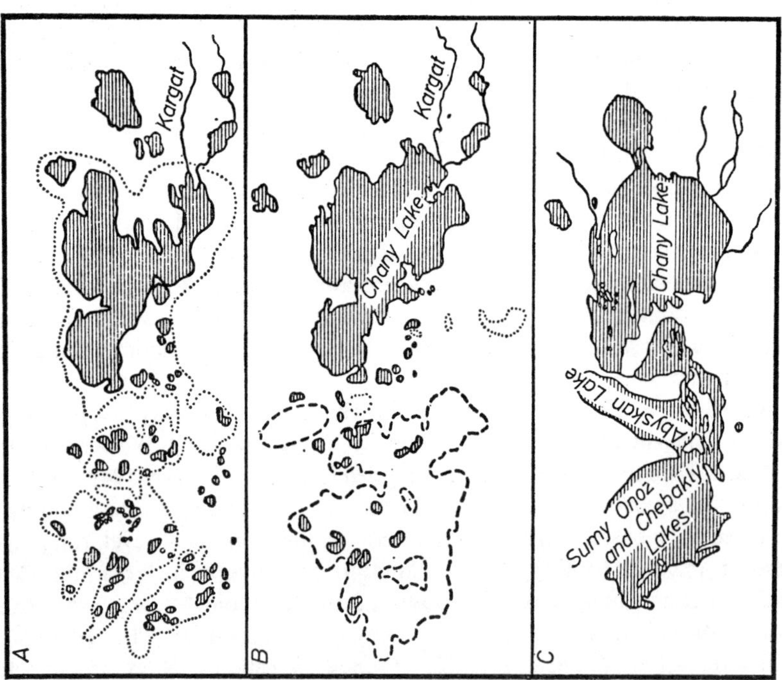

FIG. 87. Chany Lake and the adjacent lowland (from A. V. Shnitnikov). A. Geologic sketch map for the 1940's: dotted lines indicate the outer boundary of Recent flood-plain deposits and of lower terraces. B. Sketch of topography for the twenties of the twentieth century; the broken lines indicate the 100-m level, the dotted lines the 110-m level. C. Sketch map of the Kolyvan viceregency, 1787.

a layer of saline water 1·65 m thick. In 1933, when visited by M. G. Valyashko and I. I. Kornilov, the maximum depth of brine was 0·5 m and the floor was covered with a layer of mirabilite. In 1940 the lake was completely dry, and the floor was covered with halite, thenardite, and mud (A. I. Dzens-Litovskii 1959). The brine level was 8–10 cm below the level of salt. In 1946, when I visited the lake, it contained water and there were no salts on the floor, but a bed of thenardite was preserved a short distance below the mud. In 1947 there were no salts at all (A. I. Dzens-Litovskii 1959).

The cause of the periodic fluctuations in lake levels is the periodic variation in precipitation. "A comparison of rainfall cycles with fluctuations in lake level shows a well-defined dependence of the level on amount of rainfall. The lake level lags behind changes in rainfall by 2–4 years" (Shnitnikov 1950, p. 124).

Besides short-period (intrasecular) fluctuations in climate, longer changes, extending over hundreds and thousands of years, have been established, such as the 1800-year cycles of Shnitnikov and the so-called "xerothermic" phase (or phases) of L. S. Berg. The effect of these fluctuations on the behavior of a continental lake cannot be observed directly, but the phenomenon is clear from the above discussion. We would add merely that *climatic fluctuations that lasted for hundreds and thousands of years have been much sharper than short-period fluctuations, and their effects must have been felt in even the largest lakes.*

Thus, arid conditions have produced a general trend in aqueous basins toward contraction and even, in some cases, to conversion of the basins to dry lakes. Climatic fluctuations, with periods of a season, of many years, or even of a century, in combination with tectonic movements have converted this gradual trend to a varied rhythmic process. These consequent rhythmic sequences are incomparably more variable and more clearly manifest than those found in sediments of humid regions. *This rhythmic character is a specific aspect of sedimentation in arid basins generally, and is especially pronounced in basins with saline and very saline waters.*

4. HYDROLOGIC TYPES OF SALINE BASINS

Distinctive morphological features of saline basins combine with even more distinctive features of hydrology and hydrochemistry.

Two types of arid-climate basins may be recognized on the basis of hydrologic conditions. In the first type, which represents the most widespread and most extensive of present-day lakes, both small and large shallow-water bodies (Baskunchak, Elton, Ébeity, and others), *the salinity is practically the same throughout the entire water mass, horizontally and vertically.* Slight freshening may be found only locally, where ground water seeps into the lake or where streams enter. This type of lake may be called a hydrologically homogeneous saline basin. It normally forms where seepage of ground water is the principal supply.

The second type, on the contrary, is characterized by *spatial inhomogeneity of salinity in the water, the salinity clearly changing from one end of the basin to the other.* This inhomogeneity is due to the dominance of surface-water supply and to the asymmetrical disposition of inlets relative to the shape of the basin.

Lake Balkhash is a classic example of horizontal variability among lakes of continental origin. It is about 600 km long and up to 74 km wide, and is clearly divided into five segments, connected by shallow and narrow straits. The Ili River, from the south, supplies over 80% of the annual influx of water and salt to the basin. The water, evaporating and becoming increasingly salty, flows slowly to the east, toward the end of the Burlyu-Tobe segment. It is augmented, in part, by inflow from Karatal and Lepsa Rivers. All five segments of Lake Balkhash are huge interconnected evaporative basins, each receiving the mineralized solution from the preceding basin, modifying it by further evaporation and concentration over its surface, and then passing it into the next segment. This results in a steady increase in salinity through the several segments. Whereas the salinity of the Ili River is 0·01%, it is 0·02% in the southern (Ili) segment of the lake, then 0·03%, rising to about 0·5% in the far eastern segment.

Good examples of horizontal inhomogeneity of water were found in the lagoons Mertvyi Kultuk (Dead Bay) and Kaidak, when they were in existence, and also in the Kara-bogaz Gulf and the Sivash. Part of the northern Caspian before it entered the Mertvyi Kultuk was considerably freshened by the Ural River, up to 0·2–0·3%. In the strait leading to the Mertvyi Kultuk the salinity increased to 1·4–1·6%. In the Mertvyi Kultuk itself it was 3·08–3·07%, and in the Kaidak it increased rapidly to 3·8–5·97% in the extreme south.

In the Kara-bogaz Gulf in the period from 1890 to 1930, the western part adjacent to the strait was much less mineralized than the eastern half. In consequence, streams of Caspian water pouring through the strait were deflected to the south, and the freshened water in the gulf on the south reached the bays of Tyshke-Sergaz. On the north this effect was noticeable as far as the dam at the At-Chalma Wells (on the northern spit). To the east the streams of Caspian water could be seen in calm weather for distances of 10–12 km from the bar. The distribution of salinity changed sharply with strength and direction of wind, however, not only in the surface layers but also at depth. For example, at times it was possible to observe a thin layer of Caspian water even in the vicinity of the industrial plant at Sartas (on the north), where the specific gravity of the surface layer of brine (10 cm) ranged from 1·14% to 1·16%, though at a depth of 1 m the specific gravity corresponded to the summer norm of 1·175. But when northeasterly and easterly winds blew, the brine was driven directly against the sandy barrier. Some of it even passed through the strait and spilled into the Caspian. After the fall of the Caspian level (1930–1960) and the marked increase in salinity of Kara-bogaz Gulf, the saline inhomogeneity of the surface water layer was increased (Fig. 88). In the freshened part of the gulf, furthermore, vertical layering of the brine was found: less salty and lighter upward, more salty and heavier downward.

In the Sivash lagoon, fed by the Sea of Azov, a very marked irregularity in the water salinity is found in different parts of a single basin. The salinity of the Sea of Azov is about 1%. The eastern part of the Sivash, next to the Arabatskii Spit, is 3%. In the western part of Sivash, farther and farther from the winds of the lagoon, the salinity quickly increases to 7–15%, and in the

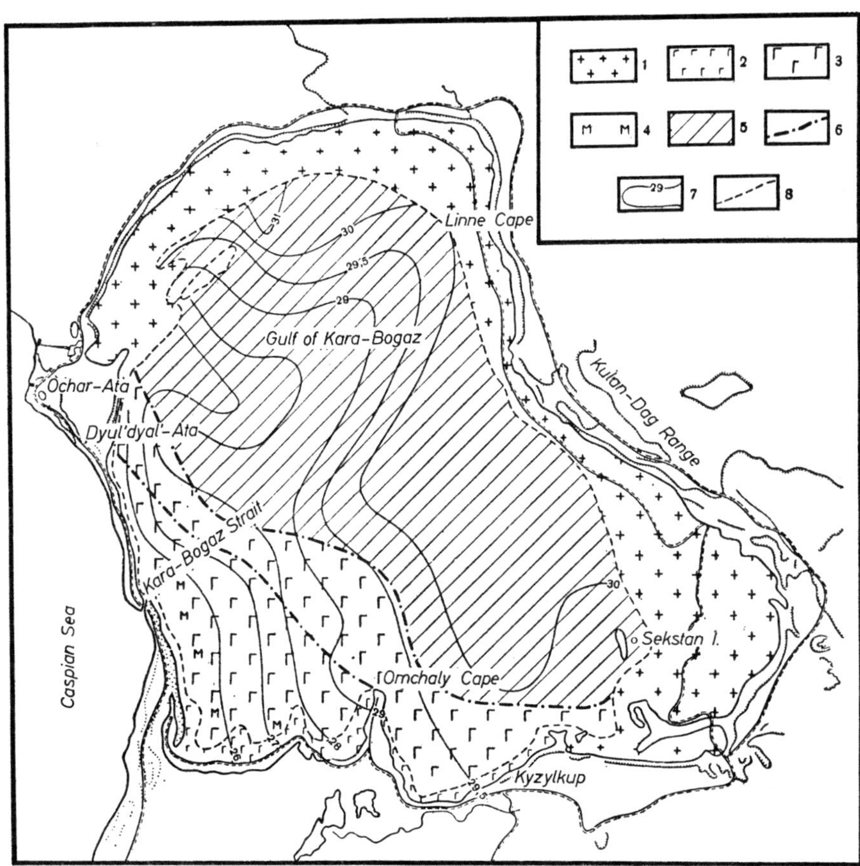

FIG. 88. Facies of the Kara-bogaz Gulf in 1956 (from M. P. Fiveg 1961). 1. Halite deposits on the dried-up part of the gulf floor; 2. dried-up part of the gulf with gypsum deposits; 3. area of glauberite precipitation; 4. mirabilite in sediments; 5. area of halite, halite + epsomite, and epsomite precipitation; 6. limit of halite precipitation in 1956; 7. isopleths of salt concentration in surface layer of brine in 1956; 8. outline of gulf.

extreme western arms it increases to 24–25%. Many of these arms become dry lagoons.

The Bocana de Virrila in the Sechura Desert of northeastern Peru is a striking example of horizontal inhomogeneity of salinity (Morris and Dickey 1957). This relict of an ancient river bed, filled with sea water is about 20 km long, about 2 km wide, and 0·6 m deep in the lower part, about 0·3 in the upper part (Fig. 89*A* and *B*). At the mouth of the gulf the salinity is

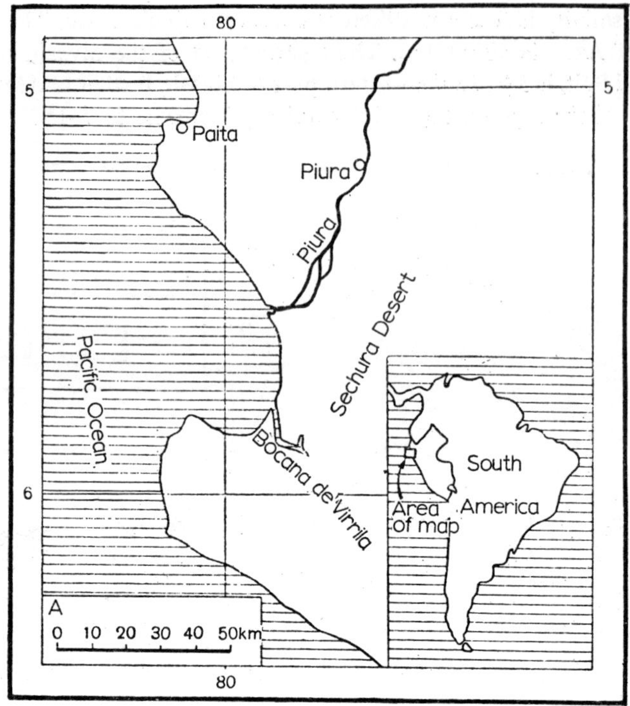

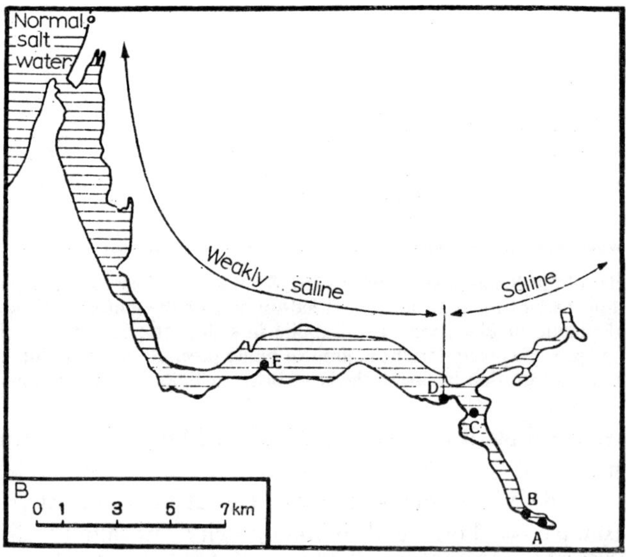

Fig. 89. Salinity distribution in the Bocana de Virrila (from Morris and Dickey). *A*. Location of the Bocana de Virrila, *B*. morphology of the Bocana de Virrila.

approximately normal marine: 3·5%. At point E it has risen to 8·88%, at D to 10·33%, at C to 13·24% on the surface and to 19·15% on the floor, at B to 34·55%, and at A to 35·49%. All determinations were made in August 1955. This basin has no shoals, submerged or above water, to restrict circulation.

Thus, in arid climates, when basins are fed from one end only, a characteristic pattern of increasing mineralization of the brines develops in a horizontal direction according to distance from the source. When the water feeding a basin progressively evaporates, salts of different composition precipitate successively according to their saturation. Since the water moves from one locality to another, from the intake in the lagoon to distant and more saline arms, *a horizontal differentiation takes place in different parts of the lagoon of the salt that is annually supplied to it.* In general little attention has been given to this circumstance, but it is clearly important in the interpretation of halogenesis in ancient saline basins, which were mostly characterized by supply of water and salt from one end of the basin, as will be pointed out below.

Vertically the salinity in saline basins is normally fairly uniform, but near any inflow of water with low salt content, a well-defined layering generally develops (Kara-bogaz Gulf, Bocana de Virrila).

5. HYDROCHEMICAL TYPES OF SALINE BASINS AND THEIR CHARACTERS

The chemical composition of water in present-day saline basins of arid regions varies greatly.

The first attempt to classify basins on this basis was made by N. S. Kurnakov (1917). He recognized two classes: (*i*) salt lakes containing $MgSO_4$ and having a metamorphization coefficient $MgSO_4/MgCl_2 > O$, and (*ii*) salt lakes with no $MgSO_4$ and with a metamorphization coefficient of zero. Unfortunately this classification does not include all highly saline basins, much less basins with low mineralization.

A broader classification was proposed by M. G. Valyashko in 1933. He wrote, "In their chemical composition, the waters of salt lakes, together with all mineralized waters of the earth, may be divided into three fundamental chemical types: carbonate, sulfate, and chloride. These types are determined by mutual combinations of the principal anions (CO_3^{-2}, HCO_3^-, SO_4^{2-}, and Cl^-) with the principal cations (Ca^{2+}, Mg^{2+}, and Na^+). The position of any lake water in this scheme determines the nature of the physicochemical processes in the lake and the equilibrium system. This permits us to understand and predict the trend of these processes. The determinative systems are the following:

For the carbonate type:

$$Na_2CO_3 - NaHCO_3 - Na_2SO_2 - NaCl - H_2O.$$

For the sulfate type:

$$2Na^+, (2K^+), Mg^{2+}, SO_4^{2+}, Cl^+, H_2O.$$

For the chloride type:

$$CaCl_2 - MgCl_2 - NaCl - H_2O.$$

The sulfate type has two subtypes: sodium-sulfate and magnesium-sulfate (chlormagnesian). Each of these subtypes may be further subdivided into two groups according to relative abundance of the ions HCO_3^{1-} (CO_3^{2-}) as compared with ions of Ca^{2+} and Mg^{2+}" (Valyashko 1952b pp. 14-15).

Valyashko's classification is simple and convincing from the chemical viewpoint; it has therefore become widely used in practice. For lithologic purposes, however, any classification should be more comprehensive. Lithologists working on arid-climate sedimentary processes are interested not only in sedimentation at high salinities but *in the entire course of successive changes in sedimentation from the least-mineralized fresh-water stage of a basin to its extreme saline state.* The classification must therefore include hydrochemical characteristics of the initial fresh-water stage of basins in each class, and it must indicate the changes this initial water undergoes as it becomes saline. Only by knowing this hydrochemical evolution of a basin can the systematic patterns in arid basins of sedimentation be determined.

Figure 90 shows the concentration of soda-rich water, up to 32%, for lakes in the Tanatar group on the Kulunda Steppe. These form an almost ideal example of the pure soda line (Strakhov 1951), because the soda ratio in the different lakes

$$\frac{Na_2CO_3}{NaCl + Na_2CO_4 + Na_2CO_3}$$

fluctuates very little, being approximately 50% ($\pm 4\%$).

It is clear that at 23-25°C all three basic salts—NaCl, Na_2CO_3, and Na_2SO_4—are steadily concentrated, but with no variation in the proportions of the constituents. Only at salinities above 28% does Na_2CO_3 begin to precipitate, while NaCl and Na_2SO_4 continue to be concentrated in the brine.

With the increasing concentration of soda, the pH tends to increase rapidly. At a salinity of 0·2% the pH has risen to 9·4, and at 1% it is about 10. Observations on pH at higher mineralizations are scattered. In Tanatar I Lake in 1945, the pH was repeatedly measured at 11·5-11·6, the salinity being about 25%. This confirmed an earlier observation by V. L. Isachenko in 1931, though the salinity was not recorded (Kulunda Expedition 1934-35). There is thus no doubt of progressive increase in pH, although it takes place at a much slower rate than during the initial stage of increasing salinity.

The behavior of calcium and potassium is entirely different. As a lake first becomes salty, at mineralization up to 0·1%, calcium is concentrated in the sodic water, attaining an average content of 1·5 mg-equiv./liter, and at times even more. As the salinity of the sodic water increases further, the calcium concentration decreases in a manner following a damping curve. At a salinity of about 1%, the amount of calcium falls to 0·3-0·5 mg-equiv./liter, and then to mere traces. Many analyses reveal no calcium at all, although analyses in the same group may show a calcium content as high as 0·5-0·7 mg-equiv./liter: but such results are probably analytical errors.

The behavior of magnesium is similar to that of calcium but with some differences. The maximum magnesium concentration is reached somewhat

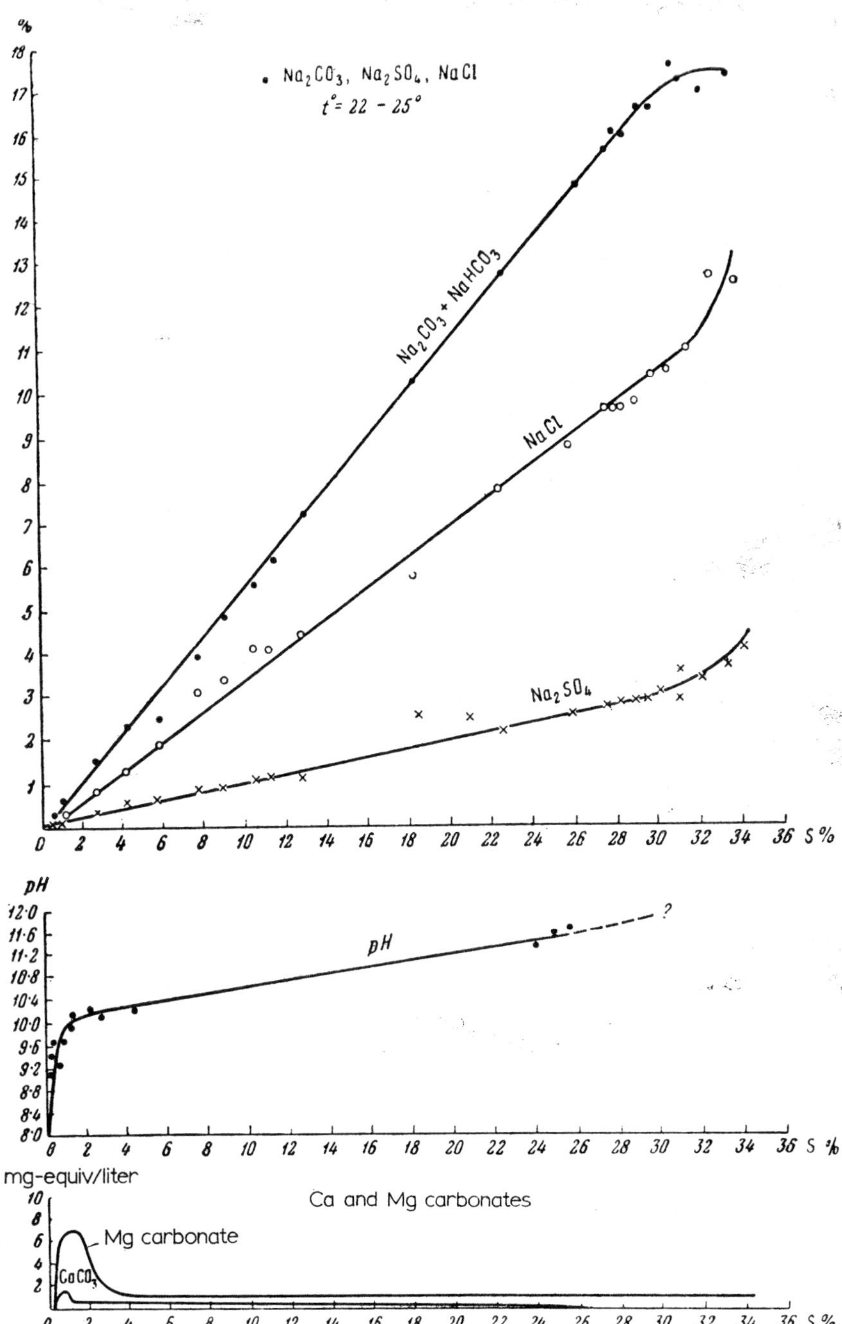

FIG. 90. Concentration in soda lakes of group Ia; Tanatar system of lakes on the Kulunda Steppe (from E. S. Telentyuk 1952, and S. Z. Makarov 1935).

later and at a higher mineralization of the sodic water than that for calcium: 0·3–0·4% against 0·1%. The ceiling of magnesium concentration is much higher on the average than that of calcium: 7–8 mg-equiv./liter instead of 1·5 mg-equiv./liter. The minimal level to which the magnesium concentration decreases during subsequent increase in salinity of the sodic water is also higher than that for calcium: about 1 mg-equiv./liter for magnesium as against mere traces for calcium.

This behavior of the alkaline-earth metals means *that during the progressive increase in salinity of soda lakes a point is very quickly reached (at approximately 1%) beyond which these constituents are expelled from solution. At salinities of about 3–4% sodic water rejects calcium (except for traces) and allows only negligible quantities of magnesium in solution (up to a limit of about 1 mg-equiv./liter).* Practically the only soluble carbonate in waters of this type is the sodium carbonate. At low mineralization this is present as bicarbonate. At low mineralization this is present as bicarbonate, but with increasing concentration it progressively changes to monocarbonate.

The "pure line" of the Tanatar Lakes demonstrates the type of sodic waters, in which soda is the principal salt and other salts are subordinate. This type is relatively rare. Usually much more soda is a subordinate constituent; and Na_2SO_4 and $NaCl$ are dominant, with variable ratios between them. Therefore, for a more precise hydrochemical characteristic of a lake, a sulfate coefficient is necessary, along with the index for total mineralization and the soda coefficient. This may be expressed by

$$K_S = Na_2SO_4/NaCl.$$

It is therefore essential to recognize three grades of soda lakes: *strongly sodic* with a soda coefficient $K_{Na} > 30\%$ (index Ia), *moderately sodic* with K_{Na} from 30 to 3% (index Ib), and *weakly sodic* with $K_{Na} < 3\%$ (index Ic). The other types of soda lakes will not be considered here, because the behavior of the principal salts does not change appreciably. All the salts (Na_2CO_3, $NaCl$, and Na_2SO_4) are concentrated in the water of any of these lakes in proportion to total salinity, deviating from this rule only locally. Sodium sulfate or sodium chloride may take the place of soda with decrease in K_{Na} and change in K_S. More substantial deviation takes place in pH and in the behavior of $CaCO_3$ and $MgCO_3$. As K_{Na} decreases, active reaction in the solution naturally declines, from 11·6 to 9·6–10 as a maximum. This allows much more $MgCO_3$ to be held in solution than previously, and $CaCO_3$ may begin to accumulate, though in negligible quantitie. All these changes in behavior in both the principal and the accessory salts, lead to a gradual approach to the sulfate class; with $K_{Na} < 3\%$, weakly sodic lakes in their general hydrochemical characteristics come close to the variety of sulfate lake that is here given the index IIa.

A characteristic feature of the sulfate class is the complete absence of Na_2CO_3 among the dissolved salts. The CO_3^{2-} and HCO_3^- ions here bind only part of the Ca^{2+} and Mg^{2+}; the other, and greater, part of the latter two constituents is bound up with SO_4^{2-}, forming $CaSO_4$ and $MgSO_4$, sometimes with Cl^- to form $MgCl_2$. *The appearance and accumulation of these salts constitute a diagnostic feature of the sulfate class.* Of the other salts, NaCl, and frequently Na_2SO_4, are always present, generally in large quantities.

The characteristic coefficients of sulfate lakes are

$$K_{Mg} = \frac{MgSO_4}{\varepsilon\ \text{salts}}; \quad K_S = \frac{Na_2SO_4}{NaCl}, \text{ and } K_K = \frac{MgSO_4}{MgCl_2}.$$

In view of the large number of salts occurring in various proportions in solution, saline lakes of the sulfate class (II) vary considerably in their hydro-chemical characteristics. Three fundamentally different groups may be recognized; groups IIa, IIb, and IIc.

The characteristic features of the IIa group are :(*a*) abundant NaCl and Na_2SO_4 with little $MgSO_4$ and no $MgCl_2$, (*b*) abundant CO_3^{2-} and HCO_3^-, which generally bind all the Ca^{2+} and part of the magnesium.

The concentration diagram of these lakes (Fig. 91) is constructed chiefly from data on the Mormyshanskie Lakes of the Kulunda Steppe. Unfortunately the line here is apparently less pure than that of the Tanatar group.

As these lakes become salty, at temperatures around 25°C, the concentration of the principal salts NaCl, Na_2SO_4, and $MgSO_4$ increases continuously to the total mineralization of 26–27%. Near this concentration Na_2SO_4 begins to precipitate. This causes a swing in the Na_2SO_4 curve toward the horizontal and in the NaCl curve toward a steeper upward slope. The $MgSO_4$ curve remains practically unchanged. Unfortunately the mutual relationships of these salts at higher mineralizations in lakes of this group are not known.

Of the accessory salts, the content of Ca and Mg carbonates varies charac-teristically. The $MgCO_3$ content, having been very low when the mineraliza-tion of the water was negligible, increases very rapidly as the salinity of the water increases, almost in proportion to the salinity increases. At a salinity of 0·4–0·8% the value reaches 14–16 mg-equiv./liter. The rate of increase then declines, and at a salinity near 20–22% (?) a sudden transition occurs, and the concentration of $MgCO_3$ is reduced. Unfortunately, there are insufficient data for reliable construction of this second part of the curve, but there can be no doubt that the mass of $MgCO_3$ generally decreases at maximal salinity.

As salinity first begins to increase, the concentration of $CaCO_3$ rises to a value of 2–3 mg-equiv./liter at a total mineralization of 0·5%. Then the $CaCO_3$ content declines, falling approximately to 1–0·75 mg-equiv./liter, but when the salinity of the water is high (over 10%), $CaCO_3$ is again precipitated, reaching a concentration of 6–7 mg-equiv./liter at a total mineralization of 28–32% in the brine. This second rise in concentration has not yet been widely traced: it is clear in some basins, but less distinct in others. We may say only that in general the $CaCO_2$ curve is a subdued reflection of the $MgCO_3$ curve.

The pH curve is very distinctive. When a body of water first becomes salty, the pH rises sharply, and at salinities of 0·4–0·6% it reaches 9·2–9·4, i.e. it is almost at its limiting value. On further increase in salinity, up to 5–6%,

the pH remains practically the same, perhaps rising to 9·4–9·5 at salinities of
1–3%. At greater concentrations of salt, judging from isolated determinations,
the pH again declines, falling to 7·8–7·6 as a consequence of the accumulation
of $MgSO_4$ in the brine.

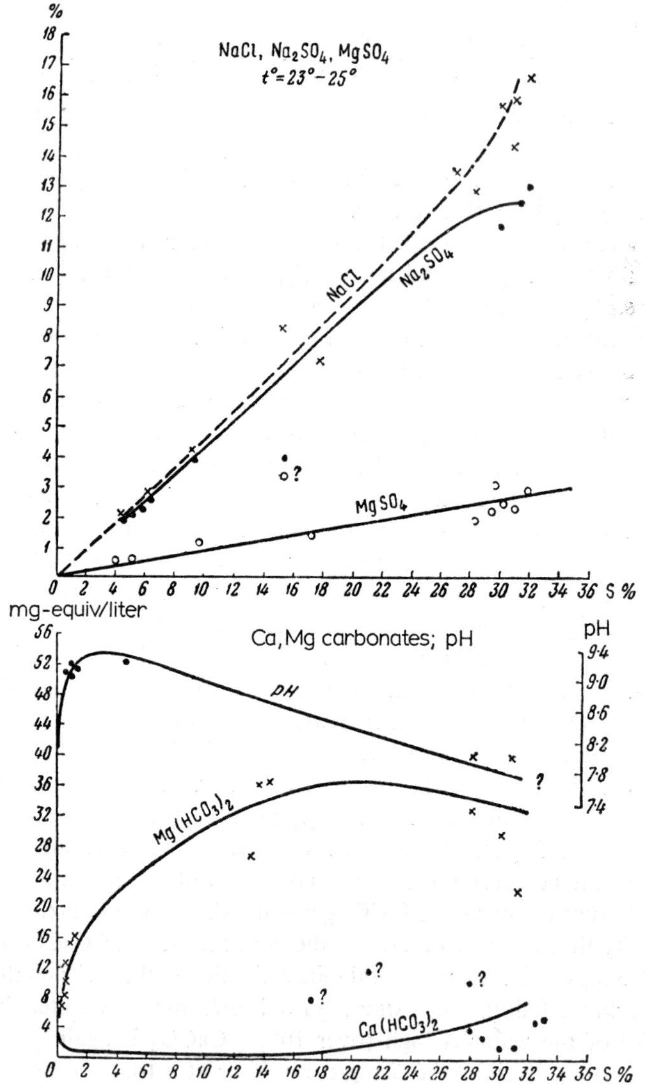

FIG. 91. Concentrations of the sulfate class, group *IIa* (from S. Z.
Makarov).

Group IIb differs from group IIa by the invariable presence of NaCl,
Na_2SO_4, $MgSO_4$, $MgCl_2$, and $CaSO_4$ in large quantities. The carbonate ion
is scarce and is associated only with some of the Ca. Magnesium carbonate

is missing. The amount of sulfate in solution reaches a maximal value for the entire sulfate class (Fig. 92).

On concentration, the NaCl, KCl, and $MgSO_4$ content increases continuously with total mineralization of the salt water, but the Na and Ca sulfates

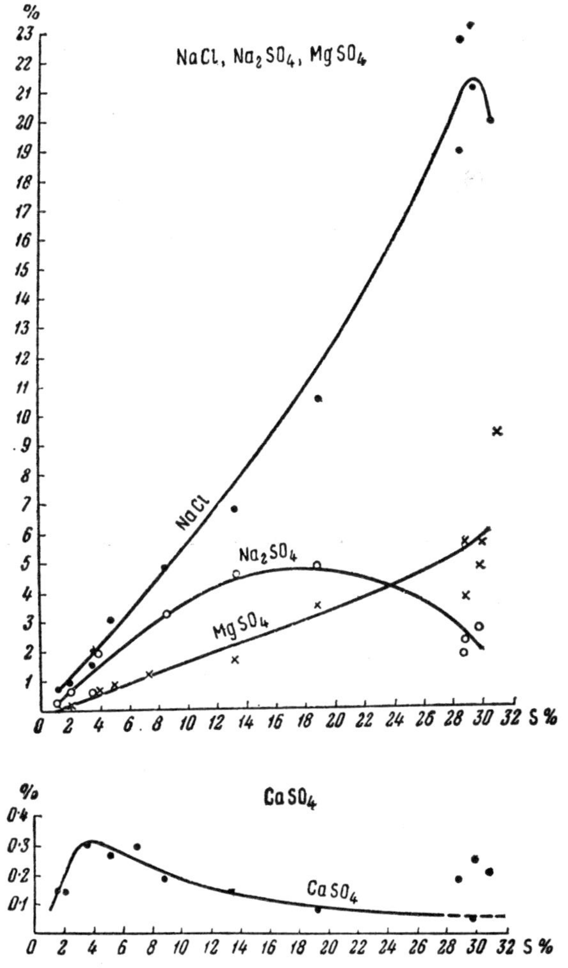

FIG. 92. Concentrations of the sulfate class, group *IIb* (from S. Z. Makarov).

behave differently. $CaSO_4$ very quickly reaches saturation (approximately at salinity of 3%), and then the content rapidly falls (precipitation of gypsum!). Later, at salinities above 13%, Na_2SO_4 follows the same course. Na_2SO_4 also ceases to concentrate in the water, and at higher salinities decreases sharply with the precipitation of mirabilite. In this group of lakes Ca and Na sulfates behave fundamentally in the same way as the Ca and Mg carbonates in soda lakes, but the inflection points on the curves are shifted sharply to the right by virtue of the greater solubility of the sulfates.

Unfortunately, few data are available on the $CaCO_3$ content and pH in lakes of this group. It must be thought, however, that the lakes of group IIb scarcely differ in this from basins of the Aral-Caspian group described below. The concentration diagram for the third group of the sulfate class of lakes (IIc) has been plotted from data on lagoons in the Black Sea and Caspian basins.

The Black Sea water was initially very similar to normal oceanic water in its salt content (Table 14). The type of saline development is therefore very similar to the mineralization history of oceanic water.

During the progressive increase in salinity in such water (Fig. 93), the NaCl content increases continuously up to a mineralization of about 28%, after which the curve flattens suddenly and then slopes downward, because halite precipitation begins. On the other hand, the curves of $MgSO_4$ and

TABLE 14

Composition of Salt in the Black Sea and the Ocean

Basin	Salt content, in %						Salinity, %	$\frac{MgSO_4}{MgCl}$
	$CaSO_4$	$MgSO_4$	KCl	NaCl	$MgCl_2$	$CaCO_3$		
Ocean	3·97	6·40	1·69	78·32	9·44	0·21	34·30	0·67
Black Sea	2·58	7·11	2·99	77·72	9·07	1·59	18·46	0·77

$MgCl_2$ rise more steeply at these salinities, reflecting the decline of NaCl in solution. The $CaSO_4$ content increases continuously up to mineralizations of about 14%, reaching the saturation point of gypsum ($CaSO_4 \cdot 2H_2O$), after which, because of the depressing effect of $MgSO_4$, the calcium sulfate concentration rapidly declines. The curve showing alkali content is a subdued copy of the $CaSO_4$ curve. It is characteristic that the solubility curve of $CaCO_3$ at 30°C is amazingly similar to that of the alkali content for the basin. Only beyond the point of inflection at 14–15% do the lines of $CaCO_3$ and alkali content diverge with increasing salinity; but the divergence is insignificant and in no way obscures the fundamental similarity of the curves.

This similarity leads to the conclusion that *as basins of the Azov-Black Sea type increase in salinity the alkali content is essentially determined by the $CaCO_3$ content, always supersaturating the solution* and changing inversely as the behaviour of $CaSO_4$. Impurities that here affect the alkali content are very small, and they accumulate only after $CaSO_4$ has saturated the water and begins to precipitate as gypsum.

The pH of the saline water at first increases slightly to 8·6, then steadily decreases to below 7 at high salinities (the effect of large amounts of magnesium sulfate and chloride).

Characteristic changes take place as waters of the Caspian and Aral seas increase in salinity. A distinctive feature is the marked abundance of Ca and Mg sulfates in the earliest stages of development of these basins (Table 15). These constituents remain abundant with increasing salinity of the water (Figs. 94, 95).

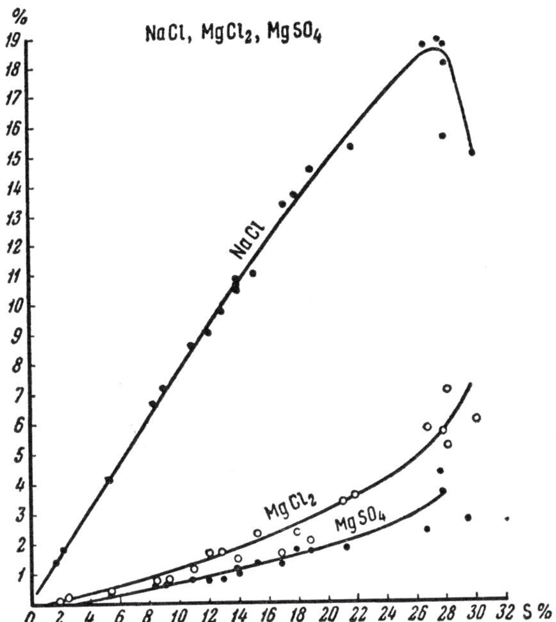

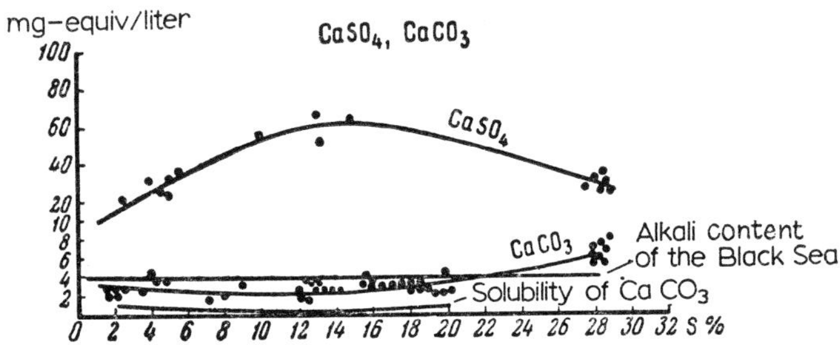

FIG. 93. Concentration of the sulfate class, group *IIc*; Azov-Black Sea basins (from N. S. Kurnakov and others).

The enrichment in sulfates shown on the concentration diagram of the Aral-Caspian basins is very characteristic. Because of its initial abundance, $CaSO_4$ reaches saturation in gulfs of the Caspian Sea and begins to be precipitated from waters having a total mineralization of about 5%—not 14% as in the Black Sea group of basins. Where salinities are high, the $CaSO_4$ concentration falls to much lower levels than in basins of the Azov-Black Sea type. Hence there is a very steep rise in the $CaSO_4$ curve in the salinity interval

N

1 to 5%, followed by a sharp drop. In lakes of the Aral region, because of their large $CaSO_4$ contents, gypsum saturates the water and begins to precipitate at even lower salinities, at only 3%. It may be recalled that some euryhaline bivalves (such as *Cardium edule*) may live in these seas at salinities up to 5%.

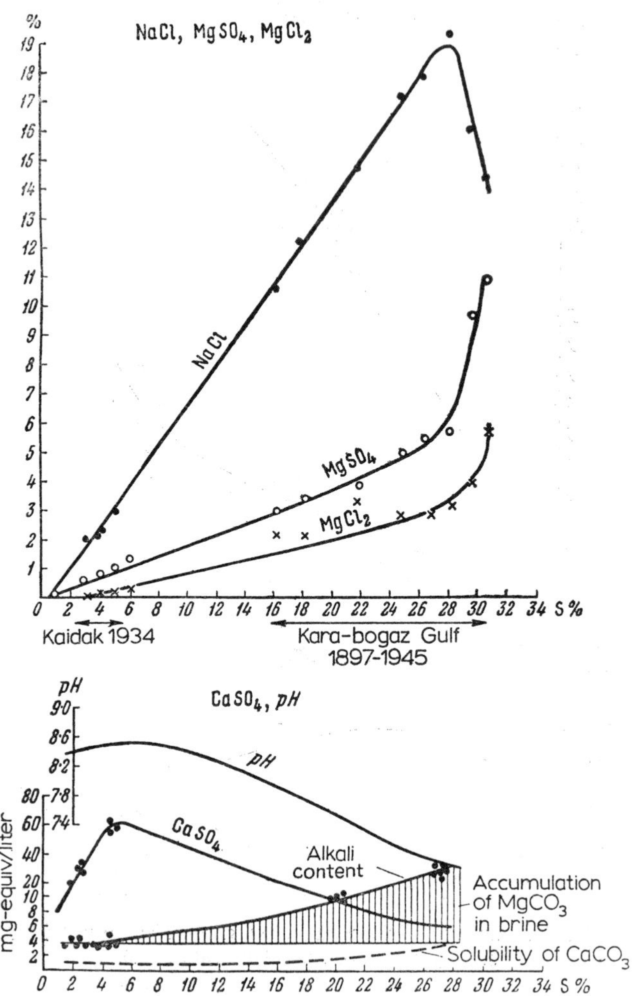

FIG. 94. Concentrations of the sulfate class, group *IIc*; gulfs of the Caspian Sea.

This leads to burial of shells of these organisms in typically gypseous or calcareous-gypseous mud. This anomaly, at first glance incomprehensible, occurred for example, in the Kaidak Gulf before that basin dried up (Makarov 1935) and also in the subfossil horizon of Kara-bogaz Gulf (Strakhov 1947).

But the enrichment of the Aral-Caspian basins in sulfates has an even more marked effect on the alkali content of the concentrating saline water. In the Caspian, Kultuk, and Kaidak, it is 3·2–4·0 mg-equiv/liter (rarely as much as 5); in the Kara-bogaz Gulf 13–14 mg-equiv/liter, and in Kuuli Lake it reaches

TABLE 15

Composition of Salt in the Black, Caspian, and Aral Seas
(from Knipovich, 1938)

Basin	Salt content, in %						Salinity, %	$\dfrac{MgSO_4}{MgCl_2}$
	$CaSO_4$	$MgSO_4$	KCl	NaCl	$MgCl_2$	$CaCO_3$		
Black Sea	2·98	7·11	2·99	77·72	9·07	1·59	18·6	0·77
Caspian Sea	6·92	23·58	2·21	62·15	4·54	1·24	12·86	5·20
Aral Sea	12·91	25·80	1·87	56·72	1·36	0·93	11·28	19 00

22–35 mg-equiv/liter. Up to salinities somewhat less than 5% in these basins, the solubility of $CaCO_3$ and the alkali content parallel each other fairly closely. At salinities above 5% they diverge sharply. *The alkali content clearly and very sharply surpasses the very slow rise in the solubility curve for $CaCO_3$.*

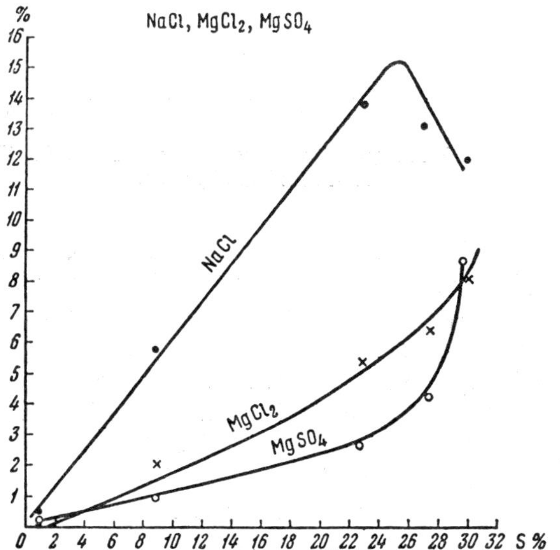

FIG. 95. Concentrations of the sulfate class, group *IIc*; lakes
of the Aral region.

This means that *in the Aral-Caspian lagoons and lakes $CaCO_3$ plays a subordinate role in the alkali content at high salinities, in contrast to the Azov-Black Sea type of basin.* The principal increase in alkali content is associated with the accumulation of some new component in the brine. It should be noted that the appearance and concentration of this supplementary component occur only after $CaSO_4$ has reached the saturation point and has begun to be deposited in the sediment. This clearly indicates that the cause of growth in alkali content of the Aral-Caspian basins lies in the formation of $MgCO_3$ according to Haidinger's reaction:

$$CaCO_3 + MgSO_4 \rightarrow MgCO_3 + \underline{CaSO_4}.$$

This reaction is effective only after $CaSO_4$ saturates the water and begins to be precipitated. The same process undoubtedly also takes place in basins of the Azov-Black Sea type, but is scarcely perceptible, and the increase in alkali is thus very weakly seen. The greater intensity of the Haidinger reaction in basins of the Aral-Caspian group is due primarily to the fact that the reaction begins to operate much earlier than in basins of the Azov-Black

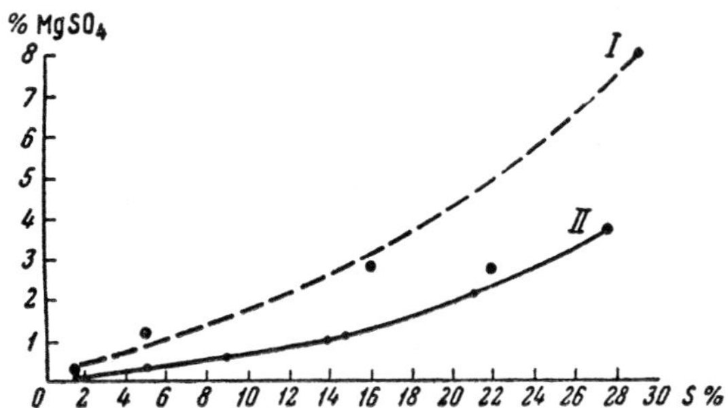

FIG. 96. MgSO₄ in lagoons and lakes of the Aral-Caspian (*I*) and Azov-Black Sea (*II*) types.

Sea group because the water becomes saturated with $CaSO_4$ much sooner. A second factor is that the $MgSO_4$ content is much higher in waters of the Aral-Caspian basins than in those of the Azov-Black Sea group. According to the laws of mass action this should strongly shift the reaction to the right (Fig. 96).

The chloride class of saline basins is distinguished from the sulfate class by the absence of magnesium sulfate; instead, $MgCl_2$ accumulates in large quantities and $CaCl_2$ appears. Brackish and weakly mineralized representatives of this group are as yet unknown. Highly saline members were described by N. S. Kurnakov (1917) from the Crimean lakes of the Perekop group, where the principal salts are $NaCl$, $MgCl_2$, and $CaCl_2$ (Fig. 97). The diagram may be clearly divided into two fields: one of salinity up to 25%, the other above 25%. Up to a mineralization of 25% all salts accumulate continuously in the solution, not mixing with each other, because they are in the unsaturated state. At a salinity of 25% $NaCl$ becomes saturated, and this at once changes all future relations. Under the suppressive effect of $MgCl_2$ and $CaCl_2$, which have an ion in common with $NaCl$, the concentration of the last rapidly decreases, and the concentration of $MgCl_2$ quickly rises. At salinities of

36–38% NaCl is practically eliminated from the brine, and $MgCl_2$ is the dominant salt. At the same time the $CaCl_2$ content also increases sharply.

It is thus clear that the hydrochemical features of basins of different types become progressively differentiated with increase of salinity. Though these

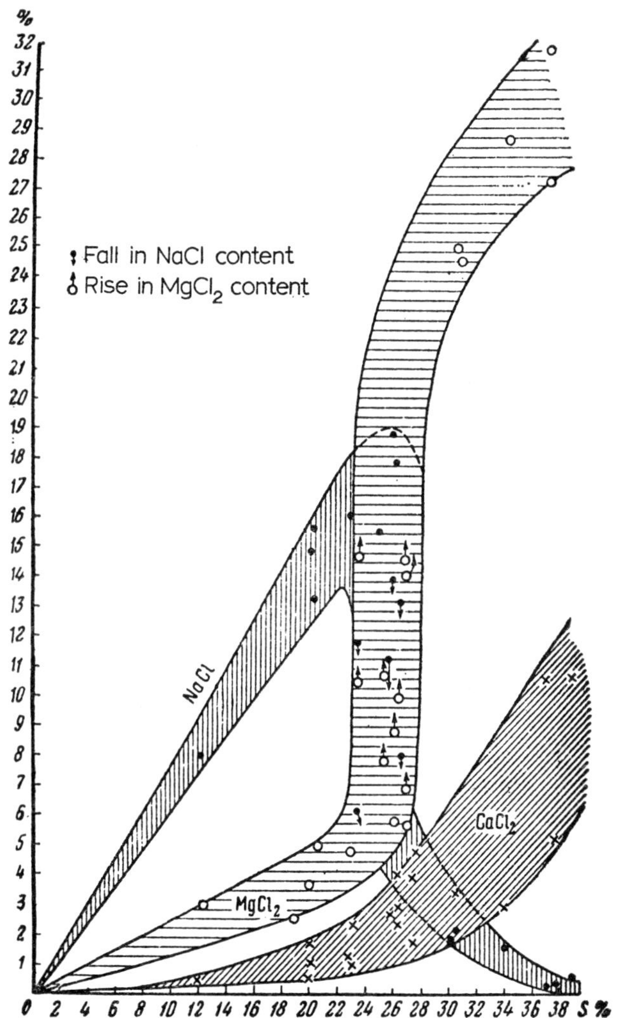

FIG. 97. Concentrations in basins of the chloride class III; lakes of the Perekop group (from N. S. Kurnakov and S. F. Zhemchuzhnyi. 1. Fall in NaCl content; 2. rise in $MgCl_2$ content.

differences are originally very slight, with practically no effect on sedimentation, they stand out sharply with increasing salinity, and strongly influence sedimentation. This is seen not only in the saline composition of the brines but also in pH. In soda lakes the progressive increase in alkali content leads to

extremely high values of pH (11·6 or higher). In sulfate lakes of the IIa group
the pH is also very high, but it is lower in other groups, especially at high
salinities. In chloride lakes the pH is low. Thus, *the increase in salinity of lakes
along with differences in composition of the salt, gives rise to great differences in
alkali-acid conditions. In some groups the alkalinity becomes higher than in any
other known sedimentary zone on the surface of the earth.*

Even all the differences in the hydrochemical features of salt lakes, the
general picture of their composition shows a well-defined pattern, especially
when basins of identical salinity are compared (Fig. 98). The total carbonate
content of the water, the alkali content, and the pH decline sharply from soda
to chloride lakes. The sulfate content increases from soda lakes to sulfate
lakes and again drops sharply in chloride lakes. Different sulfate sults appear

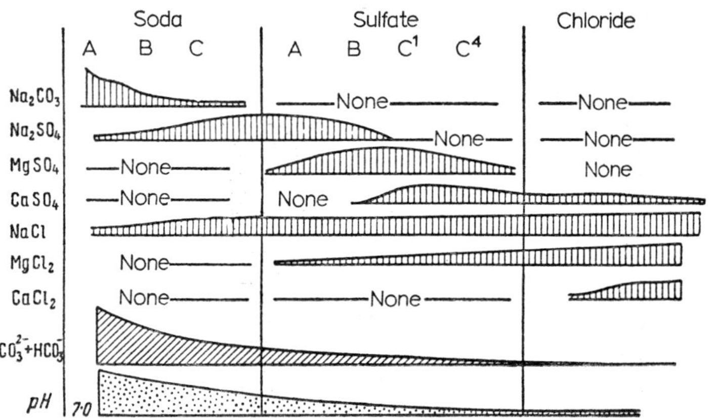

FIG. 98. Distribution of saline components for principal hydrochemical
types of basins.

in this variation: only Na_2SO_4 in soda lakes; Na_2SO_4, $MgSO_4$, and $CaSO_4$
in sulfate lakes; and only $CaSO_4$ in chloride lakes. $MgSO_4$ among the sulfates
is thus diagnostic of the class of sulfate lakes and of no others. Chlorides are
only moderately abundant in soda lakes, and only in the form of NaCl. The
chloride content increases in sulfate lakes, and $MgCl_2$ appears with NaCl.
In chloride lakes the chloride content becomes maximal, and NaCl, $MgCl_2$,
and $CaCl_2$ are found.

It may be said that in general, *in basins having a maximal content of any
group of salts—carbonates, sulfates, chlorides—the composition of this group
is very varied. As the total content of a given group declines, variety of constituent
salts in the group decreases.*

6. SOME METHODS OF GRAPHICAL ANALYSIS OF WATER COMPOSITION IN
  SALINE BASINS

The method of analysing the composition of modern saline basins at
high stages of mineralization, as applied in salt investigations during the

study of salt precipitation proper, merits attention. Basically the salt composition is represented in a diagram that indicates only the readily soluble components and excludes such poorly soluble forms as $CaCO_3$, $MgCO_3$, and $CaSO_4$, as well as potassium salts, because they are present in very small quantities in the brine. In this Fig. 99 the composition of soda lakes is represented by the triad: Na_2CO_3-Na_2SO_4-Na_2Cl_2.

The same method is used for constructing a diagram for water of chloride type. In this diagram the salt content is reduced to the sum of $2NaCl$, $MgCl_2$, and $CaCl_2$.

The diagram for water of sulfate type is more complicated, since the number of salts is greater: $2NaCl$, Na_2SO_4, $MgCl_4$ and $MgCl_2$. This may be illustrated by a square proposed by Jänecke, the corners representing the indicated salts, but in such a way that the cations vary along the horizontal and the anions

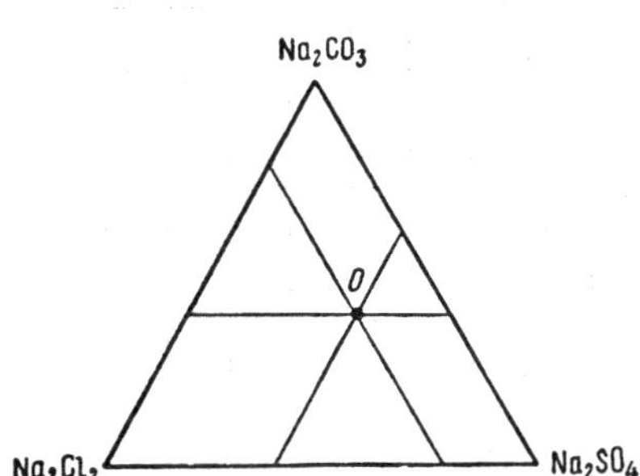

FIG. 99. Location of a point, O, representing composition within the triangle.

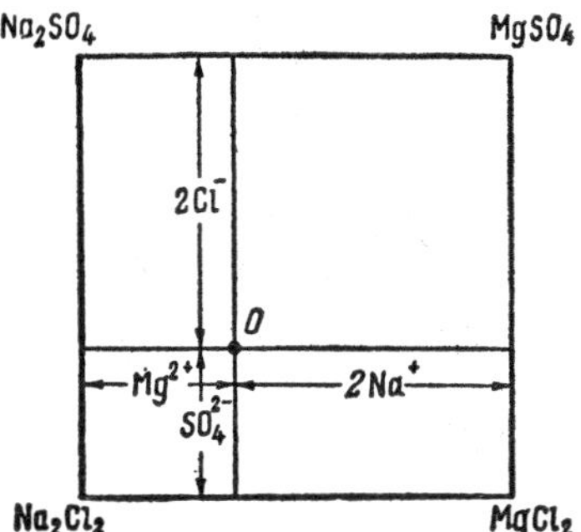

FIG. 100. Locating a point, O, representing composition within a Jänecke square.

along the vertical. After re-calculating the composition of the solution, according to ion content, to an equivalent form, we obtain: $2Na^+ + Mg^{2+} = SO_4^{2-} + Cl^-$. By taking the total of cations and the corresponding total of anions to be 100%, we may write: $2Na^+ + Mg^{2+} = 100$; $SO_4^{2-} + Cl^- = 100$.

In Fig. 100 the percentage of two ions are plotted: one cation (normally Mg^{2+}) and one anion (normally SO_4^{2-}). This is sufficient, because the other two will each be the difference between the plotted value and 100. From the Na_2Cl_2 corner the value of SO_4^{2-} is measured along the ordinate and Mg^{2+} along the abscissa to locate the point O, which represents the given solution.

It is possible to trace in these diagrams changes in the composition of the water of a basin during increasing salinity. It is necessary, however, to bear in mind the following basic aspect. If, during changes in salinity, none of the salts designated by the corners forms solid phases and is removed to the bottom sediments, the representative point of the solution on a diagram does not change position. This follows from the representative point being here but *a reflection of the salt ratios in the solution*, not a measure of absolute content as in the concentration diagrams (Figs. 91–97). Points on these ratio diagrams are displaced only when crystallization of salts occurs, because only this brings about a change in the ratios of the salts. These diagrams are therefore useful only in studying highly mineralized solutions, from which solid phases are being precipitated.

Figure 101 illustrates changes in composition of the chloride waters of Krasnoe Lake, from which NaCl is being precipitated. The representative points fall on a single straight line passing through the Na_2Cl_2 corner, indicating that NaCl becomes relatively less significant with time and that $CaCl_2$ and $MgCl_2$ become more significant. This straight line has been termed the *path of halite crystallization.*

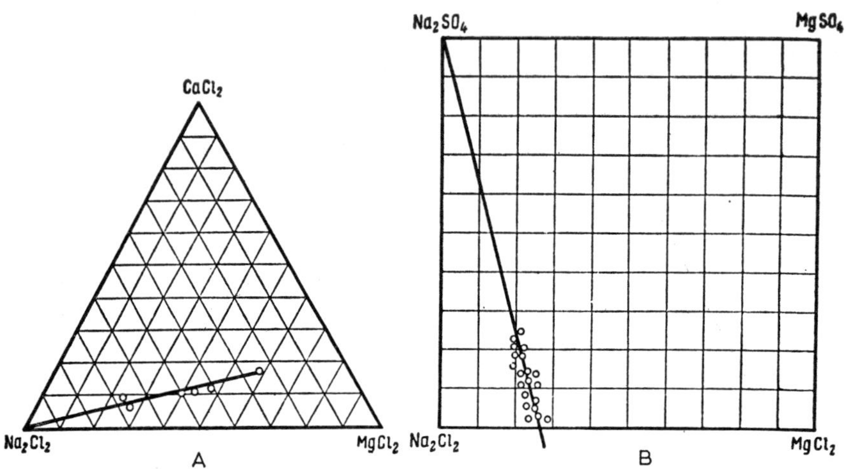

FIG. 101. Composition of salts in the brine of Krasnoe Lake (*A*) in the nineteenth and twentieth centuries, and in the brine of Kuchuk Lake (*B*) for the period 1930–44 (from O. D. Kashkarov).

Figure 102 shows paths of halite crystallization (in the lower part) at a number of sulfate lakes. The differences in these paths indicate that the crystallization of halite in the different basins begins at appreciably different salt ratios. The path of mirabilite crystallization in Mormyshanskoe Lake is also shown.

When the sequence of precipitation is complex, one salt beginning to precipitate after another, the diagram shows a broken curve consisting of several segments. This is seen in the changing composition of the Kara-bogaz brine during 1920-55 (Fig. 103). Prior to 1938 only mirabilite had precipitated, and therefore all points representing autumnal composition of the brine for this period lie on the straight line Na_2SO_4 (*A*), passing through the point that corresponds to composition of the Caspian Sea. Summer precipitation of halite began in 1938, bringing about a change in the course of crystallization, shown (for the summer period) by the straight line Na_2Cl_2 (*B*), since the mirabilite precipitated during autumn is completely dissolved by the beginning of summer. Joint summer precipitation of halite and blödite began in 1944.

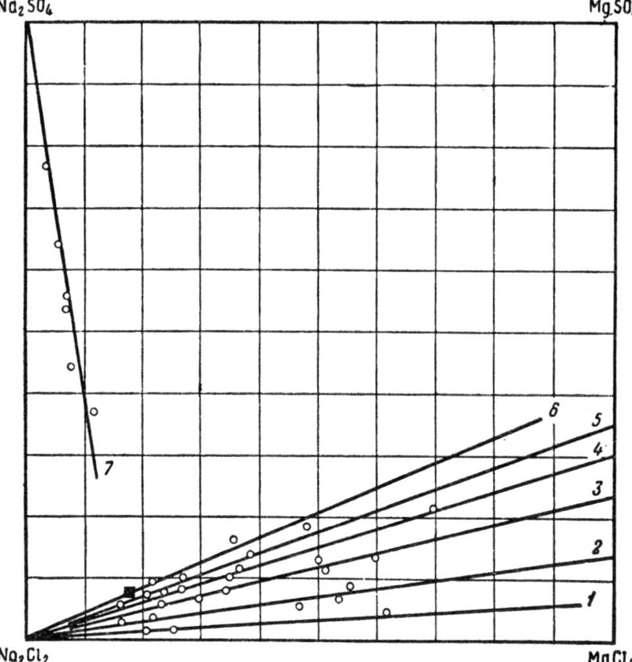

FIG. 102. Composition of salts in the brines of several lakes in the nineteenth and twentieth centuries (from O. D. Kashkarov). 1. Inder Lake; 2. Karabash and Kerleutskoe Lakes; 3. Sakskoe Lake before 1895 and Genicheskoe Lake; 4. Sakskoe Lake after 1900 and Sasyk-Sivash Lake; 5. Chekrakskoe Lake; 6. Adzhi-Baichi Lake and the Black Sea (black square); 7. Mormyshanskoe Lake.

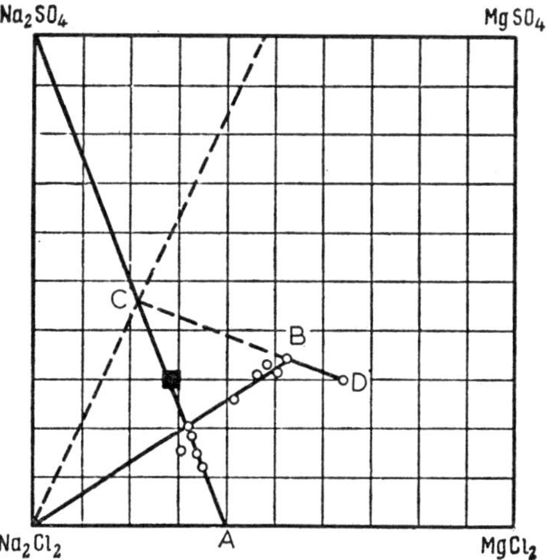

FIG. 103. Composition of salts in the brines of Kara-bogaz Gulf from 1927 to 1945, for the summer periods (from O. D. Kashkarov); the square indicates composition of Caspian water (from S. V. Bruevich).

This corresponds to point C on the diagram, and this represents a new change in the course of crystallization, shown by the straight line CBD.

It is apparent that the triangles used for graphical representation have sides in common with the square: for example, the triangle Na_2CO_3—$2NaCl$—Na_2SO_4 has the side Na_2Cl_2—Na_2SO_4 in common with the square. A generalized diagram consisting of two triangles and a quadrilateral with coincident sides, which represents all hydrochemical types of modern saline basins may therefore be used (Fig. 104). It further illustrates the entire process of chemical change in brine and of salt formation in summer and winter periods.

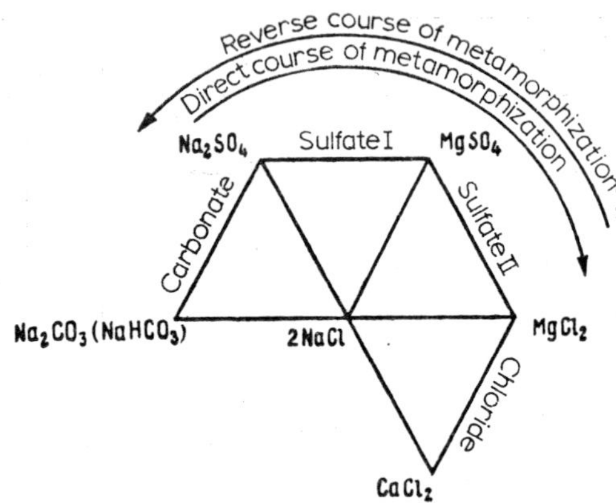

Fig. 104. Chemical composition of the principal chemical types of natural saline waters, their mutual relations, and their transitions from one type to another, (from M. G. Valyashko).

7. Physical and Physicochemical Features of Brines in Saline Basins as Compared with the Water in Basins of Low Mineralization

For a proper approach to the study of chemical sedimentation in saline basins, it is necessary to keep in mind, apart from a knowledge of their hydrochemical types, those specific physicochemical properties of the brines that distinguish them from those of low mineralization met in humid regions.

Increasingly saline brines show noticeable increase in both specific gravity and viscosity (Table 16).

In addition, the intensity of evaporation from the surface of the brine decreases as the salinity increases. Evaporation is controlled by the vapor pressure of water above a solution, which decreases in proportion to the increase in salt concentration in the solution (Fig. 105*A* and *B*). In consequence, there is a *decrease in evaporation from the surface of saline basins and also in the rate of salinity increase at high stages of mineralization.*

Some quantitative representation of this retardation may be made. In 1934

TABLE 16

Density and Viscosity of Saline Solutions at 25° (after L. L. Ézrokhi 1953)

Solution	Salt content, in %					Density, kg/m³	Abs. viscosity in centi-poises
	Na₂SO₄	NaCl+ +KCl	MgCl₂	MgSO₄	Mg (HCO₃)₂		
Brine of Kuchuk	6·64	15·55	4·39	—	—	1·218	2·33
The same	0·27	18·23	4·66	—	—	1·178	1·85
Brine of Kara-bogaz Gulf	—	0·60	27·80	2·50	0·50	1·297	9·60
Brine of Kuchuk	—	1·20	25·70	2·50	0·50	1·310	8·30
The same	—	—	32·80	1·70	0·40	1·332	11·90
Distilled water	—	—	—	—	—	1·000	0·895

I. B. Feigel'son (1936) observed evaporation from Lake Elton from April to October. The salinity ranged from 0 to 36° Bé (Baumé degrees), corresponding to a range of 0 to 34·06% (Table 17).

As the solution approached the eutonic point, evaporation of the water declined sharply and progressively. At 36° Bé it amounted to only 11% of the initial value for distilled water.

These conclusions were confirmed in 1940 at Inder Lake by Ya. I. Tychino and M. G. Valyashko (1952). They studied the evaporation of (1) water, (2) a saturated solution of NaCl, (3) brine, (4) sylvite brine corresponding to the initial precipitation of KCl, (5) carnallite brine, corresponding to the initial precipitation of KCl.MgCl₂.6H₂O, and (6) eutonic brine, corresponding to

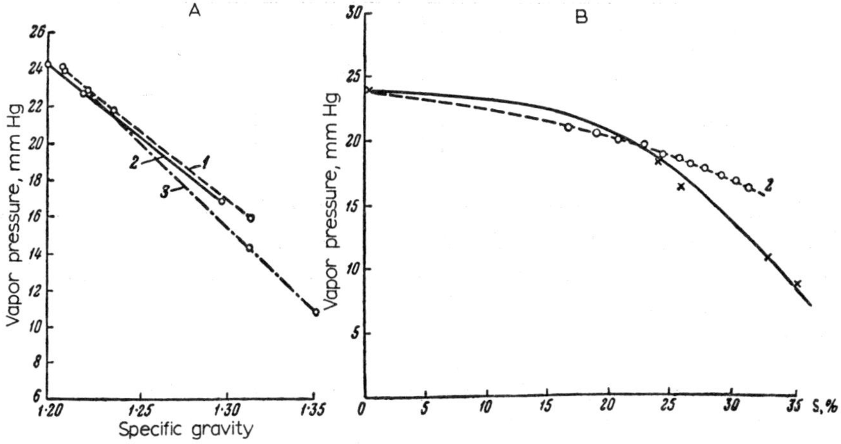

FIG. 105. Vapor pressure of solutions of different composition and mineralization. *A.* Solutions of NaCl at 35°: 1, 2, and 3 represent different compositions. *B.* Solutions of NaCl + MgSO₄ + MgCl₂: 1. Experimental solutions (from E. F. Solov'eva); 2. Kara-bogaz Gulf brine (from T. V. Rode).

TABLE 17

Evaporation of Water of Lake Elton)
(from Feigel'son)

Concentration of brine, °Bé	Coefficient of evaporation
0	1
20	0·73
24	0·63
26	0·53
28	0·38
30	0·37
33	0·24
36	0·11

the precipitation of carnallite, $MgCl_2.6H_2O$, and $MgSO_4.6H_2O$. The evaporation was studied in an evaporator having an area of 3000 cm² (Table 18).

In these experiments, as the salinity of the brine increased, the intensity of evaporation from the surface declined sharply. Basins may pass rather rapidly through the early stages of growing salinity, but the subsequent transitions become increasingly slower and more difficult.

TABLE 18

Comparative Evaporability of Brines of Different Concentrations
(from Ya. I. Tychino and M. G. Valyashko)

Solution	Total evaporability for May–September, mm	% relative to water
Water in soil . . .	1579	100
Saturated solution of NaCl .	1229	76·7
Brine of Lake Inder . .	1179	74·6
Sylvite brine . . .	971	61·0
Carnallite brine . . .	740	46·5
Eutonic brine . . .	122·4	7·8

Evaporation becomes especially small when the lake reaches the dry stage, i.e. when the summer level of brine is below the salt level and the brine becomes intercrystalline. Such a lake (Inder) was studied for evaporation from the surface of the brine and from the surface of newly precipitated sediments (Table 19).

TABLE 19

Evaporation (in mm) of Salt from the Surface of Lake Inder in 1940

	May	June	July	August	September	Total
Lake salt	0·12	1·89	1·22	1·72	0·31	5·30
Lake brine	97	187	240	277	142	896

These figures make it clear that evaporation from the surface of brine was 169 times as great as evaporation from the surface of salt. The retreat of the brine beneath the salt affords a kind of "protective resistance" of the lake against the aridity of climate. It permits the lake to become stabilized, thus increasing the concentration of bottom brines and yielding the more soluble minerals to accumulations in the bottom sediments very slowly.

All this indicates that late stages of saline development of basins have been comparatively rare in the history of the earth, and the attainment of the eutonic point with prolonged preservation of this state has been exceptional. Mineral accumulations associated with the latest stages of mineralization are therefore of limited distribution, especially accumulations of the eutonic stage (borates (?), bischofite, and others). On the other hand, deposits of earlier stages of saline development (dolomite, gypsum, and related minerals) are widespread.

Despite progressive slowing of the evaporation rate and the rate of saline development in lakes as the salinity of the brine increases, the rate of chemical precipitation during the yearly cycle not only does not diminish but actually increases sharply. The cause of this disparity lies in the sharp increase in solubility of the salts that precipitate at higher stages of salinity. Suppose, for example, that a liter of each of three solutions is evaporated, the first solution being weakly mineralized and capable of precipitating only $CaCO_3$, the second being more saline and in a state to precipitate gypsum, and the third being very salty, precipitating NaCl. Evaporation of the first solution may yield from 0·150 to 0·300 g of $CaCO_3$ from the liter, depending on the alkali content in the solution. A liter of the second solution may give about 4 g of of $CaSO_4 \cdot 2H_2O$ (according to its solubility), and a liter of the third solution may yield about 250 g NaCl (according to its solubility). It is therefore clear that chemical precipitation of substances during summer evaporation accele ates in the more saline lake or lagoon, despite the retardation of evaporation from its surface. *The ever-increasing solubility of salts not only compensates for the effect of retarded evaporation at high salinities, but it manifoldly exceeds that effect and creates a well-defined acceleration of the sedimentational processes during the course of increasing salinity.*

Significant changes occur during increasing salinity in the very structure of a solution and in the forms in which a number of the salt components are found. During the solution of electrolytes, polarly charged molecules of water attack the surface of salt crystals and remove both positively charged metal atoms (cations) and negatively charged particles (anionic atoms or complexes). A hydrate envelope of water molecules forms around the mass of cations and ions torn from the salt crystals. In an aqueous solution of salt, such as $CaCO_3$ cations of Ca^{2+} are surrounded by some number of water molecules oriented with the negative end of each molecule toward the calcium; we also find CO^{2-} anions surrounded by water molecules with the positive charge oriented toward this radical.

The strength of the bond between molecules of the hydrate envelope and the centrally disposed cation or anion differs with different substances. *The greater the charge on the cation (or anion) and the smaller its radius, the stronger is the bond of the hydrate envelope with the ion. The smaller the charge on the ion and the larger the radius, the weaker the bond is, and the more readily is the envelope of water molecules removed.* In general, cations and anions may be arranged in the following series according to the degree of hydration:

$$Li > Na > K > Mg > Ca > Sr > Rb >$$
ionic radius: 0·78 0·98 1·33 0·78 1·06 1·27 1·49

$$Cs > Ba > Cl > Br > HCO_3 > I > SO_4$$
ionic radius: 1·65 1·43 1·81 1·96 2·65 2·20 2·95

In this series the small ions with small radii, such as Li, Na, and others, are characterized by the greatest stability of the hydrate envelopes, and the large ions, Cl^-, Br^-, and I^-, have the weakest envelopes.

As the salinity of a basin increases, the distance between hydrated ions becomes progressively smaller. This leads to a progressive destruction of the hydrate envelopes; the ions are gradually stripped bare. At the same time their association with undissociated molecules increases, and crystal-like structures develop in the salt, as demonstrated by X-ray analysis. In highly concentrated saline solutions the structure of the salt mass dominates, and the molecules of the solvent (water) are distributed among the ions and molecules, forming centers for the growth of salt crystals. The structure of the solution gradually approaches a "solid-like" state (Kireev 1956; Filatov 1956).

A substantial change in the forms of individual salt components takes place with these changes in the general structure of a saline solution. This may be clearly followed in the carbonates of Ca, Mg, and Na. Table 20 presents data relative to these changes for $CaCO_3$.

TABLE 20

Forms in which Calcium Carbonate is found in Salt Solutions

MgSO_4, %	CaCO_3, mg-equiv/ liter	Form of dissolved carbonate (in % of total dissolved material)		Na_2SO_4, %	CaCO_3, mg-equiv/ liter	Form of dissolved carbonate (in % of total dissolved material)	
		Ca(HCO_3)_2	CaCO_3			Ca(HCO_3)_2	CaCO_3
				0	1·14	100	0
1	3·48	100	0	0·28	1·84	100	0
				0·52	2·14	100	0
3·5	4·41	92	8	1·17	2·74	94	6
				3·68	3·76	93	7
8	4·76	?	?	7·40	4·28	87	13
				11·61	4·56	80	20
15	4·24	88	12	18·48	4·90	70	30
20	3·57	84	16	21·27	5·04	69	31
25	3·47	72	26	25·59	5·34	65	35
According to N. M. Strakhov				According to Cameron			

These figures show that with increase in mineralization the bicarbonate form of calcium carbonate declines and, because of this, the standard form becomes more abundant. At extreme salinities the standard salt constitutes as much as one-third of the total mass of dissolved calcium carbonate.

Reconstruction in solutions of Na_2CO_3 is even more pronounced (Table 21).

TABLE 21

NaHCO$_3$ and Na$_2$CO$_3$ in Soda Lakes of different Salinities
(from E. S. Telentyuk)

Lake and year of investigation	Total mineral-ization, %	NaHCO$_3$, %	Na$_2$CO$_3$, %	NaHCO$_3$, % of total carbonate	Na$_2$CO$_3$, %
Tanatar V, 1949 . .	1·0	0·37	0·31	65	35
Tanatar VI, 1947 and 1948	1·48	0·46	0·37	55	45
——	2·79	0·57	1·04	35·5	64·5
Tanatar VI, 1948 . .	4·11	0·54	1·82	23	77
Tanatar VI, 1945 . .	7·80	0·93	3·01	23	77
Tanatar I, 1945 and 1949 .	18·46	1·74	8·47	17	83

But the greatest change in form is found in magnesium carbonate. At salinities below 7% this constituent is present chiefly in the bicarbonate form. No appreciable quantity of the basic salt is detected. At higher salinities the monocarbonate form begins to appear, and then continues to increase in quantity. According to V. I. Il'inskii (1927) the relationships of the various forms of magnesium carbonate during the increase in salinity of Sakskoe Lake are the following:

Salinity of brine (°Bé)	Form of magnesium carbonate
11·7	$2Mg(HCO_3)_2.MgCO_3$
17·3	near $MgCO_3$
28·4	$MgCO_3.Mg(OH)_2$
36·5	$3MgCO_3.Mg(OH)_2$

As the basin carbonates of magnesium form in strongly saline brine, complex compounds gradually appear, such as $MgO.nMgCl_2.pH_2O$, from the hydrolysis of $MgCl_2$. According to V. P. Il'inskii, the presence of the basic magnesium chloride indicates that the absolute value of alkalinity of the brine is a linear function of magnesium chloride and changes little with variations in content of other salts, with time, with temperature, or method of sampling (Fig. 106). Here we may again observe the formation of complexes referred to above. The presence of basic magnesium complexes explains a number of properties recorded in brines of present-day marine lagoons. For instance, the brine of highly saline lagoons is able to absorb notable quantities of chlorine, which clearly goes into the neutralization of basic substances.

With influx of solutions containing chlorides and sulfates of the heavy metals, especially $CuSO_4$ and $BaCl_2$, the basic chlorides or oxides of these metals are precipitated. When a brine is diluted with distilled water, the pH increases noticeably, clearly through hydrolysis of basic magnesium salts. As the dilute solution stands, the pH again falls to normal, because of absorption of CO_2 from the air and by neutralization of OH ions forming during hydrolysis V. A. Kovda).

It is more remarkable that, despite the clear and undoubted concentration of basic complexes in strongly saline solutions, the pH of brines in present-day marine lagoons does not increase, but rather decreases markedly (Figs. 93 and 94). This is clearly the effect of the basic salts of the brine.

It is very likely that changes in form, such as those just described, may occur among the salts of other weak acids (borates, silicates, and others) during increase in salinity; but this question has not yet been examined.

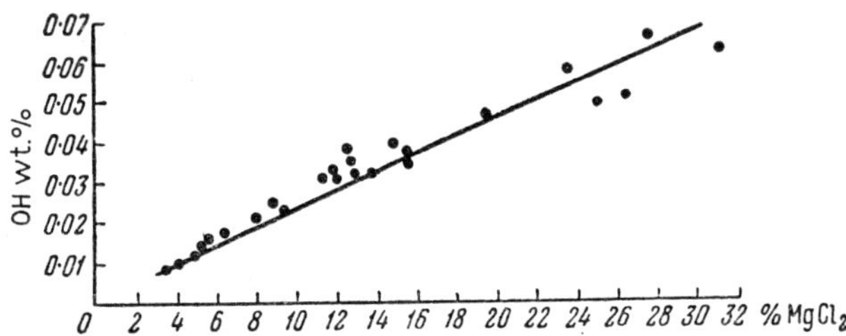

FIG. 106. Relation of total alkalinity (OH) to $MgCl_2$ in brine of Sakskoe Lake (from V. P. Il'inskii).

The most important physicochemical features shown by brines in basins of arid regions as salinity increases are: *the increase in the specific gravity and viscosity of the solution, the retardation in rate of evaporation from its surface, the gradual loss of the hydrate envelopes by ions and the formation of molecular and complex compounds, and, lastly, the transition from bicarbonate forms of calcium, magnesium, and sodium carbonates to the monocarbonate forms, even to the basic salt* ($MgCO_3$).

8. Oxidation-Reduction in Brines of Saline Basins

At the present time, the Eh determinations from maritime saline lagoons and lakes present a curious picture.

In several Taman basins (Tsokur Estuary, Solenoe Lake, Tuzlyanskaya Lagoon) the Eh is notably lower than in marine basins. "At times of freshening, a higher positive value of the redox potential is observed in these basins (Eh $= +300$ mv). At times of high salinity, normal for such basins, not only does their water have a lower redox potential, but the value may be very low and negative (from Eh $= +224$ mv to Eh $= -249$ mv in Tsokur Estuary, Solenoe Lake, and the central part of the Tuzlyanskaya Lagoon): measurements on 7 August 1947 in Solenoe Lake indicated an Eh of 12 mv, and measurements on 11 June 1948 and 6 August 1947 in the southern part of the Tuzlyanskaya lagoon and in its deeper depressions showed values of -203 and -205 mv. The redox potential thus decreases with increase in salt content. It is always higher in open stretches of water than in the near-shore zone, and in closed basins it is always low. Negative redox potentials are observed only rarely, in highly saline basins" (Savich 1950, pp. 148-149). If these rare occurrences are excluded, the boundary between oxidizing and reducing environments in saline basins normally coincides with the boundary between the bottom mud and the suprajacent water. At times it may move upward, but it never moves downward into the mud.

TABLE 22

pH and Eh of Sediments in Saline Lakes (from K. A. Ovsyannikova)

Locality No.	Locality site	Date	Depth of brine, cm	pH	Eh, mv	Density of brine, °Bé
I	Mikhailovskaya Zasukha (dry lake) near the surge basin.	14/VI 9/III	5 2	7·89 7·60	+382 +325	10·0 13·0
II	Southern shore at a distance 8 m to the left of the artesian spring where it empties into the lake.	14/VI	5	7·55	+349	10·7
III	From the barrier of the regenerating basin.	15/VI 25/VI	25 5	7·85 7·75	+274 +240	10·6 13·3
IV	At the control station on the end of the barrier.	30/VI 9/VII 25/VII	40 — 30	7·65 7·70 7·75	+258 +263 +293 (?)	11·6 13·7 18·1

Data on redox potentials in saline brine for Sakskoe Lake, given in Table 22, emphasize the following points: (a) at localities I and II, which represent the marginal parts of the lake, higher positive values of Eh were measured than at localities III and IV in the center of the lake; and (b) at each locality Eh declines with diminution in depth of water and with increase in density (localities I and III). The last conclusion agrees with the data of Savich, but the first contradicts them. The cause of this discrepancy is not yet clear.

9. The Organic Population and its Effect on the Geochemistry of Saline Basins

As basins become salty the organic population becomes impoverished in its specific composition, but it is not exterminated. A number of lower organisms live in highly saline water. "Thus, in Lake Urmia in northern Iran a few crustaceans may be observed even at salinities up to 22·4%. In Lake Bulak (or

O

"Krasnoe" or "Malinovoe"), near Fort Aleksandrovskii on Mangyshlak, Suvorov discovered at a salinity of 28·5% many *Dunaliella* flagellates, which give rise to the characteristic color of the water in this lake, together with crustaceans, rotifers, insect larvae, a type of worm, and algae" (N. M. Knipovich, 1938). Three forms of living bacteria and some indeterminate single-celled forms with a flagellum have been identified in the Dead Sea at a salinity of 28–29%. Crustaceans of the genus *Artemia* live in Great Salt Lake at a salinity of 21%; the same form is found in Lake Elton, Lake Kuchuk, Lake Yarovoe and other lakes of high salinity. The red color of such lakes is commonly due to an abundance of these crustaceans or to the presence of the microscopic algae *Dunaliella*.

The paucity of species in saline lakes is commonly accompanied by a great abundance of individuals, indicating a considerable biological productivity. This productivity begins to decline more or less rapidly only at very high salinities (above 12–20%).

Many vivid descriptions of the abundant organic life in lakes have been given. "Even on preliminary examination we were struck with the abundance of organic life in the Tanatar Lakes. The brines of these lakes literally team with life, evidence for which may be seen everywhere along the shore. Myriads of *Ephydra* larvae float on this salt water; their pupae cover every object protruding from the water, and masses of them lie on the floor of the lake. Vast numbers of *Artemia* live in the brine. Along the shore there is a layer several centimeters thick consisting of *Ephydra* pupae and bodies of various insects, among which winged ants are abundant. Many species of water birds, from gulls to shore birds feed on the abundant ants and various orthopterous insects. In Tanatar III blue-green masses of *Anabenopsis* are also present, occurring in the brine in such numbers that a plankton net is quickly choked" (Isachenko 1934, p. 155).

These observations refer to soda lakes, the salinities of which did not exceed 9%. In the sulfate Lake Kuchuk, in which the salinity was 16–19% at the time of investigation, great numbers of organisms, known only from this lake, occur. These are the *Dzensia*, described by N. N. Voronikhin. The blue-green or olive colonies of this organism appear as large masses in the zone of breakers, and they cover the lake floor near the shore with an extensive layer, partly living, partly decomposed. Where wave action is strong, *Dzensia* is cast upon the shore, where the shore sand is more or less cemented by the slimy bodies of this organism, being then sufficiently cohesive to be lifted up in large sheets. Such masses consist of individual grains of sand and algae held together by a transparent bond. Where shore deposits of *Dzensia* are very extensive, the sand is completely hidden by them (Isachenko 1934, p. 154).

Similar features are observed also in the chlormagnesian Bolshoe Yarovoe Lake, in which the salinity is 14–15%.

When we visited Lake Kuckuk in 1946, the water was red with the vast accumulation of *Artemia salina;* and at Lake Yaravoe abundant scum on the shore, cast up by the waves, consisted of eggs of this animal.

That individual species may reach such development in salt lakes may be seen also in Great Salt Lake. Eardley has pointed out that here coprolites of the crustacean *Artemia salina* constitute a major rock-forming constituent, covering several hundred thousand acres and forming masses of thousands of tons.

Significant changes in the organic community have been noted in lagoons of the Caspian Sea (Table 23). With increase in salinity the biomass of all principal groups of the organic community clearly increases at first, and then at a salinity corresponding approximately to the saturation point of $CaSO_4$, the biomass decreases.

In the much saltier Gulf of Kara-bogaz the organic community is extremely poor qualitatively: 9 species of algae and 2 species of bacteria.* But one of the

TABLE 23

The Biomass in Gulfs of the Caspian Sea

Region	Salinity, %	Biomass (number of individuals per liter)			Benthotic biomass, g/m/³
		Nanno-plankton	Phyto-plankton	Zoo-plankton	
Caspian Sea along the northern shore of Buzachi .	10·0	—	1,154,430	Rare	50·26
The pine forest zone	15·0–16·5	69,606,000	4,000,000	115,000	171·00
Mertvyi Kultuk .	31·8–35·6	506,411,000	10,850,000	140,000	200?
Kaidak . .	35·5–41·7	569,059,000	5,000,000?	208,000	257·4
Kara-Kichu . .	49·1–59·5	83,032	No data	15,000	50·8

algal species—*Aphanothece salina*—becomes extraordinarily abundant, at least locally. "At water level the shore of Kurguzul'skaya Bay in the region of Lake No. 6 is covered by a gelatinous belt in various shades of greenish olive and rose, 15–20 m wide and 0·5 m or more thick. This colloidal mass is shown microscopically to be an almost pure culture of this alga. The total isolation of this huge mass of algal material may be explained by the protection of the bay from wave action in the gulf, by the shallowness of the water, and finally by the wave action arising in the bay itself. The gelatinous algal accumulation is warmed by the direct rays of the sun to such an extent that it is uncomfortably hot to walk on top of it with bare feet. On the other hand, the bottom, protected from the sun's rays, is cold. The entire bed is in a state of active photosynthesis, judging from the emission of bubbles of O_2.

"This accumulation of algae covers part of the shore with a consolidated black mass of material, which, when dug up, gives off a sharp disagreeable odor, partly of hydrogen sulfide.

* Near the strait that connects the Gulf of Kara-bogaz with the Caspian the variety of bacteria increases to 8 species.

"In Sartasskaya Bay *Aphanothece salina* floats along the shore in rounded plate-like colonies of various hues, from rose to blue-green. The color is especially noticeable along fractures in the mass. Many of these colonies may be found, along the otherwise flat shore, behind the storm beaches. Here they assume irregular collapsed shapes and give off a disagreeable odor. These are examples of strongly saline zones. Among the rocks along irregular shores, colonies of *Aphanothece* are broken by waves into slimy masses. Those that occupy minor embayments along the shore continue to exist and develop, and, after a temporary quiescence, are again united to the general mass, as samples treated with formaldehyde show.

"In Chagala, *Aphanothece salina* locally forms abundant dense accumulations in embayments along the shore. These occur as a paste of material formed from small greenish clumps broken off from larger colonies by wave action. These migrate, depending on the prevailing wind direction, either into the open gulf, or to new positions among the small bays of the shore. Colonies of *Aphanothece salina* have been collected in the pelagic zone of the gulf" (A. D. Pel'sh, 1937).

During an investigation of the Gulf of Kara-bogaz in 1894, when the salinity was 18%, A. A. Ostroumov noted that in summer the water teemed with the crustacean *Artemia salina.* "In 1934 no crustaceans were living in the gulf, although an appreciable number of their eggs still survived" (Pel'sh, 1937).

It must be noted, however, that organic matter in marine gulfs and lagoons is not only of autochthonous origin. Much of it was transported to the sea by streams flowing into the basin. So much organic material is brought into the northern Sivash from the Sea of Azov, being deposited as it enters this northern basin, that it forms broad shoals, which serve as refuges for migratory birds. The same is true in the Gulf of Kara-bogaz; A. A. Ostroumov in 1897 made an estimate of the macroscopically visible débris chiefly of red algae and of *Ruppia* leaves transported by currents into the gulf. The amount was only moderate because the current was weak; nevertheless his reckoning indicated 200 kg of organic material per day. Under such conditions the content of dissolved organic matter in saline lagoons must be much greater than in the sea; it amounts to 4–5 mg/liter in the Gulf of Kara-bogaz, for example, compared with 3 mg/liter in the Caspian.

Despite the abundance of organic matter in saline basins, the role of organisms in the geochemistry of such basins is nevertheless very slight, much less significant than in basins of humid regions, or even during the initial stage of development in saline basins.

It is important to appreciate that all organisms living in saline basins are without skeletons of $CaCO_3$, $MgCO_3$, SiO_2, phosphate, or such mineral salts as $BaSO_4$, $SrSO_4$ and CaF_2. This means that these compounds and many others *are not extracted directly in biogenic processes* as they are in seas and lakes of low mineralization. In addition, the organic community of saline lakes is impoverished, being restricted to only a few species, of which only one normally achieved any massive development. This can only restrict the influence of living matter on the chemical processes taking place in the lake. The alkali-acid conditions in the lake are extremely important from this point of view. In seas and lakes of humid zones these conditions are determined solely by the life activity of organisms. As the salinity increases, the pH rises through a range of salinity that varies for different types of basins; it then falls more or less noticeably—weakly in the group of soda lakes, strongly in the sulfate lakes (Figs. 93-97). In these latter the pH approaches 7 at salinities of 22–25% and it even becomes acid at higher salinities. This is caused by the progressive accumulation of $MgSO_4$ and $MgCl_2$ in lakes of this type. In the daily cycle, the pH of water in saline lakes is highly stable, remaining practically constant. But in the annual hydrochemical cycle, because of general fluctuations in salinity, it may vary extensively, indicating that the *alkali-acid conditions in saline basins are determined by the salt mass and not by the effect of organic material as in seas and lakes of humid regions.*

Thus, the pH and the alkali content of brines, two of the most sensitive indicators of the activity of organisms in lake water, progressively pass from the control of the organic mass as salinity in the basin increases. In soda lakes this occurs very early, at relatively low salinities, but in sulfate lakes (group IIa) it takes place later, at higher mineralization, and in lakes of group IIc later still. The process is characteristic of all continental lakes that become progressively more saline in an arid region.

It can hardly be doubted that the above description of the organic community and its effect on the geochemistry, as observed in modern saline basins, must be valid also for saline basins of the geologic past, whether they be of lacustrine or marine origin.

10. Sedimentation at Medium and High Degrees of Salinity in Modern Basins

Progressive salinity in basins of the arid zone, up to 25–30% or more, has two important sedimentational consequences. *The first is expansion of the range of chemical precipitation as compared with the range in basins of humid regions.* New saline components are increasingly drawn into the sedimentational process, and these processes tend to approach their logical end, and commonly attain it; i.e. they culminate not only in precipitation of the mechanical suspensions introduced into the basin and of slightly soluble salts, but also of all, or almost all, the most soluble compounds—sulfates and chlorides of Ca, Mg, Na, and K, and also F, Br, and other elements.

The second consequence of increasing salinity is the development of a well-defined step-like character of the sedimentation. At relatively low salinities only the slightly soluble calcium and magnesium silicates are precipitated chemically. *At this carbonate stage, chemically precipitated material is mixed with even larger quantities of clastic material, and the resulting sediments are sands, silts, clays, or marls, with considerable carbonate content, disposed in the basin in the same manner as in basins of humid regions.* When the supply of fragmental material is low, pure carbonate sediments are formed. At a higher stage of salinity, sulfate minerals are precipitated with the carbonates. These are gypsum, thenardite, mirabilite, glauberite, and blödite. *This sulfate stage is normally characterized by a much smaller contamination of the salts by clastic material.* The clastic material is generally confined to individual beds of varying thicknesses, according to the times of freshening of the basin, when precipitation of sulfate minerals ceases. At high salinities *halite begins to precipitate,* commonly alone, but sometimes accompanied by complex

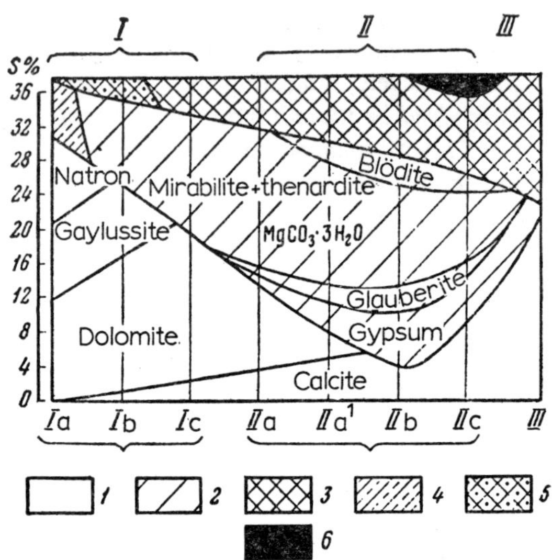

Fig. 107. General scheme of salt deposition in present-day salt lakes of arid regions. *I.* Soda lakes; *Ia.* strongly sodic ($K_{Na} > 30\%$), *Ib.* moderately sodic ($K_{Na} = 3 - 30\%$), *Ic.* weakly sodic ($K_{Na} < 3\%$); *II.* sulfate lakes: *IIa* sodium-magnesium, *IIb.* sodium-magnesium-calcium, *IIc.* magnesium-calcium; *III.* chloride lakes with NaCl, $MgCl_2$, and $CaCO_3$. 1. Carbonate stage; 2. sulfate stage; 3. chloride stage; 4. sulfate sediments highly contaminated with natron, 5. halite highly contaminated with sodium sulfate; 6. potassium salts.

chlorides and sulfates of potassium and magnesium. Minerals of the preceding stages are normally suppressed at this *chloride stage,* or are present only as impurities. As a rule clastic material is negligible, though when present, it is again confined to beds corresponding to dilution of the water in the basin and to cessation of halite precipitation.

The course of chemical sedimentation is determined by the hydrochemical type of basin (Fig. 107). The abscissa shows the hydrochemical groups of lakes, group Ia indicating highly sodic lakes, with Na_2CO_3 constituting over 30% of the salt mass; Ib is a moderately sodic lake with Na_2CO_3 ranging from 30 to 3%, and Ic is a sodic lake with Na_2CO_3 constituting less than 3% of the salt. Group II comprises the sulfate lakes. Group IIa represents sodium-magnesium lakes in which only Na_2SO_4 and $MgSO_4$ among the sulfates are present. Group IIa[1] represents magnesium lakes, IIb sodium-magnesium-calcium lakes containing Na_2SO_4, $MgSO_4$, and $CaSO_4$, and IIc magnesium-calcium lakes, in which the only sulfates are $MgSO_4$ and $CaSO_4$. Group III lakes are of the *chloride class*, which contain NaCl, $MgCl_2$, $CaCl_2$, and, of the sulfates, only $CaSO_4$ in small quantities.

In keeping with the above scheme, the sedimentation of all types of lakes passes through carbonate, sulfate, and chloride stages, but the duration of each stage and the mineralogical characteristics in basins of the different hydrochemical groups vary sharply.

The carbonate stage lasts longest in soda lakes, where the salinity ranges up to 25–27% (Fig. 108). This stage is also characterized by the greatest variety of minerals: calcite, dolomite, gaylussite, natron, thermonatrite, and trona. The $CaCO_3$ here corresponds to a negligible initial salinity ($<0.1\%$), the precipitation of dolomite to salinities up to 12–13%, gaylussite to salinities above 13%, and natron and the other minerals normally to salinities above 20%. In the transition to moderately sodic lakes, natron, thermonatrite, and trona disappear. In weakly sodic lakes gaylussite is not precipitated. Magnesian silicates such as sepiolite-cerolite accumulate in abundance during the entire carbonate stage.

Only sodium sulfate, in the form of thenardite or mirabilite, is precipitated *at the sulfate stage* in soda lakes; large amounts of natron, gaylussite, burkeite, and northupite are also precipitated. This sulfate stage in soda lakes corresponds to a state of very high salinity.

The *chloride stage* of soda lakes coincides essentially with the eutonic; NaCl is therefore strongly contaminated with sulfate and carbonate minerals that had begun to precipitate in the preceding stages.

Sedimentation in sulfate lakes takes a different course (Fig. 109). The

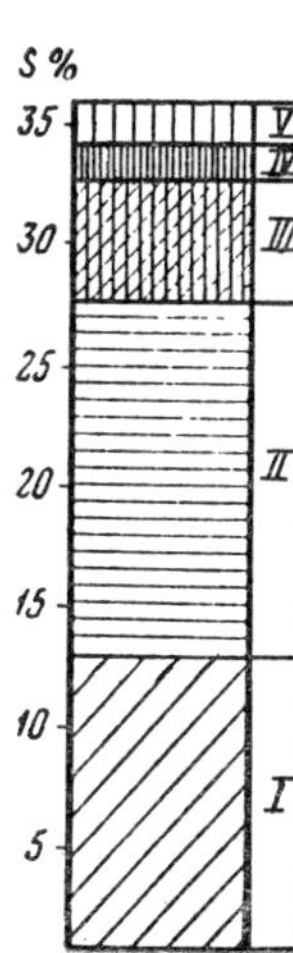

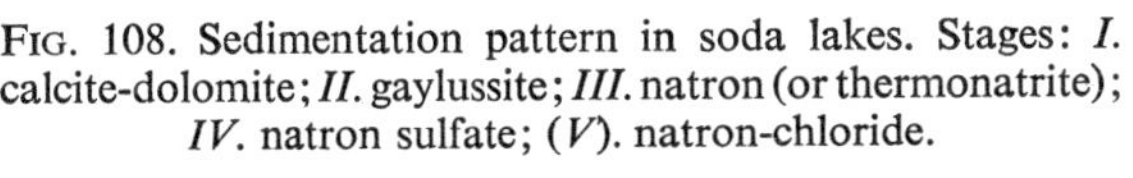

FIG. 108. Sedimentation pattern in soda lakes. Stages: *I.* calcite-dolomite; *II.* gaylussite; *III.* natron (or thermonatrite); *IV.* natron sulfate; (*V*). natron-chloride.

carbonate stage is characteristically weakly developed and is restricted to rather low mineralizations, from 12–13% (IIa) to 3·5–4% (IIc). The composition of the carbonates at this stage is uniform: dolomite is precipitated only in the left part of the field of sulfate basins (IIa) after the initial period of

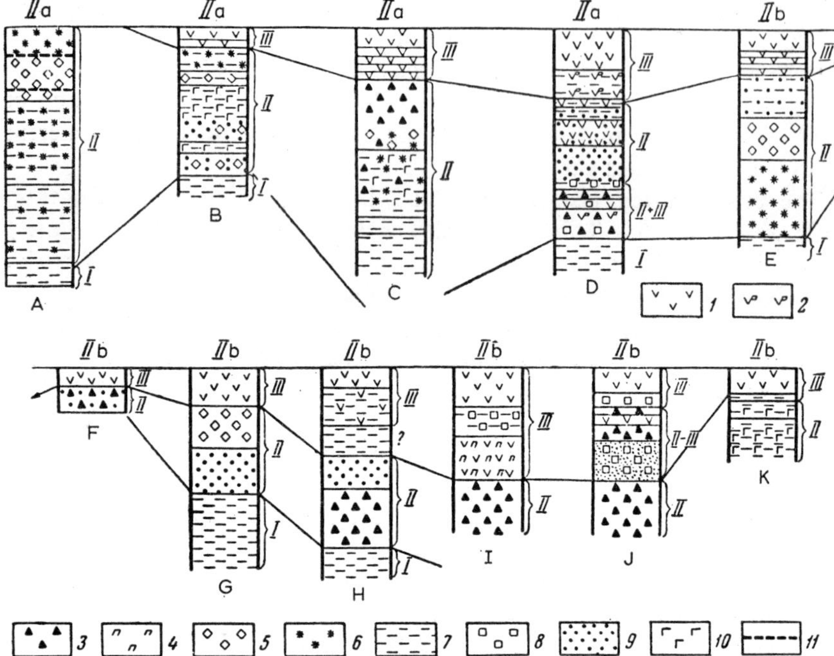

FIG. 109. The structure of salt lakes of the sulfate class. *A*. Lake Mormyshanskoe (from S. Z. Makarov), *B*. Lake Dzhamansor (from A. G. Bergman, M. G. Valyashko, and I. B. Feigel'son), *C*. Lake Tuz-Kul' (from M. G. Valyashko, A. A. Nechaeva, and T. B. Polenova), *D*. Lake Mullaly-Tuz (from A. G. Bergman, M. G. Valyashko, and I. B. Feigel'son), *E*. Lake Kok-Tyube (from N. I. Lepeshkov), *F*. Lake Aral (from N. I. Lepeshkov), *G*. Lake Karabotan (from N. I. Lepeshkov), *H*. Lake Baichunas (from N. I. Lepeshkov), *I*. Lake Maloe Korduvanskoe (from V. I. Nikolaev and N. D. Kuznetsov), *J*. Lake Bol'shoe Korduvanskoe (from V. N. Nikolaev and N. D. Kuznetsov), *K*. Lake Kurgan-Tuz (from A. G. Bergman, M. G. Valyashko, and I. B. Feigel'son). *I*. Carbonate stage; *II*. sulfate stage; *III*. chloride stage; *II* and *III* sedimentation with mixed features of the sulfate and chloride stages. 1. Halite (of current season); 2. halite (accumulated from undissolved seasonal deposits); 3. blödite; 4. epsomite; 5. thenardite; 6. mirabilite; 7. mud; 8. NaCl in crystals poorly cemented together or uncemented altogether; 9. glauberite; 10 gypsum; 11. thenardite incrustation.

The chemical type of lake is shown in Roman numerals above the columns.

calcite precipitation; only calcite accumulates during the carbonate stage in the other two groups of sulfate lakes. Magnesium silicates also form during the carbonate stage in lakes of group IIa (Lake Balkhash).

The sulfate stage in lakes of the sulfate class, on the other hand, is characterized by great range and long duration, embracing a broad salinity range. The variety of sulfate minerals increases sharply. Apart from sodium sulfate, precipitating as thenardite and mirabilite, calcium and magnesium sulfates also form, appearing partly in the simple hydrate form (gypsum and epsomite), and partly as double salts (glauberite and blödite). The variety of sulfate phases increases on the diagram from the IIa group to the IIc group. During the course of increasing salinity the least soluble phases form first: gypsum ($CaSO_4.2H_2O$), then the more soluble such as glauberite ($Na_2SO_4.CaSO_4$), mirabilite ($Na_2SO_4.10H_2O$), or thenardite (Na_2SO_4), and finally the most soluble—blödite ($Na_2SO_4.MgSO_4.4H_2O$). Carbonates are precipitated with the sulfates, but the composition is sharply different from that of the carbonate stage. Calcite and basic magnesium carbonate accumulate in all groups of sulfate lakes at medium and high salinities.

The chloride stage of sulfate lakes is characterized by prolonged accumulation of halite with relatively little admixture of sulfate minerals.

Halogen sedimentation in chloride lakes (i.e. lakes containing much NaCl, $CaCl_2$, and $MgCl_2$ with no Na or Mg sulfates and very little $CaSO_4$) is

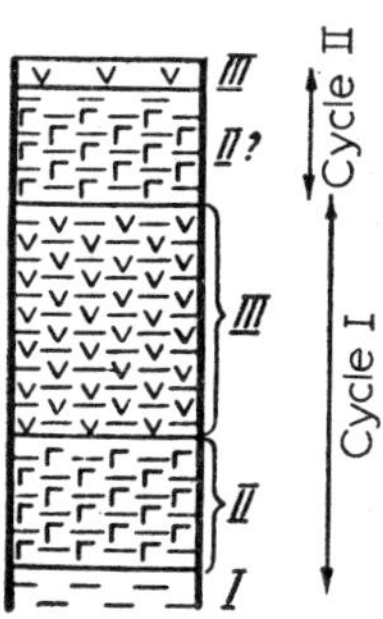

FIG. 110. Structure of deposits in chloride lakes (Lake Staroe). Stages: *I*. carbonate; *II*. sulfate; *III*. chloride.

characteristically simple (Fig. 110). The carbonate stage is again greatly prolonged through a wide salinity range and it becomes almost equal to the carbonate stage of soda lakes, but mineralogically it is poor, being represented only by calcite. The sulfate stage is sharply reduced, and is represented only by gypsum. The chloride stage, on the contrary, is prolonged, extending over a wide salinity range. It is marked by the precipitation only of NaCl; this mineral is very pure, containing few other minerals (carbonates or sulfates).

Several features should be emphasized concerning the diagram of present-day salt deposition. The boundaries that separate the carbonate stage from the sulfate stage, and this latter from the chloride stage are distinctive. On the sodic end of the diagram, the first boundary is situated at high salinities of about 25–27%. This boundary declines progressively, and in groups IIb and IIc it lies at salinities of 3–5%. In basins of chloride type, the boundary again turns sharply upward. The boundary between the sulfate and the chloride stages gradually and continuously declines from the sodic end of the diagram, (where it lies near the eutonic point) to the right, where it reaches a salinity of about 23%—far from the eutonic point. In this direction the halite sediments become progressively purer, and freer from other salt impurities.

Because of this, the total mineralization for the entire period of saline development in the various hydrochemical types of lakes varied substantially. In soda lakes it may be represented chiefly, perhaps even exclusively, by *carbonate deposition* in manifold mineral forms. In sulfate lakes, mineralization is manifested chiefly in *the deposition of various sulfates* with markedly subordinate carbonate formation and somewhat more extensive chloride formation. In chloride lakes, the process is *almost exclusively halite deposition*: or, when halogenesis is undeveloped, it is marked by the virtual absence of salt phases. In other words, *the hydrochemical type of basin is clearly reflected in the nature of the salt deposits formed in it.*

Consideration of the relationships between the various salts in basins of different hydrochemical types permits us to assess the characteristic behavior of the boundaries between mineral associations on the general halogenic diagram. The fact that sulfate sedimentation begins only at a very advanced stage of salinity in highly sodic lakes is a simple consequence of the low sulfate content of such lakes, i.e. a consequence of the low sulfate coefficient. As sulfates accumulate progressively in weakly sodic lakes, the lower boundary of the sulfate stage quickly falls to mineralizations of 3–5% in the waters of groups IIa, IIb, and IIc. Toward the chloride class the sharp decline in the sulfate ion (the disappearance of $MgSO_4$) again strongly raises the initial point of sulfate precipitation. The upper boundary of the sulfate stage is controlled not by the total amount of sulfates but by the metamorphization coefficient of the brine. For the left side of the diagram, where Na_2SO_4 is present, this coefficient is expressed by the ratio $Na_2SO_4/MgSO_4$; for the right end of the diagram, where Na_2SO_4 is absent, the coefficient is represented by $MgSO_4/MgCl_2$. The higher these coefficients the later in the scale of mineralization does pure sulfate sedimentation give way to the chloride stage. Since each of these two ratios falls to zero toward the right of the diagram, this naturally brings about a progressively lowering of the upper boundary of the sulfate stage from sodic lakes to chloride lakes.

Until recently it was thought that accumulations of potassium salts were characteristically absent in Recent halogen sediments, even those of marine origin, and that high concentrations of F, Sr, Br, B, and other halophilic elements were extremely rare.

Potassium salts and other components are concentrated in the bottom and intercrystalline brine, not precipitating in the solid phase under natural conditions, or forming only negligible deposits. Carnallite and kainite thus crystallize from lakes on the drying margin of the Gulf of Kara-bogaz. Syngenite and kainite are known in very small amounts in the salt deposits of Dzhaksy-Klych and Chumysh-Kul' Lakes. Carnallite was found in "salt bumps" forming on the dry surface of seasonal precipitates in Lake Inder. Under artificial conditions, however, such as those developed in the salt industry, precipitation of potassium salts has been observed repeatedly. Here, sylvite and carnallite have been precipitated in an experimental basin during concentration of the intercrystalline brine of Lake Inder. In lake No. 5 of the

Bekdash region, north of Kara-bogaz Cape, where the parent brine is discharged after the sulfates have been removed, carnallite crystallizes in the summer time. Carnallite in the sediment changes to kainite by diagenesis and the amount of kainite increases with depth. On an average, the content of K_2O in the bottom salt deposits is here about 13%; in the material formed along the shore the content is 12·6%. Carnallite also precipitates out of the parent brines during concentration in the experimental basins of Lake Saki. In 1927 abundant crystallization of carnallite along with hexahydrite ($MgSO_4 \cdot 6H_2O$) was observed in industrial basins containing brine prepared for production of magnesium chloride. Carnallite has been obtained in commercial quantities from basins in which Dead Sea water has been evaporated, and also from some basins in Western India and from industrial plants in the Mediterranean and Yellow Sea regions. Nevertheless, if the large volume of potassium salts obtained artificially is disregarded and only natural precipitation is considered, the impression is gained that these salts are not formed in natural conditions at the present time. Present-day halogenesis is consequently much less complete than that of earlier periods. Recent discoveries of potassium salts in the Tsaidam basin of China, however, force us to revise these views. Because of their exceptional interest, these discoveries will be considered in detail (Yuan Chien Ch'i, 1959).

The Tsaidam basin is in the southern part of central China, between the Nan Shan mountain range on the north and the Kunlun on the south. It is oval, 800 km long from west to east and about 350 km wide; its area is approximately 120,000 km². The basin lies at an average elevation of 2700 m above sea level; the bordering mountain ranges rise to above 5000 m. The basin has interior drainage and slopes from northwest to southeast. The interior part is a rocky, saline plain, traversed by low, highly dissected ranges.

The basin is now dotted with a large number of salt lakes, many of them very large. The salt phase is normally halite, more rarely mirabilite, but potassium salts have been recorded from a number of lakes as follows:—

1. *Lake Gaskule*, 117 km² in area, 1 m in depth; contains 0·8% potassium salts (western part of lake); *carnallite* is reported from the eastern and southeastern parts of the lake.

2. *Lake Makhe*, 10 km² in area; has a slight admixture of carnallite, forming 0·71% of the salts. This is a boron lake (ulexite 52·94%).

3. *Lake Detsu-Makhe* is a brine lake, 11 km² in area, 0·15–0·30 m in depth; within the basin of this lake, 6 km from the present lake, potassium salts occur in a 1 km² area in silty sediments, forming a bed up to 0·15 m thick; KCl forms as much as 12% of this bed; potassium salts are found in other parts of the basin over a total area of 6 km².

4. *Tatiner group of lakes*:

(*a*) *Eastern Tatiner Lake*, 900 km² in area, dry for the most part, brine covering an area of 167 km² with a depth ranging up to 2 m; a *deposit of halite* ranging up to 20 m in thickness and containing *carnallite* in the upper part of the bed is found in an area of about 40 km² in the southeast part of the basin, where the lake has dried. The intercrystalline brine of this area contains 0·3–0·4% K_2O.

(b) *Western Tatiner Lake* has a water surface of 90 km²; a bed of halite along the northern dried belt (35 km²) contains carnallite mixed with sand; the bed is 0·2 m thick. A similar occurrence is found in the western belt.

5. *Lake Tsarkhan* (about 3000 km²) is a huge dry halite deposit in which three brine lakes occur: Lake Saine in the west, Lake Dabusyn in the southwest, and Lake Khabusyn in the southeast. Outwardly this is a vast light-brown area, with an irregular, broken surface with local bulging segments of salt crust broken into large plates (5–10 up to as much as 30 cm in thickness) that are tilted at angles of 12–15°.

Carnallite is found over an area of 40 km² in the southern part of the deposit, within the upper layer to depths of 0·6 m; the highest percentage of KCl, of the order of 10%, is in the uppermost part, within 0·15 m of the surface. In this area the highest KCl content (from 3·5 to 15%) is restricted to an area of 25 km², outside which it diminishes appreciably. In the northern part of the lake the content is 2% or less, and in the eastern part it is generally imperceptible.

In the western part of the central zone of the lake, KCl is found on three horizons in bed *I*: 0·3–2·7 m (KCl 2 to 5·35%), over an area of 100 km²; 6–8 m (KCl 2 to 5%), over an area of 50 km²; and 9–13 m (KCl 2 to 6·4%), over an area of 150 km². In the second, the *buried layer* of halite, separated from the first by a 11-meter layer of clay, there is no appreciable KCl. Other small accumulations occur to depths of 50 m.

The amount of carnallite, where present, is very irregularly distributed both in area and in depth. Zones with high percentages alternate with carnallite-free zones. Uniformly disseminated potassium salts occur frequently in lenses in the upper parts of the section but the quantity declines downward. Near the bottom of the bed carnallite is present in nests, thin beds and irregular lenses. The KCl content here ranges from 5 to 8%. Carnallite fills the intercrystalline space in the halite but the carnallite content is generally low.

In the northern part of the Tsarkhan deposit, deposits of bischofite of considerable area occur in three beds, each 5–6 cm thick, in the uppermost part of the first salt bed. They are separated by halite.

This carnallite distribution leads to the conclusion that in the zone of the present Dabusyn salt lake there once existed a broad deep basin with brine from which halite was precipitated and with locally more concentrated brine (intercrystalline) from which carnallite was precipitated. Present-day Lake Dabusyn is apparently only a relic of this ancient basin.

Data on the accumulation of potassium salts in continental basins of China are still very scarce, but three fundamental points may be clearly seen. These salts accumulated only in huge lacustrine basins. They have been localized in only restricted parts of the total lacustrine area, and they have been associated with the dry-lake stage, forming in those parts of lake-flats that are characterized by low relief, where intercrystalline brine rich in potassium has drained, creating pools where evaporation was completed.

Thus, the stage of potassium mineral precipitation cannot be considered excluded from present-day halogenesis, but it takes place very rarely and under rather specific conditions. The cause of these relations will be discussed below.

11. THE DISTRIBUTION OF HALOGEN DEPOSITS WITHIN MODERN SALINE BASINS

The various halogen deposits that form in present-day saline basins are systematically distributed over the area of the lake, the localization of the salts being predetermined by the nature of supply to the basin of water and salts. Two types of distribution may be distinguished: *symmetrical* and *asymmetrical*.

The symmetrical type is illustrated by profiles through Saki, Elton, Mulally (in Kyzylkum), Khadzham-Sayad (in the Sundukli Desert) Lakes and by the map of Pechatnyi Island made by the author in 1945 (Figs. 111-113 and 143).

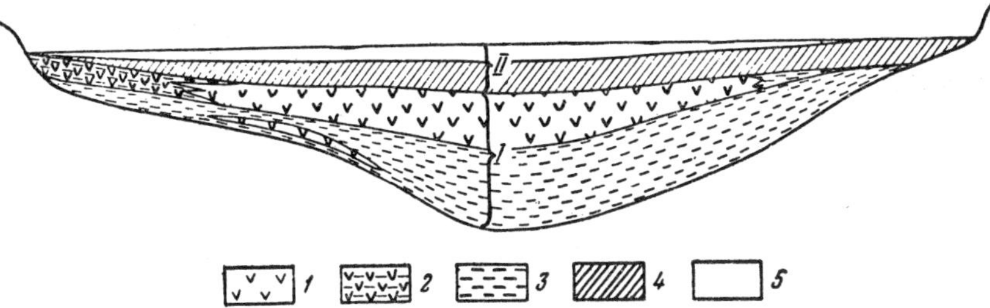

FIG. 111. Distribution of deposits in Saki Lake (modified from A. I. Dzens-Litovskii). 1. Halite; 2. highly contaminated halite; 3. clay underlying salt; 4. clay (ooze) overlying salt; 5. brine. *I*. First stage of deposition; *II*. second stage of deposition (beginning).

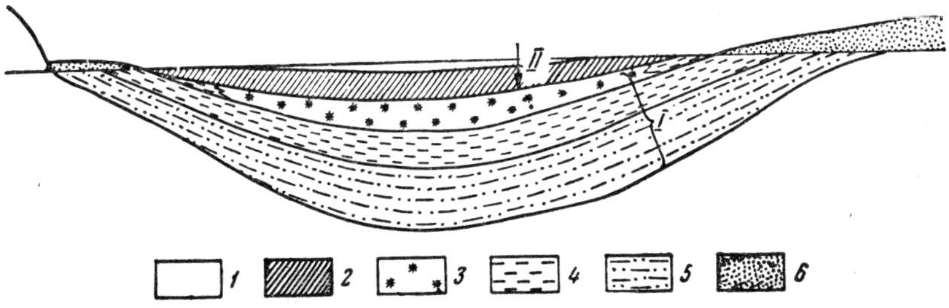

FIG. 112. Section of Lake Mulally in Kyzylkum (from A. I. Dzens-Litovskii). 1. Brine 2. black mud above salt; 3. mirabilite; 4. light grey (carbonate?) mud; 5. grey sandy mud; 6. sand. *I*. First stage of deposition; *II*. second stage of deposition (beginning).

The essential character of the symmetrical type of salt distribution is that the sediments of the highest stage of salinity—halite beds in the example considered —are deposited in the center of the lacustrine basin, and the greatest thickness is found here. The beds gradually decrease in thickness toward the margins, giving way to clay-sand sediments. These clastic sediments immediately next to the salt lens still contain single, isolated aggregates of salt (halite or thenardite, sometimes mirabilite) *or crystals of gypsum. Nearer the shore these disappear and the sediments become normal sand-silt-clay deposits with some disseminated carbonate.* Because of this distribution the salt beds have a lenticular cross-section, the thicker part corresponding to the central part of the basin, and the tapering edges marking the near-shore shallow-water zone. A salt bed thus commonly acquires a characteristic structure of many large pockets, variable in outline, with underlying muds being squeezed upward between the contacts. This results from the fact that salt beds crystallize on wet plastic muds, into which they sink irregularly, and are thus naturally broken into units that subside differentially, with upwelling ooze between the blocks (Fig. 143).

Symmetrical salt beds are characteristic only of lacustrine basins with ground-water supply and a uniform inflow around the margin of the basin.

In the asymmetrical type of distribution, *the composition of the salt phases changes from one end of the basin to the other in a manner that indicates a*

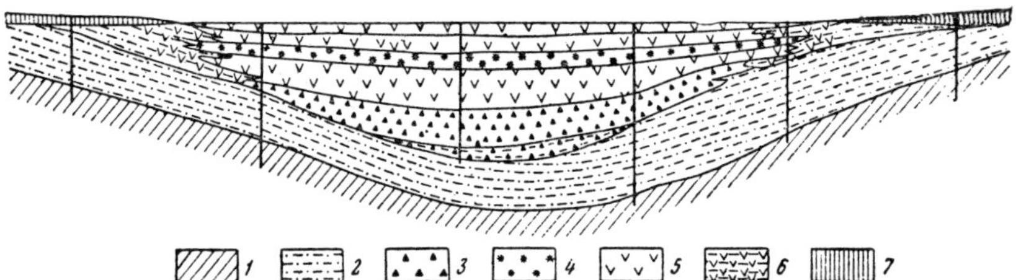

Fig. 113. Salt deposits of Khadzham-Sayad Lake in the Sundukli Desert (after A. I. Dzens-Litovskii). 1. Lake bed; 2. sandy mud; 3. blödite; 4. mirabilite; 5. halite; 6. mud with salt and sand; 7. salt marsh.

progressive mineralization of the water from which they have been precipitated. This type of distribution arises in basins supplied with water and salt from only one end. The deposits of the highest salinity stage are found farthest from the point where solutions are discharged into the basin.

An outstanding example of this type of distribution is in Lake Balkhash, which has different degrees of salinity in different parts of the basin. In the western half, the zone with lowest mineralization, argillaceous-calcareous sediments are deposited. In the eastern half, with the highest mineralization, calcareous-dolomitic sediments form. Around the margin of the lake lies a ring of salt lakes with deposits of mirabilite, thenardite, and halite (Fig. 114). "The deposits of the salt lakes are commonly separated from the principal basin by sand bars, barrier beaches, or broad transverse bars, which allow filtration of the water. Some deposits undoubtedly represent gulfs of Lake Balkhash, now separated from the major body through drying and shrinking over a period of time. This suggests that the sediments in salt lakes are a distinctive facies closely associated with modern deposition in Lake Balkhash. In spring, the water seeps through the barrier bars from the salt lakes into Balkhash; but in summer, when the lake levels drop sharply or the water evaporates altogether seepage occurs in the opposite direction" (Sapozhnikov, 1951, p. 109). From this it is seen that the salt-bearing deposits associated with Lake Balkhash *are found furthest from the point of discharge of the principal inlet stream, the Ili River.*

The distribution pattern of salt deposits associated with Lake Balkhash shows very distinctly what may occur on a different scale with greater or lesser clarity in a number of other present-day deposits.

From this point of view the Bocana de Virrila in Peru is of importance. Near the mouth of this gulf, where the salinity is 8·8–10·3%, only calcareous muds are deposited. In the farther reaches of the gulf, with salinities up to 19%,

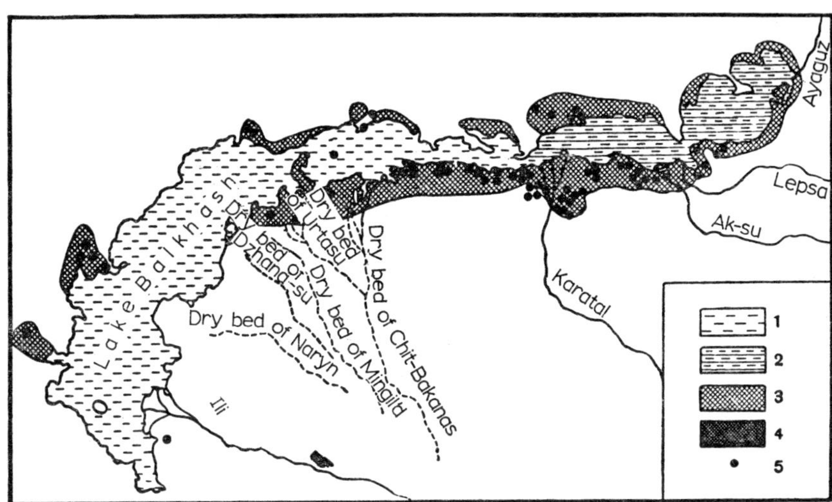

FIG. 114. Distribution of principal types of chemical sediments in the Balkhash region (from D. G. Sapozhnikov). 1. Calcareous sediments; 2. calcareous-dolomitic sediments 3. sulfate and chloride-sulfate sediments; 4. salt marsh; 5. salt lake or salt deposit of variable composition.

gypsum with carbonate impurities is deposited, and in the uppermost part, where the salinity was as high as 34·5% and more in August 1955, freshly precipitated halite is found.

A similar picture is observed in the Sivash. The eastern part of this body of water is characterized by low mineralization, and, in keeping with this, clay-carbonate muds are deposited. The western part is more saline, especially in individual arms (zasukhas), and gypsum correspondingly appears, even beds of halite at times. The lagoonal sediments are generally quite variable horizontally. *Deposits of the highest degree of salinity are found in zones farthest from the point where water and saline solutions are supplied to the basin.*

The Gulf of Kara-bogaz, before its recent phase of shrinking and saline concentration, precipitated gypsum over its entire area. In a few local zones and in depressions on the floor, mirabilite was deposited. The salinity here was 12–21%. As the Caspian began to dry up and the less water then moved from it to the Gulf of Kara-bogaz, evaporation in the Gulf increased and halogenesis developed a more advanced stage. As early as 1939, Ya. B. Blyumberg noted intense precipitation of halite, not merely periodically but as a constant bottom phase, leading to the formation of a halite bed (Blyumberg, Nikolaev, Egorov, 1940).

In 1944 A. D. Pel'sh recorded the precipitation of blödite. This mineral formed 30% of the halite bed in 1945. According to A. A. Ivanov (1953), in 1944-45 epsomite began to crystallize with mirabilite in wintertime, in addition to the summer precipitation of common salt and blödite. Lastly, glauberite has also been found in the deposits of the Gulf of Kara-bogaz (Bukshtein and Garkavi 1952), and even carnallite occurs in small pools on the higher parts of the floor (Fiveg 1955). Clearly, chemical sedimentation in the Gulf of Kara-bogaz has advanced considerably in recent decades, and the basin has become properly a salt-former. As for the spatial distribution of the salts, it has been found that the western, less saline part of the gulf adjacent to the strait is the site chiefly of gypsum-carbonate muds (Fig. 88). Farther east, glauberite accumulates in a wide belt running from northwest to southeast. Along the southwestern edge of this zone the glauberite is mixed with mirabilite. The central and largest part of the basin is the zone of halite, with considerable admixture of epsomite and small crystals of blödite and glauberite. The extreme northern, eastern, and southeastern parts of the gulf are now dry and are characterized by halite deposits. In depressions in these zones, where intercrystalline brine drains, epsomite and even carnallite are precipitated. Clearly a considerable facies variation in the chemical sediments that are strictly synchronous has developed within a single salt basin. Here, as elsewhere, the salt deposits of the lower stages of salinity shift toward the source of water supply to the basin, whereas salts of the higher stages of mineralization are displaced in the opposite direction and far from this source of water. The extreme development of this tendency is found in the localization of epsomite in Sartas Bay and of carnallite in lakelets in the dry basins already separated from the main water mass of Kara-bogaz.

Thus, the hydrochemical type of a basin directly determines the essential composition of the chemical sediments that form at each stage of its development —carbonate, sulfate, and chloride—and consequently the mineralogical-petrographic type of its halogenesis.

The localization of water supply about the margin of a basin controls the spatial distribution of synchronous halogen deposits on the floor or, to put it differently, the structure and texture of the halogen formation. This structure is symmetrical in basins with uniform water supply the entire margin and is asymmetrical in basins with water supplied from any one part.

The close connections of hydrochemistry and hydrology of present-day salt basins with the mineral composition of the sediments and with their distribution on the basin floor is important not only in itself but also from the point of view of comparative lithology, since it clarifies much concerning the composition and structure of ancient halogen formations.

12. SALINE SEDIMENTS AS A PHYSICOCHEMICAL SYSTEM

As the deposits of humid basins represent a complex system, sharply out of physicochemical equilibrium at first, slowly reaching equilibrium during the course of diagenesis, so halogen sediments of arid basins also form a system at first out of physicochemical equilibrium, reaching such equilibrium only during diagenesis. Since it will be necessary to analyze the physicochemical mechanism of halogen sedimentation and lithogenesis in the remainder of this work, it seems appropriate to end this review of modern halogen deposits by defining their characteristics as a physicochemical system.

Data on this problem come almost exclusively from deposits of average salinity stages (10–18% S), used in balneology as medicinal media. But even

TABLE 24

Moisture Content and Mechanical Composition of Sediments in Salt Lakes (from Kurnakov *et al.* 1936)

Lake	Moisture content, % of dry weight	Content of pelitic fraction, %
Dzharyl-Agach . . .	58·09	71·48
Tobechikskoe . . .	51·41	79·00
Sasyk-Sivash (white mud) .	27·09	28·09
Chokrak	46·66	76·62
Saki	41·58	42·57
Sasyk-Sivash . . .	40·87	74·62
Krasnoe	24·96	67·57
Staroe	19·83	53·19

this limited evidence allows us to explain many characteristic physicochemical features of sediments in salt basins.

Surface deposits in basins of arid zones, like those in humid regions, are characterized primarily by a high moisture content (Table 24).

The water content in the surface layer of mud ranges mostly between 40 and 50%, but there are basins with a low moisture content in the muds (20–25%). In comparison with marine muds, the clay sediments of saline basins contain appreciably less water. Thus, *increase in salinity of bottom brine leads to decrease in moisture content in the accumulating sediments.* Many investigators (S. A. Shukarev, M. I. Ravich, and others) indicate that the same comparison can be made between muds of slightly saline lakes with those of highly saline basins. "The muds of such lakes as Karadzhal, Sasyk, Dzharylgach, Donuzlav, and Mayakskoe are characterized by the highest water content when the brines in the lakes are most dilute ($d=1·02–1·08$). Muds in the Perekop group of lakes—Staroe and Krasnoe—are characterized by the lowest water content, and the brines of these lakes are the most

P

concentrated" (Ravich 1936, p. 203). It must be borne in mind, however, that these relations between moisture content of the muds and concentration of the bottom brines are maintained only where sediments of uniform grain size, lacking any solid salt phases, are being deposited. As the sediments become coarser the moisture content naturally decreases. Impurities of salt crystals characteristically have the same effect of reducing moisture content. The porosity of freshly precipitated halite, free from clastics, reaches a maximum of 30–35%. Data on other salts are lacking, but there can hardly be any great difference.

Thus, as the salinity in basins of arid regions increases, the water content in comparable sand-clay surface deposits gradually declines to 20–25% of the dry weight. Salt deposits are generally characterized by low porosity and moisture content. The porosity and moisture content decrease still more with depth in the sediment.

Because of a high water content, deposits of salt basins at the middle stages of salinity, during gypsum accumulation, abound in bacteria. Very different physiological groups are encountered: aerobic and anaerobic, sulfate-reducing and putrefying, among others. The vertical distribution of bacteria in the muds is characteristic. According to N. N. Kryukova (1956) sulfate-reducing bacteria in the muds of Lake Saki are equally developed in all samples taken from the surface of the sediments to a depth of 110 cm. No growth of other anaerobic bacteria on Tauson medium was observed by the author in tests on samples from all depths in the mud. The total number of aerobic microorganisms ranges from 200,000–250,000 per gram of dry mud down to 20,000–30,000 per gram at a depth of 110 cm. In general, then, the same relations occur as in sediments of humid regions (Vol. 2, Chap. 8).

Nevertheless, from general considerations it would appear that the number of microorganisms in sediments declines sharply as the water passes through increasingly more saline stages to the final stages of halogenesis. It is possible that near the eutonic stage, when potassium salts are being precipitated, the environment in the sediments as well as in the bottom brine may be almost if not entirely, sterile. In any case, the activity of organisms in the sediment becomes negligible.

Only the principal saline constituents of the interstitial water in sediments of saline basins have been studied thus far. It has been shown that *at the brine stage in a lake the interstitial water has the same composition and type as the bottom water of the lake, but the mineralization is generally somewhat less.* At the same time sulfate reduction and decomposition of organic material are well-defined processes in the sediment. These change the oxidation-reduction and acid-alkali conditions in the interstitial water and lead to generation of various gases: H_2S, CO_2, N_2, CH_4, and others.

From K. A. Ovsyannikova's (1951) data on the Eh and pH of sediments in Lake Saki (Table 25). It becomes clear that the oxidation-reduction potential of the surface sediments is low, –200 to –210 mv, in zones (3 and 4) where it was not disturbed by extraction of the mud. In the disturbed zones (1 and 2) it is higher, but it levels off quickly, tending toward the earlier value. The pH of the muds proves to be only slightly alkaline.

The Eh increases with increase in depth of sediment below 30 cm, fluctuating between 75 and 69 mv at depths from 100 to 110 cm. The causes of such variations are not yet completely clear.

According to V. I. Bakhman and E. F. Prokof'eva (1953), the amount of gas in the muds of Lake Saki is as follows: free H_2S 16·44 mg/kg, H_2S in hydrosulfide 17·0 mg/kg, free CO_2 263 mg/kg, CO_2 in HCO_3^- 31·4 mg-equiv/liter, CH_4 2·40 mg/kg of mud, N_2 1·28 mg/kg of mud. Similar relationships are found in other lakes. "When there is much muddy water in weakly mineralized

TABLE 25

Eh and pH in the Upper Layers (0–30 cm) of Muds in an Eastern Basin

Locality No.	Sample locality	Date	pH	Eh, mv	H_2S, mg per 100 g of dry mud
1	Locality from which mud was taken in 1948.	18/VI 20/VII	7·1 7·2	−138·3 −150·0	146·0 170·0
2	Locality from which mud was taken in 1946 from under a carbonate crust.	25/VII	7·3	−191·0	200·0
3	At the control station near the breakwater of the State Health Resort, at the front edge of the gypsum crust.	26/VII 3/VII	7·2 7·3	−202·8 −210·0	— 185·0
4	From below the gypsum crust, 150–200 m from the end of the breakwater.	18/VI 18/VII 21/VII	7·5 7·2 7·4	−170·3 −215·8 −211·0	277·0 283·0 —

muds, gases occur in the dissolved state. As the gas content increases, the gas phase appears, with the development of gas bubbles particularly when the mud is stirred or when mud is extracted from great depth. The mud may even appear to boil because of the volume of escaping gas." (Bakhman and Prokof'eva 1953, p. 150).

These features of interstitial water in bottom muds are characteristic of the middle stages of mineralization (10–18% salinity). When the salinity exceeds 25%, as the eutonic point is approached more and more nearly, the microbiological processes of desulfatizing and putrefying organic material gradually die out. Furthermore, the Eh and pH of the brine, closely associated with the biological processes, increasingly approach the values found in the bottom water, i.e. the physicochemical environment in the sediments becomes increasingly like that in the bottom water.

One of the characteristic features of sediments in saline basins, considering the sediments as a physicochemical system, is the distinctive nature of the hydrodynamic conditions in the basin. In the bottom muds of humid-climate lakes and seas the interstitial water is almost immobile, and the redistribution of compounds dissolved in it, from one point in the sediment to another, takes place only very slowly, by diffusion (Vol. 2, Chap. 8). In deposits in saline basins the process is much more complex; the specific gravity of the bottom brine increases appreciably during summer evaporation, reaching values from 1·05 to 1·33 and even more by the end of the evaporating season. Such brines are noticeably heavier than those impregnating the bottom sediments, with the result that, towards the end of the evaporating season, the bottom brine sinks into the sediment, displacing the lighter brine. In other words, the interstitial water at some depth is included in the general hydrologic cycle of water in the basin. The depth to which the heavy brine sinks into the sediment is determined by the difference in specific gravity between it and the interstitial solution and by the friction of the liquid against the solid phases in the pores. It is not yet possible to express this in figures, but the existence of water exchange in the upper part of the bottom muds in saline basins cannot be doubted.

Rising currents of water from other causes are active at the same time as the seasonal hydrodynamic microcycles in the sediments of saline basins. All modern basins in arid zones have a ground-water supply in addition to a surface supply, the amount being insignificant at times, but still important in the hydrodynamic behavior of the interstitial water of the bottom muds. Within the drainage area surrounding a lake there is usually a reservoir of ground water, flowing into the lacustrine basin and discharging its load. This is the situation on the so-called Salt Lake Steppe, which is part of the Kulunda Steppe. Ground water is discharged there along the Baklanikha River in its alluvial deposits and also from the terrace sands adjoining these deposits on the east and the west. At the foot of the sandy belt along the entire eastern margin swampy areas lie along the shores of Krivoe, Topkaragaya, and other Lakes as far as, and including, Iodnoe Lake (at the time these lakes were studied). At times, zones of reeds and rushes develop, demonstrating the intensity of ground-water discharge from the pine forest into these lakes. Along the southern and eastern margins of Malinovoe Lake one may also find a belt of fresh ground water. Similar seepages of ground water occur at many places along the western margin of the pine forest.

In draining into the lake, the fresh ground water comes into contact with the sediments both from the side of the basin, and from below. Being under hydrostatic pressure, it seeps slowly upward through the sediment, either oozing through the pores where these permit, or by breaking through special channels in the sediments, locally known as "air holes". These holes are known from both the margins and the central parts of many lakes on the Salt Lake Steppe. Externally the air holes are small swampy and muddy zones where sticks may be thrust deep into the incoherent mass. Similar air holes are found in the more central parts of Malinovoe Lake, in a salt bed. Jets of water rising upward from these holes may sometimes be seen, the water then mixing with the lake brine. The water is usually sodic, with a relatively low mineralization. The salt water of Lake Malinovoe, on the other hand, had a salinity of about 25% in 1945, approaching the group transitional between *IIb* and *IIc*. This same phenomenon has been reported from other lakes, such as Pechatnoe and Lomovoe.

Thus, on the Salt Lake Steppe, besides the sinking of heavy surface brine into the deeper layers of the sediment, rising current of less salty water is also found. The first process takes place along a continuous front, whereas the second takes place through small localized channels, disrupting the continuity of the downward flow.

There is no doubt that under different environmental conditions the strength of the upward moving ground water might be less than in the lakes of the Salt Lake Steppe. In particular, it is possible that in larger basins, with thicker deposits of sediments, the role of ground water in the hydrodynamic regime of salt accumulation is much less significant. It is clear, nevertheless, that this factor cannot be completely ignored even in dry lakes.

A third factor in the hydrodynamic regime of halogen sediments is the rising current of fresh water generated by the dehydration of minerals rich in water of crystallization: gypsum, mirabilite, epsomite, and others. This dehydration takes place at some depth, (not yet determinable), below the surface of the sediment. It becomes the source of a considerable quantity of water that is slowly squeezed upward by compaction of the sediment.

Thus the hydrodynamical system of halogen sediments is complex; it involves the sinking of heavy surface brine and the upwelling, through many small channels, of comparatively light fresh ground water and dehydration water. In lakes of low mineralization this hydrodynamical regime is scarcely noticeable, but with increasing salinity it becomes increasingly well defined.

In comparing the sediments of saline basins, as physicochemical systems, with the sediments of humid zones, one may easily detect three specific features. With increasing salinity in the basin, (1) the moisture capacity of sand-silt-clay sediments decreases and at the halogen stage proper it is generally small; (2) the microbiological processes of sulfate reduction and gas generation diminish, and the Eh and pH of the sediments approach the values found in the bottom water, and (3) a complex hydrodynamical regime is gradually established, becoming progressively better defined. This involves simultaneous downward movement of heavy water of the bottom brine and upward movement of fresher ground water and dehydration water.

CHAPTER 2

THE PHYSICOCHEMICAL MECHANISM OF PRESENT-DAY HALOGENIC SEDIMENTATION

The formation of present-day saline deposits is a very complex process. It has been determined only through experimentation into the conditions under which different mineral species are precipitated. Before setting out on an analysis of the mechanism of present-day halogenesis, therefore, it is necessary to review the basic physicochemical laws controlling the formation of solid phases of highly soluble salts.

1. LAWS CONTROLLING THE FORMATION OF SOLID PHASES OF READILY SOLUBLE SALTS ACCORDING TO EXPERIMENTAL DATA

Experimental work on solid phases of readily soluble salts was begun in the nineties of the last century by Van't Hoff and has been continued by many other investigators, both foreign (D'Ans and others) and Soviet (N. S. Kurnakov and his school). As a result, in addition to determining a great number of partial systems of physicochemical equilibrium, general laws governing the formation of solid salt phases have been defined. These have special significance in saline sedimentation, and will now be discussed.

Solid salt phases are produced from their solutions in two ways: by evaporation of the solvent (water) and by freezing.

As a solution evaporates and the concentration of salts gradually increases, the least soluble salt (A) first reaches saturation and begins to be precipitated in the solid phase; this is followed by the more soluble (B), then by (C), and so forth, in the order of increasing solubility. In this sequential precipitation, however, a certain quantity of *all salts* will be held in solution during the entire time of evaporation, a smaller quantity of each salt the lower its solubility. In the course of evaporation a point is finally reached beyond which *increase in concentration of the solution ceases and all the salts present are precipitated simultaneously*. This is the *eutonic point*, and the corresponding solution is said to be eutonic. At this point the vapor pressure of the solution is minimal. In natural solutions under natural conditions the eutonic point is normally reached at a total mineralization of 35–36 to 40% in the brine and at air temperatures above 32°. At the eutonic point the solutions simply dry up without changing in composition.

When a solution freezes, then too a sequential precipitation of one solid phase after another takes place. In contrast to the evaporative process, however, the solution is not concentrated but is diluted gradually, and the contents of all salts diminish, especially those that began earliest to separate out in solid phases. A point is finally reached at which all the salts and the solvent (water) go simultaneously into a solid phase. The solution, without changing its composition, freezes and becomes solid. This point is called the *eutectic* point. For natural solutions under natural conditions, depending on the initial mineralization and composition of the solution, this point is reached at a temperature somewhere in the interval −21 to −35°C.

Thus, *whether a salt solution is evaporated or frozen, it reaches some limiting state at which individual and sequential precipitation of solid phases gives way to simultaneous precipitation of all constituents in the solution.*

The temperature interval that separates a eutonic solution from its eutectic, under natural conditions, is 60–70°.

In this interval the precipitation of salts depends on three factors: (1) temperature of the solution, (2) total concentration of the solution, and (3) quantitative relations of the salts in the solution. Each factor will be examined in turn.

(1) *The effect of temperature is manifested primarily in the degree of hydration of the precipitating salt minerals.* Figures 115-118 (from Bergman and

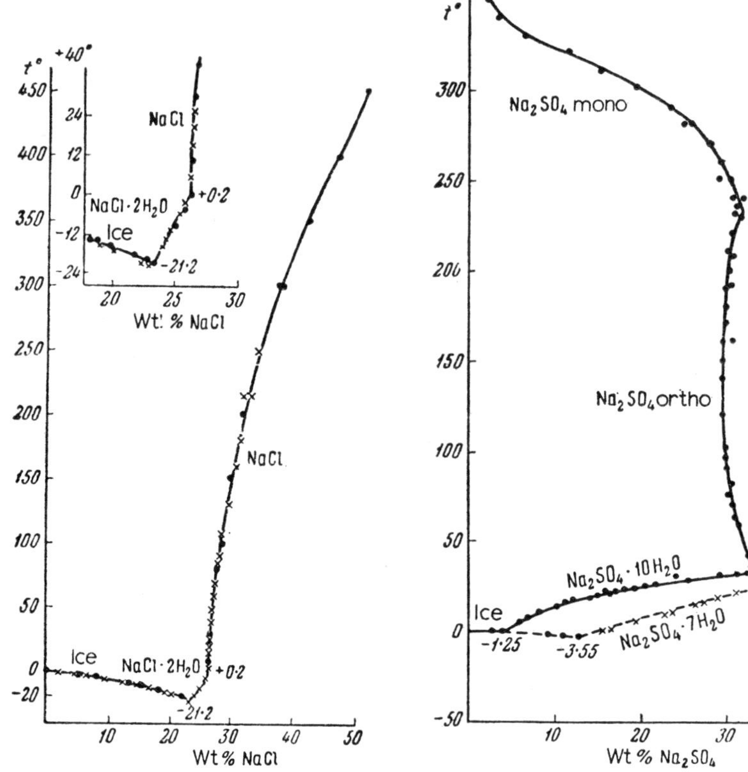

FIG. 115. Temperature-solubility curve for the system sodium chloride—water (from A. G. Bergman and N. P. Luzhnaya).

FIG. 116. Temperature-solubility curve for the system sodium sulfate—water (from A. G. Bergman and N. P. Luzhnaya).

Luzhnaya 1951) illustrate the temperature solubility relations for $NaCl$-H_2O, Na_2SO_4-H_2O, $MgCl_2$-H_2O, and $MgSO_4$-H_2O.

Figure 115 shows that sodium chloride has only one hydrated form, $NaCl.2H_2O$, which exists between the limits of $-21.2°C$ and 23.3% $NaCl$ to $+0.15°C$ and 26.3% $NaCl$. At higher temperatures and salinities, halite, the anhydrous salt, precipitates. Thus, elevation of temperature causes dehydration of the precipitating solid phase.

Sodium sulfate has two hydrated phases at low temperatures (Fig. 116): mirabilite ($Na_2SO_4.10H_2O$) and heptahydrated sulfate ($Na_2SO_4.7H_2O$). The first occurs in the interval from $-1.25°C$ and 4.08% Na_2SO_4 to $+32.38°C$ and 33.24% Na_2SO_4. The second is found correspondingly from $-3.55°C$ and 12.7% Na_2SO_4 to $+24.25°C$ and 34% Na_2SO_4. The heptahydrate is precipitated during slow cooling of a solution saturated at $32.4°C$ in the absence of any seed crystals of mirabilite or of salts isomorphous with it. A solution saturated with the heptahydrate is oversaturated relative to mirabilite, and, if a seed crystal of the latter is introduced, this mineral crystallizes instantaneously. Above $32.38°$ sodium sulfate exists in the anhydrous form, forming several (3–5) structural modifications.

Dehydration of the solid phases during elevation of temperature is graphically seen here.

Magnesium chloride is distinguished by its capacity to yield many hydrated forms: $MgCl_2.12H_2O \rightarrow MgCl_2.8H_2O$ (in modifications α and $\beta) \rightarrow MgCl_2.6H_2O$ (bischofite) $\rightarrow MgCl_2.4H_2O \rightarrow MgCl_2.2H_2O$. The stability conditions are as follows:

	Beginning		End	
	$t°$	$S, \%$	$t°$	$S, \%$
1. $MgCl_2.12H_2O$ . . .	-33.6	21.57	-16.4	30.50
2. $MgCl_2.8H_2O$ (β metastable)	-40.0	26.50	-15.0	32.70 (metastable)
3. $MgCl_2.8H_2O\alpha$ . . .	-15.0	31.85	-3.4	34.34
4. $MgCl_2.6H_2O$ . . .	-15.0	33.86	-5.0	34.24 (metastable)
5. $MgCl_2.6H_2O$ . . .	-0	34.40	$+11.6$	46.12
6. $MgCl_2.4H_2O$ . . .	$+116.0$	46.12	$+181.5$	55.80
7. $MgCl_2.2H_2O$ . . .	$+181.0$	5.80	$+300.0$	67.84

The gradual dehydration of solid phases as the temperature rises in a solution is seen even more clearly here than in the relations of sodium chloride or sodium sulfate (Fig. 117).

Like magnesium chloride, $MgSO_4$ generally occurs only in the hydrated form at temperatures below $200°$, forming no anhydrous salts. Nevertheless, even here the higher hydrates (dodecahydrate and heptahydrate of $MgSO_4$) are characteristic of lower temperatures in the solution, and the lower hydrates of higher temperatures (Fig. 118).

Thus, in all cases the more hydrous forms of solid salt phases change to less hydrous forms when the solution is heated.

Temperature also proves to have a very substantial effect on the formation of

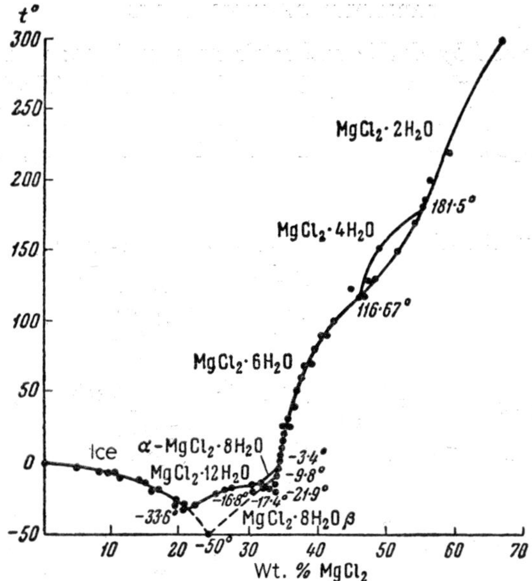

FIG. 117. Temperature-solubility curve for the system magnesium chloride—water (from A. G. Bergman and N. P. Luzhnaya).

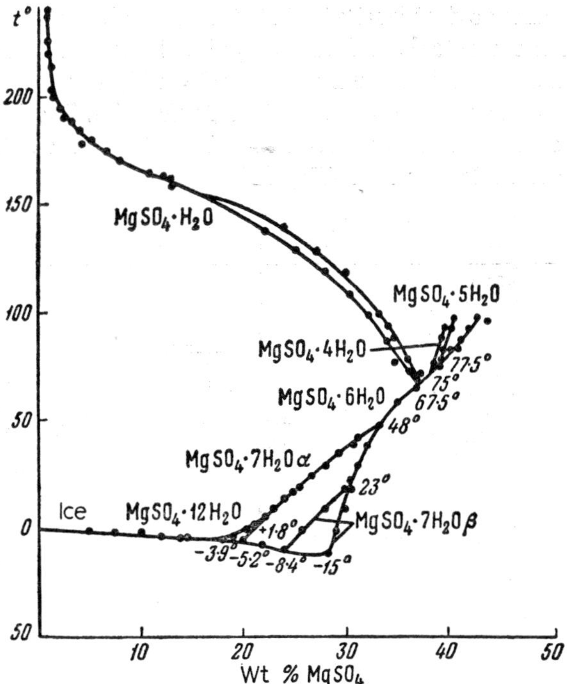

FIG. 118. Temperature-solubility curve for the system magnesium sulfate—water (from A. G. Bergman and N. P. Luzhnaya).

minerals represented by double and triple salts, such as: trona (Na_2CO_3. $NaHCO_3.2H_2O$), burkeite ($2Na_2SO_4.Na_2CO_3$), blödite ($Na_2SO_4.MgSO_4.$ $4H_2O$), glauberite ($Na_2SO_4.CaSO_4$), carnallite ($KCl.MgCl_2.2H_2O$), kainite ($KCl.MgSO_4.H_2O$), northupite ($2NaCO_3.2MgCO_3.NaCl$), and others (Fig. 119).

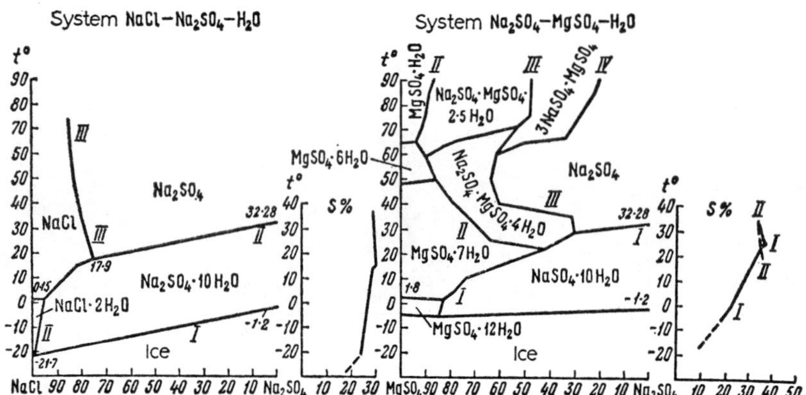

FIG. 119. Triple sulfate system.

Only in rare cases do double salts begin to form near 0°C, i.e. approximately midway between the eutonic and eutectic points in natural solutions. One example is trona. Such salts form most intensely at higher temperatures and concentrations, correspondingly nearer the eutonic point. The formation of triple salts is confined to this region. Below 0° and near the eutectic point neither double or triple salts form (Fig. 120).

Mineral	-20°C	-15°C	-10°C	-5°C	0°C	+5°C	+10°C	+15°C	+20°C	+25°C	+30 C	+35°C
1 MgCl₂ 12H₂O												
2 MgCl₂ 8H₂O												
3 MgCl₂ 6H₂O												
4 NaCl 2H₂O												
5 NaCl												
6 Na₂SO₄ 10H₂O												
7 Na₂SO₄												
8 Na₂SO₄ MgSO₄ 4H₂O												
9 MgSO₄ 12H₂O												
10 MgSO₄ 7H₂O												
11 MgSO₄ 6H₂O												
12 H₂O –Ice												

FIG. 120. The formation of minerals in the system. $2NaCl + MgSO_4$, $Na_2SO_4 + MgCl_2$ at temperatures from −20 to +35° (from A. G. Bergman and N. P. Luzhnaya).

In summarizing we note that *elevation of temperature brings about a loss of water of crystallization by the salt minerals and leads to the formation of double and triple salts. A decrease in temperature works in the opposite direction.*

(2) *The effect of concentration on the formation of minerals is complex.* With increase in mineralization in a solution it is possible that the more soluble minerals will form, and this brings about a stage-like character to halogenesis, as already stated.

Furthermore, concentrated solutions act on hydrated minerals as dehydrators, and thus lower the temperature field of stability of these minerals.

Lastly, the increase in concentration of complex natural saline solutions favors the formation of minerals represented by double and triple salts. This may be graphically shown by the formational pattern of salts during evaporation of sea water.

Thus, the effect of salt concentration on the formation of minerals is in many ways similar to the effect of temperature. An elevation of temperature or an increase in mineralization causes dehydration of solid phases and, in addition, leads to the precipitation of double and triple salts. The reverse is true with decrease in temperature and concentration.

(3) *A still more important factor in the formation of minerals during halogenesis is found in the quantitative relations among the individual salts in the solution.* Experiments by many physico-chemists (Van't Hoff, N. S. Kurnakov and his school, and others) have shown it is *precisely these relations of salts in the solution that determine the actual composition of solid phases that form from evaporation or freezing of the solution.* This is clearly shown in diagrams (Figs. 121-123) that illustrate the solid-phase composition corresponding to various salt ratios at 25°C in all three hydrochemical types of salt basins.

The diagrams were obtained as follows: a strong saline solution was prepared artificially, the salt ratios corresponding to any particular point on a diagram, of a sulfate lake, for instance. Then some quantity of known salt, such as mirabilite, was introduced, and the whole system of solid phase and solution was held for a long time and shaken until it was in equilibrium. Sometimes all the solid phase first introduced was entirely dissolved, whereupon more material was added. At other times the solid phase increased at the expense of the solution, and so forth. The establishment of equilibrium was controlled by repeated analyses, and it was considered that equilibrium had been reached when the composition of the solution remained constant. The point on the diagram represented by this solution was then calculated, and it was noted to what solid phase of salt this corresponded. If this phase was mirabilite, as in the example here discussed, it was noted that the plotted point was also a point of stability for that mineral. After numerous experiments with solutions of different composition, the stability field of mirabilite was defined, beyond which the mineral is dissolved and the solid phase proves to be a different mineral, such as thenardite or blödite. After further experiments with different compositions of the solution and by introducing solid

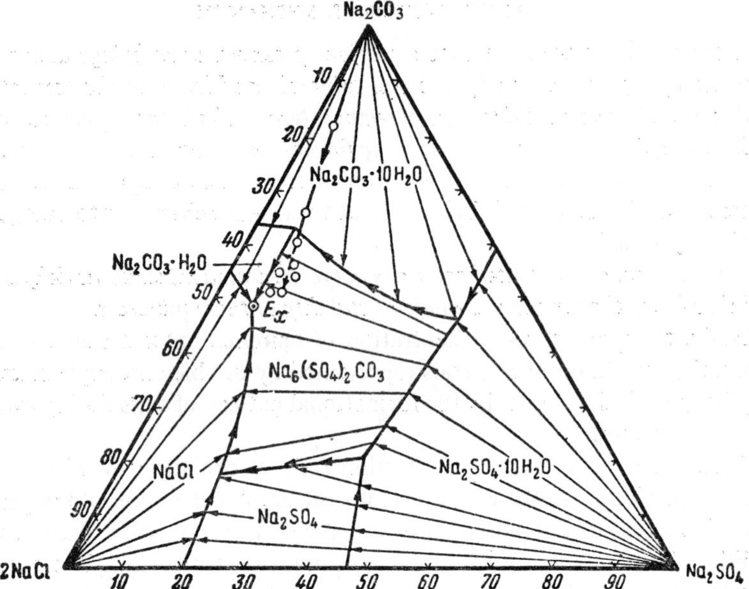

Fig. 121. The $+25°$ isotherm of the system Na^+, $CO_3^{2-}(HCO_3^-)$, SO_4^{2-}, Cl^-, H_2O (from M. G. Valyashko). E_x is the eutonic point. Arrows indicate change in composition of the brines during isothermal evaporation and crystallization of salts.

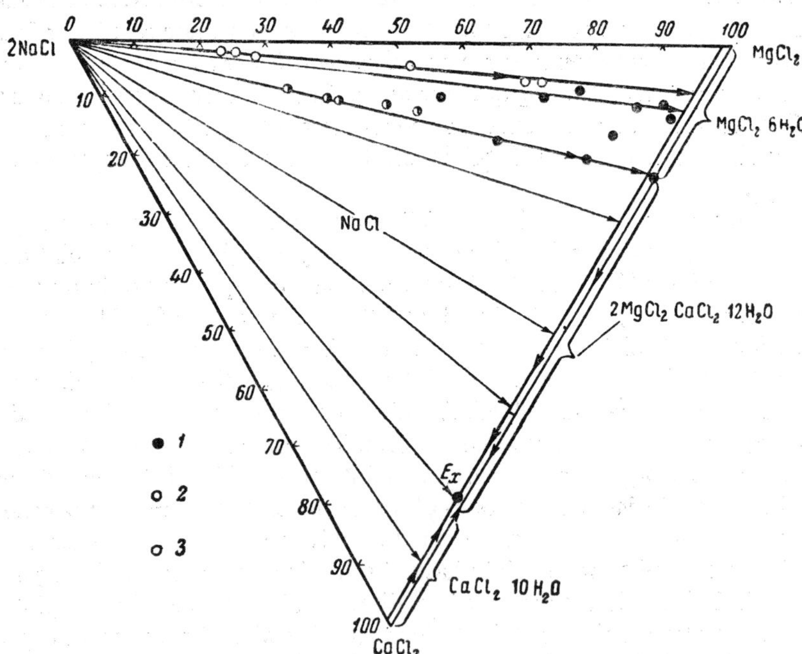

Fig. 122. The $+25°$ isotherm of the system Na^+, Mg^{2+}, Ca^{2+}, Cl^-, H_2O (from M. G. Valyashko). E_x is the eutonic point. Arrows indicate change in composition of the brines during isothermal evaporation and crystallization of salts. 1. Staroe Lake: 2. Krasnoe Lake; 3. Kiyatskoe Lake.

phases, the stability fields of the various salts over the entire area of the square in the diagram representing sulfate water were defined. There are seven such minerals at 25°C: mirabilite, thenardite, halite, blödite, epsomite, hexahydrite, and bischofite. The diagram thus obtained has a deeper significance than Figs. 101–103, in which the stability fields of the solid phases are not given. In the earlier type of diagram only the saline composition of the solution is represented; but in Fig. 123, not only are changes in composition of

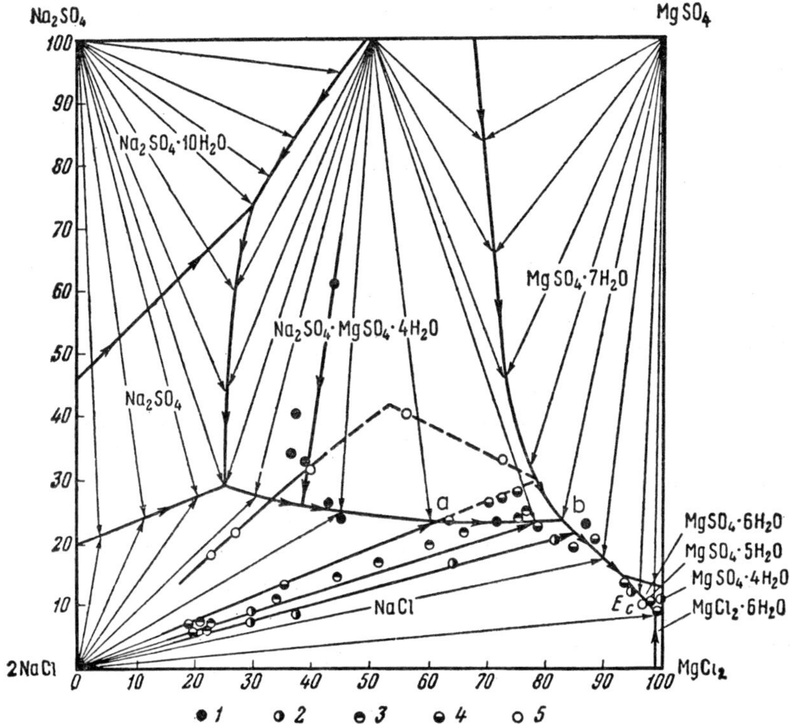

FIG. 123. The +25°C isotherm of the system Na⁺, Mg²⁺, SO₄²⁻, Cl⁻, H₂O (after M. G. Valyashko). E_x is the eutonic point. The arrows indicate change in composition of brines during isothermal evaporation and crystallization of salts. 1. Assa River and the lakes in its system; 2. Lake Saki; 3. Sasyk-Sivash; 4. Northern Sivash; 5. evaporation of summer brines from Lake Kuchuk.

the solution at different points indicated, but the composition of the solid phase corresponding to each composition of the solution is also depicted.

This is of great significance because it makes possible a forecast of the salt minerals to be precipitated first, when a solution of any given composition is subjected to isothermal evaporation, and what the general course of saline sedimentation will be. Let us assume that the water of a lake, not yet highly mineralized, is represented by a point in the halite field. This means that the first mineral to be precipitated from the highly mineralized (i.e. evaporated solution will be halite.* By connecting the representative point with the Na_2Cl_2 corner of the square the crystallization path of the lake brine is defined. Not only do we see that halite will precipitate along the extension of this path to its intersection with the halite-blödite boundary, but that the brine also undergoes a change corresponding to points along the crystallization path. Where the path intersects the boundary of the halite-blödite field (point a) the situation changes. Both halite and blödite are now in equilibrium in the solution, and both will be precipitated, and the point representing the solution will shift to the right toward the eutonic point, represented here by the symbol E_c.

At point b epsomite is added to blödite and halite, and then the precipitation of blödite ceases and only epsomite and halite will be precipitated. The previously precipitated blödite begins to convert to these salts. At point c the precipitation of epsomite gives way to the precipitation of hexahydrite, and the previously formed epsomite begins to change to this form of magnesium sulfate. This hexahydrite then gives way to the pentahydrite, then to the tetrahydrite. At point E_c, the eutonic point, bischofite begins to be precipitated, and all three henceforth are precipitated together until the solution dries up completely.

Diagrams for the fields of salt crystallization in soda and chloride lakes were plotted in a similar manner. The arrows on these diagrams indicate the course of crystallization to the terminal eutonic points of the solutions. Stable forms in the water of soda lakes at 25°C are $Na_2CO_3 \cdot 10H_2O$, halite (NaCl), mirabilite ($Na_2SO_4 \cdot 10H_2O$), burkeite ($Na_6SO_42(CO)_3)_2$, thermonatrite ($Na_2CO_3 \cdot H_2O$), and thenardite (Na_2SO_4). For chloride waters under the same circumstances the stable forms are halite, bischofite ($MgCl_2 \cdot 6H_2O$), $CaCl_2 \cdot 10H_2O$, and tachhydrite ($2MgCl_2 \cdot CaCl_2 \cdot 12H_2O$).

All three diagrams of salt water composition, as already pointed out, may be converted to a single complex diagram by combining them along common sides (Fig. 124). By so doing, a composite diagram of solutions and of solid salt phases in equilibrium with them (at 20°C) is obtained that correspond to any permissible ratios of dissolved salts. We may note that *the eutonic points in each of the adjoining diagrams remain distinctive and individual, which means that crystallization in each part of the composite diagram proceeds independently, not transferring to another part of the diagram.* This circumstance clearly reflects the independence of the three hydrochemical types of water—soda, sulfate, and chloride—throughout all stages of increasing salinity. However completely the solution is evaporated, and however changed the composition becomes through precipitation of solid phases, the hydrochemical type of solution, whether soda, chloride, or sulfate, remains unchanged. In other words, *simple evaporation cannot shift a solution from one hydrochemical type to another. Such transitions can be effected only through metamorphization*

---

* All the relatively insoluble carbonate (calcite and dolomite) and sulfate (gypsum and glauberite) minerals that precede the precipitation of halite are not considered here, of course, because these components are not plotted on the diagram.

of the brines by material supplied to the basin (bicarbonate waters + clay particles).

The diagrams thus far examined have to do with water containing no, or negligible, potassium, thus excluding the precipitation of independent potassium minerals. But when potassium accumulates in notable quantities potassium minerals begin to form, and additional diagrams illustrating potassium minerals along with the others are necessary. This is particularly so when we have to do with high salinity stages of sea water rich in potassium. Van't

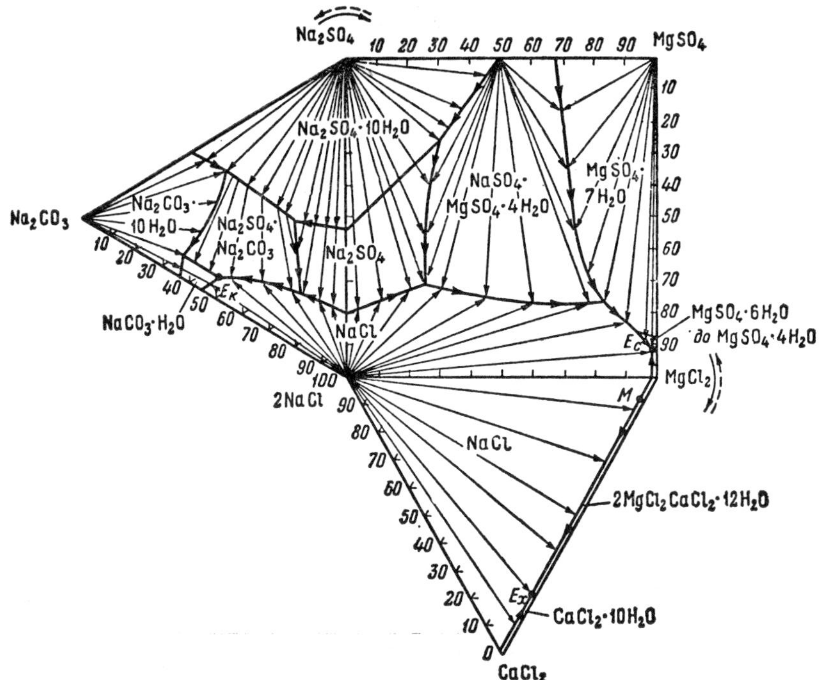

FIG. 124. Composite diagram of three principal chemical types of water. The arrows within the diagram represent crystallization course of the brines. The arrows above the diagram represent direction of forward metamorphization (——) and inverse metamorphization (– – –).

Hoff's triangular diagram demonstrated the changes in salt composition of potassium-bearing brines and the corresponding solid salt phases (Fig. 125). The complex five-component solution—$2K^+$, $2Na^+$, Mg^{2+}, Cl^-, SO_4^{2-}—is represented by three indices indicating contents of the ions $2K^+$, Mg^{2+}, and SO^{2-} in percentages of the sum of their equivalents. The components themselves are represented at the apices of the triangle: Mg^{2+} ($MgCl_2$), $2K^+$ ($2KCl$), and SO_4^{2-} (Na_2SO_4). The NaCl content in the solution is not considered here.

In keeping with the increasing complexity of the saline solutions, the number of solid phases in equilibrium with the different combinations of dissolved salts increases sharply, and at 25°C is 14: halite, thenardite, glaserite, blödite, picromerite, leonite, epsomite, hexahydrite, sylvite, kainite, carnallite, kieserite, bischofite, and langbeinite.

The point *Ok* on the diagram represents the composition of sea water, and the arrows indicate the course of crystallization from this water toward the eutonic point. We may note one characteristic feature of halite in this series of minerals. Once the solid phases have begun to form, halite continues to be

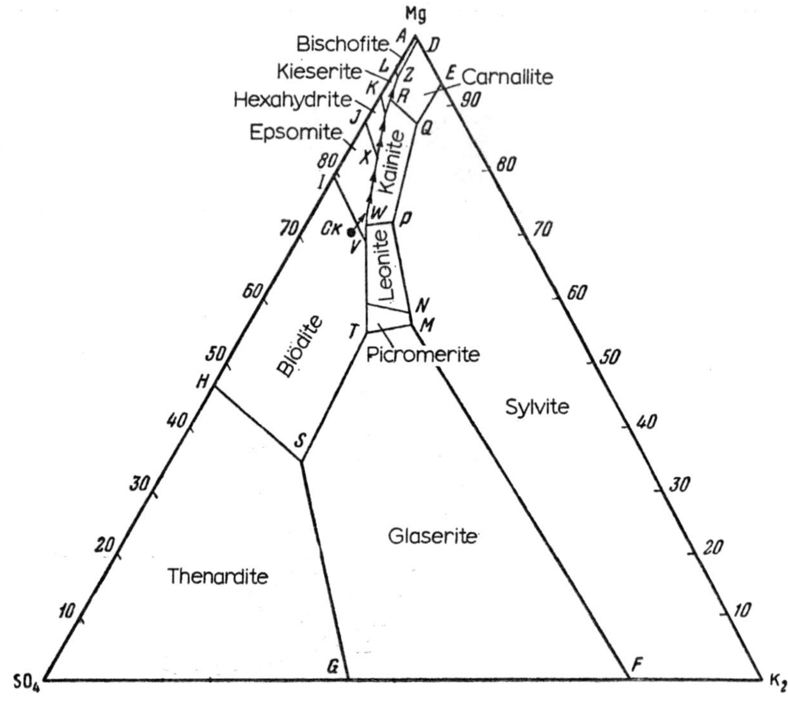

FIG. 125. Stability diagram at $+25°C$ in the system $2Na^+$, $2K^+$, Mg^{2+}, SO_4^{2-}, Cl^-, H_2O (from J. T. Van't Hoff). *Ok* is the point representing ocean water. The arrows indicate the crystallization course of ocean water. The Latin letters represent double and triple points.

precipitated throughout the entire course of crystallization from the solution, commonly forming a major part of the salt mass coming out of solution. For this reason halite is not represented on the diagram.

This assemblage of solid phases may then separate from solutions of different composition during isothermal evaporation at 25°C. At other temperatures the assemblage differs, of course, but the solid phases that precipitate still depend on the salt ratios.

Thus, in the range within which the solution changes from the eutectic point to the eutonic, the composition of solid phases is determined by the combined effect of temperature, concentration, and ratios of salts (or, more precisely, of their ions) in solution.

In order to understand the course of saline sedimentation it is very important to consider another factor which is often neglected. Despite the substantial difference in the chemical characteristics of water belonging to the different hydrochemical types, some minerals are common in saline sediments of mixed types of water on the composite diagram. NaCl, for instance, is found in sediments of all three types of water. Na_2SO_4 (in the forms of thenardite and mirabilite) is found in soda and sulfate waters. Gypsum is present in sulfate and chloride waters (at lower stages of mineralization). For such minerals with an extended range of precipitation we may establish the rule that *during the process of increasing salinity, one particular mineral may precipitate at markedly different degrees of total mineralization in a solution in different hydrochemical types of basins.* For example, gypsum in the sulfate lakes of group *IIa* precipitates at salinities of about 10–15%, but in sulfate lakes of groups *IIb* and *IIc* at salinities near 3–5%. In chloride lakes this mineral again forms at very high total mineralization in the brine. Even within a single hydrochemical group, the decrease or increase in content of any single salt, especially a readily soluble salt, may appreciably shift downward or upward the point at which another less soluble salt, one having a common ion with it, begins to be precipitated. Ocean water and water of the Black, Caspian, and Aral Seas all belong to the same hydrochemical group *IIc*, but they differ in the $MgSO_4$ content relative to the total salts: 6·4% in the ocean, 7·11% in the Black Sea, 27·58% in the Caspian, and 25·80% in the Aral Sea. When these waters are concentrated, gypsum begins to be precipitated from the first at a mineralization of about 15%, about 14%, about 5%, and at 4% respectively. As we know, the mollusc *Cardium edule* continues to survive in salinities up to 5%. Therefore, in muds of Kaidak Gulf and in a fossil horizon of the Aral Sea, shells of this form are encountered in sediments rich in gypsum. This is a primary, sedimentational association and not an epigenetic feature. Similar changes in the initial points at which precipitation begins under conditions of changing concentration are seen for other sulfate minerals (mirabilite, glauberite, blödite), but the extent of the range is markedly less with these minerals.

2. Annual Hydrochemical Cycles in Saline Basins and their Influence in Salt Deposits

Since solid phases of salts may form both during evaporation and during freezing, the actual course of salt formation in nature is highly complex. This results in the appearance of so-called *annual hydrochemical cycles* in saline basins. They first appear in the middle stages of salinity, but they become especially characteristic at high stages of mineralization. Investigation of these cycles has been made by Soviet specialists in this field, working on a number of basins. As a result, the cycles are known in considerable detail.

Figure 126 illustrates the annual hydrochemical cycles of soda Lake Tanatar I (summer salinity up to 25%). In spring, normally in April, the salt concentration in Tanatar I falls to 8–10% because of spring floods. By May,

Q

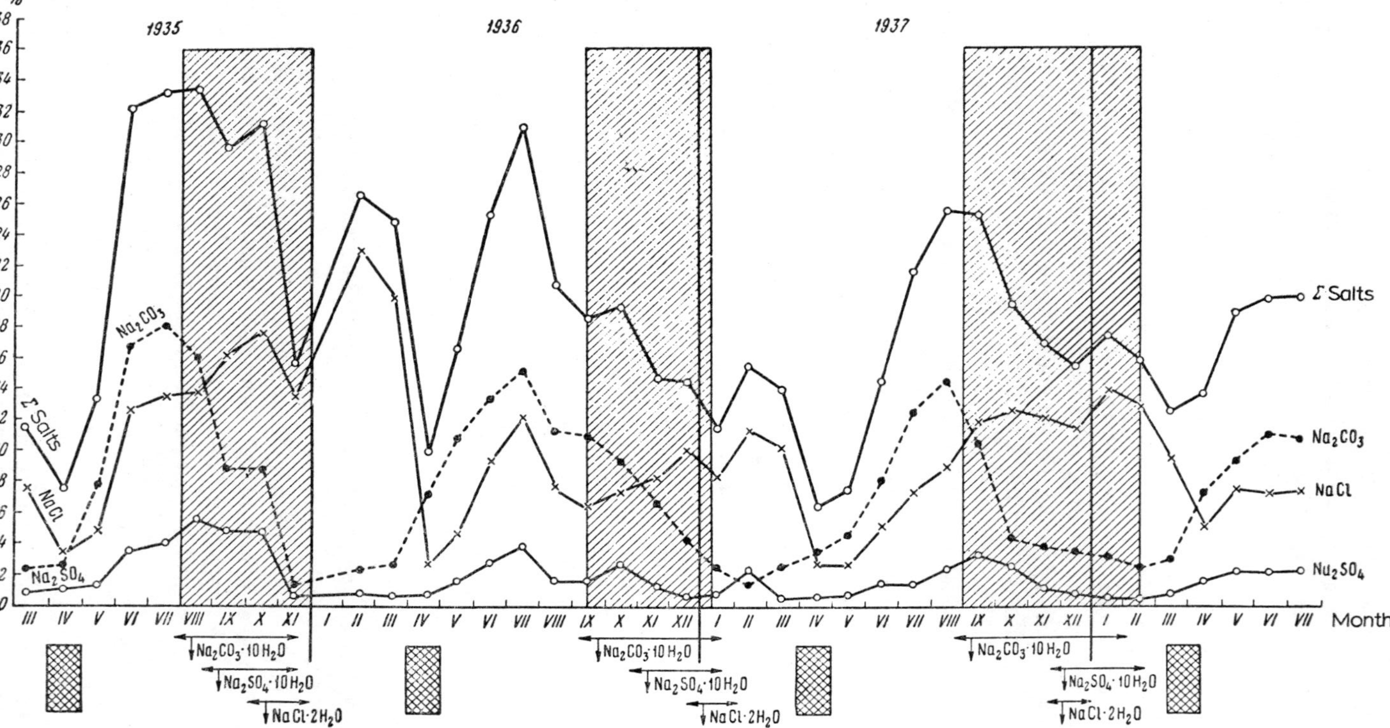

Fig. 126. Annual hydrochemical cycles of Lake Tanatar I (Kulunda Steppe). Crosshatching represents flood stage; inclined hatching indicates periods of salt precipitation.

however, evaporation is strong and mineralization increases anew, reaching a maximum in August or early September. With the advent of cold weather, sequential precipitation of minerals begins in the strongly concentrated brine. Soda (natron) is precipitated first, and this continues from September to December. Mirabilite follows (October-January), and still later hydrohalite (December-January). The concentration of the brine falls to 10–12% by virtue of this winter precipitation of minerals. In January-February, freezing of the water raises the concentration of the residual brine, and then the spring cycle begins afresh. The spring waters dilute the brine, and dissolve the winter sediments. The cycle is then repeated.

This behavior is characteristic of relatively moist summers with considerable rainfall. In very dry years the hydrochemical cycle is complicated by summer precipitation of salts, as described by S. Z. Makarov (1935). The area of the water surface on July 1 was approximately 100 hectares, its average depth about 15 cm and the volume of brine about 150,000 m^3. Intense precipitation of salt began on July 10 by the separation of trona. Precipitation began at a salinity of about 28% and continued for two days, forming a layer about 2 cm thick. When the mineralization reached 32% continuous precipitation of halite, natron, and sodium carbonate heptahydrate began. Halite crystals were formed during evaporation of the water by day and carbonate crystals at night, when the water temperature fell. Finally, on July 7, when the salinity reached 36%, a third solid phase, burkeite, began to separate out in small crystals, forming a uniform friable layer over the entire lake floor. It was subsequently cemented into very strong porcellaneous plates. The salinity of the brine increased no further. Decrease in volume of the lake—to complete dryness—took place with separation of all five phases: trona, heptahydrate sodium carbonate, natron, halite, and burkeite. The solution obviously reached the eutonic point and then slowly dried up. This limited intensity of summer salt precipitation was sedimentationally ineffective, however, because *autumn rains entirely dissolved the newly formed salt bed.*

In sulfate lakes the annual hydrochemical cycle is expressed in another series of minerals, as for example in Lake Ebeity (group *IIa*) (Fig. 127) and Lake Kuchuk (Fig. 128). In Lake Ebeity, 120 km from Omsk, the spring concentration of brine is about 20%. This lake belongs to the magnesium-sodium class, with much Na_2SO_4 and small amounts of $MgSO_4$ and $MgCl_2$. Towards the end of the evaporative season, in August, the mineralization of the brine reaches 30–31%, rarely higher, and salt precipitation begins. Mirabilite precipitates first, in August and September, but this takes place only when the highly saline brine drops to a temperature of 19°C. When the temperature reaches 0°C, the precipitated mirabilite amounts to 65–70% or the initial sodium sulfate content in the brine. At $-7°$ a new phase, ice, is added to the mirabilite, and at $-21·8°$ hydrohalite ($NaCl.2H_2O$) also begins to crystallize. Complete freezing of the eutectic solution in the experiment was observed at $-30°C$. "Under natural conditions eutectic congelation cannot be observed during relatively warm winters (the lake does not freeze). Even

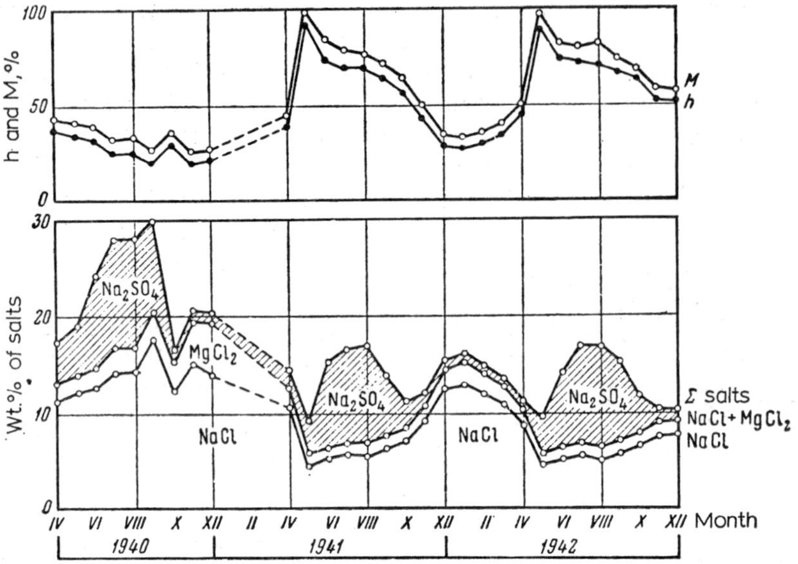

FIG. 127. Annual hydrochemical cycles of Lake Ebeity (from I. T. Druzhinin). M represents total brine, h depth of brine.

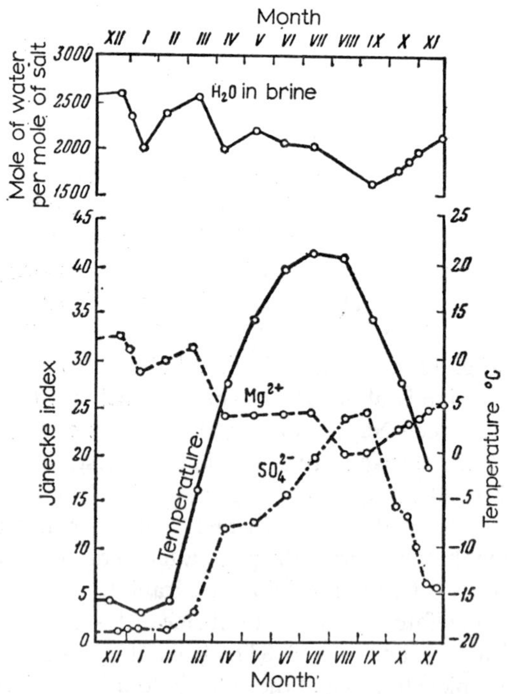

FIG. 128. Annual hydrochemical cycles of Lake Kuchuk (from S. Z. Makarov).

during the severe winter of 1940 (minimal temperature of $-42.7°C$) uncongealed brine was observed at a temperature of $-23.5°$ beneath a 25-cm layer of snow in an unconsolidated layer of mirabilite" (Druzhinin 1946, p. 119). Mirabilite precipitation by means of cooling increases and reaches a limit at $-15°$, when 98% of the sulfate present in the summer concentrated brine precipitates, and only 2% remains in solution.

In spring and early summer, most of the mirabilite returns to solution, and the bottom deposit of mirabilite increases only slightly. This monotonous alternation of precipitation and solution from year to year is varied by the formation of halite also at the end of the evaporative season in especially dry years. This phenomenon was observed, for example, in 1918, 1933, and 1939. The brine at this time (1939) had a temperature of $30.8°$ and a concentration of $32.8°$ Baumé. The absolute yield of halite was small, however, and the salt

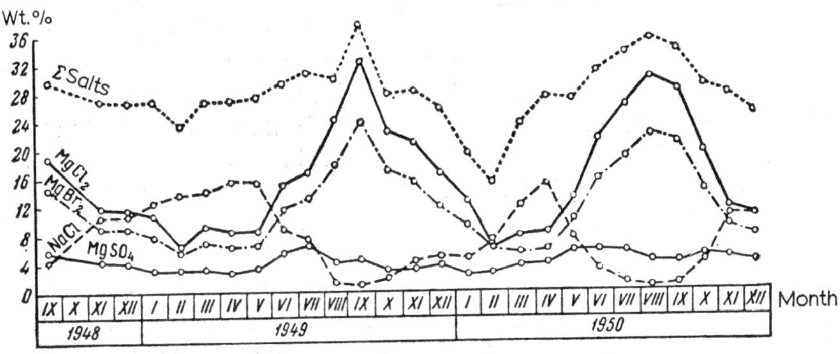

FIG. 129. Annual hydrochemical cycles in Lake Elton (from I. B. Feigel'son).

was strongly contaminated with sodium sulfate. During the next freshening of the lake the halite sediments disappeared without a trace.

Lake Kuchuk on the Kulunda Steppe is characterized by similar hydrochemical cycles.

The annual cycles of lakes near the chloride class are demonstrated by Lake Elton. This lake is at a very high stage of mineralization and the annual cycles are therefore very clear (Fig. 129).

The minimal salinity of brine in Lake Elton occurs in the winter period through March. Beginning in April the evaporation becomes strong, and NaCl is precipitated in the second half of the month, continuing to form until October. In August and September epsomite is precipitated with the halite, followed by hexahydrite. The epsomite molecule is lost from the water through the dehydrating effect of strong concentration of the brine, which contains a large quantity of $MgCl_2$ (up to 70–90% of total salts). At the end of the evaporative season the brine approaches the eutonic point, when bischofite separates out in the bottom phase during some nights of the dry summer.

This was first noted in Lake Elton by N. S. Kurnakov and then later (in August 1936) by I. B. Feigel'son. Bischofite in the bottom phase is, however, short-lived because the slightest rise in vapor pressure of the solution causes $MgCl_2$ slowly to dissolve. A considerable amount of $MgCl_2$ (up to 25%) is observed in the salts precipitated at the end of the evaporative season.

In autumn, before the brine has become diluted but after noticeable cooling has begun, mixed salts or the so-called cryo-sel-mixte ($MgSO_4 . 7H_2O + NaCl$) is precipitated. This occurs only in brines having more than 13% $MgCl_2$ and at temperatures below 5°C (Feigel'son 1936). During cooling accompanied by dilution of the brine, in November to January, mirabilite is precipitated, and this causes a general reduction in salinity of the brine. In spring the winter minerals are dissolved, and a new cycle of summer sedimentation begins.

These annual hydrochemical cycles are characteristic not only of continental lakes but also of modern saline basins of marine origin. In large basins connected to the sea by channels, the hydrochemical cycles are characterized by their great size. On the maritime Rann of Cutch (in India) 18 billion cubic meters of sea water, on average, is trapped by the summer monsoons, which corresponds to 350 million tons of salt. In winter, during the dry monsoons, this water is almost entirely evaporated; beds of salt from 10 cm to 1 m thick form in depressions, and on the slightly elevated zones efflorescent crusts of salt, up to 2–3 cm thick, form. During the subsequent period of flooding this is again dissolved. Even greater masses of salt, first precipitated and then again dissolved, were observed in the Gulf of Kara-bogaz before its recent constriction. From October to January, when the temperature of the brine fell below 5°C, mirabilite was precipitated. This was followed by a reduction in the $MgSO_2/MgCl_2$ coefficient and by a sharp fall in specific gravity. V. P. Il'inskii, G. S. Klebanov, and Ya. G. Blyumberg (1936) calculated that the mass of annually precipitated mirabilite was of the order of 10–11 billion tons on average, but this fluctuated strongly from year to year. In the winter of 1929-30, 16·6 billion tons was precipitated, but in 1931-32 only 5·6 billion tons. Precipitation depended both on the winter temperature and on the state of the gulf at the time. Because of the vast amount of mirabilite that formed, and even without it, the thick brine of the Gulf of Kara-bogaz became a pasty mass and large quantities of this salt were thrown by waves on to the shore. A thick bed of mirabilite also formed temporarily on the floor of the gulf. In spring and summer most of the salt again dissolved and only a negligible fraction remained to contribute to the formation of a mirabilite bed in the central, deeper part of the gulf.

It is clear that the sharpness of development of the hydrochemical cycles depends not only on the mineralization of the water but also on the climatic conditions. The extraordinary intensity of the annual cycles in the Soviet part of the arid zone largely depends on the broad temperature range during the course of the year. It is well known that in our latitudes (Western Siberia, Central Asia, the Caspian region) brines are heated by the sun to 30–35°C and even higher during the summer months, and in winter the temperature falls to 15–25° below zero, and even lower. Lakes in these regions that have a low salt concentration are covered by ice in the winter, but those with high mineralization do not freeze even at −25°C, and this naturally fosters winter precipitation of salts. With such large fluctuations in temperature the amount of periodic solid phases extracted annually and then redissolved must be very great. In the more southerly parts of the arid zone, where the temperature range is less and the average annual temperature is higher, the yearly hydrochemical cycles are naturally different. "The cold-loving solid phases—natron and mirabilite—become progressively less abundant and finally cease to be precipitated. The range of mineral development is restricted to a series of more distinctly warmth-loving minerals. *Saline sedimentation shows a steady shift away from polythermal processes of salt formation towards isothermal precipitation*" (Makarov 1935), although strictly isothermal precipitation of salt does not take place in nature.

Since the formation of minerals at medium and high degrees of salinity takes place in both summer and winter, the question naturally arises: which is the principal control in the development of basins in the arid zone and the formation of their sediments—summer evaporation or winter cooling?

Increasing salinity of the water, so characteristic of basins in arid zones, requires removal of the water. This is achieved only during summer evaporation, because during winter cooling practically no removal of solvent occurs.* *It follows therefore that for the general development of basins in arid zones it is the summer part of the annual cycles that is decisive.* As a rule, it determines the principal features of composition of the accumulating salts, although at times in the development of the basin or under specific conditions the winter part of the cycle does have an effect, especially during the precipitation of Na and Mg sulfates. Only because of the dominant effect of the summer part of the annual cycle on the sedimentation in the increasingly saline basin is it possible to define the basic features of halogenesis by means of the graphs for isothermal behavior in salt solutions at 25°C, as determined experimentally.

The sedimentary products of annual hydrochemical cycles, therefore, differ for different basins. In some basins these cycles do not lead to stable accumulation of salt beds because of the solution in spring and autumn of the summer and winter deposits. In other lakes a surplus of undissolved bottom phases remains, and a salt bed is formed. The question naturally arises: under what conditions does the sedimentation of solid phases in the annual cycle cease to be ineffective and start to produce a salt bed? These conditions are twofold.

In weakly mineralized lakes the dissolved salts are in a solution that is to some degree unsaturated. However, because of constant discharge into the lake, each salt ultimately reaches a point at which the water is first saturated, and *then oversaturated with that constituent on an average for the annual cycle.*

---

* This does not take account of the temporary formation of ice in lakes with relatively low mineralization.

The separation of each salt in a periodic phase—summer and winter—normally begins long before this point of average annual oversaturation is reached. Until this state is reached the salt precipitated in a single season (summer or winter) is inevitably dissolved during subsequent freshening, and the yearly cycle is ineffective. Only after the average annual oversaturation of a given salt is reached does the seasonal salt precipitation exceed the periodic solution; the foundation is then laid for the formation of a salt bed.

The second condition is the mineralization of the intercrystalline brine. M. G. Valyashko (1952a) emphasized that the degree of saturation of inter-crystalline brine with halite in various lakes varies considerably and the NaCl

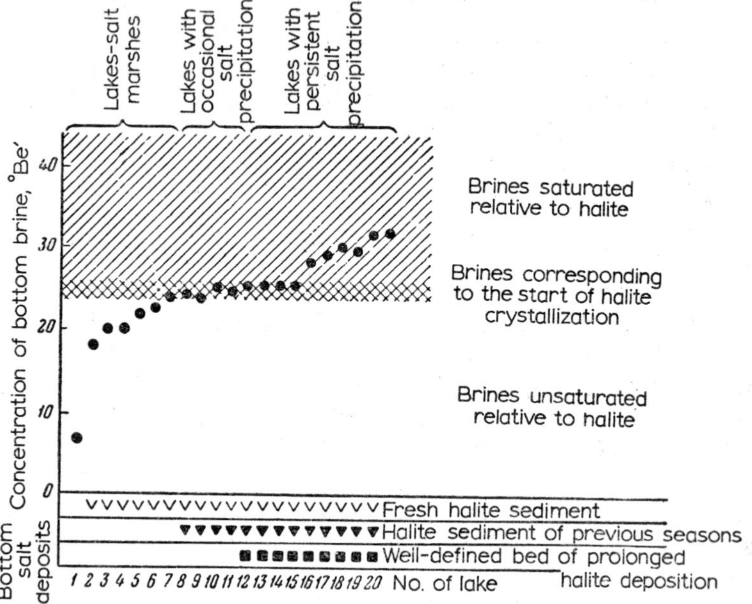

Fig. 130. Relation of halite accumulation to the degree of NaCl saturation of bottom brine (from M. G. Valyashko).

content is completely dependent on the formation of a stable and persistent halite bed, perhaps of many years accumulation. Figure 130 summarizes the data of 20 lakes that precipitate NaCl in the yearly cycle. From this diagram it may be seen that if *a halite bed forms, it persists only when the intercrystalline brine reaches saturation with NaCl or even becomes somewhat oversaturated.* If the newly precipitated halite, which may reach a thickness of 10–15 cm, is in contact in its lower part with intercrystalline brine that is unsaturated with NaCl, it will be dissolved by this brine, even if some of the newly-formed salt may remain during autumn and spring freshening. By contrast, if the intercrystalline brine is already saturated with NaCl, no solution will take place; the salt will be preserved, and covered by mud introduced by autumn and spring rains, and a salt bed will begin to form. This is the so-called "old" precipitate, contrasting with the new precipitate that appears and disappears seasonally. Furthermore, not only halite, but *any salt bed begins to form only after both the lower layer of water and the brine in the sediment become saturated with the given salt phase during the annual cycle.*

Since the mineralization of surface brine begins to satisfy this requirement before mineralization of the intercrystalline brine, the formation of each salt bed passes through some intermediate metastable period, however brief, when the bed appears (in very dry years) and disappears (in moister years), until in the end it becomes firmly established.

The annual hydrochemical cycles are reflected in saline deposits by characteristic cyclic structure. Each single annual rhythm is composed of two layers. The first corresponds to spring dilution and the succeeding summer evaporation, and it is represented by clay and silt deposition with some chemical precipitates, of the constituents with lowest solubilities. The second layer forms at the end of the summer or in autumn (sometimes in winter) and is represented by the deposition of the more readily soluble components, precipitated from the more highly concentrated brine by further evaporation in autumn or by cooling.

The composition of the chemical components in the upper and lower layers, their thickness, and their structure change during the different stages of saline development in a basin, but the basin feature of an annual unit is maintained with remarkable constancy. In Lake Saki, for example, the microrhythms consist of dark and light microzones (Yu. V. Pervol'f 1953). The dark zones are composed chiefly of clays (75%) with subordinate carbonate (12%) and gypsum (11%). Considerable organic material and hydrotroilite are present, giving the beds their dark color. The light zones are poor in clay material (less than 50%), organic material, calcite (7%), and hydrotroilite, but are rich in gypsum (40% or more). The thickness of the microzones ranges from 0·1 to 5·5 mm, averaging 1·4 mm, which means that the microrhythms are 2·8 mm thick on the average. The black laminae, stained by hydrotroilite, correspond to summer deposition, the light ones to autumn (end of August and September). In Kerleutskoe Lake clay-carbonate laminae with some admixture of organic material (0·15–0·25 mm thick) alternate with laminae consisting of a "great number of very small salt crystals," apparently gypsum (0·25–0·35 mm thick). The combined thickness of the microrhythm is 0·4–0·6 mm. The clayey laminae represent autumn-spring deposition; the gypsum is the result of summer precipitation. Similar microrhythms characterize Staroe, Krasnoe, Kiyatskoe, and other lakes.

The annual sedimentary beds at the chloride stage of modern basins consist of a clay-carbonate layer at the base, with some admixture of gypsum, and of a layer of freshly deposited halite above.

The thickness of these seasonal deposits varies considerably. The clay-carbonate layers are generally measured in fractions of a centimeter. The halite layers range from 5 to 11 cm thick. Thus, *the annual sedimentation actually observed in highly saline basins corresponds fully to the calculations previously made* (p. 193), *proving to be much more rapid than that at the carbonate or sulfate stages.*

The distinctive structure in the solid salts of the newly precipitated material is highly significant for understanding ancient saline deposits.

A freshly formed halite deposit is characterized by crystals in the form of elongated "teeth", more or less regular trihedral pyramids, perpendicular or slightly inclined to the bed (Fig. 131). In the longitudinal section the pyramids have a pinnate structure with numerous thin transparent halite plates separated by dull plates with inclusions of brine, air, gypsum, and other impurities. The interstices between groups of "teeth" are partly filled by small incompletely developed teeth and partly by fragments of these teeth and admixtures of

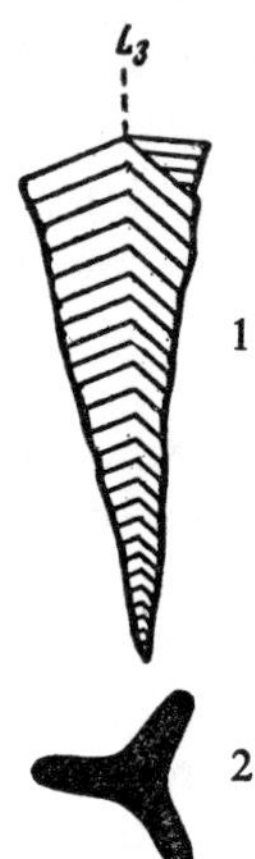

FIG. 131. "Tooth" of halite and its structure (from M. G. Valyashko). 1. Schematic illustration of a tooth of newly precipitated halite; 2. cross-sectional shape of tooth.

other minerals; alternatively they may be empty. Because of this structure in the freshly precipitated halite, the bed shows a more or less well-defined columnar structure. The total thickness of the annual layer reaches 5–8 cm. As time goes on, the newly deposited material is compacted. It commonly recrystallizes into cubic crystals or into a dense mass of halite, losing its pinnate structure. Such crystallization does not always occur, however, and the primary pinnate structure is occasionally preserved in the fossil state (M. G. Valyashko 1952).

The new deposit of halite begins by the formation of minute crystals *on the surface of the brine.* These represent morphologically the inverted plane of a pyramid, apex downward, a distinctive salt hopper, measuring from a few millimeters to 3 cm along the upper edge (Fig. 132). The temperature of the brine and the rate of crystallization during the halite precipitation are the most important factors affecting the size of the crystals. At low temperatures and with slow crystallization, the halite crystals grow big, whereas at high temperatures and rapid crystallization they remain small. The size and form of the halite crystals are also affected by the chemical composition of the solution, which contains varying proportions of other salts besides sodium chloride.

"On falling to the bottom, the hopper of halite, if it remains whole, lies either on a side or on an edge. When it lies on a side, two of the triad cube axes will be directed upwards at some angle. When it lies on an edge, one of the triad axes will be almost vertical. If the hopper breaks during falling, as must frequently happen, the fragments will fall base downward, since this end is heavier by virtue of its gradual growth, and one triad axis will be directed upward.

"As the process continues, the supply of nutrient material flows from above through convection currents in the brine that is being concentrated on the surface. Thus, the edges of the hoppers, or the fragments that are project-

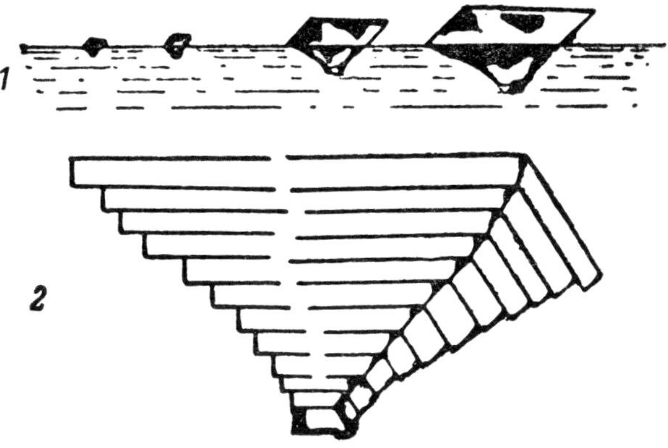

FIG. 132. "Hopper" of halite (from M. G. Valyashko). 1. Scheme of hopper formation on surface of the brine; 2. sutures on a hopper of halite.

ing upward, will receive this material first and will grow most rapidly. Only the hoppers and fragments that do not come to rest with their apices upward lie in the freshly precipitated halite and continue their growth laterally. The surfaces, along which the newly precipitated halite bed breaks, and the structure of the newly precipitated halite bed itself, show that of the four faces of the hopper only one, or rarely two, will show accelerated development, and this (or these) will be turned upward. Normally, the remaining faces do not grow.

"Further deposition of material along such an upward-turned axis takes place layer by layer on the sides of a cube that appears to be suspended by the apex on this axis. The growth is most rapid along the edges, along which the formation of skeletal shapes may be observed (Figs. 131 and 133*A*, B, and *C*). Unequal supply of material, caused by changes in evaporation rate and in temperature from day to night, creates a zonal structure in the crystal, marked by variable quantities of inclusions. More inclusions are incorporated during more rapid growth than during slow growth. Apparently the most rapid growth occurs during evening and night hours of the summer, when the temperature falls and the rapidly cooled upper layer of concentrated brine descends to the floor.

"When a hopper falls on a side and two of its edges are turned upward, equal development of both edges is possible. One may observe the emergence at the surface, not of an apex of the cube, but of an edge and two apices, i.e. the emergence upward of a diad axis, but disposed at some angle other than 90°. This situation is not common. For example, an examination of the surface of a newly precipitated halite bed, over an area of about 5 cm², has shown that 21 crystals had a triad axis turned upward, but only one had an edge emerging. Freshly precipitated halite thus represents a previous space that is gradually filling with growing hoppers of halite, the growth occurring chiefly along a triad axis pointing upward" (Valyashko 1952a, p. 45).

These structures of salt beds and of halite crystals are of considerable value in comparative lithologic analysis of ancient saline formations. *The microrhythms in these old formations, corresponding exactly in composition and thickness to the annual microrhythms of modern salts, demonstrate conclusively the presence of annual hydrochemical cycles in ancient saline basins. The pinnate structure in salt crystals and the tooth-like forms, even if these are but relict traces, prove that the formation of fresh halite followed the same laws then as in present-day saline basins.* These textural criteria may now be used for deciphering the formation of fossil salt beds.

3. Metamorphization of Brines and its Influence on Halogenic Sedimentation

A specific feature of saline mineralization, as it reaches its culmination in present-day basins, is that *it takes place in basins in which the water is undergoing continuous metamorphization.* This circumstance radically distinguishes natural salt deposition from laboratory experimentation.

The essential aspects of brine metamorphization will first be examined, followed by its relation to mineral growth.

Brine in basins that are becoming progressively more saline may be of various hydrochemical types, becoming more and more clearly distinguishable as the salinity increases. Similarly, the ground water feeding the basins varies markedly in composition. The water in rivers from humid zones is much more uniform, belonging chiefly to the sulfate type. In addition, because of the low mineralization, water in rivers and in subsurface streams generally abounds in some components, especially $Ca(HCO_3)_2$, that are very weakly represented in brine of moderate and high mineralization. This gives rise to highly complicated relationships between the brine and the solutions entering the basin.

When the hydrochemical type of water supplying the basin is the same as that of the basin water itself, and when the salinities are similar—which can be true only in the initial stages of saline development—the solutions entering the basin simply mix with the basin water and supplement its salt content. The process is the same as that in humid zones. *When the types of water are dif-*

ferent and the salinity of the inflowing water differs markedly from that in the basin, chemical reactions take place, including double exchange between the salts.

Let us imagine a soda lake with salinity of the order of 10%, with a high alkali content, and with very high pH (11·0–11·6). In such a lake $CaCO_3$ is practically insoluble and $MgCO_3$ dissolves only very slightly. Let us assume also that the lake is receiving sulfate water of two varieties: (1) magnesium-carbonate water with $CaCO_3$ (little), $MgCO_3$ (much), and $NaCl$, and (2) calcium-carbonate water with $CaCO_3$, $CaSO_4$, $MgSO_4$, and $NaCl$. When the magnesium-carbonate water mixes with the lake water the entire mass of $CaCO_3$, being insoluble under these conditions, is precipitated, and is distributed over the floor of the basin: $MgCO_3$ behaves similarly. But $NaCl$, not reaching saturation, will continue to accumulate in the lake water, increasing the mineralization.

The reaction with calcium-carbonate water is substantially different. The $CaCO_3$ will be quickly precipitated, but between the $CaSO_4$ and $MgSO_4$ entering the basin and the $NaCO_3$ in the lake water a reaction of the following type develops:

$$CaSO_4 + Na_2CO_3 \rightarrow \underset{\downarrow}{CaCO_3} + Na_2SO_4$$

$$MgSO_4 + Na_2CO_3 \rightarrow \underset{\downarrow}{MgCO_3} + Na_2SO_4$$

Since both carbonate products of the exchange reactions—$CaCO_3$ and $MgCO_3$—are insoluble in the waters of soda lakes, the reaction moves from left to right. As a result, a much greater quantity of carbonate falls to the floor of the lake than might be precipitated from the brine. The water of the lake simultaneously changes, or, as we usually say, is metamorphized. In such a lake the amount of soda decreases and that of sulfates increases. *When calcium-carbonate water discharged into a soda lake, the lake is slowly converted to a sulfate lake, particularly a sodium-sulfate lake.*

Similar relationships are found in sulfate basins, beginning with moderate mineralization, when $CaSO_4$ begins to saturate and oversaturate the water.

When any water containing $Ca(HCO_3)_2$ enters such lakes, the following exchange reactions are possible:

1. $Ca(HCO_3)_2 + MgSO_4 \longrightarrow \underset{\downarrow}{CaSO_4} + \underset{\downarrow}{Mg(HCO_3)_2}$

$$CaSO_4 \cdot 2H_2O \quad xMg(OH)_2 \cdot yMgCO_3$$

2. $2Ca(HCO_3)_2 + MgSO_4 \rightleftarrows \underset{\downarrow}{CaMg(CO_3)_2} + \underset{\downarrow}{CaSO_4 \cdot 2H_2O} + 2CO_2$

Haidinger's reaction

3. $2Ca(HCO_3)_2 + MgCl_2 \rightleftarrows CaMg(CO_3)_2 + CaCl_2 + H_2O + 2CO_2$

Marignac's reaction

4. $Ca(HCO_3)_2 + MgSO_4 + Na_2SO_4 \rightleftarrows \underset{\downarrow}{CaNa_2(SO_4)_2} + \underset{\downarrow}{Mg(HCO_3)_2} + H_2O + CO_2$

$$xMg(OH)_2 \cdot yMgCO_3$$

A. A. Verigo (1880), N. S. Kurnakov and S. F. Zhemchuzhnyi (1917), and A. E. Rykovskov (1932) attempted to test experimentally this kind of metamorphizing effect of $Ca(HCO_3)_2$. The results were generally positive, although the mineral character of the magnesium salts that were being precipitated could not always be determined. A more thorough demonstration of this process has been made by M. G. Valyashko and G. K. Pel'sh (1952). In an extensive series of experiments, a small amount of calcium bicarbonate was added to 200 g of solution saturated with epsomite, thenardite, or mirabilite. The solution was then evaporated to its initial volume during constant stirring. More bicarbonate solution was then added; the resulting solution was evaporated again, and the process was repeated until a sufficient amount of constant solid phases had accumulated. The experiments were carried out in a thermostatically controlled chamber at 25°C ($\pm0\cdot2°$) and were continued for 250 days. Systematic microscopic examinations were made each day of the solid phases precipitating out. At the end of the experiment the sediment was separated and analyzed chemically and microscopically, using immersion oils. The alkalinity of the liquid was determined by titration with $0\cdot1N$ HCl.

These experiments showed that all four reactions described above occur, but the *first* is the most important for amount of solid phases produced, giving rise to gypsum and basic magnesium carbonate. These two products make up 99% of the newly formed sediment. As for reaction 2 (Haidinger's) and reaction 3 (Marignac's), they apparently also occur, but are weak. Dolomite consequently makes up less than 1% of the solid phases, and indeed may be entirely absent.

During metamorphization of solutions containing high concentrations of Na_2SO_4, one constant phase may be detected, crystallizing as fine needles that precipitate in the sediment in great quantity. The composition is near that of glauberite, but it differs by having two molecules of water.

The hydrochemical result of this double-exchange reaction is evident. The sediments receive supplementary enrichment in calcium carbonate and sulfate (and sodium sulfate in part). *The water gradually loses sulfates, is desulfatized; i.e. it changes its hydrochemical type; it is metamorphized.*

This does not exhaust the range of exchange reactions, but no new examples are cited here, because they all repeat the same formula, which may be expressed as

$$P_1 + P_2 \rightleftarrows P_3 + \text{solid phase,}$$

where P_1 is the brine, P_2 is the introduced solution, P_3 is the resulting solution, and the solid phase is the sediment resulting from the reaction.

According to the exchange processes, these reactions are responsible for the appearance of minerals in the sediments that commonly would not form without such exchanges. They also bring about some type of metamorphization of the brine, giving it new features and ultimately changing the basin from one hydrochemical type to another.

Why do these reactions fail to take place in basins of low mineralization, and appear only as salinity increases? According to the modern theory of solutions such exchange reactions are effective only when one or both of the resulting products (Ca and Mg carbonates, Ca and Na sulfates) precipitate. With the relative solubility of some of these, especially of the sulfates, this is possible only when the water of the basin is noticeably, or even highly, saline. *The possibility of double exchange actually taking place is thus attained only as the basin becomes progressively more saline.* These reactions are therefore a physicochemical characteristic of moderately and highly mineralized basins.

The hydrochemical processes here described take place both with surface supply of dissolved salts by streams to the lake and by ground-water flow, and they may even take place in the ground-water current when this water is saline and when it interacts with water of low mineralization. To put it briefly, such processes occur when saline solutions of different compositions and degrees of mineralization come together. But the sedimentational effect of exchange reactions has proved to vary for the different conditions under which the reactions occur. When they take place at the meeting of river water and brine, in the agitated surface water of the basin, the reaction products, especially fine-grained $CaCO_3$ and, in part, $CaSO_4$, are transported by the moving water over the entire area of the basin. Being mixed with the sediment, more or less disseminated through it, these products have not been detected. When the double exchange takes place, however, during slow influx of ground water, filtering through pores or channels of the sediment, as may be directly observed in many salt lakes, the reaction products, especially $CaCO_3$ and $MgCO_3$, remain in place, and under certain circumstances they enrich the nearshore sediments with carbonates, forming areas of inshore marls. In other places distinctive calcite and magnesite concretions are formed, as noted by D. A. Vital' (1950) in a number of lakes of the Kulunda Steppe. Rivers furnish saline basins not only with dissolved substances, but also with greater or lesser amounts of suspended material, including considerable quantities of colloidal and subcolloidal clay minerals. At the beginning of saline development, when the salinity of the arid basin is still at the level of mineralization of basins in humid regions, the introduction of finely dispersed particles has no appreciable influence on sedimentation in the basin or on the composition of the water. *As the salinity of the basin increases, cation exchange between the finely dispersed clay minerals and the brine becomes increasingly strong, and this has an increasingly marked effect on the nature of the sediments and on the composition of the brine. Another type of metamorphization is developed.*

Let us recall the essential features of cation exchange and the laws that govern it. When colloidal micelles are introduced into a solution of electrolytes an interaction takes place between the constituents: some of the cations from the sorption envelope of micelles go into the solution, and other cations from the solution take their places.

This process of cation substitution on the surfaces of micelles is called base exchange. Anions do not participate in this process. The cation exchange may be schematically represented thus:

$$P_1 + M_1.\text{(c-e c)}^- \rightleftarrows P_2 + M_2.\text{(c-e c)}^-,\!^*$$
$$\text{solid} \qquad\qquad \text{solid}$$
$$\text{phase} \qquad\qquad \text{phase}$$

where P_1 is the natural solution (natural water), P_2 is metamorphized natural water, (c-e c)$^-$ is adsorbed complex, $M_1.$(c-e c)$^-$ is a colloidal complex with the M_1 cation (which it yields in the solution), $M_2.$(c-e c)$^-$ is a colloidal complex with the M_2 cation, (which it obtains from the solution in exchange for the discarded M_1).

Two factors determine the course of the exchange process: the energy of ion adsorption and the concentration of the ions in solution.

By "absorption energy" K. K. Gedroits meant the intensity with which the different cations are adsorbed by colloidal micelles. The standard of absorption energy is the number of cations (in mg-equiv) bound by 100 g of absorbent from a normal solution of the cation. Gedroits explained that *absorption energy is minimal for monovalent cations, much higher for bivalent cations, and still higher for trivalent forms. The absorption energy increases for any particular valence as the atomic number increases.* The cations most widespread in nature are therefore arranged according to absorption energy in the following series:

$$Li^+ < NH_4^+. < Na^+ < K^+ < Rb^+$$
$$Mg^{2+} < Ca^{2+} < Cd^{2+} < Co^{2+}$$
$$Al^{3+} < Fe^{3+.}$$

As for the effect of concentration, *as the content of the absorptive cation in the displacing solution increases the number of this type of cation on the micelles also increases, but in lesser degree than the rise in cation content in the solution.* Large concentrations of monovalent ions in a solution overcome the effect of high absorption energy of bivalent ions and displace them from the surface of the micelles. Both laws of ion displacement lead to the result that *the composition of the cations adsorbed on the micelles and the quantitative relations among these cations are intimately connected with the composition and degree of mineralization of the surrounding medium, and they alter with change in these latter factors.*

This basic pattern is clear in Table 26, in which data is given on the adsorption of bases of several soils and modern marine deposits. From Table 26 it is also clear that H plays an important role among absorbed cations in acid soils with low pH. It is followed by Ca^{2+} and Mg^{2+}. H disappears in slightly alkaline soils, and Ca^{2+} and Mg^{2+}, especially the former, are sharply concentrated. In all these examples, the soil water, leaching the soil, is of calcium-bicarbonate type and is very weakly mineralized. On arid steppes, where the mineralization of soil waters in the solonetz soils (dark and strongly alkaline) increases and Na is concentrated, this cation is also found among the adsorbed

---

* c-e c. stands for 'cation-exchange complex'.

bases. In marine muds, deposits from even more highly mineralized water, not only is Na abundant, but so also is K; the content of K + Na increases in the adsorbed complex to 25–27% of the total absorbent capacity, and the role of Ca and Mg decreases accordingly. The connection between adsorbed bases and the hydrochemical peculiarities of the environment is clear.

TABLE 26

Composition of adsorbed Bases in different Types of Soils, Marine Muds, and Clays

Sample	Composition of adsorbed cations, mg-equiv., %					Author
	Ca	Mg	Na	K	H	
Chernozem, Bugul'ma region .	87·2	12·8	—	—	—	K. K. Gedroits
Peat-podsol soil A_2 . .	13·9	7·8	—	—	78·5	K. K. Gedroits
The same, horizon B . .	54·6	41·5	—	—	3·9	K. K. Gedroits
Chelyabinsk podsol, horizon $A+A_1$	11·4	23·6	—	—	65·0	K. K. Gedroits
Solonets, Chelyabinsk region, horizon 0·5 cm . . .	52·9	39·7	7·4	—	—	K. K. Gedroits
The same, horizon 24–27 cm .	20·0	57·5	22·5	—	—	K. K. Gedroits
Black Sea mud, Sta. 107 . .	40·0	25·1	27·9	7·0	—	A.D.Arkhangel'-skii and É. S. Zalmanzon
The same, Sta. 6 . . .	53·0	31·8	9·2	6·0	—	The same
The same, Sta. 5 . . .	68·7	18·8	9·0	3·5	—	The same
Marine mud from the Netherlands coast . . .	24·0	49·0	19·0	8·0	—	Hissink

Since continental waters (streams and ground water) are mostly of the calcium-carbonate type, the adsorbed complex of colloidal particles removed by them are characterized by abundance of calcium, poverty of magnesium, and complete absence of sodium and potassium. On flowing into sea water, and even more into water of saline basins with an abundance of Na + K + Mg, the adsorbed complex must reorganize; Ca^{2+} will be displaced by $Na^+ + K^+$, as a result of which an excess of Ca^{2+} in the form of $CaCl_2$ will develop in the solution

$$Ca^{2+}(c\text{-}e\ c)^- + 2NaCl \rightleftarrows 2Na^+(c\text{-}e\ c)^- + CaCl_2.$$

If the brine is sufficiently salty and gypsum reaches saturation, a double exchange between the salts will take place:

$$2H_2O + CaCl_2 + MgSO_4 \rightleftarrows \underline{CaSO_4 . 2H_2O} + MgCl_2.$$
$$\downarrow$$

As a result, the brine gradually loses magnesium sulfate, and after a long period of time may become purely chloride, with a high content of $MgCl_2$. *When* magnesium sulfate has been exhausted, precipitation of gypsum ceases, and $CaCl_2$ accumulates in the solution. Thus, the sulfate basin not only loses its

R

sulfates, but acquires the distinct features of a calcium-chloride basin. This mechanism of brine alteration in basins of increasing salinity is the essence of the so-called "colloidal chemical theory of salt lakes", as advanced by S. A. Shchukarev and T. A. Tolmacheva (1930).

Many facts, and particularly the experiments of A. N. Buneev (1947), emphasize the significance of these processes. Air-dried fresh-water Quaternary clay, from the Lenin Mountains, was placed in measured amounts in artificially prepared sea water and allowed to settle. These clays, containing almost no water-soluble salts and possessing a total capacity of about 30 mg-

TABLE 27

Changes in Composition of Artificial Sea Water during Precipitation of Fresh-water Clays

Solution	Sea water	1st sample	2nd sample	3rd sample	4th sample
Amount of precipitated clay, g	0	20·0	34·8	51·9	38 ml of preceding solution + 30 g of clay
Cations, g/liter					
K^+	0·418	0·307	0·151	Not detected	
Na^+	10·795	9·970	9·385	7·531	6·841
Ca^+	0·374	0·972	1·723	3·117	3·667
Mg^{2+}	1·255	1·346	1·243	1·354	1·311
Anions, g/liter					
Cl^-	19·432	19·432	19·432	19·432	19·432
SO_4^{2-}	2·574	2·453	2·457	2·156	1·816
HCO_3^-	0·065	0·092	0·060	0·091	0·092
Total cations, mg-equiv./liter	602·6	606·6	600·1	594·4	588·2
Na/Cl	0·85	0·79	0·74	0·60	0·54

equiv. per 100 g with a predominance of adsorbed calcium, brought about profound changes in the composition of the sea water (Table 27).

With the addition of each new increment of clay the Ca^{2+} content in the solution increased sharply, the Na^+ content declined, and the Mg^{2+} concentration remained practically unchanged. Beginning with the third sample, gypsum reached saturation in the solution, and crystallites appeared on the walls of the vessel without further concentration. Potassium quickly disappeared from the solution. In a separate experiment it was established that the Br/Cl ratio did not change during precipitation of clay in sea water. The Na/Cl ratio fell from 0·85 to 0·54.

It must be borne in mind, however, that these changes in cation composition of the solution begin when relatively large amounts of clay are being added.

The same experiment was repeated with sea water having a density of 20° Bé and saturated with gypsum (Table 28).

Thus a well-defined desulfatization of the brine took place, and the brine changed to a calcium-magnesium-chloride type. This corresponds to the metamorphization theory for maritime lakes in which exchange-sorption processes take place, with the difference, however, that we observe partial desorption of Mg, not adsorption. The figures obtained give an indication of how profound the changes may be in the composition of the brine by virtue of precipitating clays, if these are introduced in very large quantities into the water.

In addition to A. N. Buneev's work experiments by M. I. Ravich and

TABLE 28

Precipitation of Clays in Concentrated Sea Water (20° Bé)

Component	Sea brine	
	prior to clay precipitation, g/liter	after clay precipitation, g/liter
KBr	0·762	0·756
KCl	4·487	2·084
NaCl	185·300	157·800
$MgCl_2$	23·860	41·669
$MgSO_4$	15·380	—
$CaSO_4$	3·190	5·059
$CaCl_2$	—	7·096
Totals	232·979	214·464
$\dfrac{MgSO_4}{MgCl_2}$	0·64	0

others have supported the view that the exchange reactions pointed out by S. A. Shukarev do actually occur.

Base exchange between the brine and introduced colloidal clay does not take place precisely as described above in all hydrochemical types of basins. It is not always so effective; but in most cases exchange does take place, and its effect increases with rise in mineralization, a fact that follows directly from the laws that govern this exchange.

Thus, *in proportion to the degree a lacustrine or marine basin in arid climate becomes more saline, double-exchange reactions between salts in the brine and introduced solutions develop with increasing intensity, as does cation exchange between the brine and finely dispersed clay.* Both types of reaction lead to supplementary precipitation of several minerals in solid masses and to a more or less well-defined *metamorphization of the brine.* This means that *the evolution of any basin in the arid zone is not merely the history of growing salinity but at the same time a history of metamorphization of the basin, at times weakly, at times very strongly manifested.*

The concept of direct and reverse metamorphization of an aqueous basin was introduced by M. G. Valyashko. Direct metamorphization involves changes in the brine that lead to replacement of less soluble components by more soluble forms, in particular the replacement of CO_3^{2-} by the sulfate ion SO_4^{2-}, and then by Cl^-, i.e. the conversion of soda waters to sulfate, and sulfate waters to chloride. The reverse process is called *reverse metamorphization of a solution.*

Let us now consider the question: what details in the mineral composition of modern salt deposits actually reflect metamorphization processes in the bottom brine? The essential mechanism of forming solid phases during increase in basin salinity indicates that salt minerals may be divided into two genetic groups: *evaporite* minerals, forming simply by evaporation of a brine of any given composition, and *metamorphogenic* minerals, owing their development to metamorphization processes.

An understanding of the cycle of metamorphogenic minerals in moderate and advanced stages of salinity has only become possible by setting up experiments under conditions resembling the natural environment. M. G. Valyashko and his co-workers (1952, 1953) obtained dolomite, basic salts of $MgCO_3$, gaylussite, gypsum, and glauberite as metamorphogenic minerals. Their distribution according to hydrochemical type of saline basin is shown on a composite diagram which combines individual diagrams of the separate types of water (Fig. 134). The solid lines indicate the boundaries of evaporite minerals; the broken lines indicate fields of metamorphogenic mineral development.

Only the single metamorphogenic mineral gaylussite forms in the soda field (Fig. 134), and only calcite in chloride waters, but the sulfate field is characterized by several varieties of metamorphogenic solid phases: glauberite, gypsum, dolomite, and the basic salts of $MgCO_3$. Within the sulfate field itself, the greatest quantity of metamorphogenic minerals lies in the $MgSO_4$ band: gypsum, dolomite, and basic salts of $MgCO_3$. This means that at advanced stages of salinity, when solid salt phases are being precipitated, the principal metamorphizing reactions are interactions between $MgSO_4$ and $Ca(HCO_3)_2$:

1. $Ca(HCO_3)_2 + MgSO_4 \rightleftarrows CaSO_4 . 2H_2O + Mg(HCO_3)_2 x Mg(OH)_3 . y MgCO_3,$

2. $2Ca(HCO_3)_2 + MgSO_4 \rightleftarrows CaSO_4 . 2H_2O + CaCO_3 . MgCO_3 + 2CO_2$

(Haidinger's reaction).

Of these two reactions the first is dominant, the second scarcely detectable. Interaction products are therefore almost exclusively gypsum and basic $MgCO_3$ salts, whereas dolomite is present in amounts ranging from traces up to 1% of the total precipitating mass of solid phases. Although these two types of reaction are possible also during interaction with $MgCl_2$, the first does not take place and the second is scarcely detectable.

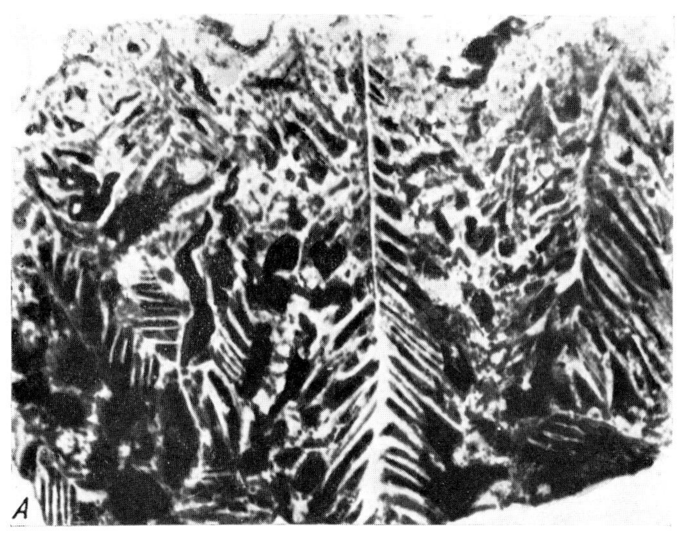

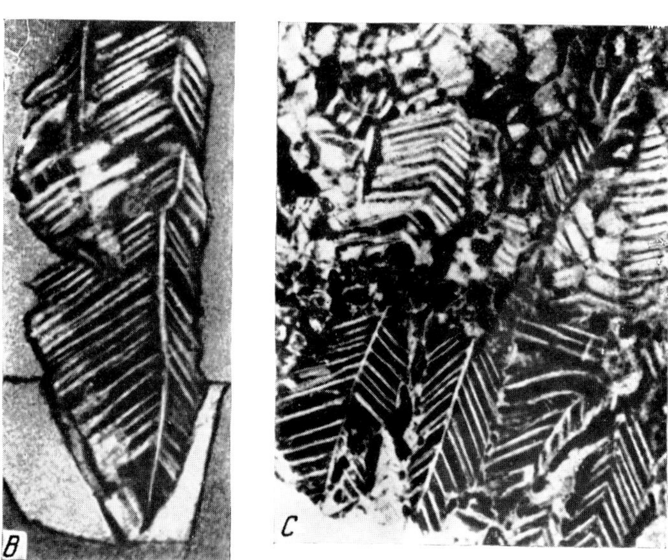

Fig. 133. Microscopic structure of freshly precipitated halite (from M. G. Valyashko). *A.* Vertical section of a bed of freshly precipitated halite (thin section 27–3b, from the Krym-Éli salt plant, salt extraction of 1946); *B.* growth inclusion, giving the pinnate form on a "tooth", reflecting the daily increment of salt (thin section of a "tooth"); *C.* vertical section of bed of freshly precipitated halite (thin section 27–5, from Krym-Éli salt plan, salt extraction of 1927).

From these facts it becomes clear that Haidinger's reaction, so popular in lithologic works, is *essentially magnesite-forming* and *not in any way dolomite-forming*, as has been erroneously maintained heretofore by V. P. Krotov (1925), who publicized this idea of N. S. Kurnakov (1917).

The observations of Valyashko relative to conditions under which the Haidinger reaction first begins to operate are very important. Valyashko set up experiments on the metamorphization of sulfate solutions ($MgSO_4$ and Na_2SO_4) not only at high mineralizations , in the presence of solid phases of

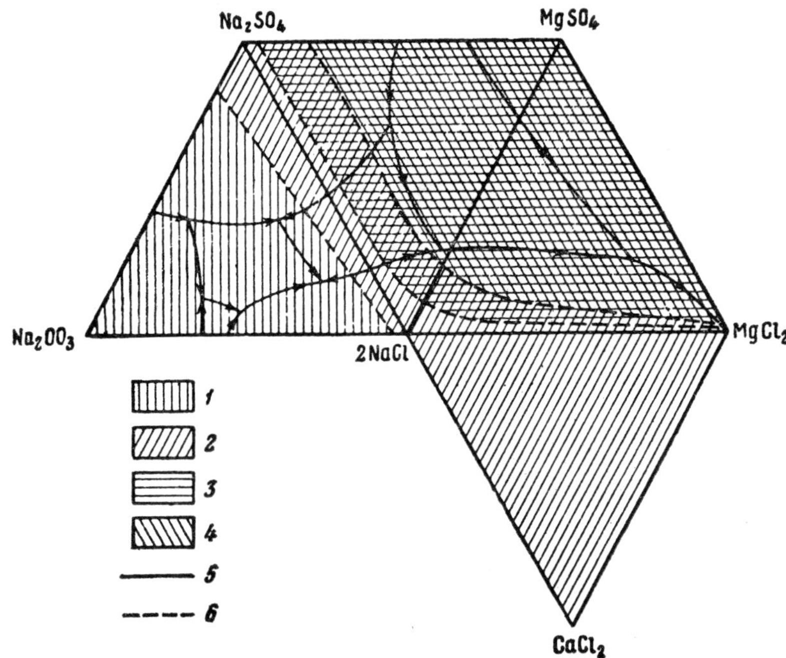

Fig. 134. Composite diagram of metamorphogenic minerals (from M. G. Valyashko). *Fields of crystallization*: 1. gaylussite; 2. calcite; 3. gypsum; 4. basic magnesium carbonates. *Boundaries of crystallization:* 5. periodic phases; 6. invariable phases.

these salts, but also at low mineralizations. He demonstrated (Fig. 135) that at mineralizations corresponding to unsaturated solutions in $CaSO_4$ metamorphization does not occur. $Ca(HCO_3)_2$ introduced into the vessel during evaporation of the solution simply precipitates as a low-solubility salt. *Metamorphization begins only when $CaSO_4 \cdot 2H_2O$, forming according to Haidinger's reaction, saturates the solution and begins to precipitate on the bottom.* This stage corresponds to the gypsum field, which is situated above the calcite field on the diagram. But $MgCO_3$, correlatively associated with gypsum in this process, is not precipitated at this time. It continues to accumulate in the solution, and $CaCO_3$ is preserved in the sediment. The Haidinger reaction clearly takes place weakly at this stage. Only later do the basin salts of $MgCO_3$

appear in the sediment. Both reaction products then contribute to the sediment, demonstrating that Haidinger's reaction is strongly operative.

Thus, there can be no doubt that *Haidinger's reaction begins to be effective only after the solution becomes oversaturated with gypsum.* For sea water, this happens only at salinities above 15%. To apply Haidinger's reaction in interpreting the origin of marine dolomites, as some lithologists still do, is incorrect.

For a proper comprehension of the formation of salt minerals, it is necessary to keep in mind that a particular mineral may be in the evaporite group or in the metamorphogenic group, depending on the actual physicochemical

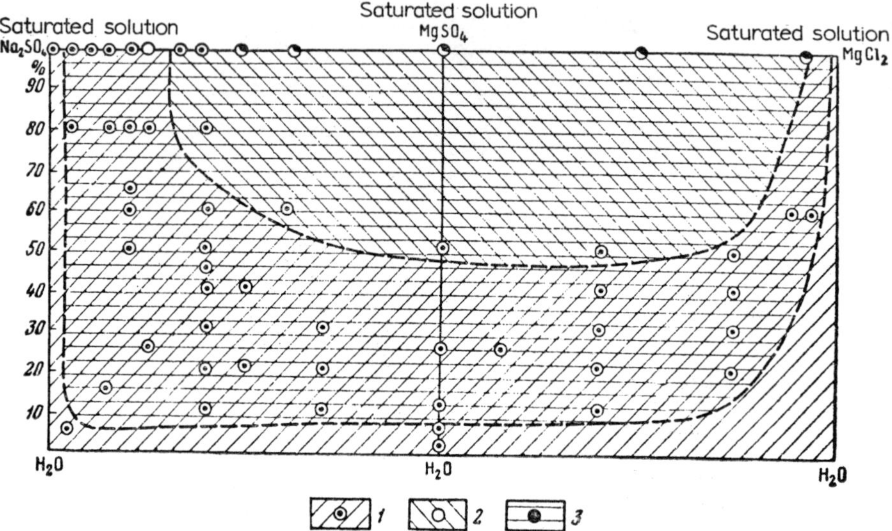

FIG. 135. Conditions for Haidinger's reaction (from M. G. Valyashko). 1. Calcite; 2. gypsum; 3. basic magnesium carbonates.

conditions. In soda lakes, when these are supplied with water of soda type, dolomite is typically an *evaporite* mineral and it precipitates from solution without preliminary metamorphization of the water. But when sulfate water is introduced into a soda lake, exchange reactions begin to take place between Na_2CO_3 and $MgSO_4$, and the precipitation of dolomite is the result chiefly of metamorphization of the water. Dolomite is then typically a *metamorphogenic* mineral. Dolomite in the sediments of group *IIa* lakes is always an evaporite mineral, but in the sediments of chloride lakes, it is always metamorphogenic. Gaylussite in soda lakes, hydromagnesite and magnesite in sulfate and chloride lakes, and glauberite in sulfate lakes are typically and invariably metamorphogenic. Gypsum, however, is typically an evaporite mineral, and only in sediments of moderate and advanced stages of salinity in basins of group *IIa* does it become typically metamorphogenic.

This means that the ability to discriminate solid phases of evaporative or metamorphogenic origin among the mineral complexes in deposits of any stage of development in basins of any hydrochemical type requires thorough analysis of the physicochemical processes in the entire developmental history of the basin. The very possibility of making reliable analysis of salt minerals is proof of the advanced stage of knowledge concerning present-day sedimentation in saline basins.

4. The Stages of Halogenesis in Modern Basins, and the Controlling Factors

We have already notes that the formation of authigenic minerals in basins of humid regions mostly took place in two stages: during sedimentogenesis the initial material accumulates, and during diagenesis this is converted to stable authigenic minerals. The same stage-like character is found in the formation of salt minerals, but also some additional features.

Indications of this stage-like development of salt minerals was obtained by N. S. Kurnakov (1938) during studies of so-called "solar crystallization" of salts, i.e. chemical precipitation under natural conditions, in lakes of the Saki group.

J. H. Van't Hoff showed that the solid phase following halite is blödite, which is then converted to epsomite. Kainite is formed later. Subsequently the epsomite is converted to hexahydrite, and the hexahydrite is then dehydrated to kieserite. In the next stage kainite is converted to carnallite, and at the eutonic point bischofite is added to the carnallite and kieserite.

This sequence of crystallization may be presented as follows:

1. Halite.
2. Halite + blödite.
3. Halite + (epsomite → hexahydrite → kieserite) + kainite.
4. Halite + kieserite + carnallite.
5. Halite + kieserite + carnallite + bischofite.

Figure 136 shows N. S. Kurnakov's comparison of the experimental data of Van't Hoff with the results of solar evaporation of brine under natural conditions. The solid lines show the boundaries of crystallization fields on the solar diagram, and the broken lines represent boundaries of the stability fields on Van't Hoff's diagram for $+25°C$.

Comparison of these two diagrams shows that the *solar diagram is considerably simpler.* It does not contain fields of blödite, leonite, or kainite, which are overlapped by the fields of magnesium sulfates (heptahydrate and hexahydrate), potassium chloride, and, in part, carnallite, the field of which is considerably enlarged. The field of kieserite ($MgSO_4 . H_2O$) and of the pentahydrate and tetrahydrate of magnesium sulfates is also absent, being replaced by the hexahydrite field.

During solar evaporation of a brine, epsomite is the solid phase following halite, not blödite, which ought to be precipitated according to the position of the representative point on the stability diagram. During further evaporation and concentration of the brine, epsomite changes to hexahydrite (dehydration of the brine!). Carnallite is then added to hexahydrite, and this mineral continues to crystallize till the eutonic point is reached, where it is joined by bischofite. As compared with the stability diagram, the "solar" eutonic point is marked by abundant magnesium sulfate.

In working out the "solar diagram" N. S. Kurnakov used data on basins with lower potassium content than that of sea water. No precipitation of KCl was therefore observed, and no relevant data are available. But if water with a potassium content normal for sea water is evaporated, following the precipitation of magnesium sulfate, sylvite separates out, up to the beginning

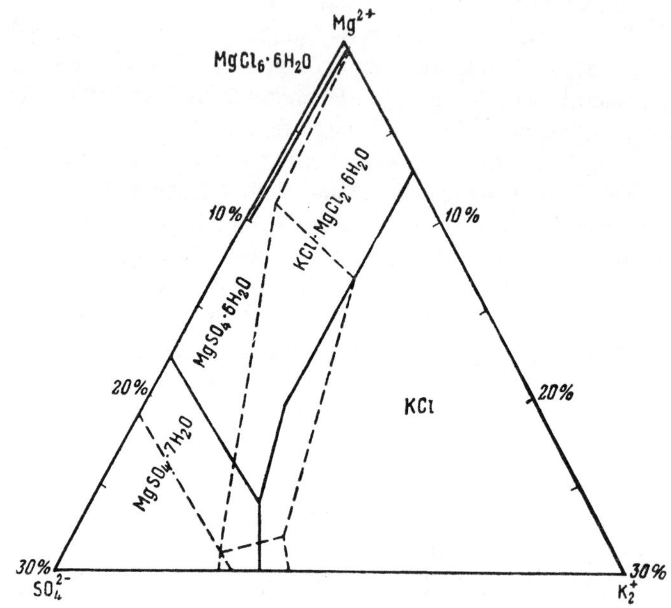

FIG. 136. "Solar diagram" from the data of N. S. Kurnakov and V. I. Nikolaev (solid lines) and boundaries of the stability crystallization fields according to Van't Hoff (broken lines).

of carnallite formation (M. G. Valyashko and E. F. Solov'ev 1953). The normal sequence of salts during "solar evaporation" of water in a saline marine basin thus proves to be as follows:

1. Halite.
2. Halite + epsomite.
3. Halite + hexahydrite.
4. Halite + hexahydrite + sylvite.
5. Halite + hexahydrite + carnallite.
6. Halite + hexahydrite + carnallite + bischofite.

The course of solar evaporation of brine in saline marine basins is substantially different from that indicated by the diagram of stability states, as proposed by Van't Hoff.

Similar differences have been observed during solar evaporation in lacustrine basins. O. D. Kashkarov (1956) constructed a composite diagram of the solid phases of salts arising in all three hydrochemical types of lakes during summer evaporation, and he compared this with the isothermal stability diagram (Fig. 137). A characteristic feature of the solar diagram is the presence of fields of mixed salts: burkeite and sulfates, burkeite and natron, sodium sulfates of various forms, halite and blödite, which occupies part of the blödite and thenardite fields, and correspondingly reduces these. These fields of mixed salts demonstrate the formation of metastable solid phases, arising during solar evaporation.

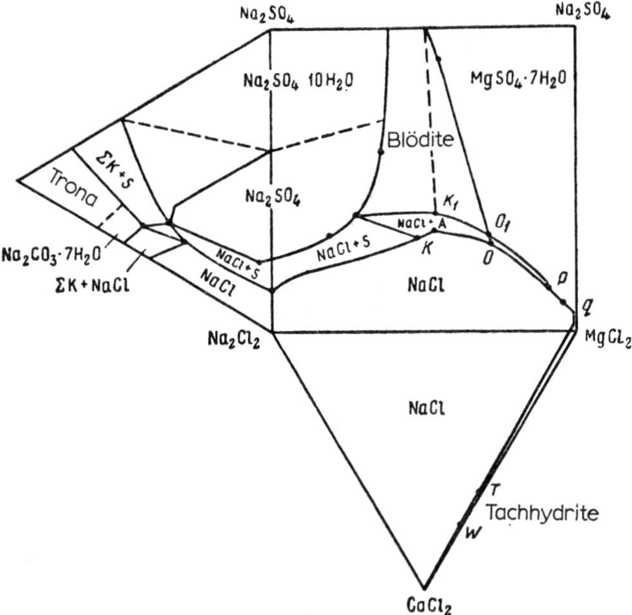

FIG. 137. "Summer diagram" of lacustrine equilibrium (from O. D. Kashkarov). *S. Sodium sulfates:* Na$_2$SO$_4$ and Na$_2$SO$_4$. 10H$_2$O; *ΣK. sodium carbonates:* Na$_2$CO$_3$. 10H$_2$O, Na$_2$CO$_3$. 7H$_2$O, and others; *A.* blödite. Compare this diagram with Fig. 124.

The cause of the divergence between solar diagrams and stability fields obtained experimentally *lies in the presence of oversaturated solutions in the brine and in the different rates of their removal.* Strong and prolonged agitation of the solutions, the introduction of seed crystals of stable phases into the solution, and other procedures employed during experimental work finally reduce the oversaturation and destroy the metastable phases. The rate of this may vary strongly for different circumstances, however.

Two situations may be clearly distinguished. In the first, agitation and seeding prove to be very effective and quick-acting, as in the example of sodium sulfate at low temperatures, when oversaturation is instantly eliminated by seeding of mirabilite. In the second, agitation of the solution begins to have an effect only a long time after initiation, and seeding of a stable salt has no effect, the seed crystals even being dissolved in some cases, as with blödite. The metastable phases NaCl and $MgSO_4.7H_2O$ (or $Na_2SO_4.10H_2O$ and $MgSO_4.7H_2O$), from which blödite is formed after many days (up to 16 and more), are maintained in juxtaposition with the "blödite solution" and are replaced by the stable phase only extremely slowly. It is equally difficult to eliminate oversaturation during the formation of kainite.

The causes of these differences in the rates of eliminating oversaturation and differences in stability of the metastable phases are not yet clear. *The most important factor in the process is probably the rate at which the stable phase crystallizes.*

When the stable phases crystallize rapidly, oversaturation is eliminated quickly and the metastable sediments are rapidly destroyed. The development of oversaturation itself requires special precautions, among which slow evaporation and absence of agitation are essential.

When the stable phases crystallize slowly, oversaturation is reduced correspondingly slowly, and special means are required to accelerate it.

Van't Hoff, D'Ans, N. S. Kurnakov, and others used various techniques for reducing oversaturation, and after holding the solutions at oversaturation for a prolonged period, they endeavoured to obtain solid phases in the sediment. The diagrams therefore received the name of "stability" diagrams, and the sequence of mineral formation indicated by them was termed the "course of stable crystallization". But with natural solar evaporation of brine there are no processes for the elimination of oversaturation, which explains why the formation of mineral salts under natural conditions differs from the course indicated by the stability diagram.

For substances that crystallize rapidly, much more rapidly than the evaporating solution changes during a summer from saturation with one substance to saturation with another, minerals are formed during the sedimentational stage, which ultimately contribute to the sediment: e.g. gypsum, anhydrite, mirabilite, halite, natron, gaylussite, epsomite, sylvite, carnallite, among others.

For substances that crystallize very slowly, the rate clearly lagging behind the rate of summer evaporation of the solution, solid (mineral) phases are not precipitated during the sedimentational stage. Instead easily crystallizing, metastable compounds are precipitated, which during slow recrystallization, change to stable minerals that could not form during evaporation of the brine because of the slowness of their crystallization. Minerals of this complex two-stage origin are thermonatrite, kieserite, glauberite, thenardite, blödite, kainite, langbeinite, and other double salts of potassium and magnesium. That the rate of evaporation is the prime factor in this relationship is demonstrated by the industrial basins of Lake Saki. Kainite and epsomite should crystallize in these basins according to the composition of the brine, but during rapid solar evaporation carnallite and hexahydrite form. Kainite is precipitated only during slow evaporation, at only half the rates encountered during normal operation of the industrial plant.

We may note that minerals passing through a metastable phase are rare during the initial stages of saline development (dolomite), are more common in the intermediate stages (thenardite, glauberite), and are abundant at advanced stages (blödite, kainite, langbeinite, and others).

From experimental work and observations on natural basins valuable data on the kinetics of formation of salt minerals have now been assembled, as the formation of blödite shows (V. I. Nikolaev and D. I. Kuznetsov 1935). During evaporation of solutions corresponding to the composition of blödite, crystals of sodium chloride are precipitated first and settle to the bottom of

TABLE 29

The Composition of Saturated Solutions and of the Precipitating
Solid Phases (in %)

Experiment	NaCl	Na$_2$SO$_4$	MgCl$_2$	MgSO$_4$	Total salts	Water
			Saturated solutions during crystallization			
VII	18·96	—	0·55	12·63	32·14	67·86
XII	14·80	6·44	—	10·73	31·97	68·03
			Precipitated solid phases			
—	92·86	3·09	—	1·25	97·20	2·8
a	33·87	17·63	—	22·77	74·26	25·74
b	8·47	36·06	—	33·06	77·60	22·40
c	0·99	39·67	—	35·69	76·35	23·64
d	1·67	40·44	—	33·96	76·03	23·97

(*a*) from surface crust after one day   ⎫
(*b*) from bottom crust after 5 days   ⎪
(*c*) from bottom crust after 15 days   ⎬ blödite+halite
(*d*) NaCl+blödite (95%); number of Na$_2$SO$_4$ ⎪
    and MgSO$_4$ molecules equal   ⎭

the vessel. Some sodium and magnesium sulfates are present, and some halite individuals succeed in growing to macroscopic size. But after one or two days the quantity of NaCl crystals in the total sediment no longer increases, but rather decreases. The content of Na and Mg sulfates in the solid phase increases, and crystals of blödite appear in dense crusts on the walls and the floor of the vessel.

The longer the vessel is left in the thermostatically controlled state (at 25°C), the fewer crystals of sodium chloride remain, the greater the quantity of sulfate salts form, and the more extensive the crust of blödite crystals, on the floor and the walls of the vessel. "After 15 days crystals of NaCl cannot be macroscopically identified on the floor or the walls of the vessel, but only a dense, crystalline crust of blödite" (Nikolaev and Kuznetsov 1935). At various times analyses were made of the precipitating solid phases (Table 29).

If the amount of water used in analysis is measured, it is possible to calculate the number of magnesium and sodium sulfate molecules combined in the blödite molecule ($Na_2SO_4 . MgSO_4 . 4H_2O$) and how many are still free (Table 30).

After 15–18 days the amount of $MgSO_4$ and Na_2SO_4 found by analysis of the solid phases becomes equal to the molecular value; the formation of blödite ceases.

These experiments with solutions of different composition, *but always corresponding to the saturation state for blödite in the brine*, have demonstrated that this mineral crystallizes at markedly different rates at different points in its field of occurrence. The entire blödite field may be subdivided into three zones (Fig. 138): the labile zone of epsomite, the labile zone of halite, and the stable zone of blödite. In the first two zones the formation of blödite is so slow that these zones approach the stable fields of epsomite and halite. In

TABLE 30

The Rate of Converting Initial Precipitates to Blödite

Time	Amount of NaCl in total solid phase, %	Blödite, % of theoretically possible amount
5–6 hours . . .	92·86	50
1 day	33·86	75
5 days	8·47	95
15 days	0·99	cir. 98

the third zone blödite crystallizes at an appreciable rate; the rate speeds up especially in a restricted area known as the "blödite spot". This emphasizes a fact of fundamental and general interest, also important for understanding the crystallization of other double salts: *The rate at which these salts form depends on the actual composition of the solution, even when the composition may correspond to the field of the given salt.* For blödite the rate of crystallization is determined by nearness to the blödite spot. For potassium-bearing minerals, according to M. G. Valyashko, the maximal crystallization rate corresponds to a composition of the solution near the crystallization of carnallite.

Temperature is also of fundamental significance. Blödite forms only above 15°C; at lower temperatures the initial phases—NaCl and $MgSO_4$—cannot be converted to blödite. Solid phases that are metastable relative to the composition of the bottom brine become stable in the sediment.

Thus, conversion of metastable solid phases, forming during solar evaporation, to stable blödite *is intimately related both to the composition of the brine penetrating the salt precipitates and to the temperature conditions; the conver-*

sion may take place only when favorable composition and favorable temperature conditions combine. If such a combination is not realized, the transformation of the initial metastable solid phases to blödite is retarded or may cease.

Generally speaking, the opposite situation is also possible. When the composition of the bottom brine falls precisely on the blödite spot and the temperature proves to be optimal for the formation of blödite, stable development of the mineral accelerates sharply, and *crystals are precipitated directly*

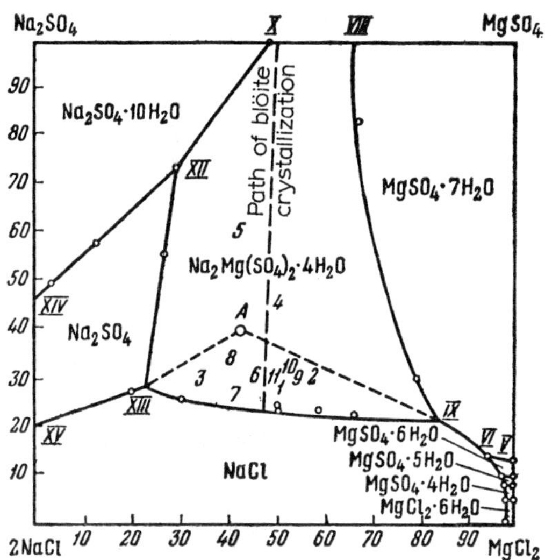

FIG. 138. Crystallization conditions for blödite with plotted compositions of lacustrine brines (from V. I. Nikolaev). *A.* Blödite spot, *A—XIII.* boundary of the metastable fields of NaCl and Na₂SO₄ . 1OH₂O, *A—IX.* boundary of the metastable fields of NaCl and MgSO₄ . 7H₂O, 2. Lake Dengiz-Kul', 1932 (from A. G. Bergman); 3. Lake Aral-Sor (north), 1936 (from I. N. Lepeshkov); 4, 5. Lake Ashi-Kul' (from M. G. Valyashko); 6, 7, 8. Lake Tuz-Kul', 1942-43 (from Valyashko); 9, 10, 11. Gulf of Kara-bogaz, 1943-45 (from A. D. Pel'sh).

from the bottom brine in the stage of sedimentogenesis. In 1942-43, freshly precipitated salt consisting chiefly of blödite was found in Ashi-Kul' and Tuz-Kul' Lakes (Valyashko 1949). In 1945, A. D. Pel'sh recognized precipitation of blödite and halite during evaporation of the brine in the Gulf of Kara-bogaz (1953). I. N. Lepeshkov noted a possible occurrence of suspended halite and blödite crystals in the brine of Lake Aralsor (Lepeshkov 1946).

Thus, what normally takes place only during diagenesis has shifted to the sedimentational stage under the above conditions. These facts demonstrate that *the dynamics of forming salt minerals is very sensitive to the physicochemical conditions in the actual basin. Any one mineral may form very slowly—in stages —under diagenetic control, but, under other circumstances, it may also form very rapidly as a sedimentational mineral.*

Another example of the stage-like development of a salt mineral is the origin of thenardite, the stages of which were observed in Bol'shoe Mormyshanskoe Lake on the Kulunda Steppe (S. Z. Makarov 1935).

"Thenardite began to form in mirabilite sediments on the nights of July 31 and August 1 because of the fall in the temperature of the brine at a time of high salt concentration. The mirabilite had precipitated in coarse crystals, forming irregular nodules of considerable size attached to objects that sink in the lake, such as logs and branches of old brushwood roads across swamps, links of stake fence and similar debris.

"The thickness of the sediments was 3–4 cm at the edge of the lake. Along the brushwood roads it reached 20 cm. During the succeeding days the entire northern and northeastern parts of the lake appeared white from a distance because of the exposed aggregates of glauberite glistening in the sun, especially when the north wind blew and drove the water to the opposite end of the lake. The layer of remaining brine in the northern part of the lake was no more than 10 cm thick. Later the $Na_2SO_4.10H_2O$ began to be dehydrated because of distinctive temperature conditions in the upper part of the freshly precipitated mirabilite. The surface of compacted mirabilite sediment, with mud, algae, $CaCO_3$ and basic magnesium carbonate salts, was heated 2–8°C above the temperature of the brine, as was determined by direct measurement. As a result of this overheating, the mirabilite lost water at the brine-salt boundary, and a thenardite crust, 0·5–1 mm thick, developed. Even where the mirabilite layer was not completely insulated, this crust created more favorable conditions for absorbing the sun's heat and for dehydrating the underlying mirabilite. The thenardite layer thus gradually grew at the expense of the mirabilite.

"It was noted that thenardite formed especially well in shallow depressions in the lake floor. The layer of glauberite salts was thicker here because of the increased thickness of the brine layer. When slumping of the newly formed salt occurs, reverse solution (with rise in temperature during the day) is retarded and does not go to completion. Compacted layers of glauberite thus readily form in low parts of the floor, and later, being dehydrated by the brine as it becomes concentrated through evaporation, this material changes to thenardite. Since these compacted masses have only a restricted distribution, the thenardite beds are patchy, the intervening zones being filled with freshly formed glauberite salt. The hollows on the floor range from 0·5 to 2·0 m in diameter, and, relatively to the floor, are rarely more than 10–25 cm deep. In such hollows (Fig. 139) a thenardite bed 2–5 cm thick is formed beneath a thenardite crust that is several millimeters thick. The bed has numerous cavities because of reduction in volume as a result of dehydration. Downward there occurs a zone of combined thenardite and mirabilite crystals, 2–4 cm thick, and below this occurs the zone of freshly precipitated mirabilite alone" (Makarov (1935).

This course of thenardite development in Bol'shoe Mormyshanskoe Lake, according to Makarov, is a model of thenardite formation in general. "A whole series of experiments, set up in the laboratory and at the lake for iso-thermal evaporation, confirmed the belief that under these conditions thenardite was not precipitated directly, but that common salt formed. On the other hand, it was possible to cause glauberite to be precipitated by cooling, and then with evaporation and elevation of temperature this salt was dehydrated. This process corresponds with the chemical relations taking place in Bol'shoe Mormyshanskoe Lake" (Makarov 1935). The experiments of V. I. Nikolaev confirm this conclusion.

The geological conditions under which thenardite forms is therefore clear. It forms only in those basins which have especially high concentrations of Na_2SO_4 and which in summer, in late July and August, precipitate mirabilite during night cooling of the brine. During the day, when the temperature rises

FIG. 139. The formation of thenardite in Mormyshanskoe Lake (from S. Z. Makarov).

above 23°C, this mirabilite is partially or completely converted to thenardite. This may occur during the same year or later. The formation of thermonatrite in soda lakes takes place in a manner similar to that of thenardite in sulfate lakes (S. Z. Makarov 1935).

Figure 140 shows salt equilibrium for lakes of the Tanatar group. Only ice forms in the solid phase when the salt concentration is below 10% and the temperature no lower than −5°C. Natron ($Na_2CO_3 . 10H_2O$) begins to form only at −5°C, and is the only salt phase down to −13°5°C. Mirabilite forms on further cooling, and then hydrohalite ($NaCl . 2H_2O$). With increase in Na_2CO_3 concentration, the precipitation of natron is shifted into the range positive temperatures, even up to 20°C or higher. However, because of the dehydrating effect of the highly concentrated brine in this temperature range, natron begins to give off water and to be converted partly to thermonatrite. This process was described by Makarov from observations on Tanatar I Lake in August 1932.

"With fall of temperature at night, natron was precipitated in nodules along the shore. These readily broke down as temperature rose during the day. After 10 August the daytime temperature dropped to 17°C, and the floor of the lake was quickly covered by a bed of natron, which extended, as acicular crystals of natron into the mud. Similar needles also formed on the surface of the bed, but only after night precipitation. By day, even when the temperature was not high, the brush-like aggregates of soda broke up, and the crystalline intergrowths, loosened in the water and settling further, crystallized to form a surface that was almost perfectly smooth. When heating was fairly rapid, as in the shallower parts of the lake, the nodules broke up and formed a friable mass of light, rose-colored crystals of thermonatrite; i.e. not only

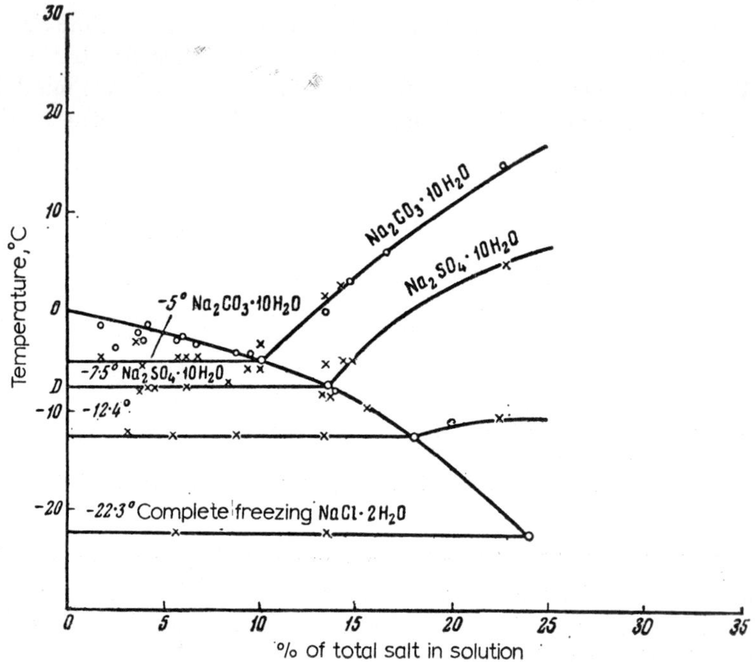

FIG. 140. Diagram of salt equilibrium in lakes of the Tanatar group (from S. Z. Makarov).

had part of the decahydrate natron dissolved, but the remaining carbonate had been markedly dehydrated. The crystals of natron that formed by night thus crystallized with the bed by day. The thickness of this bed gradually increased, reaching an average of 10 cm over the entire lake by 13 August.

"The dissolving of the precipitates and the formation of a dense bed were associated with a lower surface of reverse solution. Whereas the friable sediment went easily into the solution during the day, the dense bed dissolved very slowly, and stronger heating did not lead to reverse solution but rather to thermal dehydration of the decahydrate by the briny solution. The compacted bed of natron grew turbid, and it was penetrated to 1–1·5 mm by small crystals of thermonatrite. Small natron crystals of night-time precipitation were especially subject to dehydration by the saline solutions. The friable mass, consisting of a mixed layer of partly dehydrated salts having a large capacity for heat absorption, favored dehydration to an even higher degree within the layer. At some depth, depending on the conditions of heating, this stopped completely.

"On 22 August the structure of the bed in the lake was still comparatively simple. From the top the unconsolidated turbid layer of mixed decahydrate and monohydrate crystals extended down 1·5 mm, after which there was a continuous but very thin (0·5 mm) dull layer of 'friable biscuit', monohydrate, and below this was a layer of natron more than 10 cm thick—15 cm in places—resting on the mud.

"With time the process continued in the same direction. Along the shore of the lake, where the brine layer was only 3–5 cm deep, the entire layer of freshly deposited salt became dehydrated. In the center of the lake, where the depth of brine remained at 15–25 cm, the lower dense layer remained unaffected. It was separated from the more friable intervening layers of monohydrate and decahydrate by a very dense—up to 2–3 cm—layer of the monohydrate, strongly suggesting a white porcelain biscuit" (Makarov 1953, pp. 193-194).

On the whole, the precipitation of sodium carbonate in Tanatar I Lake was very active at the end of the evaporative season in August 1932, forming a bed more than 15 cm thick. The autumn rains freshened the lake and redissolved all the salt that had previously formed, so that from the sedimentational point of view the summer precipitation proved to be ineffective. But under other environmental conditions, such as in African lakes, this resolution does not take place and beds of thermonatrite continue to grow from year to year.

Thus the well-defined stage-like development of authigenic minerals characterizing humid lithogenesis obtains also in arid lithogenesis, but it assumes some specific features, two of which merit attention.

First, the stage-like character is no longer ubiquitous in the formation of salt minerals. Blödite, for instance, may be typically diagenetic, or it may be a sedimentational mineral; it all depends on how near the composition of the brine is to the "blödite point" within the blödite field and on how successfully the actual composition of the brine combines with the actual temperature conditions. Undoubtedly the kinetics of formation of other salt minerals is similar. It follows that *in interpreting the origin of salt minerals there can be no fixed models of uniformly applicable genetic schemes.* A genetic analysis of any saline sediments must be made individually for each deposit, and it must be based on data for brine composition (for Recent sediments), as well as on data from textural studies of the rocks themselves.

Second, even when the salt minerals form in stages, their rate of formation is many times faster than diagenetic mineral growth in sediments of humid regions. The formation of diagenetic minerals in marine sediments requires tens, hundreds, and even thousands of years for its completion. The formation of

S

salt minerals in stages may take place during one or, at most a few, hydro-chemical cycles. The difference between the two is manifest.

5. Mineral Formation in "Dry" Lakes

When a basin in the arid zone is still weakly mineralized and has only a thin layer of sediments covered by a considerable layer of saline water, the main mass of salt is concentrated in the bottom layer of water. As the water in the basin increases in salinity, however, and the thickness of the bottom deposits becomes greater, the chief location of the dissolved salts becomes more and more the intercrystalline brine. This may be graphically seen from the following calculations for several Crimean lakes (Table 31).

In Lake Sasyk-Sivash, a large basin with a layer of surface brine of the order of one meter deep, the salt content in the intercrystalline solution is almost twice that in the surface layer of water. For Lake Dzharyl-Agach and

TABLE 31

Distribution of Salt in Intercrystalline and Surface
Brine (from Kurnakov, Dzens-Litovskii, and Kuznetsov, 1936)

Lake	Salt content, %		
	in surface brine	in inter-crystalline brine	in entire lake
Sasyk-Sivash . .	35·9	64·1	100
Saki	13·8	86·2	100
Dzharyl-Agach . .	17·3	82·7	100
Karlavskoe . . .	24·5	75·5	100

Lake Karlavskoe, which also have layers of brine of the order of one meter deep (maximal depths of 1·55 and 0·75 m, respectively), the salt content in the intercrystalline brine exceeds that in the surface water by factors of 4·8 and 3·1 respectively. For Lake Saki, in which the layer of surface brine is 0·5 m deep, the salt content in the intercrystalline brine is 6·3 times that in the surface water. Similar relations are known for other basins.

Simultaneously with diminution of the importance of bottom brine as the location of salt solutions, the ratio between volume of precipitating solid phases and volume of parent brine also changes. As seen from Table 32 and Fig. 141*A* and *B, in proportion to the evaporation of parent brines, the volume continuously decreasing, the volume of solid phases precipitating from these brines increases steadily* (Valyashko, 1951.) Because of this there arrives a time in the evolution of every saline basin when the volume of parent brine is comparable with the volume of precipitated salt. From that time, the volume of brine is less than the volume of precipitated salt. *At about the time these two volumes become equal the basin reaches the dry-lake stage.* As the volume of salts continues to rise relative to the volume of remaining solution the type of dry lake becomes ever more sharply defined. This time of transition to the stage of dry lake, as experiments on evaporating sea water and brines of a number of modern soda lakes have shown, corresponds under certain circumstances to the beginning of epsomite precipitation, under other circumstances to the precipitation of potassium salts, sylvite in particular (Valyashko). Thus, *the dry-lake stage is an inevitable consequence of the system-*

TABLE 32

Changes in Volume of Parent Brine and of Precipitating Solid Salts during Evaporation (V. P. Il'inskii, 1961); Volume Computations of the Salts made by M. G. Valyashko

Lake Saki

Liquid phase		Solid phase
sp. gr.	vol., ml.	vol., ml.
1·063	1000	—
1·107	755·6	—
1·125	644·6	—*
1·141	531·2	0·65
1·179	424·8	0·93
1·224	371·0	1·48†
1·231	327·1	1·75
1·240	221·9	17·87
1·251	160·3	26·87
1·273	124·8	31·91
1·297	98·9	35·3
1·307	87·1	33·81‡
1·332	45·9	49·93§
1·361	36·4	53·29

Lake Sasyk-Sivash

Liquid phase		Solid phase
sp. gr.	vol., ml.	vol., ml.
1·066	1000	—
1·101	665·2	—*
1·108	608·4	0·44
1·162	379·1	1·46
1·220	267·6	1·90†
1·231	165·2	17·72
1·252	102·6	27·27
1·273	73·6	31·31
1·291	59·7	33·08
1·300	56·4	33·46‡
1·308	52·8	34·11
1·321	39·9	38·07
1·323	38·5	38·47
1·323	37·7	38·77§
1·336	27·1	43·64
1·344	23·4	44·22
1·350	22·2	45·05
1·361	21·3	45·30

Western Sivash Gulf

Liquid phase		Solid phase
sp. gr.	vol., ml.	vol., ml.
1·092	1000	—
1·123	769·5	—*
1·143	657·4	0·12
1·163	539·5	1·37
1·217	363·0	1·95
1·233	348·4	2·07†
1·241	179·5	28·6
1·281	85·5	41·66
1·297	74·5	43·35‡
1·307	62·0	44·40
1·325	40·9	45·87§
1·332	25·7	54·79
1·344	20·6	57·3
1·356	19·3	57·66

* Beginning of gypsum precipitation. ‡ Beginning of epsomite precipitation.
† Beginning of halite precipitation. § Beginning of carnallite precipitation.

atically changing relations between the volumes of bottom brine and of precipitating salts.

Sedimentation during the dry-lake stage is characterized by distinctive features. Most present-day lakes for example are of very small size and are confined to basins of erosional origin. *The floor of such a basin is not susceptible to active differential downwarping of one part relative to the other.* The intercrystalline brine in most present-day dry lakes therefore remains immobile between crystals of precipitated bottom sediments and clastic grains, nowhere draining off or separating from these phases. Exceptions are found only in some lakes of the Tsaidam basins (China), particularly the huge dry lake Tsarkhan, individual parts of which lie below the average level of the lake. Intercrystalline brine from the elevated area around drains to some extent into these lower basins. The history of sedimentation in dry lakes that are tectonically immobile is substantially different from that in lakes suscept-ible to differential warping in parts of the basin.

The physicochemical environment and the sedimentation in tectonically mobile dry lakes are characteristic (M. G. Valyashko and Ya. I. Tychino

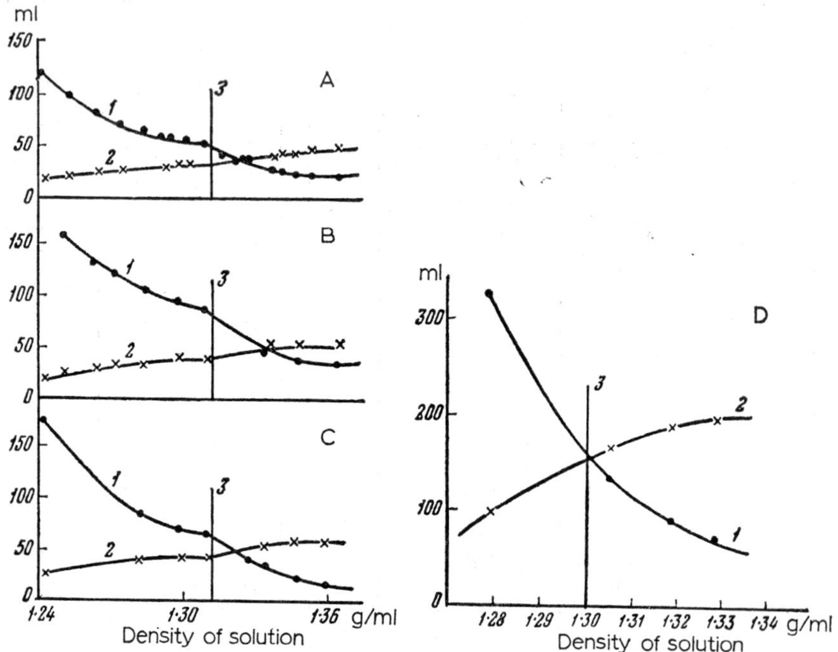

Fig. 141. Relations between parent brines and precipitation solid phases during desalinisa-tion of brine lakes (from Valyashko). *A.* Lake Sasyk-Sivash; *B.* Lake Saki; *C.* Sivash Gulf; *D.* Gulf of Kara-bogaz. 1. Volume of parent liquors; 2. volume of solid salts forming during evaporation of brines of marine origin; 3. beginning of crystallization of potassium salts (epsomite).

1952). The first distinctive feature is the marked reduction of summer evapora-tion from the surface of the intercrystalline brine. This is less than the evapora-tion from the surface of the saline basin by a factor of approximately 170. Correspondingly, the intensity of chemical sedimentation from the inter-crystalline brine is many times less, although the precipitating process by no means ceases altogether. The sharp retardation of physicochemical processes corresponds to a practically uniform composition of the intercrystalline brine, during the course of the yearly cycle, in a vertical direction through the salt deposits to a depth of 20–30 m (Lake Inder, Lake Elton). Only the very surface layer of the intercrystalline brine, being subject to dilution by rain water in the wet seasons and to evaporation during dry seasons, still exhibits annual hydrochemical cycles, and its representative point on the diagram may lie at times on the path of halite crystallization (Inder), at other times on the paths of halite-epsomite-carnallite crystallization (Elton). These cyclic processes disappear even at shallow depths in the brine. The degree of mineralization and of salinity in the intercrystalline brine in dry lakes nevertheless moves farther towards crystallization than in the bottom brine from which the salt mass containing the intercrystalline brine was precipitated. *This graphically illustrates the fact that the movement of a brine toward the eutonic point, as in the development of salt deposition, does not stop even after the bottom brine is converted to intercrystalline brine.*

The processes of mineral growth in the intercrystalline brine of dry lakes—apart from the cyclical processes in the uppermost thin horizon, which are similar to the cycles in the surface brine—remain essentially two in number. *The first is the crystallization of new evaporite minerals, corresponding to the higher stages of mineralization of the solution during slow evaporation.* If a lake becomes dry during the period when halite is being precipitated, then epsomite may be precipitated from the intercrystalline brine. If the conversion to a dry lake takes place in the epsomite stage, potassium minerals—sylvite or carnallite—may crystallize in the intercrystalline brine. *The other process involves interaction between the precipitating new phases and the earlier, leading to the crystallization of new, more stable minerals.* Thus, when epsomite precipitates from the intercrystalline brine, reaction begins between this mineral and the previously formed halite, giving rise to blödite: $2NaCl + 2MgSO_4 + 4H_2O = Na_2SO_4 . MgSO_4 . 4H_2O + MgCl_2$. In exactly the same way, the precipitation of sylvite in epsomite rock gives kainite: $KCl + MgSO_4 . 7H_2O = KCl . MgSO_4 . 3H_2O + 4H_2O$. The mineral species that grow from the intercrystalline brine are distinctive and are different from the minerals characterizing the supernacent brine.

For the formation of salts in the already accumulated bottom sediments, apart from other conditions, definite spatial requirements are necessary both for the accumulation of a sufficient quantity of brine and also for the growth of the salt crystals.

Deposits of "root"* blödite and epsomite are either confined to the muddy layers of halite beds or, more frequently, they occur beneath halite beds, at the contact with the mud. They occasionally occur in pores between crystals of stratal halite.

"The formation of lenses of such 'root' salts in the bottom deposits is possible only under conditions such that sufficiently extensive dehydration (spatially) can take place in the bottom deposits. Such processes may be displacement of the mud by crystallizing salt, replacement of incongruent minerals, and recrystallization of skeletal halite into completely formed individuals. According to our determinations, the volume occupied by halite may decrease 20–30% through recrystallization" (Valyashko 1950, p. 189).

* That is, forming from the intercrystalline brine.

The quantitative extent of this authigenic formation of minerals from inter-crystalline brine has been varyingly assessed. Valyashko, who first investi-gated these processes, was inclined to rate it high. He has stated that in Recent lacustrine saline sediments all lenticular or bedded occurrences of blödite and epsomite beneath halite mud or in the lower parts of the halite layer (Figs. 109–111) are due to precipitation from intercrystalline brine. It seems to us that this can hardly be so. In considering that the primary porosity of the saline sediments is small, up to 30%, that the salts in the brine at maximal concentration are also about 30% of the weight of the brine, and that evapora-tion of the intercrystalline is extremely slow, we believe that this *fresh* precipi-tation of minerals gives rise to only *very small amounts of salt*: individual crystals and lenses. As for bedded deposits of blödite or of large lenses beneath halite in some lakes, the origin of such deposits must be associated with precipitation from surface brine, considerably enriched in magnesium sulfate. Ordinarily, when the content of this salt in the brine is normal, if blödite actually precipitates only after halite, then, when the concentration of mag-nesium and sodium sulfate is greatly increased, the saturation point of blödite in the brine must be appreciably lowered (as indicated on p. 229 for gypsum), and blödite may begin to precipitate with halite, or even before.

However we evaluate the effect of the intercrystalline brine on the composi-tion of the salt bed, it cannot be doubted that this effect *complicates* the primary sedimentational picture of mineral growth. *In a salt bed forming at a certain mineralization of the surface brine, minerals that prove to belong to a higher stage of salinity of the brine are found.* From the fact that this super-imposition of a higher stage on an earlier stage characterizes the lower parts of the saline succession, it may be stated that as a result of the effect of intercrystalline brine a more or less well-defined *inverse stratification of the salt profile* arises.

The above discussion refers to small dry lakes in erosional basins, under-going no differential tectonic movements. The picture is substantially differ-ent for huge dry lakes, resting on floors that are capable of differential move-ments in various parts of the basin. In such basins segments warped downward more deeply than others become depressions, *lakes on halite-shores*, into which intercrystalline brine slowly drains, forming outstanding concentra-tions of the parent solution in the bottom layer of water, which continues to evaporate and yield solid phases in the ordinary way in the region near the eutonic point or agreeing with it. In other words, in huge dry lakes with tectonic movements of their floors, sedimentation of the same type as takes place in brine lakes occurs, but it is spatially restricted to local depres-sions within the "dry lake", and the source of the salts is not influx from outside but discharge of intercrystalline brine into the depression. *A peculiarity in the composition of such deposits in depressions of large dry lakes is the enrich-ment in potassium minerals.*

An example of this second course of dry-lake development is, as already pointed out, the huge dry Tsarkhan Lake (China) with its large local zones of carnallite. Lenses of this mineral correspond to sections of the dry lake warped downward somewhat farther than other parts, and therefore receiving intercrystalline brine and allowing it to evaporate further. Within Tsarkhan Lake one may still see lake segments containing brine, morphologically corresponding exactly to the above-described "lakes on halite shores". Lake Dabusyn, for instance, has an area of 330 km² (33 × 10 km); its shore and bottom are formed of halite, the very halite that constitutes the salt deposit of Tsarkhan. Fragmental halite is found accumulating here in large grains, up to 3 mm in diameter, forming a bed 3–5 cm thick. Lake Dabusyn consists of two basins, connected by a stream. A bed of salt lies on the floor of the channel and on the banks of this stream (Yuan Chien-Chin).

The differences in the behavior of dry lakes with tectonically immobile floors and of those susceptible to differential movements, interesting in themselves, are fundamentally important because *they allow us to gain a clear view of the effect of tectonic processes on salt deposition*. They show that *potassium sedimentation in saline basins is by no means always possible, but takes place only under definite tectonic conditions, where areas passing through the late stages of salt deposition are subjected to differential tectonic movements*. In other words, the deposition of salt is not only a physicochemical process but also a *tectonically controlled process*. The validity of this conclusion will be demonstrated from an analysis of modern salt deposition. Since most present-day salt lakes occupy basins of exogenetic origin, having floors that lack differential movements, potassium salts naturally do not accumulate in them. The absence of such salts, long regarded as astonishing, proves to be clearly explicable and not a fortuitous phenomenon.

6. The Relationships Between Solid Phases of Readily Soluble Salts and Solid Phases of Low-Solubility Compounds

The accumulation of authigenic minerals made up of low-solubility salts—Ca and Mg carbonates, Mg silicates, Fe compounds—continues simultaneously with the accumulation of more readily soluble compounds at advanced stages of salinity. Between these highly and slightly soluble solid phases perfectly systematic relations develop in basins of the various hydrochemical types, permitting one to formulate the hydrochemical aspect of basins that were sites of ancient salt formations.

Unfortunately, this interesting and essentially important aspect of salt deposition has been only incompletely investigated, and the known studies are concerned with the relations between solid salt phases and the carbonates of alkaline earths.

Figure 107 shows the distribution of calcium and magnesium carbonates according to different stages of salt deposition in basins of different hydrochemical types (Strakhov 1951). This is highly characteristic. Calcite crystallizes at the very beginning of chemical precipitation in basins of all hydrochemical types. In going from left to right on the diagram—from soda lakes to chloride types—the upper limit of salinity wherein calcite may exist as the only carbonate mineral increases progressively, and in the Azov-Black Sea branch of group *IIc*, as well as in chloride lakes, it is markedly dominant at all stages of salinity. Dolomite always follows calcite in the course of progressive salinity, but its distribution according to hydrochemical types is directly opposite to that of calcite. Dolomite occupies practically the entire field of soda basins. Its precipitation here begins at very low mineralization of the water and persists to the highest mineralization, combining here with other carbonates (gaylussite, etc.). Since the precipitation of gaylussite, even in strongly sodic lakes, takes place only at salinities above $\sim$13%, dolomite maintains its dominant position over a considerable range, remaining essentially the only carbonate in the clastic-carbonate muds at the carbonate stage of saline lake development. In sulfate basins dolomite is widespread only in lakes of group *IIa*. In group *IIb*, any special dolomite stage of carbonate formation following the calcite stage is missing, as in group *IIc* and in basins of the chloride class. Thus, *whereas the calcite field enlarges progressively from left to right on the diagram, the dolomite field is sharply curtailed during transition from soda lakes to sulfate lakes, and it disappears as such on the right third of the diagram.* In the intervening parts of the diagram, which include the fields of moderate and advanced stages of salinity, dolomite (but calcite in group *IIc*) gives way to a magnesite-calcite association, in which magnesite and calcite are combined in varying ratios, but generally with a sharp predominance of magnesite, especially at high salinities. It is characteristic that *the appearance of magnesite is typical of only those lakes that contain considerable quantities of $MgSO_4$ and only at that stage of development when gypsum begins to be precipitated.* These relations agree fully with the metamorphogenic origin of magnesite.

The above relations of calcium and magnesium carbonates to sulfate and chloride salt phases are so characteristic that they allow one reliably to differentiate not only the classes of soda, sulfate, and chloride lakes among deposits of the geologic past, but even to recognize groups within the sulfate class (see Part 3, Chap. 1).

The distribution of magnesium silicates has been but little studied. We may note only that in soda lakes and lakes of group *IIa*, these silicates characterize deposits of all stages of salinity (Chap. 1 above). In other groups of the sulfate class and in chloride lakes, magnesium silicates are found only in deposits of advanced stages of salinity, chiefly of the chloride stage. It is known, for example, that magnesium silicates have been found with hydromagnesite in the wavecast foam of Elton Lake at the end of the evaporative season (Vasil'ev 1956).

7. The Formation of Minerals associated with the Seepage of Ground Water into Lake Sediments

The formation of mineral salts thus far discussed is associated only with direct precipitation of solid phases from bottom water or with conversion of salts already in the sediment as these are modified by interaction with the intercrystalline brine. These are not the only processes however, at work in present-day salt basins.

Together with the surface supply, many present-day salt lakes still have a ground-water supply. Shallow ground water from the drainage areas, seeping through the bottom deposits, are discharged into the lakes. This process is especially typical of soda lakes confined to ancient alluvial deposits. Since the mineralization, and commonly the hydrochemical type of seeping ground water does not correspond to the mineralization and composition of the solutions impregnating the lacustrine sediments, a reaction takes place between the two, and some solid phases begin to separate out: most commonly calcite, dolomite, and basic magnesium salts, but also gypsum and thenardite. Essentially, however, the processes here are identical in this respect to the metamorphization of brine in the surface waters. The scene of action, however, is shifted from the supernacent brine to the intercrystalline, to the muddy bottom zone.

When working on the Kulunda Steppe, especially in the part called the Solyanoozernyi (Salt Lake) district, the author in association with D. A. Vital' (1950) succeeded in observing very effective development of minerals in lake sediments in connection with the seepage of ground water into lakes.

In the sediments along the margins of many lakes there are many "air holes", through which ground water, feeding the lake, rises and oozes out through the sediments. These are especially pronounced in Malinovoe Lake, where the seepage water is sodic but the lake brine is of sulfate type. In the blow holes themselves, and in the immediate vicinity, occur numerous calcareous concretions, disposed in groups and forming complex patterns on the surface of the lake floor, when this floor is exposed over a considerable area about the margins of the lake at the end of the evaporation season—in August and September. The concretions are of irregular form, normally pisolitic, but also with overgrown crusts. They range from 1–2 cm^3 to several cubic meters in volume, and in composition are calcite and magnesite in varying quantitative and structural relations. The concretions characteristically have numerous pores and cavities forming a system of channels along which ground water clearly flowed. On a single segment of the lake the concretions are so numerous that in time of war they have been used as quarry-stone (Fig. 142). The formation of the concretions has been due to reaction between sodic seepage water and sulfate water impregnating the sediment:

$$CaSO_4 + Na_2CO_3 \rightarrow \underline{CaCO_3} + Na_2SO_4$$
$$\downarrow$$
$$MgSO_4 + Na_2CO_3 \rightarrow \underline{MgCO_3} + Na_2SO_4.$$
$$\downarrow$$

When the seepage water and the brine are of the same sodic type, as in sodic Lake Iodnoe, differing only in concentrations—low in the ground water, high in the lake water—only the formation of $CaCO_3$ results. It accumulates in masses along the shore, where it forms a belt of high-carbonate sediments—the "white mud"—typical of lacustrine chalk or marl, in which many calcareous concretions are included. The magnesium content of the carbonates is small, but the mineral forming in the mud at this time is dolomite.

Malinovoe and Iodnoe Lakes lie along the margin of the Solyanoozernyi Steppe. In the inner parts of the steppe the lakes (Pechatnoe, Kochkovoe, Pravyi Bliznetsy, and Levyi Bliznetsy) receive ground water that is not sodic but is of sodium-sulfate type, enclosed in the alluvium of the Baklanikha River. As it seeps into these lakes under some pressure (as established by

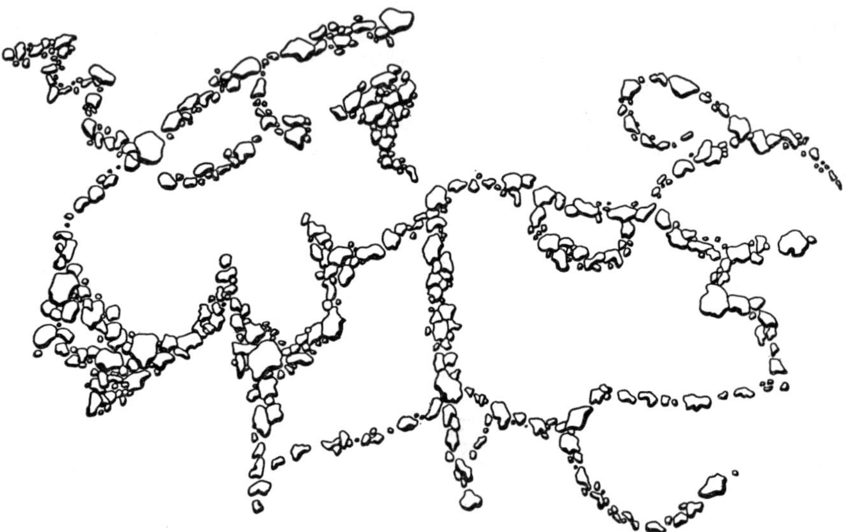

Fig. 142. Distribution of carbonate concretions in Malinovoe Lake.

drilling at Kochkovoe Lake in 1945) and in rising through a sequence of sediments including a buried mirabilite bed, the ground water becomes gradually salty, dissolving this bed from below and becoming enriched in Na_2SO_4. Upward in the lake sediments, in passing into a bed of NaCl, the water encounters intercrystalline brine saturated with NaCl, and part of its sulfate load separates, forming aggregates of thenardite, which everywhere accompany the halite bed on the bottom and which occur in black muds. The form of the thenardite segregations is characteristic (Fig. 143). They are commonly elongated in the vertical direction and are composed of well-formed and strongly bonded crystals, with vertically arranged hollow tubes, filled with mud, through which ground water has gradually seeped upward.

Thus supplementary authigenic minerals form in lake sediments by seepage of ground water of various compositions.

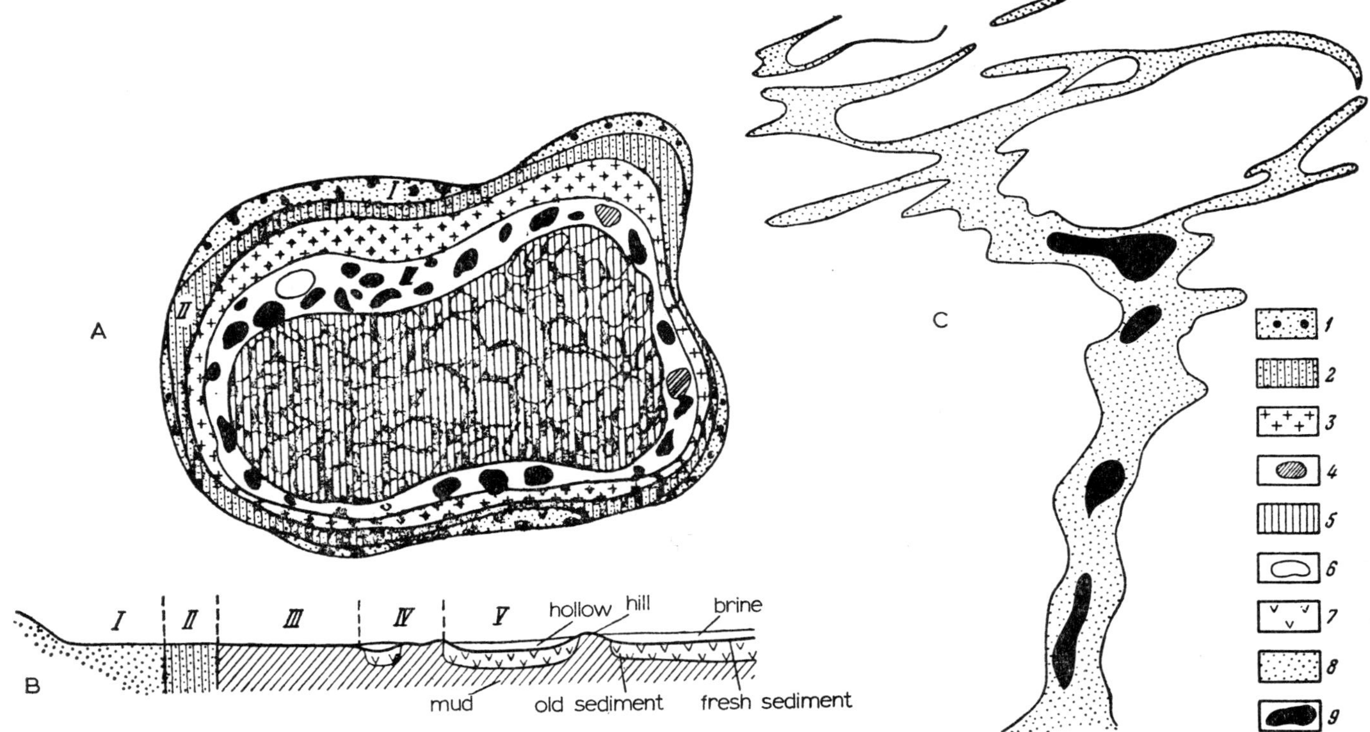

Fig. 143. Structure of the salt bed in Peschanoe Lake. *A*. Plan; *B*. cross-sectional profile; *C*. morphological details of hills and hollows. 1. Sandy beach (zone *I*); 2. muddy sands (zone *II*); 3. surface of freshly deposited mud (zone *III*); 4. hollow (zone *IV*); 5. central part of the lake, covered by brine (zone *V*); 6. hollow in central part of lake, corresponding to buried lenses of salt; 7. old salt sediments; 8. part of hill covered by fresh sediments; 9. swelling of mud with no fresh sediments.

8. Recrystallization and Lithification of Halogenic Deposits during Diagenesis

Simultaneously with processes giving rise to new solid phases of saline and other minerals, at least two other processes are at work in saline sediments: recrystallization of minerals and general compaction—lithification of the sediments.

Recrystallization of salt minerals normally leads to increase in grain size of the sediments and also, commonly, to the loss by the salt minerals of their primary pinnate structure, which is characteristic of several minerals, halite in particular. In place of the skeletal structure, the minerals develop till all the faces are more or less well defined if there is free space for growth, or they become anhedral grains when the space is insufficient. We would note that in the series gypsum—mirabilite—(thenardite)—halite—blödite, only gypsum characteristically displays very fine-grained texture; all the others are characterized normally by coarse grain, and may even form gigantic crystals.

As a rule the recrystallization of saline sediments does not affect the bedded structure, which is preserved in great detail; but at times, because of the strong seepage of ground water into the sediments of a salt lake, secondary transformations are accompanied by partial solution of the salt minerals and by fresh redeposition, and this eventually leads to a loss of the primary stratification in the salt beds and to the development of large segments of a bed (or even to entire beds) without stratification. In present-day lakes, under the bed of fresh sediments and the stratified old sediments, so-called "root salt" forms, as bed-like or nest-like lenses of halite, mirabilite, or blödite. Unstratified varieties in ancient continental salt deposits are probably correlatives of these modern "root salts".

The lithification of salt deposits takes place in two ways. Under the weight of the overlying deposits the salt grains normally come into close contact and intergrow. In this process the intercrystalline brine is entirely, or almost entirely, squeezed out. An extremely solid mass—salt rock—arises, almost devoid of pore space. Gypsum and anhydrite, the porosity of which is near zero, are also lithified in this way. The process of lithification in some rocks, however, is complicated by the fact that large crystals intergrow at points of contact, and there remains a considerable volume of pores and cavities in the rock, communicating one with another and with the surrounding intercrystalline brine. Sooner or later this brine too will be squeezed out and the rock with skeletal lithification will be converted to a dense mass. But the prolonged preservation of intercrystalline brine in some rocks naturally protracts the diagenetic transformation of the salt deposits.

The question arises, in this connection, concerning determination of the depth beneath the surface of the saline sediments at which this lithification ceases and the sediments are converted to dense salt rock. Data on a section of the Kara-bogaz deposits (V. A. Vakhrameeva 1956) are relevant. Three salt beds are distinguished here: a Recent bed and two older beds separated by gypsum-carbonate deposits.

The upper (or Recent) salt bed in Sartas Bay is 1·2–2·2 m thick and consists of halite at the bottom, halite mixed with blödite and epsomite in the middle, and epsomite at the top. It is a rather friable bed, especially in the epsomite layer. The lower part, the halite zone particularly, consists of large well-cemented grains. Only one of the 5–6 annual layers forming the bed has salt "teeth" scattered through it. It may be noted that this salt bed began to form in 1939, and the total period of its accumulation to the time when cores were taken (1951-54) was thus only 12–15 years.

In the second salt horizon, at a depth of 5–12 m, "the cementation is weak, much of the core being very friable. The rock is exceptionally cavernous. . . . Many crystals bear traces of corrosion and they commonly have interconnected narrow vertical canals about 70 cm long (the break in the core prevented tracing them farther). Cubic or cubic-octahedral crystals of halite have grown on the walls of these canals" (Vakhrameeva 1956).

Among the peculiarities in structure we may note that in Sartas Bay the bed is affected by upward currents of ground water, which are responsible for the friability of the deposit.

In the third salt horizon, which lies at a depth of 17–30 cm, "the rocks are generally dense, with few pores".

As for the gypsum-carbonate sediments beneath the salt beds, only the upper layer, separating the first and second salt horizons, is still plastic, being saturated with brine, and fluid when extracted. "The second and third mud horizons are well cemented and are brine-resistant" (Vakhrameeva 1956, p. 74). Thus, only the upper five meters of the Kara-bogaz succession, embracing the upper salt and upper gypsum-carbonate horizons, is still friable and fluid, showing only local lithification of the halite sediments. Below 5 m, lithification is more strongly developed; the sediments have been converted to compact rocks, with only individual horizons containing cavities filled with brine.

Similar phenomena are noted everywhere in present-day salt-precipitating lakes. The new deposits of the last year and of several preceding years are still distinguished by their friability. The older deposits are dense for the most part, some of them extremely solid ("chugunka").

As is well known, the lithification of sediments, the conversion of these sediments to hard, compact rock essentially signifies the end of diagenetic transformation of material. Early lithification of salt deposits is therefore an indication of the rapidity of salt diagenesis. The process runs its course to a depth of a few meters in the salt bed, taking therefore only some tens of years or perhaps a few hundred. It is easy to see that *the rapidity of diagenesis corresponds completely to the rapidity of salt sedimentogenesis, and both together attest to the ever-increasing rate of salt lithogenesis with progressive salinity of a basin.*

9. The General Physicochemical Pattern of Present-day Halogenesis

The above data lead to a brief formulation of the physicochemical pattern of present-day salt deposition. The principal features of the pattern are discussed in the following paragraphs.

(1) *The formation of salt deposits during the present epoch takes place against a background of annual hydrochemical cycles, which become progressively more pronounced as the salinity of the basin increases.* The formation of a stable salt bed becomes possible only when the intercrystalline brine as well as the bottom brine of the basin become saturated relative to the solid phase constituting the salt bed. *The annual hydrochemical cyclicity is imprinted in the microstratification of the sediments, and this becomes an essential structural feature of salt deposits.* The absence of such cyclic structure always indicates secondary recrystallization of the salts during the diagenesis. In addition to the sedimentational salt phases, pinnate structure and tooth-like crystals are also characteristic, disappearing only during secondary processes.

(2) *During salt deposition the brine in natural basins is constantly changing through the introduction of dissolved substances and finely suspended clay material; the composition is thus gradually changed, metamorphizing directly at times, losing magnesium sulfates, metamorphizing inversely at others, acquiring such salts.* When the metamorphization of salt basins is intense, even the hydrochemical type of a basin may change, shifting, for example, from sodic to sulfate or from sulfate to chloride. These changes affect the composition and quantity of evaporite minerals forming at the different salinity stages. They are also reflected in the appearace of some metamorphogenic minerals. The smallest quantity of such minerals characterizes sodic lakes, with gaylussite, and chloride lakes, with dolomite. Metamorphogenic minerals are most abundantly developed in sulfate lakes: gypsum, glauberite, basic magnesium carbonates, and dolomite. The number of metamorphogenic minerals increases as the salinity increases.

Depending on the actual physico-geographic and physicochemical conditions, a mineral of any particular composition may be either evaporite or metamorphogenic.

(3) *Generally speaking, salt minerals form in stages.* In the stage of sedimentogenesis, during solar evaporation, metastable minerals of one composition precipitate from the bottom brine, whereas stable minerals have a different composition. The conversion of metastable solid phases to stable phases involves either combination of simple salts to double, such as the formation of blödite, or dehydration of the primary solid phase, which leads to the formation of thenardite from mirabilite, thermonatrite from natron, glauberite from the dihydrate form, and hexahydrite from epsomite. Dehydration is apparently the dominant process during diagenesis of soda lakes.

The stage-like development of salt minerals always takes place rapidly, terminating within a single annual hydrochemical cycle or within a few cycles. When an appropriate composition of bottom brine combines with optimal temperatures, the process is accelerated to such an extent that stable minerals form directly in the bottom brine, and minerals that are normally diagenetic here become sedimentational. Up to the present, this has been demonstrated only for blödite, but it probably holds for other salt minerals also.

(4) *During salt deposition a progressive change occurs in the volumetric relations between precipitated solid phases and the solution from which they are precipitating: the volumes of precipitating solid phases gradually approach the volumes of parent liquor.* This leads to the conversion of a basin, previously full of water, to a dry lake, which may subsequently become covered by sand and other clastic debris.

Salt deposition is not only retarded very sharply in small salt lakes with tectonically inactive floors, but it acquires special features that distinguish it from the preceding stages. Potassium salts do not appear in them as independent horizons, but the solid phases near the eutonic stage are in the lower parts of the halite beds or beneath them. Inverse chemical stratification is developed in the sediments, or, to put it differently, a reflection of the normal profile of salt deposition is manifested.

In large dry lakes, where differential movements occur in the floor, depressed segments receive intercrystalline brine, rich in potassium, by drainage from surrounding areas, and from these newly developed briny sub-basins potassium salts are precipitated in normal fashion up through the eutonic stage. Salt sedimentation acquires its highest and most complete manifestation.

(5) *As the salinity increases not only does sedimentation increase but diagenesis also; i.e. lithogenesis as a whole takes place at an increasingly faster rate.*

CHAPTER 3

GEOLOGICAL CONDITIONS CONTROLLING THE FORMATION OF LAKES OF DIFFERENT HYDROCHEMICAL TYPES AND THEIR GENETIC INTERRELATIONSHIP

The great variety of hydrochemical types of basins in arid regions raises the problem of the geological conditions controlling their formation and the genetic relationships of the different types. These questions will now be examined as they apply to lakes of continental origin.

1. CONDITIONS OF FORMATION OF SODA LAKES

As a rule, soda lakes are confined to one of two types of arid zones: where acid, intermediate, or alkaline crystalline rocks are exposed; or where polymictic sandy deposits rich in feldspars and other aluminosilicates are developed. Examples of the first type of basin are Lakes Victoria, Rudolf, Albert, and Tanganyika in Africa; Big Soda Lake and neighboring lakes in the U.S.A.; the Tibetan lakes in Asia, Lake Sevan, and lakes of the trans-Baikal region (Doroninskoe, Gusinoe). Examples of the second type of basin are the lakes of the Kulunda Steppe, occurring in the so-called belt of pine forests growing on old alluvial sands. These sands cut across the plateau that forms the divide between the Ob and Irtysh Rivers. They are typical winnowed arkoses, with large fractions of feldspar, hornblende, and other aluminosilicates. They appear to be reworked granitic grus. It is possible that the soda lakes of the Kurgan Steppe are also related to similar sandstones.

It is natural that soda lakes should be found in regions of crystalline massifs or of their mechanically reworked products. All these rocks are poor in sulfides, their principal component being Na-rich plagioclase. They commonly carry feldspathoids also, rich in alkalis. If it is also borne in mind that during weathering of aluminosilicates the solutions are primarily enriched in cations, especially in alkalis, it may readily be appreciated that water in the region of such crystalline massifs is generally sodic in character. This water is also rich in silica, which later, during the period of saline development, leads to the formation of magnesium silicates such as sepiolite-cerolite in the sediments.

In addition to soda lakes of the origin described above, there are also those of another origin, associated with the soil-forming process, particularly with the formation of solonetz, a highly alkaline soil. It is well known that the colloids of solonetz soils are rich in absorbed Na, which they yield to solution only after the sodium salts are first leached from the soils. The absorbed Na^+ that is then given up normally replaces Ca^{2+} and Mg^{2+} in the leaching water, and thus, going into solution, is bound to CO_2 in natron. This natron, leaching out of solonetz soils, also appears to be a cause of soda lakes, according to K. K. Gedroits.

It is impossible to agree with the current view that this kind of process is essential to the formation of present-day soda lakes in general. There are a number of lakes, such as those in Africa, the trans-Baikal region, the Balkans, where solonetz soils are either entirely absent (Africa) or are negligible. Furthermore, the solonetz process is a temporary one in the history of soil development. It quickly gives way either to fresh salinization or to further leaching (solodization), and then the introduction of natron into the lake ceases. Soda lakes would have to be solely transitory features according to this point of view. They do not, however, appear to be so. In this connection it should be noted that on the Kulunda Steppe a sodic character is found not only in lakes of the *pine-forest belt but also in deep horizons of artesian water.* It is not possible to relate this regional sodic content of the water to some merely transitional episode of solonetz soil development. The sodic waters are the product of regional weathering of polymictic sands, which are widespread in this district, and not to soil processes.

Thus, *the regional distribution of soda lakes in the arid zone is due to the location of areas of igneous and metamorphic rocks and to the products of their mechanical reworking.* Furthermore, since these rocks are generally associated with complex mountain structures, *soda lakes are normally found in mountainous regions, either within the mountain systems* (the Balkan lakes, Sevan, Van, and others) *or about the periphery* (the trans-Baikal lakes, lakes of the Kulunda Steppe, lakes in the Great Basin of the U.S.A.). The soil process leading to the formation of soda lakes within areas of local solonetz soil development complicates this picture, but it does not obscure the basic pattern.

The cause of the varied soda content of the soda lakes, gradually passing to sulphate lakes, may be found in the nature of the parental crystalline rocks, the weathering of which gave rise to the sodic water. Some of these—the alkaline rocks—yield considerable Na, whereas others—ultramafic rocks— yield very little. This undoubtedly has an effect on the soda coefficient in the water. Another, much more potent factor, however, affecting this soda coefficient is the metamorphization effect of waters draining from regions of sedimentary rocks. As a rule these rocks contain notable quantities of Ca and Mg carbonates and of pyrite. During weathering, $CaCO_3$ and $MgCO_3$ pass into solution in the form of bicarbonates, and FeS_2 is oxidized first to $FeSO_4$ and then to $Fe_2(SO_4)_3$, which, being split hydrolytically, yields Fe_2O_3 and H_2SO_4. Subsequently, being neutralized by the bicarbonates of Ca^{2+} and Mg^{2+}, the water becomes enriched with the sulfates of these metals. The leaching of marine salts that were incorporated during the formation of these rocks supplies some NaCl and $MgCl_2$.

T

The processes that begin in soda lakes receiving surface waters only of the calcium-carbonate class have already been examined:

$$CaSO_4 + Na_2CO_3 \rightarrow \underline{CaCO_3} + Na_2SO_4$$
$$\downarrow$$
$$MgSO_4 + Na_2CO_3 \rightarrow \underline{MgCO_3} + Na_2SO_4$$
$$\downarrow$$
$$MgCl_2 + Na_2CO_3 \rightarrow \underline{MgCO_3} + 2NaCl$$
$$\downarrow$$

As a result, on the one hand, the bottom deposits of the basin are enriched with Ca and Mg carbonates of various mineral forms, and, on the other, a distinct change takes place in the hydrochemical characteristics of the lake: the soda content decreases and that of Na_2SO_4 and $NaCl$ increases. If the lake initially belonged to the highly sodic group with high pH, through the effect of seepage of sulfate water, the soda index decreases steadily and the lake changes from group *Ia* to group *Ib* or *Ic*. If the process continues for long, the lake may even cease to belong to the soda class, and become metamorphized to group *IIa* of the sulfate class.

Since regions of magmatic rocks commonly include considerable areas of sedimentary rocks, conditions for the above reactions are frequently realized. *The effect of water from sedimentary deposits flowing into soda lakes, in our view, is the principal factor determining the variability of the soda index.*

Simultaneously with these solutions, however, the water flowing from regions of sedimentary rocks carries into the lake a considerable colloidal clay fraction, which, interacting with the lake water, also affects the metamorphization of the water. In an alkaline soda lake, base exchange should take place according to the following scheme:

$$1. \quad Ca^{2+}(p.\ k.)^{2-} + Na_2CO_3 \rightarrow 2Na^+(p.\ k.)^- + \underline{CaCO_3}$$
$$\downarrow$$

$$2. \quad Ca^{2+}(p.\ k.)^{2-} + Na_2SO_4 \rightarrow 2Na^+(p.\ k.)^- + CaSO_4$$
$$CaSO_4 + Na_2CO_3 \rightarrow Na_2SO_4 + \underline{CaCO_3}$$
$$\downarrow$$

$$3. \quad Ca^{2+}(p.\ k.)^{2-} + 2NaCl \rightarrow 2Na^+(p.\ k.)^{2-} + CaCl_2$$
$$CaCl_2 + Na_2CO_3 \rightarrow 2NaCl + \underline{CaCO_3}$$
$$\downarrow$$

Thus, the final result of all forms of cation exchange in soda lakes is the same: *the calcium introduced in the adsorbed state is precipitated as $CaCO_3$, and the soda content of the water decreases. A highly sodic lake changes to a moderately sodic lake, then to a weakly sodic lake, and so on.* In other words, *cation exchange acts in the same direction as the double-exchange reaction with salts introduced into the basin in the dissolved state.*

In analyzing the origin and metamorphization of soda lakes, two items that frequently appear in the literature must be considered. The first associates the formation of soda lakes with the exchange reaction between $Ca(HCO_3^-)_2$

and NaCl or between $Ca(HCO_3^-)_2$ and Na_2SO_4, similar to the reaction between calcium bicarbonate and magnesium sulfate outlined above:

$$Ca(HCO_3)_2 + 2NaCl \rightarrow 2NaHCO_3 + CaCl_2$$
$$Ca(HCO_3)_2 + Na_2SO_4 \rightarrow CaSO_4 + 2NaHCO_3$$

Experimental work by M. G. Valyashko and G. D. Pel'sh (1952), however, failed to confirm these reactions. During interaction of $Ca(HCO_3)_2$ separately with saturated solutions of NaCl and Na_2SO_4, only *microaggregates of calcite* appeared in the sediment, proving merely simple decomposition of calcium bicarbonate:

$$Ca(HCO_3) \rightarrow \underline{CaCO_3} + H_2O + CO_2$$
$$\downarrow$$

and giving no indication of any exchange reaction with NaCl or Na_2SO_4. In experiments on the system Na_2SO_4—NaCl—H_2O, at 25°C, with a saturation of thenardite, finely aggregated calcite was obtained, together with glauberite in delicate, very small, acicular crystals, of the composition $Na_2SO_4 . CaSO_4$ $2H_2O$. In a solution saturated with mirabilite, the representative point of which lies near the crystallization field of thenardite, the dihydrate glauberite forms only in small quantities as an impurity in the principal mass of calcite. In both experiments the alkali contant (HCO_3^-) in the solution was negligible, being only 0·04%, i.e. 400 mg/liter, which is equal approximately to 6·6 mg-equiv/liter. *This alkali content corresponds to the solubility of $CaCO_3$ in a concentrated solution of sodium sulfate* (Strakhov 1951). *It shows that the exchange reaction between $Ca(HCO_3)_2$ and Na_2SO_4 and NaCl, with the formation of natron, does not occur.* This conclusion should be expected from the current theory of electrolyte solutions, since the reaction products of both components are more soluble than the substance undergoing metamorphization (calcium bicarbonate).

For a theory of lithogenesis this circumstance is of fundamental significance, *because it henceforth eliminates the possibility of attributing the origin of soda lakes to the effect of calcium bicarbonate on chloride and sulfate brines. Soda lakes do not form by that method.* It is clear also that they do not form by microbiological reduction of sulfates in lakes with sulfate water, although this process has repeatedly been adduced in explanation of the origin of soda lakes.

Two circumstances compel us to doubt the validity of this concept. Under the conditions necessary for desulfatizing bacteria to live, the biological reduction of SO_4^{2-} can take place only in muds with no free oxygen, not in the oxygen-bearing brine of the bottom water layer. For desulfatization reactions it is not difficult to calculate that 1·33 g of C_{org} is required to reduce 1 g of sulfur from SO_4^{2-} to S^{2-}. When the natural sulfate waters impregnating the sediments are rich in sulfates, a high content of organic matter in these sediments is clearly necessary as the energy source and base of the reducing process. One would therefore expect soda lakes, as originating from sulfate lakes by reduction of SO_4^{2-}, to be characterized by abundant C_{org} in the muds, because not all the organic matter is consumed during reduction, but only that part accessible to the bacteria. Actually, however, the muds of soda lakes differ in no way in their C_{org} content from the muds of other lakes. That is to say, it has not been found that they contain the necessary energy source and base for especially intense reduction.

The second circumstance, no less fundamental, is that *the development of desulfatizing bacteria in the deeper horizons of mud is always accompanied by the appearance of sulfur bacteria in the higher layer (purple, green), and these act in the opposite direction to the desulfatizing bacteria. In oxidizing H_2S to SO_4^{2-}, they destroy the results of the desulfatizing bacteria.* Sulfur bacteria are widely distributed in natural basins. The reddish and greenish patches in muds of salt lakes, due to their presence, are normal phenomena. These sulfur bacteria are localized at the boundary between the reducing and oxidizing zones of the mud. In contrast to desulfatizing forms, they are physiologically autotrophic, and therefore do not depend on stores of organic matter in the muds. Ecologically they are symbiotic associates of the desulfatizing forms, destroying the H_2S generated by the desulfatizers and converting it to SO_4^{2-}.

Both these circumstances compel us to be very sceptical of any view that biological agencies are responsible for the desulfatization of brines in natural basins and for the conversion of the lakes to soda lakes. Biological desulfatization does take place in salt lakes, but its intensity is probably negligible, and *it cannot be considered an important factor in the formation of soda lakes.*

Soda lakes are thus *typical of regions where magmatic rocks are widespread or where the disintegration products of such rocks have accumulated* (strings of arkosic alluvium, etc.). Lakes in drainage areas of normal sedimentary rocks containing calcite and pyrite have lower soda indices and higher sulfate content. An increase in the areal extent of these rocks leads to a gradual transition from soda lakes to lakes of the sodium-magnesium class.

2. Origin of Lakes of the Sulfate Class

In turning to the origin of lakes of the sulfate class, group *IIa* (sodium-magnesium) and group *IIb + IIc* (sodium-magnesium-calcium) will be considered separately.

Sodium-magnesium lakes represent the extreme metamorphization product of soda lakes as modified by waters originating in a sedimentary mantle in the drainage areas of these lakes. The correctness of this conclusion is confirmed by the spatial localization of sodium-magnesium lakes. *They show a distinct preference for piedmont depressions and districts of soda lakes. They extend into districts of soda lakes and form a peripheral belt surrounding such regions.* These characteristics may be observed in a number of places on the hydrochemically complex Kulunda Steppe. The same laws generally control the distribution of sodium-magnesium lakes in the arid zone.

Diversity in the hydrochemical aspect of lakes in group *IIa* (sodium-magnesium) is also due to metamorphization. This involves both exchange reactions between dissolved salts as they mix and cation exchange between brines and introduced clay material. The latter exchange takes place according to the following schemes:

1. $Ca^{2+}(p.\ k.)^{2-} + Na_2SO_4 \rightarrow 5Na^+(p.\ k.)^{2-} + CaSO_4$
   $CaSO_4 + MgCO_3 \rightarrow \underline{CaCO_3} + MgSO_4$
   $\downarrow$

2. $Ca^{2+}(p.\ k.)^{2-} + 2NaCl \rightarrow 2Na^+(p.\ k.)^{2-} + CaCl_2$
   $CaCl_2 + MgCO_3 \rightarrow MgCl_2 + \underline{CaCO_3}$
   $\downarrow$

   $\left.\right\}$ At the low salinity of the lake

3. $Ca^{2+}(p.\ k.)^{2-} + Na_2SO_4 \rightarrow \underline{CaSO_4} + 2Na^+(p.\ k.)^{2-}$
   $\downarrow$

4. $Ca^{2+}(p.\ k.)^{2-} + 2NaCl \rightarrow 2Na^+(p.\ k.)^{2-} + CaCl_2$
   $CaCl_2 + Na_2SO_4 \rightarrow 2NaCl + \underline{CaSO_4}$
   $\downarrow$

   $\left.\right\}$ At the high salinity of the lake

At low salinity, when the water is saturated only in Ca carbonate, the exchange of cations with colloidal clays according to reactions 1 and 2 leads to a gradual enrichment of the water in $MgSO_4$ and $MgCl_2$ in place of NaCl and Na_2SO_4. At high salinity, when an abundance of SO_4^{2-} creates a saturated condition for the intermediate phase $CaSO_4 . 2H_2O$, reactions 3 and 4 take place also, and sodium sulfate continues to decrease, causing a further relative increase in $MgSO_4$ and $MgCl_2$ in the brine. In other words, *through cation exchange with clay particles, the lake, while remaining within group IIa, is gradually enriched in magnesium sulfates and chlorides.* This metamorphization is further demonstrated by the deposition of some, though very small, quantities of gypsum and glauberite in the sediments of lakes belonging to group *IIa* under hydrochemical conditions that prevent the formation of these minerals directly by precipitation from the initial solution through evaporation.

A consideration of the results of cation exchange in lakes of group *IIa* leads to a more complete representation of metamorphization in soda lakes through influx of calcium-sulfate waters and colloidal clay material. The genetic relations may be illustrated in the following form:

$$I\text{---soda lake: } 1a \rightarrow 1b \text{---} 1c \begin{cases} \nearrow IIa\text{---poor in } MgSO_4 \begin{cases} \text{influx of calcium} \\ \text{bicarbonate} \end{cases} \\ \searrow IIa\text{---rich in } MgSO_2 + MgCl_2 \begin{cases} \text{influx of calcium} \\ \text{bicarbonate and} \\ \text{cation exchange} \\ \text{with clay particles} \end{cases} \end{cases}$$

The origin of continental salt lakes of groups *IIb + IIc* may be explained by the fact that *all these lakes are found in regions underlain solely by sedimentary rocks of normal composition.* For the most part these are regions of plains, but some of them (Issyk-Kul', the Great Salt Lake) are in regions of mountain structures. To put it differently, *lakes of groups IIb and IIc of the sulfate class yield hydrochemical types of saline basins that develop by the leaching of salts from the underlying sediments in the drainage areas of the lakes.*

The substantial differences in the hydrochemical aspects within these groups may be due partly to peculiarities in the petrographic composition of rocks underlying the drainage areas, and partly to differences in intensity of cation exchange between lake water and the salts or clay material transported to the lake.

A characteristic feature of lakes in group *IIb* is an abundance of sulfates, including Na_2SO_4. The enrichment in sulfates is the consequence of abundant pyrite in the sedimentary rocks of the drainage areas, yielding SO_4^{2-} during weathering. The presence of *sodium sulfate* is due in some places to leaching (solodizing) of salt marshes, where Na_2CO_3 is leached out, in other places to the weathering of feldspar, which is present in the rock and which also yields Na_2CO_3. Most commonly both these processes act together. The Na_2CO_3 that forms reacts with $CaSO_4$ and forms Na_2SO_4. This mineral may develop in the lake or in streams and ground water feeding the lake. The accumulation of $MgSO_4$ and $MgCl_2$ in the lake water results as the water and, particularly, Na_2SO_4 interact according to the above reactions with colloidal clay particles introduced into the basin.

During weathering, when the generation of soda necessary for the sodium-sulfate reaction outweighs the force of cation exchange between Na_2SO_4 and clay particles in the basin itself, conditions arise that favor formation of highly sulfate lakes with Na_2SO_4; i.e. lakes of group IIb develop. When the weathering of rocks is slight, and little Na_2CO_3 is generated but cation exchange with clay material is well developed, lakes of group *IIc* form, with $MgSO_4$ and $MgCl_2$ but without Na_2SO_4. The less Na_2CO_3 that forms during weathering in this process, the less $MgSO_4$ there will be in the brine of the salt lake (according to reactions 1–4, p. 281) and the more $MgCl_2$

3. Chloride Lakes and their Origin

The origin of lakes in the chloride group is much less certainly known. Until recently there was a tendency for these lakes to be interpreted as end products of metamorphization in lakes of the sulfate class. N. S. Kurnakov and S. F. Zhemchuzhnyi, in 1917, explained the lack of sulfate in chloride lakes (the Perekop Lakes) by the exchange reaction of Haidinger. On this assumption, the muds of the Perekop Lakes must have contained large amounts of magnesium carbonate in the basic salt form. No such accumulation has been observed, however. It follows, therefore, that if we speak of metamorphization as a cause of sulfate-free lakes this takes place, not according to Haidinger's reaction, as assumed by Kurnakov, but according to cation exchange between the brine and the introduced clay mud, as first pointed out by S. A. Shukarev (1930).

Despite the logical conclusion and the experimental support of this view concerning the origin of magnesium-calcium-chloride lakes, another, fundamentally different, method should not be ignored. Boreholes in the region of the Perekop Lakes penetrate beds containing water of the calcium-chloride class, although but weakly mineralized. But, since the Perekop Lakes are known to be fed by water from this aquifer (Dzens-Litovskii 1936), it is natural to suppose that the distinctive composition of the brines is due to the character of the inflowing water. The Perekop calcium-chloride lakes may thus represent, not a brine that has long experienced metamorphization through cation exchange with clay particles, but a brine that has developed by being supplied with distinctive calcium-chloride ground water. As for the origin of the water itself, it appears most likely that the ground water is no different from that rising along tectonic fractures in the Black Sea depression—deep, calcium-chloride ground water. In the lower reaches of the Uzboi, calcium-chloride lakes occur that are also supplied with calcium-chloride water rising along fractures from deep horizons in the earth's crust (N. V. Tageeva). Similar examples of considerable upflow of calcium-chloride waters and of their freshening at the surface are known from the Russian platform and from other widely scattered localities, such as the Saratov district of the trans-Volga region (oral communication from I. B. Feigel'son).

In summary, we may assume that the variety of hydrochemical groups and the diversity of intracontinental salt lakes are due to interaction between solutions and with clay material, the solutions being of two initial types: sodic water (*Ia*), forming in regions of volcanic rocks; and sulfate water, forming in regions of sedimentary rocks. Some of chloride water is introduced from deep horizons of the sedimentary sequence, but most of it represents the end product of metamorphization of sulfate water.

4. Some Details concerning the Effect of Drainage Areas on the Composition of Halogenic Deposits

The above supports the belief that *the type of rock in the drainage area has a decisive effect on the hydrochemical type of salt lake and on the composition of the salt deposits that form during the different stages of lake salinity.* This close connection between continental salt accumulation and the geologic environment is manifested not only in the basic features of salt deposition, but also in many of the details of chemical composition.

In soda lakes that occur in a region of volcanic rock, for example, lithium may accumulate in the brine (Searles Lake in the U.S.A.) and boron may accumulate in the sediments, in various mineral forms.

In sulfate lakes in regions of sedimentary rocks, the content of potassium and bromine may be high, especially when the drainage area is underlain by ancient marine salt deposits rich in these elements. This relationship was demonstrated by A. G. Bergman, M. G. Valyashko, and I. B. Feigel'son (1953) for lakes in the Ural-Emba and Chelkar regions, in the Ustyurt district, and in the Aral region. The average potassium coefficient for the Ural-Emba region is 8·12; for Ustyurt it is 1·30, the Amu-Darya region 8·75, Chelkar 3·58, and the Astrakhan region 1·75. Thus, the highest potassium content is found in lakes of the Amu-Darya and Ural-Emba regions. In the latter the enrichment is undoubtedly due to erosion of Permian potassium-bearing salt deposits. In the Amu-Darya region the cause is probably weathering of Jurassic potassium salts in the Zeravshan system along the upper reaches of the Amu-Darya.

Bromine is most strongly concentrated in waters of the Ural-Emba region, less so in the Amu-Darya region (0·625%). The other regions are poorer in bromine, although the average content is not very different from that in the Amu-Darya lakes (0·49% in the Chelkar lakes, 0·52% in the Ustyurt lakes). Thus, although bromine is generally found in the same regions as potassium, in its detailed distribution it deviates appreciably from that of potassium and exhibits individual peculiarities.

Thus, *whether the principal features in the composition of salt deposits of continental, lacustrine origin or the fine details of this composition are considered, their derivation from the rock type underlying the drainage area in which the salt deposits form is proved convincingly.*

This derivation from the underlying rock type should be strongly emphasized; in salt deposits of marine origin the characteristics are markedly different. The effect of the rock composition in the drainage areas on deposits of marine origin is incomparably less, perhaps undetectable in many examples. Furthermore, even when such an effect may be observed, traces of it appear much later, chiefly at advanced stages of salinity. *The principal, decisive features of marine salt deposition as well as the finer details of the composition of marine salt deposits are due to the composition of the sea water. The result is a much greater uniformity among geochemical types of marine salt deposits.*

Since the study of salt has hitherto been based primarily on marine sequences, on which the effect of rock composition in the drainage areas is small, this factor has usually been ignored in analyzing salt deposits in general and continental deposits in particular. Nevertheless, this approach is fundamentally incorrect for the continental deposits. *Salt deposits of continental lakes, as a whole and in detail, reflect the rocks underlying the drainage basins in which the lakes occur.*

5. The Distribution of Continental Halogenic Deposits within Arid Regions

In conclusion, some consideration may be given to the localization of continental salt deposits in arid zones of the present day. There are two fundamentally different aspects to this matter: (*a*) distribution within arid zones of *different stages of salinity in the lakes* and of salt deposits characterizing these stages and (*b*) localization of basins of *different hydrochemical type.*

In the first part of this monograph, where soil salinization in arid regions was described, we noted that this depends on the nature of the climatic regime and consequently changes from the margins of arid zones toward their centers. The margins are characterized chiefly by carbonate salinization, which passes inward to sulfate (gypsum) salinization, and in the central part to chloride (halite) salinization. Extreme development of arid climate is characterized by nitrate salt flats.

The distribution of *stages of chemical sedimentation* in lakes conforms to this scheme in its general outlines. On dry steppes at the margins of arid zones are lakes with minimum mineralization and with sediments in which only the least soluble salts are deposited: calcite, dolomite, and magnesium silicates. In the semi-arid zone the lake water is more strongly mineralized and the sediments therefore contain, in addition to carbonates, calcium and sodium sulfates: gypsum, mirabilite, thenardite, and glauberite. In the desert belt proper are lakes of the highest mineralization. They are commonly dry, and at times covered with layers of sand. Sedimentation is in the chloride stage: halite is precipitated, with a greater or lesser admixture of blödite and epsomite.

Maps of arid zones show, however, that this type of zoning among lacustrine salt deposits is most clearly displayed for lakes of small and intermediate sizes: up to 10–15 km in length, 5–7 km in width, and a few tens of square kilometers in area. Large lakes such as Balkhash, even more the Aral Sea, do not fit into this zonal pattern, because they represent a much lower stage of development than the zone in which they occur. This is natural, because the rapidity with which a basin passes from one stage of chemical development to the next is basically determined by the size of the basin and by the water feeding it. *Large salt lakes, therefore, lag in their development behind small and medium-sized lakes.* On a map showing the stages of chemical sedimentation in lakes, this is reflected in unzoned scattering of the low stages of chemical sedimentation within a zone of a much higher stage of development. The map is therefore more patchy in appearance than one showing soil salinization.

The distributional pattern of different hydrochemical types of salt lakes is completely different from that of corresponding salt deposits within arid zones. *Hydrochemical types of basins depend strongly not on the climate but on the rock type underlying the drainage areas, and this means that their geographic localization is due to the geological characteristics of the various parts of the arid zones.*

On the whole, the development of continental lakes in arid regions is controlled by two factors: the climatic conditions and the type of rock in the drainage areas. *The climate determines in general the stage of mineralization of the lake and the corresponding stage of chemical sedimentation:* carbonate, sulfate, and chloride. *The type of rock in the drainage area controls the hydrochemical class of arid lake basin, i.e. the particular composition of the water at each stage of salinity and, consequently, the specific individual features of chemical sedimentation at these stages.* Since climate is zoned within arid belts, it follows that the distribution of sedimental stages of salt within them is also zoned, repeating in general outlines the zones of soil salinization. The rock types in drainage areas do not, however, depend on climate. This creates variations in chemical sedimentation from one lake basin to another within any particular belt of one stage of development.

PART THREE

HALOGENESIS IN PHANEROZOIC TIME

CHAPTER 1

FACIES CORRELATIVES OF RECENT DEPOSITS AMONG ANCIENT SALINE FORMATIONS

The formation of saline sediments in ancient basins, as in present-day basins, was characterized by development in stages, which led to a systematic change, in the course of time, from carbonate sediments to sulfate and from sulfate to chloride. As a result, salt deposits of the early epochs are now seen as groups of rocks systematically related to each other both vertically and in areal distribution. This association of intimately related rocks (physico-chemically) is a classic example of a formation in the facies-genetic sense, as described in Vol. 1 of this monograph.

This means that a study of ancient halogenesis at our present state of knowledge is reduced to the recognition and description of the types of halogenic deposits, of their essential constitution and structure, and of their development and distribution in time and space. Examination of these questions constitutes the present part of this monograph.

Before considering the description and analysis of the formations, some preliminary questions may be raised. In the history of sedimentation, salt formations follow deposits that accumulate at the stage of low mineralization. What type of rock is most favorable to begin a salt formation and what is the relation of the salt formation to the deposit of low mineralization? What facies types of salt formations should we recognize? In other words: How do we classify these formations?

A. PRINCIPLES OF CLASSIFICATION OF SALINE FORMATIONS

The total volume of salt formations may be calculated only approximately. Such formations consist of deposits of readily dissolved salts: $CaSO_4$, NaCl and K and Mg chlorides and sulfates. Thus, in a succession of rocks in the arid zone, salt formations are regarded as beginning with the appearance of gypsum or anhydrite.

Despite the simplicity of this principle, its practical application meets with considerable difficulty, necessitating supplementary criteria for specifying the limits of salt formations in each particular example. The following example illustrates this.

The salt content of the Middle Miocene basin of the southern U.S.S.R., despite the fact that it undoubtedly lay in the arid zone, was low, about 2·1–2·2% (Fig. 144). This is demonstrated by the similarity of its fauna to that of the present Black Sea. Nevertheless, gypsum accumulated in various parts of this Middle Miocene sea. Furthermore the character of the deposits differed in different places. Two types may be recognized.

In the first type the gypsum occurs in small lenses 1–3 km long, and ranging in thickness from a few centimeters to 2–3 m. Normally only one lens will appear in any succession. In places the gypsum merely impregnates the sands and clays, forming centimeter-thick seams. At El'dar-Eliiskii, 5 km southwest of Kerch, a single bed of gypsum 3–5 m thick may be traced for about 2 km. Nearby, the gypsum in the Chekurkoyashkoe deposit forms two beds with a combined thickness of 1·8–3·3 m, but it is strongly contaminated with sand and clay, to as much as 30%. In the northwestern Caucasus, at the villages of

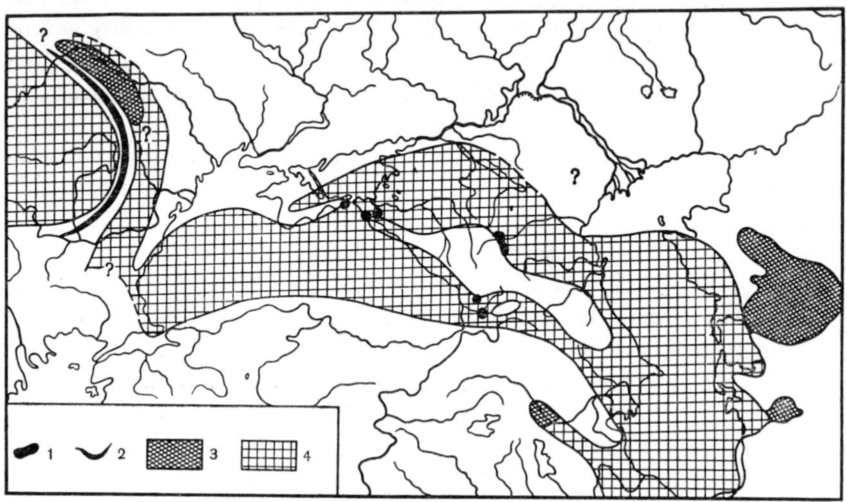

FIG. 144. Paleogeography of the Middle Miocene basin (modified after V. P. Kolesnikov).
1. Salt facies; 2. mountain chains; 3. salt formations; 4. marine deposits.

Krymskaya and Varenikovskaya, a gypsum lens in the Karagan horizon ranges up to 5 m in thickness and may be traced for more than 2 km. Along the Sunzha River, 35–40 km east of Mount Ordzhonikidze, a unit 25 m thick occurs in the middle part of the Datykh Series, in which gypsiferous and saliniferous clays are interbedded with thin beds of gypsum (0·5–1 cm thick). At the base of this unit a bed of argillaceous gypsum, 0·5 m thick persists for several kilometers, though its total extent has not been determined. It may indeed be several tens of kilometers long, forming a continuous bed, or it may be a chain of individual gypsum masses. Local lenses of halite up to 1 m thick also occur. In the Rion valley, north of Kutaisi, the Karagan horizon includes gypsum beds, many as thick as 1 m, forming an economic deposit. About 20–25 km south of Kutaisi, several lenticular deposits of gypsum are found, tens of meters long, the largest being 10 m thick. *A characteristic feature of all these deposits of saline rocks is that spatially they do not form isolated bodies, but are subordinate to the general mass of normal marine sediments.* They formed during very short intervals of time, their development being purely episodic, after which salt deposition completely ceased. *Salt deposition of this type did not develop as a significant process, independent in space and time. It appeared momentarily, against a background of normal marine sedimentation and then ceased.*

The second type of gypsum is very different. Examples are found in the salt deposits of the Carpathian foredeep, the trans-Caspian region, and Nakhichevan. All of these reflect a process that *lasted for a long time over an area isolated from the remainder of the Middle Miocene basin.* As a result a special rock association formed, clearly different from a marine association in its composition and texture. Thus, the gypsum-carbonate Upper Tortonian group of the Ciscarpathian region extends without interruption from the banks of the San and Vistula in Poland to the northern border of Moldavia, forming a belt more than 330 km long and ranging in width from 1·5 km at Nemirov in the northwest to 40 km in the region of Zaleshchikov in the southeast. Throughout this area the gypsum-anhydrite rests on an irregular erosion surface that causes considerable thickness variation in the gypsum-anhydrite. The distribution of this horizon is systematic. The purest varieties of sulfate rocks were deposited far from the rising Carpathian mountain chain, on the platform in the region of the present Dniester. Towards the Carpathians clastic material appears in this horizon, first in the form of impurities in the gypsum and anhydrite, then as thin beds of clay. Near the Carpathians the thickness of the sulfate rocks decreases. Beds of sandstone (Khaluchi) as well as clay, with pebble- and coarser conglomerates are present. Gypsum is then reduced to lenses or thin beds. Thus, the removal of material during the formation of the sulfate horizon took place with the growth of the Carpathians. Fluvial agents transported large quantities of clastic material into the Carpathian foredeep, freshening the water and preventing chemical precipitation of sediments near the shore. Gypsum was precipitated only at some distance from the Carpathian chain, and reached maximal development in that part of the basin on the platform (the Dniester region). The thickness of the sediments also changes regularly. In the inner zone of the Carpathian foredeep, near the edge of the Carpathians, the thickness of the horizon, represented here by purely clastic rocks, is about 100 m. In the inner part of the depression, the thickness falls to 35–40 m, and on the platform, where purely chemical sedimentation took place, it decreases to 25–35 m (Glushko 1956).

The gypsiferous deposits of Middle Miocene age in the trans-Caspian region were formed in two well-defined gulfs: the Ustyurt in the north and the Turkmen in the south. The area of the first was 5–6 times that of the present Gulf of Kara-bogaz. The area of the second was about one-fourth to one-third the area of that body. The deposits consist of alternating gypsiferous, calcareous, and sandy rocks, the total thickness amounting to a few tens of meters. On the western scarp of the Ustyurt Plateau, the thickness is 50 m; in the vicinity of the Tuarkyr anticline it is about 30 m, and in outcrops to the east it does not exceed 10 m. The thickness of individual gypsum beds, according to Yu. B. Aizenberg, ranges from 10–12 m on the eastern shore of Karabogaz to 1 m at the eastern scarp of the Ustyurt Plateau. Details of the distribution of the sediments within the gulf are not yet known. The thickness of the gypsum in the Turkmen gulf is 0·5–1·5 m near Kyzyl-Arvat. In the southeastern part of the Nakhichevan basin, which in the Middle Miocene (Karagan time) was at the head of a long marine gulf, not only were gypsiferous clays deposited, but also rock salt, to a thickness of 46–85 m.

The basic difference between the processes giving rise to the Middle Miocene deposits of the first and second types is, therefore, clear. The first type represents transient episodes, during which small lenses of gypsum and, sometimes, of halite formed, *subordinate to the total mass of deposits accumulating from a weakly mineralized marine basin in the arid zone.* The second type represents prolonged formation of a rather thick complex, *clearly isolated spatially from marine deposits proper with their characteristic sequence.*

The question naturally arises: How may the salt formations of each of these origins be distinguished? In order to answer this question it is necessary to recall that by a formation we mean the association of sedimentary rocks that accumulates during prolonged persistence of uniform environmental and tectonic control. It is clear that *salt formations constitute only the gypsum-anhydrite horizon of the Ciscarpathian region, the gypsiferous sequence of the trans-Caspian region and Turkmenia, and the saline beds of Nakhichevan, but the small gypsum lenses found individually in near-shore marine deposits cannot be regarded as formations. These saline lenses are essentially local facies in a formation of an entirely different type.*

Thus, halogenic rocks found in arid-climate deposits may form an independent salt formation or may be members of formations of weakly mineralized basins. *They are distinguished as formations only where they formed as a result of prolonged persistence of definite environmental and tectonic conditions that favored halogenesis and gave rise to a group of rocks having special characters in time and space, being isolated from deposits forming in basins of low mineralization.* When these conditions do not obtain, saline rocks are but members of arid-climate formations in the early stages of mineralization.

The distinction between saline rocks as formations and as facies (members of a formation in the early stages of salinity) has not hitherto been made. (cf. A. A. Ivanov 1960, M. P. Fiveg 1960a). Nevertheless, it appears that the distinction is of fundamental significance. *It helps the correct understanding of the conditions under which salt deposition in lithogenesis develops in the growth of formations.*

The classification of saline formations, though they are so variable in facies, constitution, and texture, is based on the same environment-tectonic principles employed for classifying formations of humid zones (Vol. 1, Chap. 2). From this point of view, all ancient saline formations may be reduced to five well-defined types.

The first includes *continental formations* that accumulated on plains of arid regions or in areas of intermontane basins. The salt formed in fairly large lakes similar to modern salt lakes, within basins that may be either erosional or tectonic. The second type of saline formation is *lagoonal*. It is made up of a complex of deposits that form in areas adjacent to the sea, in small saline lagoons morphologically similar to the present-day estuaries along the Black Sea coast, the Sivash, and also the irregular chain of inter-island straits and gulfs along the eastern shore of the present Aral Sea. Tectonic structures here correspond to no single estuary or lagoon, but rather to a series of them. The third type of saline formation has accumulated during *prolonged persistence of large salt basins that are clearly enclosed gulfs, communicating with the open sea only through relatively narrow straits.* These were approximately the facies analogues of the present Gulf of Kara-bogaz and the Bocano de Virrila, though they differed in size. Each such basin corresponded to a single large syncline. The fourth type of saline formation includes deposits in highly saline marginal segments of huge open seas. These segments were more or less vast *open gulfs*. They corrrespond to syneclises about the margins of epicontinental seas. There are no existing correlatives of such salt-producing basins. Lastly, the fifth type of saline formation includes marine complexes proper, accumulating in tremendous elongate intracontinental seas, usually of irregular outline, set in the middle of a broad continent and communicating with the ocean at only one end. The other end terminated in the arid zone of the continent. Tectonically the saline formations of intracontinental seas corresponded to structurally complex regions with numerous syneclises and anticlises. Salt-producing seas of this type also are non-existent at the present time.

Thus, of the five formation-producing environments of ancient time, only three have approximate analogues in the present epoch. The other two are entirely absent and must be considered specifically ancient features. These facies types of saline formation will each be considered in turn.

B. HALOGENIC FORMATIONS OF CONTINENTAL TYPE

1. GENERAL REMARKS

Continental saline formations are rarely encountered, their distribution being confined to the latest episodes in geological history. They may be observed in the process of formation at the present time, and their development in post-glacial time may be defined. Our views on modern saline formation, already discussed, are based in considerable measure on the study of these deposits. They are much more rarely encountered among the Lower Quaternary and Neogene deposits that are known to us. Examples of these are the halite-glauberite formations of the Tien Shan intermontane depression (V. N. Shcherbina 1952, 1956a). Analogues are known from the Pliocene of the Tsaidam basin of China, and it seems likely that many others may occur in parts of Asia, Africa, and North America. There is a sharp decrease in the number of continental saline formations among Paleogene deposits.

U

Older continental saline formations are extremely rare, and those that are known are very small, in the Lower Carboniferous of the Dzhambul region, and of uppermost Devonian age in the Teniz basin, among others. *Continental saline formations are thus geological rarities, limited in their distribution almost exclusively to the very latest moments of geological history.* Despite its rarity, however, this type is of great theoretical importance, since it permits us to appreciate some essential aspects in the accumulation of saline formations not apparent in present-day salt deposition.

2. The Miocene Halogenic Formations of the Tien Shan Intermontane Depression and their Origin

In the region of the Tien Shan lie a series of intermontane basins of varied size filled with Neogene and Quaternary continental deposits (Fig. 145). Some of them even appear to have late Paleogene strata at the base. Their characters are summarized in Table 33.

TABLE 33

Sizes of Intermontane Basins and Thicknesses of their Saline Deposits
(from A. A. Ivanov and Yu. F. Levitskii 1960)

Basin	Length along greatest axis, km	Width, km	Thickness of Miocene deposits, m	Thickness of saline deposits, m	Remarks
Eastern Chu .	—	—	4000–5000	>1000	Lower horizons of salt-bearing series not exposed.
Kochkorka .	40	>20	>2000	Several hundred m	Because of deformation, the thickness of the saline series was not determined
Dzhumgol .	70	from 5–10 to 25–30	—	75	—
Kekemeren .	60	100	100	Tens of m	—
Ketmen'-Tyube	50	10–30	?	260	—
Naryn . .	250	from 10–15 to 55 on the west	>2000	500	—
Kegen' . .	140	from 10–15 to 40–45	1500–2000	500	—
Ili . . .	—	—	up to 1500	100	—

The length of these basins normally ranges from 40 to 250 km, and the width from 10 to 30 km. The thicknesses of the continental deposits that floor them are tremendous, generally exceeding 1500 m in places as much as 4000–5000 m. The salt-bearing (or gypsiferous) formations proper normally have thicknesses of the order of many hundreds of meters.

V. N. Shcherbina distinguished three interrelated facies among the saline

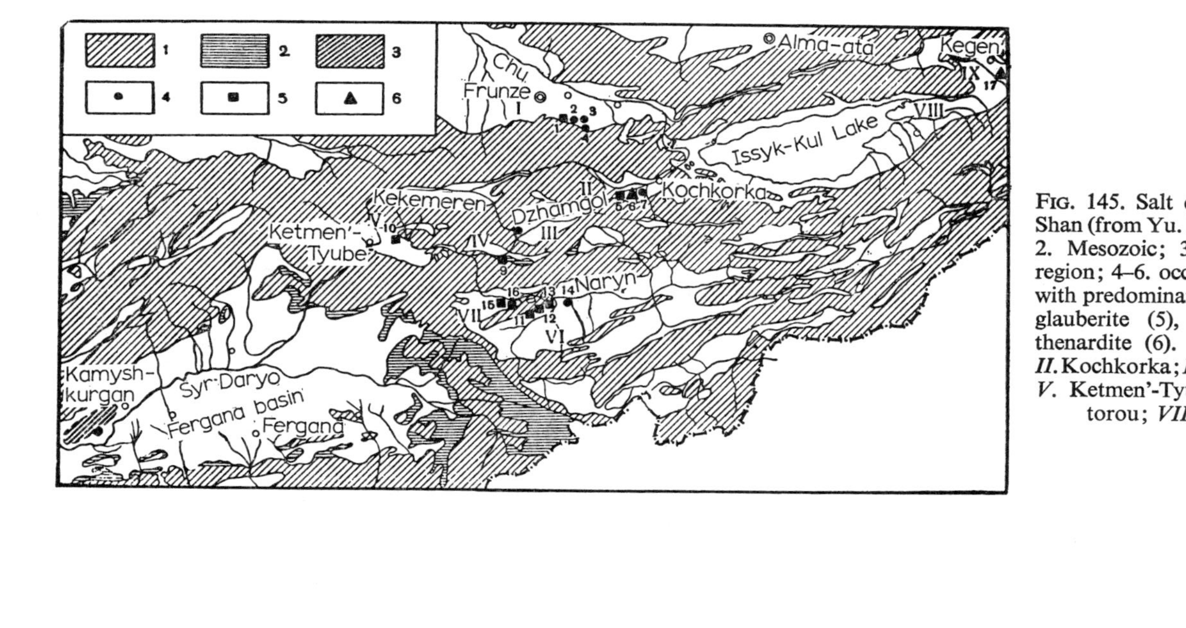

FIG. 145. Salt deposits in basins of the Tien Shan (from Yu. F. Levitskii 1960). 1. Paleozoic; 2. Mesozoic; 3. Samgar-Kamyshkurgan salt region; 4–6. occurrences of salt mineralization with predominance of glauberite (4), halite and glauberite (5), and halite, glauberite, and thenardite (6). *Intermontane basins: I.* Chu; *II.*Kochkorka; *III.*Dzhamgol; *IV.* Kekemeren; *V.* Ketmen'-Tyube, *VI.* Naryn; *VII.* Toguz-torou; *VIII*, Issyk-Kul; *IX.* Kegen'.

formations of the Tien Shan: clastic, intermediate, and salt-bearing. The clastic facies is developed chiefly in the peripheral parts, especially the eastern parts, of the formations. It consists of red sandy conglomerates with subordinate argillaceous rocks and with calcareous or, locally, gypsiferous cement. In the eastern parts of the formations the thickness of this facies reaches hundreds of meters, but in other parts of the marginal zone it is measured in tens of meters, or perhaps a few hundred. It consists of *slope-wash and fan material, and partly ordinary stream deposits.* This facies gives way towards the inner parts of the formations to the so-called intermediate deposits. These are chiefly clays and marls, greenish, dark brown, light brown, or reddish grey, with subordinate sandstones and pebble conglomerates with calcareous cement. One may find sporadic lenses and seams of gypsiferous rock and gypsiferous clays and, in places, disseminated halite. The thickness of this facies unit is measured in tens and hundreds of meters. *In facies this is chiefly a near-shore lacustrine deposit with some members of fan and deltaic material.*

The salt-bearing deposits are found mostly in the central and western parts of the formation, where they accumulated at considerable distances from the zone of erosion and denudation.

The salt-bearing succession always begins with clastic rocks, which grade upwards first into gypsiferous and then into salt rocks. The section is always capped by a bed of gypsum. These deposits represent a lacustrine facies, forming chiefly in the central part of the lake basin. Changes in composition through the section represent its inception as a basin, its gradual change to a saline basin, and then its conversion from a saline basin to a fresh-water basin. *Along both the depositional strike and dip the saline deposits change systematically and gradually to gypsiferous deposits and then to sporadically saline gypsiferous-argilous deposits.lace These in turn also give way gradually to deposits with disseminated gypsum, and then to non-gypseous clay and sandy-clay deposits.* Rocks of the saline-gypseous facies are mostly greenish, green, or greenish grey, but in some basins they are red (especially the gypseous horizons, but in the Kekemeren basin the saline beds also). The thickness of this facies is measured in hundreds of meters.

Four distinctive features characterize the composition of the saline formations of the Tien Shan. One of these is the abundance of clastic material. Even in the purest varieties of gypsum and salt rocks, the clastic fraction ranges from 1 to 15%. Normally it ranges from 15 to 35%, and in the argillaceous varieties is more than 35%, even reaching 80–95% in the rocks with disseminated salt and gypsum. Fragmental rocks, normally argillaceous but rarely marly, are interbedded with the saline accumulations. *The saline deposition in the Miocene intermontane basins took place against a background of intense mechanical sedimentation.* "Mechanical analyses have shown that the clay material in the salt and gypsum rocks are very finely dispersed. The fraction smaller than 0·01 mm constitutes 75–90% of the rock, and the fraction over 0·1 mm is essentially absent, and never above 3%. The content of the silt fraction (0·1–0·01 mm) ranges from 1 to 25%. The same distribution holds also for most of the argillaceous rocks interbedded with the gypsum and salt rocks. Among them, however, varieties rich in the silt fraction (up to 40–50%, and poor in clay (30–40%) are known. In these the content of the fraction greater than 0·1 mm increases to 10–12%, and even some of the fraction 0·25–0·40 mm may be present" (Shcherbina 1956a, p. 119).

Another distinctive feature of the continental salt deposition of the Tien Shan is the abundance of sulfate minerals. In the early stages of salt deposition this sulfate is gypsum, later converted in part to anhydrite. As salinity increases, glauberite, thenardite, and, in places, blödite form. Halite is almost the only other salt mineral that appears. In their quantitative distribution, the sulfate minerals and halite may be clearly distinguished in the saline formations of the Tien Shan: in one the sulfates are sharply dominant among the salts. Gypsum deposits are found about the margins of the formation and these give way towards the center to glauberite and halite-glauberite. Pure halite beds are lacking, or poorly developed. In these rocks halite is but a minor impurity. Thenardite deposits are also found in this association. This well-defined sulfate salt formation is found in the eastern Chu, Kochkorka, Dzhumgol, and Naryn basins. In the other variety of saline formation halite is the principal mineral. Gypsiferous deposits in the marginal parts of the formation give way towards the center to halite, either pure or with small impurities of glauberite. No independent beds of glauberite are present, or they are few and small. A feature of these halite salt formations is that the rare beds of blödite found in the region are confined to them (in the Ketmen'-Tyube basin). Deposits of thenardite are also found. This halite type of continental formation is clearly represented on the west of the Fergana depression, and in the Kekemeren, Kegen', Ketmen'-Tyube, and Toguztorou basins.

A feature common to both varieties of saline formation is the absence of any appreciable potassium minerals. Similarly, with rare exceptions (a few lenses of blödite rocks in the Ketmen'-Tyube basin), *deposits of MgSO$_4$ are also essentially absent in the continental saline formations of the Tien Shan* (maximum 1·2%).

A fundamental feature of the saline formations of the Tien Shan is the abundance of carbonate. In gypsum-free and salt-free clays, the carbonate content ranges from 3 to 30%, in places to as much as 40–50%, marls thus occurring along with calcareous clays. In the gypsiferous clays and gypsum rocks carbonates make up from 1 to 22%, in the glauberite rocks from 3 to 8%, and in the halite rocks from 0·02 to 10%. Furthermore, "the content of microgranular carbonate is generally inconsiderable in the fraction smaller than 0·005 mm. Most of the microgranular carbonate is concentrated in the fraction 0·05–0·005 mm, falling sharply in the fraction larger than 0·05 mm, and is essentially absent in the fraction larger than 0·1 mm. This general pattern may be traced through all varieties of rock. In the 0·1–0·05 fraction, microgranular carbonate normally occurs as masses having aggregate polarization, indicating a finely granular state. On the other hand, in the fraction finer than 0·005 mm, in which the smallest quantity of scattered micro-granular carbonate is found, the carbonate is always present in individual small grains" (Shcherbina 1956a, p. 119).

The mineral composition of the carbonate varies. In red and grey calcareous clays the carbonates are calcite, manganocalcite, ferrocalcite, and dolomite. Calcite is always sharply dominant, quantitatively ranging from a few percent to 40, and even 70%. The remaining carbonate minerals, particularly dolomite, are not invariably present, and when present, the amounts are small. Two varieties of gypsum may be distinguished: one, in which calcite is the dominent carbonate, manganocalcite and dolomite forming only negligible admixtures, and the other, in which magnesite is the chief carbonate. Most of the microgranular carbonate proves to be magnesite. Calcite may also appear in this form but always in subordinate amounts, as may also microgranular dolomite and ankerite. In the salt rocks—of glauberite, thenardite, and halite—magnesite is always the dominant carbonate. Calcite is not always present, and when present it occurs in but small quantities. There is even less ankerite.

Thus, *the outstanding feature of chemical sedimentation in the Miocene basins of the Tien Shan is that the carbonate association of pure calcite, which is typical of basins with low mineralization, gives way to magnesite at the gypsum-accumulating stage, and this holds to the highest stage of salinity.* The formation of dolomite is negligible at all stages of chemical sedimentation, and the dolomite association intermediate between calcite and magnesite does not form.

For the sake of completeness it may be added that spectrographic analysis has established that strontium is invariably present in small quantities in glauberite rocks (hundredths and, rarely, a few tenths of a percent). Barium has been occasionally detected in thousandths of a percent.

The field relations of the several types of saline rocks in the Tien Shan formations have been much less studied than their composition. Exposures of the salt-bearing rocks are generally confined to anticlinal structures, where the saline deposits have participated in complex salt tectonics and have lost their primary features. Furthermore, these exposures are of only limited size, measuring up to 10–15 km. The salt rocks form beds from 1–1·5 m to many meters thick, extending for many hundreds of meters, up to 1–1·5 km. These salt beds are separated by beds of gypsum, gypsiferous and salt-bearing clays. Salts of different compositions appear repeatedly, in the succession, but it has not yet been possible to define any macrorhythms in the succession. Neither are there generally any traces of seasonal microrhythms nor of the specific pinnate structure of crystals, so clearly manifest in present-day halite deposits. All these distinctive negative features of the saline rocks are apparently the result of intense recrystallization.

The Tien Shan saline formations originated in large, salty, internal-drainage lakes, apparently like the terminal lake of the Chu River, situated among drainage areas of plains type and undergoing prolonged, slow subsidence.

The mineral compositions of the salt phases and of the included carbonate reliably indicate the hydrochemical type of basin in which the deposits accumulated. The abundance of sulfate rocks points to lakes of the sulfate class (*II*). The absence of dolomite among the carbonates means that magnesium-carbonate lakes (group *IIa*) were entirely absent from these sulfate basins, as were perhaps calcium-carbonate lakes (*IIb* and *IIc*). The discovery that massive accumulation of magnesite took place as early as the gypsum-forming stage leads to the conclusion that the lake waters were rich in $MgSO_4$, and that they probably corresponded to the Aral-Caspian branch of sulfate basins, as was first pointed out by V. N. Shcherbina (1956a). The rarity of blödite in the sediments, a mineral usually characteristic of this group of lakes, is explained not by the small amount of $MgCO_3$ in the water, as Shcherbina stated, but by the fact that this basin very rarely reached the degree of salinity necessary for blödite precipitation. The same circumstance is of primary importance in explaining the absence of potassium salts.

"In these basins, which are found in a hot, dry climate and are supplied by streams, isochoric or near-isochoric evaporation took place over a prolonged interval of time, leading to the accumulation of large quantities of Ca, Mg, Na, and K sulfate and chloride salts in the lakes and to the precipitation of some of these salts in solid form. The lakes were very distinctly zoned hydrochemically. The most saline parts of the lakes were at some distance from the source of supply and from the shore-line. This fact is supported particularly by the very fine dispersion of the clay material in the salt-bearing and gypsiferous deposits. These parts of the lakes were characterized by poor water exchange with other parts of the lakes and by stagnation of their water, apparently caused by topographical irregularities of the lake floors. This was pointed out particularly by S. S. Shul'ts in his analysis of the conditions under which the Tertiary deposits of the Kochkorka basin formed. Anticlines, in which the Tertiary rocks of the Kochkorka basin were folded, developed simultaneously with the formation of the Tertiary lakes. Shoals forming in these lakes in this way cut off individual segments. The lake margins, adjacent to the shore lines and near the sources of water supply, were characterized by low salinity, and gypseous deposits formed in them. The greatest freshening occurred in zones near stream mouths where marly, clayey, sandy, and conglomeratic rocks, free from salt and gypsum, were deposited synchronously with the salt-bearing and gypsum-bearing rocks" (Shcherbina 1956a).

The deposition of salt took place in stages. The Tertiary basins were fresh at first, when red and grey clays with some carbonate were laid down. As salinity increased, gypsum-bearing sediments accumulated, and continued to do so for a comparatively long time. When mineralization became greater than 20%, salt deposits began to form, containing glauberite, halite, and, at times, thenardite and blödite.

Some of the lakes show evidence that the degree of salinity remained approximately constant for prolonged intervals, as indicated by thick sequences composed solely of salt rocks (halite-glauberite deposits). Some show evidence that the mineralization fluctuated periodically and widely, as a consequence of which glauberite, halite, and thenardite alternated with gypsum. In some places, salt deposition ceases entirely. This latter is seen in sections of interbedded salt and gypsum rocks with clayey-marly rocks in which no salts of any kind occur. The pattern of such changes has not yet been studied.

The deposition of saline formations ceased with gradual freshening of lakes in all the basins. This caused a change first to gypseous rocks and then to clay-marl deposits with remains of fresh-water faunas.

3. Continental Halogenic Formations of Soda and Chloride Types

In the section of this monograph devoted to present-day salt deposition it was pointed out that, in addition to sulfate lakes, there are also soda and chloride lakes with salt deposits differing sharply in composition. In the example of the halite-glauberite formations of the Tien Shan we have discussed saline complexes that formed in basins of sulfate type. The question arises: Are there in the sedimentary sequence saline formations that have formed in basins of soda or chloride type?

In principle, of course, the answer can only be positive, because soda lakes are known to have existed in earliest times. They were probably more extensively developed then than in more recent time, in view of the magmatic rocks that must have underlain them, such rocks having covered much greater areas of the earth's surface in early geologic times than now. But in practice, except the post-Quaternary formations of soda lakes, which may still exist, no single example from the geologic past of any soda complex as an independent saline formation can be recognized. At best either traces of soda lakes during the early stages of mineralization are met (see the section on dolomite in Chap. 2), or negligible occurrences of sodic salt deposition. We know even less about formations of the chloride type, because no saline deposit related to a basin of this hydrochemical type has been recorded.

It should be stated that this gap in our knowledge is due partly to incompleteness of the geologic record, and partly to the lack of thorough analysis of the available continental deposits. It may be recalled that the halite-glauberite formations of the Tien Shan became known only very recently, and the salt deposits of the Tsaidam basin were investigated for the first time only in the last few years. Undoubtedly further work on the continental deposits of Asia will define saline formations of sodic and chloride types.

4. The Restricted Distribution of Continental Halogenic Formations; Conditions controlling Potassium Content.

Even considering the incompleteness of the geologic record and the restricted study of continental formations, two cardinal facts have been established: (*a*) *continental saline deposits have a very restricted distribution among geological deposits*, and (*b*) *the normally observed incompleteness of the saline process is such that it always appears to be represented by the stage of halite deposition, with almost complete absence of potassium accumulation.*

Potassium accumulation is rarely found, and with certainty only in the Tsaidam basin.

The cause of the extreme rarity of intracontinental saline formations was pointed out by V. N. Shcherbina (1956). In order for any considerable thickness of saline sediments to accumulate, beginning in basins with moderate salinity and continuing to high salinity, it is necessary that the basin maintain a constant volume of water for a long period of time, and for the floor to subside. The first condition requires evaporation from the surface of the water in amounts that equal the annual supply of water to the basin, or exceed this but slightly; i.e. evaporation must be isochoric or very nearly so. When this condition is not met, and evaporation greatly exceeds the water influx, the basin quickly dries up. Saline formations do not develop, therefore, but under the best conditions a thin salt member may be developed in a clastic-carbonate formation. When the annual influx of water exceeds the evaporation, the lake becomes freshened and the deposition of salt ceases altogether. When tectonic activity is weak the effect is also negative. The floor of the basin then subsides very slowly, sediments fill the lake and the water overflows the banks, forming a very thin layer over a broad area. This increases evaporation, the isochoric process fails, and the accumulation of saline formations is terminated. *The deposition of continental saline formations has thus required the combination of isochoric, or near-isochoric, evaporation and an appreciable tectonic activity that has preserved the basin even when salt deposition has become rapid.* But isochoric evaporation, taken alone as a specific process, could not have been either frequent or widespread. Its combination with the necessary tectonic activity must have been especially rare, and in consequence continental saline formations were exceptionally rare in the past, being the exception rather than the rule in arid lithogenesis.

The incompleteness of continental salt accumulation, in particular the general absence of potassium-salt deposition, follows from the specific supply of saline constituents to the lake basins. This took place entirely by stream discharge, which carries only a negligible content of potassium salts. The potassium content was also low in the water of the basins in which the saline formations were deposited. For even small lenses of potassium salts to be deposited in such a saline formation, as from the lakes in the Tsaidam basin, tremendous volumes of inflowing fresh water must have evaporated. Not only the prolonged existence of such lake basins would be essential but also very large size, because during isochoric, or near-isochoric, evaporation the volume of water that evaporates, and consequently the volume of inflowing water, is proportional to the area of the water surface in the basin. It follows that *only vast intracontinental lake basins can become sites of complete development of the salt-accumulating process, culminating in the precipitation of potassium salts.* Medium-sized, and especially small, basins result in incompleteness of the salt-forming process, at best reaching only the halite-stage of precipitation. Exceptions are found only in lakes supplied with water having a markedly high content of potassium, but these are extremely rare.

The great rarity of continental saline formations and the incompleteness of development of their salt-accumulating process results from the realization the coincidence of rare phenomena, following a strict pattern: it is not a fortuitous circumstance.

For this reason saline formations of the geologic past are almost exclusively marine, not continental, and the salt-forming process in them is found frequently to have reached completion. Nevertheless, under some facies conditions the process developed very distinctively, and this led to considerable variation in the facies of types of marine saline accumulation. These will now be discussed.

C. HALOGENIC FORMATIONS OF LAGOONAL TYPE, THEIR COMPOSITION AND DEPOSITION

The term "lagoonal" formation refers to that complex of sediments that accumulate in saline lagoonal basins, i.e. in gulfs of restricted size separated from the sea by bay bars or barrier beaches. The shape of the lagoons and their connections with the sea change considerably with time. In the initial stage a lagoon connects with the sea through straits. Later, barrier beaches completely cut off the lagoon from the open sea, and supplementary spits within the lagoon itself subdivide it into several lakes. When these become filled with sediments the history of the lagoon comes to an end. Lagoonal formations are among the most widespread of salt-bearing complexes, and are characterized by the following well-defined typical features.

1. THE UPPER CAMBRIAN UPPER LENA FORMATION AND ITS CORRELATIVES ON THE SIBERIAN PLATFORM

The Upper Lena (Verkholenskaya) formation occurs in a wide belt along the southern margin of the Siberian platform, extending over a great area from the lower reaches of the Lower Tunguska River to the Irkutsk amphitheater and thence to the lower reaches of the Olekma River (Fig. 146). The belt ranges in width from 200 to 300 km and reaches a length of 2000 km. The thickness of the formation ranges from 20 to 100 m. Throughout this entire region the saline formation rests on an eroded surface of Lower Cambrian rocks and represents a sequence that accumulated during transgression of the sea. Stratigraphically it forms the lower part of the Upper Lena series. The rocks in which the saline formation occurs are variable, being chiefly red argillites, marls, clays, sandstones, and, to a lesser extent, dolomites. The saline deposits consist of gypsum, in places restricted to the base of the formation, elsewhere occurring at higher levels and separated from the base by a unit of clastic rocks 20–30 m thick, locally more (up to 95 m). In some successions both these horizons are present. Saline sediments are known from practically the entire area of the formation, but locally they are meager, occurring as gypseous clastic rocks or as thin, or very thin, seams of gypsum; elsewhere they are distinct, forming beds up to several tens of meters thick. Higher gypsum contents are observed in the Irkutsk amphitheater (the Tyret'-Balagansk region, the upper reaches of the Lena and Kirenga Rivers) and in the extreme northeastern part of the formation (Olekma-Chara region). The intervening districts are characterized by low gypsum content. The gypsiferous sequences and the units in the regions of high gypsum content are spread over tens and hundreds of square kilometers, but the thickness is generally small, mostly below 20–30 m. This gypsum occurs in bedded and

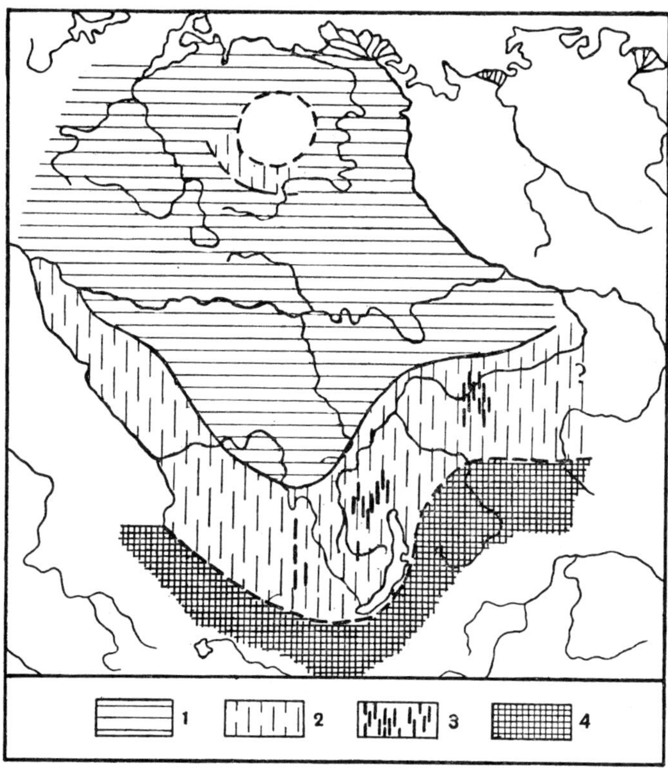

FIG. 146. Facies of the Upper Lena formation. 1. Normal marine deposits; 2. lagoonal saline formations; 3. region of maximal gypsum development; 4. land.

lenticular deposits ranging in thickness up to several meters, very rarely reaching 25–30 m. The gypsum normally contains sand and clay impurities, and it is also interbedded with sandstone, siltstone, and clay. Facies transitions have been repeatedly observed from gypsum to gypsiferous dolomite and then to pure dolomite, and from gypsum to gypsiferous and pure clay and siltstone. In the host rocks of the gypsum cross-bedding, ripple marks, filled mud cracks, crystal molds, with other shallow-water structures are widespread. Fauna are absent, and algae are only rare. No large-scale erosion is observed.

The topographic and tectonic environment in which the Upper Lena formation accumulated may now be specified. The zone of accumulation in the south was in an uplifted segment that supplied the fragmental material constituting the formation. An open but shallow sea lay to the north. The area where the formation accumulated was a broad plain, barely above the level of the sea and sloping towards it. At its other end this plain gave way to a more or less elevated continent. The plain was characterized by some slight variation in relief, with the topographic lows occupied by sea water and the highs acting as divides or underwater shoals. As always with very low relief, the outlines of the tongues of water were extremely irregular, commonly branching, and the basins themselves were very shallow, ranging in depth up to a few meters. Some lagoons were separated from the sea by bay bars and barrier beaches, others communicated freely with the sea. In front of the lagoons there probably occurred a string of islands rather than an open sea. The present-day Sivash, the gulfs of the Anapa Peninsula, the eastern coast of the Aral Sea are modern analogues of the Upper Lena lagoonal landscape.

Even in arid climates where plains adjoined uplifted regions, some water succeeded in crossing the plains, carrying fragmental material with it, normally rather fine, which gave rise to red or variegated beds. Extremely shallow water, bars in some places, island zones in front of estuaries in others, made water exchange with the open sea difficult. Increased evaporation all through the lagoons and the inter-island gulfs gave rise to a continuous flow of water from the sea into the lagoons, which were continuously filled from the sea, at times augmented in part from the drainage divides. These conditions inevitably led to increase in salinity of the lagoons and to salt precipitation. The completeness and intensity of the process were greatly limited by two circumstances, however. One was that the lagoons were not tectonic depressions that might continue to exist for a long time through progressive downwarping. Both lagoons and inter-island depressions developed through exogenic processes, near-shore wave action and currents that redistributed the clastic material, with the construction of barrier beaches. It is this type of process that was responsible for the shallowness of the lagoonal basins. The maximal depth of lagoons is generally twice the height of marine waves, (see Part II). The second important circumstance was the tectonism, the great slowness and low amplitude of epeirogenic movements. As a result the rate of sedimentation markedly outstripped the rate of total downwarping of the sloping plain. The sediments quickly filled the lagoons and the inter-island lows, and sedimentation ceased at the sulfate stage, failing to reach the chloride stage. *Thus, the tectonic regime sharply limited the extent of saline accumulation, cutting it off in the initial sulfate stage of development.*

The tectonic regime affected sedimentation on the seaward-sloping plain in still another way. The low relief of the plain was exceptionally sensitive to crustal movements. Even local movements of low amplitude (1–2 m) had profound effect by causing increased communication with the sea or eliminating it altogether, or spilling water from one low area into another.

All this gave rise to a considerable irregularity in the lagoonal saline formations, particularly in the relationship between the gypsum and clay rocks, and the subdivision of the sulfates by wedging beds of clastics. These features of the Upper Lena lagoonal formation demonstrate the extraordinarily great variability in the depositional environment. *This variability resulted from the especially intense activity of various physico-geographic agents and to the extreme sensitivity of the flat maritime plain with its microrelief to these agents.*

Two additional features in the processes that took place on this plain may be noted. The very slight dissection of the divide segments between lagoons in combination with arid climate resulted in practically no erosion on the positive topographic elements. Hence, erosion surfaces within the formation were absent, or, at best, were extremely rare. *Sedimentation, varying in space and in time, took place without any redistribution of the previously deposited material. Furthermore, a traverse from the shore through the formation passes by imperceptibly gradual transition from continental deposits of the marginal zone through lagoonal deposits proper to normal marine deposits.*

The impression is given that the lagoons were not separated from the open sea, although in fact they were, and the separation was well defined. The exceptional flatness of the maritime plain on which the Upper Lena legoonal formation was deposited in the distant geologic past destroyed the morphological features of the separation. These features, like the shore-line features along the lagoon, were masked, which now makes it impossible to plot the individual lagoons and inter-island water belts

The same type of lagoonal saline formation continued to accumulate on the Siberian platform during Ordovician and Silurian time, the site changing continually, giving rise to at least three monotypic, but spatially differentiated, formations (Nikiforova 1955). The first, which formed at the end of Early and the beginning of Middle Ordovician time, extends in a wide belt along the southern margin of the Anabar massif, reaching from the basin of the Moiero River to the lower course of the Vilyui River. In many places here (the Moiero River, Markoko River, middle course of the Vilyui River) gypsum locally thick, occurs. The second lagoonal formation accumulated in Middle and Late Ordovician. It lies in what is now the middle course of the Vilyui River, extending in tongues to the Anabar massif and to the middle course of the Lena River. The third formation accumulated at the end of Ordovician time in approximately the same area as the second, but it is separated from that formation by a thick marine sequence. From the meager published data (chiefly of O. I. Nikiforova), all these formations closely resemble the Upper Lena formation. Of saline deposits proper only gypsum is represented. Rock salt and potassium salts are absent. The only difference between these Ordovician-Silurian lagoonal formations and the Upper Lena formation is the presence of a fauna in the fragmental rocks, most commonly eurypterids, more rarely brachiopods and even corals. This indicates a more complex facies structure in the Ordovician-Silurian lagoonal formations and, in particular the presence of normal marine deposits.

2. The Lower Famennian Lagoonal Formation of the Principal Devonian Field

The Cambrian and Silurian lagoonal formations of the Siberian platform were characterized by their vast size, though elsewhere much smaller complexes of the same type are known. One of these latter is the saline complex in the middle of the Shelon' beds in the principal Devonian field (Fig. 147). It extends in an east-west direction from the meridian of Porkhov on the east to the Baltic shore on the west. East of Porkhov the lagoonal formation gives way to continental sand-clay deposits. This zone corresponds to the axis of the Baltic syneclise, which passes eastwards into the axial part of the Polessk salient. In consequence of this tectonic control, the thickness of the Shelon' sequence as a whole, as well as of the saline formation, varies systematically. The thickness of the Shelon' beds at Porkhov is 16 m, on the

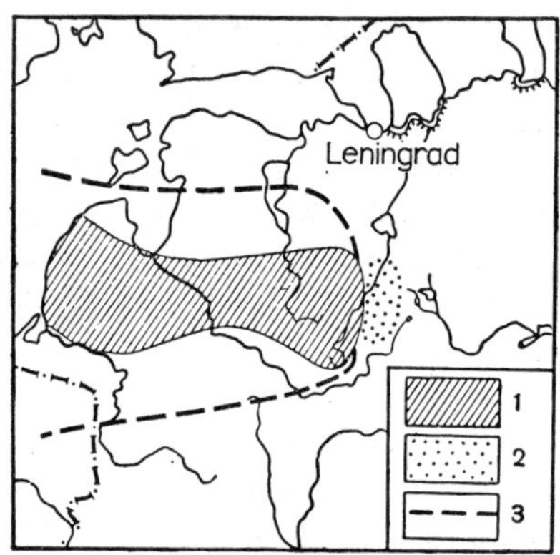

Fig. 147. The lagoonal formation of Middle Shelon' time. 1. Lagoonal formation; 2. continental deposits; 3. inferred boundary of formation before erosion.

Kudeb River 13 m, at Izborsk 15–17 m, at Plavinas 12 m, at Baldone 20 m, at Riga 11 m, at Jaunjelgava 25 m, and at Northern Litva 40 m. The thickness of the saline formation at Porkhov is 1·5–1·8 m, to the southwest 0·7–2·8 m, 30 km northeast of Riga (Baldozh) 5–12 m, and in the northern part of Latvia (Birzhai) 3·2–6·3 m. It is thus a very thin formation for the most part.

This formation is characterized by its variable lithology, both vertically and horizontally, and also by the incompleteness of the salt-depositing process: only gypsum deposition is represented. Beds and lenses of gypsum of different thicknesses alternate with beds of clay, marl, dolomitic marl, and dolomite. All these facies occur in lenticular bodies, giving way to one another within short distances. The gypsum members are discontinuous, forming bedded and lenticular bodies ranging in area from a few square kilometers to insignificant lenses (Gekker 1934). The gypsum deposits give way to gypsiferous dolomites and clays, but at a number of points gypsum may be entirely absent. Where gypsum is present, the structure of the beds is characteristically complex, of which the Birzhai succession may serve as an example. Here the gypsum bed is as follows, from the base upward: 0·5–2·3 m of gypsum, 1·5–1·8 m of dolomite that is gypsiferous in some beds 0·7–1·0 m of gypsum with seams of dolomite at the base, and 0·5–1·2 m of dolomite with insignificant laminae of gypsum and with grey marl at the top (Ivanov 1960).

The present distribution of the Shelon' formation is determined to a considerable extent by subsequent erosion. On the whole, the formation accumulated on a small seaward-sloping plain which coincided with the northern, eastern, and southern margin of the Baltic syneclise, merging partly also into the adjacent segments of the positive zones. The paleogeography corresponded fully to that of the Upper Lena formation. The period of accumulation was short and the initial stage of development gave rise to saline deposition. In the general history of the Frasnian basin in European U.S.S.R., this formation corresponds to a regression phase.

3. SOME COMPARISONS FROM HISTORICAL GEOLOGY

It is clear that lagoonal formations resemble in their nature the epeirogenic formations of humid regions, largely because of the similarity in the topographic-tectonic conditions. In both, sedimentation occurred on a plain sloping toward the sea, adjoining a more dissected part of the continent at its upper edge. In both, the transition from land to sea was gradual, and the contacts between the formation and marine deposits were diffuse. In both, tectonic movements were scarcely noticeable, with cyclic sequences occurring throughout the entire formation, though rhythmic tectonic movements were absent. Both types of formations are irregular. It appears that only the difference in color—grey in humid zones, red in arid—the replacement of coal by gypsum, and of carbonaceous shales by carbonate and clastic rocks with disseminated sulfatization are the principal, climatically controlled differences. This example clearly demonstrates that a single topography and its corresponding tectonic regime, developing in different climatic belts, may give formations that differ sharply in composition.

In Table 34 the principal data on the stratigraphic position of saline formations of lagoonal type are presented with their distribution relative to structural elements of the crust, and the occurrence of gypsum and rock salt in them. It is clear that lagoonal formations are spread through practically the entire stratigraphic column, from the Cambrian to the present. In this respect they are the most stable type of saline formation from the general viewpoint of historical geology. Furthermore, they are characterized by a wide range of tectonic conditions, since they have formed in geosynclinal zones during normal development, in foredeeps, and, lastly, on platforms. Nevertheless, platforms represent the principal sites of their accumulation, particularly the slopes flanking the elevations and syneclises, partly even the central parts of these latter basins. In geosynclines they are localized partly in the marginal zones, at the edge of the platforms, and partly in intrageosynclinial deposits. Relative to the dynamic conditions in the sedimentation areas, saline formations of lagoonal type accumulate with practically equal readiness either during subsidence and marine transgression or during uplift and regression.

TABLE 34

Stratigraphic Distribution of Lagoonal Saline Formations

Geologic age	Platforms, old and young	Geosynclines		Site of formation
		during normal stage	during closing stage	
Ng	●2 ●3 ■4		■1	1. Sicily; 2. Algeria; 3. Tunisia; 4. Egypt.
Pg	●5 ●2 1● ●4		●3	1. Florida; 2. India; 3. California; 4 and 5. Central Asia.
K	●1 ●2 ●3			1. Taukyr; 2. Kuba-dag; 3. Kugytang-tau.
J	●2 ●3	●1 ●4 ●5 ●6 ●7 ●8 ●9		1. Carpathians; 2. Moldavia; 3. Donbas; 4. Northern slope of Caucasus; 5. Southern slope of Caucasus; 6. Little Caucasus; 7. Kuba-dag; 8. Balkhany; 9. Konstdang.
T	●2 ●3 ●1 ■9	●7 ●8 ●5 ■6 4●		1, 2 and 3. China; 4. Algeria; 5. Italy; 6. Balearic Islands; 7 and 8. Yugoslavia; 9. England.
P			●1	1. Dzhezkazgan.
C	●4 ●2 ●3 ●1 ●5 ●6 ●7			1, 2, 3 and 4. Different parts of Russian platform; 5. Sarysu dome; 6. Bedpakdala; 7. Dzhambul.
D	●1 ●5 ●3 ■4	●2		1. Teniz basin; 2. Kuznetsk basin; 3. Minusinsk basin; 4. Tuva basin; 5. Shelon' formation.
S+O	●5 ●4 ●3 ●2	●1		1. Williston basin; 2 and 5. Siberian platform; 2. Lower Ordovician; 3. Middle Ordovician; 4. Upper Ordovician; 5. Silurian.
E	●1			1. Upper Lena formation.

●　gypsum;　　■ rock salt

A characteristic feature of these lagoonal formations is their restriction to the deposition of sulfates, as a general rule, at the beginning of the salt-forming process. Only rarely did the process reach the precipitation of NaCl, and more rarely the precipitation of potassium salts. The chief cause of this was the weakness of tectonic movements, a weakness which led to rapid filling of the basin with gypsum and to disruption of the salt-forming process.

Despite the wide distribution of lagoonal formations in time and in space, the mass of gypsum concentrated in them—to say nothing of halite—is small, and not to be compared with those vast masses typical of other types of saline formations.

D. THE DEPOSITION OF HALOGENIC FORMATIONS IN LARGE CLOSED GULFS

The second group of marine saline formations are the deposits in large gulfs isolated from the sea. *Such gulfs have each occupied a large structural depression during sedimentation in a tectonically active region. This restriction to isolated synclines distinguishes formations of this group from those of lagoonal type deposited on plains gently inclined toward the sea, occurring during periods of relatively weak tectonic activity.* The conditions under which the gulf type of formation accumulates are extremely variable, resulting in a wide-ranging facies series, perhaps the most varied of all facies associations in saline formations.

In morphology and internal structure, two series of gulf saline formations may be clearly recognized: one may be called the Kara-bogaz type after those deposits accumulating in the present Gulf of Kara-bogaz, and the other may be called the Virrila type for the similarity to the deposits in the present-day Bocana de Virrila (South America).

1. Lower Givetian Halogenic Formations in the Moscow Syneclise

A good example of the Kara-bogaz type of formation is the carbonate-sulfate Lower Givetian complex of the Moscow syneclise. The platform structure in Middle Devonian time was already marked by the large Voronezh salient of the crystalline basement, bordering the Moscow syneclise on the south and southwest (Fig. 148). The Volga-Kama arch, equally large, bordered the platform on the east and northeast. The Baltic shield and the Central Byelorussian massif were much less pronounced. These uplifts, creating positive elements of relief, surrounded the broad Moscow syneclise, open to the southeast. The difference in height between the uplifted zones and the depressed central part was of the order of 1000 m (M. M. Tolstikhina 1952). Against this background of macrorelief were found smaller variations in height, amounting for example to 50–70 m on the Voronezh massif.

Deposition of the saline formation was preceded by the accumulation of a clastic sand-silt-clay complex with characteristic distribution. On the crests of the basement uplifts (Voronezh massif, Volga-Kama buried massif), and

X

also on the upper slopes, deposits of this first phase are absent (Voronezh, Pavlovsk, Baranovka, Syzran', Mar'ina Posada, and elsewhere), or are sporadic and very thin (Staraya Russa, Veimarn, and Valdai). It is likely that the uplifts were at that time predominantly zones of erosion of clastic material. Thin sandy deposits, commonly red, accumulated on the slopes of

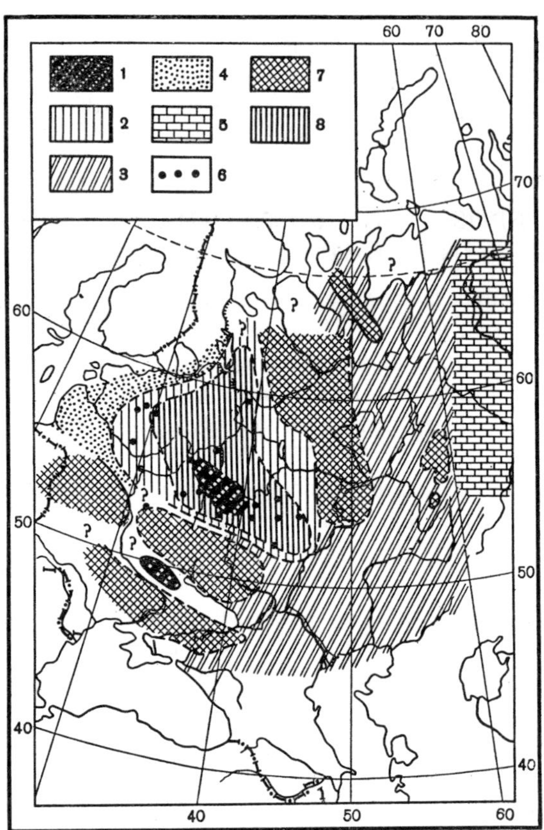

Fig. 148. Facies of the sulfate-carbonate Givetian complex of the Russian platform. 1. Salt-bearing deposits; 2. carbonate deposits of different thicknesses; 3. interbedded clastic and carbonate marine deposits; 4. clastic continental-lagoonal deposits; 5. marine carbonate deposits; 6. boreholes cutting the sulfate-carbonate complex; 7. segments of land in the platform sea; 8. carbonate-gypsum lagoonal deposits.

the uplifts. In the greater part of the syneclise clayey deposits with some sand accumulated, with rare local units of carbonate-bearing clays and salt-bearing rocks (Baryatin, Plavsk). The thickness of such deposits locally reached 40 m. Deposits in the deepest part of the basin were chiefly clays, although some sandy sediments are found even there. Between the southern slope of the Baltic shield and the northern slope of the Central Belorussian massif occurs a sequence that is predominantly sandy, some tens of meters in thickness. All these rocks are dominantly of continental facies: deluvial, alluvial, and lacustrine, but brackish-water deposits are known as indicated by the presence locally of numerous fish, lingulids, and estherians. Dwarfed estherians, with very thin valves, are characteristic. These brackish-water deposits are most widespread in the central part of the syneclise, in the region of Tula and Kaluga. According to E. G. Burova (1956), "all these rocks are impregnated to some extent with rock salt and have been notably pyritized". However, since the above fossils are preserved in them, halite must be considered as epigenetic, formed during katagenesis. At the beginning of Early Givetian time marine incursions reached the region of the Moscow syneclise from the southeast through the Ryazan-Pachelma downwarp, giving rise to small lagoonal gulfs, somewhat saline, but only to such a degree that euryhaline forms—fish, lingulids, and estherians—inhabited them.

In the succeeding phase of Givetian sedimentation sulfate deposits accumulated widely in the Moscow syneclise (Fig. 148). In the central part, embracing an area about 300 km long and 25–75 km wide (Ivanov 1960), the thickness of this complex reached its maximum, over 50 m, mostly 75–84 m. Beds of rock salt are also found here. From this area of maximum deposition the thickness of the complex rapidly decreases from 50 to 20 m; here the rock salt disappears from the succession, and the complex consists only of alternating beds of gypsum (anhydrite) and dolomite. In the marginal zone, the sulfate complex gradually decreases in thickness to zero. There is a simultaneous change in another feature. In the area of its maximal development (Serpukhov, Tula, Zubtsov, Borovsk), the sulfate complex is characterized by a remarkable freedom from clastic impurities. At some distance from this zone (Red'kino, Kaluga, Plavsk) some contamination of clastic material appears: thin and relatively rare beds of clay and siltstone. About the margins of the marine basin (Balakhna, Gorki, Kostyukevichi, Valdai, Porkhov, Staraya Russa) the dominant rocks are clay, marl, dolomite, and locally sandstone. "Locally (in the Leningrad district), these rocks are characterized by evidence of desiccation on the surfaces of individual beds, by pseudomorphs after rock salt, by variegated, commonly red colors in the rock, and by similar features. Moreover, in the westernmost part of this region, in the Leningrad district and northeastern Byelorussia (Drissa, Gorodok), beds of red and grey sandstone, commonly cross-bedded, are found. These must be considered continental, running water having been essential in their origin. On the large platform uplifts (Voronezh massif, Volga-Kama massif) deposits of the second stage of Givetian sedimentation are entirely missing. We know of no deposits here that may be considered stratigraphic correlatives of this sedimentary complex. It must be assumed that sediments did not accumulate on the Devonian surface of the platform in the regions of the large uplifts. These regions remained zones of erosion and denudation as during the earlier stages of development" (Tolstikhina 1952, p. 85).

These changes in thickness and composition of the sulfate complex demonstrate beyond doubt that the saline basin in which the sediments accumulated increased continuously in size. It first occupied the central part of the Moscow syneclise, in a relatively restricted area (Plavsk, Serpukhov, Tula, Bobrovsk, Zubtsov), but then gradually encroached upon the slopes of the uplifts, though never covering their crests. We have here a rare transgression of an *excessively salty marine basin*. In this process the salinity of the water decreased, as is shown by the vertical changes in lithology, especially well seen in the central part of the complex. In the Tula district two members may be distinguished in the complex: "(*a*) a carbonate-sulfate-halite member, composed of rock salt with thin seams of anhydrite and dolomite, commonly argillaceous, and (*b*) a sulfate-carbonate member, represented by dolomite with subordinate gypsum. The sulfates gradually give way upward to dolomites" (Burova 1956, p. 41). These members are succeeded by a sequence of argillaceous-carbonate deposits, in which dolomites pass progressively upwards to limestone. Similar changes may be traced through all the other bore-holes records. "Deposits of the Narova beds (Givetian) are characterized by great persistence of the lithology, which may be clearly separated into two parts. The lower part is throughout composed of anhydrite, gypsum, and dolomite. The upper normally consists of limestone, dolomite, and marl, with corals, brachiopods, and ostracods" (R. M. Pistrak and V. A. Sytova 1951, p. 55). A gradual decrease in salinity of the huge body of salt water from its initiation to its end, i.e. in proportion to the degree of transgression, cannot be doubted.

A characteristic feature of the chemical sedimentation in the Early Givetian saline gulf of the Moscow syneclise is the sharply defined cyclicity of the sulfate-carbonate complex. This is observed first in the lower salt-bearing member and it culminates in the change from halite beds proper to anhydrite and dolomite. "Rock salt forms beds that range in thickness from several millimeters to 35–40 cm, and are separated by beds of anhydrite and dolomite normally ranging from a few millimeters to 15–18 cm thick" (Burova 1956, p. 41). Outside the limits of salt deposition all sections of the complex show an alternation of carbonate and sulfate rocks with units of various thicknesses in cyclic sequence. Unfortunately, the lateral persistence of these cycles has not been investigated.

Thus, the sulfate-carbonate formation of the Moscow syneclise accumulated in a huge and progressively growing gulf that was connected to the open sea by a relatively narrow strait. The length of the gulf along its major dimension was about 1800 km, and along the least 1000 km. The connecting strait lay in the region of the Ryazan-Saratov downwarp. The existence of this strait is proved at Mosolov, Morsov, and Pachelma. "In the region of Mosolov, the closest to Moscow, all the lithofacies peculiarities of the Moscow succession are preserved. This attests to a uniformity of depositional conditions, the sequence having accumulated apparently in a single basin. In the Morsov region, lying like Mosolova on the axis of the Ryazan-Saratov downwarp but somewhat to the southeast, the saline deposits are thinner; and still farther to the southeast, between Pechelma and Serdobsk, pure saline deposits disappear from the section and marine deposits of various lithic types become abundant" (Tolstikhina 1952, pp. 84-86). Despite the huge size of the gulf, its general morphological similarity to present-day Gulf of Kara-bogaz is obvious, and I therefore class it with that type.

In considering the sedimentation in the salt-producing gulf of the Early Givetian sea, it is necessary to discover (*a*) what caused the deposition of rock salt at the very beginning of the basin's development, and (*b*) what controlled the cyclic character of the salt deposition.

In order to answer these questions it is necessary first to survey the supply of water and salt to the basin and the principal features of hydrology in the basin. The supply was apparently from two sources. The change from sulfate and carbonate rocks to clastic rocks about the margins of the basin indicates that surface water discharge undoubtedly took place, partly from precipitation in the arid part of the drainage basin but chiefly from precipitation in the humid part of that area. These waters brought with them the clastic material. This source can hardly have played any significant part in supplying the basin with salts, however, in view of the low mineralization of the surface waters. The chief supply of salt and water to the basin came through the Ryazan-Pachelma strait from the sea on the eastern part of the platform.

The influx of tremendous masses of water through this strait must have inevitably caused the specific hydrological features of the saline basin. The inflowing surface water, as usual, was squeezed against the shore, probably the southwestern shore, and to some extent decreased the salinity of the basin. Farther from shore, however, this water mixed with the highly saline basin water, thus losing its individuality. The current generated by inflow of sea water into the basin gradually died out. The largest—central, western, and northern—parts of the basin were characterized by more or less constant high salinity and were either free from steady currents or had only weak currents, extensions of the surface discharge current from the strait. In the deeper water levels in the vicinity of the strait, where the heavy salt water of the gulf came in contact with the lighter sea water, the difference in density caused a reverse current along the bottom from the gulf toward the sea. The strength of this current, however, was determined by the depths of the strait; any deepening of the strait increased the outward flow, and so caused some decrease in salinity of the basin. This hydrological mechanism must have become more effective as the water of the gulf became saltier.

The hydrological scheme described here makes it possible to understand much of the history of the Early Givetian saline gulf of the Moscow syneclise. At the beginning of the Givetian epoch, when continental deposition was still dominant, a tongue of the sea began to extend into the Moscow basin, clearly demonstrated now by the fossil content of the lower clastic group of Tula, Kaluga, Borovsk, and elsewhere. But the water had not yet become appreciably salty. Some expansion of the tongue then occurred, and rock salt was simultaneously precipitated. The increase in the water volume in the gulf and the simultaneous sharp increase in salinity is apparently contradictory. In point of fact the two events quickly succeeded one another: *first*, the sharp expansion of the strait caused a great influx of water into the gulf; and *second,* the shoaling of the gulf stopped the outward flow from the gulf to the open sea. As a result the gulf quickly became salty and salt was precipitated. The example of the present Gulf of Kara-bogaz shows that when no outward current is present, the gulf becomes increasingly salty within a period of several decades, of the order of 50–150 years. Geologically speaking this is instantaneous. This explains why saline deposits abruptly replace the basal fossiliferous clastic beds in the Lower Givetian succession. With renewed subsidence of the crust, the strait widened and deepened, which led to the progressive decrease in salinity of the gulf, partly from the influx of only slightly mineralized sea water from the east, but chiefly because of increased outflow of water along the bottom. This mechanism operated spasmodically; subsiding movements of the crust, causing transgression and decrease in salinity, ceased at times, and even gave way to weak upward movement. When subsidence ceased, and especially when uplift occurred, the outward flow of water became weak or stopped altogether because of shoaling in the connecting channel; the water in the basin became increasingly salty and gypsum was precipitated. When subsidence became appreciable, events were reversed; the basin decreased in salinity, gypsum was no longer precipitated, and carbonates began to form.

Such tectonic behaviour in the sedimentation zone, through changes in water discharge in the strait, caused the characteristic cyclicity of chemical deposition over the huge saline Moscow gulf. During its history this gulf was progressively converted to a platform sea in its marginal zone.

2. THE HALOGENIC FORMATION OF THE LATE SILURIAN MICHIGAN GULF

Another example of a saline formation accumulating in a gulf of a large platform depression is the Upper Silurian salt deposits of the Michigan basin on the North American platform. This basin was a large, slightly oval depression, elongated from southwest to northeast. The length along its major axis was about 800 km, and the total area was about 500,000 km². The beds dip toward the center of the basin 6–7 m per kilometer. The principal salt sequence is therefore about 3200 m deep in the central part of the basin, whereas it is exposed at the surface about the margins. On the west the Michigan basin is bounded by the Wisconsin and Kankakee arches and on the east by the Findlay arch. East of the basin, in the Appalachian marginal depression, another, much less extensive, saline formation accumulated in Late Silurian time. This now occurs in the states of New York, Ohio, and Pennsylvania. The salt deposits of the Michigan basin are upper Upper Silurian (Salina formation), forming the lower part of the Cayugan series. The total thickness of the sequence ranges from 30 m at the margin to 1350 m in the center of the basin.

The structure of the formation is complex. It is divided into three similar cycles having the following composition in the marginal sections (from the base upward):

1. *Dolomite*, brownish, with subordinate limestone and anhydrite, locally interbedded with one or two units of rock salt. The thickness is generally 260 m, with a maximum of 330 m.

Rock salt, 90 m thick, almost pure, with thin beds of dolomite.

2. Green-grey *dolomitic clay*, or argillite-dolomite with layers of clay. The thickness is 18–48 m.

Rock salt, almost pure, locally with thin beds of orange-yellow dolomite. The thickness is 10–20 m.

3. Dolomite, but in southwestern Ontario clay with a few beds of dolomite. The thickness is 15–30 m.

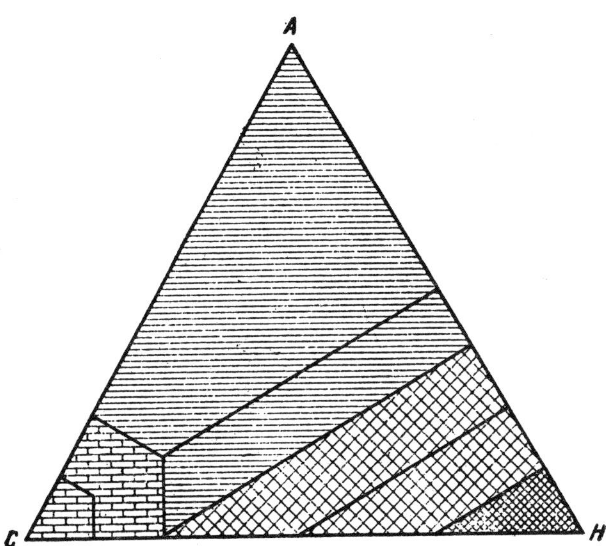

FIG. 149. Compositional diagram of evaporite rocks in the 56 boreholes of the Michigan basin (from Briggs). *A.* anhydrite; *C.* carbonate rocks; *H.* halite.

Rock salt, the thickest horizon is interbedded chiefly with grey, green, and red, anhydritic clays of variable thickness and with argillaceous dolomite. Anhydrite is present throughout almost the entire succession. The thickness of the unit is normally 90 m, the maximum 370 m.

Red *dolomite* and green clay.

The composition of the Michigan salt deposits is distinctive in the extensive development of dolomite and rock salt and the weak development of anhydrite. There are no thick beds of anhydrite underlying or overlying the halite sequences. In this respect the Michigan salt deposits closely resemble the Lower Cambrian saline formation of the Siberian platform, described below (Chap. 2).

Briggs (1958) has shown the weighted-mean compositions of the evaporate rocks, computed for 56 boreholes (Fig. 149). The area of the diagram is divided into four fields. The first field (at apex *C*), marked with the brick pattern, represents limestone and dolomite, pure or containing small admixtures of anhydrite and halite. The second field, horizontally ruled, is the field of anhydrite-dolomite and dolomite-anhydrite, without NaCl in the upper half and with admixtures of NaCl in the lower part, reaching large quantities in the lowest part. The third field (at apex *H*), shown by the coarse grid, represents mixed halite-anhydrite-dolomite rock. The fourth field, fine grid,

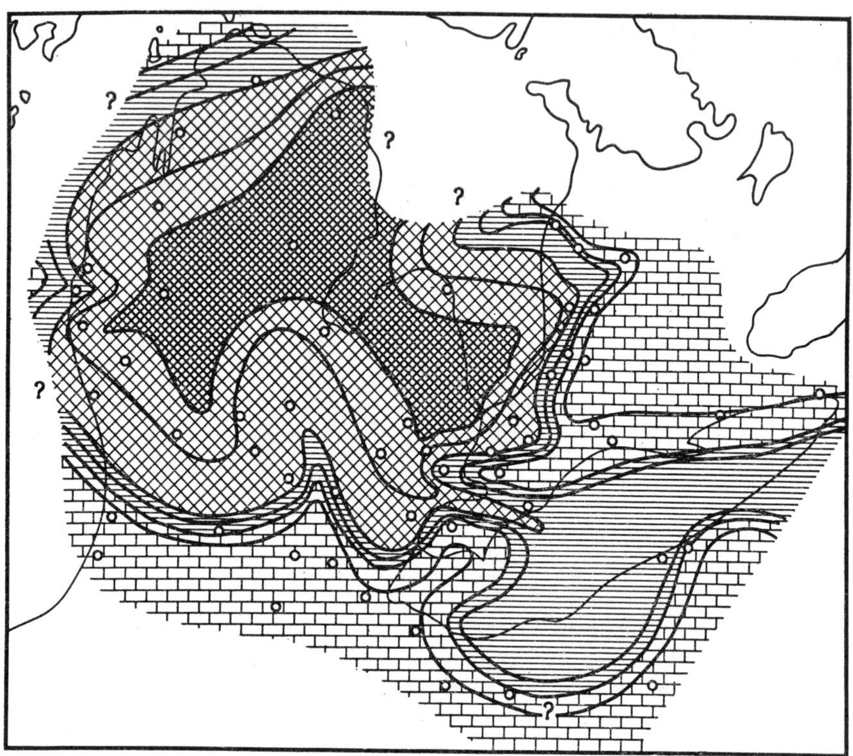

FIG. 150. Facies of the evaporite rocks in the Michigan basin and the Ohio-New York basin (from Briggs). Symbols as for Fig. 149.

represents pure halite. The weighted-mean compositions of the evaporate thus computed were then plotted on a map, and isopleths representing the four groups of the diagram were drawn. A systematic pattern of salt deposits was thus obtained (Fig. 150). About the margin of the formation, where it is thinnest, the section consists chiefly of carbonate rock—more particularly, almost exclusively, of dolomite, with minor layers of limestone. This marginal carbonate facies is especially broad on the south, where it forms two characteristic tongues: one wider, along the Findlay arch, the second narrow, acute-angled, west of the arch. A second carbonate field is found on the northeast, with a large salient on the southwest along the Findlay arch. On the north of the field a second salient of the carbonate facies is noted, extending to the southwest. Very small islands of carbonate deposits are found along the western margin of the formation, at the northern tip of Lake Michigan, and approximately in the middle of Lake Michigan. These occurrences are of fundamental significance. They prove that initially the formation was surrounded on all sides by a broad zone of dolomitic rocks, though subsequently reduced by erosion. Within the carbonate zone a second isopleth has been drawn to separate pure dolomite from dolomite somewhat contaminated with anhydrite and halite (corresponding to the subdivisions of the carbonate field in Fig. 143). Pure dolomite is found along the outermost part of the carbonate zone: dolomite with anhydrite and halite occur in the inner zone. A narrow band of anhydrite rock parallels the belt of carbonate rocks everywhere. The narrowness of this band is due to the restricted distribution and the thinness of this group of rocks in the succession of the saline formation.

The inner part of the formation is composed of halite rock of variable composition. The outer part of the halite zone consists of mixed halite-anhydrite-dolomite rock, the inner part of relatively pure halite. It is clear that halite deposits accumulated much more intensely in the Michigan basin, at the time of saline sedimentation, than in the Appalachian trough. The two basins were connected by only a narrow strait at what is now Chatham, where a tongue of mixed halite-anhydrite-dolomite rocks extends from the Michigan basin to the Appalachian. Within the Michigan basin the zone of pure halite deposition has irregular margins where mixed halite-anhydrite-dolomite rocks invade the field of pure halite. These indentations correspond to similar indentations in the marginal carbonate zone into the inner parts of the formation.

On the whole the pattern of the formation proves to be both simple and systematic. *The marginal thin zones consist of carbonate rocks, pure dolomites at the very edge, dolomite with anhydrite inward into the formation. Farther toward the center of the formation the thickness increases sharply and layers of mixed halite-anhydrite-dolomite rocks and of pure halite rock become more abundant. Halite rocks are especially characteristic of the innermost parts of the formation.*

The petrographic features of the saline rocks are extremely curious (Dellwig 1955). In the Glen Bradley borehole, in the central part of the formation, the rock salt shows well defined interbedding of halite and anhydrite-dolomite rock. The salt layers range in thickness from 1·5 to 8 cm; the anhydrite-dolomite layers never exceed 5 mm in thickness. The beds are strikingly uniform, with smooth contact surfaces, and persistance of thicknesses. The halite layers are composed partly of transparent and partly of dull salt. The dull variety has a microscopic pinnate structure, forming those boat-like forms and "teeth" previously noted as developing during sedimentogenesis. These zones of dull salt correspond to individual skeletal crystals and to groups of crystals in salt crusts that settled to the bottom, having formed on the surface of evaporating brine as a result of intergrowth of individual "boats".

The dullness of this pinnate salt is due to the great number of fine inclusions of brine between thin layers of halite, forming a framework of crystals. The transparent zones of the halite beds form by recrystallization of the pinnate skeletal forms, and inclusions disappear. According to Dellwig, however, outlines of the square upper surfaces of the once-floating "boats" may be preserved within crystals. The quantitative relations between the dull (pinnate) and transparent (recrystallized) varieties of halite within the halite beds vary strongly: in some places one is dominant, in other places the other. An interesting mineralogical feature of the halite beds is the presence of polyhalite and carnallite. The polyhalite is generally found at the base of the halite beds, in idiomorphic crystals (about 2μ on a side), commonly as pseudomorphs after anhydrite. The carnallite is found either in rounded grains or as idiomorphic crystals. In some thin sections very rare individuals of langbeinite occur in the form of idiomorphic crystals to about 6μ.

The thin layers of anhydrite, separating the halite layers, are extremely fine-grained. Over 8% of the grains have the major dimension less than 2μ, and the remainder have diameters ranging up to 4μ. These anhydrite layers contain anhydrite, dolomite, celestite, quartz, polyhalite and organic matter. No systematic change in grain size in a vertical direction is observed within the layers, nor is there any segregation of individual minerals into independent aggregates. The lower surface of the anhydrite beds is locally even, because of truncation (solution) of the underlying halite bed (especially well seen in truncation of the pinnate crystals), but is mostly irregularly serrate because of protruding halite crystals.

The form of the iron is highly distinctive. In the halite beds it occurs exclusively as powdery or dendritic hematite, disseminated in halite crystals; in anhydrite it appears only as well-developed pyrite. This clear separation of the forms of iron is striking within the seasonal bedding of the halite rocks. It is undoubtedly due to the fact that the anhydrite beds are rich in organic matter.

The structure of the halite rocks about the margins of the formation is the same as in its inner parts, but some new features of fundamental importance for understanding salt-forming processes are also present. One of these is the notable thickening of the anhydrite layers in the marginal zone and its complex structure. They are thin-bedded alternations of almost pure dolomite with pure anhydrite microlayers. Crystals of pinnate halite are occasionally found. This complex structure of the anhydrite about the margins of the formation clearly indicates that this represents not a seasonal variation, but rather the deposits of many years. Either seasonal layers of halite did not form, because of insufficient salinity in the marginal parts of the basin, or the halite remained in the embryonic stage of development, leaving individual crystals of pinnate halite. It is possible that thin beds of halite forming at the end of the evaporative season subsequently returned to solution in the humid part of the year when the basin decreased in salinity, leaving no traces behind. In other words, halite accumulated much less readily in the marginal parts of the basin than in the central part, many years going by without any permanent deposition. This was a normal occurrence in the marginal part of a basin. A still more curious phenomenon is the uneven bedding of the rocks in many places. The anhydrite layers commonly rest on a ripple-marked surface of the halite (Kaufmann and Slawson 1950). We consequently find undoubted traces of extensive shallow water and of the effect of actively moving water on the salt beds. It is hardly accidental, therefore, that crystals of dull (pinnate) salt are always broken in this marginal zone (according to Dellwig). An interesting feature in the structure of rocks of the marginal zone is that the anhydrite crystals in the anhydrite layer are *commonly pseudomorphs after gypsum.* These

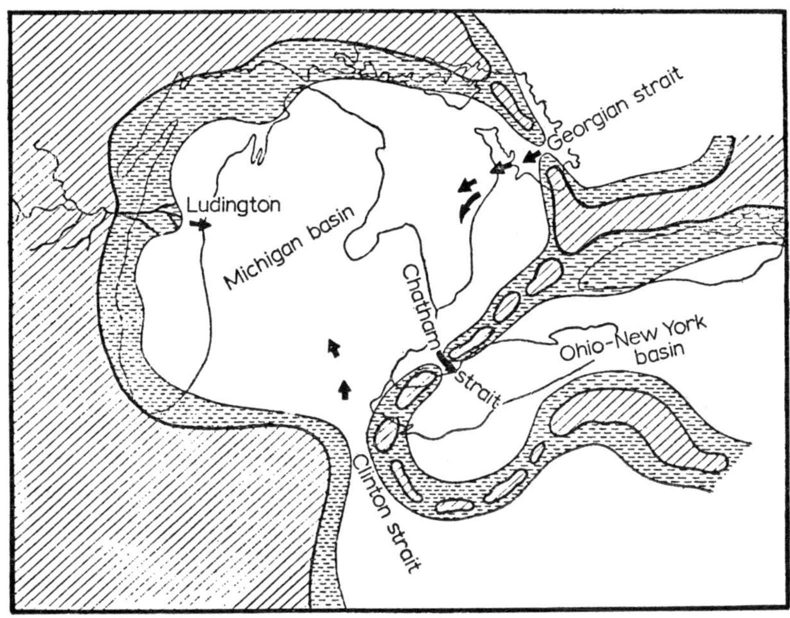

FIG. 151. Paleogeography of the Michigan basin (from Briggs).

crystals generally range from 4 to 8μ in length, though some are larger. They are found chiefly at the contacts between anhydrite and halite layers, but they may occur within the latter. Where this is true, the anhydrite crystals show signs of corrosion by the halite. These observations are of great significance in understanding the primary forms of sulfate sedimentation. In this example it proves to be *gypsum.* We may note, finally, that dolomite grains in the anhydrite layers are commonly rounded. They have the aspect of purely fragmental grains, transported into the salt basin from the drainage area which was underlain at that time by carbonate rocks to a considerable extent.

These facts furnish a general view of the paleogeography of the Michigan basin in which the saline formation accumulated.

It was deposited in a huge gulf of the open Late Silurian sea that occupied the region of the Michigan basin (Fig. 151). Coarse cyclicity characterized sedimentation in the gulf, being expressed in a threefold repetition of a mono-typic course of paleogeographic changes. The initial moment of each macro-cycle represented an epoch of accentuated subsidence and was characterized by large size of the gulf and by relatively broad connections between the gulf and the open sea of the Appalachian trough. Connection with the northern basin existed through the broad Georgian strait. Connection with the marginal Appalachian basin was through the Findlay arch, which was flooded for the most part, and through the Clinton strait to the southeastern basin. The broad connecting passages with the surrounding basins were zones of low salinity in the waters of the Michigan gulf, very near normal sea water. Thick beds of richly fossiliferous limestone and dolomite, interbedded with clays, were deposited in the gulf at this time. They may be observed at the base of the first, the second, and, in part, the third cycles. In the second half of each cycle the drainage areas of the gulf were uplifted somewhat, decreasing the area of the gulf and, significantly, making communication with the open seas more diffi-cult. The Georgian and Clinton straits grew narrow and shallow. The Findlay arch became dry land for the most part, separating the Michigan gulf from the neighbouring segment of the Ohio-New York basin and cutting off this latter from the Appalachian sea to the south of the Findlay arch. Restricted circulation resulted from the fact that bottom outflow through the strait ceased and only currents in the upper water remained. These hydrodynamic conditions led inevitably to progressive salinity of the water in the Michigan gulf and to a change from carbonade to sulfate deposition, and then to halite. The saline horizon characteristic of each cycle was formed.

The development of carbonate rocks, particularly dolomites, along the margins of the formation and the deposits of anhydrite and halite toward the interior may be interpreted only as an indication of a lack of uniformity in the salinity of the basin brine. About the margins the mineralization was somewhat less than 15% (beginning of gypsum precipitation), whereas it was 25–27% in the central part (beginning of halite precipitation). There was, consequently, a difference of at least 10–12% in salinity between the two zones. The variable mineralization within a single mass of water was main-tained by the influx of sea water through the straits and of fresh river water (Fig. 150). Carbonate rocks now extend much farther into the interior of the formation at places where water entered the basin than at other parts of the margin. It is obvious that the influx of slightly mineralized water here freshened the brine in the gulf over a considerable area and retarded the precipitation of sulfates and halite. Briggs has used an additional method to explain regions of freshened water within the basin. From drill cores he studied the relations between thickness of calcareous beds and thickness of the total mass of car-bonate rocks in the Salina formation (Fig. 152). He started with the sound idea that the calcareous content of carbonate rocks is proportional to the freshen-ing of the brine at any given point. It may be clearly seen from the map that the largest freshening current entered through the Georgian strait, penetrating far into the central part of the gulf. The existence of this strait may be traced through the entire Salina time. The Clinton strait was effective chiefly during the second saline cycle, since the greatest thickness of the carbonate sequence having the highest percentage of calcareous material is found in the lower part of this cycle. The fact that fresh water flowed into the basin here is demonstrated by the large embayment in the isopleths of 30–40% calcareous content.

The presence of ripple marks and traces of mechanical crushing of primary halite crystals in the Saline formation indicates shallow depth at least about the margins of the gulf. The depth could hardly be much greater in the central

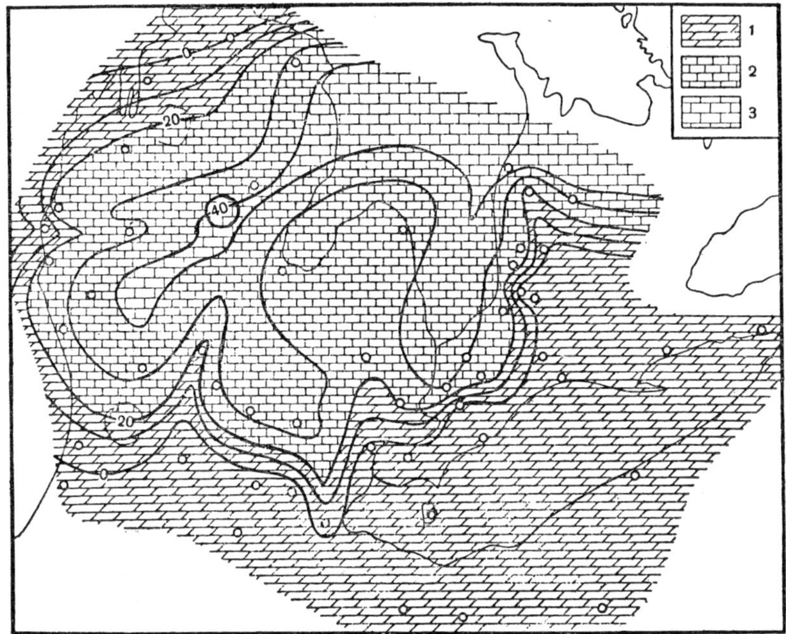

FIG. 152. Calcareous content of rocks of the Salina formation (from Briggs).
1. 0 – 20%; 2. 20 – 40%; 3. 40% and more.

part of the basin, because it is hard to imagine crusts, made up of many intergrown halite "boats", passing through great thickness of water without being damaged.

Judging from the successions of the saline beds in each of the three saline cycles, the structure in these horizons is complex. There are irregular alternations of halite rock proper with anhydrite-dolomite, indicating that the salinity during the formation of each of the three halite horizons varied, apparently as straits into the gulf varied in depth.

The deposition of salt in the Michigan gulf was characterized also by a well-defined seasonal character: annual hydrochemical cycles, manifested in a change from anhydrite-dolomite layers to halite layers. The presence of these cycles proves that the salinity of the surface brine fluctuated markedly during the year. During the summer evaporative season, the intensity of evaporation became so great that not only did the volume of water entering through the straits and by inflowing streams evaporate completely, but the total water mass of the basin itself was reduced. The concentration was markedly increased and halite was precipitated. During the cold and probably moist winter season, the basin was again filled by inflow of sea water through the gulf and of fresh water from streams. Rainfall also contributed directly. The weakening of evaporation led to the accumulation of water of low mineralization and the salinity of the upper horizon declined. At the beginning of the evaporative season dolomite was precipitated from this diluted brine, and then anhydrite. The mixture of these two gave rise to the anhydrite-dolomite units that separate the halite beds.

Thus, the accumulation of each evaporative horizon in the formation was a complex cyclic process, in which both tectonically caused macrorhythms and mesorhythms combined with seasonal microrhythms.

Since the thickness of each pair of annual layers ranges from 2 to 10 cm, with an average of 6 cm, and the total thickness of salt in the central part of the formation is 600 m, the total length of the salt-forming process must have been about 10,000 years, i.e. a negligible period when compared with the time it took the entire formation to accumulate. The greatest part of this time was occupied in the accumulation of dolomite and anhydrite. This example indeed demonstrates the rapid course of salt accumulation.

3. Miocene Halogenic Formations of the Carpathian Foredeep

In addition to saline formations that form in circular or oval gulfs like Karabogaz, others form long, narrow basins, communicating with the sea only at one end and resembling, even if in more complex form, the present gulf of Bocana de Virrila. Unfortunately our knowledge of sedimentation in these gulfs is very incomplete. We know only about salt-producing zones, and commonly we can reconstruct the features of the basin as a whole only hypothetically. Despite the incompleteness of our factual knowledge, the paleographic type of salt-producing basin may be clearly defined. Two examples will be given.

The Miocene history of only the Soviet part of the potassium-rich Carpathan marginal basin will be considered; it extended from the Polish border on the northwest to the Rumanian border on the southeast. The depression extended for 280–300 km within this zone. The breadth of the downwarp during its growth and filling with sediments ranged from 35–40 to 60 km and more. A. A. Bogdanov (1949) divided the downwarp into two sharply different zones: an inner zone on the southwest and an outer zone on the northeast (Fig. 153). "The inner zone of the downwarp lies directly on the northern limb of the Carpathian fold zone and extends in a narrow belt along the margin of the Carpathians. It was characterized by extensive downwarping during Lower and Middle Miocene time. This part of the downwarp is the zone where thick salt beds have accumulated; it is now complicated by intense folding. The outer zone formed on the platform, migrating within the limits of the platform during Tortonian time."

The succession of Tertiary deposits containing the salt beds is shown in Fig. 154. Four saline formations are recognized, separated by thick clastic sequences with no salt or with only disseminated salt in the cement.

At the base of the section is the Polyanitsa series (80–400 m), found only in the inner part of the downwarp. It consists of dark grey, locally slightly greenish, calcareous argillites, grading into flaggy grey or greenish grey marls,

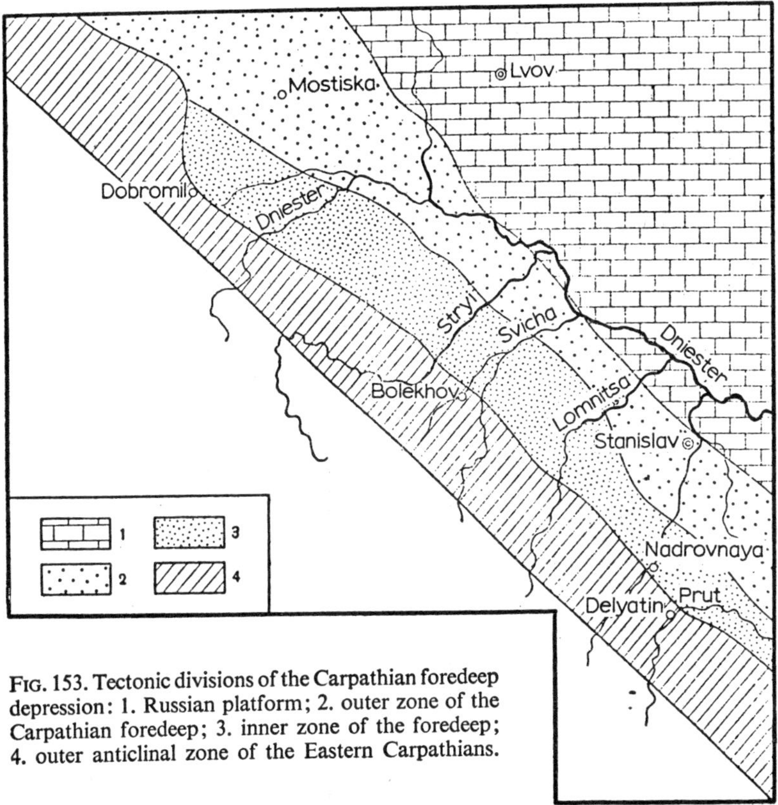

FIG. 153. Tectonic divisions of the Carpathian foredeep depression: 1. Russian platform; 2. outer zone of the Carpathian foredeep; 3. inner zone of the foredeep; 4. outer anticlinal zone of the Eastern Carpathians.

with layers of light grey siltstone and calcareous sandstone. Locally, on the Prut River, thick lenses of conglomerate occur. Even disseminated salts are absent in the Polyanitsa series.

The *first saline formation*, called the Lower Vorotyshcha series, rests conformably on the Polyanitsa series. Its distribution, like the two overlying formations, is restricted also to the inner zone of the downwarp. For the most part the formation consists of gray and dark gray clay, siltstone and sandstone, with some conglomerate. Salts are disseminated through the rocks, contributing to the cement, but they also form independent beds and smaller units. In some localities, such as at Dolina, Ninyuv and elsewhere, the section

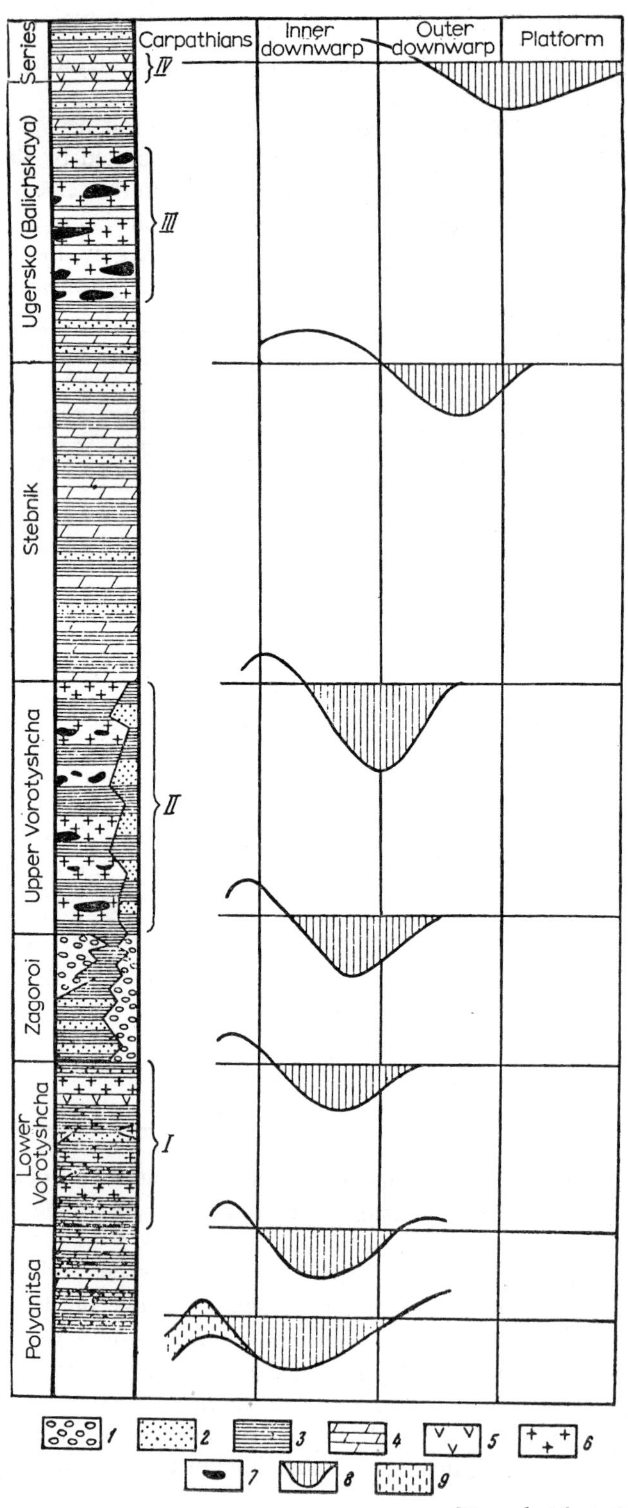

[*Legend at foot of next page*

contains sequences of rock salt up to several hundred meters thick. At Bolekhov the upper part of the Lower Vorotyshcha series has a sequence of anhydrite-halite rock over 100 m thick, containing aggregates of polyhalite (Korenevskii, 1956a). Consequently, as early as the accumulation of the first saline formation, chemical sedimentation had reached the threshold of the stage of potassium-salt precipitation, and at times even entered this stage for brief periods. The thickness of the Lower Vorotyshcha series reaches 350–400 m in localities of its greatest development.

The overlying Zagoroi series, in the central part of the Soviet sector of the Ciscarpathian region, between Bolekhov and Nadvornaya, consists chiefly of clay and argillite with disseminated salt (in the cement) and gypsum. To the northwest and southeast, more extensive areally, many sandstones and conglomerates also occur. The conglomerates are especially thick in the vicinity of Truskavets (the so-called Truskavets conglomerate) and Sloboda Rangurskaya (the so-called Sloboda conglomerate).

The conglomerates were derived chiefly from the folded Paleozoic rocks in the Carpathians. Weak Cu-Pb-Zn mineralization has been observed in the the Truskavets region (see Part 1, Chap. 2).

The *Upper Vorotyshcha* or second saline formation contains two facies: the salt facies proper and the so-called Dobrotov facies. The former is composed of dark clay, siltstone, and sandstone, in which beds of clayey rock salt are intercalated. At a number of localities potassium salts are also present. The Upper Vorotyshcha salt sequence differs from the Lower Vorotyshcha in its much greater salt content, and also by the presence of well-defined beds and lenses of potassium salts. The Upper Vorotyshcha series of this lithic type extends from the upper reaches of the Dniester on the northwest to the basin of the Bystritsa River at Nadvornaya on the southeast. In this region, according to S. M. Korenevskii (1956e, p. 243), "a reliably established zone of potassium-salt deposits extends from northwest to southeast from Modrych through Stebnik, Dobrogostov, Ulichno, Girne, Ninyuv, Morshin, Smolyanoi, Trostyanets, Svarychev, Rasbul'nu, and Dzvinyach to Starun and Lanchin". The Dobrotov facies of the Upper Vorotyshcha series is characterized by the absence of salt deposits and by the appearance of variegated coloration in the upper part. It is composed of alternating fine-grained sandstone, dark grey and greenish grey argillite and marl. The sandstone is dominant, especially in the lower part of the series. Upward layers of rose-colored clay and marl appear. The sandstone and clay alternations resemble the rhythmic character of flysch. The thickness of the Upper Vorotyshcha series is about 600 m.

The overlying *Stebnik series* (600–1000 m) is also variegated. It consists

---

Fig. 154. The Tertiary deposits of the Carpathian foredeep; 1. Conglomerate; 2. sand; 3. clay; 4. marl; 5. gypsum; 6. halite; 7. potassium salts; 8. site of sedimentation; 9. initial stage of downwarping; *I-IV*. Halogenic formations.

Y

chiefly of carbonate-bearing clay shales and marls of grey, greenish, brown, rose, and reddish colors. Friable, calcareous sandstones are present in subordinate amounts. Disseminated salt and gypsum are characteristic of only the lowermost part of the series.

The *third saline formation* is confined to the middle part of the Balichskaya (Ugersko) series, the lower and upper parts of which consist of complex alternations of sand-clay-marl rocks. This saline formation, sometimes called the Kalush formation*, is petrographically very similar to the Upper Vorotyshcha, and is frequently considered part of that sequence. It includes a series of potassium-salt beds and lenses.

The upper saline formation is the *Tiras* or gypsum-anhydrite series. It occurs chiefly in the outer zone of the downwarp, on the platform, only part occupying the inner zone. As the name itself indicates, the evaporite cycle is incomplete, having stopped at the sulfate stage. Furthermore, controlled by the tectonic character of the substratum, the thickness of the formation is very small, only 25–40 m.

Above this series lie the Upper Tortonian and Sarmatian marine deposits.

In general the Miocene deposits of the Carpathian foredeep show a well-defined cyclical character. Each clastic sequence with its adjacent saline formation makes up a coarse cycle, beginning with sandy clays that are essentially free of salt and ending with saline sediments, more or less contaminated with fragmental material. There are four macrocycles in all (Fig. 154). The first is characterized by an extensive but incomplete development of saline sedimentation: the accumulation of halite without potassium salts. The second and third cycles are most complete and contain potassium salts. The fourth cycle is the least complete, stopping at the stage of calcium sulfate precipitation. The region of saline accumulation, in connection with the general displacement of the axis of downwarping, slowly migrated from the southern part of the inner zone of the depression through the central part, and finally to the outer zone.

Some of the characters of composition of the saline formation in the Ciscarpathian region, especially the abundance of clastic sand and clay, should be emphasized. Beds of mudstone, siltstone, sandstone in the saline formations make up at least 50% of the total thickness. But even the remainder of the saline rocks is strongly contaminated with clay to a minimum of 10–15% of the saline rocks and most commonly 25–30% or more. The salt beds are thus strongly argillaceous, and they grade into clays with disseminated halite. Thus the saline formations in the Ciscarpathian region contain more clastic material than salt proper. *Fundamentally, the formations are not saline, but rather clastic-saline.*

Furthermore, the *small development of specific beds of gypsum or anhydrite during the initial stage of saline sedimentation is striking.* Excluding the fourth, gypsum-anhydrite formation, the thick Miocene succession contains only

* Opinions concerning the stratigraphic position of the Kalush formation are contradictory. We here follow the scheme of A. A. Ivanov (1960).

individual occurrences of isolated, notably thick gypsum members. S. M. Korenevskii (1956a, p. 241) thus points out as a great rarity that the upper part of the Lower Vorotyshcha series "contains a sequence of anhydrite-halite rocks with nodules of polyhalite (at Bolekhov), the thickness being more than 100 m. The $CaSO_4$ content in the rocks of the Upper Vorotyshcha series is small; only two or three units of anhydrite up to 0·5–1 m thick are present. During Balichskaya (Ugersko) time the saline process in the Kalush-Golyn part of the basin began with the deposition of gypsum in the upper part of the clays in the Kalush series and with the formation of a basal anhydrite, ranging in thickness from 0·5 to 12 m, at the top of this series," but precipitation of sulfates ceased abruptly after this. This essentially represents all the independent deposits of $CaSO_4$ thus far recorded. For the most part gypsum or anhydrite merely constitute minor impurities (of the order of 5–10%) in the rocks, contributing to the cement. This lack of independent gypsum-anhydrite beds in the succession is probably due to intense clastic sedimentation, taking place simultaneously with chemical precipitation. Actually, the absolute rate of gypsum sedimentation from brine is low, and this mineral may form independent beds only when the introduction of clastic material into the basin is small. When this introduction of clastic material increases sharply, as in this example, the gypsum that accumulates on the floor of the basin becomes so strongly contaminated with clay that it is but a secondary constituent in the total mass of sediment. This mechanism is accentuated still further by the fact that, judging from the composition of the potassium salts, direct metamorphization in the basin brine giving rise to the saline formations of the Ciscarpathian region was practically absent. As a result, the sediments were deprived of the supplementary $CaSO_4.2H_2O$ that always appears during metamorphization of brine. The gypsum in these basins was almost exclusively evaporative; metamorphogenic gypsum did not form. As a result, the total $CaSO_4$ was clearly less than normal.

The third characteristic feature of the saline formations discussed here is the distinctive occurrence and composition of the potassium accumulations. Detailed studies on the occurrence of the potassium salts have been made in only two regions: Stebnik, where the Upper Vorotyshcha formation is potassium-bearing, and Kalush-Golyn, where the potassium salts are confined to the Balichskaya (Ugersko) series (saline formation III). The occurrence is very similar for both, which may therefore represent the essential aspects of the phenomenon (Fig. 155).

"The most extensive deposit of potassium salts (in the Stebnik region) may be traced for 2·5–3·0 km along the strike and 750 m or more down-dip. However, there is a more common distribution of potassium-salt deposits that extend along the strike for distances up to 1 km, and extend down-dip 200–300 m. The thickness of these deposits does not depend strictly on the length along the strike. It ranges from a few meters to 100–150 m, including beds of rock salt and saliferous clays and breccia" (A. A. Ivanov 1959, p. 318). The same relationships apply in the Kalush-Golyn region, in a different stratigraphic horizon. "Deposits of potassium salts, which include beds of saliferous clays and rock salt, range from 1–2 to 25–30 m in thickness, some even reaching 45–50 m. The thickness of the beds, between the potassium-salt members, is 10–15 m and consists commonly of brecciated clays and rock salt. The total range in thickness is from a few meters to 50–60 m. The most extensive deposits may be traced for 2·5–3·0 km and down-dip for 0·8–1·2 km. Most of the deposits are much smaller, however" (Ivanov, p. 321). In keeping with the very restricted linear dimensions, the areas of the potassium-salt lenses are very small. In the Stebnik and Kalush-Golyn deposits the area of the largest deposits is 6–10 km². Most are much smaller. In plan and in section the distribution of the potassium-salt deposits is irregular. The deposits are more concentrated and more widely developed in the lower parts

Fig. 155. Occurrence of potassium salts in the Upper Vorotyshcha series in the Solets-Stebnik-Smolyanoi district (from O. P. Gorkun). Upper Vorotyshcha series: 1. salt-bearing sand-clay breccia; 2. dense dark gray clays with small halite content; 3. salt-bearing sandstone; 4. lenses of potassium-salt rock; 5. Zagoroi series—carbonate-bearing clays with polymictic sandstone having halite-filled joints.

of the Vorotyshcha and Balichskaya formations. In the middle, and especially the upper, parts of the formation, deposits become rarer, discontinuous, lenticular, and small. The zones where the deposits are crowded together have a total area of 30–50 km², a very restricted distribution.

The mineralogy of the potassium-salt deposits of the Ciscarpathian region is complex and is characterized by an abundance of K and Mg sulfates. The main primary rock-forming minerals are: halite, sylvite, carnallite, kainite, and langbeinite. Minor minerals are: polyhalite, anhydrite, kieserite, and glauberite. In some parts of the deposits, where weathering has occurred, the secondary minerals gypsum, picromerite, mirabilite, epsomite, glaserite, and others are found. Different combinations of primary minerals give rise to a series of potassium-bearing rocks, including sylvite, carnallite, kainite, langbeinite, langbeinite-kainite, and polyhalite rocks. "In all the rocks invariant constituents are halite and clay (clastic-authigenic), the halite ranging between the same limits (10–40%) in all varieties, whereas the content of clay is greatest in sylvite and carnallite rocks (up to 25–30%) and least in langbeinite rock (1–10%). In all the sulfate rocks, and also in one sylvite variety, polyhalite is always present (2–10%). Its place in the other variety of sylvite rock and in carnallite rock is taken by anhydrite (1–8%)" (Ivanov, 1959). The contents of the remaining minerals in the rocks range between wide limits, and this gives rise to many transitional rock-types. Kainite and langbeinite-kainite rocks are the most widespread. The langbeinite rock occurs as lenses, rather thick members, and beds in association with other transitional rocks. "Sylvite rock is found chiefly in the Kalush-Golyn deposit, where it forms independent lenses. In other deposits sylvite rock occurs in small lenses and beds in the upper and lower parts of deposits of potassium sulfate salts. Carnallite rock is found only in a few districts of the Kalush-Golyn deposit and is not of commercial interest. Sylvite was generally precipitated in shallow-water, commonly near-shore belts of the salt-producing basin, where much clay was being introduced. The precipitation of sylvite commonly preceded the deposition of slightly saliferous sand-clay members. The sylvite rock of the Ciscarpathian region is normally the potassium-salt rock containing the most clayey material. The weighted-mean content of clay in the sylvite lenses generally ranges from 18 to 33%, whereas under other conditions, potassium-salt deposits contain no more than 20%" (S. M. Korenevskii, 1956a). In the series of sylvite—kainite—langbeinite rocks (with all the transitional rocks between) the amount of clay gradually declines, and langbeinite rock contains the least clay of all varieties. "It was apparently precipitated from clarified, stabilized, concentrated brine. It is possible, therefore, that langbeinite is less abundant in the kainite lenses of the near-shore part of the Kalush basin, where the brine was frequently, though temporarily, freshened and where great quantities of clay were introduced.

"Polyhalite rock generally forms the bottom and top of sylvite and potassium-sulfate deposits. It represents the beginning and the end of potassium-salt precipitation. Beds of polyhalite are observed locally in saliferous clays, attesting to incomplete, interrupted phases of brine concentration. Polyhalite precipitated with other potassium salts (sylvite and kainite), marking the presence of local freshened zones and the introduction of calcium carbonate into the basin. According to weighted-mean contents of polyhalite and anhydrite in sylvite deposits it may be established that these contents are inversely proportional and the total quantity of both is less than in other potassium-salt deposits" (Korenevskii, 1956a).

The mineral composition of the potassium-salt deposits in the Ciscarpathian regions is characterized by a sharply defined spatial variability. This is seen in the Kalush-Golyn region, where the third saline formation is developed. Two basins existed here: the Kalush and the Golyn, separated during sedimentation by a low uplift parallel to the present Kalush anticline.

"In the near-shore Kalush sub-basin of the larger salt-producing basin sylvite lenses become increasingly abundant. Potassium-sulfate salts are present chiefly in lenses of kainite rock. In the section of salt-bearing rocks, lenses of potassium salts are distributed in the following manner: (1) lower sylvite, (2) middle sylvite, (3) kainite lens, locally with seams of langbeinite, and (4) upper sylvite, commonly interbedded with kainite and thick members of carnallite rock. In the Golyn basin the number of potassium-salt lenses increases to thirty, of which three thick lenses of potassium and magnesium sulfates are the largest, and the lower and middle sylvite lenses are smallest. Thus, conditions developed in the near-shore Kalush basin that were very favorable to the precipitation of sylvite deposits and halite-kainite rocks of rather simple composition, whereas in the Golyn basin precipitation gave rise to mixed sulfate salts of more complex composition" (Korenevskii 1956, p. 246).

"In deposits of potassium salts changes and transitions among rocks of different chemical and mineral composition are extensive. Many deposits

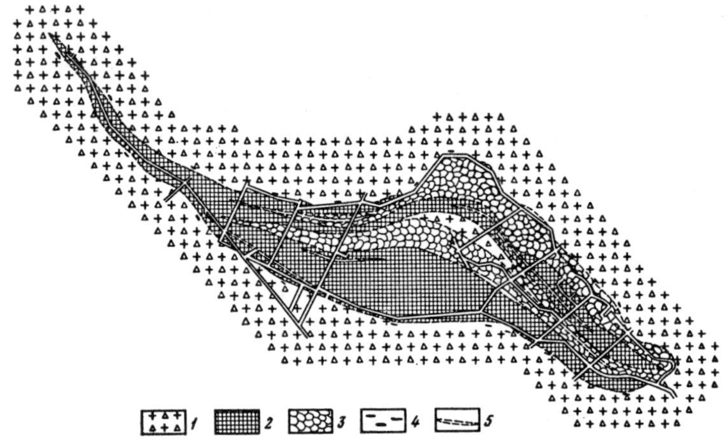

FIG. 156. Map of the Sigmund lens (from A. A. Unkovskii 1947). 1. Saliferous sand-clay breccia; 2. langbeinite-kainite rock; 3. langbeinite rock; 4. polyhalite rock; 5. rock salt.

consist of different rock-types alternating in the section. For example, the so-called upper sylvite bed of the Kalush mine is composed of sylvite, carnallite, and kainite members. The Sigmund bed in the Stebnik mine is composed of langbeinite-kainite, kainite, and langbeinite members alternating with rock salt and saliferous breccia" (Ivanov and Levitskii 1960, p. 316).

Thus, *variability in mineral content of the potassium-salt rocks even within a particular small lens is a characteristic feature of the potassium-salt deposits in the Balichskaya and Upper Vorotyshcha saline formations* (Fig. 156). Whether this variability is due entirely to primary sedimentation or whether it is katagenetic (or epigenetic) is not yet clear. M. P. Fiveg is inclined to believe that the variability is in great measure a feature associated with kata-genic alteration of the salt lenses.

These, then, are the basic features of the essential composition of the saline formations in the Carpathian downwarp. The paleogeography of the time is generally very clear but the question of the general facies type in the salt-producing basin and of the nature of the communication between this basin and the open sea should be considered. Critical data on this question have been presented by V. V. Glushko (1956).

According to this author, "the Early Vorotyshcha basin was bounded on the southwest by the rising Carpathians and on the northeast by the Paleozoic

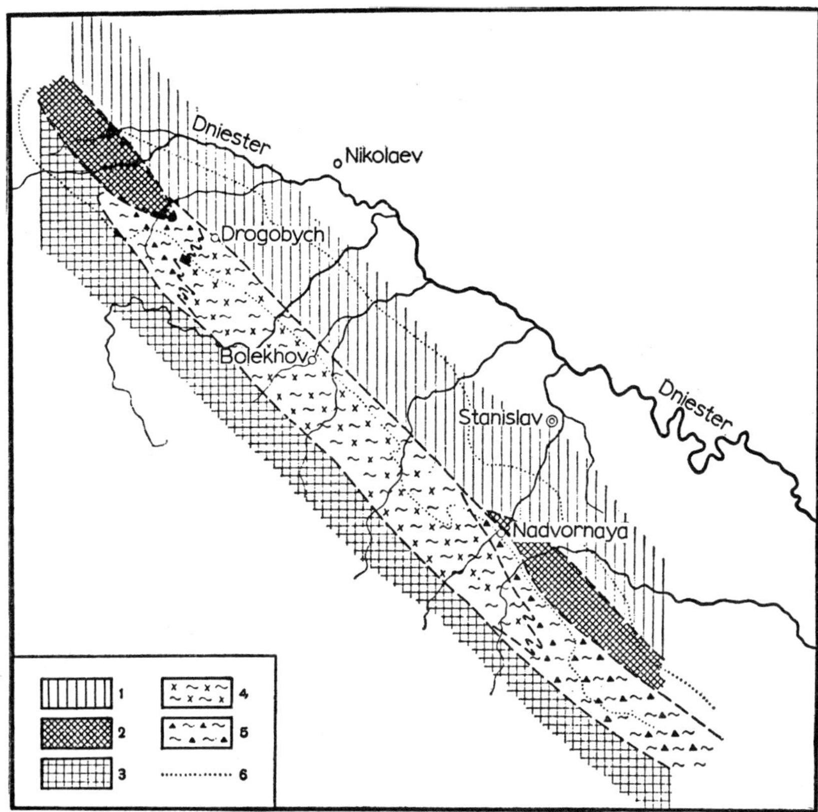

FIG. 157. Paleogeography of Early Vorotyshcha time (from V. V. Glushko. Regions of erosion: 1. Russian platform; 2. Paleozoic fold zone; 3. the Carpathians. Regions of sedimentation (Early Vorotyshcha sea): 4. saliferous clay; 5. saliferous clay containing cobbles of Carpathian and Paleozoic rocks; 6. present margin of the inner zone of the Carpathian marginal downwarp.

fold structure and the adjacent part of the platform (Fig. 157). Since there is no reliable indication that the Lower Vorotyshcha series is present in the north-western part of the depression (in the Cherkhava-Vyar interfluve), we are not without grounds for supposing that the basin was closed in the region of Monastyrets and Cherkhova. On the southeast the Early Vorotyshcha basin probably connected with the open sea, which lay in the region of present Rumania (Walachian basin)." If this was so, the Early Vorotyshcha basin was a distinctive, sharply elongated gulf, in the zone of the present Carpathian foothills. The width of the Early Vorotyshcha gulf was apparently no more than 25 km, and the length within the Soviet part of the depression was 200–250 km (Glushko 1956, p. 127). It is clear that this was an estuary-like gulf of the Virrila type, but much larger than the present Bocana de Virrila. Without knowing the entire gulf, there is some difficulty in determining sites of salt deposition, but by analogy with the present Bocana de Virrila it cannot be doubted that the facies in the Early Vorotyshcha gulf lay in the upper end, farthest from the influx of sea water. In considering the limited areal dimensions of the salt deposits proper, it may be thought that the Early Vorotyshcha basin (in the upper part, at the head) was broken into a series of small isolated lagoons in which the precipitation of salts was especially intense.*

A similar picture is seen also for the time of accumulation of the Upper Vorotyshcha saline formation. "On the northeast the Vorotyshcha basin was bounded by an elevated, greatly denuded upland of the Paleozoic fold zone and the adjacent part of the Russian platform. Along the northwestern part of the basin the Carpathian flysch was deposited (Fig. 158). There was land also on the northwest, as is proved by the absence of Vorotyshcha and Stebnik sediments northwest of Peremyshl'. Consequently, water of the Vorotyshcha basin was in contact with the open sea only on the northeast. It appears probable that there was a strait on the northeast in Vorotyshcha time, in the region of the Pokutsk-Bucovina Carpathians, bordered on the northeast by the upland of Paleozoic rocks and on the southwest by the Carpathians. It is possible that an underwater ridge prevented the free exchange of water between the Carpathian marginal depression and the open sea" (Glushko 1956, p. 132). From this description and from a comparison of Figs. 158 and 89 we see that the Late Vorotyshcha basin was also an elongated gulf of the Virrila type. But the salinity of this water was considerably greater than the salinity of the Early Vorotyshcha gulf, because potassium salts were precipitated. According to S. M. Korenevskii (1956, p. 244), these salts pre- cipitated "in a rather small area, only a few kilometers long. They were deposited only during the most extensive drying of the Late Vorotyshcha salt basin, in the deepest depressions on the floor. The positions of these depressions, which formed because of the constant instability of the floor in these parts of the basin, were commonly inherited. The formation of different horizons of potassium salts within a single area and the multi- serial character of these horizons within the field of their occurrence are due to this circumstance."

Because of the rather indeterminate stratigraphic position of the Balich- skaya (Ugersko) saline formation, which is varyingly interpreted, the paleo- geographic reconstruction is not well established. In considering the general similarity of composition of the Balichskaya and Upper Vorotyshcha forma- tions, however, and the uniformity of their potassium-salt accumulations, one can hardly doubt that the paleogeographic conditions were the same for both.

---

* In Rumania there are salt accumulations of Lower Vorotyschcha age, indicating that supplementary branches of the main gulf existed in this region.

The upper gypsum-anhydrite formation (Fig. 159) also formed in a gulf, but one which was more of the Kara-bogaz type than of the Virrila type.

Setting aside this Trias formation and considering only the three older formations, we must recognize that *they reveal a special type of salt-producing gulf, differing sharply from the Kara-bogaz type. It is a long, narrow gulf in*

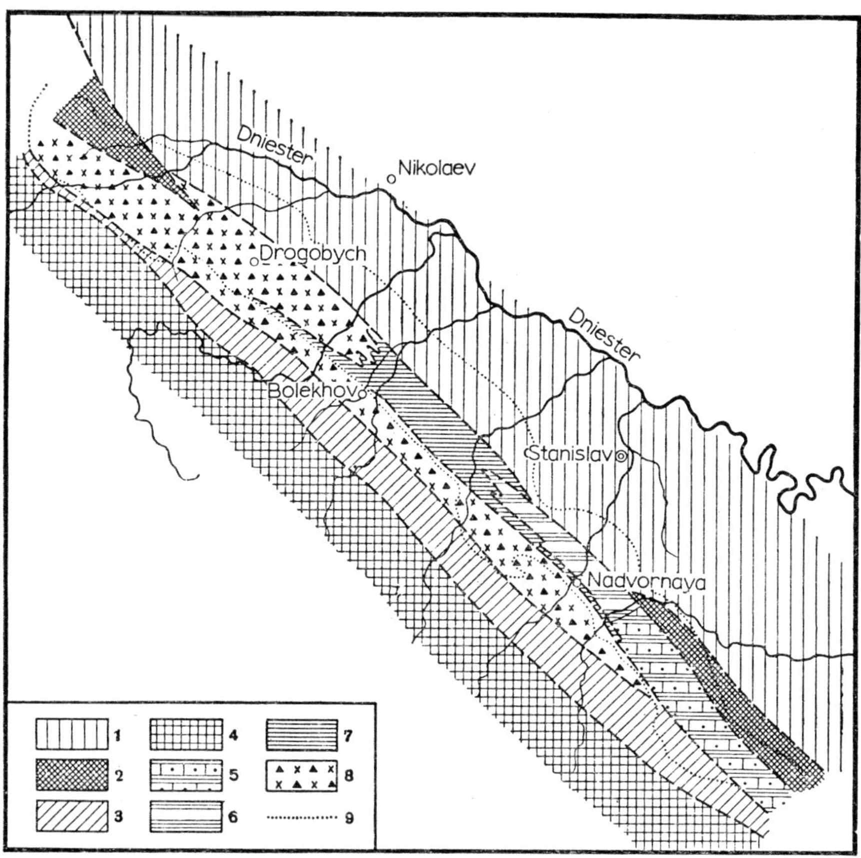

FIG. 158. Paleogeography of Late Vorotyshcha time (from V. V. Glushko). Regions of erosion: 1. Russian platform and post-Hercynian Predkeletsko-Sandomir depression; 2. Paleozoic fold zone; 3. Carpathian marginal depression; 4. Carpathians. Regions of sedimentation (Late Vorotyshcha sea). 5. alternation of sandstone (dominant) and mudstone; 6. alternation of mudstone (dominant) and sandstone; 7. alternation of mudstone and sandstone, the rocks containing small quantities of gypsum and salt; 8. salt-clay breccia, 9. present margin of inner zone of Carpathian marginal depression.

which the supply of salts comes almost exclusively from one end and progressive salinity develops away from the mouth where this supply is introduced. The saline deposits accumulate in the upper end of the gulf, and the structure of the formation as a whole is thus characterized by well-defined asymmetry, as in the present Bocana de Virrila. The morphology of salt basins of the Virrila type is thus quite different from that of basins of the Kara-bogaz type, corresponding with the sharp difference in structure of the formations. At the same time there are features in common. *Basins of both morphological types have developed in tectonic depressions that have remained active for a prolonged period but that have been characteristic of each type and that have determined the history of sedimentation.*

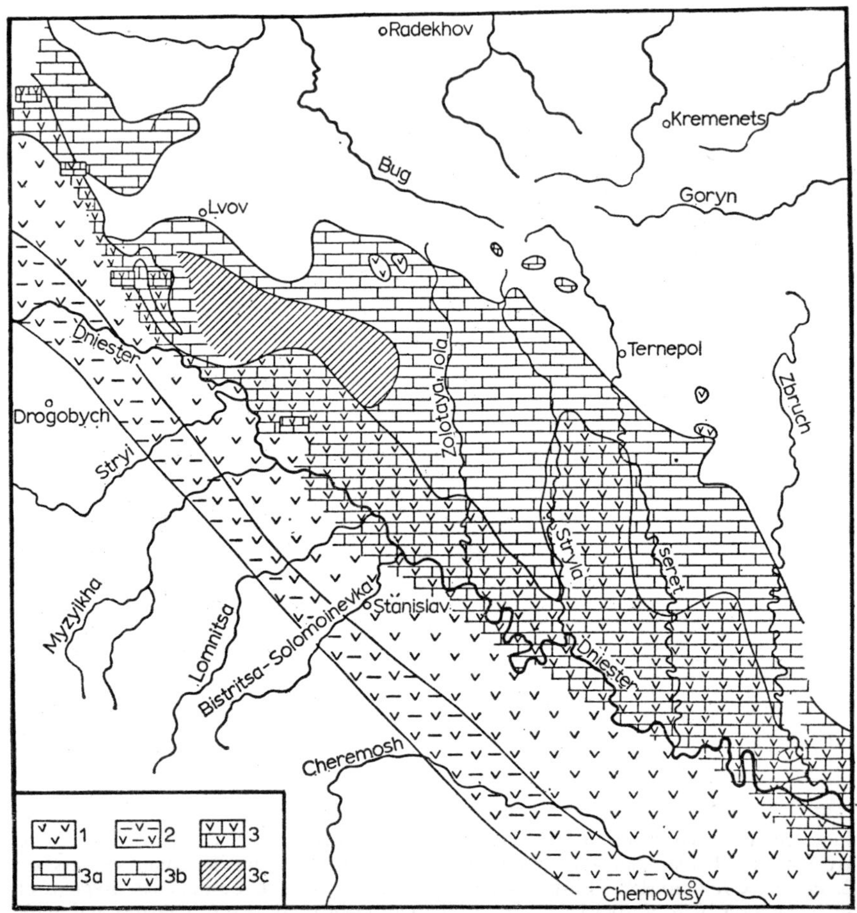

FIG. 159. Paleogeography of Trias time (from L. N. Kudrin). Gypsum-anhydrite facies: 1. gypsum and anhydrite; 2. gypsum and anhydrite with intercalations of clay; 3. gypsum and overlying chemically precipitated (Rata) limestone. Facies of the chemically precipitated Rata limestone: 3a. chemical Rata limestone underlain by gypsum; 3b. chemical Rata limestone, 3c. gypsum interbedded with chemical Rata limestone.

4. HALOGENIC FORMATIONS OF THE RHINE GRABEN

Another example of formations that accumulate in gulfs of the Virrila type is that of the Cenozoic deposits of the Rhine graben (Fig. 160). Three saline

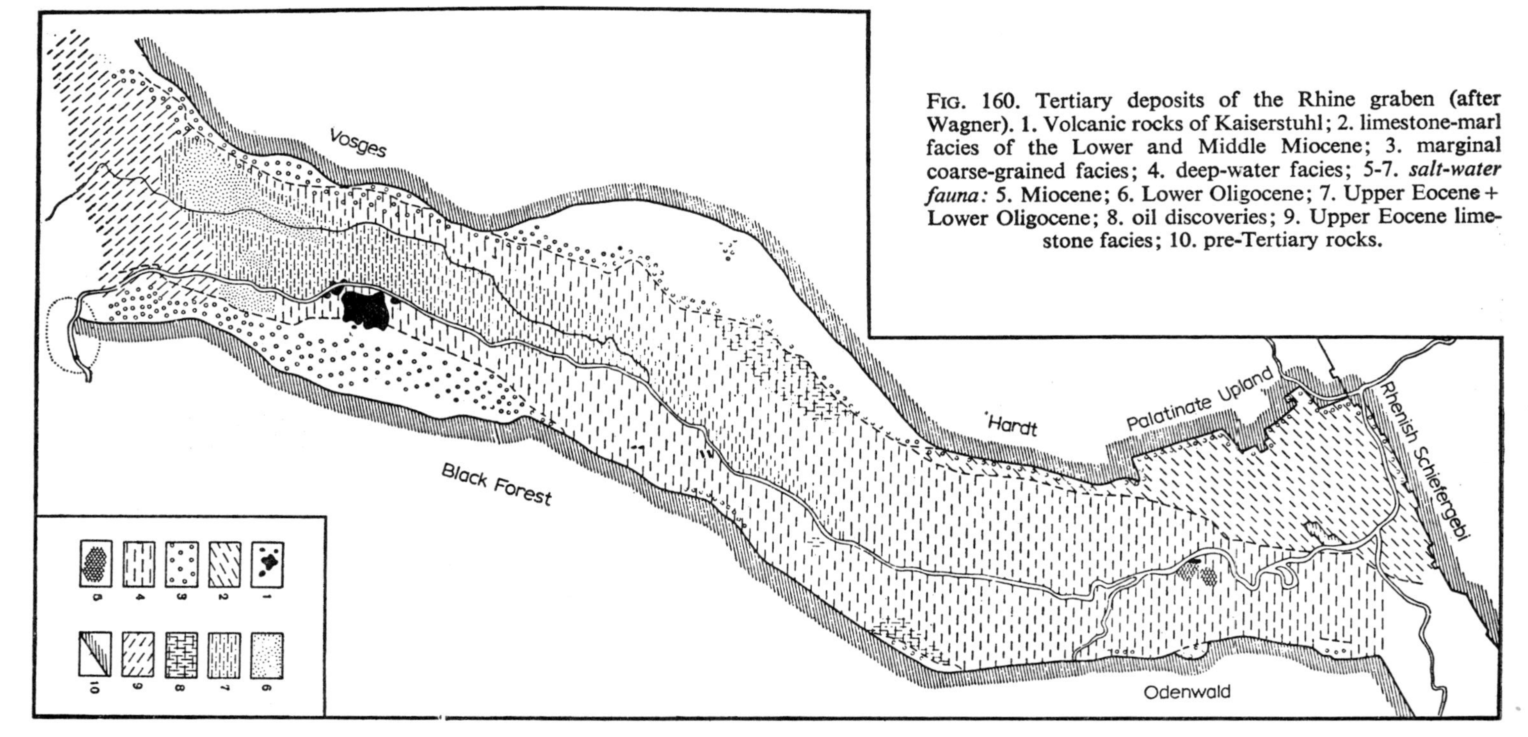

FIG. 160. Tertiary deposits of the Rhine graben (after Wagner). 1. Volcanic rocks of Kaiserstuhl; 2. limestone-marl facies of the Lower and Middle Miocene; 3. marginal coarse-grained facies; 4. deep-water facies; 5-7. *salt-water fauna:* 5. Miocene; 6. Lower Oligocene; 7. Upper Eocene + Lower Oligocene; 8. oil discoveries; 9. Upper Eocene limestone facies; 10. pre-Tertiary rocks.

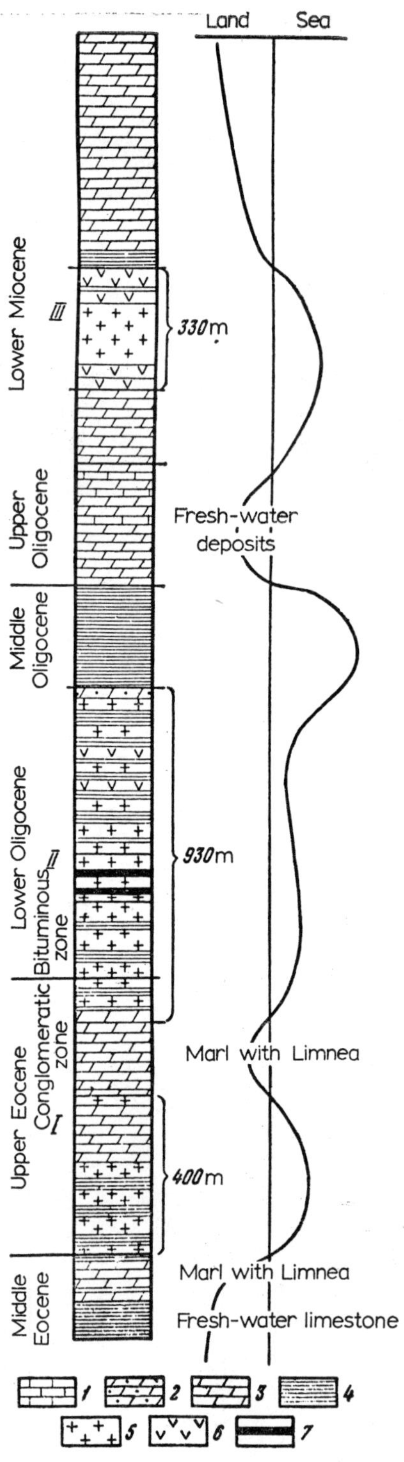

[*Legend at foot of next page*

formations are found here (Fig. 161). The first, up to 400 m thick, is Upper Eocene in age. Its lower two-thirds consists of rock salt interbedded with grey-green, very calcareous marl (Kalkmergel). The halite forms about 74% of the total thickness of the sequence. Marl and dolomite are dominant in the upper third, however, and NaCl constitutes but 11·5%. The second saline formation, 930 m thick, Lower Oligocene in age, is separated from the first by a 150-meter section of marl with shells of the fresh-water mussel *Limnea*. This formation consists of an alternation of rock salt and clay. It is bituminous, thin-bedded, and richly fossiliferous. The halite content varies from place to place but is generally 35–46%. A member with markedly more NaCl (up to 65%) is found almost precisely in the middle of the succession. Two beds of potassium salts are found in this member, which will be described below. Above a 750-meter sequence of Middle Oligocene clays and Upper Oligocene marls occurs the third saline formation, Lower Miocene in age, made up of anhydrite at the bottom (65 m), rock salt in the middle (150 m), and anhydrite again at the top (60 m). The section is capped with Lower Miocene bituminous marl with fresh-water fauna of *Corbicula* and *Hydrobia* (Wagner 1955).

This succession is found in the central parts of the Rhine graben, where the deposits represents relatively deep-water sedimentation. Toward the flanks of the graben these deposits give way to shallower-water sediments of the marginal facies (Randfacies): siltstones, sandstones, conglomerates, and organic limestones (Fig. 160). This facies has, however, been preserved only patchily in the southern part of the graben.

The structure and distribution of potassium salts in the Lower Oligocene formations are of considerable interest. The potassium salts accumulated in three individual troughs in the Rhine graben: Wittelsheim, Münchhäuser, and Buggingen (Fig. 162*A* and *B*).

Two horizons of potassium salts may be distinguished: a lower, or main, horizon and an upper (Fig. 163*A* and *B*).

The main horizon is found in all three troughs over a total area of approximately 250 km². The thickness is 4·30–5·12 m, and the structure is complex. A lower bed consists of alternating layers of halite and sylvite with rare interbedded dolomitic marl. The total thickness is 1·95 m. The middle bed is of the same composition, and is 1·75 m thick. The upper bed also consists of alternating halite and sylvite at the base, with carnallite rock above, and sylvite again at the top. The total thickness of the upper bed is 0·60 m. Two dolomitic marls are interbedded between the potassium salts. The lower one, between the first and second salt members, is 0·15 m thick, and the upper is 0·10 m thick.

Fig. 161. Section of Tertiary deposits in the Rhine graben (from Wagner). 1. limestone; 2. bituminous marl; 3. marl; 4. clay; 5. halite; 6. anhydrite; 7. potassium salt.

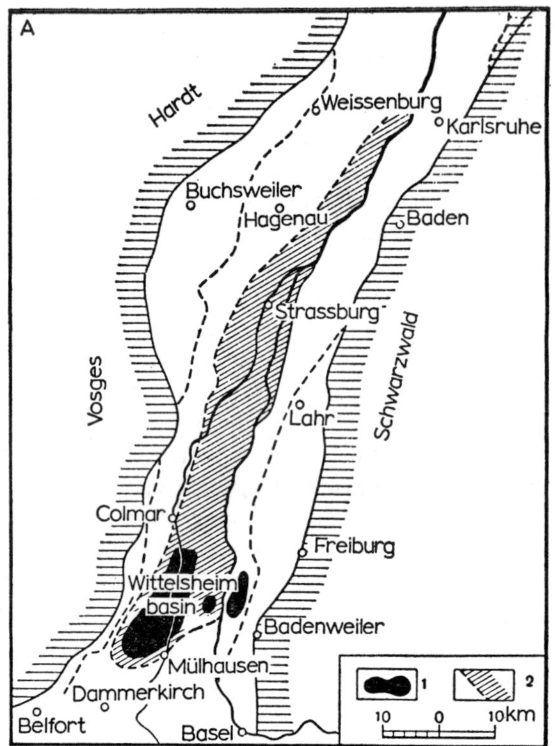

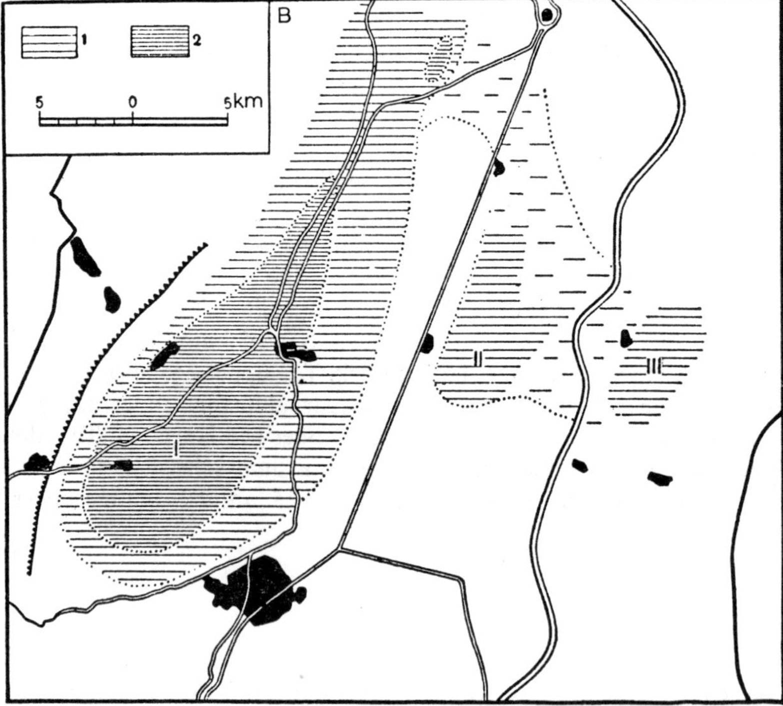

Fig. 162. Potassium-salt deposits of the Rhine graben (from Wagner). *A*. Distribution of potassium-salt (1) and halite (2) deposits. *B*. Distribution of the lower sylvite bed (1) and the upper sylvite bed (2): *I*. Wittelsheim basin; *II*. Munchauser basin; *III*. Buggingen basin.

It is of interest that the lower dolomitic marl contains a vast number of insect remains (737 species and 40 genera) and plants. These represent tropical and subtropical fauna and flora, living where the average yearly

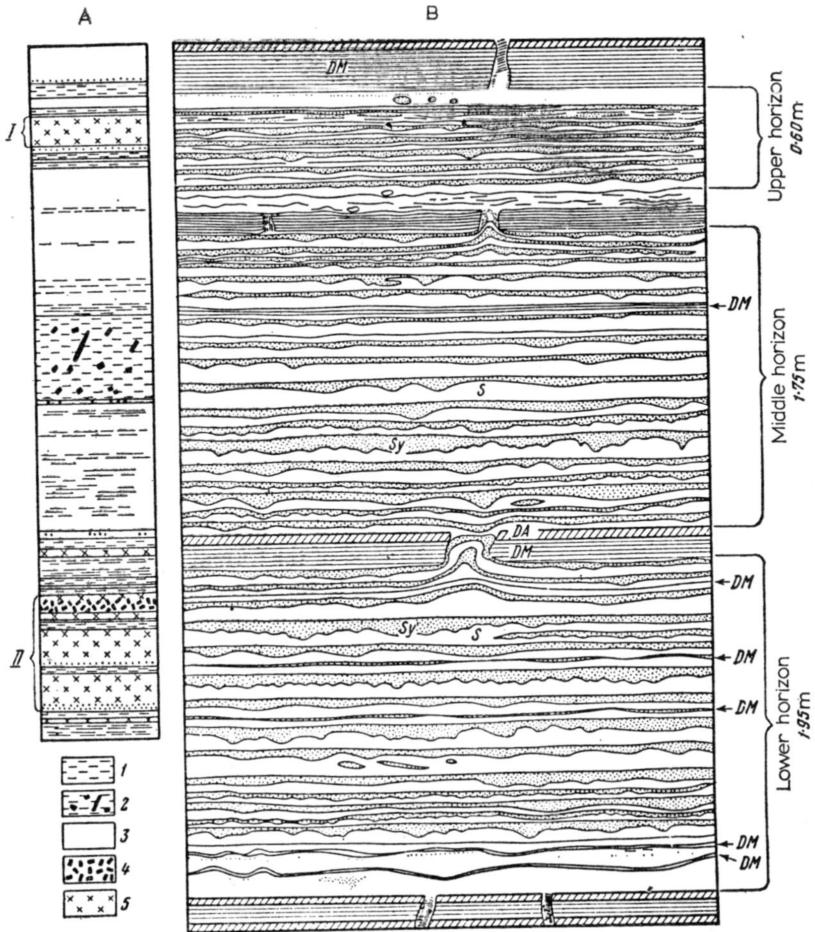

FIG. 163. Section through the potassium-salt strata (from Wagner). *A. Entire section:* 1. dolomitic marl with anhydrite; 2. deformed rock salt; 3. rock salt; 4. carnallite; 5. sylvite; *I.* upper bed, 1·5 m thick (27·5% K_2O), *II.* lower (main) bed, 5·0 m thick (22% K_2O). *B. Structural details of the main bed:* S. rock salt; *Sy.* sylvite, *F.* fibrous salt; *DM.* dolomitic marl; *DA.* dolomitic anhydrite.

temperature was 19°C, and blown into the salt-producing basin during dust storms.

The main potassium-salt horizon is separated from the upper by a 20-meter sequence of alternating dolomitic marl, saliferous clay, dolomite-anhydrite, and salt. The dolomitic marl is dominant in the lower part of the sequence, rock salt toward the top.

Two impersistent intercalations of sylvite, negligible in thickness, are also found in the lower half of the sequence.

The upper potassium-salt horizon only 1·15 m thick is found in the central part of the Wittelsheim trough and is absent elsewhere. A characteristic feature of this horizon is that rock salt and potassium salts are more sharply isolated here than in the main horizon. "It is typical to find light, milky, coarse-grained sylvite crystals up to 4 cm across with a red border. The milky color of the sylvite results from subangular crystallites of NaCl" (Wagner, 1955).

The history of Paleogene sedimentation in the Rhine graben is now known in its general outline. In the Mesozoic the Vosges and the Black Forest were a single Hercynian massif invariably rising above sea level and, therefore, forming a zone of erosion. Sediments began to accumulate in the Middle Eocene: fresh-water lacustrine limestones and oil shales. The same conditions continued to the very beginning of the Upper Eocene, represented now by marls with *Limnea*. The sea then transgressed, quickly became highly saline, and the *first* saline formation was deposited. The accumulation of the formation ended in a freshened basin with deposition of marl containing *Limnea*. Marine conditions were restored in the Early Oligocene and continued to Middle Oligocene time. The sea again quickly became very salty, and a second very thick (930 m) saliferous formation accumulated. At approximately the middle of the depositional period of the formation the basin water became sufficiently salty for potassium salts to be deposited. At the beginning of the Middle Oligocene the saline basin was again destroyed by the opening of broad communications with the sea. Ordinary carbonate-clastic sedimentation replaced saline deposition, Rupelian clay and marl with foraminifera accumulating first and then the Meletta clays. In Late Oligocene time marine conditions gave way to continental for the third time, and marls and limestones with fresh-water molluscs accumulated. Transgression was repeated once again in the Early Miocene, and the third saline formation was deposited. The period of deposition was relatively short, and the salt-producing basin became a fresh-water lacustrine basin, in which marl with *Hydrobia* and *Corbula* accumulated.

Thus, within a short period from the Middle Eocene to the Early Miocene the sea transgressed the region of the Rhine graben and saline formations were deposited. In other words, *the sea water that penetrated the graben quickly lost its free connection with the open sea, and the basin became a highly saline marine gulf in which saline deposits were abundantly deposited, sometimes to completeness, as during deposition of the second saline formation.* Unfortunately, the actual shape of the gulf and the location of the strait communicating with the open sea are not yet known. "The ingression of this Sannoisian sea in Alsace poses an unsolved paleogeographic problem. The transgression does not seem to have come by way of the Mainz basin in the north, nor by the Jura Mountains in the south, since marine deposits of Sannoisian age are unknown in either place. A direct communication with the Paris basin by way of the Phalsburg depression has been suggested (van Werveke). Perhaps more likely would be a connection with the lagoons of the Rhône and Saône (Dollfus). For in this direction, the Oligocene ends against the Jurassic of the Belfort pass, with lagoonal-lacustrine formations lying directly on the Jurassic. . . . Between this region and the Alsace plain the Oligocene may have been removed by erosion" (Gignoux 1952, p. 460).

Whatever paleogeographic variant of this communication with the sea is adopted, the fact remains that the saline formations accumulated in a gulf and at its farthest end. This circumstance is distinctive and important in our classification. In considering the shape of the Rhine graben, one must recognize that morphologically the basin is of the Virrila type and that its sedimentation was of a markedly asymmetrical character. Like the present Bocana de Virrila, the Rhine gulf at the time of its existence received normal clastic-carbonate sediments near the strait that connected it with the sea. Farther from the strait calcium sulfate precipitated from the progressively saltier water, and at the extreme end of the gulf halite and double potassium salts formed.

One additional important factor should be noted: as in the Carpathians, *the potassium salts were formed only in parts of the gulf that warped downward most rapidly, where the sequence accumulated to its greatest thickness.* Similar relationships are common and will be discussed later.

There is yet another fundamental control: *sedimentation in the Rhine graben took place in parallel with rapid uplift of the Vosges and Black Forest,* which supplied clastic material. Great thicknesses of Paleogene deposits thus accumulated, reaching about 3761 m, in a relatively short time between the Middle Eocene and the Early Miocene. Of this thickness only 1256 m, or 43%, are of saline rocks; the remainder is clastic-carbonate material, especially marl and dolomitic marl. Furthermore, the saline formations themselves contain large percentages of clastic-carbonate material, at least 50%. Excepting this clastic-carbonate admixture, the solid salt phase proper makes up only about 20% of the Paleogene rocks. This aspect relates the saline formations here to those of the Carpathian marginal depression, although the tectonism was substantially different.

5. Other Halogenic Formations deposited in Gulfs

A few of the better known formations deposited in large gulfs and connected with the sea by means of narrow straits have been described: they correspond to syneclises that warped downward for prolonged periods, to foredeeps, or to grabens. In addition to these formations, there are still others, of much smaller volume, that have accumulated in gulfs. These include the potassium salts and halite beds in the Eocene of the Ebro basin, Upper Cretaceous gypsum of a large gulf in Tunisia, Upper Jurassic rock salt in northwestern Germany, Upper Jurassic gypsum and, in part, halite deposits of the Caucasus, Lower Triassic saline sequence in the northern and southern foothills of the Pyrenees, Eifelian salts of Kara-Tuz and also the Aquitanian

z

basin, the Devonian saline sequence in the Dnieper-Donets basin, Pliocene salts of Uzun-su, and Cambrian rock salt of Pakistan. This list of gulf-type formations will undoubtedly be increased. The facies character of gulf formations is, however, more difficult to establish the smaller the formation both in area and thickness. Even more so, if the saline formation was incompletely developed, depositing only gypsum for example, it becomes increasingly difficult to distinguish gulf type from lagoonal type. The situation is complicated still more by the fact that the small salt deposits have been studied least.

Despite this incompleteness, three fundamental conclusions may be drawn. *Firstly*, formations of the gulf type were deposited as early as the Early Cambrian, in Pakistan (Salt Range), and then continued to accumulate through subsequent periods to the present. The history of these formations is as long and as complete as that of lagoonal formations. *Secondly*, formations of the gulf type commonly display fuller development of the saline process than lagoonal formations. Pure gypsum beds are few. Halite is normally precipitated, and potassium salts are common, both sulfates and chlorides. *Thirdly*, formations of the gulf type extend over a broad tectonic range. They accumulate in geosynclinal zones during normal tectonic conditions (rarely) and in the closing stages (most commonly), and also on platforms, both old and young. *The main bulk of gulf-type formations, especially the large formations, clearly occur on platform areas.* In this connection gulf-type formations represent a step toward more stable structures when compared with lagoonal formations.

E. ANALOGIES OF ANCIENT AND RECENT TYPES OF HALOGENESIS AND SOME RELATED PRINCIPLES

A number of types of ancient saline formations have now been described, similar in form to deposits now accumulating. In such comparisons, however, questions always arise concerning the accuracy of the analogies, the limits of possible and permissible deviations from similarity, the significance of differences that lie still within the limits of analogy, and so forth. It is therefore necessary to consider these questions.

Attempts to find analogies between recent and ancient saline deposition have recently come to be considered inadequate and even reprehensible in some geological circles. These analogies appear to be typical manifestations of uniformitarianism, and, since this concept is regarded as antiscientific, a negative attitude is usually taken regarding any parallelism between ancient and modern saline deposition; such comparisons are under suspicion even before they are examined. It is considered that modern salt deposition is one thing, ancient deposition a different matter, and that it is better to avoid methodological "sins" and to refrain altogether from comparing them. Such an approach has indeed been set forth repeatedly by A. L. Yanshin.

It must be stated that to some extent the negative attitude toward a comparison between recent and ancient saline depositions is valid, because the comparisons themselves normally have been oversimplified and followed rigidly. From the present epoch the so-called "lagoon" has been taken as the standard of comparison—a salt basin with very indeterminate features and properties, and all ancient salt-producing basins have been placed in the lagoonal category without attempting to examine them critically (see the monograph of A. A. Ivanov and Yu. F. Levitskii 1960). It is not possible to agree with this. It is doubly erroneous: in the first place, present-day saline basins are by no means all of one type. As pointed out in the second part of this book, there are at least three morphological types: *intracontinental lakes*, with no communication with the sea; *lagoons*, small saline basins cut off from the sea purely by exogenic means, by the formation of bars through the action of surf and longshore currents; and *saline gulfs* isolated from the sea by tectonic activity, such as a tectonic depression filled from the sea but cut off from the sea by some slight uplift, through which a narrow connecting strait may pass. Another defect in the comparisons previously made has been the failure to classify ancient salt-producing basins and their corresponding formations according to composition and facies type. At least five facies types of formations, as indicated above, are known in the geologic record: *continental formations*, accumulating in basins having no connection with the sea; *lagoonal formations*, accumulating in areas adjacent to the sea and experiencing only weak tectonic movements, generally with the saline process only imcompletely expressed because of the dominant exogenic factors that give rise to sand bars, isolating the lagoons; *gulf formations*, accumulating in relatively large gulfs, each corresponding to a large syneclise, downwarp, or graben; *formations of marginal parts of the open sea*, each one corresponding also to a single tectonic depression or a series of small depressions; and *formations of huge intracontinental salt-producing seas*, accumulating on a complex tectonic substratum. From this list of facies types of recent and ancient saline deposition it becomes clear that *the first by no means completely spans the range of the second, but that, at the same time, analogies between the two undoubtedly exist*. It is very probable that present-day saline deposition in intracontinental lakes is the closest analogy to the deposition that gave rise to our older continental saline formations, or that present-day deposition of salts in lagoons may have its analogy in the conditions and processes obtaining during the accumulation of ancient lagoonal saline formations, or that saline deposition in present-day gulfs such as Kara-bogaz and the Bocana de Virrila give us a picture of the large salt-producing basins in which gulf-type saline formations accumulated in the past. It is just as probable that the present has formations that accumulate in the marginal parts of the sea or in intracontinental seas, analogous to those known from the past. Thus, *among ancient formations there exist more or less close analogies of different facies types of present-day salt deposition and also undoubted specific formations that have no current counterpart*. With these complex relations it has become necessary to develop a classification of both present-day and ancient salt deposition. *The concept of facies types of saline deposition has become one of the central ideas in the science of salt deposition in general.*

The recognition among ancient saline formations of analogues of recent salt deposition by no means suggests that the two are identical. It implies merely a fundamental similarity of the main features, admitting divergence of details. Continental saline formations of the past and present are similar in that salts accumulated in them outside the influence of marine basins, deriving their salts only from streams and ground water entering the sedimentational basin. And since streams throughout geologic history have carried greater or lesser amounts of clastic material, both modern and ancient continental saline sediments are strongly contaminated with sand, silt, and clay. Another fundamental similarity is the incompleteness of the saline process, the general absence of potassium-salt deposits. Along with these important similarities, however, there are also differences: differences in size, the ancient lakes being much larger than present lakes, and the duration of the depositional process being longer in earlier times, commonly resulting in great thicknesses of saline residues.

Ancient and recent lagoonal deposits are similar not only in their deposition in very small semi-isolated basins but also in the general interruption of the saline process at the beginning of the gypsum stage and by the thinness of the sulfate accumulations themselves. In both, these features follow from the slight mobility of the seaward-sloping area on which the lagoons formed. The smallness of the basins has led to the accumulation of clastic sediments from transported material and beach abrasion and to contamination of the saline sediments themselves with this material. Differences between ancient and recent lagoonal saline deposits are found in the greater size of the older lagoons, the greater number of lagoons in the coastal zone, differences in thickness of the saline beds and sequences, and in other details of the internal fabric of the formations.

Recent and ancient gulf formations are similar in that each formation corresponds to a specific tectonic depression that underwent subsidence for long periods and then became more or less filled with a sequence of rocks. They are also similar in the principal morphological features of the basins, whether rounded and equidimensional like the Gulf of Kara-bogaz or long and narrow like the Bocana de Virrila. The basins are small and the saline deposits are argillaceous and subordinate to a sequence of clastic or carbonate-clastic host rocks. With increase in size of basin the clastic material becomes increasingly a marginal feature, and the saline deposits, localized chiefly in the central parts, become freer of contaminating admixtures. This pattern is illustrated by comparing the formations of the Rhine graben and the Carpathian foredeep with the sulfate-carbonate complex of upper Lower Devonian age in the Moscow basin and of Upper Silurian age in the Michigan basin. Differences between ancient gulf formations and their modern counterparts are chiefly the larger size in general of the older gulfs and the occasional greater completeness of the saline process, yielding potassium salts.

It is therefore clear that *the recognition of analogies between recent and ancient saline deposition, resting on a preliminary discrimination of natural facies types among both* allows the prediction in ancient formations of several features that direct observation has not yet established. Knowledge of chemical types of modern intracontinental lakes, for example, permits us to predict reliably that, among ancient deposits, representatives of soda lakes as well as of sulfate lakes of groups *IIa* and *IIb* may be found. Knowledge of the physicochemical mechanism of modern salt deposition and, in particular, of yearly hydrochemical cycles and of the structures in the salt beds reflecting these cycles, knowledge of brine metamorphization and the reflection of this in the mineral content of the salt phases, and knowledge of "solar crystallization" of brine—all these make possible the reliable interpretation of compositional and structural features in ancient saline formations.

CHAPTER 2

HALOGENIC FORMATIONS IN THE MARGINAL ZONES OF OPEN CONTINENTAL SEAS

Together with formations in which the facies types of modern saline deposition are readily recognizable occur a number of which *the depositional environment has no known correlative among modern deposits. They are specific products of past geologic time.* For a long time this character was not recognized because of lack of sufficient data. Only now, after extensive borehole information in the U.S.S.R. and the U.S.A. has become available, have the true nature and distinctive character of many marine saline formations become clear. As has been already pointed out, there are two such specific ancient types of saline formations. One accumulated in the marginal zones of open epicontinental seas, the other in huge intracontinental salt-producing seas resting on a complex tectonic structure. Both types will now be described in detail.

A. LOWER AND MIDDLE TRIASSIC HALOGENIC FORMATIONS OF GERMANY

The Röth (Upper Bunter) and Middle Triassic deposits of Germany are an example of halogenic formations deposited in marginal zones of epicontinental seas. The tectonic base upon which both these formations accumulated was the Central European depression, which first began to form at the end of the Early Permian time. This depression occupied a large part of the present-day North German basin and the Polish-Lithuanian syneclise. It was bounded on the west by the Caledonian and Hercynian massifs of England and France, including the Ardennes, on the north by the Baltic shield, on the east by the Polesian salient, and on the south by the Bohemian massif, from which a narrow uplifted zone extended to the southwest, separating the Central European depression from the Tethys geosyncline.

The Röth (Upper Bunter) halogenic formation lies at the top of the Lower Triassic and it conformably overlies the so-called Bunter sandstone, a clastic continental arid-climate red bed. On the south, along the mountainous border of the depression, coarse clastics occur in this formation. To the north, in the Magdeburg-Halberstadt basin of Saxony, clays are dominant, chiefly red, more rarely light bluish or green. Still farther north the upper part of the succession contains carbonates. In thickness the continental formation of the lower Bunter normally ranges from 200 to 300 m, rarely more.

In contrast to the arid-climate deposits already described, the Röth saline formation is marine. It occurs over the entire area of the Central European depression, having been deposited in a marine basin that connected with the normal geosynclinal sea through the Silesian-Moravian strait to the east of the Bohemian massif.

The formation changes systematically in composition from the area of marine ingress to the basin. In the strait, along the southern and western slopes of the Schwientoch mountains, the deposits are chiefly clastics, especially in the lower part, where flaggy sandstones with interbedded clay are abundant. Upward in the succession limestones become increasingly

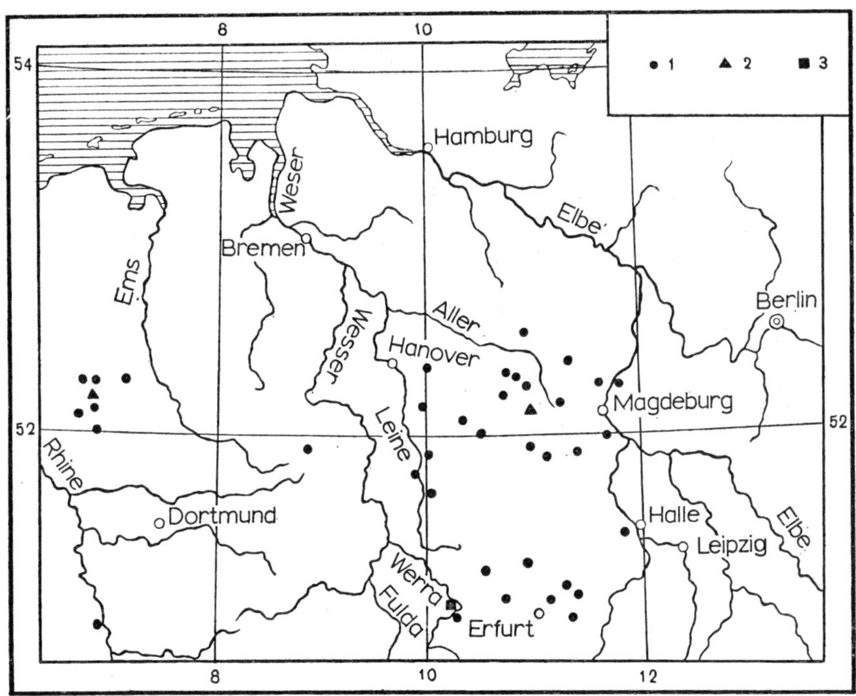

Fig. 164. Localities of rock salt in the Röth (according to E. Fulda, after M. P. Fiveg). 1. Rock-salt localities in upper Bunter (Röth); 2. salt wells; 3. soda recovery plants.

common. The thickness of the Röth here exceeds 100 m. To the north, in the region of the Polish lowland (at Szubin), it is made up of clays with numerous thin fossiliferous limestones and marls. West of Magdeburg a broad zone is marked by saline deposits. They are concentrated in three areas: a small area between the Rhine and the Ems, between the Elbe and the upper reaches of the Ems, and north of Erfurt (Fig. 164).

The succession of saline deposits in the largest of these areas (west of Magdeburg) is as follows: a thin bed of anhydrite (2 m) at the base, an overlying bed of halite interbedded with anhydrite (40–80 m), and an upper anhydrite layer to cap the formation (2–14 m). This is succeeded by 30 m of grey marl interbedded with dolomite and with nodules and beds of anhydrite. Rock salt in the lower part of the horizon is grey or watery clear, commonly coarse-grained. In the upper part of the horizon it is reddish or yellow. Some salt horizons have scattered aggregates of red polyhalite. Grey crystals of glauberite are also common, and anhydrite is invariably present. Beyond the borders of rock salt accumulations, anhydrite is found over a considerable area. To the west and south, near the shores of the ancient basin, it gives way to grey sand-clay rock.

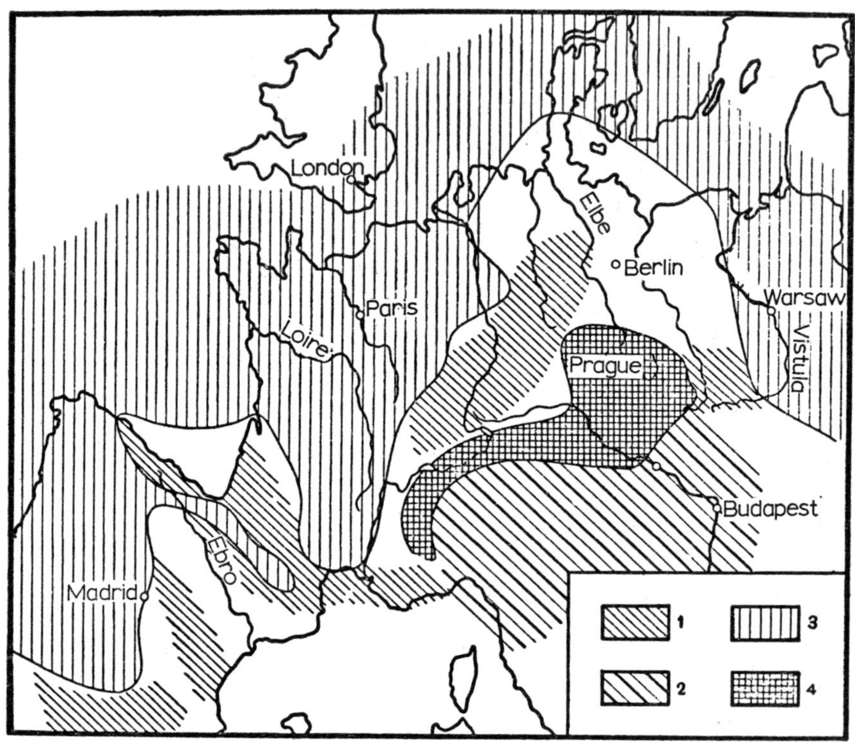

FIG. 165. Paleogeography of Early Triassic time for the Aquitanian basin (modified by Fiveg from Gignoux). 1. German type Triassic; 2. geosynclinal Triassic; 3. continental deposits (land); 4. Vindelician island.

The thickness of the Röth in the western part of its occurrence ranges from 130 m, in successions lacking gypsum and salt, to 240 m, where rock salt is present. The thickness of rock salt in the northwestern corner of the salt-bearing area exceeds 100 m.

It is therefore clear that *the saline deposits were laid down in those zones of the epicontinental sea of the Central European basin farthest removed from the influx of sea water.* This gives to the formation a *well-defined asymmetrical structure.* Saline deposits are subordinate in the formation as a whole. The salt phases are accompanied by large quantities of fragmental material, attesting to the dissected character of the adjacent continent.

The Middle Triassic saline formation occupies the same area as the Röth, resting conformably on this formation (Fig. 165). The Middle Triassic formation is much less contaminated with clastic material than the Röth, however, and the distribution of salt deposits within the formation is substantially different. It would therefore appear that the Middle Triassic formation is an independent body, entirely separate from the Röth formation.

Figure 166 (from Richter-Bernburg 1953b) shows the distribution of sedimentary types within the saline formation. Marine deposits proper (limestones with crinoids, sponges, and oysters) are found in very restricted areas

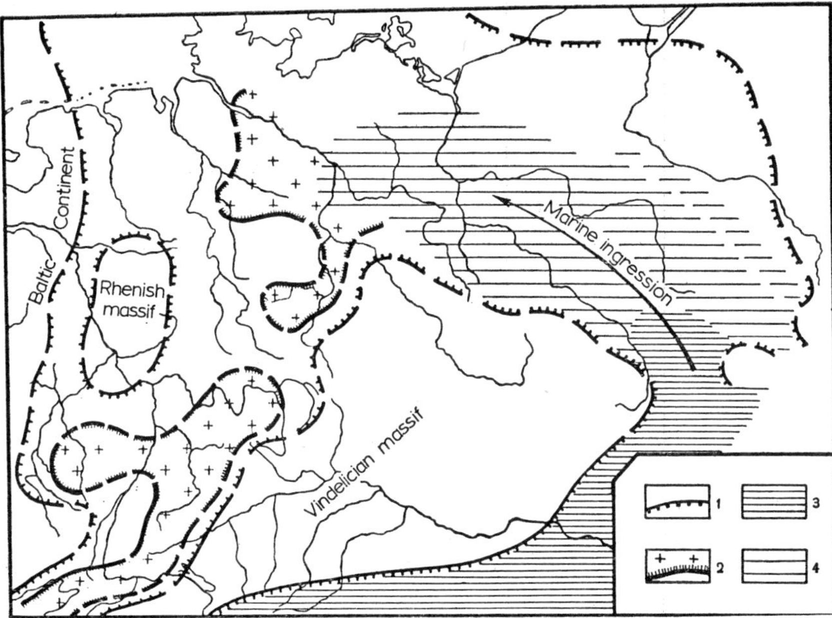

FIG. 166. Middle Triassic facies of Germany (from Richter-Bernburg). 1. Inferred boundary of sedimentation zone; 2. original distribution of rock salt; 3. distribution of $CaSO_4$-free marine deposits; 4. distribution of $CaSO_4$-bearing marine deposits.

in the southeastern corner of the basin between the Bohemian massif and Lysa Gora—the Silesian gap—where water of normal salinity entered the basin. In the broad adjoining zone on the north and northwest Diplopore dolomite was deposited first, followed by alternating beds of dolomite and gypsum (anhydrite). Toward the west the amount of anhydrite increases and at Berlin the Middle Jurassic is 80 m thick, of which 12–16 m is anhydrite and the remainder is dolomite. West of the Elbe, rock salt appears and may be traced through southwestern Germany to Switzerland. It occupies two isolated areas of irregular outline, each bordered by a broad zone of anhydrite, and is confined to the lower part of the formation (Fig. 167). The vertical sequence demonstrates the classical paragenetic series of salts: at first dolomite (bituminous), then a thin basal anhydrite, followed by rock salt with a total thickness of 8·3 m, but with a bed of anhydrite only 0·12 m thick. Despite the thinness of the anhydrite layer, it is persistent over almost the entire area of the basin and is a good marker horizon. Above the rock salt

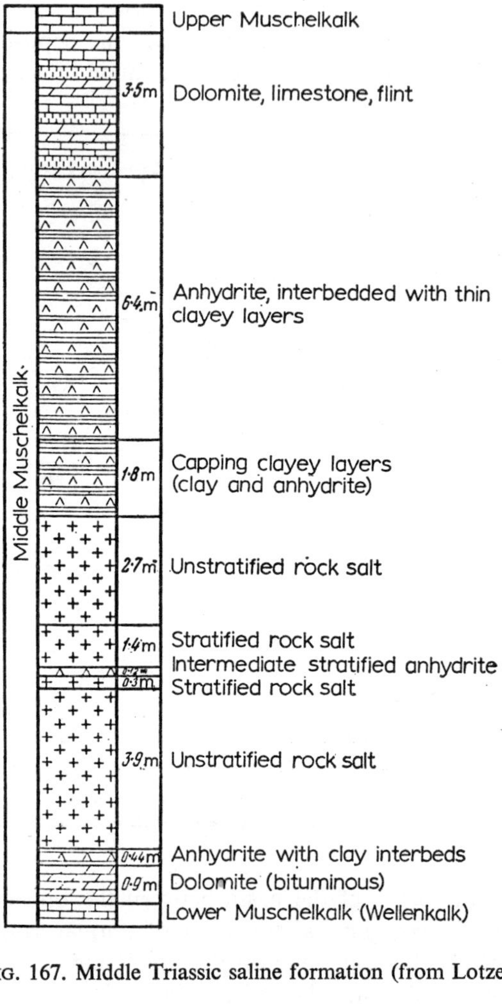

Fig. 167. Middle Triassic saline formation (from Lotze).

there is the capping anhydrite, 8·2 m thick, interbedded with clay, and then dolomite and limestone. The total thickness of the Middle Triassic saline formation in Germany is thus small, but it locally reaches 40 m. Inshore deposits developed in northwestern Germany and also in the vicinity of Würzburg and farther to the southwest; these are represented by red marls.

Thus the structure of the Middle Triassic saline formation as a whole shows, *even more clearly than the Röth, both distinct asymmetry and also a strictly systematic distribution of the saline deposits.* Near the influx of sea water pure carbonate deposits accumulated. Farther away, over a broad area, carbonates and sulfates were interbedded; and still farther from the influx of sea water, in the western half of the basin—essentially that part cut off entirely from the open sea—halite accumulated. This change in deposition clearly points to progressively greater salinity away from the conncction to the sea. Let us consider the complex configuration of this part of the basin and the large Rhenish island within it. It is probably that this part of the basin was characterized by complex relief of the floor and in addition, as shown by the 50-m isopach, by the most intense downwarping.

F. Lotze (1938) termed the Middle Germanic salt-producing basin the "German Gulf of Kara-bogaz". It appears preferable, however, to consider this basin a marginal part of Tethys rather than a gulf. It occupied a tectonically and geomorphologically complex depression cut off from the normal basin by a large mountainous island, and it contained islands within it. The complexity of the form in the flooded region was responsible for the difficult connections between the marginal part of the sea and the central open part, a relationship that causes increase in salinity in arid climates.

The Röth and Middle Triassic saline formations are not the only representatives of formations that formed in marginal zones of a sea flooding tectonically complex regions. The Upper Jurassic saline formations of Tadzhikistan and the Permian rocks of the Delaware basin in North America probably are also of this type. If this is correct, a new essential feature of saline deposition in marginal parts of open seas is brought to light: the greater completeness of the sedimentary process, leading to the deposition of potassium salts. In this respect, the Jurassic potassium salts are of the sulfate-free type, whereas the deposits of the Delaware basin belong to the sulfate group. In essence marginal zones of open seas have a complex tectonic structure, being developed on young epi-Hercynian platforms that have broken up tectonically.

B. THE UPPER FAMENNIAN SULFATE-CARBONATE FORMATION OF THE MOSCOW SYNECLISE

Not only marginal zones of the sea that occur on a tectonically complex substratum, but other zones of the sea have flooded a basement of simple structure, such as a broad syneclise bordered by anticlises. Here the morphology of the marginal zone of the sea is also simple, and restricted connections with the main water mass of the sea are less obvious, but discernible by careful investigation. This relationship is known for saline formations of old Precambrian platforms, as described below.

The Dankov-Lebedyan beds of the Moscow syneclise are, in this regard, of exceptional interest. Until recently these beds were known almost solely from outcrops in the central Devonian field, but they have now been extensively drilled in European U.S.S.R. V. G. Makhlaev (1956) first showed the true nature of these rocks in the Moscow syneclise and the relations between these rocks and Upper Famennian sequences lying farther east. In order to understand the specific features of this formation, the general aspects of Late Famennian paleogeography in the western half of the U.S.S.R. will first be considered.

1. GENERAL PALEOGEOGRAPHY OF LATE FAMENNIAN TIME

At that time a large basin lay in the western half of the U.S.S.R. (Fig. 168), occupying the Ural-Siberian geosynclinal zone, the eastern half of the Russian platform, and the Crimean-Caucasian geosynclinal zone.* Within the Ural-

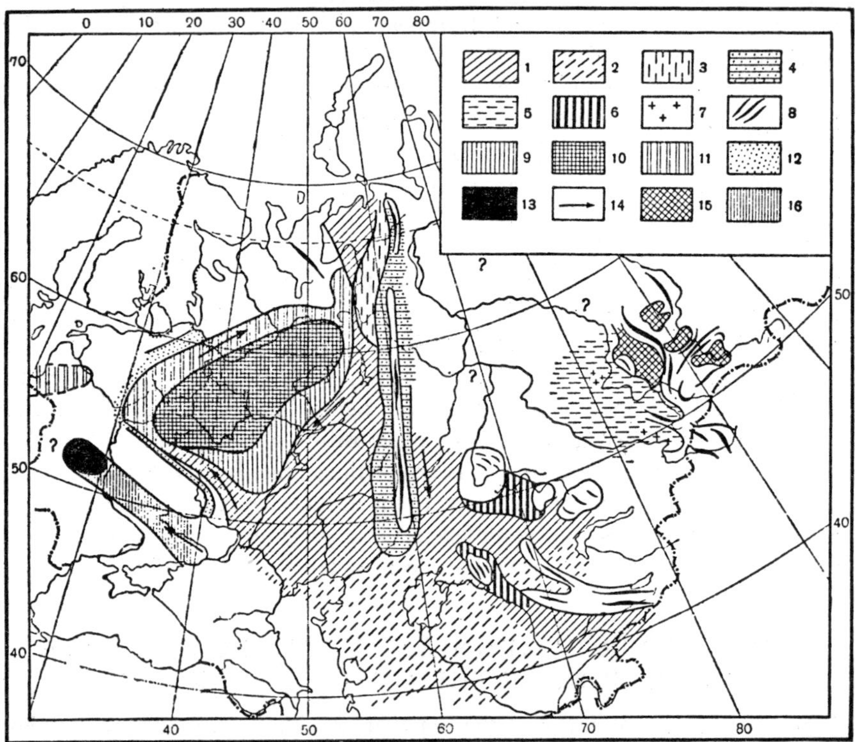

FIG. 168. Paleogeography of Dankov-Lebedyan time. 1. Chiefly limestone; 2. the same, inferred; 3. bituminous deposits of Domanik type; 4. sand-clay deposits; 5. the same, inferred; 6. dolomite of Central Kazakhstan; 7. volcanic rocks; 8. mountain chain; 9. normal dolomite; 10. gypsum interbedded with dolomite; 11. dolomite interbedded with gypsum; 12. in-shore (partly continental) sand-clay deposits; 13. potassium salts + NaCl, 14. direction current; 15. continental deposits; 16. rock salt.

---

* In preparing this map, use was made of the paleogeographic maps of S. M. Andronov for the Urals, Martynova for central Kazakhstan, T. N. Bel'skaya for the Kuzbas, V. G. Makhlaev for the Moscow syneclise, and T. I. Kushnareva for the Timan and the Pechora basin.

Siberian zone, accessible only in its southern half, the basin was studded with islands. Along the eastern Urals there apparently extended a continuous high mountainous cordillera, under active erosion and supplying polymictic sand, silt, and clay to the adjacent parts of the sea, especially near the Southern Urals, where the thick Zilair sequence was deposited. Islands were also numerous in Central Kazakhstan and Central Asia,* but they were low and flat. Only a narrow, thin, inshore margin of sand-clay deposits accumulated around the islands, therefore, giving way abruptly to carbonate deposits in the inter-island zones. Organic and organo-detrital limestones were dominant among the carbonate rocks. Dolomitic sediments (patchy, metasomatic, but locally sedimentational) framed the islands and even extended into the inter-island zones. Thin lenses of gypsum, negligible in area, are found only south of Lake Teniz and are undoubtedly lagoonal deposits. Independent saline formations did not accumulate. In the eastern half of the geosynclinal zone the islands and the adjacent land areas were strongly dissected. To the east of the Irtysh, therefore, clastic sediments are dominant. Locally (the Rudnyi Altai, the Novosibirsk region) volcanic eruptions occurred. In the region of the western Siberian Caledonian fold zone, marine sedimentation gave way to accumulation of a thick continental sequence in the intracontinental basins between the Kuznetsk Alatau, Gornyi Altai, and Sayan Ranges (the Minusinsk and Tuva basins). Gypsum accumulated only in the northeastern part of the Kuznetsk basin in late Late Famennian time, forming minor platy beds about the margin of the marine basin. There is no doubt that this gypsum is also lagoonal, including deposits of maritime lakes that developed from the lagoons. There are no independent saline formations.

Late Devonian sedimentation on the Russian platform was substantially different. Here lay a huge flat-bottomed shallow-water basin, forming two large arms. One of these, narrow and long, was the site of the Dnieper-Donets basin, in which the saline sequence of the Pripet depression and the Dnieper-Donets syneclise accumulated in Late Famennian time.† The second arm lay in the region of the Moscow syneclise and was very distinctive. It was a huge gulf-like body, very broad, grading imperceptibly into the open sea, which was to the east. Morphologically it resembled the Bay of Biscay, or that part of the Arctic Ocean called the Laptevykh Sea. The broad front along which the Moscow syneclise graded eastward into the open sea will therefore be called a marginal zone of the sea cut into the continent, and not a gulf.

Sedimentation in this zone was distinctive. Around the margin of this embayment continental red beds accumulated. They now form a wide belt on the northwestern border, where they lie next to the Baltic shield; and they form a very narrow belt, finally wedging out, along the northeastern slope of the Voronezh massif. This difference is undoubtedly due to the difference in steepness of the slopes on these salients of the Precambrian basement. The

* The map of Martynova shows a series of small island uplifts that could not be plotted on the scale of our map.

† Frasnian saline deposits were also deposited here, but they formed in independent formations and are not considered in this discussion.

continental zone is bordered by a belt of clastic-carbonate deposits, chiefly marine, also broad on the northwest margin of the syneclise and narrow along the southwest. Dolomites are widely developed among the carbonates in the belt, especially on the northwest, and thin beds of gypsum appear. The central part of the Moscow syneclise is a zone of dolomite, interbedded with very extensive but generally thin beds of gypsum and anhydrite (up to only a few meters thick). In the Moscow, Tula, and Uglegorsk regions, however, the thickness of gypsum may be as much as 20 m, and the total number of gypsum beds is greater. The total area of this dolomite-anhydrite field in the syneclise is 700,000 km². On the southeast the field is bordered by a comparatively narrow zone of dolomite interbedded with limestone. The number of dolomite beds steadily decreases toward the east and the succession finally becomes chiefly limestone, with only local and subordinate patches of metasomatic dolomite. Limestone occurs on the eastern half of the platform in a broad north-south belt from the shores of the Barents Sea to the northern Caspian and extending to the southeast into Central Asia, merging with a belt of limestones of Central Kazakhstan.

We are thus confronted for the first time with the paradoxical example of a basin that is essentially a marginal part of an open sea, broad and with no apparent obstacles to connection with the water of its central part, giving rise to saline deposits usually regarded as typical of deposits accumulating in basins isolated from the sea. Is our interpretation of the basin correct? And if so, by what mechanism were the saline accumulations deposited? In order to answer these questions, it is necessary to examine closely the structure of the Dankov-Lebedyan beds in the Moscow syneclise, following the work of V. G. Makhlaev.

2. Lithologic Types and Structure of the Dankov-Lebedyan Deposits in the Moscow Syneclise

The lithologic and facies types in the Dankov-Lebedyan deposits are extremely variable (Table 35). They include sandstone and pebble conglomerate, mudstone, calcareous sandstone, many types of limestone (oolitic, organic, pelitomorphic, stromatoporoidal, algal, and others), limestone breccia and conglomerate, dolomite and dolomitized limestone with structures similar to those of the limestone, gypsum-dolomite rock, gypsum, and anhydrite rocks. In facies these correspond in part to segments of the sea with normally saline water, in part to those with high salinity, and in part to those with very high salinity. Makhlaev has recognized the first two facies types in sediments of the near-shore zone to a depth of 10–15 m, of the shallow-water zone 10–60 m deep, and of the moderately deep zone more than 60 m deep.

In composition the section of Dankov-Lebedyan beds may be clearly divided into two similarly constituted cycles. The first embraces the Lebedyan, Mtsensk, and Kiselevo-Nikol'skoe beds. At the beginning of Lebedyan time red sand-clay inshore deposits, in part, and continental (?) beds, almost completely unfossiliferous, accumulated in a zone along the western margin of the basin (Fig. 169). On the northwest this zone is broad; on the southwest, at the Voronezh massif, it is narrow. To the east of the sand-clay zone a belt of clastic-carbonate deposits formed. The accumulation varied sharply from place to place. On the northwest and west only unfossiliferous dolomitic rocks or dolomites with rare fish remains are found in combination with the sand-clay beds. Individual horizons have nodules and very thin beds of

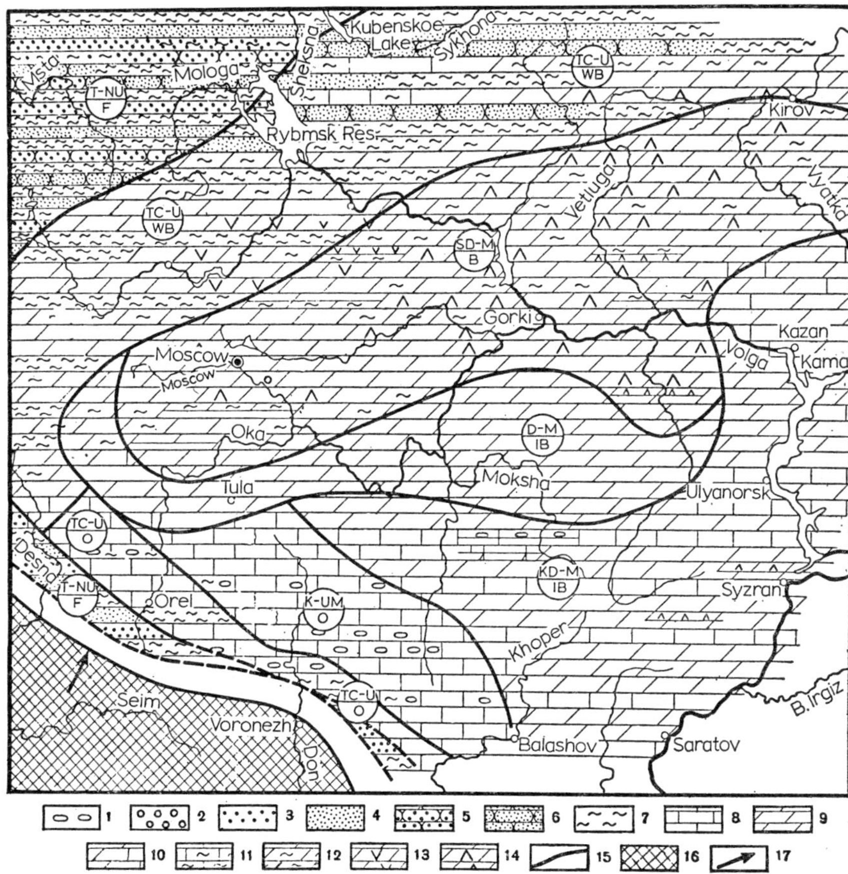

FIG. 169. Facies at the beginning of Lebedyan time (from V. G. Makhlaev, modified), 1. Carbonate pebbles and cobbles; 2. carbonate sand; 3. quartz sand, 4. silt; 5. sandstone. 6. siltstone; 7. clay; 8. limestone; 9. dolomite; 10. limestone-dolomite; 11. calcareous marl and argillaceous limestone; 12. dolomitic marl and argillaceous dolomite; 13. gypsum-dolomite rock; 14. anhydrite-dolomite rock; 15. boundary between facies zones; 16. zone of denudation; 17. direction of drainage. *Fractions in circles indicate: (above the line, facies first)*: T. clastic; C. carbonate; K. calcareous; D. dolomite; TC. clastic-carbonate; KD. limestone-dolomite; SD. sulfate-dolomite; (*and second, after the dash, the depth zone of the basin*): N. near-shore; U. shallow water; M. moderately deep water; (*and below the line, salinity of water in the basin*): F. freshened; O. marine; WB. somewhat briny; IB. moderately briny; B. very salty or briny; L. lagoonal (freshened or somewhat briny).

gypsum. These features indicate an appreciable increase in salinity of the water. On the southwest, near the Voronezh massif, gypsum disappears from this zone, dolomite gives way to limestone with a rather varied fauna, including not only euryhaline pelecypods and gastropods but also stenohaline brachiopods and crinoids. These features attest to normal salinity of the basin.

The greater part of the Lebedyan basin is characterized by carbonate sedimentation. In the basin of the River Don, i.e. near the Voronezh massif, a broad belt extends from northwest to southeast in which various limestones occur (oolitic, organic, clastic, biomorphic, and others), relatively rich in fauna, including stenohaline groups (individual corals for one). This part of the sea was clearly characterized by salinity closest to normal sea water. In the Volga and trans-Volga regions rather thick beds and macrolenses of sedimentational-diagenetic dolomites are frequently found with the limestones, but the fauna is still somewhat varied. To the north both limestone zones give way to a narrow and irregularly shaped belt of primary sedimentational dolomites that are barren or contain only meager remains of euryhaline forms. The central part of the formation, generally occupying the axial part of the Moscow syneclise, consists of a sulfate-dolomite series made up of dense barren microgranular and fine-granular dolomite interbedded with beds of gypsum. This part of the basin clearly contained water of maximal salinity.

This sedimentational picture applies only to the beginning of the Lebedyan epoch. In late Lebedyan time clays became relatively more abundant along the western border of the basin and accumulated closer to the center of the basin than in early Lebedyan time. At the same time the zone of episodic sulfate accumulation in the central part of the syneclise broadened, extending far to the east. The zone of dolomite accumulation enlarged correspondingly. The zone of pure limestone accumulation, by contrast, became markedly restricted and the fauna inhabiting these zones was markedly impoverished. Of special importance is the fact that almost all stenohaline forms disappeared. *All these features agree in indicating some reduction in size of the Lebedyan basin and also a notable increase in salinity. This increase in salinity was apparently due to more difficult communication between the water of the basin and the normally saline sea in the trans-Volga region.* The inflow of sea water into the Lebedyan saline basin occurred along the northeastern edge of the Voronezh massif, through a zone that had become even more narrow.

Regression, characteristic of the second half of Lebedyan time, soon gave way to transgression. *Mtsensk time was characterized by a new expansion of the marginal zone of the sea and by simultaneous decline in salinity* (Fig. 170). Proof of this transgression is found in the shift of the belt of sand-clay and clastic-carbonate deposits toward the west and in the general expansion of the area of pure carbonate accumulation. Proof of declining salinity is seen in the almost complete disappearance of gypsum in the central part of the basin. In Mtsensk time only normal sedimentational dolomite accumulated in this zone. The fauna of Mtsensk time characteristically became abundant and varied. Many stenohaline groups were represented.

TABLE 35

Facies Types of Deposits in the Dankov-Lebedyan Basin (from V. G. Makhlaev, with some modifications)

2 A

Zone	Region of normal-marine and slightly freshened water (limestone sediments)	Region of high salinity (dolomitic sediments)	Region of greatest salinity (sulfate sediments)
Near-shore to depths 10–15 m	1. Limestone breccias, coarse turbizone breccia. Limestone conglomerates. 2. Calcareous sandstones, oolitic and organic shell limestones (pelecypod, brachiopod and ostracod, serpuline, crinoidal), and also detrital limestones containing coarsely fragmental calcareous material or associated with limestone conglomerates. *Stromatoporoidal limestones.* 3. Quartz sandstones and pebble conglomerates, locally calcareous, locally non-calcareous. Kaolinitic clays in sandy rocks. Gradual transitions occur between groups (2) and (3) through pelecypod shell limestone with abundant quartz and grains.	1. Dolomitic conglomerates, breccias, coarse turbizone breccias, and dolomitic sandstones associated with dolomitic conglomerates and breccias. 2. Dolomitic shell limestones (pelecypod, serpuline, ostracod, and others); detrital and oolitic dolomites, either with coarsely fragmental dolomitic material or associated with coarsely fragmental dolomitic rocks. *Stromatoporoidal dolomites.* 3. Quartz sandstones and sandy pebble conglomerates associated with dolomites.	No sediments
Shallow water with depths from 10–15 to 50–60 m.	1. Fine-clastic turbizone breccia. 2. Calcareous sandstone and siltstone, microgranular limestone (detrital and chemical) without coarse calcareous material and not associated with coarse clastic limestone. 3. Quartz sandstone and siltstone, carbonate-free mudstones or calcareous mudstone with or without sand admixtures.	1. Dolomitic, fine-clastic turbizone breccia. 2. Dolomitic sandstone and siltstone, detrital oolitic and granular, crystalline dolomite with no coarse-clastic dolomitic material and not associated with coarse-clastic dolomites. Banded dolomites with quartz siltstone interbeds. 3. Quartz sandstones, siltstones, and mudstones associated with dolomites.	1. Gypseous-dolomitic and gypseous sediments. 2. Quartz silty and clayey rocks associated with gypsum.
Moderately deep water with depths greater than 50–60 m.	1. Fine-grained limestone (Elets type) with small quantities of brachiopod shells and crinoid stalks or without fossils. *No algal remains.*	1. Thick persistent members of massive dolomite without clastic quartz or fragmental dolomitic material (the dolomitic correlatives of Elets limestone). Banded dolomite without quartzose silty material.	Dense, massive, pure gypsum and anhydrite. Bedded sulfate rocks.

During Kiselevo-Nikol'skoe time the paleogeography and facies conditions changed once again. The area of the sea was reduced, causing a migration of the belt of clastic-carbonate deposits toward the east, to a zone near the center of the Moscow syneclise. The clastic material became coarser, sandy rocks appearing along with clays. Over a broad area in the center of the syneclise sulfates again accumulated, and in considerable quantities. The belt of calcareous sediments along the northeastern border of the Voronezh massif is very thin. Organisms were sharply restricted, represented only by well-defined

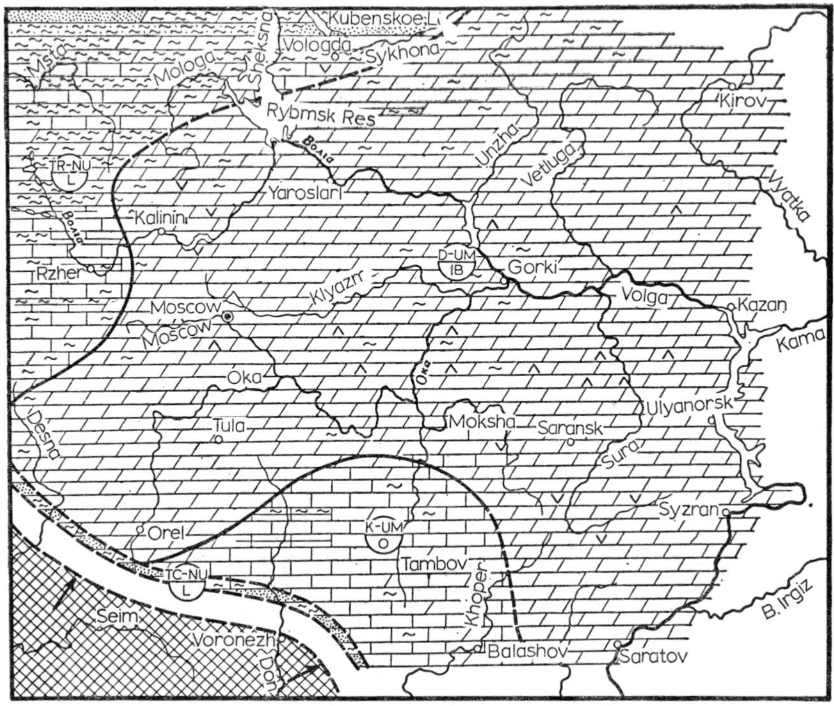

FIG. 170. Facies of Mtsensk time (from V. G. Makhlaev). Symbols are the same as for Fig. 169, with this addition: *R.* marl.

euryhaline groups—ostracods, pelecypods, and worms—but of these groups only six species are known, belonging to six genera, though individuals were abundant. There can be no doubt that the basin became much saltier with regression.

The second cycle in the history of the Dankov-Lebedyan basin developed in Turgenevo time. The general dimensions of the basin and its facies zones very closely resemble that of Kiselevo-Nikol'skoe time (Fig. 171). The same marked shift of the marginal clastic deposits to the east, the same marked development of sulfate-dolomitic rocks, the same suppression of calcareous sediments occur. The fauna, however, became richer, with 15 genera and 16 species. Nautiloids and foraminifers appeared, indicating a lower salinity in the gulf, at least in the marginal part near the Voronezh massif. A new feature in the lithofacies map of Turgenevo time is the isolated patch of gypsum somewhat west (and southwest) of Samarskaya Luka.

The succeeding Kudeyarovka epoch was characterized, like the Mtsensk, by expansion and marine freshening. Transgression is indicated by a sharp displacement of the clastic-carbonate zone to the west and by dominance of clay in the marginal zones of the basin. The freshening of the water is demonstrated by the disappearance of sulfate deposits from all parts of the syneclise,

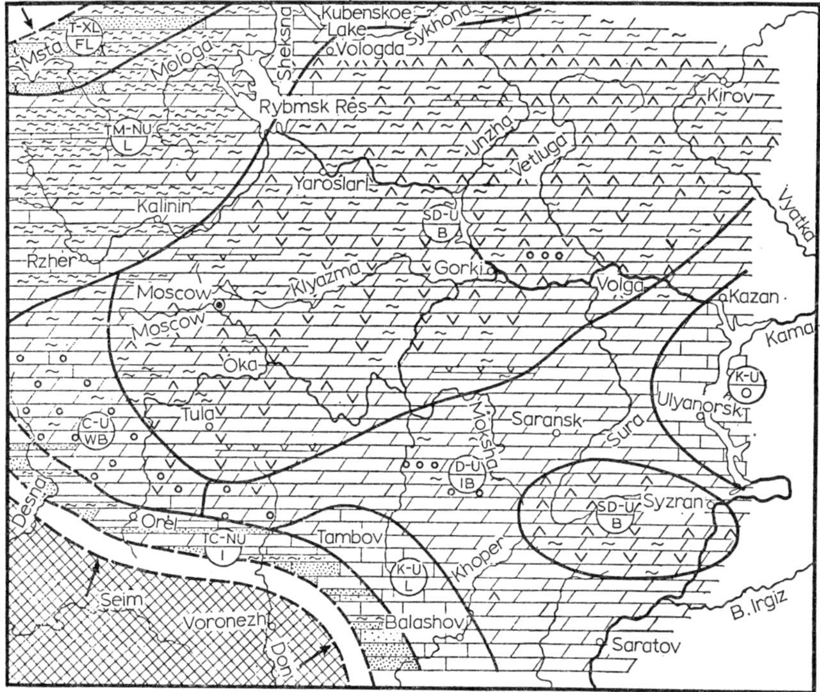

FIG. 171. Facies of Turgenevo time (from V. G. Makhlaev). Symbols as in Fig. 169, with additions: *R* stands for marl; and *X* continental.

particularly from the central part. Only dolomites are found here, with fossils, although they are scarce and stunted. In the marginal part of the basin next to the Voronezh massif, and also in the trans-Volga region, where limestone was deposited, organic forms were varied: 27 species belonging to 15 genera. These include several stenohaline groups: five species from four genera of brachiopods and one species of echinoderm. In hydrochemical conditions and in paleogeography the basin in Kudeyarovka time was very similar to the basin in Mtsensk time.

During the final moments of the second cycle, in Ozerki time (Fig. 172), the marginal zone of the sea was again reduced in size, and the belt of clastic deposits shifted inward into the syneclise. Sulfates again accumulated in the center of the basin, very abundantly this time, reaching a maximum for the entire Dankov-Lebedyan interval. The thickness of the gypsum layers is greatest in the southwestern part of the basin, where individual layers are as much as 20 m thick. The fauna over most of the area of the syneclise died out entirely; even in the least salty belt along the Voronezh massif, only 5 species of 5 genera survived, one each of foraminifers, worms, ostracods, brachiopods, and pelecypods.

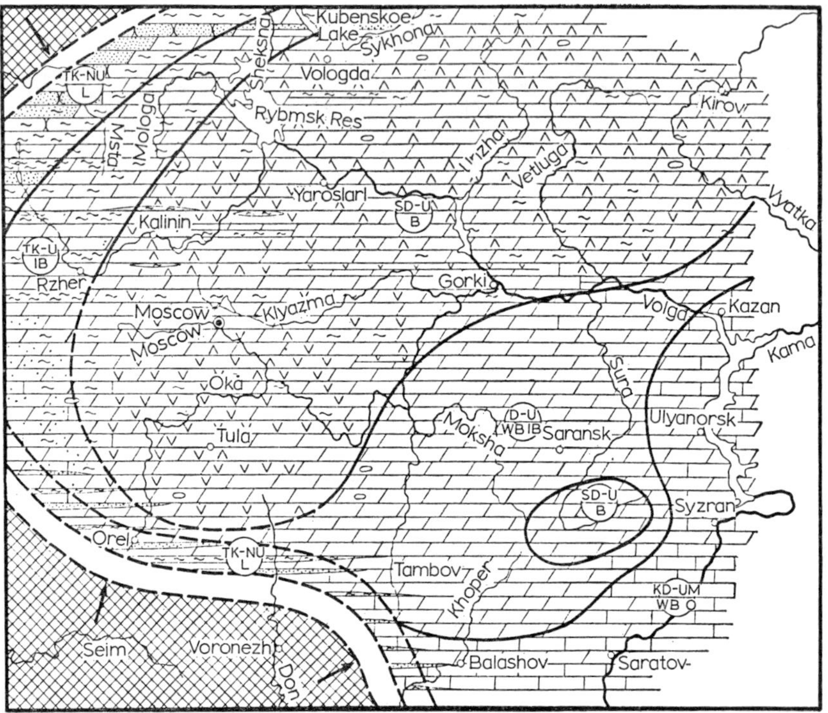

FIG. 172. Facies of Ozerki time (from V. G. Makhlaev). Symbols are the same as for Fig. 169.

In Khovanshchina time the basin, preserving its earlier size, became strongly freshened, to the extent that gypsum deposition entirely ceased and even dolomites were replaced by limestones. The fauna expanded noticeably: 19 species belonging to 19 genera, including four species of brachiopods and one of coral. This time the freshening of the basin was not associated with transgression but resulted from climatic change. Arid climate gave way to humid. A new plan of climatic zoning began to appear, the Hercynian, with broad increase in humidity of the climate at the beginning of this time interval (Vol. 1, Chap. 5).

Setting aside Khovanshchina time and its specific features, we see a strict pattern in the course of events in the two preceding cycles of the Dankov-Lebedyan epoch. At the beginning of each cycle, corresponding to the smallest size of the sea, extensive development of clastic deposits characterized the western margin and the water became markedly salty in the central part, where not only unfossiliferous dolomite accumulated but also beds of gypsum, some of them notably thick. The basin then expanded, shifting the zone of clastic sedimentation to the west, in part beyond the limits of the map, and causing freshening of the water in the central part of the basin, where gypsum ceased accumulating and only dolomite was precipitated, some of it barren but some containing fauna. In the parts of the basin with least salinity, along the northeastern border of the Voronezh massif, fauna abounded at this time, including stenohaline groups: brachiopods, echinoderms, and corals. The last stage of each cycle was characterized by a new reduction in size of the sea, in the appearance of sand-clay clastic sediments along the western near-shore zone and of gypsum in the central part. The fauna died out entirely in the center and only the near shore segments with lowest salinity preserved a meager and stunted fauna of stenohaline groups.

"Comparison of sequences from the lower and upper sedimentary cycles—the Lebedyan and Turgenevo, the Mtsensk and Kudeyarovka, and so forth—reveals a distinct evolutionary trend in the Dankov Lebedyan basin. The basin was progressively reduced in size, although with interruptions, became increasingly shallow and salty, and its fauna disappeared. For example, deposits of the Lebedyan sequence are found in the central field farther to the southwest than their correlatives of the Turgenevo sequence that formed at comparable distances off shore. The first include comparatively deep water members, chiefly of limestone in the central field, with a varied fauna. The second is entirely shallow-water, chiefly of dolomite, in the same region, with a much poorer fauna. The water in the central field of the Kudeyarovka basin was saltier than the Mtsensk, and the first basin lay within the boundaries of the second (though the variety of organisms found in the two sequences indicate the two were more or less identical). . . . The lower sedimentary cycle contains almost twice as many species, and many more genera, of organisms as are found in the upper cycle. This fact convincingly demonstrates the worsening conditions for life in the Dankov-Lebedyan basin during the second half of its existence. The scarcity of fauna appeared as early as Elets time and continued with some interruptions to the end of the Devonian period, since the fauna of the Zadonsk beds are richer in fauna than the Elets beds, and these in turn are richer than the lower sedimentary cycle of the Dankov-Lebedyan beds. The decrease in number of species in the upper cycle is apparent in almost all groups of fauna, except the nautiloids and gastropods, which were more numerous in this part of the upper cycle than in any sequence of the lower cycle" (Makhlaev 1961).

Thus, the cyclic development of the Dankov-Lebedyan basin of the Moscow syneclise occurred as *an irreversible evolutionary trend toward reduction in size and toward increase in salinity of the basin. In this process the accumulation*

*of gypsum coincided with regression, phases of reduced access of the water to
the open sea, which then occupied the eastern part of the Russian platform.*

The gypsum sediments of this distinctive basin and the facies conditions of
their accumulation merit further attention.

3. LITHOLOGIC TYPES OF GYPSUM DEPOSITS AND THEIR FORMATION

Deposits of sulfates are widespread in the Dankov-Lebedyan beds, but the
largest masses are found in the upper part of the Lebedyan sequence, in the
Kiselevo-Nikol'skoe sequence, in the upper part of the Turgenevo sequence
and in the middle part of the Ozerki sequence. Makhalaev (1961) calculated
the relative amounts of sulfate rocks, in percentage of total thickness, for the
respective beds of the Tula-Serpukhov succession and the Povarovo borehole
as follows:

	Tula-Serpukhov section	Povarovo section
Khovanshchina sequence . .	none	none
Ozerki sequence	60·0	30·0
Kudeyarovka sequence . . .	?	5·0
Turgenevo sequence . . .	10·4	9·2
Orel-Saburovo sequence . .	1·7	0·0
Kiselevo-Nikol'skoe sequence .	14·0	3·2
Mtsensk sequence . . .	0·0	0·0
Lebedyan sequence . . .	0·4	0·7
Dankov-Lebedyan beds as a whole .	23·4	10·5

The sulfate minerals are gypsum and anhydrite. There are reliable indications that the gypsum is always derived from the anhydrite by secondary
hydration in the uppermost horizons of the succession. The previous mode of
occurrence of the beds is preserved, however, and we shall therefore consider
the sulfate forms identical for further discussion.

The sulfate deposits occur as beds, lenses, concretionary masses and impregnations in dolomitic rock.

(1) *Beds and lenses* range in thickness from a few centimeters to several
meters of either massive or thin-bedded gypsum and anhydrite.

The stratified gypsum and anhydrite consists of thin beds of sulfate interbedded with clayey marly dolomite. The thickness of the interbedded laminae
ranges from parts of a millimeter to 2 cm. The thickness of the thin-bedded
gypsum members may reach several meters.

The massive gypsum is made up of aggregates of grains, flakes, and plates,
varying in dimension. "Scales, plates, and needles are normally combined in
sheaves of differing orientation. In places they form spherulites, semispheru-
lites, or rosettes. Some beds and sequences of gypsum have zones that are
almost free of inclusions exhibiting macroscopic differences. But mostly they
contain beds, lenses, and aggregates of clayey, microgranular dolomitic, or
mixed composition, and films of organic material. Microscopic, occasional
individual crystals of barite, celestite, fluorite, quartz, chalcedony, and scattered organic material in the gypsum also occur. Relict granules of anhydrite
are frequent, and individual granules of dolomite occasionally occur"

Makhlaev 1961). The inclusions give rise to variations in structure in the massive gypsum, resulting in massive, lenticular, concretionary, irregularly wavy, and pseudobreccia forms. The last four are characterized by the presence of clay or dolomite veinlets in the gypsum. When dolomitic inclusions are abundant, the structure is patchy, and the rock becomes transitional in composition between gypsum and dolomite.

"Texturally, the anhydrite is made up of prisms ranging in length from a few hundred millimeters to 0·5 mm, arranged in mutually parallel and horizontal positions that may form streamers of variable direction or may show no alignment. Idiomorphic prismatic crystals of anhydrite are occasionally seen in finer-grained masses of the same material or in microgranular dolomite" (Makhlaev 1961). The anhydrite contains the same macroscopic and microscopic inclusions as the gypsum.

(2) *Sulfate concretions* in the massive dolomites are a few millimeters to 10 centimeters and more across. "They are circular, elongated in a horizontal, vertical, or other direction, or they are irregular. Their outline is smooth or embayed and normally very clear. The concentration of concretions may be highly variable; in some places they are rare, in others they may almost entirely displace the host rock, restricting it to narrow zones between closely crowded concretions. The reverse spatial relations are also observed between gypsum and dolomite, in which the latter merely forms residual aggregates in the former. Where the concretions are more abundant, the gypsum forms only a network in a mass of dolomite" (Makhlaev 1961). Genetically the concretionary occurrence of gypsum probably developed from redistribution of sulfates in mixed sulfate-dolomite sediment that formed the transitional zone in the basin between the central sulfate zone proper and the purely dolomitic marginal zone. The different proportions of sulfate and dolomite in the original sediment predetermined the petrographic features of the final rock.

(3) *Sulfate impregnation of dolomite* appear macroscopically to be completely homogeneous rocks, but they represent fine mixtures of dolomite and gypsum. "The gypsum forms numerous small aggregates in the dolomite, irregular in form and differently oriented optically. Texturally the rock is distinctive, but it is related through transitions to dolomite with intergrown and idiomorphic crystals of gypsum. It forms independent beds and thin units interbedded with dolomite or marl that is free of gypsum. The dolomite enclosing concretions of gypsum is also impregnated with gypsum in greater or lesser degree" (Makhlaev).

The horizontal distribution of the sulfate rocks is characterized by restriction to the central part of the Moscow syneclise. Toward the shores and also at the eastern and southeastern edges of the basin the amount of gypsum declines. It may be noted that *the dolomitic rocks of the central part of the syneclise, being interbedded with sulfate rocks, are characterized by aspects of relatively deep-water precipitation—they contain no algae or fragmental dolomitic material* (Table 35). All this proves that the accumulation of gypsum took place in relatively deep water but in a basin of the syneclise with a relatively flat bottom. Toward the margin, in the near-shore zone, gypsum was not deposited.

The form of the gypsum deposits changes in different parts of the syneclise in accordance with quantitative variations. In the central parts of the syneclise the gypsum is found chiefly in beds and lenses; concretions and impregnations are markedly subordinate. Along the margins beds of gypsum disappear, and only lenses, concretions, and impregnations remain. At the very edge even the lenses disappear. "In the central Devonian field cavities formed from solution of gypsum crystals occur, chiefly in the Kiselevo-Nikol'skoe and the Ozerki sequences, more rarely in the Turgenevo, and still more rarely in the Kudeyarovka. These cavities are tabular, idiomorphic, and well defined. They range up to several centimeters in major dimension and are confined to primary sedimentary dolomite, either bedded or massive, lacking inclusions or having cavities from the solution of rock salt crystals. Fragmental material is locally present. Mostly the tabular cavities are scattered through the beds, commonly forming lenticular and concretionary masses that may be traced for appreciable distances in some part of the section. This distribution is observed in the upper part of the Kiselevo-Nikol'skoe sequence at Niznee Shchekotikhino near Orel, in the middle of the Turgenevo sequence at Korsakovo on the Zusha River, and in the upper part of the Kudeyarovka sequence at Yur'evo on the Plava River. They are undoubtedly diagenetic" (Makhlaev).

We should emphasize that all modes of occurrence of primary gypsum deposits in the Dankov-Lebedyan basin are associated only with dolomite and not with limestone.

The composition of the upper Upper Devonian saline formation again leads to the question: how can gypsum accumulate in the marginal part of an open sea in which there exists a broad connection with the central part of the basin? This question will be examined with reference to the connection between the basin of the Moscow syneclise and the open sea.

4. The Nature of the Connection between the Basin of the Moscow Syneclise and the open Parts of the upper Famennian Sea

The essence of the problem is as follows: at the time of gypsum accumulation, was the marginal zone of the sea in any measure cut off from the open sea by shoals or exposed islands? Or was there no isolation, and did the water of the gulf communicate freely over a broad front with the open basin on the eastern half of the Russian platform? Both opinions have their adherents. D. V. Nalivkin and M. S. Shvetsov maintain that islands in the Dankov-Lebedyan basin of the syneclise produced some isolation. L. M. Birina and M. S. Petrov favor the view that no islands restricted movement, and that the marginal zone of the sea communicated freely with the normal marine open sea to the east. It appears to me that the first view is essentially correct, because the restriction of water circulation by islands and shoals is supported by a number of factors.

The thickness distribution of the Dankov-Lebedyan beds, from a few tens of meters to 175-200 m, supports this view (Fig. 173). The maximum thickness is found in the axial part of the Moscow syneclise, the minimum on the southwestern and northwestern margin. A characteristic and very distinct minimum (100 to 25 m and less) is found on the Tokmovo arch, separated from the Voronezh massif by a belt of greater thickness. In the northern part of the syneclise, toward the Tokmovo arch, an area of reduced thickness is found at Kotel'nich. It is no less noteworthy that east of the line through Borisoglebsk,

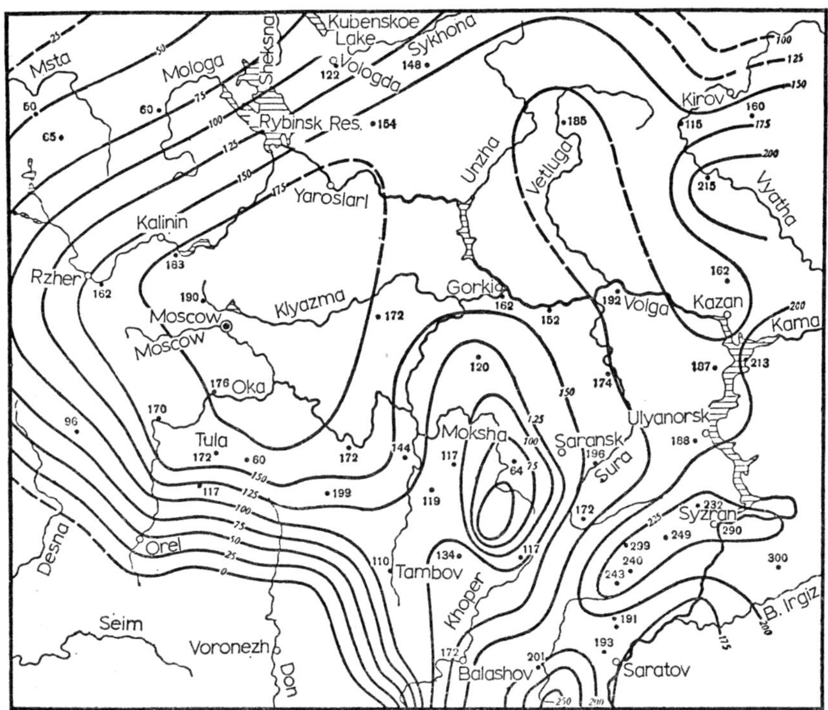

FIG. 173. Isopach map of the Dankov-Lebedyan deposits (from V. G. Makhlaev).

Serdobsk, Ulyanovsk, Kazan, and Kotel'nich the thickness of the Dankov-Lebedyan deposits increases rapidly, exceeding 200–225 m, whereas within the eastern margin of the syneclise it is commonly less than 150 m. Two fundamental conclusions may be drawn from these relationships. *In Dankov-Lebedyan time the area of the Moscow syneclise, i.e. the marginal zone of the sea as a whole, subsided more slowly than the area of the open sea on the eastern part of the Russian platform. Furthermore, there were zones along the eastern margin of the syneclise—the Tokmovo arch and the district about Kotel'nich— where subsidence and sedimentation were especially slow. These were undoubtedly areas of shallow water, converted during uplift to shoals or even to exposed island zones of varying size, hindering communication between the basin of the syneclise and the open sea.* Such are "the islands on the southern part of the Tokmovo arch at the beginning and the end of Dankov-Lebedyan time and the islands (or peninsulas) in the Lebedyan region at the beginning of Lebedyan time. Other zones, not necessarily appearing as islands, may well have existed during this time as also during Orel-Saburovo and Khovanshchina times. The grounds for this statement are the shallow-water character of the basin during these time intervals and the presence of local clay members in the central part of the basin" (Makhlaev 1959a). To this may be added that in the vicinity of Timan at the extreme northeastern part of the syneclise there were invariably large islands during all of Dankov-Lebedyan time, separating this part of the Moscow syneclise from the open sea.

Thus, there undoubtedly existed a number of underwater shoals and exposed islands along the eastern and northeastern margin of the sea, restricting communication and creating conditions for prolonged preservation of highly saline water in the central part of the syneclise. This was especially true during periods of uplift, regression and increasing salinity, with the accompanying precipitation of gypsum. During transgression, the general subsidence led to a considerable freshening of the gulf and to the cessation of calcium-sulfate precipitation. Only dolomite was deposited, but apparently at the lower limit of salinity necessary for this process, because the fauna in the gulf grew rich and spread over almost the entire floor.

In the Dankov-Lebedyan deposits, the belt along the northeastern border of the Voronezh massif differs sharply from the remaining part of the basin by clear indications of normal marine salinity. In this zone limestone or, at times, dolomitized limestone was deposited throughout the epoch. The richest and most varied fauna lived here, corresponding most nearly to normal marine conditions. At times the fauna was sufficiently abundant to give rise to shelly limestone. Along the margin, this limestone belt grades into dolomitized limestone, then to dolomite. The fauna becomes impoverished, loses its marine aspect, and finally disappears.

The zone between the Voronezh massif and the Tokmovo arch was, throughout Dankov-Lebedyan time, a belt along which a strong current of normal marine water flowed northwestward from the sea, along the edge of the massif. With the change in salinity of the basin water, pure limestone precipitation gave way to dolomitic limestone, and the fauna died out progressively.

In considering these processes it may be assumed that by the time the water reached the zone of the modern upper Dnieper its original constitution had been lost, the salinity corresponding there to that of the water along the western margin of the Moscow basin. This does not mean, however, that the water ceased moving northwesterly along the coast. The current continued, then turned to the northeast and moved along this shore to the region of the present southern Timan, where it entered the open sea. What the situation was along the eastern margin of the Moscow basin, along the shoals and islands, may be only surmised. It appears that along the outer side of this shallow-water zone water must have moved out of the basin at the southern end of Timan. In the process it must have become continuously less saline. If this hypothesis is true it would appear that the saltiest central zone of the Moscow basin represented a huge halystasis. During halystasis, streams of water poured into the basin, as surface waters, through the broad straits between islands and underwater shoals, and heavy salt water flowed out of the syneclise along the floor. As elsewhere, the intensity of the water exchange between the marginal zone of the sea and the remaining open part was controlled by the tectonic activity in the zone of underwater shoals and islands. Subsidence of this shallow-water zone increased the outflowing current of salt water from the syneclise to the open sea, whereupon the Moscow basin became less salty. Uplift of the shallow-water zone retarded the elimination of salts, and the basin became saltier. These fluctuating movements of the crust in the shallow-water zone took place synchronously with subsidence and uplift of the Moscow syneclise as a whole. Regression of the sea from the basin was thus always accompanied by increasing salinity of the sea and transgression by decreasing salinity, with the formation of the characteristic macrocyclic structure of the Dankov-Lebedyan formation.

C. THE LOWER CAMBRIAN SALINE FORMATION OF THE SIBERIAN PLATFORM AND ITS DEPOSITION

A still larger example of saline formations that have formed in the marginal zone of a platform basin is found in the Lower Cambrian sequence along the southern rim of the Siberian platform. It has become known only recently through deep drilling in connection with oil exploration. The paleogeography of Early Cambrian time on the Siberian platform will first be considered.

1. The Location and Setting of the Saline Formation in the Lower Cambrian Deposits

Saline deposits are found here in the Irkutsk amphitheater in a broad belt that stretches from the lower reaches of the Angara River to the upper course of the Lena. On the northeast it reaches the Kempendyai and Suntar Rivers, and on the east the lower reaches of the Olekma and Amga Rivers. This belt of saline deposits is about 3000 km in length; the width locally reaches 600–700 km, and the total area is about 2,000,000 km².

The mountain structures of the Sayan and Transbaikal now lie south of the southern edge of the formation. The saline formations were not deposited in this area, to judge from the composition of the adjacent parts of the platform rocks. In Early Cambrian time the Sayan-Transbaikal region was terrestrial and bounded the area of evaporate deposition. In the central and northern parts of the platform the saline formation gives way to a sequence of calcareous rocks, either pure or somewhat dolomitized, representing deposition in a sea of normal salinity. Gradual transitions occur between these normal marine sediments and those of highly saline water (Arkhangel'skaya, Grigor'ev, and Zelenov 1960). Thus, *the Lower Cambrian saline formation was confined to a broad marginal zone of a sea that made a broad embayment into the continent open to the main area of the sea.* The essential similarity to the disposition of the Dankov-Lebedyan basin of the Moscow syncline relative to the broad expanse of the Late Famennian sea is clear. The tectonic framework of both formations is also similar. On the Russian platform the Dankov-Lebedyan sulfate formation covered the floor of the Moscow syncline. On the Siberian platform the Lower Cambrian saline formation filled the deep Sayan-Lena marginal depression, being confined on the north by a submerged uplift along the middle course of the Stony Tunguska.

There are substantial differences, however. The Lower Cambrian saline formations of the Siberian platform differ from the Dankov-Lebedyan formation by much greater thickness, by complete development of the saline process, including the deposition of halite, and by very complex structure.

2. The Initial Stages of Early Cambrian History on the Siberian Platform and the Development of Conditions for Halogenesis

At the very beginning of the Cambrian purely clastic deposits accumulated in the region of the Irkutsk amphitheater: sandstone, siltstone, mudstone of the Ushakovka series and lower part of the Moty series, the two series interdigitating in part (Fig. 174). The fragmental rocks are finer grained in the

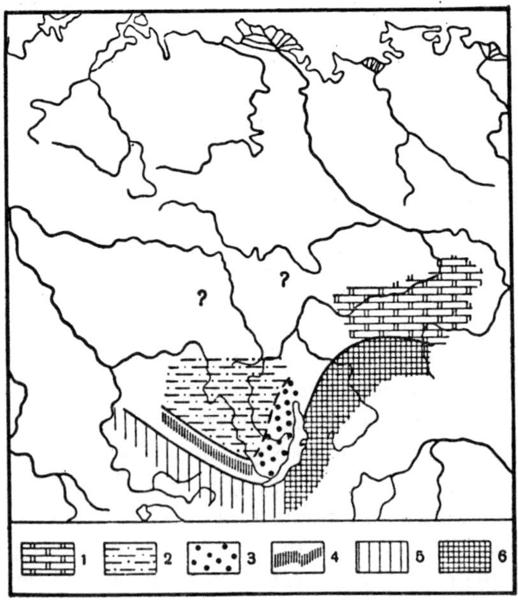

Fig. 174. Formational environment for sediments of Ushakovka and early Moty time (from Ya. K. Pisarchik and K. K. Zelenov). 1. Limestone-dolomitic deposits; 2. sand-clay marine deposits; 3. thick land-conglomerate marine deposits; 4. continental red beds; 5. land with relatively minor relief; 6. land with mountainous topography.

inner parts of the amphitheater and coarser toward the margins. This is especially pronounced in the southeastern Baikal zone, where the Ushakovka series is characterized by coarse graywackes. Over most of the area the Ushakovka series is characterized by regular bedding and is undoubtedly a marine basin sediment. In the narrow zone near the Sayan, however, cross-bedding appears in the sandstone and the color becomes red. This was a zone of continental, channel, and deltaic deposition (Ya. K. Pisarchik 1958b). Stratigraphically these beds correspond to the upper part of the Ushakovka and lower part of the Moty series. The clastic complex shows a characteristic thickness distribution. In the central part of the amphitheater the Ushakovka series is about 120 m thick, but in the narrow Sayan belt it reaches 400 m, and in the Baikal region 900–1000 m. It is clear that both these latter zones were marginal depressions at the time the clastic sediments were deposited. They rimmed the central inner part of the Irkutsk amphitheater and, on the outer side, lay next to the uplifted continental zones of the Eastern Sayan and Transbaikal.

To the northwest of the Irkutsk amphitheater, in the lower reaches of the Stony Tunguska and Lower Tunguska, grey and red clastic deposits similar to the Ushakovka series also accumulated during this time. In the Berezovskii downwarp, along the Lena, Amga, and Maya Rivers, the correlative of the Ushakovka series is the lower part of the Tolba series. This rests on deeply eroded Precambrian rocks and has a conglomeratic sandstone at its base. In the Berezovskii depression the thickness of this horizon increases to 50 m, and the conglomeratic rocks are coarser. On the platform the thickness of the horizon decreases to 10 m and less, and the rocks are finer-grained. The remaining overlying part of the Tolba series that corresponds to the Ushakovka series consists of dolomite with algal and oolitic horizons. The dolomite shows local signs of sulfatization, especially in the Berezovskii depression. In the central and northern parts of the Siberian platform, correlatives of this clastic-carbonate formation are not definitely known, but, as subsequently, carbonates, chiefly limestones, accumulated.

During the initial stage of Early Cambrian history on the Siberian platform as a whole, a broad marine transgression quickly flooded the northern and middle part of the platform, reaching the Irkutsk amphitheater, the Berezovskii downwarp, and the Aldan massif, stopping before the uplifted parts of the Eastern Sayan, Baikal region, and Transbaikal. The purely clastic sedimentation in the Irkutsk amphitheater indicates that the continental framework about the amphitheater was appreciably rugged and was being strongly eroded, especially the Baikal region. To the east, however, where the carbonate Tolba series was accumulating, the surrounding continental areas were flat. The areas flooded by the sea subsided at different rates and to different depths. Subsidence was particularly intense in the relatively narrow zones next to the Eastern Sayan and Baikal uplifts and in the Berezovskii downwarp. It was much less within the Irkutsk amphitheater and on the Aldan massif. *This combination of more rapid subsidence in the marginal depressions of the platform and slower subsidence in the inner parts is a characteristic feature of the tectonic regime during the initial stage of Early Cambrian sedimentation.* It is noteworthy that at the time the clastic deposits were accumulating in the Irkutsk amphitheater there was no evidence of increasing salinity of the water in the basin, but on the Aldan massif dolomite was widely deposited and, what is especially important, show some sporadic lenses of anhydrite and gypsum, as in the Berezovskii downwarp.

In middle Moty time the facies-genetic conditions changed substantially (Fig. 175). The marginal Sayan, Baikal, and Berezovskii depressions ceased

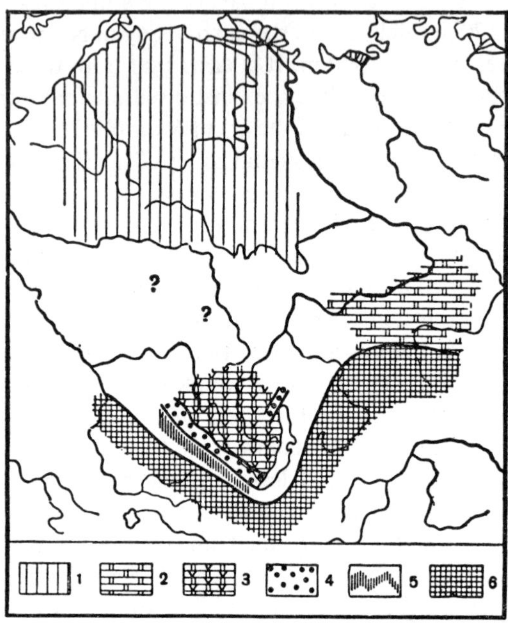

FIG. 175. Formational environment for sediments in middle Moty time (from Ya. K. Pisarchik, K. K. Zelenov, and others). 1. Calcareous deposits of a normally saline sea; 2. dolomite; 3. dolomite interbedded with anhydrite from hypersaline parts of the sea; 4. sand-clay marine deposits; 5. continental deposits; 6. land.

their sharp downward warping and subsidence became uniform. The transport of clastic material into the sedimentational zone in consequence declined sharply the decline becoming progressively more noticeable through middle Moty time. The succession consists of chemical precipitates alone, with only slight contamination of clastic material. These changes were due in part to the further expansion of the basin and in particular to reduction in relief of the drainage areas. A last specific feature of the middle Moty complex in the Irkutsk amphitheater is the widespread development of anhydrite among the chemical precipitates, either pure or contaminated with clay and carbonates, as in the Berezovskii depression. To the east, however, in the region of the

ldan massif, the appearance of sulfates was again sporadic and weak.
ormal microgranular, oolitic, and algal dolomites are dominant.

The composition of the middle Moty rocks indicates without doubt that
e salinity of the basin increased markedly and at times became high (above
5%), making it possible for calcium sulfate to be precipitated (primarily
gypsum). "Sedimentation took place under variable conditions, being
companied by frequent emergence of the sediments from the water and by
eir consequent reworking and redeposition. This is indicated by the nature
f the bedding in the rocks (thin, lenticular, undulatory in the marginal parts
f the basin, commonly cross-bedded in the sands and silts), by the presence
f numerous rather coarse platy fragments of shale, and also by rounded
agments of dolomite and dolomitic marls, crystal molds after rock salt,
pple marks, mud cracks, rain (?) prints, worm tracks, traces of local
osion, and so forth" (Ya. K. Pisarchik, 1958a). These structural features are
und most abundantly in zones near the Sayan and are less well developed in
e inner parts of the Irkutsk amphitheater. This results from the proximity
f the first zones to the shore line. Miniature slide phenomena are also
equent, indicating episodic tectonic activity.

Sedimentation took place continuously during a gradual decrease in the
ansport of clastic material from the land to the basin during the later stages
f middle Moty time. This may have been due to the increasing distances from
e zone of denudation or to topographic reduction through erosion. Both
ctors were apparently important. Shallow depth in the saline basin is in-
icated by algal remains in the dolomites. In places the algae represent the
rincipal rock-forming agent. Preparation for the accumulation of saline
eposits was thus completed, and salts then accumulated on a grand scale.

. GENERAL STRUCTURE OF THE SALINE FORMATION; THE USOL'E COMPLEX, ITS COMPOSITION, AND DEPOSITION

The Lower Cambrian saline formation of the Siberian platform is charac-
erized not only by its wide areal distribution but also by great thickness and
y complex structure (Fig. 176). It is made up of three macrocyclothems:
he *Usol'e*, embracing the upper horizons of the Moty series, all the Usol'e
eries, and the lower part of the *Belaya* series; the *Belaya*, including the middle
nd upper parts of the Belaya series and the lowermost part of the Angara
eries; and the *Angara*, embracing most of the Angara series. Each macro-
yclothem is characterized not only by a distinctive combination of rocks
ut also by areal distribution.

One characteristic feature of the Usol'e macrocyclothem, as pointed out
y Ya. K. Pisarchik, is the distinctive distribution of thicknesses, indicating
completely different plan of tectonic movement from that prevailing earlier
Fig. 177). In the Sayan zone the macrocyclothem is only 120–150 m thick,
ut in the inner part of the Irkutsk amphitheater it reaches 800–1350 m. In
he Berezovskii depression the thickness of the equivalent deposits reaches
700 m; on the Aldan salient it is 170–180 m. The distribution of rapidly and

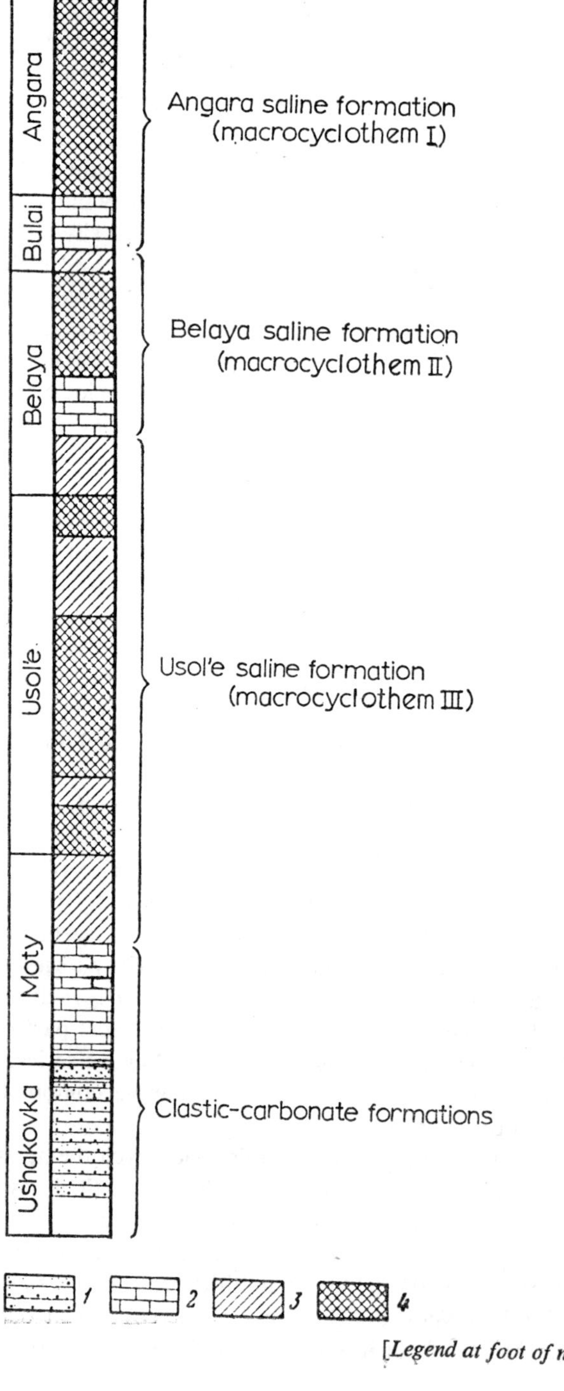

[*Legend at foot of next page*

slowly subsiding regions during deposition of the saline formation thus follows an entirely different plan, in places directly opposite to that at the beginning of clastic-carbonate accumulation: *the marginal parts of the platform underwent minimal subsidence, the central part of the amphitheater maximal.* The belt of maximal subsidence on the northwest extended to the site of the present Yenisei ridge, on the northeast to the Berezovskii downwarp and the Kempendyai salt domes. There are no reliable data for the region farther north, but indirect evidence from the modern structure of the southern end of the Tunguska basin (P. E. Offman 1959) indicates another sharp reduction in

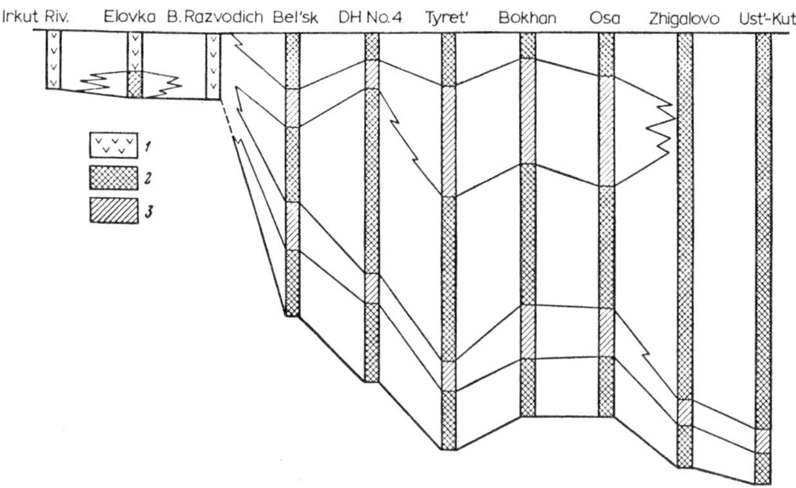

FIG. 177. Sections of the Lower Cambrian Usol'e sequence along a north-south profile (from Ya. K. Pisarchik). 1. Anhydrite-dolomite complex with subordinate halite; 2. rock salt with thin beds of dolomite and anhydrite; 3. anhydrite-dolomite complex.

thickness. If this is so, then *a huge depressed zone formed on the southern part of the Siberian platform when the Usol's saline complex was forming, occupying the site of the Irkutsk amphitheater and extending northeastward to the region of the Kempendyai domes and the Berezovskii downwarp. The depressed zone was bordered on all sides by belts that subsided at rates but one-fifth to one-eighth that of this zone.*

Differences in composition of the deposits are very closely related to these different rates of subsidence in the different areas (Fig. 178). In the narrow, weakly subsiding Sayan and Baikal belts, the saline formation is represented

---

FIG. 176. Composite section of the Lower Cambrian saline formation of the Siberian platform (from Ya. K. Pisarchik). 1. Clastic rocks; 2. limestone-dolomite; 3. sulfate-carbonate with indications of salt; 4. halite interbedded with sulfates and pure anhydrite.

2 B

by alternating dolomite and anhydrite and by transitions between the two. Clastic admixtures are insignificant, and clastic rocks proper are entirely absent. Only along the Urik River are dolomites and limestone-dolomites interbedded with unsorted, cross-bedded, red sandstone, similar to that in the underlying sequences. These sands are probably continental alluvial or deltaic

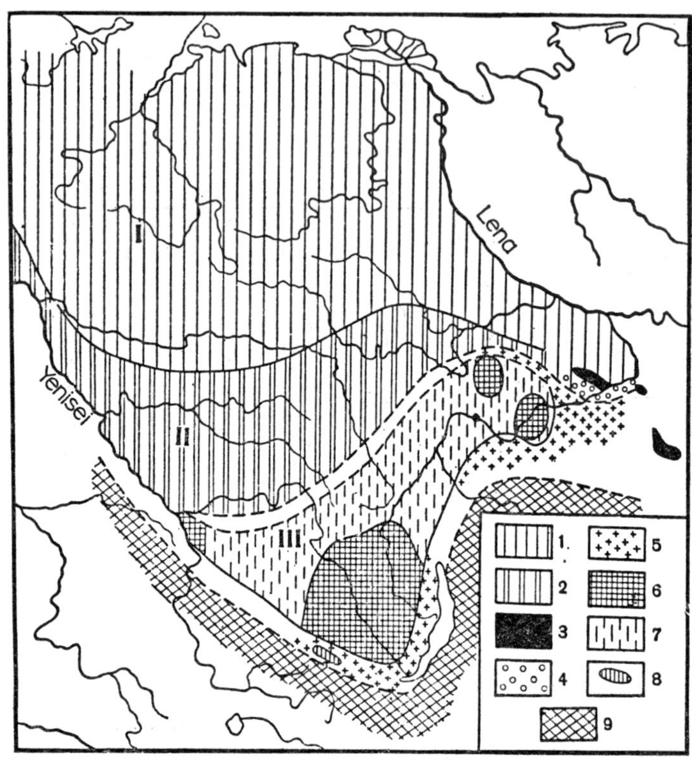

FIG. 178. Formational environment for the Usol'e sequence (from Ya. K. Pisarchik, N. A. Arkhangel'skaya, V. N. Grigor'ev, A. A. Ivanov, and others). *I.* Zone of open sea, *II.* transitional zone, *III.* marginal salty part of the basin. 1. Limestone of normally saline sea; 2. dolomite of a slightly more saline sea with underwater shoals and islands; 3. algal and archeocya-thid bioherms (according to K. K. Zelenov); 4. oolitic dolomite; 5. anhydrite-dolomite; 6. salt deposits, reliably determined; 7. salt deposits, inferred; 8. continental red beds; 9. land.

deposits. On the northeast, the sulfate-carbonate zone may be traced eastward from the Berezovskii downwarp, where it grades into the Aldan massif (K. K. Zelenov).

The entire zone of deep subsidence to the north of this belt of weak subsidence was filled with saline deposits, rock salt interbedded with dolomite, anhydrite-dolomite, and anhydrite. Argillaceous variants of these rocks are found only as thin interbedded units, and clastic admixtures in the other rocks are negligible. The thickness of the beds of rock salt ranges up to 60–70 m, and the thickness of the salt-free rocks ranges up to 40–50 m. A salt sequence of this composition has been recognized in four rather widely scattered areas. The westernmost section is in the Taseevo region along the lower reaches of the Angara River, where a borehole proved 2207 m of salt without reaching the base of the saline sequence. It is however not clear if all this sequence belongs to the Usol'e series, or if the lower part is of Moty age. The second large area of salt accumulations has been found by deep drilling in the inner part of the Irkutsk amphitheater. It embraces the upper reaches of the Angara and Lena Rivers and their tributaries. The third area is on the extreme northeast, in the basin of the Kempendyai and Suntar Rivers, where

TABLE 36

Salt Content of the Usol'e Series (from A. A. Ivanov and Ya. K. Pisarchik)

Region	Drill hole	Thickness of rock salt, m	Relative salt content, %
Irkutsk amphitheater	Kamenka 1-r .	3	0·4
	Elovka 1-r	39·5	29
	Usol'e-Sibirskoe 4-r	433	63
	Bel'skaya 1-o	351	59
	Polovina 4-rs .	470	63
	Tyret' 1-r	545	61
	Bokhan 1-r (supplemented by 2-r and 4-r) .	520–540	61–64
	Osa 1-r (supplemented by 2-r and 4-r)	506–559	60·5–66
	Zhigalovo 1-o (supplemented by 2-r).	606–730	64–67
	Ust'-Kut	405–420	42–43·5
Berezovskii downwarp	Russkorechenskaya	19	10
	Namana	40	18·6
	Olekminsk	34	12
Taseevo region		1400	63

the salt has formed a number of domes. The fourth and last area is in the Berezovskii depression. These four areas are separated by zones in which Usol'e salt beds have not been definitely recognized. Indirect evidence, however, particularly salt springs, makes it seem entirely probable that the Usol'e salt deposits are continuous between the areas where they have been reliably identified.

The total thickness of rock salt in the central part of the Irkutsk amphitheater reaches 420–730 m, and the relative salt content is remarkably constant over a great part of the region, ranging from 59 to 75% (Table 36). The thickness of the rock salt decreases sharply in only a very narrow belt transitional to the sulfate-carbonate zone, and here the salt content drops to 42–49% and lower. The thickness of the salt beds is very great in the Taseevo area, and the relative salt content is also high (63%). In the Berezovskii depression the relations are substantially different. The thickness of salt is relatively low (19–40 m) and the salt content is also low (10–18%). This material was clearly deposited in the marginal zone of salt accumulation. There are no data for the region of the Kempendyai faults, but it can hardly be doubted that this zone also belongs to the marginal belt of salt accumulation.

To the north of the zone of intense subsidence and salt accumulation lies a broad belt of dolomitic sediments, including the lowest reaches of the Lower Tunguska River, the entire basin of the Stony Tunguska, the middle course of the Vilyui River, and extending thence to the Lena and Amga Rivers (N. A. Arkhangel'skaya and V. N. Grigor'ev 1960). Between this belt and the halite zone there is undoubtedly a narrow strip of the dolomite-sulfate facies, similar to that of the Sayan and Baikal regions, but it has not yet been definitely recognized. North of the dolomite zone occurs a broad belt of limestones, including the present area of the Anabar massif and the adjacent parts of the Tunguska and Lena-Vilyui basins.

In most places the Usol'e macrocyclothem may be clearly subdivided into five lithic members according to the rock salt content. In the first, third, and fifth members, rock salt is dominant, but in the second and fourth, dolomite, anhydrite, and rocks transitional between these are dominant. If the upper Moty dolomite-anhydrite series is added, the resulting six units present a well-defined pattern. Each dolomite-anhydrite unit with its overlying salt forms a first-order cycle, with increasing salinity of the water, and three such cycles may be recognized. Mineralogical analysis of the clastic material indicates that the deposits of each cycle were synchronous in the various parts of the salt basin (Ya. K. Pisarchik 1958b).

As in other saline formations, second- and third-order cycles within the first-order cycles may also be recognized. They are of more local development, however. The finer cycles represent seasonal bedding. In rock salt this is manifested by the presence of fine laminae (parts of a millimeter) of fine-grained and microgranular anhydrite or dolomite anhydrite, or in places clay. The halite layers separated by these laminae are 1 to 7 cm thick. The coarse rhythms (first to third orders) are undoubtedly associated genetically with tectonic processes, particularly tectonic movements of the basin floor, which, at times, widened the connection between the marginal salt-producing part of the sea and the part with normal salinity and, at other times, restricted it. The finer cycles were probably the result of periodic fluctuations in climatic conditions up to annual hydrochemical cycles.

With the range of fluctuation in salinity, the composition of the halitic beds is of fundamental significance, and the question of whether potassium salts appear or not requires an answer. The rock salt has been generally recrystallized to grains 0·5 to 4 cm in diameter. Extremely small crystallites of sylvite (hundredths of a millimeter) within halite grains commonly occur (Voronova 1954). The KCl content in the rock salt is normally measured in tenths of a percent, however, rarely reaching 1% in rock having an NaCl content of 78–99·6%. Only occasional records of carnallite and sylvite are known, as in the Bel'skaya section in the upper part of the Usol'e series, where they occur in small grains and thin beds in the rock salt (A. A. Ivanov 1950b). Grains of sylvite to 1–1·5 cm in diameter occur in rock salt from the upper half of the Usol'e series at Polovina. In the Zhigalovo borehole gamma logging has also revealed potassium salt. Thus, the salinity of the marginal salt-producing part of the sea occasionally reached very high values, at least the lower level of potassium-salt precipitation, i.e. more than 30–32‰. The times when such salinity was attained clearly corresponded to the times of most restricted connection between the marginal highly saline and the normally saline parts of the sea.

There are no reliable criteria for direct determination of the depths of the marginal part of the basin at the times of rock salt accumulation, though indirect and convincing data are available. These are the composition and structure of the dolomitic members that alternate with the halite and anhydrite. "The dolomites of the Usol'e series are similar to those of the Moty series. Dolomites that have been but slightly recrystallized generally have a microgranular texture and very commonly contain algal growths. Apart from *Collenia* these rocks, much more extensively than the Moty series, show fine undulatory banding, in what are probably algal dolomites. Clotted, oolitic, or oolitic-fragmental texture occurs along with uniform microgranular structure in the dolomite, as in the algal Moty series. Anhydrite crystals are common, and salt is frequent" (Ya. K. Pisarchik 1958b). The algal dolomites are clearly shallow-water deposits; and, since they are interbedded with anhydrite and halite, commonly forming numerous thin alternations, it is obvious that the salt beds are also shallow-water deposits. This appears even more obvious when we consider that halite accumulates very rapidly, and this must inevitably cause shoaling in the basin.

Although most of the halite beds are completely recrystallized, traces of primary sedimentational structures may still be found. These appear as relict zonal structure in the halite crystals. They have been observed in several parts of the Bel'skaya, Tyret', Osa, and Zhigalovo sections. Normally they are macroscopic, in individual halite crystals of 0·5–4 cm diameter, but pinnate structure is especially obvious under the microscope. Growth zones in the crystal are normally marked by numerous small gas bubbles and by cubic cavities filled with brine. In places anhydrite and dolomite are concentrated along these growth zones, probably having been incorporated at the time the halite crystal formed.

Usol'e time as a whole presents a lithofacies picture on a grand scale. The great northern part of the platform was undoubtedly a flat shallow-water marine basin having normally saline water and receiving calcareous sediments. South of this was a broad belt of sea with water of a somewhat higher salinity, from which dolomitic muds were precipitated. The southern part of the platform, corresponding to the Irkutsk amphitheater and its extension to the Berezovskii downwarp and the Kempendyai faults, was characterized by water with high, and commonly very high, salinity, reaching the stage of anhydrite precipitation above 15% and halite precipitation above 25%. In the extreme south, in the narrow near-shore zone, in the Sayan and Baikal regions and further to the northeast, the salinity of the water again decreased, but not greatly, since intermittent precipitation of dolomite and gypsum took place. The salt-producing zone of the Early Cambrian sea in this situation proved to be the marginal part, which had the form of a broad open gulf.

The connection between this salt-producing part of the marine basin and the remaining incomparably larger part of the sea, in which the salinity was normal (deposition of limestone) or only somewhat elevated (deposition of dolomite), is not well defined. Evidence is restricted to two localities: in the lower reaches of the Angara River on the extreme west and along the Lena between Olekminsk and the Sinyaya River on the extreme east. In both places transition is very gradual between deposits of normal or slightly over-normal salinity and the salt deposits of the marginal part of the sea (N. A. Arkhangel'-skaya and V. N. Grigor'ev 1960). No trace of a boundary separating the salt-producing part of the basin from the normally saline part has been found. From this fact it is concluded that, everywhere else in the basin, deposits of the normally saline zone probably grade transitionally into the saline accumu-lations. There were no barriers between these two radically different zones of the sea; on the contrary, water of the shallow, normally saline northern part of the basin, becoming salty through evaporation, gradually and imperceptibly changed to very salty water in the extreme south, where salinities exceeded 25%, with no underwater or subaerial barriers. This is supported by calcula-tions using the concept of turbulent diffusion. It is impossible to agree with the reasoning of these authors, however. Calculations of increasing salinity become convincing only when they are founded on well-based views con-cerning the interaction between the water masses of the normal and the highly mineralized zones of the sea. The authors did not consider this aspect. Their calculations are based on arbitrary assumptions which detract from their value. In order to solve this problem of the relationship between normally saline and highly saline parts of the sea, it is necessary first to draw an analogy from a well-founded example of similar kind. It has been shown that salt-producing basins of the gulf type may be separated from normally saline or somewhat more saline entities. The marginal part of the Dankov-Lebedyan sea had restricted connection with the remaining normally saline part of the basin at the time gypsum was being precipitated. This relationship was even clearer in the Triassic of Germany. From these examples we conclude that *there must have also been some restricted connection between the normally marine northern part and the salt-producing margins of the Early Cambrian sea of the Siberian platform.* There are some indications that shoals and islands existed. Immediately north of the salt-producing zone are segments of the Siberian platform characterized by relatively thin early Paleozoic deposits, because of their emergent tendencies. These structural elements favor the development of shoals and islands, which might act as barriers to the free exchange of water between the two parts of the sea. The form of the transition between limestones of the normal salinity zone and gypsum-dolomite rocks on the extreme east, along the Lena, Amga, and Aldan Rivers, is also very significant. A zone of oolitic dolomites and archaeocyathid-algal limestones occurs between the gypsum-dolomite rocks and the limestones (K. K. Zelenov 1957). The oolitic dolomites are characterized by a high degree of sorting and by cross-bedding. All this indicates very shallow water. There is no doubt that the oolitic grains of dolomite, being agitated by water movement formed shoals, possibly even emergent ridges and islands. The archaeocyathid-algal bioherms, considered singly, were not high (up to 1·2 m) and could not have retarded water exchange individually. But an entire zone of them would undoubtedly form a large shoal in comparison to those parts of the sea lying to the east, where no biohermal masses developed, and this zone would to some extent retard water exchange. All these facts favor a restricted water exchange between the more saline and the normal parts of the single basin, and they suggest that the restriction was precisely in the zone where N. A. Arkhangel'skaya and V. N. Grigor'ev pictured free and unobstructed passage.

Thus, *although direct data for solving the problem are few, various kinds of indirect evidence suggest that the connection between the marginal salt-producing zone and the normal zone of the Early Cambrian Usol'e sea was obstructed by underwater shoals, and at times even by islands of accumulated calcareous oolites. These shoals occupied a belt transitional between dolomitic sediments and sulfate-dolomite deposits.*

4. THE BELAYA AND ANGARA MACROCYCLOTHEM OF THE LOWER CAMBRIAN SALINE FORMATION.

Deposits of the Belaya macrocyclothem occupy practically the same area as those of the Usol'e macrocyclothem (Fig. 179). There was therefore no appreciable expansion or contraction of the general marine area on the southern Siberian platform, in Belaya time, although the lithology of the macrocyclothem changed substantially. The lower quarter of the Belaya macrocyclothem (the Lower Belaya beds) is a limestone-dolomite unit containing the trilobite *Bulajaspis* along the southern edge of the basin, a form completely absent in older rocks of the same localities. The succeeding major part of the Belaya macrocyclothem is a saline complex—alternating layers of rock salt, dolomite, anhydrite, and transitional anhydrite-dolomite rocks. The uppermost part of the macrocyclothem consists of a thin sulfate-dolomite unit, belonging stratigraphically to the lower Angara series. *Thus the basin, while preserving an essentially uniform area, underwent successive changes in salinity. It was first freshened to the point that sulfate precipitation ceased almost completely; it then became highly saline to the extent that halite formed, before finally undergoing renewed freshening.*

From this discussion it becomes clear that the cause of salinity change in the water was the fluctuation in intensity of water exchange between the marginal zone of the sea and the zone of normal sea water. During the first third of the macrocyclothem, because of subsidence of the floor along the northern margin of the Irkutsk amphitheater, the water exchange increased. During the second and more extensive interval, exchange was restricted because of uplifts in the northern marginal zone, and conditions similar to those prevailing during halite precipitation in Usol'e time were restored. During a brief closing moment, the exchange of water again increased and the basin freshened.

We have said that during halite precipitation conditions were similar to

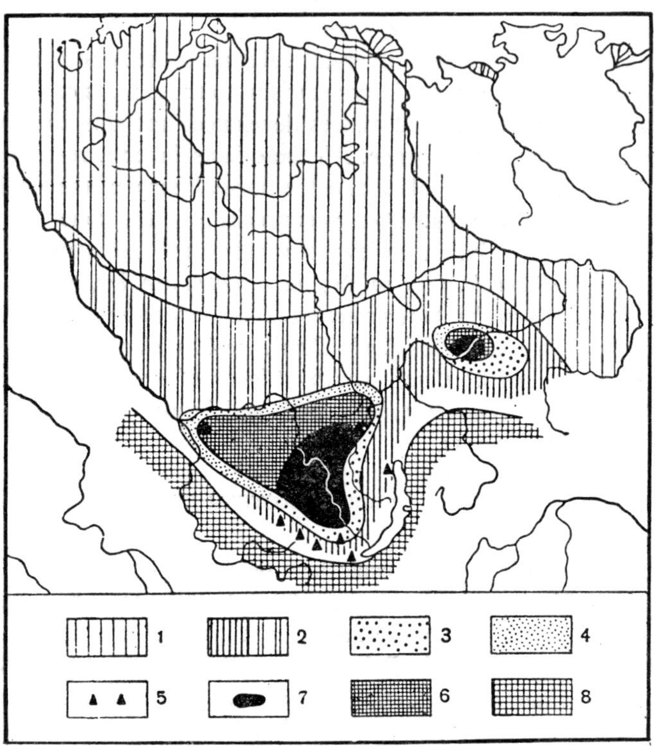

FIG. 179. Formational environment for the Belaya sequence (from Ya. K. Pisarchik *et al.*). 1. Limestone of normal sea water; 2. dolomite of the transitional zone of slightly more saline water in the zone of shoals and islands; 3. anhydrite-dolomite, proved; 4. anhydrite-dolomite, inferred; 5. localities with *Bulajaspis*; 6. salt deposits, inferred; 7. salt deposits, proved; 8. land.

those of Usol'e time; they were, however, by no means identical. This is shown in the lithofacies zoning of Late Belaya time (Fig. 179). *Instead of a single area of salt deposition, which in Usol'e time included the Irkutsk amphitheater and extended northeastward to the Berezovskii downwarp and the region of the Kempendyai salt domes, there were two areas of halite accumulation in Late Belaya time: one in the amphitheater proper and the other in the Berezovskii downwarp.* Between them lay a zone of dolomite deposition, corresponding to water only slightly saltier than normal, along the present Lena River. The total area of salt accumulation was therefore reduced. *It is obvious that some of the earlier barriers that separated the marginal zone of the sea in Usol'e time were not restored during the Belaya macrocyclothem.* The total thickness of salt and the relative salt content during this time were also lower (Table 37).

The relative salt content in the Berezovskii downwarp, as before, was especially low.

No indications of potassium salts have been found in the rock salt of the Belaya macrocyclothem. It is obvious that the salt-producing part of the marginal marine zone in Belaya time was insufficiently saline for this, as also during Usol'e time.

Discoveries of the trilobite *Bulajaspis* in the near-shore belt of Belaya dolomites along the Eastern Sayan and in the Baikal region are very significant in this connection. These indicate that the southern margin of the marine zone

TABLE 37

Salt Content of the Belaya Macrocyclothem (from
A. A. Ivanov and Ya. K. Pisarchik)

Borehole	Total thickness of salt, m	Relative salt content, %
Bokhan 1-r . .	26·5	51·0
Osa 1-r . . .	43·4	43·4
Tyret' 1-r . .	61·0	51·2
Ust'-Kut . .	164·0	52·7
Namana . . .	67·0	18·0
Solyanka . .	68·0	?
Olekminsk . .	112·0	30·0

was then only slightly above normal salinity, allowing even highly developed euryhaline marine forms to live. The cause of the decline in salinity is not yet known. Ya. K. Pisarchik (1958b) and N. A. Arkhangel'skaya and V. N. Grigor'ev (1960) have stated that it was initiated by the breakup of the previously single Eastern Sayan continent into a chain of islands with intervening straits. The influx of marine water through these straits from the neighboring southern geosynclinal zone brought about the freshening. Without denying this possibility, I would suggest as an alternative, that a freshening current of marine water may have come from the northern normally saline part of the platform sea through the same broad straits between the salt-producing zones of the Irkutsk amphitheater and the Berezovskii downwarp. This current then moved along the Baikal zone to Irkutsk before turning northwest along the Sayan. Here it probably became sufficiently salty to lose its individuality. The fact that *Bulajaspis* was a genus of the platform basin and is unknown from the neighboring Sayan geosyncline supports this view. If all these concepts of inflowing diluent sea water are correct, we find here the same system of circulating currents as occurs in the saline marginal zones of other seas (see above). The accumulation of salt in this case too was due to halystasis.

On the whole saline deposition during the Belaya macrocyclothem was much weaker than that during the Usol'e cyclothem, in mass of accumulated salt, in area of deposition, and in degree of mineralization attained in the salt-producing zone.

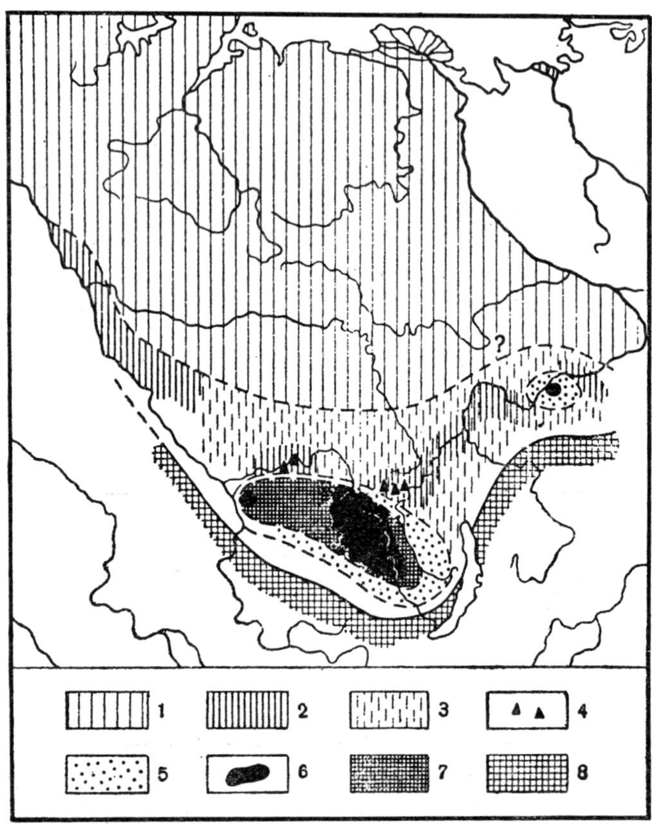

FIG. 180. Formational environment for the Angara series (from Ya. K. Pisarchik and others). 1. Limestone of a normally saline sea; 2. dolomite, proved; 3. dolomite, inferred; 4. localities of *Bulajaspis* and other forms; 5. anhydrite-dolomite; 6. salt beds, proved; 7. salt beds, inferred; 8. land.

These features are developed even more strongly in the Angara macrocyclothem (Figs. 180 and 181). The area over which this cyclothem occurs is smaller than that for the previous two cyclothems chiefly because of subsequent erosion of the Angara rocks. Although the basin remained stable through Angara time and occupied the same area originally, the composition of the deposits changed substantially. The lower part of the macrocyclothem, the Bulai series, consists of dolomitic rocks interbedded with anhydrite-dolomite and dolomitic marl in the basal beds. Sulfates become very rare in the upper

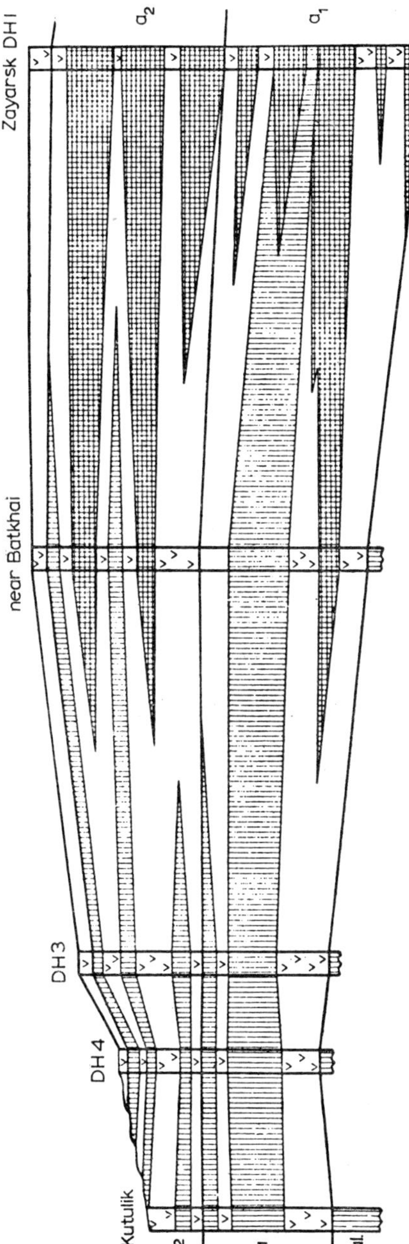

FIG. 181. The Angara saline series (from Ya. K. Pisarchik). 1. Rock salt with interbedded anhydrite; 2. anhydrite-dolomite group; 3. limestone-dolomite group: *Series: Bul*—Bulai, a_1. lower Angara, a_2. upper Angara.

horizons. Remains of the trilobites *Tungusella* and others are present. Clearly, a considerable freshening of the basin occurred at the beginning of the macrocyclothem, caused by increasing subsidence in the northern "barrier" zone of the Irkutsk amphitheater and by improved water exchange with the northern normally salty part of the sea. Under these conditions 120–160 m of carbonates of the Bulai series accumulated. Later, connection with the open sea was again restricted and the saline Angara series was deposited: beds of rock salt interbedded with dolomite and anhydrite, an association that was normal for the Early Cambrian. The total thickness of this series ranges from 260 to 400 m. The great reduction in the salt-producing zone, even when compared with Belaya time is striking (Fig. 180). Marked thicknesses of rock salt (up to 270 m) are found only at Zayarsk on the Angara River, in the middle of the western salt district, where the relative salt content is about 40% (Ivanov and Levitskii 1960). In the Berezovskii downwarp, salt of the same age is only 34 m thick at Olekminsk (relative salt content 9%), 78 m at Ust'-Biryuk (drilling incomplete), and 171 m at Del'geiskaya (28% salt).

Like the preceding saline macrocycles, the Angara shows a well-defined cyclicity of the second and higher orders. *In the saline beds of the Angara macrocyclothem a much greater fraction of clastic material is found than in the preceding macrocyclothem together with independent beds of sandstone and siltstone. This proves rejuvenation of the drainage areas, the beginning of the uplift that led to general marine regression in Middle Cambrian time.*

5. The Type of Early Cambrian Saline Formation on the Siberian Platform

For a full appreciation of the Lower Cambrian saline formation the type of formation must be determined—transgressive, stable, or regressive.

Different opinions are expressed on this subject. N. S. Zaitsev and Ya. K. Pisarchik consider the formation to be transgressive; A. A. Ivanov, on the other hand, is inclined to believe it to be regressive. The former view is apparently based on the characteristic change in the succession from basal clastic to chemically precipitated rocks, carbonates at first, then saline, with a progressive decrease in fragmental material. This is characteristic of sedimentation in a transgressive basin. Interpretation of the Lower Cambrian saline formation as transgressive, however, contradicts the evidence of constant area of deposition from the very beginning of Moty time. The diminution in amount of clastic material in the rock may be explained by progressive lowering of the continental drainage area. These same facts do not support the view that the Lower Cambrian saline formation is regressive. Actually, as appears from the above analysis, *the formation is typically a stable type. It accumulated while the marginal saline zone of the sea maintained an essentially constant area, but its connections to open normally saline sea changed. This was caused by episodic deepening and shoaling of the transitional zone, causing improved or restricted water exchange.*

D. THE MIDDLE DEVONIAN ELK POINT POTASSIUM FORMATION OF NORTH AMERICA AND ITS ORIGIN*

Another example of saline formations of marginal-sea type is the Elk Point formation in the Williston basin of North America. This basin is one of the syneclises on the Canadian platform, plunging toward the Cordilleran geosynclinal zone. The Williston basin includes the western part of North Dakota, the eastern part of Montana, and the southern part of Saskatchewan. Estimates of its area have ranged from 244,000 to 500,000 km², depending on the horizon used for the calculation.

The tectonic structure of the Williston basin in Devonian time, according to A. Baillie (1955), is shown in Fig. 182. A great depressed zone lay in the

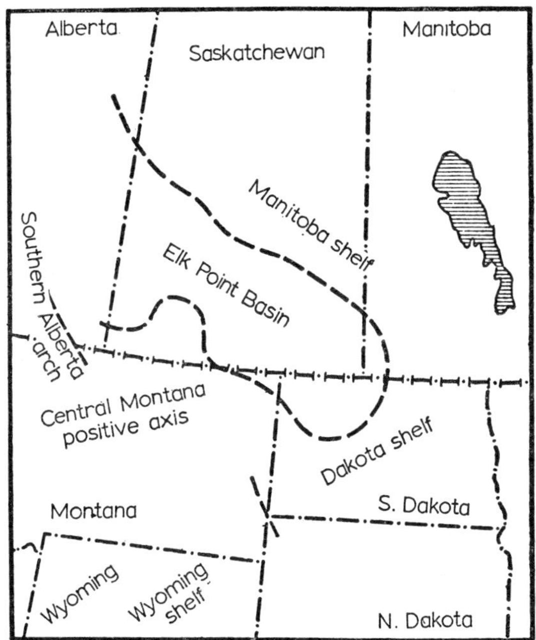

Fig. 182. Tectonic plan of the Devonian Williston basin
(from Baillie).

middle of the basin, in the deepest part of which the Elk Point saline formation was deposited. This zone cut diagonally across southern Saskatchewan and extended northwest into Alberta. The area in which the Elk Point formation accumulated was surrounded by less mobile segments: the Manitoba shelf on the north and northeast, the Dakota shelf on the south and southeast, and the elevated belt of Montana on the southwest, grading southward into the Wyoming shelf. On the west the Elk Point basin was bounded by the

* The following description is based chiefly on the summary of A. P. Bogdanova made in 1959 at the Institute of Halurgy.

uplift of southern Alberta. The northwestern boundary was probably the Sweetgrass uplift which, according to Sloss, began to form in Middle Devonian time and continued to the end of the Devonian. North of the Sweetgrass anticline lay the central Alberta basin, and to the south the Williston basin, the two regions of maximal subsidence in Devonian time. One was located in the southern part of Saskatchewan and is now outlined by the 600-m isopach; the other lay in the central part of Alberta. The greatest thickness of Devonian rocks is more than 1200 m in these basins. On the shelf zones, which are essentially the borders of the syneclise, the thickness is only one-third to

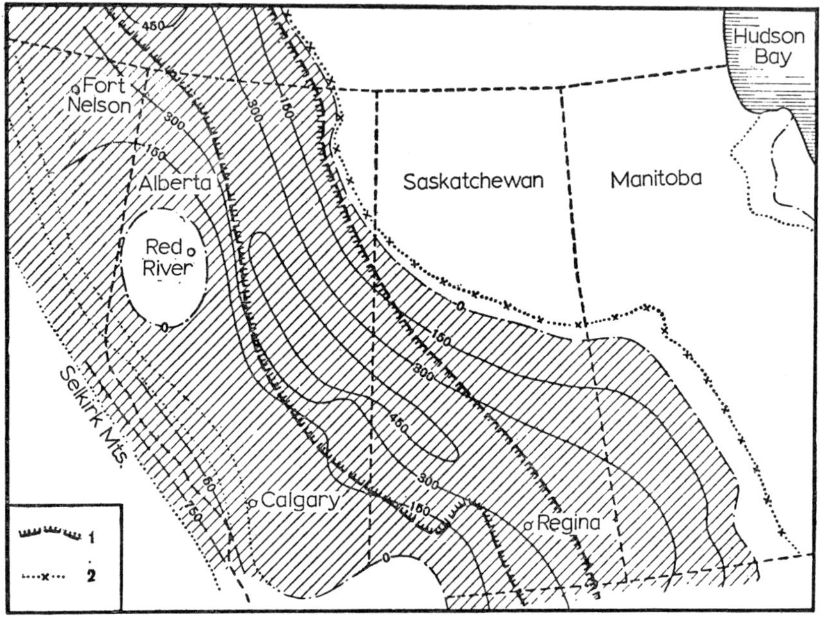

Fig. 183. Isopachs and present distribution of Devonian and Silurian rocks in the Williston basin (from A. P. Bogdanova). 1. Boundary of salt distribution; 2. edge of the Precambrian shield.

one-fourth that in the central part. To the east the thickness decreases and the formation wedges out against the Canadian shield.

There is a major break everywhere at the base of the Devonian: the upper part of the Silurian and all the Lower Devonian are missing. The Middle Devonian is represented chiefly by the Elk Point sequence, with brick-red argillaceous dolomite interbedded with red shales (15 m) at its base. Above this occurs light-colored fine-grained laminated limestone (22 m), succeeded by a thick sequence (115 m) of reef dolomite that is saccharoidal, straw-yellow to light brown. The Elk Point salt formation, 180 m thick, rests conformably upon the marine carbonates. It consists of alternating beds of rock salt and anhydrite with a few beds of dolomite. Three horizons of potassium salts are found in its upper part. The formation is succeeded by a 60-m sequence of red and green shales interbedded with micritic stromatoporoid limestone. Nodules and beds of anhydrite are found locally in the upper part of this unit.

The salt extends along a wide belt from southeast of Saskatchewan to the northwest, cutting across Saskatchewan and Alberta and extending into the northern provinces (Fig. 183). Its northern and southern boundaries are not yet defined, but it is known to be 1600 km long (Webb 1951), and dominated by halite. Interbedded anhydrite is found in the upper part of the formation, the thickness locally reaching 30 m in the northeast. Near the edge of the formation anhydrite overlaps the salt and replaces it along the margin, extending

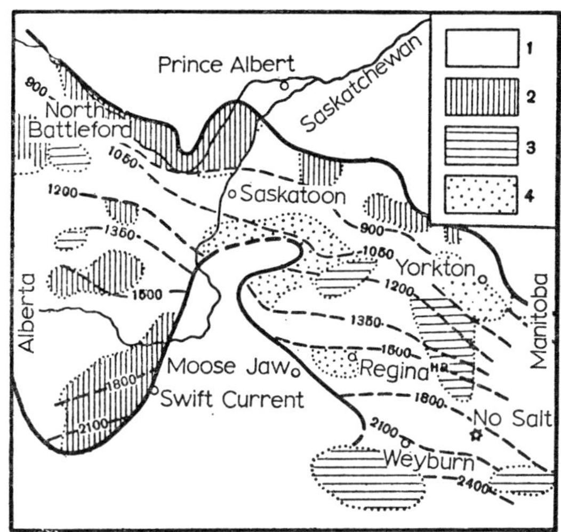

FIG. 184. Depth (in meters) to potassium salts and their K_2O content (from Tomkins, taken from A. P. Bogdanova). 1. No information; 2. less than 15% K_2O; 3. from 15 to 25% K_2O; 4. more than 25% K_2O.

also on to the shelf. The thickness of the rock salt ranges up to 192 m. A huge irregular area with no salt extends from the south into the salt-bearing zone and almost cuts it in two. Salt isopachs approach the salt-free area at right angles and could be easily linked across the gap. It is therefore concluded that the salt has been leached from the zone where it is now absent. Erosional features and leaching of salt have been noted elsewhere. Locally the upper part of the saline sequence has been removed completely.

Potassium-rich rocks occur in the upper part of the formation. Three potash horizons have been recognized. Sylvite is the main potassium salt, with small quantities of carnallite. The thickness of the potash horizons ranges from 0·75 to 4·2 m; the areal distribution has not yet been defined, but from borehole evidence would appear to be extensive. The highest potassium content ($K_2O > 25\%$) is found in the central part of the basin, where the rock salt is thickest (Fig. 184).

The paleogeography for the time of Elk Point salt accumulation may be reconstructed only in its general outlines. At the end of the Late Silurian and in Early Devonian times the region of the Williston basin was raised above sea-level and became a zone of denudation. At the beginning of the Middle Devonian the basin was again flooded to become a zone of shallow-water, chiefly carbonate deposition, at times dolomitic, at times calcareous, and latterly a zone of reefs. In Middle Devonian times the environment changed substantially: the broad sea of the Williston basin, hitherto connected with the open geosynclinal basin, apparently narrowed markedly, which led to a sharp increase in salinity of the water remaining in the syneclise. The Elk Point saline formation was deposited. The uplifts that caused the restriction in water circulation with the open sea cannot be found along the northeastern border of the syneclise, inasmuch as the broad continent of the Canadian platform extended to the east, probably embracing a large part of the shelf area. Uplifts might have occurred only along the southwestern edge of the basin, in a belt extending from southern Alberta to central Montana and thence into Wyoming and South Dakota. The Sweetgrass anticline also was probably elevated at that time. It is unlikely that the uplifts formed a continuous elevated ridge, because the saline formation has no notable clastic fraction, except in the upper horizons of the formation. The restricted connection with the open sea on the southwest must therefore have been through a chain of shoals and islands. Because the water in the straits was shallow, water circulation was restricted to an inflow of seawater; the highly saline brine could not readily flow in the opposite direction. This led to a rapid rise in salinity and a complete development of the saline process, culminating in the precipitation of potassium salts. The brine was also deprived almost completely of magnesium sulfates, and yielding therefore only potassium and magnesium chlorides.

Thus, the paleogeography and the tectonic conditions prevailing during the deposition of the Elk Point formation were similar to those controlling the Dankov-Lebedyan and Usol'e formations: they are indeed of the same facies type and differ only in greater completeness of the saline process in the Elk Point formation, which led to potash formation. This was probably caused by the greater isolation of the salt-producing margin of the open sea than prevailed during the deposition of the Dankov-Lebedyan and Usol'e formations.

E. THE DEVELOPMENT AND STRUCTURE OF MARGINAL SALINE FORMATIONS AND THE CONDITIONS CONTROLLING THEIR ACCUMULATION

The characteristic morphological feature of the basin that gives rise to a marginal saline formation is *a broad indentation into the continent, comparable to the modern Bay of Biscay in its general relations.* Where the structural basement was a young post-Hercynian platform with considerable tectonic differentiation and notable relief, the morphology of the marginal part of the sea was also complex. It was characterized by islands, some large, some isolating the margins from the main sea, sharply restricting the circulation of the water. The marginal basin then closely approached the gulf type, and when connected to the open sea by a single strait became a gulf of the Kara-bogaz type. Normally, however, there have been two or more passages for the entrance of sea water, which clearly distinguishes the marginal part of a sea from a gulf. Such were the Röth (Upper Bunter) and Late Triassic basins of Germany, and the Late Jurassic basin of Tadzhikistan, among others. But when the flooded basement was part of an ancient platform and was only weakly differentiated tectonically, with a low relief, the morphology of the marginal part of the sea has been simple. *The marginal part of the sea has been free of islands and of underwater uplifts in general, and it has graded gently over a broad front into the open sea, appearing as a single water mass. Actually, however, even in such seas, submarine elevations between the marginal zone and the open sea have formed shoals, which have restricted water circulation.* Surface water has always passed from the open sea into the marginal zone, where it has evaporated, and salts have passed into the marginal zone of higher mineralization. The heavy salt water of the marginal zone has flowed toward the open sea, but only through individual, deep straits. *The weaker this inverse flow, the more rapidly the marginal zone has reached high stages of salinity and the more near to completion has the saline process gone.* The Dankov-Lebedyan, Usol'e, and Middle Devonian Williston basins, illustrate this type of marginal basin, which outwardly appears to communicate broadly with the open sea but was in fact partially isolated by shoals. At times reef structures on shoals have further isolated the marginal zones of such seas.

Sedimentation in the marginal zones of a sea in many ways differs substantially from that in gulfs, as is indicated by the composition and form of the saline formations. That the areal dimensions of the marginal zones of a sea have generally been much greater than those of gulfs, even large gulfs, has been manifested from the degree of contamination of the saline deposits by terrigenous material. This has been deposited as an inshore zone of sand, silt, and clay of varying width of which only negligible amounts reached the interior of the basin. Thus the saline formations that formed in such basins have been exceptionally pure for the most part. Furthermore, the shoals have contributed no clastic material, and the islands only very little. On the whole, therefore, the distribution of clastic material is concentrated only along the inshore edge of a formation, being absent elsewhere. In formations of the Röth and Middle Triassic type, having large islands that separate the marginal zone from the main part of the sea, the distribution of clastic particles becomes more complex, although the asymmetry at no time disappears completely.

The localization of saline deposits precipitated from brines at various stages of salinity follows a simple rule: *the salts requiring highest mineralization for the production of the solid phase are confined to the deepest parts of the basin floor and to the more central parts of the formation having the greatest thickness. Salts precipitating from brines with the lowest salinities are concentrated in*
2 c

the zones of shallower water, in the marginal parts of the formation, having the least thickness. Areas of rock salt are therefore localized in the central and thickest parts of formations. Gypsum and anhydrite encircle these zones, occurring about the margin. Potassium salts are always found within the zone of rock salt, occupying a smaller zone and being restricted to the upper horizons of the formation.

This simple scheme is normally subject to various complications two of which are especially fundamental. It is a matter of fact that no single solid salt phase shows continuous development at any single point in a marginal marine formation, but rather shows an alternating series of beds of mixed composition. In the center of the formation potassium salts, when present, alternate with rock salt, and beds of the latter are interbedded with dolomite and occasionally with clays and marls. On closer examination these alternations prove to be subordinate to cycles of various durations, covering different areas of development as the marginal part of the sea becomes alternately fresher or saltier. The freshening takes place when the straits connecting the marginal zone with the open sea become deep and the outflow of heavy saline waters is accelerated. Increase in salinity follows shoaling in the straits, which restricts the outflow of heavy saline waters. This cyclic sequence in marginal-sea saline formations is a feature common to all. Another complication follows from the characteristic exchange of water between the marginal and open zones of the sea. When the connection occurs along a broad front, i.e. when the isolating uplifts are submarine, the distribution of the various salt phases in a cross-section through the formation is symmetrical, as in the Dankov-Lebedyan and Usol'e formations. But when the water enters the marginal zone through a single narrow passage, as in the Röth and Middle Triassic basins of Germany, the localization of salt phases within the saline formation becomes markedly asymmetrical. Dolomitic sediments are deposited near the strait, gypsum farther away, and halite at the greatest distance. This results in a well-defined chemical differentiation of the salts.

Saline formations of marginal seas are characterized by a distinctive transition to deposits of the open sea. In the zone of shoals separating the margins from the open sea shallow-water limestones of many types accumulate continuously. These shallow-water facies grade into salt deposits in the marginal zone and into open-sea sediments on the other. This generally results in a *complete transition from marginal saline formation to normal marine deposits,* such as found in the Dankov-Lebedyan and Usol'e formations. Absolutely no visible isolation of the marginal zone from the open sea appears in the formation, and this has led to completely erroneous conclusions. (Arkhangel'skaya and Grigor'ev 1960). Actually the uplifts that restrict free exchange of water between margins and the open sea are generally submarine features (Röth and Middle Triassic of Germany). No complete break exists between deposition in the two zones and no belt of denudation. Even where such breaks might exist, deposition in the connecting strait is transitional from the marginal facies to the open-sea facies.

The tectonic prerequisite for the development of marginal-sea formations is the existence of a depression or syneclise in front of an advancing sea, with the axis of this depression lying more or less parallel to the transgressive basin. During flooding, the syneclise inevitably forms a gulf-like indentation into the continent, like the modern Bay of Biscay, but separated from the main sea by shoals or islands.

Where the size is not great, the outline is more or less equal and the connection with the open sea is by only a single strait, a gulf of the Kara-bogaz type is formed. Where the depression is very large, occupying a platform syneclise, and where connections with the open sea are broad, a marginal sea is formed. Clearly transitions occur between the Kara-bogaz and the marginal-sea type. In areal development, but not in thickness, formations of marginal seas are generally much larger than gulf deposits of the Kara-bogaz type. Gulf formations of the Virrila type also grade into larger, specifically old, saline formations, but of different character. These will be examined in the following chapter.

In the stratigraphic column, saline formations of marginal seas are met much more rarely than lagoonal and gulf types. Indeed their number is almost limited to those discussed in this chapter. The great bulk of them are of Paleozoic age, and no post-Jurassic examples are known. It is characteristic that marginal formations are completely absent in geosynclinal zones at all stages of their development. They are restricted to platforms, especially Precambrian platforms. This tendency, already noted in the distribution of gulf formations, reaches its logical conclusion *in the formations of marginal seas, which accumulate only on platforms*, because the vast syneclises, in which such marginal formations are deposited, only develop there.

F. THE POSSIBILITY OF SALINE ROCKS ACCUMULATING IN THE PELAGIC ZONES OF THE OPEN SEA

In describing the initial stage of arid-climate lithogenesis it has been pointed out that dolomite is characteristic of more saline waters. It is confined, in some places, to gulfs and marginal zones of open seas, elsewhere to the pelagic zones of open seas, here occupying great areas. Under these latter conditions it is absent in the marginal zones. At higher salinities, when salt deposits are formed, gypsum and rock salt occur only in gulfs and marginal zones of some seas. The occurrence of these salt phases in the pelagic zones of epicontinental seas is a matter for enquiry. Should we speak of pelagic saline formations as well as marginal-sea formations?

It would first appear that the answer should be positive. According to A. B. Ronov (1956), very high $CaSO_4$ content has been recognized in Middle and Upper Carboniferous rocks of the Russian platform: e.g. 33·67% SO_3 from Middle Carboniferous at Kel'tom, 7·67% at Kotel'nich, and 5·64% at Soligalich; for the Upper Carboniferous, 11·18% SO_3 was recorded at Vologda,

17·66% at Gorki, 8·61% at Kotel'nich, 13·05% at Krasnokamsk, and 8·77% at Soligalich. Such high SO_3 content, *averaged for a horizon*, cannot be attributed to secondary sulfatization, but rather to beds of gypsum.

In interpreting these data from the formational point of view, it must be borne in mind that *the localities where beds of gypsum or anhydrite are found in the pelagic zones of Carboniferous seas are widely separated by zones that are known to consist of pure carbonate rocks*. This indicates not only an isolation of the sulfate deposits within the pelagic zones but also their very small total area. In applying the principles for distinguishing saline formations, as already discussed, we cannot speak of pelagic saline formations but only of *pelagic saline facies of some type, arising in the open sea*. The mechanism by which these deposits accumulated is probably as follows: in the pelagic zones of the shallow Middle and Late Carboniferous seas there were belts of low embayed islands and shoals, separated by labyrinthine straits. At times, during restricted water circulation, such zones quickly became saltier than the surrounding sea, and gypsum accumulated in small areas. Such conditions in pelagic zones were anomalies, and prevailed for only short periods of time. Gypsum hence accumulated here only occasionally and was never thick.

It therefore appears that *gypsum lenses in the Middle and Upper Carboniferous deposits of the Russian platform by no means represented normal pelagic sediments, but formed in specific lagoons within island segments of pelagic zones of the sea*. Even so, it is clear that at higher stages of salinity the distribution of saline sediments associated with epicontinental seas was fundamentally different from the distribution during the initial stage of arid lithogenesis. Masses of dolomite could accumulate not only in gulfs and marginal seas, but over great parts of the pelagic zones also. *With the beginning of the saline process the accumulation of salt phases was excluded from the pelagic zones of epicontinental seas, but was localized only in lagoons, gulfs, and the marginal zones of some seas. No pelagic saline formations accumulated, and in principle they could not form*. Lagoonal deposits within pelagic zones sometimes formed minor deposits, but only under specific conditions, anomalous for the pelagic zone, comparable with the modern Bahama shoals and islands.

CHAPTER 3

FORMATIONS OF INTRACONTINENTAL MARINE SALINE BASINS

Saline formations of intracontinental halogenic seas are among the most imposing phenomena of ancient salt deposits. They are remarkable not only for their great areal extent and great volume, and for the completeness of the halogenesis, which may reach its culmination in the eutonic point, but also for their complete restriction to the Permian period. The halogenic formations of intracontinental salt-producing seas may indeed be spoken of as specific Permian features.

Spatially and temporally these Permian formations belong to two maxima of halogenesis: one Early Permian and confined to the eastern part of the Russian platform and the Ural foredeep, the other Late Permian and localized in the Central European basin (Zechstein deposits) and along the eastern margin of the Russian platform (Kazanian deposits). The saline deposits for each of these maxima will be considered separately.

A. LOWER PERMIAN SALINE FORMATIONS OF THE RUSSIAN PLATFORM AND THE URAL FOREDEEP

1. GENERAL CONDITIONS OF HALOGENESIS

It is well known that the Ural-Siberian geosynclinal zone entered a stage of intense folding at the beginning of Carboniferous time. The orogenesis that began in the central part of the zone (Altai, Kazakhstan, Central Asia) gradually shifted toward the platforms, the Siberian and the Russian. By the end of the Carboniferous not only were the regions of the Turgai Plateau and the eastern slope of the Urals, in the western part of the geosynclinal zone, involved in the folding, but the region of the Central Urals and, in part, the Western were also affected. At the beginning of Permian time, therefore, a broad zone of Hercynian folding lay to the east of the Russian platform. The zone consisted of a system of moderately elevated and dissected mountains, which were being intensely eroded. A similar situation characterized the southern margin of the platform. As a result of Sudeten and later Carboniferous folding, the fold structures of the Donbas developed, extending eastward to the lower reaches of the Volga, to Mangyshlak, and thence along the southeastern edge of the platform to join the Ural fold system. The zone of deformation here was very broad, including the Stavropol plateau and its continuation into eastern Ciscaucasia.

Apparently the two fold structures, the Ural and the Donets-Mangyshlak, effectively separated the Russian platform from the marine basins that persisted in the Ural-Siberian and, especially, Mediterranean geosynclinal zones.

Between the platform and the folded structures, in a long ribbon from the extreme northern to the southern end of the Urals and through the lower reaches of the Emba and Ural Rivers to Askrakhan, lay an active downwarping foredeep, grading into the Donets foredeep on the west.

Sea water displaced from the geosynclinal zone by the ever expanding fold structure became restricted to the foredeep and the adjacent parts of the platform in Early Permian time. It occupied a north-south basin, 3000 km long and 1400 km wide. On the east and south the basin was closed by the folded mountain structures, on the west by the uplifted parts of the Russian platform, particularly the Baltic shield and the Voronezh and Azov-Podolia massifs. Connection with the open sea existed only on the north, through the Pechora basin and in a fairly broad belt in front of Timan Ridge. Thus, *the Early Permian basin on the eastern part of the Russian platform was a true intracontinental basin, lying between broad land areas with low, weakly dissected relief on the west and with mountainous country on the east and south.* In its form it *was a distinctive sacklike marine basin, somewhat similar to gulfs of the Virrila type but incomparably greater*. This latter circumstance distinguishes the Early Permian basin of the Russian platform from gulfs and leads to its recognition as an intracontinental sea.

The tectonic complexity of the floor of the basin should be noted. Even among the larger structures, the Pechora depression on the north and the northern part of the Moscow syneclise are outstanding. The Tokmovo, Tatar, and Bashkir arches lay to the south, with intervening depressions. Still farther south lay the Caspian syneclise; and, lastly, the Ural and Donets foredeeps bounded the eastern and southern edge of the platform. Nor were these downwarps themselves by any means uniform. They were essentially a series of subsiding segments of varying development. The Cherdyn'-Solikamsk, Chusovaya-Sylva, and Belaya zones, the last with its extension into the Aktyubinsk zone, were among the more rapidly subsiding segments. The anticlises were also affected by different degrees of tectonic activity. As a result, halogenesis in the basin was very complex, especially in the spatial distribution of the various types of salt deposits. The resulting saline formations were therefore not only complex, but also different from any of the saline formations already described.

A distinctive tectonic feature of the Early Permian basin in Eastern Europe was the clear upward tendency of the surrounding land areas. Continuing fold movements in the Urals and along the southern border of the platform reduced the size of the basin, forcing it both westward and northward. The uplifts of the Baltic shield and the Tokmovo, Tatar, and Voronezh arches displaced the water from the western margin toward the east. Of special importance, the connection of the basin with the open sea became progressively more restricted. During the *Schwagerina* and Sakmarian-Artinskian epochs for example, the connecting strait was operative on both the eastern and western sides of Timan, whereas during the Kungurian it was open only on the eastern side, and here too the basin finally contracted and ceased to exist. *The shrinking basin changed during the Early Permian into a closed, excessively saline intracontinental sea.*

These tectonic aspects in arid regions have but one consequence—progressively increasing salinity in the basin and the continued development of the saline process towards a more advanced stage of completion. This is precisely what is indicated in the composition of the Lower Permian deposits of the Russian platform and of the foredeep. Four stratigraphic stages are distinguished in this sequence: *Schwagerina*, Sakmarian (the Tastuba and Sterlitamak horizons), Artinskian, and Kungurian. From the viewpoint of lithofacies, however, the subdivision must be different. Three successive formations may be distinguished, of different lithology and spatial arrangement: *dolomitic*, corresponding to the *Schwagerina* stage and the Tastuba horizon of the Sakmarian stage; *sulfate-saline*, including the Sterlitamak horizon and the Artinskian stage; and *potassium-halite*, comprising the Kungurian deposits. These will be examined in turn.

2. THE LIMESTONE-DOLOMITE FORMATION

The paleogeography and lithology of the basal limestone-dolomite formation during the first half of its history, in *Schwagerina* time, are shown in Fig. 185. Although the basin as a whole already displayed features of an intracontinental marine basin, supplied with water from the open sea only from the north, it was still broad, and, in addition, it included a large gulf along the Dnieper-Donets syneclise. Over the greater part of the sea sediments of the initial stage of arid lithogenesis accumulated, and only in the restricted area of the Donbas did evaporites form: gypsum and salts. On the whole, then, the basin was in only an early stage of halogenesis.

The localization of petrographically different sediments within the principal north-trending part of the basin shows a characteristically well-defined asymmetrical zonation.

A comparatively narrow belt of terrigenous clastic deposits from the eroding Ural mountain chain extended along the Ural foredeep. In the narrow eastern part of this belt the material was commonly coarse, with conglomeratic tongues at the southern end. In the central part of the deep, where downwarping was most intense, as in the Belaya depression on the south and Cherdyn'-Solikamsk-Chusovaya depressed zone on the north, relatively deep-water, bituminous, argillaceous-marly and micritic limestones with ammonoid, radiolarian, and other fauna were deposited. A discontinuous chain of reefs formed along the western margin of the deep, being most massively developed at Chusovskie Gorodka, on the Ufa Plateau, and in the southern part of the Belaya depression, where they now appear either as monadnocks or as buried reef masses in the vicinity of Ishimbai. South of Sterlitamak the reefs generally increase in size, extending to the Orenburg region of the Urals. The belt of sediments in the deep is joined on the west by a broad zone of platform sediments, almost exclusively carbonate, generally free from clastic material. These sediments were initially limestone with negligible admixture of dolomite. The limestone is generally organic or bioclastic, with a varied warm-

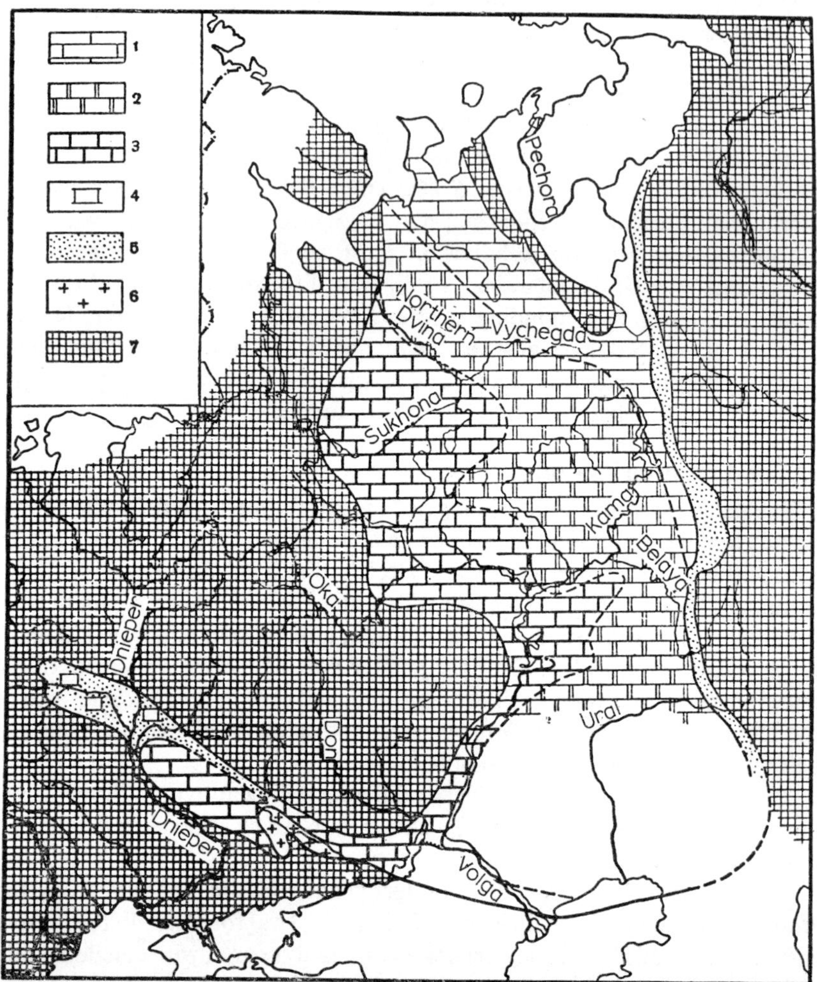

FIG. 185. Paleogeography of *Schwagerina* time (from M. P. Fiveg, 1960). 1. Limestone of normally saline sea (fusulinid and related types); 2. zone of carbonate rocks with subordinate dolomite; 3. zone of dominant dolomite; 4. terrigenous rocks with interbedded dolomite and limestone (Dnieper-Donets basin); 5. terrigenous rocks of the Ural foredeep and other depressions; 6. saline sequence; 7. land.

water fauna (brachiopods, bryozoans, crinoids, and the like) and with algal debris, clearly indicative of marine water of normal salinity. Farther west dolomite increases and the fauna becomes more monotonous. Fusulinid limestones gradually give way to stafellid limestones. Organogenic carbonates decrease in volume, and chemically precipitated calcite and dolomite become increasingly significant, alternating with organogenic limestone. In the western third of the basin dolomite becomes dominant, locally making up the entire succession; the fauna is poor and stunted. These are clear indications of the increasing salinity of the water. A distinct asymmetry of salinity from nearly normal on the eastern edge to manifestly high on the west is apparent. "Along the western margin of the basin lagoons formed, and dolomite, gypsum, and possibly, halite were deposited in them, but were not preserved" (Fiveg 1959).

Thin sand-clay rocks are interbedded in the carbonate succession along the margin of the western third of the basin. It is possible that this increasing clastic fraction was introduced chiefly by wind from the adjacent arid continent.

If the dolomite content in the rocks reflects to some degree the salinity of the water, the limestone and dolomite distribution may be used in reconstructing the hydrodynamic regime of the sea. The influx of normally saline water from the ocean on the north clearly took place through the Pechora depression and in a broad belt to the west of Timan. The water thence moved southward, gradually becoming more saline by evaporation. This process was retarded somewhat by the stream inflow from the Ural landmass. Because of this the water remained at normal salinity along the eastern margin of the platform and in the foredeep, even far to the south. But even here, the belt of such water narrowed to the south, and at the extreme southern end became appreciably more saline. Along the western third of the basin, where dolomite accumulated, the water was notably saline (at least 4·5–5%) and, consequently, was heavier. This inevitably led to a countercurrent from the basin to the open sea, which must have flowed along the floor in the western part of the connecting strait. The oblique position of the boundary of the dolomite facies in plan is obviously supporting evidence for the existence of this bottom current. Tongues of limestone, projecting into the western dolomite from the east, are apparently the result of currents splitting off from the main eastern meridional stream of normal sea water. Inasmuch as the basin was huge and lay entirely within the arid zone, the water filling the Caspian syneclise may be expected to be more saline than the water on the north or even in the central part of the sea. Unfortunately, data for testing this view are not available.

The only zone where saline sediments accumulated in *Schwagerina* (and Tastuba) time was the northwestern Donbas and the adjoining parts of the Dnieper-Donets basin.

A characteristic feature of the saline deposits in this zone is the repeated alternation of rocks of different petrographic type. In the lower anhydrite-clay sequence beds of clay and anhydrite alternate, with the clayey units constituting at least 60–70% of the sequence. In the middle sequence—the saline sequence proper—beds of rock salt are interbedded with anhydrite and, more rarely, saliferous clays and carbonates. The most complete of these saline sequences is found near Artemovsk and Slavyansk. In the Artemovsk region this sequence is 300–350 m thick; the succession contains 22 to 26 beds of rock salt each ranging from parts of a meter to 38–40 m in thickness, and totalling 144–185 m. In the Slavyansk region the thickness of the salt sequence ranges from 266 to 386 m, averaging about 325. The total thickness of salt is approximately 190 m, corresponding to a relative salt content of 60–70%. The thickness of the salt sequence diminishes in all directions away from these two localities, and the beds of rock salt gradually decline in number and thickness. The salt sequence may be traced from southeast to northwest for at least 130–140 km, and from southwest to northeast for 80–90 km. Its actual areal extent is probably greater, but it is quite possible that at the very edges of this area the rock salt wedges out or gives way to clay-anhydrite rock. A thin saline sequence is found in the upper sand-clay gypsiferous series. It consists of alternating beds of clay, gypsum, anhydrite, and, commonly, argillaceous sandstone and carbonate rock.

The deposition of rock salt was greatest and, at the same time, the least contaminated in the Artemovsk and Slavyansk regions, where the NaCl content in the halite is about 97–98% and the insoluble residues constitute only parts of a percent. Toward the margin of this zone the halite rock becomes contaminated with $CaSO_4$ and clay. A potash deposit in the upper parts of the saline sequence in the Slavyansk dome is low grade (2–3% K according to A. A. Ivanov). A well-defined annual layering of thin clay-anhydrite layers (1–5 mm) and thicker layers of pure rock salt (8–12 cm) is seen, but this has been obliterated for the most part by recrystallization.

The paleogeography of the halogenetic region may be reconstructed in general outline. It may be recalled that the Kal'mius-Torets and Slavyansk-Bakhmut depressions now lie between anticlinal axes of the Donbas. It is very likely that in *Schwagerina* time these axes extended to the northwest and affected the relief over a much greater distance than at present. The salt-producing basins also lay between these axes, being essentially gulfs with restricted communication with the open sea. Complex rhythms of tectonic movements repeatedly modified these connections, enlarging them at times and terminating halogenesis, or sharply restricting them and re-establishing salt accumulation. It may be noted that halite accumulated in the zones of maximum downwarping.

3. The Sulfate-Dolomite Formation

The area in which the sulfate-dolomite formation accumulated corresponded closely to the *Schwagerina*—Tastuba basin (Fig. 186). Like the latter, the Sterlitamak-Artinskian basin had a well-defined north-south zoning, but the constitution of the zones were substantially different. The smallest differences appeared on the east. Along the eastern border of the Ural foredeep, as before, purely terrigenous rocks accumulated in great thickness. In the central part of the deep these deposits give way to thin bituminous marls, clays, clayey limestones with radiolarians, ammonites, and occasional fusulinids carried out from the shallower water of the western zones. This thin sequence of rocks corresponds stratigraphically to the complete and thick succession of Sterlitamak-Artinskian deposits of the near-shore zones, and, following Shamov, is considered the deep-water deposits of the foredeep. This sequence did not develop throughout the foredeep, however, but only in the deepest parts: the Chardyn'-Solikamsk and Chusovaya-Sylva depressions on

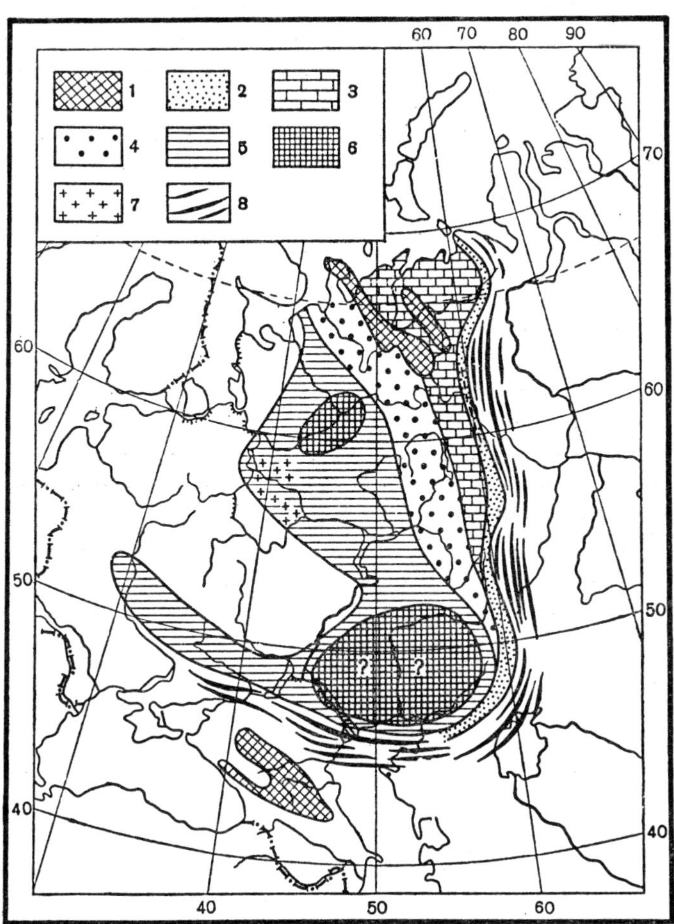

Fig. 186. Paleogeography of the Sakmarian and Artinskian epochs (from Nalivkin). 1. Positive zones; 2. near-shore deposits (fragmental); 3. limestone; 4. dolomite with gypsum; 5. dolomite interbedded with gypsum; 6. gypsum and rock salt; 7. local lenses of NaCl; 8. mountain chains.

the north and the Belaya depression on the south. Along the western border of the deep, as before, there extended a chain of reefs, faulted into blocks, found at Chusovskie Gorodka, on the Ufa Plateau, at Shikhany, and forming buried masses at Ishimbai and on to the south. Organic and detrital limestones with brachiopods, bryozoans, numerous fusulinids, sponge spicules, and algae still survived in the marginal belt of the Russian platform nearest the deep, but even here notable quantities of dolomite appear. This limestone belt attained its greatest width on the north, where it occupied the entire area of the Pechora basin between Timan and the Urals. South of the 60th parallel the belt narrowed sharply, and it wedged out quickly south of Ishimbai.

From the nature of the sediments and the organic remains it is clear that these zones are of normal or nearly normal salinity.

Immediately west of the limestone belt lies a broad zone of dolomite interbedded with calcareous dolomite, locally with limestone and with gypsum. Gypsum increases progressively toward the west. The fauna shows signs of being stunted and is restricted in number of species: they are chiefly euryhaline forms of pelecypods and gastropods. Brachiopods are scarce; stafellids replace fusulinids. It is clear that the salinity of the water in this zone was greater than that of normal sea water, and it was commonly above the limit allowing life and the development of bottom forms.

In the large western half of the basin the nature of sedimentation changes once again (Fig. 187); anhydrite and gypsum become the principal constituents, and dolomite is ubiquitous, forming beds of varying thickness but normally less thick than those of the sulfates. Gypsum-anhydrite rocks are especially well developed along the western margin of the middle part of the basin, and also in the southern part, corresponding to the trans-Volga region. Locally the succession in these districts is almost completely sulfate, with hardly a trace of dolomite.

I. N. Tikhvinskii (1960) attempted to define the controls of the variation in the anhydrite-dolomite ratio in the trans-Volga region. He showed that the principal factor was the tectonic structure of the floor, which was reflected in the floor relief during sedimentation. The greatest concentrations of gypsum and anhydrite are found in the structurally depressed segments: dolomite, with subordinate sulfate beds, accumulated chiefly on the intervening elevated areas. It is very probable that this pattern has a wider significance.

A fundamentally new factor in the sedimentation of Sterlitamak-Artinskian time was the accumulation of rock salt in different parts of the basin and in relatively large masses. Three such areas may be considered. The first is in the north, in the Dvina-Mezen basin, where the broad Dvina-Sukhona saline basin occurs. This is 300 km long from southwest to northeast, and 200 km wide. The rock salt deposits in this basin are remarkably thin (0·1–10 m), and the 5–7 beds make up 35–40% of the sequence. The rock salt is usually coarsely crystalline, and is either pure and transparent or carries inclusions—thin veinlets and laminae of red-brown and greenish grey clay and anhydrite—which locally are more abundant than the salt. In places the salt is rose-colored, though normally it is grey and brown. The NaCl content in salt samples ranges from 50 to 95%, the $CaSO_4$ content from parts of a percent to 8–10%, and the insoluble residue from 1–1·5 to 25% and more. No highly soluble salts, specifically potassic, were detected in chemical analyses.

It is very significant that *the zone where the rock salt of the Dvina-Sukhona basin accumulated* (Fig. 188) *coincides with the zone of greatest thickness of the saline formation. To put it differently, the salt accumulated in zones characterized by maximal mobility.* Still, the rock salt of the Dvina-Sukhona basin is clearly found in the central part of the salt-producing intracontinental sea (Ivanov and Levitskii 1960).

A second zone of rock salt accumulation lies at the western edge of the formation. In contrast to the area of rock salt described above, this marginal

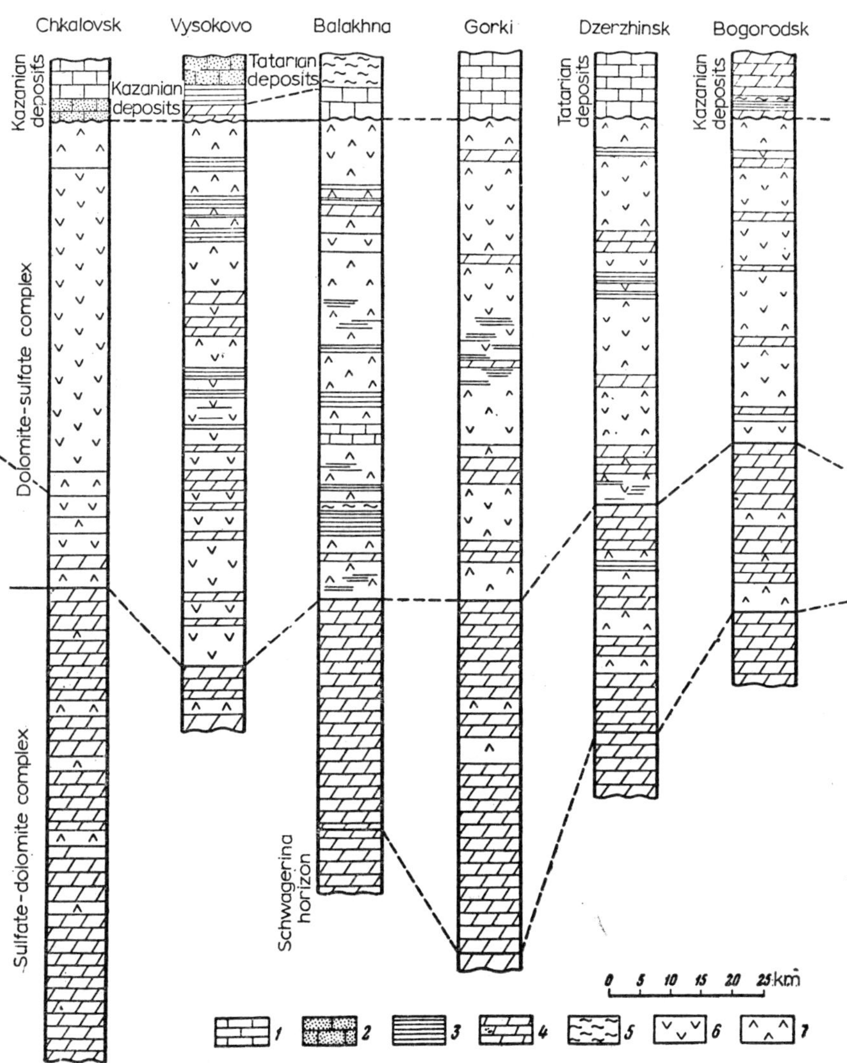

FIG. 187. Composite sections of Sakmarian-Artinskian deposits of the Russian platform (from Ivanov). 1. Limestone; 2. sandstone; 3. clay; 4. dolomite; 5. marl; 6. anhydrite; 7. gypsum.

occurrence is not continuous, but takes the form of small individual patches, genetically related to the marginal lagoonal zone, which shifted from east to west, following the episodic shifting of the shore line. "In the northern part of the Zalesskii area of the Gorki region and the adjoining Makar'ev area of

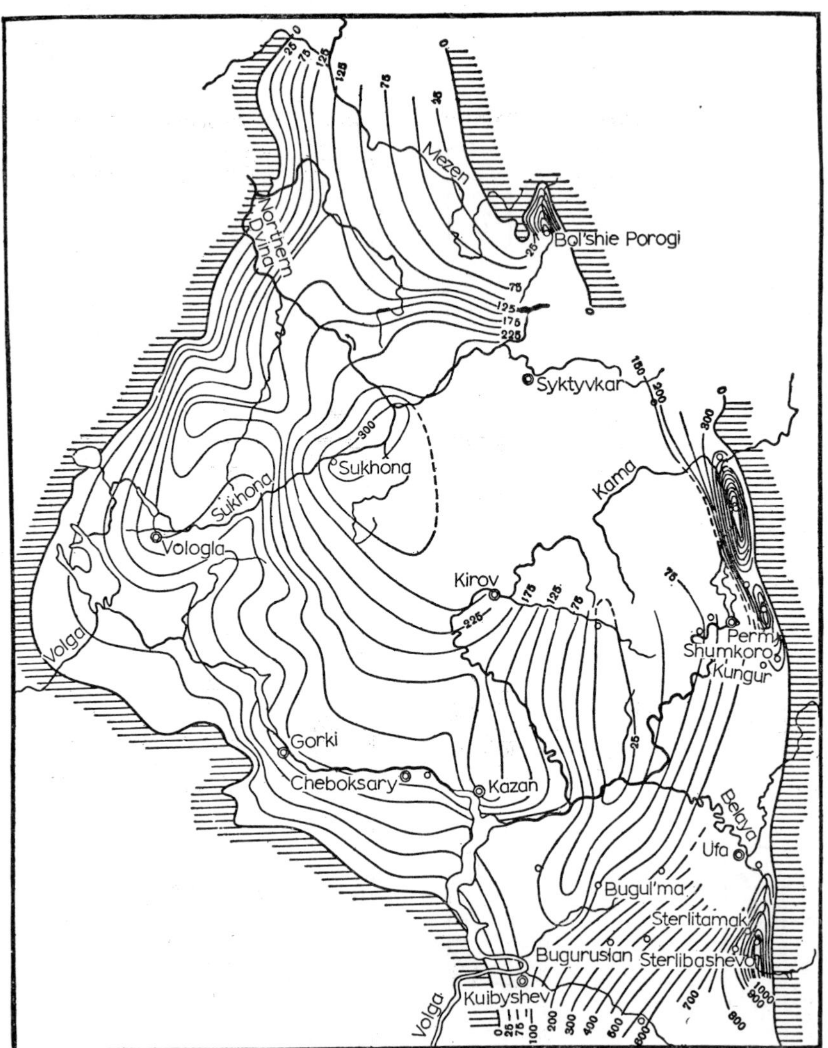

Fig. 188. Thickness (in meters) of the Lower Permian saline formations (from Ivanov). The zone of zero thickness is hachured.

the Kostroma region S. K. Nechitailo has established the presence of a salt sequence 10–50 m thick. In the Ryabov region, five beds of rock salt 1·8–7·6 m thick were found in this sequence. It appears likely that the numerous salt springs known in the Yaroslav, Ivanov, and Gorki regions and in the

Mari and Chuvash Republics are connected with this sequence" (Ivanov 1960, p. 212). Rock salt has also been found in the Kazan region and in the trans-Volga region, in the basin of the Ik River near Tuimazy. These deposits are 6–7 m thick.

The third rock salt zone is the broadest. It occurs in the Caspian basin. The largest salt masses here are undoubtedly of Kungurian age, but, as M. P. Fiveg suggested, the lower part of this vast rock salt sequence began to accumulate at the end of Artinskian time; i.e. it belongs to the above dolomite-sulfate-halite formation, as illustrated on the map. In general outlines this represents the composition and structure of the Lower Permian dolomite-sulfate-halite formation and the paleogeography of the basin in which it accumulated.

By comparing this with the composition, structure, and paleogeography of the underlying limestone-dolomite formation, it is easy to find both similarities and differences. Similarity is found in the dimensions of the basins in which the two formations accumulated, in the almost identical spatial localization, and in the uniform pattern of north-south zoning of the deposits, indicating an increase in salinity from east to west. However, the two differ substantially in composition of the sediments. The limestone-dolomite formation represents the initial pre-saline stage of arid sedimentation. The dolomite-sulfate-halite formation is the true saline formation, though the saline process did not reach culmination. From this it appears that, *while preserving the general paleogeographic pattern, the Sterlitamak-Artinskian basin differs from the Schwagerina-Tastuba basin by much higher salinity of the water, generally above 15% but at times reaching 24–27% in places where halite precipitated.* This salinity, at least at the end of the epoch in which the formation accumulated, was maximal in the blind southern end of the basin, and from here generally decreased toward the northern connecting strait. At the same time, the salinity in the western half of the sea was much higher than in the eastern half. This salinity pattern within the basin leaves no doubt that the general water circulation during the deposition of the dolomite-sulfate-halite formation was the same as during that of the limestone-dolomite. It is very significant that the character of the northern connecting strait was essentially the same during both intervals of deposition. This means that *the general increase in salinity in Sterlitamak-Artinskian time was not the result of decreased water exchange between the intracontinental sea and the ocean, but rather was the simple consequence of the constantly changing salt content of the sea, i.e. of the constant excess of salt introduced over salt removed through the strait by the bottom countercurrent.*

A characteristic feature of the saline deposition in the Sterlitamak-Artinskian basin is *the connection between composition of the saline phase and the tectonic structure and mobility of the different parts of the floor.* Not only did the thickness of saline deposition increase, but solid phases representing more saline solutions accumulated in tectonic basins of different kinds and sizes. Calcium sulfate accumulated in depressions within the zone of dolomite deposition. Rock salt accumulated in greater or less degree in the zone of sulfate deposition. In other words, *saline deposition in tectonically more actively depressed areas outstrips the deposition in more elevated and stable segments.*

Richter-Bernburg (1953) and Fiveg (1960) account for this distribution by the flow of heavier salt solutions into the tectono-geomorphic depressions from the surrounding higher areas. As a result, more soluble salts are pre-cipitated in the depressions than on the surrounding prominences. In our opinion this explanation is unacceptable. It is clearly thought by its pro-ponents that a gravity-controlled stratification of the water results from the flow of heavy solutions into the depressions; from this heavier lower water the more soluble salts are precipitated. But how, precisely, do the salts precipitate from this heavier water, even if it is saturated with some component? To accomplish this and to maintain permanent supersaturation of the solution a constant removal of the solvent—water—would be necessary. This is clearly impossible when lighter and less mineralized water overlies the heavy solu-tion. Because evaporation unavoidably takes place only from the upper light layer, having no effect on the lower layer, solid phases that collect on the floors of the depressed zones should correspond to the composition of the upper layer, not of the lower. The mechanism of precipitating heavy solid phases of salt from a bottom-flowing solution is therefore not explicable in this way. Fiveg called on the factor of cooling in the deep heavy layer of brine as a mineral-producing process, but at the sulfate stage, as at the halite stage this factor is ineffective in mineral formation. Another interpretation is clearly necessary.

The Sterlitamak-Artinskian basin on the whole was an ordinary platform, shallow-water sea. The relatively elevated stable areas were especially shallow. Deeper water was found in the actively subsiding depressions between these positive segments. Chemically precipitated phases are generally very fine-grained, particularly calcite and dolomite in the initial stages of saline develop-ment, as well as gypsum. This inevitably means that part of the chemically precipitated carbonate was agitated on the positive areas by water movement and was transported to the adjoining depressions. The thickness of carbonate sediments in these depressions thus increased at the expense of the deposits on the elevated zones. At higher stages of salinity this process of redistribu-tion led to an extensive accumulation of gypsum and anhydrite in the depressions at the expense of the higher areas. In this case the effect of another, supplementary factor is added: the seasonal, or longer-periodic, freshening of the sea. Both affected the relative thicknesses of the water layers. On the higher areas the entire mass of water was freshened, causing partial and, at times, complete re-solution of the accumulated $CaSO_4$. In the zones of greater depth, however, freshening of the water did not reach the floors. In the deeper zones gypsum was not redissolved, and in consequence thicker gypsum-anhydrite beds accumulated here than on the adjacent highs. The same pro-cess was repeated at even higher stages of salinity, when rock salt was precipi-tated. A variation of the mechanism involved wave erosion of halite grains, but the coarseness of the grains prevented any effective transport of such material from the elevated to the low areas. The unequal solution of halite, however, during freshening on the elevated zones and in the depressions, because of the great solubility of NaCl, became especially effective. Thus, halite that fell to the floor of the elevated areas was almost completely dissolved soon after its precipitation; in the depressions it was preserved. As a result, the halite deposits were restricted to the depressed zones, especially to rapidly subsiding segments, and they did not form on the elevated and more stable areas. Thus, even when the average salinity of the surface water remained uniform throughout the year, the depressed zones received more saline deposits than the shallow-water and stable zones.

4. The Kungurian Potassium-Halite Formation

Additional features appear in the composition, spatial localization, and paleogeography of the Kungurian saline formation. The principal factor governing the general character of this formation was the sharply expressed positive movements of the crust, beginning at the end of Artinskian time and continuing through the Kungurian. As a result, the sea drained from the entire region between the Baltic shield and Timan, from the entire area of the Moscow syneclise and the Tokmovo and Tatar arches. Water remained only in the Ural foredeep, the Caspian basin, and a narrow belt along the eastern part of the platform. A long, narrow, north-trending basin resulted, distinctly bottle-shaped, closed at the south end and connecting with normal marine water only through a strait in the Pechora depression, in the zone between Timan and the Urals. But even here the connection became progressively more restricted during Kungurian time. "The history of this strait derives from the following facts: known marine deposits of Sakmarian and Artinskian age occur on the western and eastern slopes of Timan; Kungurian strata are absent from the western slope, and only thin Filippovskoe beds [lower Kungurian] are known from the eastern slope. The abundance of continental rocks increases eastward over the broad area of the platform basin east of Timan. The upper part of the Vorkuta series are here continental coal-bearing rocks. Judging from these data we may conclude that the northern strait, broad at the beginning of Permian time, gradually narrowed and finally ceased to exist in the second half of Kungurian time.

"These features of the strait led to a dominantly unidirectional movement of the water into the basin, essentially compensated by loss through evaporation in the epicontinental basin. Because of the shallowness of the strait, any reverse flow of brine was small.

"With the gradual narrowing of the strait, the inflow of oceanic water decreased, practically ceasing in the second half of the Kungurian epoch. The basin on the eastern part of the Russian platform henceforth became a continental basin of the broad platform depression, isolated from the sea" (Fiveg 1960a).

These changes in paleogeography of the Kungurian basin (Fig. 189) were

2 D

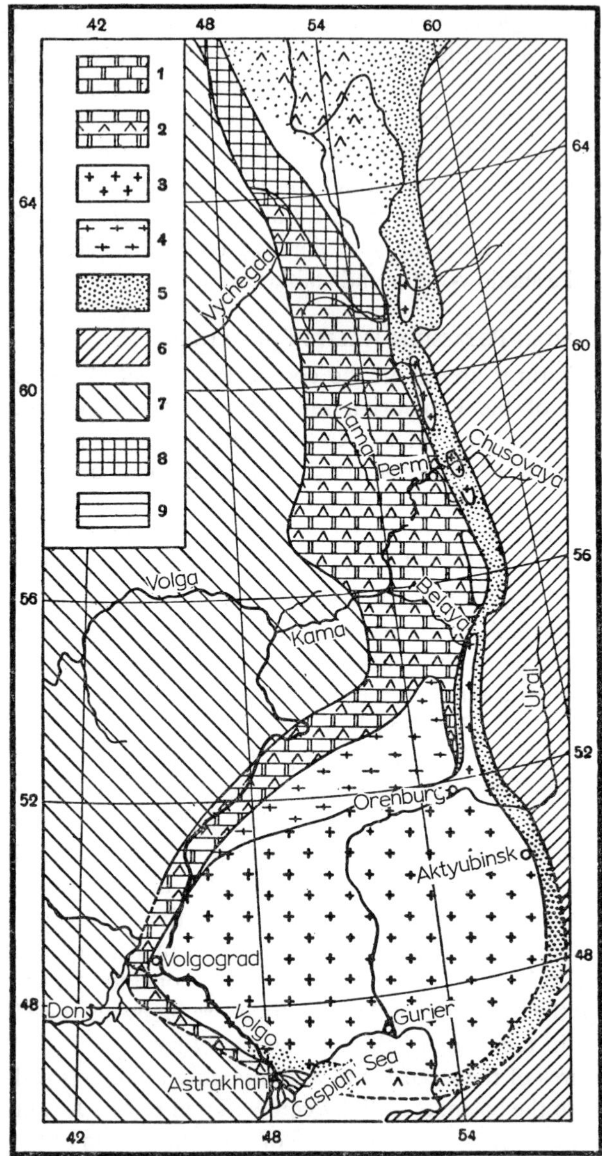

FIG. 189. Paleogeography of Late Kungurian (Irenian) time (from Fiveg). 1. Dolomite; 2. alternating beds of anhydrite and dolomite; 3. rock salt in the Ural foredeep and the Caspian depression; 4. rock salt on the southeastern slope of the platform; 5. terrigenous rock, 6. western coast of the Angara land mass; 7. western arid plain; 8. Timan island; 9. land on the eastern part of the Bashkir arch.

clearly reflected in the composition of the resulting saline formation. *Its distinctive feature as contrasted with the Sterlitamak-Artinskian formation is the sharp increase in abundance of rock salt, which now becomes the dominant rock type both in areal extent and in thickness. Furthermore, the saline process, especially in the second half of Kungurian time, went farther toward completion, forming not only potassium salts but also deposits of the eutonic stage.*

In Kungurian times, clastic sediments, sandstone, siltstone, and clay, generally grey and polymictic, with carbonate and clay cement were deposited in the east, in the Ural foredeep. Minor beds and lenses of dolomite and anhydrite are present, several meters or a few tens of meters thick. In places (e.g. the Belaya basin) gypsum forms beds of considerable thickness, up to 120–190 m. In the Aktyubinsk sector of the Ural region the clastic strata include rock salt, locally of appreciable thickness. This same facies zone, but without the rock salt, extends along the southern margin of the Caspian basin into the Ural foredeep on the southwest. In the far north, above 64° N lat, the clastic-saline assemblage gives way in the Vorkuta basin to coal-bearing deposits, corresponding to the transitional zone between arid and humid.

A huge zone of rock salt, broken into separate segments, runs parallel to the belt of clastic-sulfate deposits. The greatest area of rock salt is found in the Caspian syneclise. A complete succession of the saline formation is known from only one locality in this region, at Krasnoyarka, 40 km southeast of Orenburg. Underlying the salt are pure gray dolomitized limestones with an Artinskian foraminiferal fauna, including stafellids. Four horizons may be distinguished in the Kungurian saline formation, which rests conformably on the Artinskian. The lower unit, about 187 m thick, consists of anhydrite, which passes upward into a thick bed of salt. The next unit, 755–1500 m thick, is rock salt. Above this occurs an intermediate sulfate-saline unit, 185–350 m thick, composed chiefly of alternating beds of rock salt and anhydrite. The middle part of this unit contains a layer of halite enriched with polyhalite. In places polyhalite forms 50–85% of the bed, but most commonly it is 25–30% or less. Small amounts of sylvite, up to 4–6% also occur; kainite was found in one sample (up to 8%), and glauberite has also been recorded (Ivanov 1960). The upper horizon of the saline formation is chiefly anhydrite and gypsum (50–210 m thick).

The saline deposits that floor the Caspian syneclise are known only in their upper part, corresponding to the upper part of the Krasnoyarka succession. Rock salt is dominant; anhydrite is markedly subordinate and dolomite almost completely absent. "Rock salt in most of the cored sequences is relatively uniform. It is crystalline-granular and in most places exhibits well-defined banding with alternating, annual, light (5–6 cm) and dark (1–1·5 cm and more) layers, contaminated by anhydrite and argillaceous material. The salt is most commonly light grey and pale in tint. In places, e.g. the Azinskii structure, most of the salt is rose-colored, and in some localities it is yellow, red, and even dark grey and white. Occasional pockets of rock salt are composed of large colorless and transparent halite crystals. The NaCl content in the salt is above 90%. The content of the principal impurity, anhydrite, ranges up to 4%, and even more, in varieties of salt transitional to saliferous anhydrite. Insoluble residues (carbonate-clay material) normally range from traces to 3% and more in varieties of argillaceous salt. $CaCl_2$, $MgCl_2$ and $MgSO_4$ are either absent or are less than 1%" (Ivanov 1960, p. 176).

The Kungurian saline deposits of the Caspian depression are strongly deformed by salt tectonics. The primary bedded salts have developed diapiric salt domes, either folding or fracturing the overlying beds. In the Volga-Ural-Emba salt-dome field the domes number several hundred, of which only ten have as yet been drilled. Geophysical data indicate that the thickness of the salt in the cores of the domes ranges from 2200 to 8900 m in the southeastern part of the Caspian syneclise, increasing from southeast to northwest. The greatest thickness of salt is found in the vicinity of Lake Inder. These facts emphasize the vast amount of rock salt that accumulated in Kungurian time in the Caspian syneclise.

Potassium salts occur with the rock salt in the upper parts of the saline formation of the Caspian basin, in the Krasnoyarka, Greben', Nezhino, Dzhuaktyubinsk, Zhilyanskii, Bol'she Sulak, Ilets, Ozinskii, Chelkar, Inder, Ashchebulak, Andzhar, Sagiz, Novobogatinskoe, Akat-kul', and other structures. Although most of the discoveries have been made in the northern, eastern, and southeastern parts of the syneclise, it is likely that potassium salts are also present in the western part of the syneclise (Lakes Baskunchak and Elton) and, consequently are characteristic of the entire syneclise. Since the salt in the drilled cores of the salt domes is highly deformed, it is not possible to detect the stratigraphic positions of the potassium horizons, much less to correlate the successions. It is, however, clear that the potassium horizons occur in the upper parts of the formation. The potash-bearing rocks are mineralogically complex and include sylvite, locally in considerable quantities, carnallite (much more rarely), polyhalite, kainite, and kieserite. The last mineral occurs frequently as a secondary replacement. In the Azgir vicinity (Chapchachi), thenardite has been found in the upper parts of the salt sequence.

Potassium beds in the Caspian syneclise thus contain abundant sulfate minerals. The distribution of the polyhalite rock is worthy of note. Thick beds and lenses of polyhalite are known to occur in the salt deposits in the northeastern and eastern part of the Caspian basin, the polyhalite content ranging up to 60–80% and more. In the central part of the basin (Inder) polyhalite occurs only in veinlets of secondary origin. Some structures show a vertical change from one type of potassium rock to another. In the lower part of the succession in the Zhilyanskii structure thick beds of polyhalite rock occur, succeeded by rock salt and then sylvinite. Elsewhere the sequence may be different, even reversed.

North of the Caspian depression the saline deposits form two tongues: Sterlibashevo and Ishimbai (Belaya). The first is very wide and occurs in the marginal, most strongly downwarped, southern part of the trans-Volga region. The saline deposits, here 700–750 m thick, are complex. At the base, resting nonconformably on the Artinskian beds, occurs 40–60 m of dolomite with thin interbedded anhydrite. This is succeeded by a thick sequence (280–290 m) of anhydrite with thin layers of dolomite and clay, followed by an anhydrite-salt sequence 300–400 m thick, consisting of thick beds of rock salt alternating with thinner beds of anhydrite. The succession is capped by a clay-anhydrite sequence 32–59 m thick. Of the total thickness of this formation, rock salt makes up 25–35%; the rest is anhydrite. In this respect the Sterlibashevo tongue differs fundamentally from the saline deposits of the Caspian syneclise.

The rock salt in this tongue contains considerable amounts of clay and anhydrite. The NaCl content ranges from 50 to 98%, averaging 77–91%. $CaSO_4$ ranges from 2·4 to 7·5%, and the clay from 1·5 to 14·5%. Potassium salts are present in the upper part of the rock salt sequence, and are represented chiefly by sylvite and, in possible rare occurrences, by carnallite. The potassium content is normally very small.

Since rock salt appears to be confined to the higher horizons in the Sterlibashevo tongue, *they correspond stratigraphically to the upper part of the saline sequence in the Caspian syneclise.* In other words, at the end of Kungurian time the area of the rock salt facies at the south of the basin clearly grew larger. This is strongly supported by the increasing isolation of the basin and by the progressive increase in salinity.

The second tongue of saline deposits corresponding to the rock salt of the Caspian syneclise is in the Belaya basin (Figs. 190 and 191), where three horizons are recognized: salt, gypsum-anhydrite, and transitional (Strakhov 1947). The salt horizon in the buried parts of the downwarp consists chiefly of rock salt and interbedded anhydrite. Locally saliferous grey clays occur. The thickness of this horizon ranges from 500–600 to 1000–1200 m. The largest deposits of rock salt reach a thickness of 500–600 m, and they are separated by anhydrite beds, which may reach a thickness of 140 m. The relative salt content in the central parts of the downwarp ranges from 48 to 98%, averaging 68%. On the western border it decreases to 25–35%. Polyhalite appears in grains and small aggregates, disseminated through the salt sequence at several isolated beds of lenses. One such occurrence is in the Yarbishkadak district. The polyhalite is rose-colored, brownish red, orange, or rose-grey. It ranges from a few percent to 48% in the salt samples. No other potassium minerals have been identified.

The gypsum-anhydrite bed overlying the salt sequence is characterized by great uniformity in composition, being made up exclusively of anhydrite, or of anhydrite and gypsum. The latter occurs in the upper parts of the bed, and is epigenetic. In thickness the bed ranges from 80 to 548 m, depending on the tectonism of the site.

The transitional sequence is a red argillaceous series, the lower horizons of which contain abundant sulfate.

Since the rock salt in the Belaya basin occurs in the lowest part of the succession, it clearly began to accumulate earlier than in the Sterlibashevo

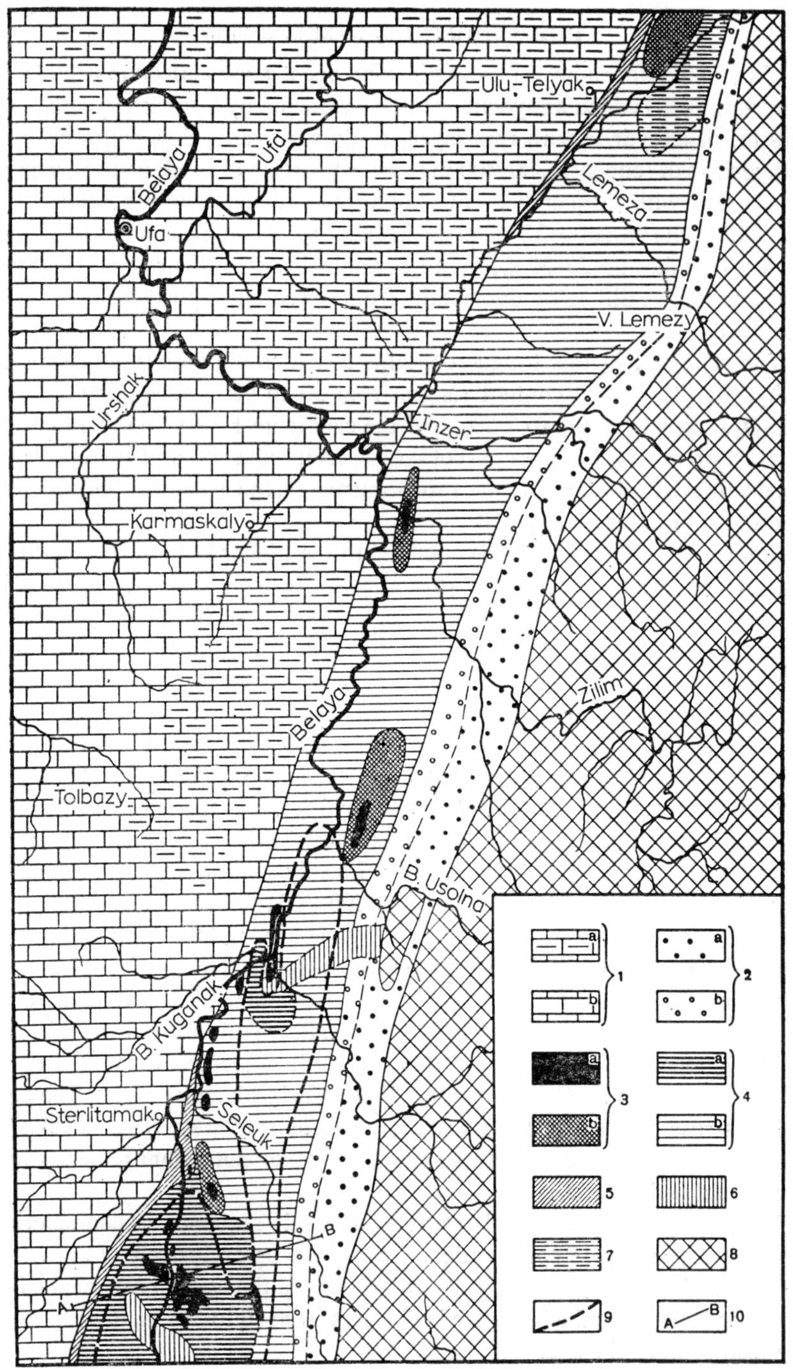

FIG. 190. Principal types of Kungurian deposits in the Bashkir segment of the Ural region. 1. Central (western) type: (a) outcrops, (b) inferred distribution; 1. eastern type: (a) outcrops, (b) inferred distribution; 3. supra-massif type: (a) outcrops, (b) inferred distribution; 4. salt-bed type: (a) outcrops, (b) inferred distribution; 5. transitional Allaguvatskii type; 6. Allakaevskii type; 7. Kazyakskii facies of saline type; 8. pre-Kungurian basement 9. boundary of thick salt bed; 10. locations of cross section.

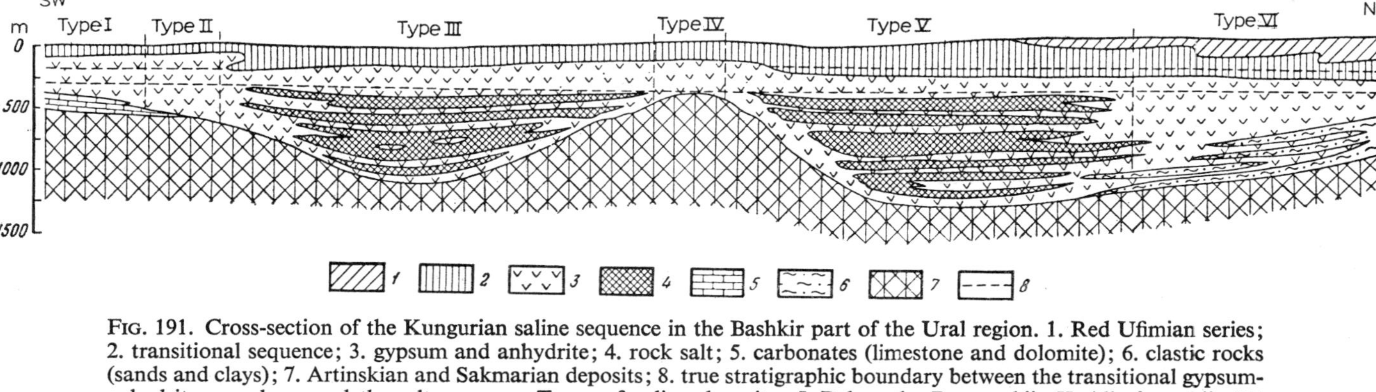

FIG. 191. Cross-section of the Kungurian saline sequence in the Bashkir part of the Ural region. 1. Red Ufimian series; 2. transitional sequence; 3. gypsum and anhydrite; 4. rock salt; 5. carbonates (limestone and dolomite); 6. clastic rocks (sands and clays); 7. Artinskian and Sakmarian deposits; 8. true stratigraphic boundary between the transitional gypsum-anhydrite complexes and the salt sequence. Types of saline deposits: *I.* Pokrovsko-Ryazanskii; *II.* Allagkaevskii; *III.* salt; *IV.* supra-massif; *V.* eastern.

tongue, possibly at the same time as the beginning of halite deposition in the Caspian syneclise or only slightly later.

Several other small but widely separated salt deposits are found in the northern part of the Ural foredeep. The largest occupies the Cherdyn'-Solikamsk depression and extends southward into the Sylva-Chusovaya basin; the smallest lies in the Pechora basin (Ivanov and Lebitskii 1960).

The deposits of the Cherdyn'-Solikamsk depression have been most thoroughly studied (Figs. 192–193). Here the Kungurian saline formation begins with a clay-anhydrite sequence 300–380 m thick, consisting chiefly of clay-carbonate rock with subordinate anhydrite, though locally these proportions may be reversed. Thin beds of rock salt appear in the uppermost part of this sequence. This clay-anhydrite unit is succeeded by the so-called lower rock salt 260–460 m thick, above which lie the potassium salts, 100–110 m thick.

According to Fiveg (1948), the rock salt is a uniform light grey throughout. Only in the uppermost beds, near the contact with sylvinite, is it locally light yellow, or rose-colored. The entire mass of salt is banded by alternating annual layers, one of granular or cryptocrystalline halite (6–10 cm), the other thinner, of anhydrite-clay-carbonate rock (1–2 mm). The salt is characterized by pinnate structure, the product of crystal growth of the halite, which points to a primary sedimentational origin (Dubinina 1951a).

The potash salts occur on two horizons: sylvinite below and carnallite above.

The sylvite horizon is composed of a series of beds of red sylvite, normally 5–7 in number, alternating with beds of gray or colorless rock salt. Each sylvinite bed is itself complex: grains of sylvite and halite are commonly isolated in independent beds of annual accumulation, with a thin clay-anhydrite layer at the base, followed by halite and then by sylvite. This sequence is most clearly developed in the upper part of the sylvite bed. Facies change along the strike from sylvinite to rock salt is found throughout the sylvinite zone. This is manifested by the persistence of some beds of sylvinite which remain practically unchanged in thickness or structural features but change in composition to rock salt with large fractions of anhydrite and carbonate-clay material. According to Ivanov and Levitskii (1960), these transitions are probably related genetically to primary sedimentational conditions of salt crystallization in the basin. They have a regional character, but their areal distribution has not yet been defined. The thickness of the zone of red sylvite rock is about 25 m on average. The thickness of individual sylvite beds ranges from 0·5 to 10 m, which are separated by rock salt in beds ranging from 1 to 3–4·5 m in thickness.

The carnallite zone consists of alternating beds of carnallite rock and halite. The zone is sharply deformed tectonically, and commonly brecciated, which makes interpretation difficult. The principal components of the carnallite rock are carnallite and halite, the former occurring in red, orange or brown crystalline grains up to several centimeters in diameter. The halite

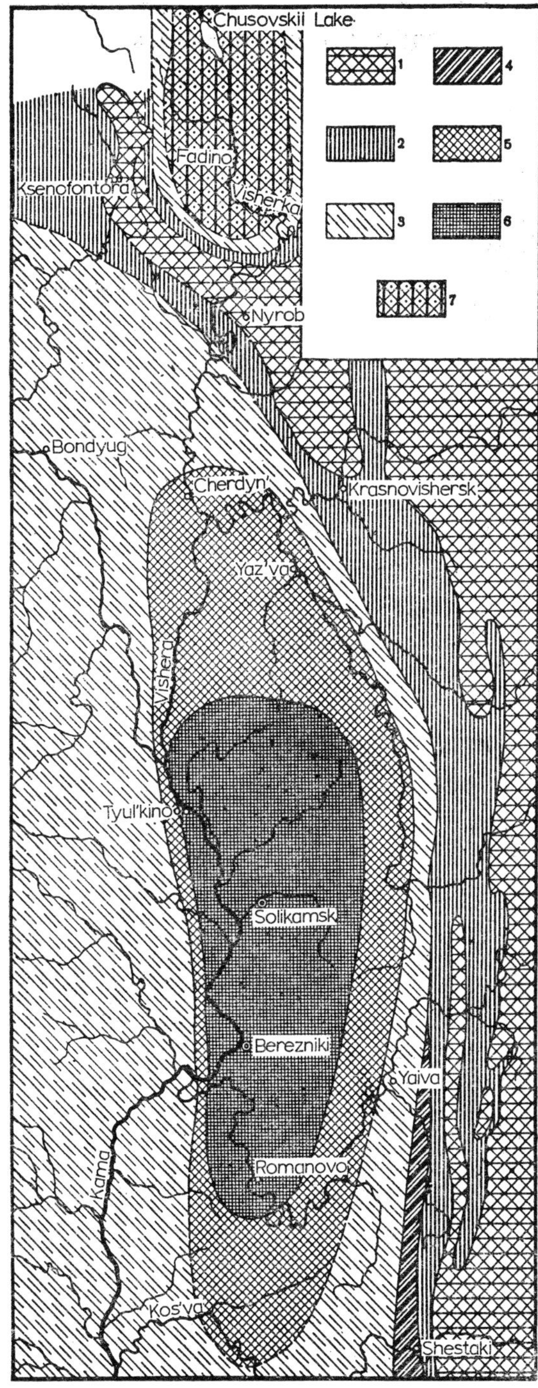

FIG. 192. Distribution of the Solikamsk potassium sequence (from Ivanov).
1. Devonian, Carboniferous, and Lower Artinskian (carbonate) deposits;
2. Upper Artinskian (terrigenous) deposits; 3. clay-anhydrite sequence;
4. facies correlatives of the clay-anhydrite sequence (marginal marl and
clay-sand facies with gypsum and anhydrite; 5. underlying rock salt;
6. sequence of potassium salts; 7. salt sequence in the Visherki basin.

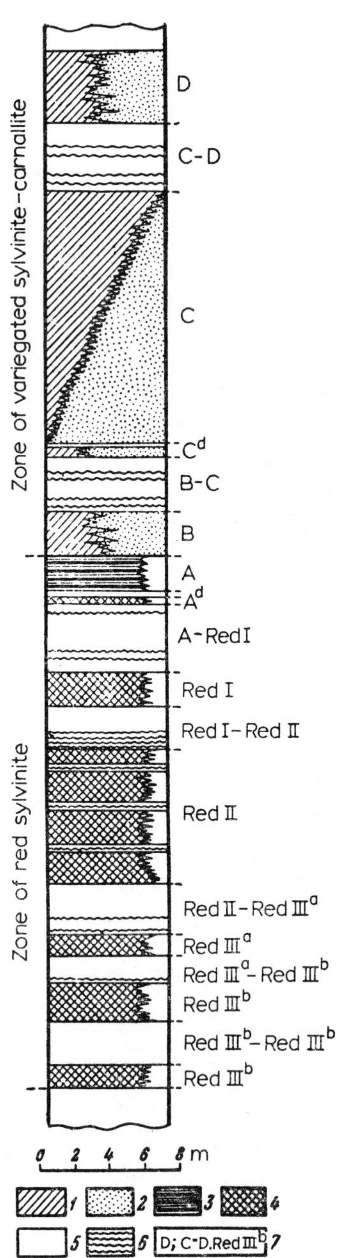

Fig. 193. The potassium-salt horizon of the Upper Kama deposit (from Ivanov). 1. Carnallite rock; 2. variegated sylvite; 3. banded sylvite; 4. red sylvite; 5. rock salt, 6. rock salt with inter-bedded clay: 7. marker beds.

grains, colorless or dirty grey, occur in beds and laminae in the carnallite rock, and also in isolated aggregates of various sizes. Seasonal bedding is usually ill-defined.

The total thickness of the carnallite zone is 75–85 m. In the lower parts of the zone it normally gives way to the so-called variegated sylvinite. All the features of the carnallite rock, seasonal bedding in particular, are generally preserved in this transition. According to Ivanov, Dubinina, Vakhrameeva, and others, this transition is a primary sedimentational facies change from one type of bottom precipitate to another. In the uppermost carnallite zone the so-called upper sylvinite is found, and is widely preserved at the highest horizons of the formation, though the thickness of the overlying rock salt is small. In contrast to the variegated sylvinite, "the upper sylvinite is chiefly epigenetic and was formed by hydrometamorphic processes through the introduction of unsaturated magnesium chloride solutions on the carnallite rock" (Ivanov 1960, p. 137).

The saline formation is capped by rock salt and a transitional sequence. The latter is mainly anhydrite and carbonate, with abundant clay, rock salt occurring only in thin beds.

The most extensive horizon of the saline formation is the basal clay-anhydrite sequence (Fig. 192). It extends far beyond the border of the Upper Kama basin, over the Russian platform to the west and the Ural foredeep to the north and south. The overlying rock salt covers a much smaller area, 6000–6500 km^2, and only within the foredeep. Potassium-salt deposits occupy a still smaller area, only 2800–3000 km^2 (Ivanov and Levitskii 1960). In addition to these features, the saline formation as a whole exhibits notable changes along its margin. On the northeastern and eastern edges the thickness of the formation decreases and the saline formation is replaced chiefly by clay-marl and clay-sand sediments, with only local development of gypsum and anhydrite deposits. Along the northern and western edges of the formation, the underlying rock salt decreases in thickness and wedges out. The content of potassium salts in the potash horizons is reduced in some parts of the marginal zone, though the average thickness may be preserved. Elsewhere the thickness may decrease considerably though the high content of potassium is preserved. In still other places, either the carnallite zone or the sylvinite zone may disappear completely.

South of the Cherdyn'-Solikamsk basin the saline deposits continue into the Sylva-Chusovaya depression, but they become thinner and lose their potassium-salt horizons. A recently discovered upper Pechora saline basin, north of the Upper Kama basin, is very similar in its makeup to that of the Cherdyn'-Solikamsk depression. Although the salt facies proper throughout the northern basins appeared rather later in the sequence than elsewhere they are younger than the rock salt sequence in the Caspian syneclise.

On the whole, Kungurian halogenesis summarized as follows. Salt deposits first accumulated in the Caspian syneclise at the end of Artinskian time, and continued into the Kungurian. Somewhat later, salt deposits were deposited in the Belaya basin; and considerably later they formed in the Sterlibashevo tongue. At this time, or slightly later, salt deposits accumulated in the northern parts of the Ural foredeep. In other words, the area of salt accumulation gradually expanded from south to north. West of the salt zone, where the Kungurian basin was most restricted in size, the saline formations are of substantially different form. Anhydrite accumulated here, with alternating beds of dolomite, in some places locally barren, in other places containing an appreciable fauna, numerically large, specifically small. The total thickness of these deposits is now 100–150 m.

The Kungurian saline formation as a whole is highly distinctive. First, *the rock salt facies is outstandingly well developed, in area and in total mass, among the other sediments.* At the same time, the distribution of types of sediments shows *a very clear asymmetry in the facies zone, both from east to west and from south to north.* This complex asymmetrical distribution of the deposits is supplemented by *a well-defined connection between the different facies types of saline deposits and the tectonism of the floor.* The dolomite-anhydrite (western) facies formed on a relatively stable, only slightly down-warped floor, and the beds are therefore thin. The halite deposits developed over tectonically more mobile structures—the Cherdyn'-Solikamsk basin in the Ural foredeep and similar depressed zones—and the deposits are consequently much thicker. Potassium-salt rocks accumulated in the most mobile and most intensely downwarped zones. Finally, *the saline process reached a very advanced stage at times, yielding eutonic sediments such as bischofite in some zones, especially in the Caspian syneclise.*

5. EVOLUTION OF THE LOWER PERMIAN SALINE DEPOSITS ON THE RUSSIAN PLATFORM AND IN THE URAL FOREDEEP

The compositions and sequences of the limestone-dolomite, dolomite-anhydrite-halite, and potassium-halite formations provide a clear demonstration of the evolutionary trend of Early Permian halogenic deposits in the basins of the Russian platform and of the Ural foredeep. The *limestone-dolomite formation* represents the initial stage of arid-climate lithogenesis, when the relatively insoluble carbonates calcite and dolomite were precipitated almost exclusively, partly chemically and partly biologically. The *dolomite-anhydrite-halite formation*, of similar distribution, was characterized by the precipitation of easily soluble compounds : gypsum and halite. This was therefore the initial stage of the saline formation, although the saline process was still far from complete; the precipitation of potassium salts had scarcely begun (the Donbas), and there were no eutonic phases. *The sedimentational transition to the saline stage took place with no perceptible change in the shape of the basin or the nature of the connection with the open sea.* The increase in salinity resulted from prolonged salt increment in the vicinity of the strait, which took place when the inflow of surface water from the ocean introduced more salt from year to year than was removed by outflowing bottom currents.

The progressive accumulation of these salts in a basin that did not change in size, shape, or depth led to a change in sedimentation.

The potassium-halite formation represents more complete evaporation. This is expressed first in the much wider development of rock salt than in the preceding formations, and secondly in the greater completeness of the saline process, which reaches the stage of massive deposition of potassium salts and, in places, to the formation of eutonic minerals (bischofite and others). The change to this higher saline stage took place during shrinkage of the basin, involving narrowing and shoaling of the gulf, which led to sharp decrease or even to complete cessation of effluent bottom currents. This could only lead to an even more pronounced increase in the salinity of the water in the basin. *The marine basin of essentially normal salinity in Schwagerina time thus changed progressively to a salt-producing intracontinental sea with steadily concentrating brine, a type of basin that is entirely unknown in the present geological epoch.*

Despite the well-defined evolutionary trend of the saline sedimentary process, some features that each succeeding epoch inherited from the preceding may be noted, which consequently appear in the structure of all three Lower Permian formations. One of these features is the *variously expressed distributional asymmetry of the different types of sediments.* In the limestone-dolomite formation this involved a change from clastic deposits along the eastern margin of the formation to limestone in the central parts and to dolomite along the western margin, reflecting the increasing mineralization of the water from east to west. Asymmetry of sedimentation along the length of the basin was much less strongly developed. In the dolomite-anhydrite-halite formation, not only is asymmetry pronounced from east to west, but it is also well defined along the length of the basin, from south to north. It is especially striking where the initial deposition of rock salt in the Caspian syneclise gives way northward to synchronous gypsum and dolomite, the gypsum first becoming especially abundant in the trans-Volga region. In the potassium-halite Kungurian formation both types of asymmetry, north-south and east-west, reach their maxima.

In the structure of the Lower Permian saline formations the connection between different types of saline sediments and tectonic activity of the floor constantly appears. This is most pronounced in the restriction of rock salt and potassium salts to zones that were most actively downwarped. Within these zones the potassium salts are confined to segments that were outstandingly mobile. This is an expression of the tectonic control over the hydrodynamic behaviour of the water and the seasonal, or perhaps longer, episodic freshening of the water in the basin.

The combination of a well-defined evolutionary trend in halogenesis with a complex asymmetry in distribution of sedimentary types and also with a constant influence of different tectonic mobility in individual parts of the basin has given the Lower Permian saline sedimentation the complexity apparent even on tye paleogeographic maps of the various stages of the basin's development.

B. UPPER PERMIAN SALINE FORMATIONS OF EUROPE

1. CONDITIONS OF HALOGENESIS

The Late Permian salt deposition in Europe was much more complex. It was restricted to two isolated regions. One lay in Western Europe, in the Central European (North German) syneclise, and the second was on the eastern part of the Russian platform, slightly west of that area where salt accumulated in Early Permian times. In both these regions of Late Permian deposition the salt-producing seas developed as a result of large-scale transgression of a boreal basin. They had the form of huge shallow-water marine basins that were supplied with water from the open sea from only the north, and were closed far to the south (Fig. 194). These were intracontinental basins,

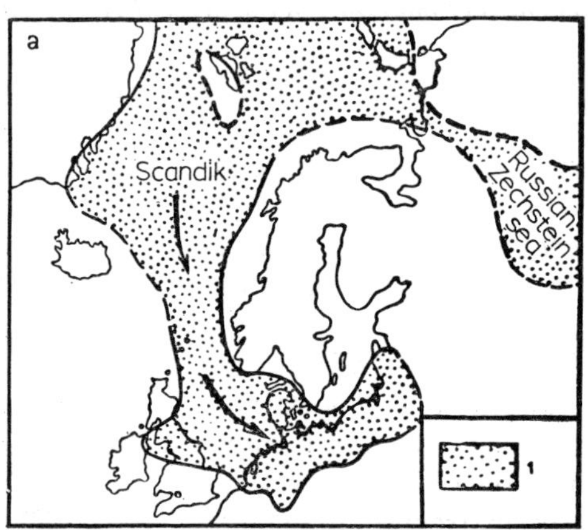

FIG. 194. General paleogeography of Zechstein time (from Stille, 1932). 1. Sea. Arrows indicate direction of sea advance.

having all the principal features of the Early Permian salt-producing basins of the Russian platform.

The North German depression, containing the Zechstein salt producing basin, developed in part on Hercynian structures, partly on Caledonian (England), and, to some extent, on the marginal zone of the Russian platform. Along the northern and eastern edge of the basin lay a low arid platform-continent, which supplied neither marked volumes of water nor clastic material. Newly uplifted mountain chains framed the basin on the south and southwest. Here a humid or semihumid climatic zone developed at greater altitudes, and from here the supply of water and of fragmental material was therefore much greater.

The basin of the Russian platform was also characterized by a difference in its drainage areas. A high Hercynian folded region lay to the east, from which there undoubtedly came abundant water and clastic material. On the west lay the same flat arid platform-continent that framed the Zechstein basin. The supply of water and clastics from this region was slight.

The two intracontinental basins developed at the beginning of the Late

Permian and they continued to exist throughout the remainder of the Permian, though their outlines were modified by tectonic movements, and their salinities varied. These changes were reflected in the character of the sedimentation. Because of these conditions each basin gave rise not to one but to several formations, differing from each other in composition and, partly, in structure. Four such formations occur in the Zechstein basin, termed by Richter-Bcrnburg (1955) Z_1 (Werra formation), Z_2 (Stassfurt formation), Z_3 (Riedel formation), and Z_4 (Upper Zechstein formation). Two formations accumulated in the basin of the Russian platform: Lower Kazanian limestone-dolomite and Upper Kazanian dolomite-sulfate-halite. These will be examined in turn.

2. ZECHSTEIN SALINE FORMATIONS OF WESTERN EUROPE

Figure 195 shows a generalized succession of the Zechstein saline formations of Western Europe (Richter-Bernburg 1953a). The lower (Werra) formation begins with a thin Zechstein conglomerate resting on underlying rocks that range in age from the Rothliegende (Lower Permian) to the Silurian. This is succeeded by the Kupferschiefer, already described in this volume (Part 1, Chap. 2), which gives way to the so-called Zechstein limestone, rich in brachiopods, bryozoans, pelecypods, and other forms. This in its turn is succeeded by anhydrite and rock salt, with two beds of potassium salts. An upper anhydrite caps the section.

A distinctive feature of the Werra formation as compared with the overlying formations is the distribution of its saline rocks over the area of the basin (Fig. 196). Along the northern and southern margins of the basin occur zones of variable width, in which sedimentation was restricted to carbonate and sulfate deposition. The central part of the basin, of very irregular outline, was characterized by the dominant accumulation of anhydrite, but with a thin bed of dolomite at the base. The sulfate exceeds 150 m in thickness. Lenses of rock salt are frequently encountered in the anhydrite zone, as in the vicinity of Sangerhausen in Saxony, Wschowa on the Oder (Poland), and elsewhere. The thickness of the rock salt in the Wschowa area reaches 101 m. It is interesting to note that the carbonate rocks and anhydrite that accumulated along the margins and in the center differ lithologically. The carbonate rocks along the margin are dense, light-colored, bioclastic and locally oolitic. Off-shore the rock passes to thin-bedded bituminous marl, with the disappearance of shell débris. Apparently these bituminous marls formed in somewhat deeper water than the bioclastic limestones. The anhydrite exhibits similar changes. The thickest anhydrite beds, light-colored, massive and locally veined lie in the inshore belt. Thin-bedded bituminous varieties, that are banded and locally include interbedded dolomite, cover most of the central part of the basin and were also apparently deposited in somewhat deeper water than the marginal type. Potassium-salt rocks formed during Werra time along the southwestern margin of the basin, within the Lower Rhine (southern Ems region), Werra, and Fulda depressions.

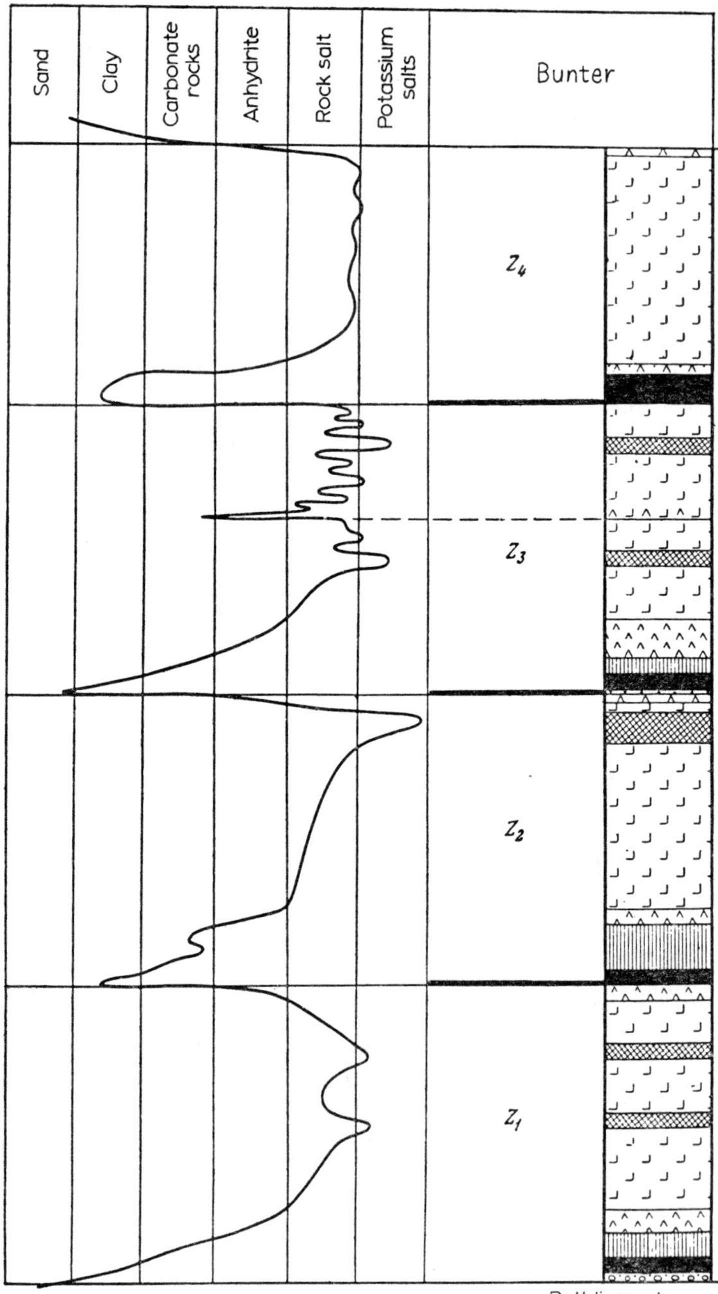

FIG. 195. The Zechstein saline formations of Germany (from Richter-Bernburg). *Formations*: Z_1. Werra; Z_2. Stassfurt; Z_3. Riedel; Z_4. Upper Zechstein.

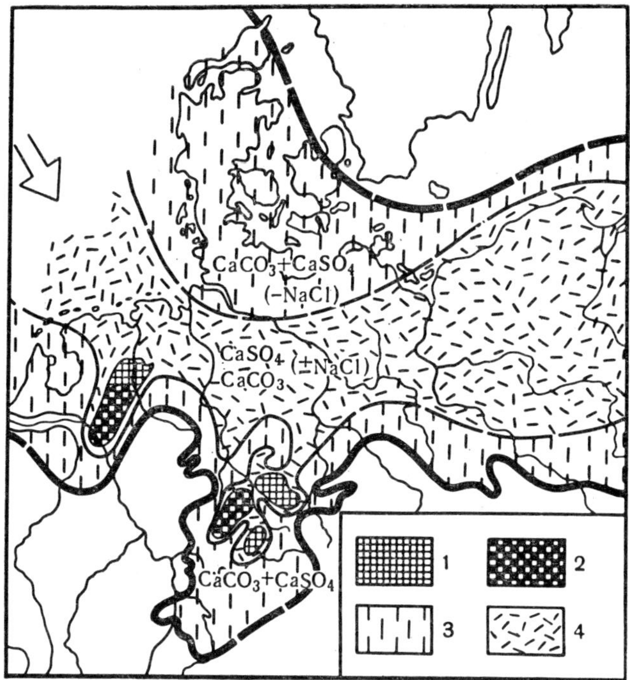

FIG. 196. Paleogeography in Werra times (from Richter-Bernburg).
1. NaCl; 2. K-Mg salts; 3. $CaCO_3 + CaSO_4$ (−NaCl); 4. $CaSO_4 +$
(±NaCl)−$CaCO_3$.

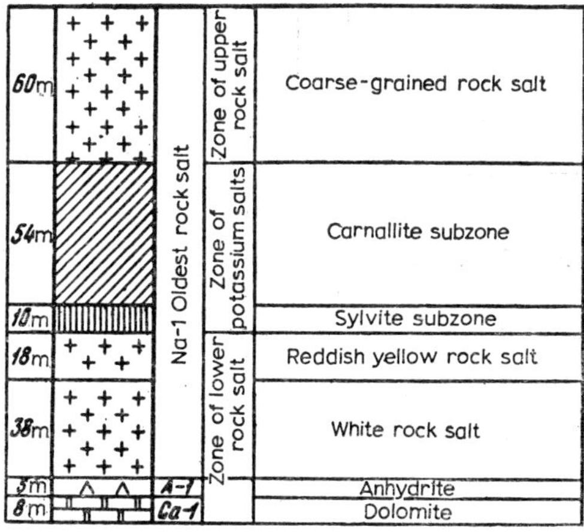

FIG. 197. The saline sequence of the Werra formation in the
Lower Rhine basin (from Fiveg).

2 E

The saline sequence in the Lower Rhine depression (Fig. 197) occurs over an area of 320 km², and is up to 476 m thick. At the base of the succession is a white inequigranular rock salt with disseminated water-clear crystals of halite (98% NaCl). Above this lies reddish yellow salt, succeeded by a zone of potassium salts with a total thickness of 64 m. This includes subzones of so-called "Hartsalz" and carnallite. The first consists of two beds of Hartsalz (a mixture of 20% sylvite, 37% kieserite, and 43% halite), each 1–2 m

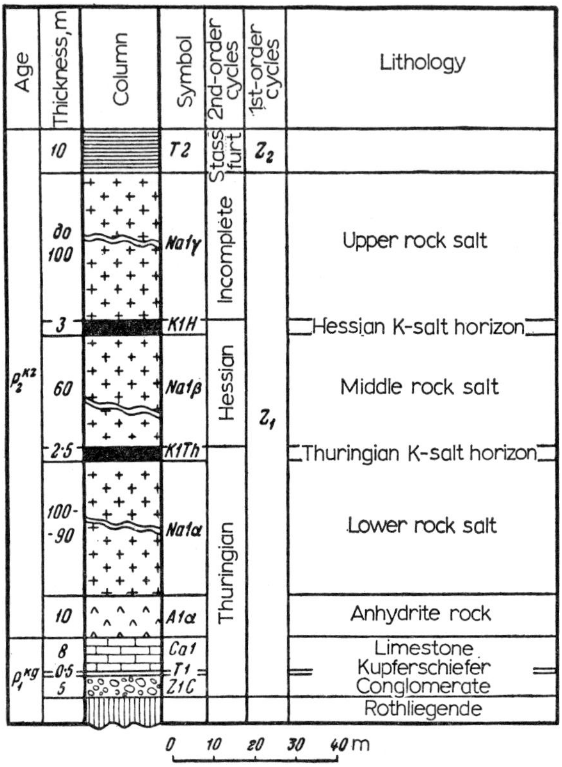

FIG. 198. The salt sequence of the Werra formation in the Werra region (from Fiveg).

thick. The carnallite subzone consists of rock salt with numerous layers of carnallite rock (90% carnallite, 4% kieserite, 1% anhydrite, and 5% halite). The succession is capped by coarse-grained rock salt. The potassium-salt rocks are developed only in the central part of the Lower Rhine basin.

In the Werra basin the saline succession is different (Fig. 198). It is 250–260 m thick and consists of three units: lower, middle, and upper rock salt. Between these occur two beds of potassium salts, the lower 2·5 m thick, the upper 3 m.

Thus, the Lower Rhine basin and the Werra and Fulda basins were traps, accumulating highly soluble salts on their floors. It should be emphasized that the thickness of the Werra formation in these basins was maximal relative to the other parts of the main Zechstein basin. *The accumulation of salt took place only in the parts of the basin with greatest mobility, particularly with the most intensely subsiding floor.*

The general structure of the Stassfurt formation resembles that of the Werra formation. A basal thin bed of red-brown saliferous clay is followed by a sequence consisting predominantly of carbonates, known as the *main dolomite* (Hauptdolomit). At the margins of the formation this is a thick series of massive, locally oolitic dolomite, but in the center the thickness diminishes, through displacement by anhydrite. At the same time the dolomite is enriched in organic material, becomes finely banded, and grades into a fetid bituminous dolomite up to 4–6 m thick. Upward the carbonate horizon gives way to the sequence of so-called *basal anhydrite* (Basalanhydrit). About the margins of the basin this ranges up to 100 m in thickness, decreasing to 3–5 m in the centre because of its displacement by halite. Furthermore, the massive marginal character changes to a thin-bedded structure, and the rock becomes rich in organic material. It is characteristic that thin-bedded bituminous varieties, both of the main dolomite and of the basal anhydrite, almost coincide in areal distribution, and cover a greater area than similar facies in the underlying Werra formation. If these deposits are regarded as being of relatively deeper water origin, it follows that the area of deep water increased somewhat in Strassfurt time.

The so-called older rock salt, up to 500 m thick, rests on the basal anhydrite. Both in areal distribution and in localization within the basin, the Stassfurt salt differs sharply from the rock salts of the Werra formation. *This salt is not restricted to small depressions around the margin, but it covers the central part of the basin, over an area of about 350,000 km², extending beyond the limits of the bituminous facies of the underlying horizons* (Figs. 199 and 200).

The rock salt deposits of the Stassfurt formation are surpassed in area only by the rock salt zone of the Lower Cambrian Usol'e basin and the Kungurian deposit in the Caspian syneclise. The older rock salt is predominantly grey and usually shows a well-defined seasonal bedding. Impurities are chiefly anhydrite, which, with the insoluble residues, increases toward the margin of the basin.

The older rock salt varies in thickness chiefly by local redistribution as a result of salt tectonics. The greatest thicknesses are found in the middle part of the formation, i.e. in the central part of the depositional basin.

The upper part of the Stassfurt formation is the Stassfurt horizon of potassium salts, with a total thickness of 25 m (Fig. 201). The eastern and northern boundaries of this horizon are not clearly defined, but the other boundaries are known. The known area occupied by this horizon is 80,000 km². It is highly probable, however, that potassium salts were precipitated over an area of the order of 100,000 km² (Lotze 1938, p. 486), i.e. approximately

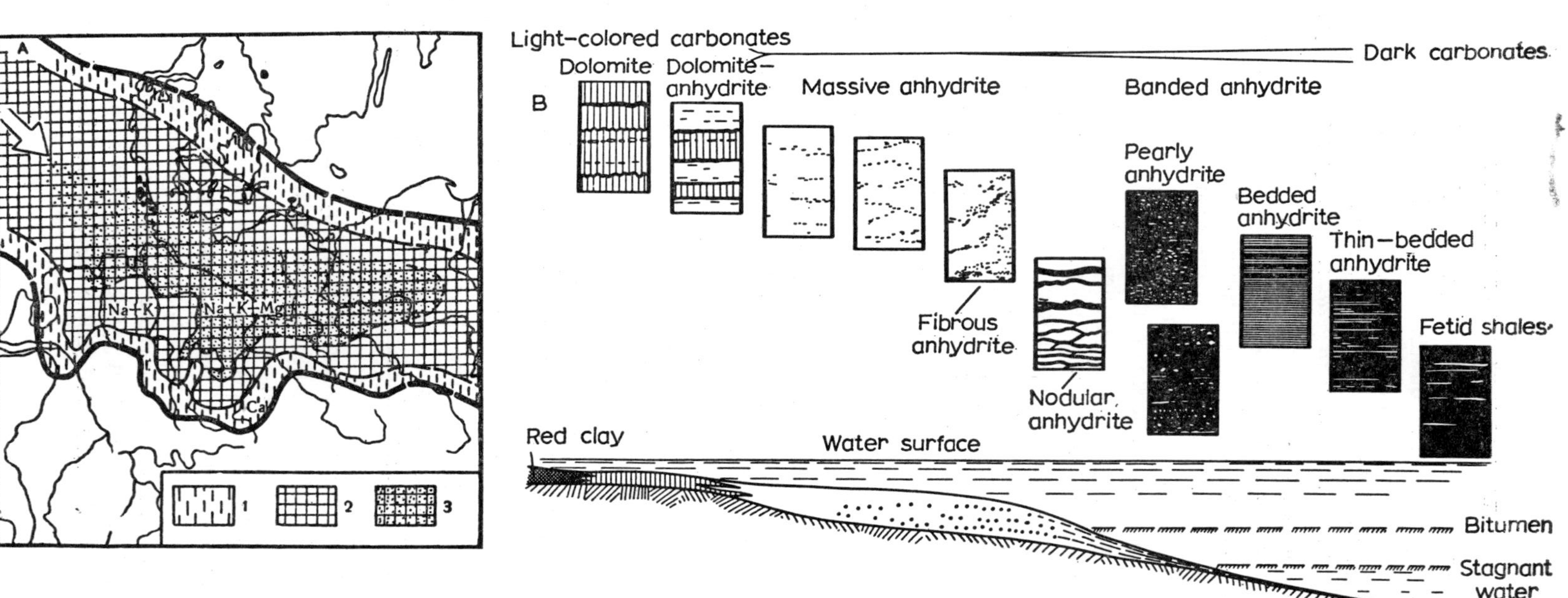

Fig. 199. Depositional environment of the Stassfurt formation (from Richter-Benburg). A. Paleogeography; 1. Ca; 2. Na+K; 3. Na+K+Mg. B. Facies types of anhydrite and their paleogeographic relations.

one-third to one-fourth the area covered by the older rock salt. *This is the largest known bed of potassium salts.* It consists of carnallite rock with some kieserite (15–20%) and of sylvite also with kieserite (18–20%). In some varieties of sylvinite there may be significant amounts of anhydrite (15–20%) and langbeinite. The carnallite and sylvite rocks are interbedded with rock salt, the latter showing well-defined lamination. *A distinguishing feature of the Stassfurt potassium horizon is the variability of its composition; zones of*

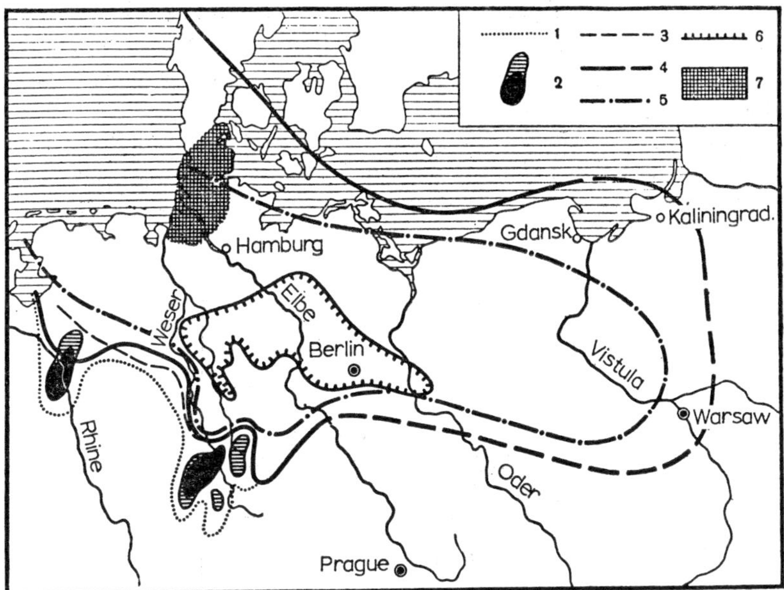

FIG. 200. Relative positions of halogenic areas during Permian time in the North German basin (from Fiveg). 1. Southern limit of the oldest (Werra) rock-salt formation (Z_1); 2. rock salt with a total thickness greater than 100 m, lined area representing distribution of potassium-salt rocks; 3. northern limit of the oldest rock salt (Z_1); 4. limit of older rock salt of Stassfurt formation (Z_2); 5. limit of potassium-salt horizon of the Stassfurt formation (Z_2); 6. limit of potassium-salt horizons of the young (Riedel) rock salt formation (Z_3); 7. rock (lacustrine) salt in the Rothliegende (Lower Permian).

carnallite rock change abruptly and in different directions to zones of kieseritic sylvinite (hartsalz), and both these rocks give way to rock salt. These variations occur at different horizons. Along with this variability in composition, several distributional patterns of potassium salts within the Stassfurt horizon as a whole are clearly shown. Carnallite rocks are dominant in the center. Hartsalz (kieseritic sylvinite) here occurs in irregular lenses of different sizes. This zone is surrounded by a belt in which the number of hartsalz lenses increases markedly at the expense of the carnallite rock. Furthermore, at the southern edge of the predominantly hartsaltz belt, kieserite becomes less abundant than anhydrite. Here a distinctive variety of anhydritic hartsalz occurs, with an average composition of 30–35% sylvite, 50–60% halite, 15–20% anhydrite, and less than 1% kieserite. This distribution of salt rocks, according to Lotze, is a reflection of primary zoning of salt deposition in the basin. We may add that alternations of rock salt and potassium-salt beds within the Stassfurt horizons can be traced throughout the entire distributional area of the horizon, indicating that fluctuations in salinity were simultaneous throughout the huge salt-producing basin.

The third saline formation of the Zechstein (Z_3) is to some extent a repetition of the two preceding (Fig. 202). Saliferous clays 3–3·5 m thick occur at the base. These are subdivided into three horizons, the lowest with a high anhydrite content (60%), the uppermost with abundant magnesite (35%). At the margin of the formation the clay gives way to bedded sandstones, locally yielding a marine fauna. In a belt 30–50 km wide around the margin of the basin, platy dolomite up to 20 m thick rests on the clay. In the center of the basin the clay is overlain by the main anhydrite (1–1·5 m) with lenses of

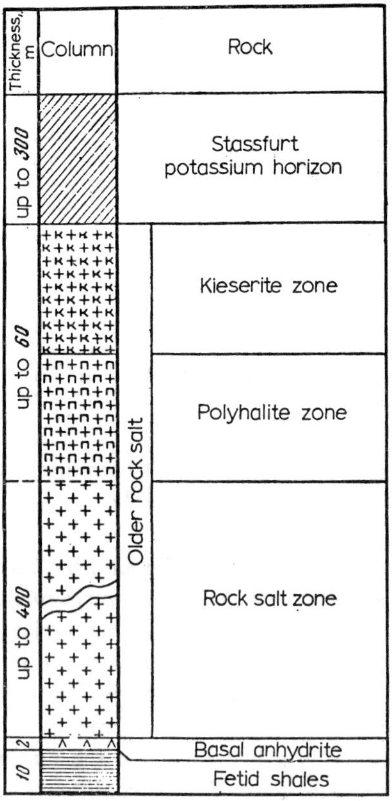

FIG. 201. The Stassfurt formation (Z_2) at Stassfurt (from Fiveg).

dolomite at the base. Above this, forming the main part of the formation, lies the so-called younger rock salt. It differs from the saline sequences of the older formations in occupying a much smaller area than the rock salt of the Stassfurt formation, and in occurring in only the western part of the North German depression. It is much more variable than the Stassfurt salt, with a number of horizons of differing color and texture.

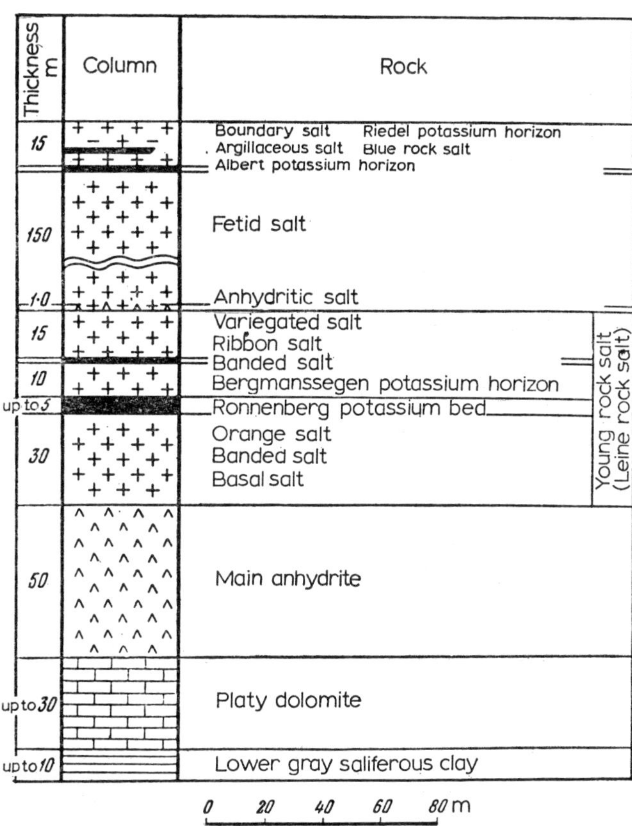

FIG. 202. Composite succession of the Riedel formation (Z_3) (from Fiveg).

The rock salt sequence has two subordinate horizons of potassium salts: the Ronnenberg (4–5 m) in the middle of the sequence and the Riedel (6–10 m) in the upper part. The first is predominantly milky white and light grey high-quality sylvinite. Kieserite appears in the sylvinite in some localities, forming up to 8·5% of the rock, and carnallite is also present locally. The sulfate content in this horizon is generally small. The Riedel potassium horizon is also chiefly sylvite. As the map shows, the potassium deposits of the Riedel formation are of smaller areal extent than those of the Stassfurt horizon.

The fourth or upper Zechstein saline formation begins with red clay (c. 20 m), which gives way to a bed of anhydrite with a distinctive pegmatoidal texture, in turn overlain by the youngest rock salt with a total thickness up to 100 m. The rock salt changes appreciably from the base to the top of the succession (Fig. 203), the upper part being rich in clay, and with interbedded clays. At one mine the argillaceous salt is overlain by a bed of hartsalz 0·7 m thick. This is kieseritic sylvinite, with 17·8% sylvite, 30·8% kieserite, and 50·7%

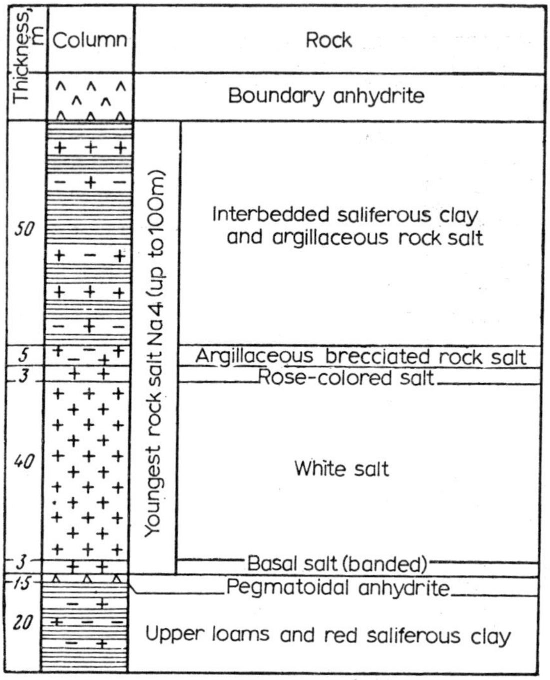

Thickness, m	Column	Rock
		Boundary anhydrite
50		Interbedded saliferous clay and argillaceous rock salt
5		Argillaceous brecciated rock salt
3		Rose-colored salt
40		White salt
3		Basal salt (banded)
1·5		Pegmatoidal anhydrite
20		Upper loams and red saliferous clay

Youngest rock salt Na4 (up to 100 m)

FIG. 203. Composite section of the Upper Zechstein formation (Z_4) (from Fiveg).

halite. It occurs over a considerable area, corresponding to the inner zone of potassium salts of the Riedel formation.

The distribution of the Zechstein saline formations of the North German basin, within the known paleogeography (Fig. 196) indicates that they accumulated in that part of the intracontinental halogenic basin farthest from the strait. Two questions arise: what type of sedimentation took place near the strait when the saline formations were accumulating, and what governed the change from one saline formation to the next? Unfortunately the part of the basin near the strait and the zone of the strait itself are now below the North

Sea. Answers to both questions can therefore be based only on general considerations and analogies.

It is accepted that the Zechstein basin of Western Europe was morphologically a typical basin of the Virrila type, but of vast size. All the characteristic patterns of sedimentary distribution found in basins of that type should therefore occur. In particular, it must be assumed that from the head of the gulf on the south the halitic sediments, corresponding to the water of maximum salinity, must generally give way northward to anhydritic sediments; these latter must give way near or in the strait to dolomites, and these, in turn, to organic limestones in the open sea. The latter are actually found in the Permian of Spitzbergen (Svalbard). Of course, the tectonic movements that gave rise to the complex morphology of the floor must inevitably have resulted in a more varied facies distribution in the northern part of the Zechstein sea than is outlined here, but with no fundamental change.

The cause for the change from one formation to another lies in the tectonic-morphological controls in the vicinity of the strait. At the beginning of each formation, the strait was deep, and the outflowing saline current dominated the marine surface current in its transport of salt causing so marked a decrease in salinity that principally carbonates were precipitated over the entire area. Differential subsidence in the vicinity of the strait, particularly within the basin itself, corresponded to uplift in the surrounding drainage area, and this led to an increase in denudation and a greater influx of clastic material into the basin. This is expressed in the succession by the appearance of clays at the base of the formation and by marls in the deeper parts of the basin. Subsequent uplift near the strait caused shoaling. The outflowing current from the basin to the open sea was sharply restricted, stopping altogether at times, and this led to a rise in salinity in the basin and to progressive precipitation of the more soluble salts, up to potassium salts. Small-scale fluctuations in the currents at the strait gave rise to cyclicity within each saline formation.

3. UPPER PERMIAN FORMATIONS OF THE RUSSIAN PLATFORM

Late Permian sedimentational history on the eastern part of the Russian platform is essentially the same as in the Zechstein basin, but with some differences.

The Lower Kazanian limestone-dolomite formation is much more widespread, covering almost the entire area of deposition of the Lower Permian formation except for the Ural foredeep, which was emergent in Kazanian time.

The Lower Kazanian formation is in composition and structure similar to the *Schwagerina*-Tastuba formation (Fig. 204). Its asymmetrical form is clear. According to Forsh (1955) a ribbon of sand-clay rocks that accumulated in a narrow lagoonal zone forms its eastern margin. Some fossils indicate distinctly freshened lagoons, but others point to lagoons with normal salinity. Insignificant lenses of gypsum occur at some places among the sand-clay rocks; at other places coals are found. These facts indicate a great instability of the physical and geographical conditions in this transitional belt between continental and marine conditions. A broad zone of carbonate deposits is found to the west of the lagoonal zone. These are usually normal, pure

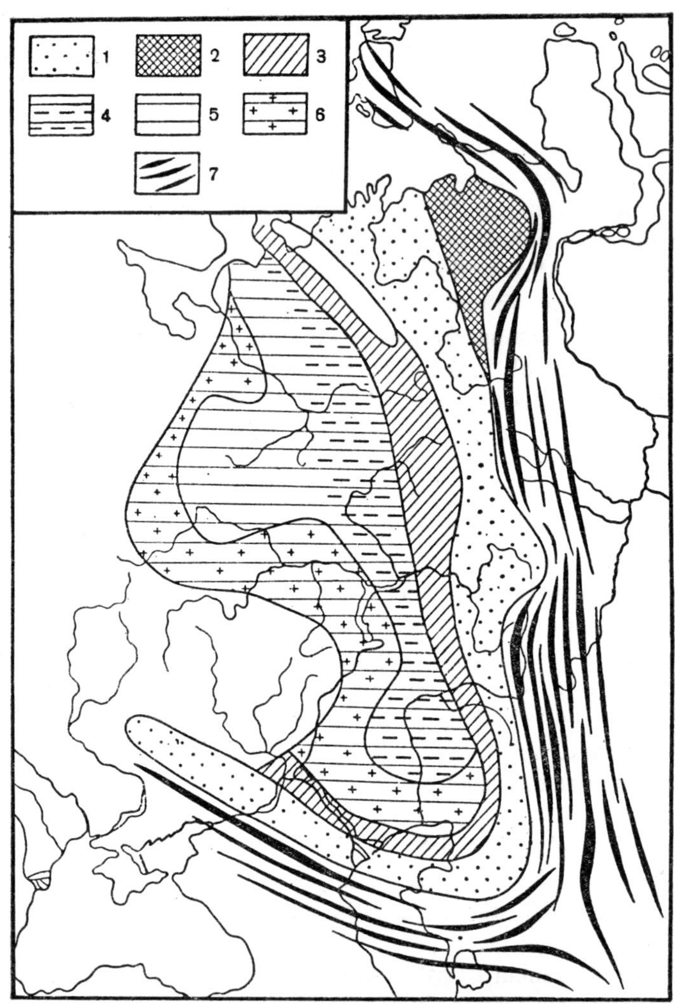

FIG. 204. Paleogeography of Early Kazanian time. Continental red beds; 2. coal-bearing deposits; 3. near-shore-lagoonal zone of sand-silt deposits; 4. limestones with abundant fauna; 5. dolomitized limestones with impoverished fauna; 6. dolomites with poor, stunted fauna; 7. mountain chains.

limestones, with appreciable dolomite only in the upper third of the succession, increasing upward. The limestones are mostly bioclastic or micritic. Many are oolitic, some biomorphic, and some pelitomorphic. They contain an abundant and, for the Kazanian, varied fauna of brachiopods, e.g. *Spirifer rugulatus,*

Dielasma, Productus, bryozoans, pelecypods, gastropods, and others. The fauna points to a near-normal marine salinity, deviating only slightly toward higher salinity. The composition of the carbonate rocks is notably different in the western half of the basin, where pure limestones are replaced by calcareous and normal dolomites. The fauna here is poor; brachiopods almost die out and euryhaline pelecypods and gastropods are more abundant, forming coquinoid banks. Organic remains disappear entirely from many parts of the succession. Oolitic and pelitomorphic dolomites are prominent, forming major units in the succession and extending in tongues far to the east. All this indicates that the salinity in the western half of the Early Kazanian sea was greater than in the east, and that it increased progressively toward the west. It is understandable, therefore, that small lenses of gypsum appear in the upper part of the formation in some localities near Samarskaya Luka.

Forsh has shown that the Lower Kazanian deposits, at least in the trans-Volga region, may be divided into three similar cycles: Baitugan, Kamyshla, and Barabashina. The lower part of each cycle is more calcareous, the upper more dolomitic. In addition, the dolomite content in the succession increases upward. On the whole the Lower Kazanian limestone-dolomite formation is free of saline deposits. It corresponds to the initial stage of arid-climate sedimentation.

At the end of Early Kazanian time, according to Forsh, the trans-Volga region emerged from the sea to become part of the continent. Denudation modified the surface, forming large-scale erosional features one of which developed in the southern part of the trans-Volga region, from Buguruslan to Buzuluk (and somewhat south). This was a large trough, open to the south, from which the Barabashina, Kamyshla, and upper part of the Baitugan beds were eroded. The uplifts in the trans-Volga region divided the original single basin into two: a northern basin connecting through a strait with the open sea, and a southern basin, occupying the Caspian syneclise. The northern basin maintained normal marine salinity, whereas the southern, becoming something of a closed basin within an arid climate, became much more saline.

The history of the Upper Kazanian formation begins with slow subsidence of the trans-Volga region and with renewed marine flooding (Fig. 205). During this transgression the Buguruslan-Buzuluk erosional trough was covered with water and a distinctive saline sequence was deposited, 200 km long and 100 km wide, and reaching a thickness of 216 m. A 20-meter anhydrite bed at the base is succeeded by a bed of rock salt 133 m thick. A second anhydrite, 20 m thick, overlies the rock salt, and in its turn, is overlain by a second rock salt, 19 m thick. A unit of anhydrite-gypsum, 24 m thick, caps the sequence (Fig. 206). "The gypsum overlying and underlying the salt is much more widespread than the salt. It is found where the Kamyshla and even the Barabashina beds have been preserved from erosion. The salt is restricted in its distribution to areas where these beds are absent. The basin in which the salt was precipitated was consequently smaller than the basin in which the gypsum was deposited, and, apparently, it was not as deep" (Forsh

1955, p. 127). According to Forsh, the saline sequence accumulated as follows. "The water in the southeastern basin, with a very high saline concentration, quickly filled the lower reaches of the erosional valley, forming a large gulf

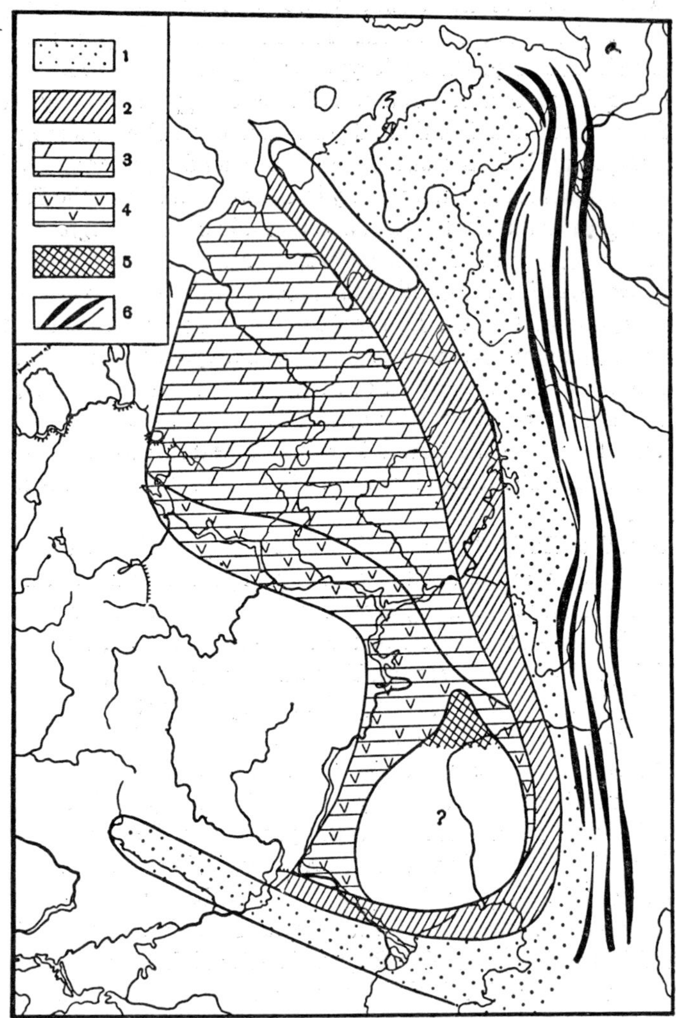

FIG. 205. Paleogeography of Late Kazanian time. 1. Continental red beds; 2. near-shore-lagoonal sand-clay deposits; 3. dolomite, containing some sulfates in the upper horizons; 4. gypsum and anhydrite with subordinate interbeds of dolomite, the gypsum becoming sharply predominant upward; 5. rock salt accumulations at the base of the Upper Kazanian beds; 6. mountain structures.

in which the lower anhydrite of the Buguruslan section accumulated. At the time the salt was deposited, the size of this gulf had been greatly reduced, becoming markedly narrow, apparently occupying only the deepest part of the valley" (*ibid.*).

During further transgression, not only the southern basin but also the northern basin in the trans-Volga region were covered by the sea, and the Late Kazanian intracontinental basin was ready for deposition of the upper carbonate-sulfate formation.

The general shape of this basin (Fig. 205) was similar to that of the Early Kazanian basin, but the sedimentation was notably different. Two zones may be generally distinguished. One includes the broad northern zone between the mouth of the Dvina River and Yaroslavl, extending along the eastern half of the basin. Its boundary passed somewhat east of Balakhna, Ioshkar-ol, and Bugul'ma and farther, to the eastern shore line of the basin. Sedimentation within this zone varied from north to south. In the central part of the Sukhona-Mezen' basin, according to Ivanov and Levitskii, the lower horizon of the Upper Kazanian deposits (50–55 m thick) consist chiefly of carbonates:

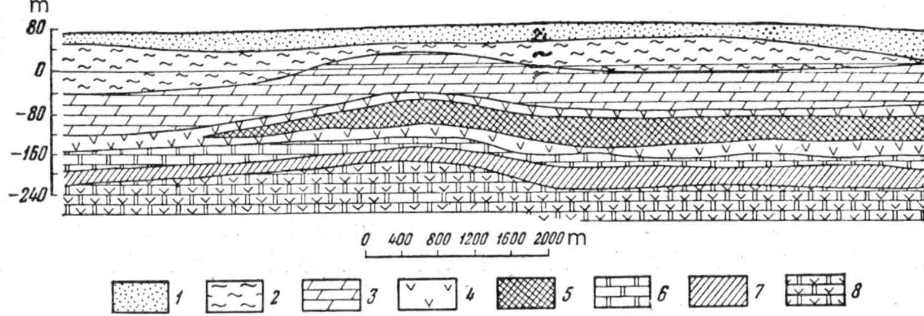

Fig. 206. Structure of the Upper Kazanian saline deposits in the Buguruslan region (from Ivanov). 1. Quaternary loam and gravel, 2 – 5, Upper Kazanian subseries; 2. red clay, sandstone, dolomite, and marl; 3. dolomite with interbeds of marl, clay, and sandstone and with some gypsum and anhydrite; 4. anhydrite, 5. rock salt; 6. Lower Kazanian subseries—dolomite, limestone, and marl with interbedded clay and sandstone; 7. Ufimian series—variegated clastic rocks; 8. Kungurian series—anhydrite-dolomite complex.

marls, dolomitized limestone, and calcareous clays and siltstones, with an abundant fauna. Gypsum is rare, occurring only as small nodules and rims about grains. The upper horizon is composed of dolomitized limestone and dolomitic marl, barren of fossils. Gypsum is generally present as lenses that range in thickness from 0·5 to 6 m. The thickness of the entire horizon is 16–34 m, increasing in the zones where gypsum is abundant. The gypsum content in the upper part of the succession decreases to the northeast of the Sukhona-Mezen' basin. To the south, at Vologda, Soligalich, Shar'ya, Kotel'nich and in the Kirov region, the deposits maintain their character and the gypsum content of the upper sequence is inconsiderable. From here the belt of weakly gypsiferous carbonate rocks, gradually growing narrower, extends to the south in the trans-Volga region, occupying the eastern part of the region and reaching to the upper course of the Ural River. Limestones interbedded with the dolomites contain a few poor species of pelecypods and gastropods.

This zone corresponds to a part of the basin in which the water was generally variably saline, at times with a bottom fauna and yielding a limestone precipitate (dolomitized) at other times excluding organisms and yielding dolomitic rock and, locally, even lenses of gypsum. Toward the south, the high salinity increases with increased dolomitization and gypsum formation.

The composition of the Upper Kazanian formation is substantially different in the southwestern belt that extends approximately from Gorki to Kazan, the mouth of the Kama, Bugul'ma, Kuibyshev, and farther to the southeast. The total thickness is here 30–80 m, rising to 100–120 m in the trans-Volga region. In both the lower and upper horizons, dolomite completely displaces limestone, and the interbedded gypsum lenses and beds increase sharply, becoming especially numerous in the upper horizon. In an area bounded by Vasilsursk, Cheboksary, Mariinskii Posad, Urmary, Ulema, and elsewhere, the thickness of gypsum beds ranges from 0·5 to 8–9 m, locally 16 m, in a gypseous horizon ranging in thickness from 10 to 37 m. The gypsum content of the sequence is 15–60%, being especially high in the vicinities of Karly, Ulema, and Umary. In the region of Kazan, the mouth of the Kama, and farther to the southeast, gypsum is found throughout almost the entire formation, reaching its greatest development in the upper half. Nevertheless, the gypsum does not form individual continuous beds that may be traced throughout the region; it occurs in large lenses ranging from 3 to 11 m in thickness and in length 7–8 km, rarely more. Such features in the southwestern zone attest to considerable salinity in the corresponding part of the basin, at times exceeding 15%.

The Upper Kazanian as a whole is undoubtedly a saline formation, but the saline process at the time of its deposition remained at the early stage of gypsum precipitation, which was quantitatively only poorly developed. Only in the Buguruslan-Buzuluk district did a large lens of rock salt accumulate.

4. A Comparison of the Zechstein with the Kazanian Saline Process

From comparison of the saline process of the Zechstein formation with that on the Russian platform, it is clear that the former was much greater than the latter in area, thickness, and completeness of its development. *The Zechstein halogenesis took place on a grand scale and reached an advanced stage of development, whereas the Kazanian process remained at the initial stage.* The cause of this difference in sedimentation between the two intracontinental basins, which had a single source of salt supply and a similar disposition relative to surrounding country within the same arid zcne, lay in the nature of the connection between the basin and the open sea. *The Kazanian basin undoubtedly had a broader connection and a more intense exchange of water because of greater depth in the strait. The connection of the Zechstein basin was more restricted because of the shallower depth in the strait, and the water exchange was weak.*

CHAPTER 4

THE MORPHOLOGY, HYDROLOGY, AND DISTRIBUTION OF SEDIMENTS IN ANCIENT MARINE HALOGENIC BASINS, AND THE INFLUENCE OF THESE FACTORS ON THE CONSTITUTION OF SALINE FORMATIONS

We have reviewed many marine saline formations and their depositional basins, and have found that the paleogeography of the basins can usually be reconstructed fairly accurately. This leads to a consideration of pattern in the morphology, hydrology, and distribution of sediments in the marine halogenic basins of the geologic past and of their influence on the constitution of the saline formations.

1. MORPHOLOGIC-TECTONIC TYPES OF HALOGENIC MARINE BASINS AND THEIR INTERRELATIONSHIP

The morphologic-tectonic types of halogenic marine basins may be classified in two parallel series:

Recent and ancient   Specifically ancient

2. Gulfs of Kara-bogaz type→3. Marginal zones of epicontinental seas

1. Lagoons

4. Gulfs of Virrila type   →5. Intracontinental salt-producing seas

The initial link in both series, the lagoon, represents purely exogenetic, atectonic basins, resulting from the action of exogenetic factors, particularly wave action which produces sand spits across gulf-like indentations in the shore lines of a land mass. Lagoons have consequently always been small and shallow, generally having depths from a few meters to a few tens of meters, this depth being approximately equal to twice the wave height. Morphologically, lagoonal basins are unstable ephemeral features, being easily converted to maritime lakes by the same exogenetic processes that produced them.

All the succeeding members of the series are morphologic-tectonic rather than exogenetic, forming in a depressed structure or structures. *The type of structure essentially determines the morphological type of salt-producing basin.*

Gulfs of the Kara-bogaz type have formed where the tectonic depression invaded by sea water has been more or less equidimensional, round or subangular, and slightly elongate in one direction. The size of such basins has ranged from very small—such as the trans-Caspian Middle Miocene gulf, the present Gulf of Kara-bogaz, and the Akchagylian Uzun-Su—to huge, as the Michigan and Moscow basins of early Givetian time. The shore line was generally simple. Communication with the sea has been through a single narrow strait. Gulfs of the Kara-bogaz type always appear as independent, isolated basins attached to a continental sea by a narrow strait. These gulfs are traditionally regarded as shallow, but it should be borne in mind that shallowness in such large basins as the Michigan and Moscow gulfs surely means depths of tens of meters (up to 50–60 m). It may be recalled that ripple marks have been observed in salt in the marginal part of the Salina formation of the Michigan basin, and crystals with pinnate structure are broken to pieces. Furthermore, clastic grains of dolomite are found in the anhydrite deposits. All this indicates that at least the marginal part of the salt basin has been shallow, comparable to modern salt lakes. The central part of the basin has been undoubtedly deeper, because the above features in the salt disappear in this zone; but there are no objective criteria for determining the actual depth. The morphology of the basin floor containing the gulf is always simple. It is represented by a single depression, with no subsidiary basins. Although lagoons and gulfs of the Kara-bogaz type differ fundamentally in origin of the basin, being endogenetic and tectonic respectively, it is not always possible to differentiate between them when the latter are small. Classification of saline deposits as lagoonal or gulf accumulations must therefore be at times tentative, as has already been pointed out.

Salt-producing basins in the marginal parts of epicontinental seas formed when there existed a broad syneclise in front of a transgressing sea, with its axis parallel to the shore of the sea and with its seaward edge subsiding more strongly than the opposite. The sea readily transgressed the near edge of the syneclise and flooded it. A marginal segment of the sea developed, appearing on the surface to have free communication with the open sea but actually being more or less cut off from it by a series of underwater shoals, as in the Dankov-Lebedyan basin of the Russian platform, the Usol'e basin of the Siberian platform, and the Elk Point basin of the Canadian platform. When the relief of the flooded segment was relatively strong, however, the uplifts were above water, appearing as a chain of more or less mountainous islands, as exemplified by the Röth and Middle Triassic basins of Germany. These salt-producing basins generally had simple outlines. The syneclise that formed the basin was a single depression, but one occasionally developed with islands and subsidiary basins within the major depression, as the Middle Triassic basin. The depths of these marginal salt-producing basins would be of interest, but this problem cannot ordinarily be solved by lithic analysis, and remains a matter of speculation.

Until recently there have been no pertinent data. Makhlaev (1959a), who studied the dolomite-sulfate formation of the Dankov-Lebedyan basin, demonstrated that the marginal zones of the salt-producing basins in which organic limestone, dolomitized limestone, and dolomite accumulate are mainly shallow. Algae, oolites, desiccation cracks, and wave-zone breccia disappear in the dolomites of the central parts of the basin and in the anhydrites interbedded with them, indicating a greater depth for deposition of these sediments. Makhlaev considered that deposits with algae accumulate at depths less than 50–60 m, and deposits without algae at depths greater than 60 m. Depths in the central parts of the Moscow syneclise in Dankov-Lebedyan time were thus greater than 60 m; but there is no evidence that they reached 100–200 m.

Data on the Lower Carboniferous saline sequence of the Siberian platform are of significance. Pisarchik (1958) succeeded in establishing the fact that the dolomites interbedded with rock salt in the Usol'e series contain algal remains, and these rocks are therefore shallow-water, having been deposited at depths not exceeding 50–60 m. The interbedded rock salt also could not have been deposited in deep water; at most, it formed at depths measured in tens of meters or less. This example, despite the fact it is unique at this time, is of fundamental significance. It shows that thick saline deposits may form in relatively shallow zones in the marginal parts of salt-producing basins, that the accumulation of rock salt is not merely the filling of a ready-made depression, and that simultaneous subsidence of the floor of the basin is a primary factor in the development of rock salt deposits.

The Virrila type of salt-producing gulf has formed in a long narrow depression, possibly a graben, which may have existed in some segment of a geosynclinal zone during the stage of normal development (Late Jurassic in the Caucasus), in a marginal depression (the Carpathian foredeep), or on a platform (Rhine graben, Dnieper-Donets syneclise in Middle and Late Devonian time). Accordingly, the salt-producing basin was long, terminating in a narrow estuary, perhaps rectilinear, perhaps somewhat bent, withdrawn from the open sea. The outline of the basin was generally simple, and the floor was a single entire unit, though it is true that subsidence was not perfectly uniform over the entire length of the basin. Depths of gulfs of the Virrila type have not been investigated in great detail, but some indications of the relative depths are known. "The depth of the Carpathian foredeep did not remain the same over the entire duration of sedimentation on the floor. During terrigenous sedimentation the basin was probably deeper than during accumulation of predominantly saline sediments. But even during accumulation of clastic sediments the basin was relatively shallow, as is indicated by wave-cut features, ripple marks, mud cracks, and fossil debris in the sandstone beds of the Zagoroi and Dobrotov series" (Ivanov and Levitskii 1960, p. 322). The thin, lenticular character of rock salt accumulations in the fore-Carpathian gulf might have arisen in small basins, already very shallow when they formed intermittently in the salt-producing part of the gulf proper. There are no data on depths in other gulfs of the Virrila type, but one can hardly conceive of their being different from the preceding.

Intracontinental salt-producing basins of the Permian period are essentially extensions of the Virrila type of gulf, but on a huge, incomparably grander scale. It is precisely this vastness that makes their tectonic framework substantially different from that described. *Intracontinental salt-producing marine basins are broken into a number of individual depressions with intervening uplifts, in*

2 F

contrast to the single, narrow, though elongated zone of subsidence in gulfs of the Virrila type. This was true in the Lower Permian basins of the U.S.S.R. and in the Upper Permian Zechstein and Kazan basins. It may be stated, therefore, that *intracontinental halogenic basins form where the transgressive sea encounters a broad linear zone of subsidence, a zone that embraces complex tectonic movements and that has a long axis of subsidence normal to the shore line of the sea, not parallel as in marginal halogenic basins.* This difference in direction of the axis of subsidence relative to the shore line is responsible for the growth of the two large, specifically ancient types of salt-producing basins indicated above.

The outlines of intracontinental salt-producing seas, those that have been reliably reconstructed, such as for the Werra formation, the *Schwagerina* horizon and some others, are characteristically complex. The configuration of the floor also was probably complex, uplifts of positive tectonic structures alternating with low areas of tectonic downwarps.

The intracontinental salt-producing seas of Europe are probably the only examples for which the depths are discussed specifically in the literature. I. P. Gerasimov, P. I. Lunin, M. P. Fiveg, and others have maintained that the rock salt of the Kungurian deposits of the Ishimbai part of the Ural region was deposited in a synclinal basin with depths up to 1000 m. Kühn (1955), basing his conclusions on determinations of bromine content in the Zechstein rock salt, suggested that these deposits formed in a basin with depths no greater than 380 m, and these depths decreased with time. It should be noted, however, that these determinations are neither convincing nor consistent. The view concerning the depth of the Kungurian basin at Ishimbai is based on an unjustified assumption of stratigraphy of the Kungurian deposits in the Belaya depression. It is a matter of fact that the saline horizon is transgressive at the margins of the basin, which means that it was initiated only in its lowest part, shifting toward the margins as the basin subsided. Such a mechanism of salt accumulation naturally excludes the possibility that the basin was initially 1000 m deep, becoming shallower with time. Actually the salt-producing basin was generally shallow, gradually transgressing over a broader area as the foredeep subsided more extensively. Kühn's estimates are tentative because they are based on the assumption that the halite was precipitated during winter cooling of the water mass, not during summer precipitation from the surface. This concept of the process does not agree with the pinnate structure of the halite in the annual cycles. This structure indicates that halite precipitation took place during summer evaporation and not during winter cooling. The disposition of the Zechstein basin along the very southern edge of the arid zone, immediately adjacent to the moist tropical belt, also casts doubt on the possibility of any considerable winter cooling, and that this process could be effective. In conclusion, we must recognize that the existence of deep intracontinental salt-producing seas has not yet been proved. All present information leads to the belief that these seas were no deeper than marginal salt-producing basins.

Marine salt-producing basins of the geologic past were thus distinguished by extreme variability in size and shape, but by much less variability in depth, which ranged from a few meters to several tens of meters, probably not exceeding 100–120 m. *In all morphological types of basins, except lagoonal, the outline of the basin and the topography of the floor have been completely determined by the tectonic structure—downwarp or syneclise—that gave rise to the basin.* An invariable feature of all salt-producing marine basins has been the restricted connection with the open sea, the restrictions being perhaps barriers of islands or continuous continental segments, perhaps underwater shoals. The necessity of the barriers follows from the nature of water exchange between salt-producing basins and the open sea. Two opposing currents usually exist in the strait (or straits): *the surface current flows from the open sea into the salt basin,* bringing in water and salt; *the bottom, outward-flowing current removes salt from the basin.* The accumulation of salt has been physically possible only when the salt balance has been positive, i.e. when the introduction of salt by the surface current has been greater than its removal by the bottom current, or when the latter has been absent altogether. *Barriers in the zone of the strait, either above water or below the surface, have been necessary for realization of positive salt balance in salt-producing basins. They have retarded the removal of salt and, by so doing, have caused the basin to become progressively saltier.*

2. DEPOSITIONAL DISTRIBUTION IN SALT-PRODUCING BASINS AND THE CONSTITUTION OF SALINE FORMATIONS

Like basins of the humid zones, salt-producing basins in arid regions are characterized not only by an assemblage of deposits of definite composition but also by a systematic areal distribution of the deposits on the floor. When localization of deposits has been preserved in its basic outlines for a prolonged period, varying only in detail, several structures appear in the saline formation, reflecting the distribution of sediments for the separate time intervals. This circumstance compels us to examine carefully the types of localization among sediments in the salt-producing basins of the different morphologic-tectonic types.

The distribution of deposits in lagoonal basins, taking each lagoon separately, has been relatively simple, like that observed in modern lagoons and maritime lakes. Sand-silt-clay rocks, more or less carbonate-bearing, were deposited about the margin of the lagoon. At low salinities carbonate rocks were deposited in the center of the lagoon, at high salinities gypseous deposits. But this type of localization has not been invariable. Hydrodynamic conditions might make rapid changes. A lagoon might be readily converted into one, two, or several maritime lakes, some perhaps large, each with its own distribution of sediments. It is essential to note also, that lagoons rarely formed in isolation, but generally developed in groups, sometimes in large numbers. At times, also, these lagoons might form complex branching basins.

In the distribution of saline deposits along such lagoonal coasts as a whole, whether modern or ancient, it is impossible to detect any systematic pattern, either along the shore line or at right angles to it. *On the whole, the distribution of saline deposits on lagoonal coasts is very irregular and patchy.* When this patchy pattern of deposition persists for a long time, an occurrence that is fostered by the morphological conversion of individual lagoons into maritime lakes, by drying up or silting up, it is natural that the *structure of lagoonal formations also proves to be patchy.*

Typical lagoonal saline formations that accumulate on rather broad plains inclined toward the sea are characterized by thinness of beds, by lack of persistence of the saline and other constituent rocks, and by the absence of any clear pattern of spatial distribution. Gypsum and anhydrite rocks are most frequently found in lagoonal formations, in lenses tens and hundreds of meters long, rarely persisting for a few kilometers. The thickness of such lenses ranges from a few tens of centimeters to several meters. The lenses normally occur at different horizons in the formation, some completely isolated, some clustered, with the marginal parts somewhat overlapping. In some parts of such formations two or three separate gypsum or anhydrite horizons occur, but these are local and impersistent features. Nor is there any well-defined preference of saline deposits for any particular part of the formation. The same lack of definite pattern also characterizes the other constituents of the rock—the clays and carbonate deposits. Thus the structure of the formation is *patchy to the spatial pattern of the constituent rocks.*

The distribution of sediments of Kara-bogaz and the closely related marginal marine types of salt-producing basins is substantially different. Here each individual salt basin corresponded to its particular tectonic structure, in the form of a syneclise, and the spatial distribution of the saline rocks was controlled by the morphology and evolution of the syneclise. The distribution may be termed symmetrical, although the actual manifestation of symmetry differs for each stage of saline development. In the initial, pre-saline, stage a broad border of sand-silt sediments was deposited at the margin, dolomitic muds in the central, deeper, part. At the sulfate stage clastic and dolomitic deposits accumulated at the margin, gypsum in the central, deeper part. Periodic or episodic changes in salinity led to repeated alternation of sulfate and dolomitic muds in the center. At the chloride stage carbonate deposits accumulated at the margin; these gave way to gypsum about the inner part of the marginal zone, and rock salt formed in the central, more rapidly subsiding part. Again, periodic and episodic freshening led to complex rhythmic interbedding of dolomite and anhydrite about the edges of the basin and of anhydrite and rock salt, perhaps of dolomite also, in the central parts. Potassium salts accumulated always within the field of rock salt, generally occupying a small part of this field and occasionally interfingering with the rock salt. This is the general distributional plan of saline sediments within gulfs of the Kara-bogaz type or within marginal salt-producing basins. In the latter, some features of asymmetry, though minor, also appear, such as accumulation of clastic deposits normally on only one side, the shore side, of the basin. No border of clastic material on the submerged ridge develops.

From the foregoing it is easy to understand that the structure of the terminal accumulations in formations of the Kara-bogaz type, and in formations of marginal marine basins, must inevitably be symmetrical, as appears in the Lower Givetian formation of the Moscow syneclise, the Salina formation of the Michigan basin, the Dankov-Lebedyan formation of the Russian platform, the Lower Cambrian formation of the Siberian platform, and others. In keeping with the outlines of the syneclises, these formations are rounded bodies, more or less equant in plan, thin at the margin and thick in the center, In the central part thick accumulations of rock salt alternate with beds of anhydrite and dolomite. This symmetrical structure is somewhat disturbed by shifting of limestone-dolomite rocks farther into the interior of the formation in the zone where the basin was freshened by incoming water. Similar indentations may be noted at other localities about the margin of the formation where streams entered, but these represent exceptions rather than the rule.

Sediments in salt-producing gulfs of the Virrila type, however, show a markedly different and asymmetrical pattern of distribution. This asymmetry, furthermore, differs for the different stages of salinity. Initially a very large part of such basins was covered by sand-silt, clay, and carbonate deposits of normal marine character, and only at the head of the bay did dolomite accumulate. At the sulfate stage the zone of normal saline deposits was considerably reduced. At some distance from the zone where marine water entered the basin, dolomite was deposited, and still farther away, at the head of the bay, gypsum accumulated. At the rock salt stage the zone of normal marine sediments was confined to the strait itself and its immediate vicinity. Away from the strait this zone of normal marine sediment gave way to a zone of dolomitic mud, then to gypseous deposits, and finally to rock salt. At this stage, the headward part of the gulf apparently began to split up into smaller, individual bays and lakes, in which halite precipitated, and into intervening segments, where sedimentation had ceased. Similar phenomena took place also in arms of the gulf in its middle and lower parts. At the stage of potassium-salt accumulation, the headward part of the gulf and its arms had already dried up for the most part, but brine was preserved in the more actively downwarped segments. Intercrystalline brine flowed into such zones from the surrounding uplifted areas. Potassium salts were concentrated in these lakes in zones of the dried-up rock salt field. This represents the areal distribution of sediments during accumulation of the Rhenish Eocene-Oligocene saline formation, the Devonian formation of the Dnieper-Donets and Pripyat basins, the Miocene formation of the Carpathian region, and others. This history of accumulation also predetermined the asymmetric form of the saline formations. The parts of a formation corresponding to the entrance of the gulf consist of dolomitic rocks interbedded with anhydrite. Farther inside the gulf, away from the strait, the formation consists chiefly of anhydrite with interbedded rock salt. At the closed upper end of the gulf massive rock salt is found, with local deposits of potassium salts. *A well-defined longitudinal asymmetry in composition of the formation is combined with a more or less well-defined transverse symmetry.* Along both long margins of such formations the saline rocks give way in narrow belts to clastic-sulfate, clastic-carbonate, or, in places, to coarsely clastic sediments. This facies transition is beautifully displayed, for example, in the Eocene-Oligocene formation of the Rhine graben. This structure must also apparently characterize other asymmetrical formations, but confirmatory evidence is lacking.

Where a long, narrow, salt-producing gulf lies along a marginal depression (such as the Carpathian foredeep) or in a local synclinal trough of the geosynclinal zone (such as the Late Jurassic zone of the Northern Caucasus), the complex structure of the floor has been reflected in complex sequences in the saline formations that form upon it. Although longitudinal asymmetry is generally preserved in the formation, it is disturbed by a number of local deviations—sand-clay or limestone-dolomite rocks which intrude for a considerable distance into the inner part of the formation.

The sequence of the saline formation that formed in *the Permian intracontinental halogenic seas of the Western German plain and the Permian seas of the Russian platform* is generally similar to the sequence of gulf formations of the Virrila type, but they differ in their complexly asymmetrical character. In a longitudinal section through such formations as a whole, asymmetry is always distinct, following from the fact that the part corresponding to the mouth of the gulf (where the saline basin connected with the ocean) shows local development of limestone-dolomite rocks, whereas sulfate rocks interbedded with dolomite accumulated farther inside the basin, with rock salt and potassium salts at the upper end. This longitudinal asymmetry combines in these formations with a transverse asymmetry. This latter results from one side of the basin lying next to a flat arid continent while the other is next to a mountainous district belonging to the culminating stage of orogeny. Such relationships naturally cause asymmetrical accumulation of fragmental material, which is absent, or almost absent, near the continental plain, but forms thick deposits on the mountainous coast. In some places, such as in the Upper Permian formation of the Russian platform, continental clastic sediments accumulate in a broad belt, forming continental molasse between the salt-producing marine basin and the mountains. The transverse asymmetry of the clastic material is complementary to a similar asymmetry in distribution of saline deposits: these latter show a strong preference for the low-lying coast, the less saline rocks forming on the mountainous coast. This form of distribution is clearly seen in the Lower and Upper Permian formations of the Russian platform. It is possible that this is also true for the Zechstein formations of Central Europe, but incomplete preservation prevents precise interpretation.

The complexity of structure in saline formations of intracontinental seas increases where such formations accumulated in basins made up of a series of tectonic depressions between uplifted segments. In such basins, thicker sequences of rocks are deposited in the depressed zones than on the uplifted segments, and these depressed zones are distinguished by a wider development of saline rocks and higher stages of the saline process. In particular, when dolomites and limestones are deposited on the uplifted segments, gypsum (anhydrite) is deposited in the depressions. When dolomite-sulfate deposits form on the uplifts, rock salt, and occasionally potassium salts, accumulate in the depressions. The scattering of such units in the formation upon the established asymmetry makes the distributional picture increasingly complex.

The distribution of saline sediments in halogenic basins and saline rocks within saline marine formations has been controlled by prolonged preservation of the tectonic structure of the floor upon which the salt-producing body of water lies. In atectonic basins, i.e. lagoons, patchy and variably distributed sediments prevail. In gulfs of the Kara-bogaz type, and in marginal marine basins, forming simple, circular or oval depressions, the sediments are symmetrically deposited, and the resulting formations have symmetrical structure. Gulfs of the Virrila type, occupying long, narrow depressions, become the sites of asymmetrical accumulation of sediments and of asymmetrical structure in the resulting formation. In intracontinental saline seas that occupy elongated zones with complex tectonic structure, the sedimentary accumulation and the resulting formation are complexly asymmetrical.

Tectonic conditions are thus seen to have a clear effect on the essentially physicochemical processes of halogenesis.

3. Complete Development of Halogenesis; the Relative Thickness of the Salts in Saline Formations and the Controlling Factors

In most types of marine saline formations halogenesis stopped at the initial sulfate stage, sometimes it went on to the later halite stage, and at times went through all possible stages to reach complete development. On the other hand relative thicknesses of the beds of dolomite, $CaSO_4$, $NaCl$, K and Mg chlorides and sulfates in no formation corresponds to the weight proportions of these salts in sea water. In some places the basal carbonate horizon of the formation may be anomalously thick, elsewhere the sulfate layer, and in still other places the rock salt bed. In other formations the carbonate bed may be entirely missing, or sulfate horizons are lacking, with only minor development of rock salt. The relationships here are extremely variable (Fig. 207).

It is natural to ask what has controlled the degree of completeness in the development of halogenesis in the accumulation of a formation, and what has determined the thickness relations of the individual types of rocks in the succession.

From the paleogeography of marine saline formations it follows that the necessary condition for the development of these formations is the restricted connection between the saline basin and the open sea. The water exchange must have been such that for long periods the *outflowing bottom current was weak or was absent altogether.* But this condition can be realized only when tectonic movements isolate some particular segment of the crust and continue to preserve this isolation. *The first tectonic prerequisite for complete development of saline formations is thus prolonged stability of tectonic and topographic conditions in the vicinity of the strait, securing a positive salt balance in the saline basin.*

The second condition involves tectonic activity in the salt basin itself: the necessity of intense and prolonged subsidence. This is seen by comparing the

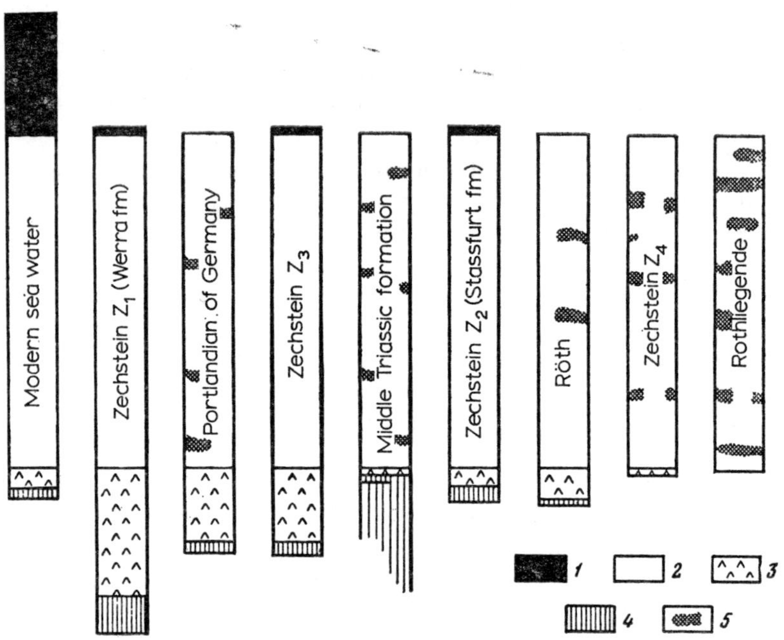

FIG. 207. Relative thicknesses of the major evaporite lithologies. (from Richter-Brenburg). 1. K and Mg chlorides and sulfates; 2. NaCl; 3. CaSO₄; 4. Ca and Mg carbonates; 5. clay.

rates of sedimentation at the different stages of salinity with the rates of subsidence. The accumulation of carbonates is extremely slow, parts of a millimeter per year. If the rate of subsidence is very slow, however, it can compensate only a small increment of carbonates. The basin will be preserved for a protracted interval, and this will make possible a strong increase in salinity of the basin. Sedimentation will shift to the sulfate stage. At this stage, however, sedimentation will increase sharply to 1–2 mm per year. If the floor subsides more rapidly than this, the basin will be preserved despite the rapid accumulation of sulfates and this will make possible continued advance in salinity till rock salt will be precipitated. But if the rate of subsidence of the basin floor proves to be less than 1–2 mm per year, the basin will fill up to the edge with sulfate after some interval of time, and the saline process will be terminated at the initial, the sulfate, stage. When the basin is preserved beyond the sulfate stage, and halite is precipitated, the rate of sedimentation reaches very high values: 3–10 cm per year. Again essentially there are two possibilities. The rate of subsidence may be less than the rate of sedimentation, in which case the basin will fill up more or less rapidly with rock salt and sedimentation will culminate earlier than the stage of potassium-salt

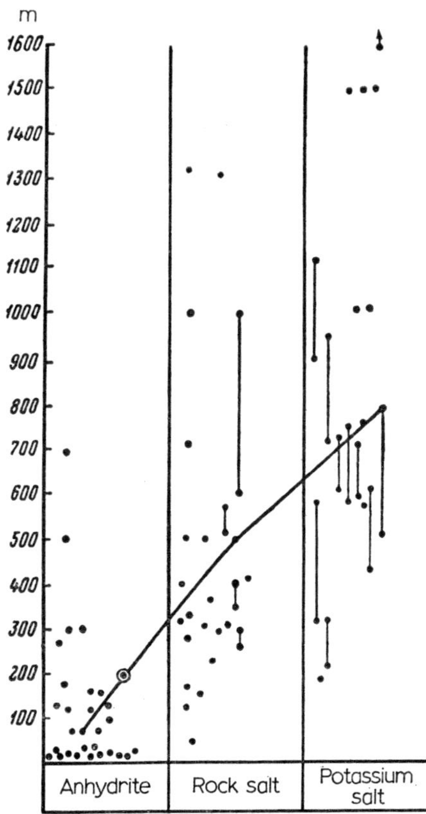

Fig. 208. Thicknesses of saline formations representing various stages of completeness of the saline process (data from Krumbein 1951, Lotze 1938, Ivanov 1960 and Fiveg 1960).

precipitation. If, on the other hand, the rate of subsidence proves to be greater than the annual rate of halite precipitation, the basin will continue to be preserved, and it may become possible for potassium and other salts to be precipitated, reaching even the eutonic phase. It is therefore clear that *complete development of the saline process has taken place only in basins in which the rate of subsidence of the floor at each stage has exceeded the rate of* *synchronous sedimentation.** How this condition is actually realized is demonstrated below.

That this condition has been actually realized in the geologic past is shown in Fig. 208, on which are plotted the thicknesses of marine saline formations representing various degrees of completeness of the saline process. Three groups of such formations are distinguished: gypsum-anhydrite, rock salt, and potassium salt. It has been shown that the points representing thickness for each formational group are not together. For the gypsum-anhydrite group, in which the saline process was interrupted at the very beginning, thicknesses are mostly less than 50 m, very rarely 60–180 m, in only three cases 280–300 m, one about 500 m, and one 700 m. The two latter, however, are completely atypical. The great thicknesses of sulfate rock here were due, possibly, to salt tectonics (more accurately anhydrite tectonics). The thicknesses of rock salt formations show a different pattern, corresponding to the greater degree of completeness of the saline process. Only one formation less than 100 m thick is recorded. Three are 100 to 200 m thick, and most are 250 to 700 m thick. Thicknesses up to 1300 m are known but are rare. Thus, *the thickness of a rock salt formation is generally much greater than the thickness of anhydrite formations.* The thicknesses of potassium-salt accumulations generally are from 600 to 1500 m; i.e. *their thicknesses for the most part are clearly greater than those of rock salt formations. There is thus a well-defined increase in formation thickness from the anhydrite group to the potassium-salt group.* The thickness of a formation is, however, a reflection of the intensity of subsidence of the sedimentary basin during accumulation of the sediments. The diagram consequently shows that the saline process terminated quickly in basins with weakly subsiding floors (group *I*), reached a more advanced stage in basins with more active subsidence (group *II*), and attained the final stage in basins with the most actively subsiding floors.

In summarizing the foregoing we note that *a combination of two conditions has been necessary for complete development of the saline process in marine saline basins: (a) prolonged and stable preservation of the tectonic regime in the vicinity of the strait, guaranteeing a positive salt balance in the basin, i.e. a predominance of salt over removal, and (b) preservation of a rate of subsidence in the basin in which the salts accumulate such that at each stage of saline precipitation this rate exceeds the sedimentation rate.* When either of these conditions is not fulfilled, the saline process does not go to completion but stops at some early or intermediate stage.

Even in formations that represent a complete, or almost complete, development of the saline process, however, the thicknesses of horizons of different compositions are highly variable; and this lends individuality to each particular formation. The cause of such variation in composition lies in the *character of the positive salt balance of the saline basin.* It is necessary to keep in mind

* Very rarely the transition of sedimentation in a saline gulf from one saline stage to another has been controlled by climatic conditions, such as in the present Gulf of Kara-bogaz (see below) and, perhaps, some of its ancient analogues (the Karaganda gulf in Transcaucasia, the Werra basin).

that, while remaining positive, this balance may vary markedly in a quantitative sense. It may be only slightly positive, i.e. with only a small excess of introduced over removed salt, or it may be moderately positive, strongly positive, or very strongly positive. The quantitative value of the salt balance has naturally had fundamental influence on the course of sedimentation. When the positive balance was slight, the basin increased in salinity very slowly. The transition from saturation with dolomite to saturation with gypsum, and then with halite, extended over a very long time interval, and this favored the accumulation of thick carbonate and sulfate beds in the lower part of the saline formation. When the positive balance was strong the attainment of saturation, first by $CaSO_4$ and then by $NaCl$, took place quickly; in consequence the beds of carbonate and sulfate were thin. With any change in the positive salt balance with time, e.g. by altering the tectonic regime in the vicinity of the strait, the development of any particular petrographic unit of the saline formation would also be altered.

Thus, it may be stated that *the physicochemical factor determined the general trend of the saline process and its stage, based on the solubility differences among the various components of the brine. The tectonic factor controlled, through the salt balance in the vicinity of the strait and through the character of subsidence within the basin, both the completeness of the saline process and the thicknesses of the individual beds of the saline formation.* One cannot fail to be astonished by the striking effect of the tectonic regime on such a process as saline accumulation, a process which one would expect to be purely physicochemical. The extent of this effect, together with other manifestations, will be discussed below.

4. The Cyclical Structure of Saline Formations and its Origin

A characteristic feature of saline formations is the varied cyclicity of rock associations in vertical sequence, expressed as repetitions of large units (macrocycles) and as thin alternations (microcycles).

As a rule, saline formations consist of three members: an *underlying* (infrasalt) *member,* made up of clay-carbonate-anhydrite rocks, corresponding to a successive saturation with dolomite and calcium sulfate; a *salt member,* made up of rock salt and potassium salts and corresponding to salinities above 25%, up to the eutonic stage; and a *suprasalt member,* corresponding to the period of general freshening, when anhydrite and carbonate-clay rocks cover the salt beds. The infrasalt, salt, and suprasalt members, according to Fiveg, together form a complete first-order cycle in the development of saline basins. Most deposits are characterized by several such complete first-order cycles. Thus, the Lower Cambrian saline formation of the Siberian platform consists of three cycles: the Usol'e, Belaya, and Angara. The Polessk formation consists of two cycles.

Each member is subdivided into second-order cycles. The lower part of the infrasalt member is characterized by an alternation of carbonate-bearing clays with marl and dolomite, the upper part consists of alternating clays and carbonate rocks with anhydrite. In these macrocycles the clay rocks correspond to times of freshening of the basin and introduction of terrigenous material. The anhydrite rocks correspond to time of increasing salinity of the water.

The macrocycles in the rock salt unit are similarly constituted: thin layers of fine-bedded anhydrite or carbonate- and salt-bearing clay at the base, a much thicker bed of halite above with more or less well-defined annual stratification and pinnate structure. According to Fiveg and Ivanov, these macrocycles are present over considerable areas in a deposit, perhaps throughout the entire area where the rock salt occurs, demonstrating the remarkable uniformity of conditions over the area. The presence of beds of clay at the base of macrocycle indicates that the precipitation of halite in the saline basin was interrupted many times by freshening of the water.

The structure of potassium-salt horizons is more complex, however. "The cyclic unit of the Upper Kama deposit begins with a saliferous clay 5–10 cm thick. Despite its comparative thinness, this bed marks a rather prolonged stage of dilution, during which time the upper water of the basin reached saturation with sodium chloride. Above, there occurs a layer of rock salt, 2–4 m thick, which accumulated over a period of 25–60 years, during which time the upper water of the basin became saturated with potassium salts. The cyclic unit culminates with seasonal precipitation of potassium salts, with the formation of a bed of potassium salts 1·5–12 m thick. The total duration of this cyclothem was 100–170 years. It should be noted that the precipitation of potassium minerals was not continuous throughout the accumulation of the thickest beds of potassium salts, but was usually interrupted every 5–15 years. For example, the red second bed (Red II) consists of seven layers, the odd layers consisting of sylvite, and the even layers of rock salt. The same accumulation, representing several years of sedimentation without potassium-salt precipitation, is shown in horizons C and D of the Upper Kama deposit, in the Stassfurt deposit, and in many other localities. Halite layers remain and do not change their position in the succession as the carnallite rock is displaced by sylvite. They are consequently used everywhere as marker horizons for correlating individual potassium-salt beds" (Fiveg 1955, pp. 7-8).

Prolonged and progressive increase in salinity, in both modern and ancient basins, has been a highly complex and cyclic process, involving many phases of freshening by inflowing water to interrupt the process of saline concentration. This freshening may have been brief and seasonal, or it may have lasted for decades or even centuries.

In interpreting the mechanism that controls the varied cyclical sequence in saline formations, let us recall that the connection between a marine salt-producing basin and the open sea is maintained by currents flowing in opposite directions: the surface current from the sea to the gulf, bringing water and salt into the basin, and the bottom current, removing part of the water and salt. Depending on the depth of the strait and other factors, the quantitative relations between inflowing and outflowing currents fluctuate considerably, and the hydrodynamic system accordingly varies sharply.

Fundamentally, three situations should be distinguished: (*a*) the increment of salts to the basin exactly equalled the discharge, (*b*) the increment of salts appreciably exceeded discharge, and (*c*) the increment was less than the discharge. In the first case, the salt basin preserved a uniform salinity for a long time and one particular type of salt was precipitated; and, by continuous tectonic subsidence of the floor, a thick sequence of essentially homogeneous sediment accumulated, such as carbonate mud, gypsum, and the like. In the second case, the saline basin gradually became more salty and changed from one stage of sedimentation, such as clay-carbonate, to another, a higher stage, such as sulfate and chloride. When the process was long continued, a more or less complete cycle of saline development resulted. In the third case, the basin freshened and the inverse course of saline sedimentation took place, from higher stages of salinity to lower.

Tectonism is by far the most significant factor in changing the hydrodynamic regime in the vicinity of the strait. Increased subsidence of the crust there led to outflowing currents becoming stronger, the saline basin being flushed out, with decrease in salinity and ultimate cessation of salinity saline deposition. Uplift of the crust acted in the opposite direction; it restricted the channel, and the outward current on the bottom became more and more confined till it disappeared. The basin then became saltier, and saline deposits were formed.

This mechanism of uplift and downwarping of the crust in the vicinity of the strait inevitably caused development of a series of typical macrocycles, each of which began with deposits at the stage of relatively low mineralization and ended with some stage of higher salinity.

There were also purely exogenetic factors, however, at times giving rise to macrocyclic structure in marine saline deposits. One of these was climatically controlled fluctuations in the level of the parent basin, which was immediately reflected in supply to the halogenic basin connected with it. A classic example of this is the present Caspian Sea (Fig. 209). As already pointed out (Part II, Chap. 1), since the thirties of the twentieth century the level of the Caspian has progressively fallen. This has caused shoaling in the strait connecting the Caspian with the Gulf of Kara-bogaz, and has led to a sharp reduction in the annual influx of water to the gulf—from 23 km^3 to (approximately) 7–9 km^3. This has been accompanied by increased evaporation in the gulf, and as a result the salinity increased annually 1–1·5% during the forties. The increasing salinity quickly reached the state of saturation with NaCl, and halite was precipitated. During the interval from the fourteenth to the eighteenth centuries, on the other hand, because of increased inflow of stream water, resulting from episodic occurrence of moister climate, the level of the Caspian was so high that the water is known to have overtopped the lower parts of the bar separating the Gulf of Kara-bogaz from the sea. This caused freshening of the gulf and deposition of clastic-carbonate muds, a freshening which could still be detected at the beginning of the nineteenth century (Mikhalevskii 1932). This is a well-defined macrocycle of purely exogenetic origin, ultimately climatically controlled.

This exogenetic mechanism for the development of macrocycles in saline formations has apparently operated but rarely in the past, only in large closed gulfs such as are found in the present Caspian, and such gulfs are not common. It is possible that gulfs of the Neogene seas in the Crimea-Caucasus region were sites of exogenetic development of cyclic structure.

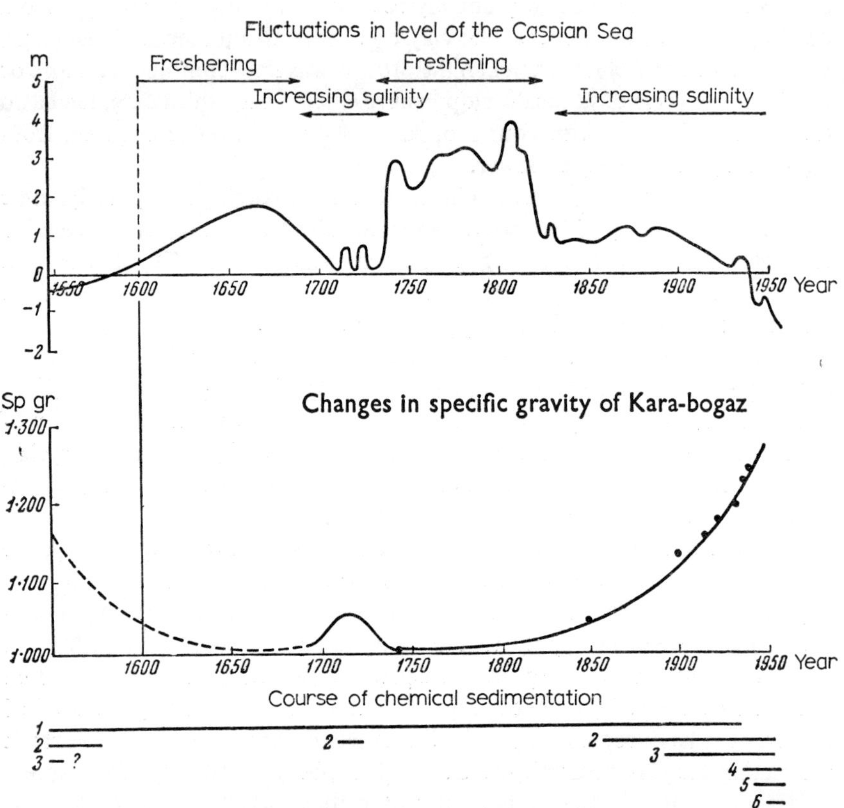

FIG. 209. Changes in salinity and sedimentation in the Gulf of Kara-bogaz in historic time. 1. Calcite; 2. clay (and $MgCO_3$); 3. mirabilite; 4. halite; 5. blödite; 6. epsomite.

5. RATES OF ACCUMULATION OF SALINE FORMATIONS AND THE TECTONIC FACTOR

After having established the absolute rates of annual sedimentation for recent sediments and ancient formations, attempts have been made to define the duration of the accumulation of saline formations. Without attempting to give an exhaustive summary of such determinations, some will be considered, particularly those that essentially give a general picture rather than a specific one.

The duration of the accumulation of the Upper Kama saline formation, reckoned by Ivanov (1953) and Fiveg (1954), is a case in point. According to Ivanov, the clay-anhydrite sequence, with a thickness of about 380 m, accumulated in 150,000–200,000 years. The time for the underlying rock salt, 250–300 m thick, was only 13,500–15,000 years, for the zone of potassium salts, 100–110 m thick, 1500–2000 years, and for the capping rock salt 400–600 years. The total duration for accumulation of the saline formation would then be 165,000–217,000 years, the clay-anhydrite sequence occupying 150,000–200,000 years, the salt sequence requiring a total of 15,000–17,000 years. This means that the salt accumulated at a rate 10–14 times that of the underlying clay-anhydrite sequence.

The duration of the interval for accumulation of the Stassfurt formation (according to Fiveg) was determined as follows. If it is assumed that 1 m of fetid dolomite required 20,000 years to accumulate (according to Richter-Bernburg), and the same thickness of bituminous, thin-bedded anhydrite required 2000 years, then the entire lower infrasalt part of the Stassfurt formation accumulated in 108,000 years $(20,000 \times 5 + 2000 \times 4)$. The time required for accumulation of the salts is approximately 10,000 years (20×550). Here again the rate of halite precipitation is approximately 10–11 times that of the underlying rocks.

The same relationships have been obtained in computing times required for the accumulation of other saline formations.

Two cardinal facts thus appear: (1) *even thick saline formations accumulated very quickly, and* (2) *sediments at advanced stages of the saline process accumulated especially rapidly.*

Both these facts again show that *the accumulation of saline deposits took place only in syneclises and depressions characterized by a rapid rate of subsidence.* But how do we interpret the marked increase in sedimentation rate at the advanced stages of the saline process? Was it due to a marked but short-lived increase in the rate of tectonic movement? Or did the rate of subsidence attain its maximum value at the very beginning? If the second view is correct, it is then clear that during the accumulation of the basal members of each formation (dolomite and clayey anhydrite) a huge depression must have formed, which was later filled with salts. The approximate depth of such a depression may be determined if the rate of subsidence from the beginning was the same as that during the accumulation of rock salt and potassium salts. Since the saline horizons accumulate at a rate 10–14 times that of the underlying clay-carbonate-anhydrite units, and the thickness of the infrasalt member in the Solikamsk deposit is approximately 380 m, the depth of the depression floor, on the basis of this assumed subsidence rate, must have been 3800 (380×10) or even 5320 m (380×14), i.e. some 4 or 5 km. *Such a figure is inconsistent with the subsequent history of the basin, and also with the processes of potassium-salt accumulation. Furthermore, even in the clay-anhydrite sequence itself there are no indications of deep-water accumulation. We must therefore reject the view that the rate of subsidence during accumulation of the Upper Kama deposit was the same as that during the salt deposition. This means that during saline deposition the actual rates of subsidence altered, being slower at first and becoming more rapid as time went on.* In exactly the same way, conditions were created that permitted the complete development of the saline process, as already discussed. Without increase in subsidence rate the saline process would have broken off during one of the early stages.

The tectonic conditions necessary for halogenesis, especially its fullest form, may now be redefined. *Halogenesis has developed, in general, only in syneclises that have been characterized by rapid subsidence. Complete development of the process has been possible only in depressions in which the rates of subsidence over short periods of time could change sharply and reach very high values. These times of accelerated movement corresponded to periods of accumulation of salt phases belonging to high and terminal stages of the saline process.* From this point of view, petrographic changes in saline formations along a vertical section are records of the changes in rate of subsidence of the depression or syneclise.

The history of the Lower Cambrian saline formation of the Siberian platform (Chap. 2) clearly illustrates such a course of accelerating subsidence in the Irkutsk amphitheater from Ushakovka time to Usol'e time.

The questions naturally arise: is this variability of subsidence rates with time a specific feature of depressions in arid zones, or does it characterize other climatic regions as well? Why has it not been detected elsewhere? The answers to these questions are clear. Variation in rates (commonly within short periods) is undoubtedly a characteristic of tectonic movements in all kinds of tectonic structures. If this phenomenon has not been detected in depressions of humid regions it is because rock indicators are absent in these zones: *the composition of humid-climate deposits in no way indicates the rates of accumulation. In arid zones, by contrast, the constitution of the different rocks is uniquely related to rates of accumulation*

The study of modern halogenesis has demonstrated the physical chemistry of the saline process, but little has been revealed concerning the effect of the tectonic factor on the process. Ancient saline formations, on the other hand, show the varied and very delicate effect of tectonic movements on saline deposition. Combined observations of recent and ancient deposits have permitted a more detailed and wider understanding of the saline process as a natural phenomenon than was possible by limiting ourselves to a single category. This simple, indeed self-evident, conclusion, repeatedly reached by examination of other sequences of rocks, is not applied by the opponents of the comparative lithologic method, with consequent damaging result.

CHAPTER 5

THE COMPOSITION AND ORIGIN OF THE
ROCKS OF THE HALOGENIC FORMATIONS

A discussion of facies types of saline formations is a necessary preliminary to the main objective of this part of the book: *the demonstration of systematic patterns in the composition of rocks and their distribution within saline formations and in ancient arid zones in general.*

In this chapter only rock constitution and distribution within formations will be considered, leaving remaining questions to the following chapter.

A. DIFFERENCES IN COMPOSITION BETWEEN
CONTINENTAL AND MARINE SALINE FORMATIONS

The outstanding feature in the essential constitution of saline formations is the well-defined difference between continental and marine members of the group. This is expressed in very diverse ways. It is manifested primarily in the distinctly *different proportions of clastic material contaminating the salt.* In both modern and ancient continental saline deposits the admixture of sand and clay is very large, and consequently the salts themselves—halite, glauberite, thenardite, and others—not only carry large fractions of insoluble residues, but they frequently alternate with clastic beds, most commonly with silts and sandy clays. Clearly, continental saline formations are essentially terrigenous-saline deposits.

In marine saline formations the rock associations are fundamentally different. Large fractions of clastic material are found only in some of these formations, particularly lagoonal, or in those deposits forming in gulfs occupying small basins of the Kara-bogaz or Virrila types (the Rhine graben or the Carpathian basin). In formations that accumulate in large gulfs, even more in the marginal zones of epicontinental seas and in great intracontinental seas, clastic material is found only along the periphery of the formations, usually only along one edge, and is essentially absent elsewhere. Saline rocks are here of great purity, the clastic component generally forming only tenths of a percent and rarely exceeding 3%. Exceptions to this are such isolated occurrences as the Starobin deposit. Still greater differences between continental and marine formations are seen in the composition of the solid salt phases.

Judging from modern continental saline deposition and its fossil analogues, continental saline deposits include deposits of natron, gaylussite, thermonatrite, northupite, lazulite, and other minerals that form in basins of soda type.

2 G

In marine saline formations these minerals are entirely lacking. Continental formations are also characterized by the widespread development of calcium and sodium sulfates, rarely of magnesium sulfates, chiefly in the form of gypsum, glauberite, thenardite, mirabilite, and blödite. In marine formations only gypsum or anhydrite are widespread, the remaining minerals in the list occurring rarely, and only in individual deposits of the so-called sulfate marine branch, i.e. enriched in sulfate. At the same time the overwhelming majority of continental saline formations contain no sylvite, carnallite, kainite, bischofite, or any other double salts of K and Mg chlorides or sulfates. In marine formations, accumulations of potassium-magnesium chlorides and sulfates are widespread and thick. They are present in 26 deposits, with a total combined reserve of K_2O of 50–70 billion tons (Ivanov 1959). Only halite is widespread in both continental and marine saline formations, which thus forms a connecting link between the two.

From these characteristics of the solid salt phases in continental and marine saline deposits, it appears that *the two types supplement one another in essential composition and, to some extent, may be considered complementary.* These differences are entirely predetermined by the facies conditions.

In order to appreciate the difference in clastic fractions as between continental and marine formations, it is necessary to keep in mind two circumstances.

(1) Continental saline formations, to the extent they are now known, are areally small, usually of tens or, at most, a few hundreds of square kilometers. They form in only small basins occupying relatively minor depressions. The formation of salts in these basins is associated entirely with the supply of water and dissolved salts by streams, which inevitably have also introduced suspended clastic material. This material has been deposited along with the dissolved fraction and has contaminated it to some extent. Because the salt-producing basins were small, clastic material has accumulated in them not only along the margins but also in the center, with contamination of the salts throughout the formations, although not to the same extent. The rapid rate of this saline sedimentation, especially at the halite stage, has in some measure reduced, but not eliminated, the clastic contamination, which is much greater than that in marine salt deposition.

(2) The greater supply of water to lacustrine basins is characteristic. In general, it has undoubtedly favored the more active accumulation of salts. In arid regions, however, increased inflow of river water has been possible only where uplands surround the basin. Similar relationships also obtained when water was brought from neighboring humid zones. Nevertheless, the increased discharge of stream waters and salts, according to the law of denudation (see Vol. 1, Chap. 1), led to the increased introduction of clastic material. Thus, the forms of water supply to lacustrine basins in arid zones, by their very essence, predetermine contamination of the salts with clastic material. This is why abundance of clastic material, *or their terrigenous-saline composition is a characteristic facies feature of continental saline formations.*

The accumulation of marine formations has been essentially different. They are always incomparably larger. The smallest of them occupy hundreds and thousands of square kilometers. Normally they cover tens or hundreds of thousands, not uncommonly even millions, of square kilometers. The Paleogene formation of the Rhine graben occupies an area of several thousand square kilometers. The Michigan Salina formation covers 500,000 square kilometers. The Lower Givetian sulfate-carbonate formation in the Moscow syneclise covers approximately 700,000 square kilometers, the Upper Permian Stassfurt sequence in the German basin about one million square kilometers, and the Kungurian sequence on the Russian platform several million square kilometers. The difference is clear; basins in which marine deposition of salt has taken place have been occupied by immense volumes of water, hundreds and thousands of times larger than those of continental saline lakes. The nature of the supply of fragmental material was also very different. Stream discharge is always accompanied by transport of clastic material; whereas this is only an insignificant part of the supply to salt-producing marine basins. In them most of the salt has come from the open sea and as such is practically free from terrigenous impurities. As a result, the saline formations of marine origin are usually free from contaminating detrital impurities. These clastics occur only along the margins of the formations, particularly where the depositional area was near mountainous terrain and stream discharge was large. Abundant fragmental material is characteristic of marine saline formations only if they have been deposited in small basins where the topography is highly dissected, such as lagoons and long narrow gulfs lying in marginal depressions and in the bordering zones of geosynclines, perhaps even on platforms (the Lower Miocene basin of the Ciscarpathians, the Rhine graben, the Tuva basin, the Nakhichevan basin, and others). In huge salt-producing seas and in the marginal zones of open seas, even if considerable clastic material is introduced locally (such as in the Lower Permian basins of the Russian platforms), it is spread over such a vast area of the basin that it provides only trivial contamination of the saline phases in most of the formation.

The causes of the substantial differences in mineral composition between continental and marine saline formations are clear. As pointed out in the second part of this volume, the hydrochemical types of intracontinental lakes of the arid zone vary considerably and include soda, sulfate, and chloride basins, the first two having at least three subtypes. The solid salt phases include such specific groups as natron and minerals with Na_2CO_3 in their composition, as well as thenardite (mirabilite) and double salts with Na_2SO_4 in their composition (glauberite, blödite). Furthermore, since continental lakes are supplied by fresh water derived from the surrounding humid zones, which carries only negligible quantities of potassium, there are no significant potash deposits in lacustrine sediments. Only in very large lakes that persist for a long time, as in the Tsaidam basin, have potassium salts occasionally accumulated, and then only in very limited quantities. The absorption of potassium by clay micelles apparently proves to have a restricting effect.

Sea water supplied to basins in which marine evaporites were deposited was hydrochemically uniform throughout Phanerozoic time, belonging to group *II* in our classification. With increasing salinity the sea water underwent some change, either losing or being enriched in $MgSO_4$. These changes were generally secondary, however, and had much less effect on the composition of the saline sediments than changes in hydrochemical types of intracontinental salt lakes. Marine saline formations, in consequence, lack such minerals as natron, minerals derived from Na_2CO_3 (thermonatrite, lazulite, and related species), thenardite, and derivatives of Na_2SO_4. At the same time, sea water in the more saline marine basins carried a hundredfold greater concentration of potassium salts and $MgSO_4$ than did stream waters. As a result, basin brines of marine origin frequently reached saturation in potassium and magnesium salts, and potassium-bearing horizons were deposited in the saline sequences.

It is therefore clear that the *fundamental differences in composition between continental and marine saline formations, in both the salt phases and the contaminating clastic material, are systematic, and essentially they result from facies differences.*

B. THE MACROCOMPONENTS OF MARINE ARID-CLIMATE FORMATIONS AND THEIR ORIGIN

In arid formations, as in humid, it is necessary to distinguish the principal formation-producing components (the rocks that make up the main bulk of the formational complex) and the microcomponents (minor constituents that are scattered throughout the complex).

The macrocomponents in marine formations are the carbonate and sulfate rocks, rock salt, potassium-bearing rocks (sylvinitic, carnallitic, kainitic, etc.), and mixed terrigenous-saline rocks. Accessory components are carbonate minerals in the saline rocks, as well as disseminated strontium, fluorine, bromine, iodine, boron, iron, manganese, phosphorus, trace elements (V, Cr, Co, Ni, Cu, and others), and organic material. The composition and deposition of these components in saline formations will be examined.

1. CARBONATE ROCKS OF SALINE FORMATIONS

Carbonate rocks are necessary components of marine saline formations. They underlie saline deposits proper to a greater or lesser extent, interfinger with them, replace and succeed them.

During the deposition of saline sediments carbonate rocks occur at the point of inflow of waters to the basin and also along its margins, being mostly inshore sedimentary facies. These sediments also commonly grade into marls and carbonate-bearing clays. Such a distribution of carbonate rocks may be observed in all saline formations, the form of which has been described in some detail. It is particularly well developed in the Michigan basin, in the Zechstein formations of Germany, in the Kungurian and Sakmarian-

Artinskian deposits of the Russian platform, in the Dankov-Lebedyan deposits of the Moscow syneclise, in the saline formations of the Siberian platform, in the Middle Devonian Williston formations of America, and elsewhere. When the saline basin is diluted and saline deposition ceases in its central parts, carbonate rocks form pelagic deposits.

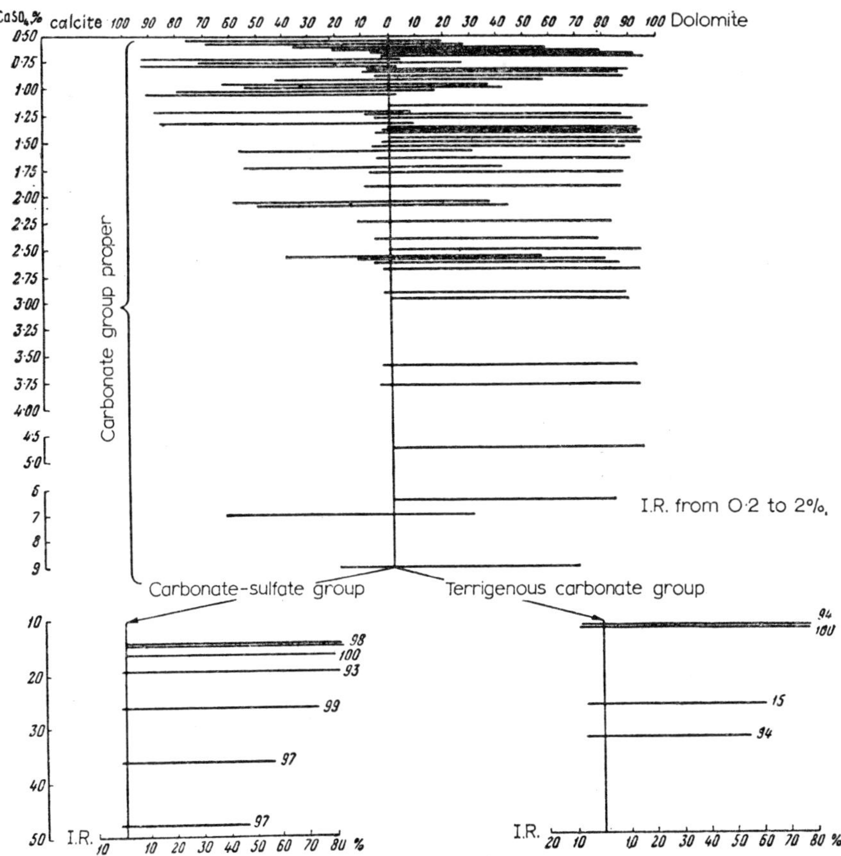

FIG. 210. Composition of carbonate rocks in saline formations (as found in the Kungurian deposits of the Ishimbai region of the Urals). The figures to the right of the straight line give the dolomite content of the carbonates (in %); *I.R.* insoluble residue.

In composition, the carbonate rocks in the Kungurian of the Ishimbai region of the Urals (Strakhov and Tsvetkov 1945) fall into three groups: (a) *carbonate rocks proper,* in which the clay component makes up less than 5%, (b) *carbonate-sulfate rocks,* containing some gypsum or anhydrite in addition to the carbonate, and (c) *terrigenous-carbonate rocks,* with much clay material (Table 38 and Fig. 210). Of these the carbonates proper will be first considered.

TABLE

Composition of Kungurian

Sample locality		Insoluble residue	R_2O_3	NaCl
		Normal dolomites, dolomitized		
Basal western sequence				
Borehole 5/11	Depth 536·7–548·2 m . .	1·64	0·44	0·16
Borehole 7/11	Depth 493–504 m . .	0·24	0·20	—
Borehole 6/11	Depth 549·3–555·9 m .	1·25	0·09	1·06
Ryazanovka 2	Depth 581·6 m . . .	0·13	0·05	—
Ryazanovka 2	Depth 601 m . . .	0·21	0·07	—
Ryazanovka 2	Depth 606–614·5 m . .	0·40	0·19	—
Borehole 5/11	Depth 567–578 m . .	0·57	0·09	1·29
Borehole 5/11	Depth 519·9–528·5 m . .	0·10	0·04	1·40
Borehole 5/11	Depth 528·5–536·7 m . .	0·46	0·16	1·78
Borehole 5/11	Depth 536·7–547·2 m . .	0·18	—	2·52
Beds in the anhydrite and salt sequence				
Borehole 5/11	Depth 409·6–411·9 m . .	0·74	0·18	0·59
Borehole 10/20	Depth 1043–1052 m . .	1·01	—	0·24
Beds in the terrigenous and sulfate sequence *along the eastern border of Kungur*				
Timashevka, Upper Kungurian . . .		6·02	1·80	None
Timashevka, Upper Kungurian . . .		3·46	1·62	None
Timashevka, Lower Kungurian . . .		18·20	2·86	None
Timashevka, Lower Kungurian . . .		0·64	0·16	None
		Anhydritic		
Basal western sequence				
Ryazanovka 1	Depth 546 m . . .	0·42	0·09	—
Ryazanovka 2	Depth 571–584 m . .	0·23	0·04	—
Borehole 7/11	Depth 481–493 m . .	2·36	0·64	—
Borehole 5/11	Depth 512–518 m . .	6·24	0·12	0·23
Borehole 5/11	Depth 480–492 m . .	1·28	0·22	1·53
Anhydrite sequence in the western margin *of Kungur*				
Borehole 3/4	Depth 466–484 m . .	5·73	0·98	0·86
Borehole 5/11	Depth 375·4–383·1 m . .	6·46	1·46	1·16
Borehole 3/4	Depth 593·0–595·7 m . .	2·78	2·16	0·71
Anhydrite and salt sequence in the central *part of the formation*				
Borehole 10/20	Depth 971–978 m . .	5·86	—	0·16
Borehole 42/2	Depth 658·4–663·4 m . .	3·26	—	Trace
Borehole 10/20	Depth 1033–1043 m . .	2·95	1·35	—
		Anhydrite-		
Borehole 3/4	Depth 527–532 m . .	1·30	0·40	—
Borehole 3/4	Depth 527–532 m . .	1·78	0·50	—
Borehole 3/4	Depth 542–547 m . .	4·52	0·92	—

38

Carbonate Rocks (in %)

$CaSO_4$	Carbonates	Of total carbonate			Σ	Dolomite content
		Dolomite	Calcite	Magnesite		
limestones, limestones						
0·77	96·75	3·04	93·71	None	99·75	3
1·32	97·0	9·43	87·57	None	98·76	9
0·54	96·30	20·23	76·07	None	99·22	20
2·07	97·74	38·64	59·10	None	99·99	39
2·06	98·80	43·70	55·10	None	101·14	44
2·55	97·32	58·00	39·22	None	100·41	59
0·97	96·69	42·88	54·01	None	100·79	43
0·75	98·51	90·43	8·08	None	100·89	91
0·71	96·46	86·34	10·12	None	99·50	90
1·46	94·82	94·38	0·44	None	98·98	99
0·66	96·09	83·00	19·59	None	98·26	86
0·56	96·80	5·05	93·75	None	97·61	3
1·63	59·67	29·66	58·91	None	—	33
Trace	85·52	84·50	11·02	None	—	88
Trace	77·69	7·02	70·67	None	—	9
1·84	96·94	87·30	9·64	None	—	90
dolomites						
14·65	83·87	82·52	1·35	None	99·03	98
6·59	92·84	39·28	53·57	None	99·70	40
13·34	76·96	74·06	2·90	None	93·80	96
18·31	80·04	69·74	5·30	None	94·87	93
1·84	90·86	89·49	4·37	None	95·73	98
5·71	89·98	82·73	3·5	None	99·31	97
10·15	76·56	76·15	0·41	None	95·79	99
2·87	87·54	87·54	—	None	96·06	100
26·33	62·23	61·87	56·64	None	94·68	15
17·86	75·70	71·45	4·25	None	96·82	95
2·55	90·06	83·78	6·28	None	96·91	93
dolomites						
48·57	46·31	45·05	1·26	None	96·58	97
36·23	57·90	56·25	1·65	None	96·58	97
31·99	54·24	51·10	3·12	None	94·37	94

The carbonates are limestones, to some degree dolomitized, calcareous dolomites, and normal dolomites. The first two groups are of a characteristically patchy texture with irregular areas of dolomite and pure limestone. This normal type of sedimentational-diagenetic dolomite was described in Chapter 3 of this work. Normal dolomite is generally massive, microgranular, more or less finely stratified, and commonly oolitic. The localization of carbonate rocks of different composition within saline formations has been strictly systematic, as may be seen in the calcareous content of carbonate rocks of the Michigan basin (Fig. 152).

The highest lime content is found in carbonate sediments at places where weakly mineralized marine and stream waters discharge into the basin. Here patchy calcareous dolomites and dolomitized limestones have formed. Farther away from points of discharge the lime content decreases and normal dolomites appear, in places sulfate-bearing. If the current entering the saline basin was strong and its effect was felt far into the basin, a great tongue of highly calcareous carbonate rock was deposited, as in the Michigan basin. Similar tongues occur in the Dankov-Lebedyan formation (Figs. 169-172) and also in the Angara and Belaya macrocycles of the Lower Cambrian saline formations of the Siberian platform (Figs. 179-180). When the supply to the saline basin is markedly from one side, the carbonate rocks show a *sharp asymmetry*: highly calcareous rocks mark the entry of weakly mineralized water to the basin, whereas dolomitic, or even sulfate-dolomite, rocks are deposited on the opposite side of the basin. Such a distribution may be seen in the Dankov-Lebedyan formation of the Moscow syneclise and in the Lower Cambrian formation of the Siberian platform. In both these formations, many limestones and patchy calcareous dolomites, in addition to dolomites, are found along the border next the sea. On the opposite side normal dolomites are found, locally sulfate-bearing.

Carbonate rocks are singularly saline formations. Some of them contain organic remains. Algae of various types are most common, and vary from small aggregates to major reef structures. Foraminifers, bryozoans, gastropods, pelecypods, and nautiloids also occur, and are generally restricted to limestones. They consequently abound at places where marine waters enter the salt-producing basin, and become less abundant with increasing dolomite deposition. This distribution of fauna and flora was demonstrated by Makhlaev (1959) for the Dankov-Lebedyan formation, where the richest and most varied fauna thrived in the narrow gulf at the southeastern edge of the syneclise. It decreased rapidly, and finally disappeared toward the north and northwest in the interior saline part of the basin. A similar distribution of fauna occurred in Belaya and Angara times.

It is important to note that *the disappearance of the fauna takes place more slowly than the change from calcareous sediments to normal chemical dolomitic sediments*. In several formations, including the Dankov-Lebedyan, the Kungurian and the Kazanian, the rock passes to a normal sedimentational dolomite with organic forms still surviving, though clearly impoverished and stunted; only where dolomite precipitated during markedly high salinity did the fauna disappear entirely. It is characteristic that plants persisted in the saline waters much longer than the fauna: algal structures are found even in sulfate-bearing dolomites.

For an understanding of the environmental conditions in the basins, the occurrence of organic forms, especially algae, is exceptionally important. Such forms show that *those parts of the saline basins where carbonate rocks were deposited were shallow, not more than 50–70 m deep, permitting sufficient light to penetrate to the floor for the development of algae.* Traces of erosion, desiccation cracks, and sedimentational breccia also indicate shallow water—as is invariably observed in lithologic investigations (Makhlaev 1959 and Pirsarchik 1958a)—and are characteristic chiefly of the margins of saline basins.

The terrigenous carbonate rocks are rocks in which the carbonate makes up more than 50% of the total. They include argillaceous limestones, dolomitized limestones, and marls. This group of carbonate deposits tends to occur along the margins of saline formations, in near-shore zones and especially where such zones lie along a mountainous coast or near the entrance of a stream. The sharply varying, and generally low, dolomite content in the carbonate component is characteristic.

In the sulfate-bearing dolomites the sulfate ranges from hundredths to several tens percent, with all stages in the transition from pure dolomite to pure sulfate rock. In those varieties which have only small proportions of sulfate, forms of the sulfate differ markedly. Most commonly platy crystals of gypsum lie in the plane of the bedding, either as isolated individuals, or in groups. Together with the well-formed crystals, zones of anhedral gypsum, thin replacements of dolomite by gypsum, and thin lenses of gypsum also occur. The presence of gypsum-bearing dolomite demonstrates that the salinity of the basin where such rock occurs must have exceeded 15% for at least in some years or during some evaporative seasons. The precipitated gypsum was first fairly uniformly distributed throughout the sediments, but during diagenesis it was segregated into individual crystals and lenses. The mineralogy and texture of the sulfate-bearing dolomites thus arose in complex fashion—partly during sedimentogenesis, partly during diagenesis.

It is necessary to keep another possibility in mind, however. When gypsum is precipitated from the bottom water and carbonate sediments are succeeded by sulfates, strongly mineralized brine, of high specific gravity, penetrates the upper part of the dolomitic sediments and may precipitate some gypsum there, admittedly in small volume. This selective crystallization has led to the formation of at least some of the above sulfate-bearing dolomitic rocks. To evaluate the effectiveness of this type of deposit, it is necessary to keep in mind that it could only begin after the bottom water passed from the saturation stage in dolomite to saturation in gypsum, when gypsum began to accumulate on the depositional surface, i.e. simultaneously with conditions that made possible the precipitation of gypsum by the sedimentary process. It is this fact that makes it extremely difficult to distinguish primary sedimentational sulfatization of dolomitic rocks from secondary processes. In considering that the opportunities for the latter are much more limited than for the primary process, however, we are forced to conclude that the role of secondary sulfatization is of only minor importance and may well be negligible.

2. SULFATE ROCKS—GYPSUM AND ANHYDRITE—AND THEIR ORIGIN

Sulfate rocks—gypsum and anhydrite—are the earliest representatives of saline rocks proper in marine saline sequences, and they will therefore be considered in some detail, beginning with their genesis. There are numerous proofs that gypsum is secondary after anhydrite in saline formations. Three facts are decisive. In surface exposures, sulfate rocks are always represented by gypsum, which changes progressively to anhydrite away from the surface. The gypsum shows all transitions to anhydrite, with clear traces of replacement of anhydrite by gypsum. Gypsum develops first along the cleavage planes in anhydrite: the individual grains become increasingly replaced until only relics of anhydrite remain in gypsum and, finally, the anhydrite disappears. The gypsum-anhydrite boundary is highly irregular; tongues of gypsum may extend deep into the anhydrite and *vice versa*; unaltered relics of anhydrite may occur in the gypsum. This is indeed the typical form of the boundary of a weathering crust.

Repeated attempts have been made to determine the depths at which the hydration of anhydrite begins. In the Ishimbai region of the Urals, where the top of the chemical sequence is higher than 150 m.OD., only gypsum occurs. When the top is between 150 m and -450 m.OD. both anhydrite and gypsum are found, and below -450 m.OD. only anhydrite occurs. Although some variations occur, gypsum is usually associated with elevations and anhydrite with depressions on the Kungurian surface (Strakhov 1947).

It is worth noting that the degree of hydration in anhydrite is controlled by the amount of deformation: the greater the faulting and jointing in the anhydrite the thicker is the gypsum zone, reaching 340–700 m over salt domes. By contrast, where the beds are almost horizontal the anhydrite is only 80–170 m thick (Strakhov 1947). Over the domes the development of gypsum is very irregular; in the relatively undeformed beds the gypsum forms an essentially flat body, suggesting a primary deposit.

Since gypsum in saline formations has formed as a secondary product by hydration of anhydrite, both minerals will be considered together. Sulfate rocks occupy substantially different positions in the different formations. In underdeveloped formations, where the saline process stopped with the precipitation of $CaSO_4$, the sulfate rocks occupy the central parts of the basin, wedging out toward the margins and grading laterally into dolomite. The individual beds of anhydrite are also separated by dolomitic horizons. In completely developed saline formations, i.e. those with salts in the central area, sulfate rocks are displaced toward the peripheral zones, forming a transition between carbonate rocks and rock salt. On the other hand, when the water in the basin is freshened and the precipitation of salts ceases, sulfate rocks again accumulate in the center. On the whole, the area of sulfate accumulation in saline formations is always much less than the area occupied by carbonates.

The $CaSO_4$ of anhydrite rocks is mixed to some degree with Ca and Mg carbonates, terrigenous material and organic substances. Quantitatively these components can be subdivided into four groups of closely related anhydrite rocks (Strakhov and Tsvetkov 1945). The first is anhydrite rock proper, with more than 95% $CaSO_4$. The second is dolomitic anhydrite with 50–95% $CaSO_4$. The carbonate-pelite-anhydrites also have 50–95% $CaSO_4$, but with approximately equal fractions of carbonate and pelitic material. The final, the anhydrite-clastic group, also with 50–95% $CaSO_4$, has clastic material as the principal impurity (Table 39 and Fig. 211). Each group will be considered in turn.

In the *anhydrite proper* the carbonate component is generally dolomite, which ranges up to 3·48% of the rock. Calcite is usually untraceable, but magnesite may be present. In exceptional cases calcite and magnesite are found together. Massive, patchy, and bedded anhydrite varieties are distinguished according to the distribution of the components. Massive anhydrite is dark blue, light blue, or light grey, generally coarsely crystalline, rarely fine-grained and homogeneous; the edges of fragments are clearly transparent. Large crystals occur in felted unoriented aggregates, but may form radial spherulites 0·75–1 cm in diameter.

The carbonates, in contrast to the main bulk of $CaSO_4$, are pelitomorphic, with grains about 0·01 mm in diameter. They are uniformly distributed throughout the rock, normally finely disseminated throughout the anhydrite crystals. In the patchy blue and grey varieties of anhydrite, the carbonate is concentrated in fine (parts of a centimeter) amorphous and irregular patches throughout the anhydrite matrix. The boundary between a patch and the anhydrite mass is normally rather sharp, but it may be transitional. The absence of any close relationship between the patches and fractures in the rock points to a primary origin. The patches developed during early diagenesis of the sediments and were connected with the unusual course of anhydrite recrystallization, during which the growing crystals of $CaSO_4$ displaced the finely dispersed and more or less uniformly distributed carbonate, giving rise to the variable texture. In the bedded anhydrite the carbonates occur in laminae (0·1–0·2 mm thick) separated by thicker anhydrite (1–2 mm). Fine clayey material is here mixed with the carbonate.

In the *anhydrite-carbonate group* the carbonate content grades up to 50%. From 95 to 100% of the carbonate is dolomite, the remainder being calcite. Magnesite is about as common in this group as in anhydrite proper, but occasionally it is abundant, as in the Kungurian rocks of the Ishimbai region, where it reaches 10–32%, and in the Kuibyshev and Saratoc districts of the Volga, where it constitutes up to 75% of the carbonate (Frolova 1955). In these rocks magnesite thus becomes a rock-forming mineral. These carbonates

TABLE

Composition of Anhydrite

Sample locality	Insoluble residue	R_2O_3	NaCl
			Anhydrite
Borehole 6/11 Depth 371–376 m; light blue massive anhydrite	Trace	None	Trace
Borehole 5/11 Depth 411·9–418·9 m . .	0·04	None	None
Borehole 3/4 Depth 553–558 m; anhydrite with carbonate veinlets	0·18	None	None
Borehole 48/20 Depth 460 m, the same .	0·42	0·10	Trace
Borehole 48/20 Depth 479 m, the same .	Trace	Trace	Trace
Borehole 7/5 Depth 713–738 m, the same .	0·64	0·14	Trace
Borehole 10/20 Depth 738–748 m, the same .	0·95	0·10	None
			Anhydrite-
Anhydritic dolomites			
Borehole 7/5 Depth 566–567 m; grey anhydrite with carbonate patches . . .	1·88	0·20	Trace
Borehole 48/20 Depth 485 m, the same .	0·88	None	Trace
Borehole 48/20 Depth 548 m, the same .	1·10	0·56	None
Borehole 42/2 Depth 157–180 m, the same .	0·35	0·05	Trace
Borehole 5/11 Depth 346–349·5 m, the same	1·72	0·26	None
Borehole 10/20 Depth 588 m, the same .	1·60	0·36	—
Dolomitic anhydrite rocks			
Borehole 5/11 Depth 464·5–472·9 m; thin interbeds of anhydrite and dolomite . .	0·48	0·44	0·31
Borehole 7/5 Depth 464·5–476·2 m; grey patchy anhydrite rock	3·48	1·08	0·16
Borehole 7/5 Depth 544·7–552 m, grey patchy anhydrite rock	3·25	0·15	Trace
Borehole 74/3 Depth 162–170 m; grey massive anhydrite rock	0·98	0·26	Trace
			Anhydrite-dolomite-
Borehole 28/20 Depth 460 m . . .	9·70	1·64	Trace
Borehole 28/20 Depth 479 m . . .	17·42	1·54	0·26
Borehole 28/20 Depth 510 m . . .	2·50	0·52	Trace
Borehole 28/20 Depth 534 m . . .	1·13	0·26	—
Borehole 42/2 Depth 432·5–440·9 m; thin-bedded anhydrite rock	3·27	—	—
Borehole 74·3 Depth 331–339 m . .	7·24	1·08	—
			Anhydrite-
Borehole 42/2 Depth 410–416 m . .	22·73	—	Trace
Borehole 48/2 Depth 465 m; sandy dolomite	25·03	3·82	Trace
Borehole 48/2 Depth 500 m . . .	14·30	0·88	Trace

39
Rocks (in %)

$CaSO_4$	Carbonate	Of total carbonate			Σ	Dolomite content
		Dolomite	Calcite	Magnesite		
rocks						
99·76	0·53	0·53	None	None	100·01	100
99·65	0·59	0·59	None	None	100·28	100
98·18	0·54	0·54	None	None	98·90	100
97·67	0·70	0·70	None	None	99·01	100
98·56	1·91	1·51	None	0·40	99·54	100
95·29	3·73	3·48	0·29	None	99·80	92
96·66	1·72	1·22	0·50	None	99·43	67
carbonate group						
78·51	22·66	20·02	2·64	None	103·85	89
86·22	11·20	11·20	—	None	98·30	100
82·43	12·47	8·94	3·53	None	96·56	72
93·51	4·86	4·86	2·00	None	98·77	100
94·05	5·17	2·91	1·26	None	101·74	75
92·43	6·82	—	3·57	3·25	94·39	None
69·53	29·39	29·39	None	None	100·65	100
56·62	36·21	36·21	None	—	96·95	100
64·44	28·96	28·96	None	None	96·80	100
84·48	13·23	13·23	None	None	98·15	100
clastic group						
69·16	16·12	14·02	2·08	None	96·50	88
51·83	27·84	24·32	—	3·52	97·14	100
89·79	5·95	2·75	3·20	None	98·76	45
91·57	5·46	3·98	1·48	None	98·44	73
93·58	2·55	—	2·46	None	99·35	None
78·48	11·73	7·73	4·00	None	98·53	66
clastic group						
63·43	9·93	4·13	5·40	None	96·19	43
55·50	9·00	8·51	0·49	None	93·35	100
80·96	1·88	0·47	1·44	None	98·39	47

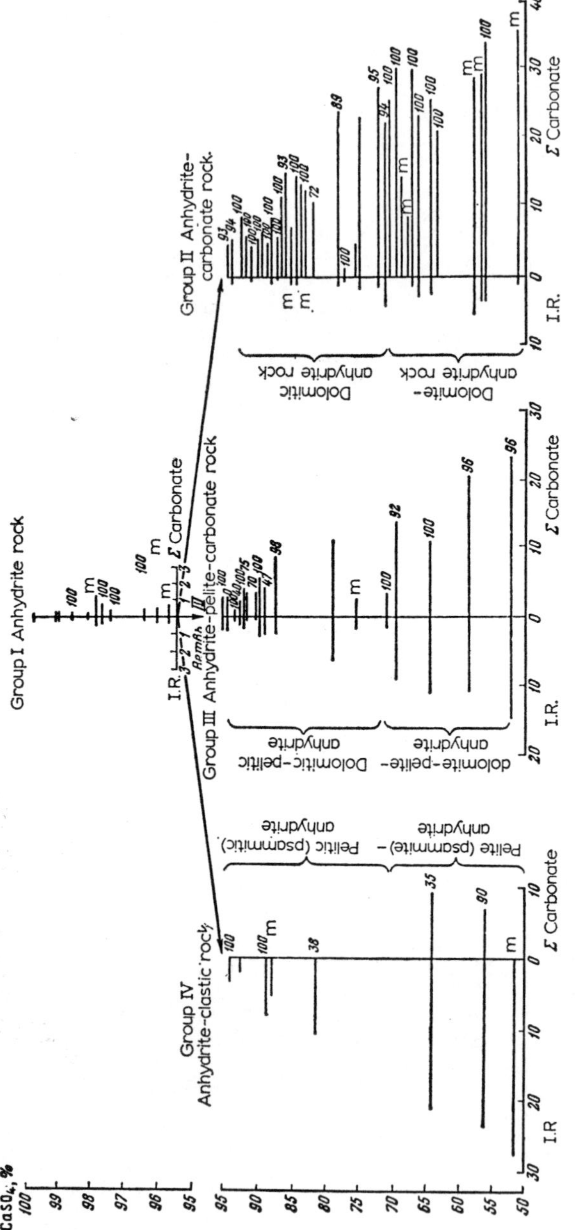

Fig. 211. Composition of sulfate rocks in saline formations (at Kungur in the Ishimbai region of the Urals). The ordinate indicates the CaSO$_4$ content in %; the abscissa indicates the insoluble residue (*I.R.*) on the left and the total carbonate on the right. The figures at the ends of the lines indicate dolomite content in the carbonates of a given sample (in % of total carbonate). *M* indicates that magnesite is present.

are micritic, showing no recrystallization. Anhydritic dolomites and dolomitic anhydrites are massive, texturally patchy, and bedded. Patchy varieties are especially abundant.

The other two groups of anhydrite rocks are distinguished by the substantial fraction of terrigenous (pelitic and silty) material, which is usually associated with a change in the carbonate component. When the pelitic material increases, magnesite disappears and the dolomite content becomes highly variable (80 to 45%).

In considering the carbonate components of anhydrite rocks, two facts are outstanding. Magnesite was first observed in 1940 by G. I. Bel'kov in anhydrite rock in the Kungurian rocks of the Ishimbai region of the Urals, but only in very insignificant quantities. Strakhov and Tsvetkov (1945) showed

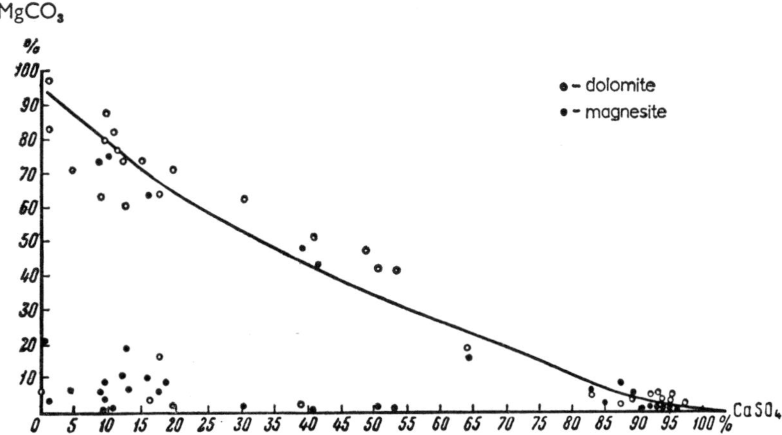

FIG. 212. Relations between MgCO₃ and CaSO₄ in saline rocks of the trans-Volga region (after Frolova).

that the mineral is widespread in the sulfate rocks of this region and that the content is locally high (up to 32%). The full significance of magnesite in sulfate rocks was revealed in the work of Frolova (1955) on the Lower Permian deposits of the Kuibyshev, Orenburg, and Saratov regions. In the first two regions, magnesite was found in 126 of 539 samples analysed, and ranged from 42 to 75% of the rock, i.e. from 2 to 100% of the total carbonate. Moreover, the magnesite fraction is clearly independent of the CaSO₄ content in the rock (Fig. 212). Indeed, the maximum magnesite values appear to lie in the initial transitional zone between dolomitic and anhydrite rocks. This, however, has still to be proved. A second fact is fundamental: *the magnesite-bearing rocks show no enrichment in chlorine, which ranges from* 0·25 *to* 0·75%. The same amounts of chlorine also occur in anhydrite rocks that lack magnesite.

All gradations exist between dolomite and anhydrite rock, as shown above. In order to explain the relative distribution of the members of this continuous series, deep drilling data of the Lower Cambrian saline formation in the

Irkutsk amphitheater, collected by Ya. K. Pisarchik, have been used. Some 325 analyses, each containing less than 3% terrigenous material, were used for the diagram, ensuring a reliable coverage (Fig. 213). The extreme members of the series, dolomites with little sulfate and anhydrite rock with small amounts of carbonate, are most abundant. Transitional members of the series, particularly those with large fractions of both components, are much less abundant. Dolomitic and anhydritic rocks thus tend to be isolated from each other to the same degree as pure limestones are separated from pure dolomites.

Accessory constituents in anhydrite rocks normally include celestite,

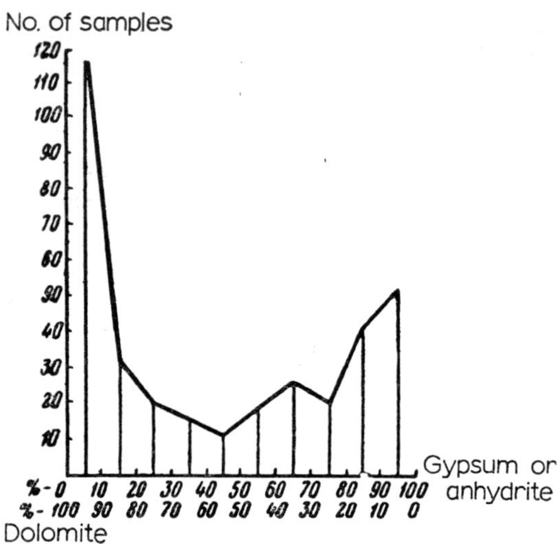

FIG. 213. Frequency of mixed rocks in the dolomite-gypsum (anhydrite) series of the Lower Cambrian saline formation of the Siberian platform (from the data of Ya. K. Pisarchik).

fluorite, and, less commonly, barite. Celestite and fluorite form occasional isolated crystals or aggregates disseminated throughout the rock.

Unusual local deposits of salts occur in the Lower Cambrian anhydrite-dolomite sequences of the Irkutsk-Usol'e part of the Siberian platform. Contrary to normal occurrences, these rocks are associated directly with halite beds. "Halite, making up 22–32% of the total rock, does not appear in any well-defined form, but occurs in a rounded salt-rich zone (from 1 to 8 mm in diameter). The salt aggregates result from the selective crystallization of halite, apparently having formed before the dolomitic sediment enclosing the anhydrite had completely lithified" (Yarzhemskii 1938, p. 93).

In addition to the mineral constituents of the anhydrite rocks, organic material is also locally present. In the anhydrite rocks of the Ishimbai region, apart from oil films, which are clearly secondary, a number of anhydrite samples were found to contain inclusions, up to 1–2 cm in diameter, consisting apparently of carbonized plant remains. At Aktash, on the Zilim River the gypsum was found to include a horizon rich in carbonized plants transported from the neighboring Ural land mass.

Since anhydrite rocks are interbedded with dolomites, the shallow-water origin of which is undoubted, sulfate deposits should also be considered as of shallow-water origin. In the marginal zones of salt-producing basins, the sediments may be re-worked and re-deposited as anhydrite cobbles and sands. Lenses of such fragmental anhydrite occur in the Kungurian of the Ishimbai region and in the Lower Cambrian of the Siberian platform. Eolian anhydrite sand is also recorded, forming dune deposits.

Near the centre of the basin, where depths are of the order of tens of meters, homogeneous unstratified anhydritic sediments are known to have accumulated within the zone of turbulence, which has led to carbonate and pelitic contamination. Thin-bedded anhydrite deposits have accumulated below the zone of turbulence, with the preservation of seasonal micro-stratification.

It was pointed out above that in present-day saline formations gypsum is secondary and has formed by hydration of anhydrite. From these considerations it might appear that the primary precipitation of sulfate from the brines of ancient basins was in the form of anhydrite. This, however, was not so. *The precipitation of sulfate from modern saline lakes in the mineralization range of 15–27%, corresponding to that of fossil sulfates, takes place exclusively as gypsum.* All experiments—from the ocean, the Black Sea, Lakes Sasyk-Sivash and Siki, the Sea of Azov, and others—on the evaporation of sea water at a temperature between 20 and 25°C have invariably yielded gypsum, not anhydrite. Experiments on metamorphizing brines, conducted by M. G. Valyashko, invariably yielded gypsum also. It may be added that individual, though very rare occurrences are known of the primary precipitation of gypsum rather than anhydrite. Dellwig (1955) recorded pseudomorphs of anhydrite after gypsum, in the seasonal anhydrite layers of rock salt in the Michigan basin. Similar observations have also been made in the Rhine graben.

This indicates apparently the removal of sulfates from brines of salt-producing basins by the formation of gypsum, and that anhydrite, now the only representative of calcium sulfate in unweathered saline formations, is a secondary mineral, the product of dehydration of gypsum, probably during katagenesis.

3. Halite Rocks, their Constitution and Formation

Halite rocks (halitites) are the earliest representatives of accumulations of highly soluble salts in marine saline formations. In symmetrical marine saline formations the halite rock occupies the central parts, giving way to anhydrite toward the margins. In asymmetrical formations the rock salt occurs in areas farthest removed from points of entrance of weakly mineralized sea or stream waters, forming in the upper closed end of the saline basin. Where the saline formation has a complex asymmetrical structure, the accumulations of halite

2 H

are found in various segments of the formation, corresponding to the most intensely downwarped depressions, but the greatest accumulations are still made in the closed terminal zone of the basin. Regardless of type of localization, rock salt forms beds of variable thickness within the formation (from tens of centimeters to several meters), alternating with anhydrite or anhydrite-dolomite. Taken as a whole, rock salt within each saline formation occupies a much smaller area than the anhydrite or, particularly, the carbonate rocks.

In addition to halite (the principal constituent), rock salt beds also contain anhydrite or gypsum, the carbonates dolomite, calcite, ankerite, and magnesite, pelitic material, and, locally, admixtures of potassium minerals (sylvite, langbeinite, and others). It is characteristic that magnesite is almost always the dominant carbonate. Pyrite is generally present, and sometimes hematite,

TABLE 40

Mineral Composition of Halite-Anhydrite Rocks, in % (from borehole 7/5)

Depth, m	Halite	Poly-halite	Anhy-drite	Σ carbon-ates	Dolo-mite	Calcite	Mag-nesite
Unknown	94·42	5·67	—	0·31	—	0·31	—
870–884	88·03	9·14	0·70	0·86	0·86	—	—
972–993	87·57	—	11·20	0·19	—	0·19	—
1042–1118	85·17	11·40	2·50	0·56	—	—	0·86
Unknown	81·56	16·31	1·21	0·25	—	—	0·25
810–839	80·35	—	13·66	0·56	—	0·96	—
840–889	60·34	28·48	6·38	1·63	1·63	—	—
889–890	11·77	84·03	—	4·98	4·37	0·61	—

giving the red color of some rock-salt deposits. In addition, iron is present in some measure in magnesium and calcium carbonates.

Three groups of halite rocks, similar to the above groups of anhydrite rocks, may be distinguished, on the basis of quantitative relations: (a) *halite rocks proper*, in which NaCl makes up over 95% of the rock, (b) *halite-anhydrite rock*, in which NaCl makes up more than 50%, the remainder being anhydrite with small admixtures of carbonate and pelitic material, and (c) *halite-carbonate-pelitic* rock, in which NaCl makes up over 50% but the carbonate-pelitic component is more abundant than the anhydrite (Table 40).

Halite rocks proper (Fig. 214) are transparent, turbid, coarsely crystalline rocks with grain sizes up to 1 cm; the grains are variable in form, generally anhedral. Commonly the grains are elongated in some preferred direction through tectonism in the salt; at times subprismatic elongation of crystals is found with geniculate angles up to 60°. Contaminating clastic material forms obscure, cloudy aggregates; but anhydrite forms individual, isolated crystals within salt grains. Furthermore, either component may form laminae (from parts of a millimeter to 2 mm in diameter), giving the rock a more or less well-defined cyclic bedding.

In some cases the rock salt beds are clearly subdivided into transparent and cloudy parts. Under the microscope the latter show well-defined pinnate structure—a relic of the primary sedimentational structure. Occasionally not only individual funnel-shaped or boat-shaped crystals occur in these beds, but also lenticular intergrowths of these forms—salt films or crusts that first floated on the surface of the brine and later settled to the floor. The transparent zones between the turbid zones are composed of recrystallized halite.

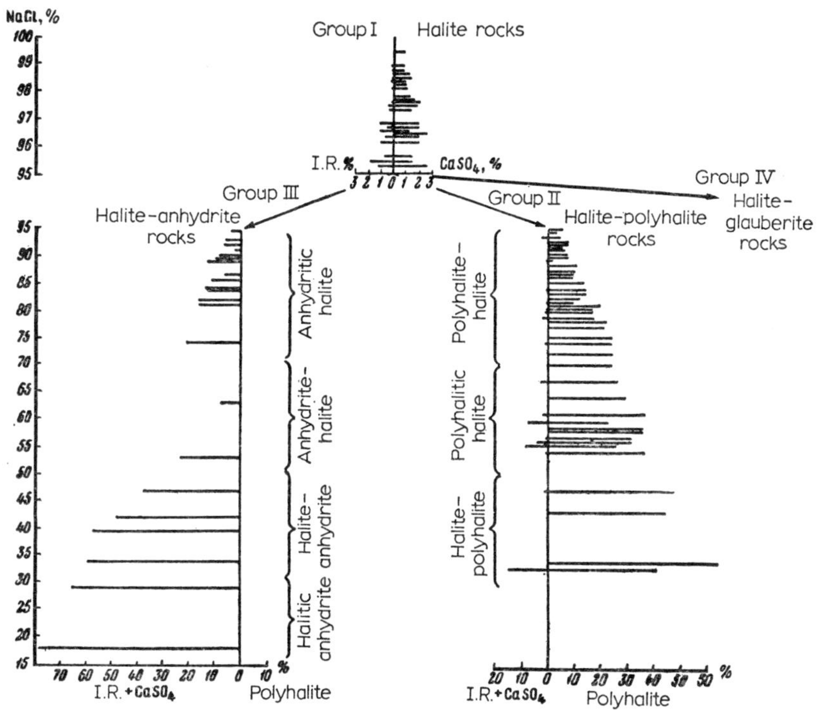

FIG. 214. Composition of halite rocks in saline formations (as observed at Kungura in the Ishimbai region of the Urals.)

The *anhydrite-halite rocks* are characterized by a great variety of petrographic types, caused not only by the variations in the principal components but also by the different distribution of the contaminants in the halite rock. Three petrographic varieties are most frequently recognized: (1) unstratified recrystallized anhydritic halite rock with cloudy poikilitic structure; (2) unstratified halite rock with the impurities distributed chiefly between halite grains; and (3) anhydritic halite rock with seasonal bedding. The varves are normally intensely crumpled. Occasional large (secondary?) grains of halite occur, either extending from one layer into another or lying between them.

Halite-carbonate-pelitic rocks are distinguished from the preceding groups by a sharp reduction in the anhydrite content: its place is taken by pelitic material with variable amounts of carbonate. The petrographic types are texturally similar to the preceding groups.

In some saline formations, especially in the Lower Cambrian of the Siberian platform and the Salina formation of the Michigan basin, distinctive rocks are found in which considerable quantities of dolomite, anhydrite, and halite are combined. These are dolomite-anhydrite, anhydrite-dolomite, or dolomite-anhydrite-halite rocks with practically equal parts of the constituent components. At low salinities in the precipitating basin, macroscopic grains of halite are not formed; halite delicately impregnates the rock through its pore-spaces and effectively cements it. At high salinities, halite forms discrete nodular concretions, and also fills cracks. The origin of these three-component rocks is not fully understood. According to Pisarchik (1959), they are probably secondary diagenetic emplacements, resulting from the downward percolation of a highly mineralized brine into dolomite-anhydrite sediments.

The high solubility of $NaCl$ results in the precipitation of all groups of primary halite rocks taking place at high salinities, at least 25–27%. As observations on present-day basins have shown, direct metamorphization of brine, such as the improverishment of the solution in magnesium sulfates, lowers the saturation point of $NaCl$; by contrast, any enrichment of the solution in $MgSO_4$ raises the point of crystallization of halite. On analogy with present-day precipitation again, the precipitation of halite took place in the past at the end of the evaporative period, especially on warm days. The solid phase of halite has always formed at the surface, where evaporation has produced a dense film, in the form of intergrowths that then settle to the bottom. In the sediment, some of the skeletal crystals are then recrystallized into transparent solid grains, but some are preserved in their incipient form. These primary structures of halite are, however, commonly lost.

Opinions differ widely on the depths of basins in which rock salt has accumulated. Some maintain that the basins were very shallow, as those of modern salt lakes in which rock salt is precipitating, while others believe that the water was very deep, up to 1000–1200 m in the Ishimbai region, or 400 m in the Stassfurt deposits of the Zechstein sea. As yet, however, these figures do not rest on any reliable evidence. In the few examples where dolomites have been found interbedded with rock salt it has been possible to find organic remains that clearly point to shallow water in the halite-producing basins (the Lower Cambrian Usol'e series of the Siberian platform). *The rapid accumulation of rock salt has been possible, not because it filled pre-existing deeps, but because the floor of the basin subsided rapidly during deposition, allowing a thick accumulation of halogenic sediments to form in a shallow basin.*

4. POTASSIUM-BEARING ROCKS, THEIR COMPOSITION AND DISTRIBUTION

Potassium-bearing rocks form the culmination of the process of marine halogenesis. This circumstance determines two characteristic features of their development. Although small inclusions of potassium and magnesium salts are sometimes found in halite rocks, large accumulations are very rare, having been recorded from only 26 areas (Ivanov 1959). Economically valuable deposits of potassium salts have been found in only eight of the many hundreds of halite deposits. This degree of rarity clearly indicates that highly specific conditions are required for their development and that these are rarely met in nature. Fossil accumulations of potassium salts are *always localized within areas of halite rocks*. Furthermore, the area of potassium salts is always less than that of halite deposits, usually representing a very small fraction of the total area. Moreover, potassium salts always appear in the uppermost part of a normal sequence of saline deposits, capping a considerable thickness of halite rocks.

In contrast to the saline rocks already described, the composition of potassium-magnesium deposits is characterized by an abundance of mineral species. The most important primary forms, i.e. those lying below the zone of weathering are :—

1. Chlorides: sylvite, KCl; carnallite, $KCl \cdot MgCl_2 \cdot 6H_2O$; bischofite, $MgCl_2\ 6H_2O$.

2. Chloride-sulfates: kainite, $KCl \cdot MgSO_4 \cdot 3H_2O$.

3. Sulfates: kieserite, $MgSO_4 \cdot H_2O$; blödite, $Na_2SO_4 \cdot MgSO_4 \cdot 4H_2O$; langbeinite, $K_2SO_4 \cdot 2MgSO_4$; glaserite, $3K_2SO_4 \cdot Na_2SO_4$; and polyhalite, $K_2SO_4 \cdot 2CaSO_4 \cdot MgSO_4 \cdot 2H_2O$.

Inherited from preceding stages of the saline process are anhydrite, $CaSO_4$; glauberite, $Na_2SO_4 \cdot CaSO_4$; and halite, $NaCl$. Halite is especially important because it forms a considerable part, even the major part, of potassium-bearing rocks.

Two features in the mineralogy of the potassium stage of marine halogenesis are outstanding: the abundance of double salts and the distribution of crystal hydrates. Both are typical of minerals forming from highly mineralized solutions (Part 2, Chap. 2), but the latter has particular physicochemical significance. Highly mineralized solutions of electrolytes are distinguished by low vapor tension above the surface; this causes retardation of evaporation, which in turn leads to retardation of further concentration. The development of solid crystal hydrates of the salt phases, bringing about the extraction of considerable H_2O in their precipitation, contributes to the loss of water as a solvent and may in some measure eliminate the retarding effect of salt precipitation caused by decrease in vapor tension above the solution. This means that *the formation of potassium crystal hydrates, fixing the solvent, is a counter process, speeding saline deposition to its final stage.*

The precipitation of potassium-magnesium salts begins at a general mineralization of about 32%, and it ends at the eutonic point at $S = 38$–40%. Thus, *the salinity range for the potassium-magnesium stage of halogenesis is* smaller than that for the dolomite and gypsum stages; in length it corresponds approximately to the halite stage. However, the abundance of solid phases that arise brings about a subdivision of even this short interval of halogenesis into a number of clearly defined stages, to each of which corresponds a mineral-petrographic zone. The number of stages varies in different saline formations because of the variations in composition of the evaporating brine. In basins with so-called sulfate brine, i.e. brine containing appreciable quantities of $MgSO_4$, the normal sequence of precipitation has been as follows:—

First (after the halite stage) was *the stage of magnesium-sulfate* deposition, represented by kieserite in combination with halite, polyhalite, and blödite in various quantitative relations.

After this followed the *sylvite stage (zone)*, characterized by sylvite in combination with kieserite and polyhalite, inherited from the preceding stage, and freshly formed kainite and langbeinite. The magnesium-sulfate zone grades into the sylvite zone with no perceptible break.

This is succeeded by the *carnallite zone*, the chief member of which is carnallite in combination with halite, sylvite, kieserite, kainite, langbeinite, and polyhalite. The transition from the underlying zone is again gradual.

Finally, the succession is capped by potassium rocks of the *bischofite zone*. This zone corresponds to a eutonic state. The characteristic mineral is bischofite, but representatives of the preceding stages are intermixed, especially halite and kieserite.

Magnesite, anhydrite, and clay minerals, introduced into the basin in various ways, occur as minor impurities throughout the entire section of potassium salts.

In basins with so-called nonsulfate brine, i.e. brine containing no $MgSO_4$, the mineral composition and zonation of potassium deposits is greatly simplified. Three zones instead of four have developed: *sylvite*, in which sylvite and halite occur in various proportions; *carnallite*, made up of a combination of carnallite and halite; and *bischofite*, in which bischofite, halite, and carnallite are present. All complex potassium minerals in combination with $MgSO_4$ are absent. Although the mineralogical differences of potassium between sulfate and sulfate-free basins are typically very marked, transitional occurrences are known, characterized by a lesser development of K-Mg sulfate minerals.

The different quantitative combinations of minerals have given rise to a great number of potassium-bearing rocks, though no generally accepted classification of such rocks has yet been proposed. Figure 215, based chiefly on the data of Yarzhemskii, shows variations in petrographic type characterizing the different stages of potassium-sulfate halogenesis. Halite is shown in the center of the cyclogram and the principal minerals of each potassium zone lie along the radii. As distance from the center increases the amount of halite decreases and the amount of the potassium minerals increases, reaching 100% at the circumference. The radii divide the cyclogram into segments (from the center) of 5, 25, 50, 75, and 95%, showing the degree of halite replacement by the potassium mineral. In order to include the observed complex relationship among the potassium minerals, supplementary radii are constructed at each radius of the cyclogram in planes perpendicular or inclined to the cyclogram. Each such radius represents a mineral and indicates the standard quantitative values.

The cyclogram shows that the number of potassium rocks, according to compositional differences, is generally large, though the distribution is extremely variable. The principal types are *sylvite rocks*, composed of halite and sylvite, locally with marked addition of kieserite (hartsalz) or anhydrite

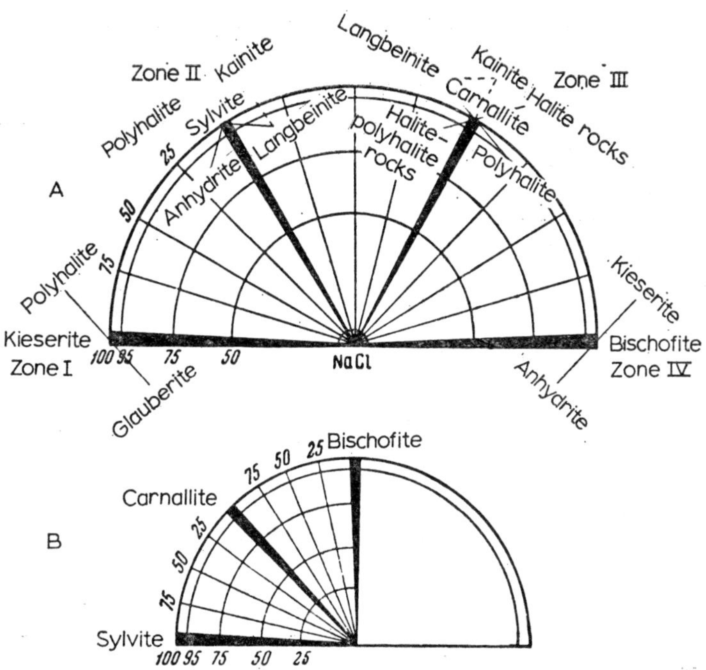

FIG. 215. Types of potassium-bearing rocks in the sulfate (*A*) and chloride (*B*) groups.

(anhydritic hartsalz), *kieserite rock*, the chief components of which are halite and kieserite, *polyhalite rock*, locally almost monomineralic but more commonly mixed with halite and frequently with kieserite and sylvite, *kainite rock*, composed of kainite and halite with langbeinite and kieserite, *langbeinite rock*, with the reverse relations of impurities, rocks that are transitional between kainite and langbeinite rocks, and, finally, *bischofite rock*, a mixture of bischofite with kieserite and halite, and locally with carnallite.

The spatial relations of potassium rocks in saline formations follow a rigid pattern. *These rocks are located at the maximum distance from the points of marine ingress to the salt-producing basin and also from the mouths of streams where they are present.* In salt basins of the Virrila type or of the group of intracontinental seas, the potassium salts have been confined to the farthest end of the basin. This may be observed in the Starobin deposit, the Zechstein deposit of Germany, and the Miocene rocks of the Carpathian region. Where more or less isolated successions occur about the margins of such pocket-like basins, potassium deposits have accumulated, as in the Kungurian region of the Russian platform. The position of potassium deposits in salt basins along the marginal zones of open seas has been less clearly defined (those in the Williston basin in Canada and the Upper Jurassic of Central Asia), because the paleogeography of these basins is not yet known in detail. It is unlikely, however, that any marked deviations from the established pattern will be found, because the very nature of the zone of potassium deposition, being the zone of highest salinity in the salt basin, requires maximal distance (or maximal isolation) from the point of entry of the weakly mineralized sea water.

Potassium-bearing deposits differ considerably in their mode of occurrence within the body of a saline formation and four types may be distinguished (Table 41).

The *first type* is represented by a single example, the Stassfurt horizon of the German Zechstein. The distinguishing feature of this deposit is its tremendous size and unique character. The total area of potassium salts amounts to 100,000 km², over which, however, only a single horizon of sulfate-type potassium salts, 6–20 m thick, was deposited. The horizon shows a well-defined annual sedimentary rhythm: rock salt beds 3–5 cm thick alternating with potassium salt beds, 5–7 cm thick. This seasonal rhythm is laterally very extensive, extending for many tens of kilometers, but the composition of the potassium minerals within the seasonal deposits varies, at times considerably. Richter-Bernburg has noted that sylvite and kieserite (hartsalz) occur about the margins of the potassium bed; carnallite and kieserite occur in combination in the more central parts of the bed.

Considering the great thickness of the annual beds at the stage of potassium accumulation (6–10 cm), the duration of the process by which the potassium bed was deposited was very short, amounting to about 100 years. This potassium stage was, however, preceded by a much longer period of halite sedimentation, during which up to 400 m of rock salt accumulated.

The *second type* of potassium deposition includes those evaporites in which the precipitation of potassium salts was very intense, but for which the space and time relationships were substantially different from those of the first type. The areal extent over which these potassium deposits accumulated is generally measured in several thousands of square kilometers, but the potassium beds are numerous, from 5 to 20 (most commonly 7–12), and the thickness of each is considerable, ranging from 0·5 to 7–15 m. Halite beds, of approximately the same thickness as the potassium beds, occur between the latter. As a result, the potassium-bearing horizon is thick, reaching 60–100 m, and the potassium content is high. In the Upper Kama horizon it amounts to 45–55%, in the Catalonian 35–50%, in the Upper Pechora 25–30%, and about the same for the Delaware basin. It has been noted that *the upper potassium beds generally lie within the boundaries of the lower; they are not only less extensive but also thinner.* Mineralogically the potassium horizons of this type belong to both sulfate and sulfate-free lines of descent. The beds are characterized by well-defined annual bedding. In the sylvite zone of the Upper Kama deposit, a basal carbonate-clay-anhydrite bed of negligible thickness is succeeded by a halite bed, in which pinnate structure is common. This is followed by sylvite in which relics of pinnate structure are also common. The thickness of an annual layer in this horizon is 3–7 cm. In the carnallite zone the upper member of the annual layers is carnallite instead of sylvite. The thickness of the bed is here 10–25 cm. In deposits of the sulfate group in other countries, the upper part of the annual layer may be polyhalite or kieserite or hartsalz, a mixture of sylvite, kieserite, halite, and other salts. The thickness of the annual layer is 7–10 cm. Pinnate structure may be present in halite, sylvite, kainite, and other minerals, or it may be destroyed by recrystallization.

The presence of many potassium-salt beds in a succession, separated by halite beds, attests that during the formation of potassium horizons of this second type prolonged and rhythmically repeated decrease in concentration of the brine, lasting for several years, occurred in conjunction with small seasonal fluctuations. If we assume the annual precipitation of halite and potassium salts to average 6–8 cm, the deposition of each bed of potassium salts with its overlying rock salt extended from 10 to 200 years. In the Delaware basin, however, where great differences in thickness between the potassium-salt and the rock salt beds are recorded, the period of potassium salt accumulation lasted for 10–40 years and that of the rock salt for 30–140 years (Ivanov 1959). The total time required for accumulation of the potassium horizon amounted to 10–12 thousand years.

The *third type* differs from the two preceding types in representing a more moderate intensity of potassium accumulation, as manifested in the smaller areal extent of the potassium beds, their smaller number, and their lesser thickness. The area covered by these beds is generally of the order of hundreds, or a few thousands, of square kilometers. The thickness is generally 2–3 m, rarely ranging down to 1 m and up to 5 m; the beds are generally 2–3 in number. The total thickness of the potassium horizons, however, is large (75 m, locally ranging up to 250 m). This thickness is largely due to increase in thickness of the intervening beds of rock salt, which commonly ranges up to several tens of meters. This indicates that the breaks between stages of development of potassium-salt beds occupied long intervals of time—250 to 2500 years. The phases of potassium salt deposition were much shorter, usually ranging from 25 to 75 years and only occasionally reaching 15–300 years. The total length of time occupied by the formation of the potassium horizons was also on the order of 10–15 thousand years.

Finally, the *fourth type* of potassium deposition is characterized by its low intensity. It is represented chiefly by potassium beds of small areal extent, from parts of a square kilometer to a few square kilometers, as in deposits of the Carpathian area. The beds may be few (1–2) or fairly numerous (8–13).

TABLE 41

Types of Potassium-Bearing Sequences (modified from Ivanov)

Structural type	Deposit	Composition and thickness of saline rocks underlying the potassium horizons (from base upward)	Structure of potassium-salt horizon				Composition and thickness of sequence of saline rocks overlying the horizon of K salts
			Total thickness, m	No. of beds of K salts	Thickness of beds of K salts, m	Thickness of interbeds of NaCl, m	
I	Zechstein Stassfurt formation (Z_2)	Anhydrite and dolomite, 1–3 m; rock salt, 40–500 m	—	1	6–20	—	Rock salt and anhydrite, 1–3 m
II	Upper Kama	Clay-anhydrite-dolomite rock, 300–380 m; rock salt 250–300 m	100–110	15–20	0·5–15	0·2–8	Rock salt, 50–70 m; rock salt with marl layers, 10–100 m
	Upper Pechora	Clay-dolomite-anhydrite, 300–250 m; rock salt, 70–170 m	22–40	4–7	0·5–0·75	0·5–9	Rock salt, 8–50 m
	Catalonia	Anhydrite, 2–12 m; rock salt, 200–500 m	60–80	12–20	0·5–7·5	0·5–7·5	Rock salt with layers of marl and anhydrite, from 40–50 to 150–200 m
	Delaware Basin	Anhydrite with layers of rock salt 200–300 m; rock salt with an-hydrite, up to 280 m	75	11–12	0·5–3·0	2·5–10	Rock salt, 50–75 m

III	Gaurdak	Anhydrite, 300–400 m, rock salt, 160–170 m	70–75	3–4	1·5–5	2·5–35	Rock salt, up to 55 m
	Tyubegatan	Anhydrite (?); rock salt, 170–175 m	35–70	2	1 and 3	50–60	Rock salt, 30–85 m
	Starobin (Pripyat basin)	Dolomite-anhydrite rock, 40–50 m; rock salt with marl layers, 250–300 m	200–260	3	20·3 and 5	150–190	Rock salt, 5–100 m
	Upper Rhine	Salt-bearing marl and rock salt, up to 1000 m; rock salt, 20 m	25	2	1·5 and 4·5	20	Rock salt, 60 m; anhydrite, 400 m
	Werra-Fulda formation and the lower Rhine depression (Z_1)	Anhydrite and dolomite, 8–15 m; rock salt, 90–120 m	60–75	2	2–3	55–60	Rock salt, 60 m; anhydrite, 400 m
	Hanover region (Z_3)	Anhydrite, 35 m; rock salt, 40–50 m	75–170	2	5–10	70–150	Rock salt, 15 m
	Yorkshire	Anhydrite-dolomite-salt sequence, 400 m; rock salt 85 m	55	2	9·5 and 5·5	20	Rock salt, 24 m; anhydrite and clay, 3 m
	Sterlibashevo	Anhydrite, 100–200 m; rock salt, 250–3000 m	—	1	2–4	—	Rock salt, 40–50 m; anhydrite 30–75 m
	Zhilyan	Rock salt, anhydrite, beds of polyhalite, more than 500 m	140–240	2	1·5 and 7	60–180	Rock salt, 50 m and more; anhydrite, clay, sandstone, 280 m and more
IV	Ciscarpathian basin: Stebnik	Clay-salt breccia, 500–800 m	up to 1000	1–8	0·5–100	20–300	Clay-salt breccia, 200–250 m and more
	Kalush	Clay-salt breccia, 5–200 m	150–300 up to 600	2–13	1–40	10–60	Clay-salt breccia, 10–80 m and more

The thickness is variable, ranging from parts of a meter to a few tens of meters, the latter being commonest. Furthermore, the mineralogy and structure of the beds, unlike those of the preceding types, are characteristically highly variable, both areally and vertically. The total reserves of potassium in deposits of this type are comparatively small; many deposits are negligible. The time involved for accumulation of these deposits has generally been short. The duration of potassium precipitation proper has ranged from tens of years to a few hundred.

Two features characterize these type of potassium deposition. *They together form a continuous series, from large deposits of potassium to negligibly small quantities.* The intensity of potassium sedimentation fluctuates widely in space and time. The transition from intense precipitation of K-Mg salts to very small deposition is seen chiefly in the decrease in area of accumulation, from many tens of thousands to only tens of square kilometers, or even only parts of a single square kilometer. The decrease of potassium accumulation has been effected partly by decrease in number and thickness of individual potassium beds. *Despite these wide variations in intensity of potassium deposition, the process has invariably been rapid.* The length of time required for the accumulation of the potassium-magnesium beds ranges from a few tens to a few hundreds of years. Among the various sedimentary processes occurring on the surface of the earth, the precipitation of K-Mg salts in saline basins is undoubtedly one of the most rapid, being exceeded by only a few others—the deposition of tuffs around volcanic centers, the deposition of mud flows and, perhaps, the deposition of sediments in tidal marshes.

5. The Paleogeographic Controls upon Potash Deposition

In considering the conditions of deposition of potassium salts three questions are of interest: (1) the physico-geographic conditions at the stage of potassium sedimentogenesis as compared with the preceding stages, (2) the mechanism determining the development of the sulfate and chloride lines of descent in potassium sedimentation, and (3) the mechanism of formation of complex potassium minerals.

Until recently, the existence of some specific sedimentogenic feature in zones of potassium sedimentation, other than the markedly high mineralization of the brine, has not been considered. Most geologists working on potassium deposits maintain that *there are no specific physico-geographic conditions in potassium-accumulating zones of saline basins.* These zones are apparently shallow-water areas with rapidly subsiding floors. From the bottom brines in these zones halite normally precipitates in some seasons (at the beginning of evaporation) and potassium-magnesium salts in others (at the end of the evaporative period). Cooling of the brine during cold periods also appears to have played an important role. Neither have there been any specific tectonic conditions prevailing during times of potassium sedimentation. The floor continued to subside, as in the preceding stages of salt accumulation, and as in other less concentrated parts of the saline basin. This concept of the stages of potassium deposition has been developed by Ivanov (1960), Lotze (1938), Borchert (1959), and others.

In 1949 Valyashko advanced another interpretation. In his opinion there is a very substantial physico-geographic condition that is specific for the stage of potassium sedimentation as compared with the preceding stages. This view is supported by a study of the relationship between the volumes of the solid salt phases precipitated from highly concentrated brine and the volumes of the remaining liquid phase—the mother liquor. In the second part of this volume, concerning modern basins, it was shown that, as the mineralization of the brine approaches the point of epsomite precipitation or even closer to the saturation point of potassium salts, the volume of the solid phases approximately equals the volume of the remaining mother liquor. It follows that, as the sediments are very friable and porous, most of the brine lies between the crystals of the solid phases, and the basin approaches the character of a dry lake.

According to Valyashko, at the beginning of this sedimentation, zones of potassium deposition in a saline basin are converted to shallow-water basins, such as modern dry lakes. But the fundamental difference between potassium basins and modern dry lakes is the high tectonic mobility of the floor. In very large saline basins differential subsidence subdivided the floor into zones that subside rapidly and zones that remain stationary, or even rise appreciably. The intercrystalline brines then flow from the relatively uplifted areas into those of relative subsidence. The former, being deprived of bottom brine, cease to be sites of further salt precipitation. Indeed, during moist seasons the salts in these zones may to some extent be destroyed. In the downwarped zones more or less large basins have developed with rock-salt shores and with highly concentrated brine, rich in K, Br, B, and Mg, from which potassium salts up to the eutonic stage are precipitated.

Thus, when active differential movements of the floor occur in huge dry salt lakes, local zones of especially intense downwarping appear, which form secondary basins with high concentrations of bottom brine. Potassium salts result from the evaporation of this bottom brine. The size of such potassium lakes with rock salt shores has been extremely variable. Some have been rather large, but shallow. Valyashko determined this as follows: the main bulk of sylvite precipitates in the autumn, during cooling of the brine, when, as was shown experimentally, a layer of sylvite about 1 cm thick is deposited from each meter of brine as the temperature falls to zero or slightly above. Since the annual beds of sylvite in the Upper Kama deposit range from 2 to 5 cm, the depth of the potassium basin was clearly 2–5 m. These same depths also correspond to measured thicknesses of annual sylvite layers in the Alsace deposits and elsewhere. Depths of 5–6 m were apparently common in basins precipitating potassium salts. The persistence of potassium basins among slightly uplifted areas of rock salt, and the total thickness of deposits in the potassium complexes, have been controlled by the duration of local downwarping, and also by the dimensions of the upwarped areas and the content of intercrystalline brine in them.

In any assessment of the two diverse viewpoints concerning potash deposition, due regard must be paid to one of the cardinal laws of saline deposition, established by Valyashko: *the volume of precipitated solid phases tends to approach the volume of the mother liquor as the brine approaches the saturation point in* KCl. Only hypotheses based on this law may be considered adequate to interpret the specific features of the potassium stage of sedimentation.

Nevertheless, accepting Valyashko's viewpoint, it must be borne in mind that *in detail the potassium basins in the past might have varied considerably.* One variant of the environmental pattern is that the deposits in which potassium rocks are found occur in small lenses as in the Ciscarpathian potash basin, and in the Zhilyan, Inder, Ishimbai, Sterlibashevo, and some other deposits. The smallness of these lenses and their great variation in composition make it difficult to treat these deposits on the basis of the first of the above explanations—as local patches in a broad saline basin from which halite is being precipitated. The features become readily understandable, however, as deposits in local pond-like hollows in a dry halite lake, into which inter-crystalline brine has flowed from the rock-salt bed. Changes in outline of the basin, accompanied by overflow of parent brine from one depression into the next after salts of the initial stages of potassium accumulation have begun to precipitate, may explain the complex composition of the potassium lenses in the Carpathian region, if it is assumed that the salt tectonics were initiated in the region in Miocene time. This also applies to other representatives of minor potash deposits, but in regions of larger accumulations, especially in the first and second types mentioned above, the pattern has clearly been different.

In the Upper Kama deposit, for example, the total area of potassium-bearing rocks is about 2800–3000 km^2, i.e. approximately one-sixth that of the present-day Gulf of Kara-bogaz. Since the principal potassium beds occupy essentially the entire area, it must be assumed on the basis of Valyashko's thesis that the brine basins within the halite field of deposition were vast as compared with the brine "pools" of the Carpathian region. The same may be said of the Starobin and Williston potassium basins.

It is possible that still larger potash-bearing basins formed elsewhere. The Zechstein potash deposits in Germany is in this regard of outstanding interest. The potash beds of the Werra formation were laid down in small local depressions, forming only a small part of the rock-salt field. These basins represent a type intermediate between the Carpathian and Upper Kama types. This applies also to the potash-depositing third of the Ridelian (or Lower Saxonian) saline formation. It is however possible that the situation was different during the Stassfurt salt deposition. The potash horizon here covers an area of about 100,000 km^2 in a field of rock salt about 350,000 km^2 in area. This raises the question: what conditions favoured the accumulation of a potassium deposit of such great areal extent? Was the potassium deposited in a single aqueous basin, as the German investigators on salt deposits maintain (Borchert, Lotze), or was it formed in a series of basins separated by "halite drainage areas"? Undoubtedly the Stassfurt horizon at present takes the form of a large number of relatively small lenticular potash bodies (carnallite rock and hartsalz) separated by rock salt.

German geologists believe that this lenticular form of the potassium beds is entirely secondary and caused by salt metamorphism, local leaching of the K salts, and conversion of potassium beds to rock-salt. Without denying the significance of such processes, their effectiveness in controlling the well-defined facies variability of the Stassfurt deposit may be doubted. On analogy with other large deposits, where primary facies variations among potassium horizons are always observed, i.e. primary sedimentational wedging of the potash beds, it may be affirmed that no single potassium bed over an area of 100,000 km² in the Stassfurt deposit was ever laid down but that the deposit consisted of a large number of more or less large potash lenses separated by a field of rock salt. Secondary processes of salt metamorphism merely intensified this primary variability; they did not create it. The individual depressions were probably large, measured in hundreds and thousands of square kilometers, as in the Upper Kama deposit.

Thus, an examination of individual potassium deposits, among them the largest, shows that *the view of Valyashko may be applicable even though the dimensions of individual salt basins amid tremendous "halite dry lakes" vary considerably.* But, if this is so, it follows that *the sedimentation of the potassium stage is not at all a simple continuation of preceding salt deposition but is a specific feature; in its general environmental aspect it represents a process by which brines form and persist in depressions and there yield potassium salts.*

6. The Origin of the Sulfate and Chloride Groups of Potassium-bearing Rocks

Rykovskov (1932) advanced the view that the cause of nonsulfate potassium deposits of the chloride group was direct metamorphization of brine, taking place intensively in initial states and being accompanied by loss of $MgSO_4$. This ideas was successfully applied to the Upper Kama deposit by Morachevskii (1940) and Urazov (1932). As applied to the entire range of nonsulfate deposits, this theory of origin became acceptable only after experimental confirmation by Valyashko and his co-workers.

Let us analyze on the salt diagram the displacement of a point representing the composition of sea water as direct metamorphization by the calcium ion affects it, according to the Haidinger reaction.

It has been established experimentally (Valyashko and Solov'eva, 1953) that during evaporation of unmetamorphized oceanic water the halite zone is followed by:

(1) a zone of normal magnesium sulfates (blödite, epsomite, polyhalite;

(2) a zone of normal sylvite rock with epsomite and sylvite;

(3) a zone of normal carnallite rock with hexahydrite, carnallite, and kainite);

(4) a zone of bischofite.

To determine deviations from this scheme let us select in Fig. 216*A* several points on the line Ok-5, such as 1, 2, 3, 4, and 5, defining an ever-increasing degree of metamorphization of the oceanic water. The course of crystalliza-

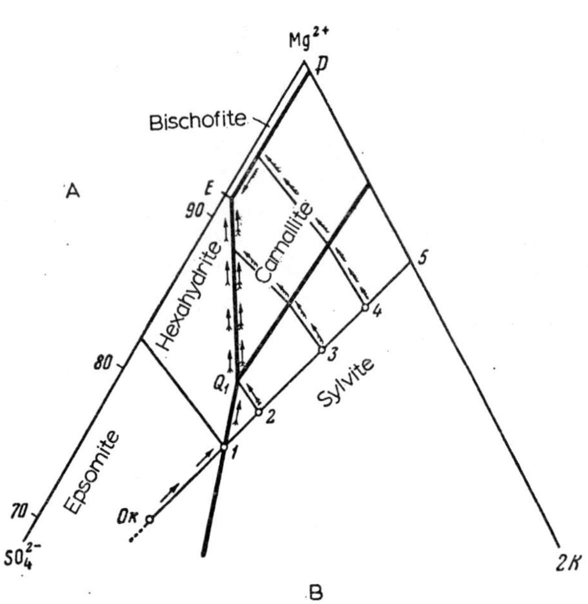

Fig. 216. Solar diagram and course of crystallization of oceanic water metamorphized to different stages (from Valyashko). *A*. Solar diagram: heavy solid line represents boundary between crystallisation fields; arrows indicate direction of change in composition of oceanic water during metamorphization (loss of SO_4^{2-}); fine line represents course of crystallization of normal (Ok) and metamorphized oceanic water at various stages of the process (1, 2, 3, 4, 5). *B*. Crystallization zones; ph represents polyhalite, *A*. anhydrite.

tion for each point is shown by the arrows. The terminal points of crystallization for compositions Ok-1, 2, 3, and 4 will be point E, the solar eutonic point; for composition 5 the terminal point will be D, corresponding to a brine without $MgSO_4$.

This diagram shows that at a low stage of metamorphization of the brine, corresponding to point 1, the precipitation of minerals during evaporation will differ little from the crystallization from normal sea water. The thickness of the zone of normal magnesium sulfates will be merely reduced because the course from Ok to 1 will be eliminated. During crystallization of brine of composition 2, a zone of sulfate-free sylvite forms after halite; the remainder of the section is as before, potassium and magnesium sulfates being preserved. During crystallization of brine of composition 3, a zone of sulfate-bearing sylvite rock forms after halite, and then a considerable thickness of sulfate-free carnallite rock. But magnesium sulfate persists in the upper parts of the carnallite zone and throughout the bischofite zone. During crystallization of brine of composition 4, the entire sylvite and carnallite sequence proves to be sulfate free and part of the bischofite zone is also free of sulfate. Lastly, brine of composition 5 characterizes completely sulfate-free potassium basins.

During progressive direct metamorphization of brine and the expulsion of $MgSO_4$ from it, the succession of salt deposits is simplified first by removal of the zone of magnesium sulfates and then by the disappearance of magnesium and potassium sulfates from the remaining zones. The progressive loss of $MgSO_4$ from the brine during metamorphization is directly and immediately reflected in the loss of the corresponding potassium and magnesium sulfate minerals from the salt succession. It therefore follows that gradual transitions must exist between the sulfate and chloride groups of potassium deposits (Fig. 216*B*).

Such are the changes in mineral groups at the terminal stages of the saline process during direct metamorphization of brines in saline basins. During reverse metamorphization of such brine, when the brine is enriched in magnesium sulfate, changes in composition of the deposits are reversed. Enrichment in magnesium and sodium sulfates leads to the appearance of glauberite during pre-halite sedimentation. After precipitation of halite, enrichment of the brine in $MgSO_4$ leads to development of a blödite zone and, during precipitation of potassium minerals, to a marked enrichment of the zone in kainite, langbeinite, kieserite, and polyhalite. This type of potassium-salt deposition from inversely metamorphized brine is found in the Neogene potassium deposits of the Carpathian region (Kalush, Stebnik, Golyn').

It is noteworthy that both sulfate and chloride types of potassium deposits existed as early as the Lower Cambrian, i.e. as early as saline deposits in general are known to occur (Fig. 217). Furthermore, potassium chloride sediments have the wider distribution: of the 26 recorded K-Mg deposits, 14 are chloride, 12 sulfate. Thus, *direct and long-continued metamorphization of brine in saline basins has been widespread in the past, and certainly more extensive than reverse metamorphization, which was only sporadic* (the Pripyat

21

System	Series, stage		Salt basin and deposit	Composition and type of potassium deposit		
				Small lenses or zones rich in K	Larger beds of K salts	Commercial deposits of K salts
Quaternary			Dallo deposit (Ethiopia)	———	———	
Neogene	Miocene		Ciscarpathian basin	- - - - -	- - - - -	- - - - -
			Maman deposit (Iran)	- - - -		
			Deposit on Sicily	———		
Paleo-gene	Oligo-cene		Upper Rhine basin (France, Germany)	———		———
	Eocene		Catalonian basin (Spain)	———		
Jura-ssic			Eastern Turkmen (Gaurdak basin)	———		
Tria-ssic	Middle		Adour basin (France)	———	———	
Permo-Triassic			Basin in Austrian Alps	- - - -		
Permian	Upper	Zechstein	North German basin	- - - - -	- - - - -	- - - - -
			Yorkshire basin	- - - - -	- - -	
			Polish basin	- - - - -	- - -	
		Salado	Delaware (New Mexico, U.S.A.)	- - - - -	- - - - -	- - - - -
	Lower	Kungurian	Upper Pechora basin	———	———	
			Upper Kama basin	———	———	
			Sterlibashevo deposit	- - - - -	- - -	
			Sterlitamak–Ishimbai deposit	———		———
			Zhilyan deposit	- - - - -	- - - -	
			Inder deposit	- - - - -	- - - -	- - -
			Ural–Emba salt-dome deposits	- - - - -	- - - -	
Carbon-iferous	Upper		Deposit in Utah	———	———	
	Lower		Deposits of Nova Scotia and New Brunswick (Canada)	———		
Devon-ian	Upper		Pripyat basin	———		
	Middle		Saskachewan	———		———
Cam-brian	Lower		Angara–Lena basin	———		
			Deposits of the Salt Range (Pakistan)	- - - - -	- - - - -	- - -

———— *1*   - - - - - - - - *2*

[*Legend at foot of next page*

deposit). Finally, it should be noted that salt basins have existed in which both sulfates and chlorides were precipitated simultaneously, or at slightly different times, in various parts of the basin. Thus, the Upper Kama deposit is free of sulfate, but sulfate salts do occur in the Ural-Emba salt-dome region. In the latter region, however, chloride deposits (Sterlibashevo) also occur in some localities. The same relations have been observed in the Carpathian region and in the Delaware basin. Where the two types of potassium-salt deposit are far apart, the existence of zones with different degrees of direct metamorphiza-tion is readily understood; but where chloride and sulfate units occur in the same region and even within a single deposit (Zhilyan in Kazakhstan and in the Carpathian region), the ordinary mechanism of metamorphization is inadequate to explain the relations, and some different process is necessary, such as displacement of brine from one local basin into another.

7. The Role of Diagenesis in the Mineral Composition of Potassium Deposits

The origin of minerals that are double salts of K and Mg is of outstanding importance: we want to know whether these are of primary origin, forming directly by crystallization from the bottom brine, or are a product of very early diagenesis, derived from primary metastable solid phases. It is well established that saline sedimentation in modern basins follows the so-called "solar" or metastable path (Kurnakov 1938). The fundamental problem is: did potassium sediments accumulate in ancient basins along the same meta-stable "solar" path as they follow in modern basins, or did they form differently? Unfortunately, no direct indication of the path followed in precipitation of the salt phases in ancient basins has been found, but general considerations furnish a completely definitive answer to the question.

If the rate of evaporation from the surface of the brine in ancient basins was similar to the present rate, then the only course of precipitation of solid salt phases must have been "solar", or metastable. Now, the rate of evaporation is

---

Fig. 217. Distribution of sulfate and chloride deposits of potassium in a composite geologic section of the lithosphere (from Ivanov). 1. Potassium and magnesium chlorides; 2. potassium and magnesium chlorides and sulfates.

controlled by climate, by aridity, by wind conditions. Present-day arid climate, according to all indications, bears the impress of the glacial epoch, both in its low temperatures and relatively high precipitation. During ancient periods of earth history, when there were no polar ice caps, the climate of arid zones was apparently drier and the temperatures at all latitudes were somewhat higher than at present. For some epochs—Late Cretaceous and Paleogene—these facts may be verified directly by measurements of paleo-temperatures (Vol. 1, Chap. 5). For other epochs, such as the Miocene, values are obtained similar to those of the present. *In any case, there are grounds for stating that the intensity of evaporation from the water surface in arid basins of ancient times was by no means less and was at times even greater than at present.* This is sufficient ground for stating with some confidence that sedimentation in the brine basins of ancient times took place as at present, along the metastable, "solar" path. It follows that diagenesis played a prominent role in the formation of the complex K-Mg minerals. The formation of the potash minerals was thus a twofold process: during sedimentogenesis the initial metastable material was precipitated; during diagenesis it was converted to stable phases. From experimental data and observations of the solid phases that form in nature during solar evaporation, as well as the mineral composition of the different zones of potassium deposits, the course of the potash-forming processes in saline basins may be plotted with high probability (Table 42).

In summarizing our analysis of the environmental conditions and the mechanism of potassium deposition, we obtain the following scheme.

Potassium sedimentation has been localized in parts of saline basins farthest removed from the sites of influx of sea water. Paleogeographically the basins have been huge, dry, rock salt regions, with individual, somewhat downwarped segments forming as local basins, supplied by inflow of inter-crystalline brine from the surrounding elevated rock-salt areas. Some such segments have been large (measured in hundreds or a few thousands of square kilometers), others have been small lakes bordered by rock-salt shores. Modern examples of such basins may be found in the present potassium lakes of the Tsaidam basin of China (Part 2 of the present volume). The water of potassium-accumulating zones was shallow but subsidence of the floor was rapid, and this has led to relatively long life of the basin despite the high rate of sedimentation. Such lakes have existed for tens to a few hundreds of years.

Whether the potassium deposits are of sulfate or chloride types has been determined by the intensity of direct or reverse metamorphization of the brine by the introduction of $CaCO_3$ and finely dispersed clay material by stream discharge. Processes of metamorphization themselves, however, did not take place in potassium-accumulating basins but rather in the larger and less saline basins of the preceding stages of development.

The potassium minerals, as in present basins, formed according to the "solar" or metastable path; that is, very early diagenesis played an important part in the development of the solid salt phases.

TABLE 42

Genetic Components of Salt Rocks (from Valyashko 1956)

No. of zone in order of precipitation	Zone (normal)	Primary rock-forming mineral	Secondary and accompanying minerals—diagenetic and epigenetic
6	Bischofite	Eutonic borate ($MgO . B_2O_3 . xH_2O . yMgCl_2$) Bischofite Magnesium sulfates ($MgSO_4 . xH_2O$, where x is 6–4) Carnallite Halite Gypsum	Boracite Kieserite Anhydrite
5	Carnallite	Carnallite Magnesium sulphates ($MgSO_4 . xH_2$), where x is 6–4) Kainite (in incipient stages of the zone's development) Halite	Kieserite Langbeinite Kainite (in lower parts of the zone Polyhalite (in lowermost parts, bordering the sylvite zone) Anhydrite
4	Sylvite	Sylvite Hexahydrite (sakiite) Epsomite Polyhalite Halite	Kainite Langbeinite Kieserite Polyhalite Leonite (?)
3	Magnesium sulfate	Epsomite Hexahydrite (sakiite) Polyhalite Halite	Kieserite Polyhalite Blödite
2	Halite	Halite Gypsum	Anhydrite Polyhalite (only in uppermost parts of zone)
1	Gypsum-anhydrite	Gypsum	Anhydrite

8. TERRIGENOUS-AUTHIGENIC SALT DEPOSITS AND THEIR COMPOSITION

Since saline basins have always received, in addition to sea water with its dissolved salts, some stream water with its suspended load of terrigenous material, saline formations have inevitably included some terrigenous-chemical deposits, at times in considerable volume, together with pure chemical rocks.

In humid zones the terrigenous component causes hardly more than a reduction in the chemical component of the sediment, but in arid climates the matter is more complex. A purely diluent effect of the clastic particles is observed only where such material is introduced by the wind or by the mechanical abrasion of the fragmental rocks of the shore areas. These mechanisms probably had merely secondary significance near ancient salt basins. As in humid zones, the introduction of clastic material by stream waters was the dominant process of contamination, and this circumstance strongly complicated the picture.

Since the turbid river water was only slightly mineralized, its effect on the sedimentation in the saline basin proved to be twofold. Its clastic load contaminated the saline sediments, forming more complex, multicomponent,

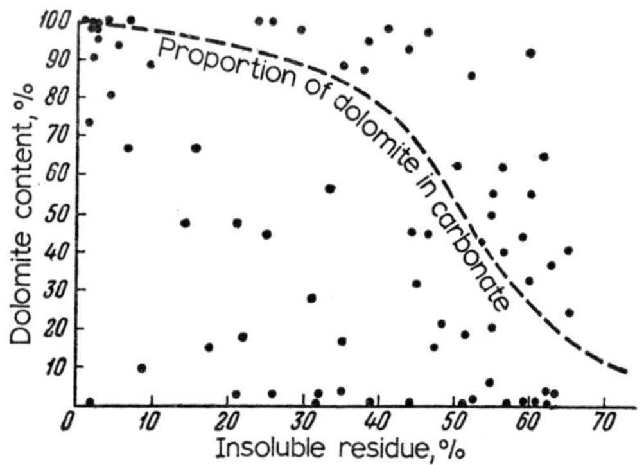

FIG. 218. The relation of dolomite content to insoluble residues in carbonate rocks of the saline formation in the Bashkirian region of the Urals.

terrigenous-chemical deposits. The fresh water also diluted the brine in the area where such deposits were accumulating. It may be assumed, as a first approximation, that the clastic fraction of the rock was proportional to the degree of freshening of the water, although there may have been no strict quantitative relationship. This freshening effect also disturbed the physicochemical equilibrium, markedly influencing the resulting chemical sedimentation. The significance of these changes may be appreciated in specific examples. In the so-called Allakaev facies of the Lower Kungurian saline sequence in the Ishimbai region (Strakhov 1947) tongues of terrigenous deposits represent traces of stream currents, bringing fresh water into a highly saline lagoon. Many analyses were made to study the relations between insoluble terrigenous residues of the Allakaev rocks and the content of $CaSO_4$, total carbonate, and proportion of dolomite in the carbonate material (Fig. 218). From Fig. 219 it is clear that as the clastic fraction increases from 0 to 40% the sulfate content of the rock declines sharply, from 93–95 to approximately 20%. Dolomite genesis remains unchanged within this range of variation: the entire mass of carbonate is precipitated as normal, or weakly calcitic dolomite. In the range of clastic material from 40 to 65%, the sulfate content falls to a few percent, and even completely sulfate-free rock may occur. Normal dolomite is here found only exceptionally; the dolomite proportion ranges from 80 down to 6%, indicating on average a sharp decline in the intensity of the dolomite-forming process. When the terrigenous fraction is above 65%, sulfate is reduced to mere traces and dolomite genesis is even more markedly reduced. Finally, above 80% clastic material, only normal calcite is deposited, or calcite with slight admixture of dolomite.

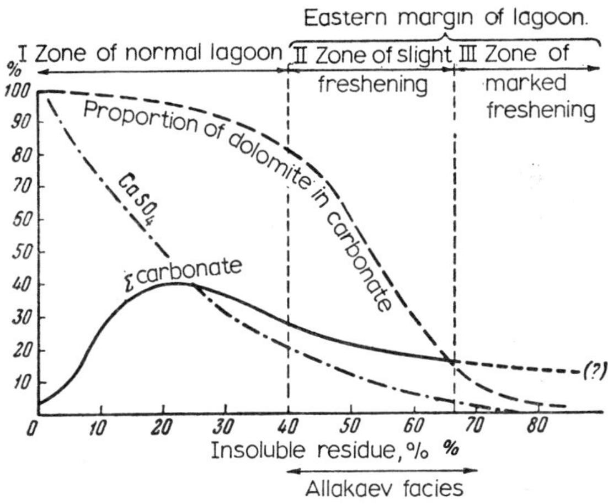

FIG. 219. Changes in the deposition of chemical sediments in the Kungurian lagoon due to discharge of fresh stream water.

A second example is found in the Lower Cambrian saline sequence of the Siberian platform. From the many rock analyses by Pisarchik (1959), about 600 with essentially no halite and with marked variation in content of sulfate and insoluble residue were selected and plotted (Fig. 220). When the content of insoluble residue is less than 5%, the $CaSO_4$ content ranges from zero to almost 100%. The remainder of the rock is dolomite. With increase in the insoluble fraction, the $CaSO_4$ value decreases steadily, and at about 60%, sulfate practically disappears, leaving only carbonate. The sharp fluctuations in the relationship of dolomite and $CaSO_4$ to insoluble residues below 5% clearly indicate that the rocks were deposited at very different salinities, from low salinity of the order of 15%, when gypsum was just beginning to be precipitated and dolomite almost entirely suppressed, to high salinity of the order of 27%, near the point of halite precipitation, when gypsum completely suppressed dolomite. *The ordinate consequently shows, in addition to the*

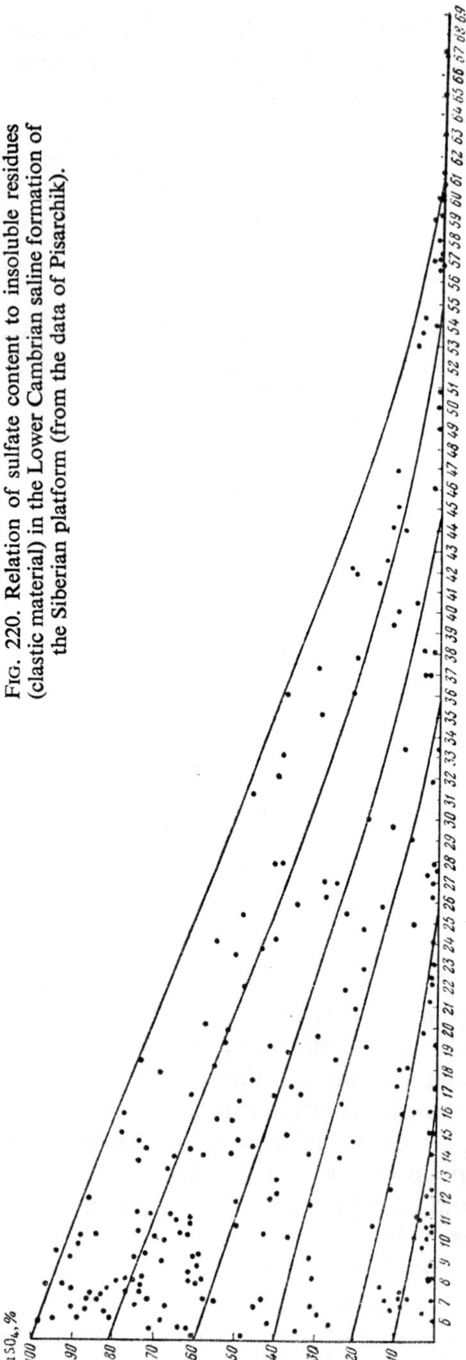

FIG. 220. Relation of sulfate content to insoluble residues (clastic material) in the Lower Cambrian saline formation of the Siberian platform (from the data of Pisarchik).

chemical component, an increase in salinity of the basin from 15 to 27% at 100% $CaSO_4$. Let us now take this upper point of maximal salinity and maximal $CaSO_4$ content in the rocks and move along the diagram from left to right, toward increase in clastic component, i.e. in a direction of gradually freshening of the brine. The value of maximal $CaSO_4$ content gradually declines, approaching zero. This process is shown by the upper curve. The introduction of fine clastic material and fresh water makes the precipitation of gypsum increasingly difficult, and this mineral entirely ceases to form when the amount of insoluble material in the rock reaches 60%, indicating a considerable freshening of this zone by fresh water. Another starting point on the ordinate corresponds in composition to the original chemical rock, 80% $CaSO_4$ and 20% dolomite. It is obvious that this rock could have been deposited at a lower value of mean yearly salinity in the initial water than that in the selected example. The gradual increase in amount of insoluble material in this rock leads to a progressive decrease in $CaSO_4$ to zero, but this latter point is reached earlier than in the above example. This is shown by curve 2 reaching the abscissa axis before curve 1. In taking a starting point on the ordinate with still less $CaSO_4$ in the rock and moving to the right along the diagram, i.e. toward an increase in amount of insoluble material in the rock, the zero value for gypsum is reached at a still lower value of clastic material (curve 3). By continuing this procedure to ever lower levels on the ordinate, i.e. in basins of still lower initial salinity, we find that a simultaneous decline in amount of inorganic material also follows and this means a smaller amount of fresh stream water in order to reduce the amount of $CaSO_4$ in the rock to zero. Finally, at the very lowest ordinate level, at a total amount of 10% of $CaSO_4$ in the initial rock, i.e. at a salinity of the brine near 15–17%, the insoluble material must make up no more than 16%, corresponding to a small quantity of introduced fresh water, in order for complete cessation of gypsum precipitation to occur.

These examples indicate that *at salinities in a basin corresponding to the precipitation of dolomite and gypsum, the influx of fresh water, bringing terrigenous material with it, may lead to the termination of precipitation of either salt. The lower the salinity of the initial water, the less fresh water and terrigenous material need be introduced to stop the precipitation of gypsum and dolomite.*

The situation is substantially different in basins of high salinity, in basins yielding halite and potassium salts.

In Fig. 221 (*A* and *B*), based on 44 analyses (Yarzhemskaya 1954), is reproduced the relationship between the amount of terrigenous material and the halite or sylvite content in rock salt and potassium salts. It is clear that the nature of the relationship is the same as that of the gypsum contamination: *the halite content decreases as the terrigenous material increases.* But there is also a basic difference: halite, forming in very salty water (curve 1), does not entirely disappear from a rock even at a very high clastic content, ranging to nearly 100%. Halite, which is precipitated at the beginning of saturation by sodium chloride, continues to form in rock with high silt and clay content.

Thus, *at the initial and middle stages of mineralization, when dolomite and gypsum are forming or when halite has just begun to be precipitated, the introduction of fresh water and clastic material leads not only to contamination, but may cause cessation of precipitation of these salts in localities, where the amount*

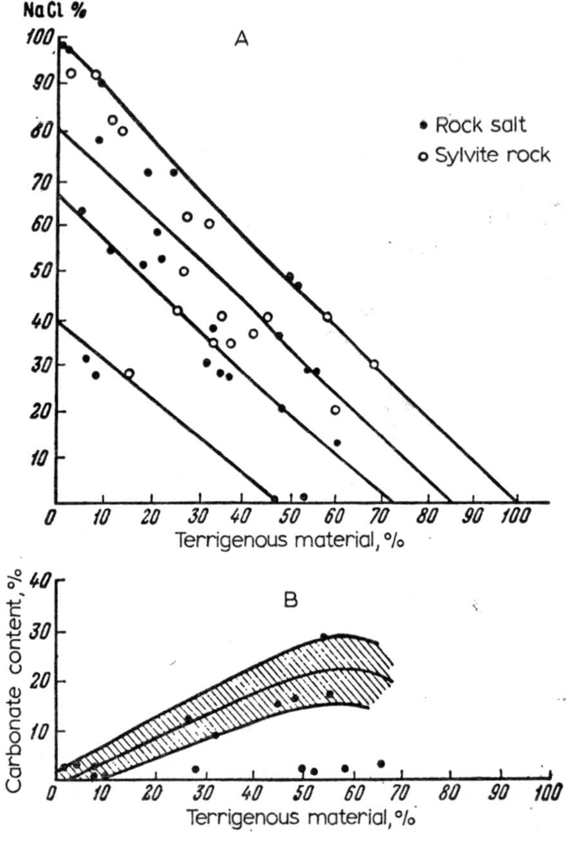

FIG. 221. Effect of insoluble material on the authigenic constituents of rock salt (from Yarzhemskaya). *A*. Relations of halite and sylvite to insoluble material. *B*. Terrigenous admixture and carbonate composition.

of clastic material is large. In other words, in sub-basins at the beginning and at the middle stage of mineralization, at localities where considerable clastic material is being deposited, *chemical deposition lags behind the general level of such sedimentation in the basin.* But when the mineralization is high, the effect of introducing fresh water and terrigenous material becomes less pronounced and is confined to extensive contamination of the solid phases.

The addition of terrigenous material to gypsum and rock salt is always accompanied by an increase in the carbonate content, normally dolomite or a mixture of dolomite and calcite (Fig. 221B). The increase clearly results from the introduction of Ca and Mg bicarbonates into the water.

Any increase in the clay fraction in these rocks is invariably accompanied by an increase in content of iron, manganese, and other components, as well as of organic material. During diagenesis, ferric iron and manganese dioxide are reduced and $FeCO_3$ and $MnCO_3$ are formed. $MgCO_3$ forms through reduction of $MgSO_4$ by interstratal water. When the primary carbonate admixture during sedimentation (i.e. $CaCO_3$, dolomite) is generally small in rocks, the appearance of diagenetic $FeCO_3$, $MnCO_3$, and $MgCO_3$ substantially changes the final composition of the carbonate. Reactions with sedimentational calcite and dolomite lead to the formation of ankerite and ferruginous dolomite, and excessive amounts of $FeCO_3$ form a siderite with more or less $MnCO_3$ and $MgCO_3$. In other words, the composition of the carbonate components in a rock strongly contaminated with clay material is characterized by high iron content and by complex mineralogy. With decrease in the amount of clastic material, the iron content of the carbonate is reduced. At the dolomite and gypsum stages, highly ferruginous ankerite, with admixtures of $FeCO_3$ in rocks rich in clay, gives way in slightly argillaceous rocks to ferruginous and then normal dolomite. At the halite and potassium stages, ferruginous ankerite and dolomite in rocks with high clay content give way to pistomesite; in rocks with low clay content, they give way to magnesite. These changes in composition of the carbonate components can now be verified (Shcherbina 1956a). They are theoretical rather than proved; nevertheless, the composition of carbonates in rocks of humid zones (Strakhov, Glagoleva, and Zalmanzon, 1959) supports this view. The content of C_{org} in a rock is of fundamental significance for actual carbonate paragenesis; the more C_{org}, the greater the amount of iron passing into FeS_2 and the less the contamination of the carbonate by iron carbonate. The purification of calcite and dolomite from ankerite, siderite, and ferruginous dolomite takes place more quickly in this manner.

The terrigenous components, while having a marked influence on the composition of chemical phases accumulating in arid-climate rocks, themselves react with the saline environment and are altered as a result. The process is most clearly seen in the clay minerals.

Yarzhemskaya (1954) has established the presence of hydromica and chlorite in the fine fraction of rock salt and potash salts. The hydromica commonly shows enrichment in K, Na, and Mg. The MgO content may be as high as 7–8%, clearly indicating a magnesian silicate. Sepiolite minerals have been identified in autumnal sediments of the present Lake Elton (Vasil'ev 1956). Pisarchik (1958b) has recognized hydromica and magnesian hydrochlorite in the fine clay fraction of the Lower Cambrian evaporites of the Siberian platform. The quantity of these minerals increases sharply with salinity. These few observed facts indicate that in highly mineralized arid basins hydromicas are enriched in alkali metals and magnesian silicates are simultaneously deposited. The abundance of these minerals varies markedly in different saline sequences.

In order to interpret these phenomena, it is necessary to recall that clay material in arid saline basins is chiefly allogenic, having been transported by streams from neighboring humid zones. Any alteration in the composition of the clay particles, at sharply defined horizons of the saline basin, indicates an interaction of the allogenic clays with the solid salt phases. The effects of the salt phases on the clay particles and of the clay particles on the chemical precipitates make saline rocks a distinctive entity, such as observed in the group of humid-climate rocks (Vol. 2, Chap. 7).

9. The Effect of Stream Water on a Saline Basin

Two specific forms of the effect of stream water on the saline deposition will now be considered: the massive accumulation of magnesite and of polyhalite in saline formations.

TABLE 43

$MgCO_3$ Content in Kungurian Rocks, in % (from Frolova (1955))

Area	Sample No.	$CaSO_4$	$CaMg(CO_3)_2$	$MgCO_3$
Krasnaya Polyana	543	61·91	16·73	7·60
Krasnaya Polyana	547	16·17	67·57	7·20
Rakhmanovka	42a	64·85	7·27	25·00
Rakhmanovka	42	37·15	16·36	43·45
Dergunovka	131	39·30	3·59	48·52
Dergunovka	134	16·06	3·94	64·60
Dergunovka	137	10·37	1·18	75·54

As already pointed out, when saline waters reach saturation with gypsum the Haidinger reaction begins to operate, as a result of which magnesium carbonate is precipitated. After some period of accumulation of this constituent, prior to the stage of oversaturation, magnesite begins to be precipitated and appears in the saline sediments. As shown by Bel'kov and later by the present author, together with Tsvetkov (1944), magnesite appears as early as the sulfate stage, but is especially abundant in rock salt and potassium rocks. As a rule, the accumulation of magnesite in saline deposits is slight, probably because of the small amount of stream water carrying dissolved $CaCO_3$ into the saline basin. But when conditions are favorable and where stream discharge becomes intense, the accumulation of $MgCO_3$ in the sediments increases sharply, and magnesite becomes a major rock-forming mineral. In the Kungurian deposits of the Kuibyshev segment of the trans-Volga region, and in the adjoining northeastern part of the Saratov region, the magnesite content of sulfate-bearing argillaceous dolomites at the base of the Kungurian series rises to 75·5% (Frolova 1955), i.e. the rock becomes an argillaceous, sulfate-bearing magnesite rock (Table 43).

The high concentration of magnesite is caused by two factors: (*a*) the zones of magnesite deposition were situated at the western shore of the saline basin; and (b) the high magnesite content is accompanied by a large terrigenous clay fraction. Both circumstances indicate that streams entered the basin somewhere near the present Samarskaya Luka, and these streams carried not only terrigenous clay but considerable $Ca(HCO_3)_2$ in solution. Interaction between calcium carbonate and magnesium sulfate in the brine, according to Haidinger's reaction, led to the formation of relatively large amounts of $MgCO_3$, which was precipitated with the clay, giving rise to argillaceous, sulfate-bearing magnesite rock. The presence of dolomite along with magnesite in sulfate rocks is explained by a change in chemical precipitation during the annual cycle. During the spring (or autumnal) freshening, when the water becomes temporarily undersaturated with gypsum, the Haidinger reaction does not operate, and dolomite is precipitated. By contrast, during the evaporative period, when the water is saturated with gypsum, the Haidinger reaction does function, and magnesite is precipitated. The sedimentary products of the different seasons are mixed in the sediment, and a distinctive dolomite-magnesite paragenesis develops (Strakhov 1951).

Whereas the massive deposition of magnesite in saline basins has been due to the reaction of the basins to the abundant discharge of stream water at various stages of the saline process, the accumulation of polyhalite, at least at a number of places, has been due to a similar reaction during very late stages of saline deposition. Thus, in the Kungurian saline sequence in the Ishimbai region polyhalite-halite rocks with a polyhalite content up to 35·6% are abundant, though not ubiquitous. The relationship of this material to pure halite deposits is seen in the Yar-bishkadak zone (Gerasimova 1956). In part, the polyhalite and halite sediments exhibit a zonal arrangement. The central and largest part of the deposit, elongated in a northwesterly direction, consists of rock salt, essentially pure (Fig. 222). Toward the southeast the thickness of the rock salt decreases as wedges of anhydrite-halite and clay-anhydrite rock appear. Toward the northeast and southwest, polyhalite aggegates appear in the rock salt and become progressively more abundant, ultimately forming a polyhalite-halite rock. At the same time anhydrite increases, and magnesite appears, the amounts of each increasing steadily toward the margins of the deposit.

In analyzing the origin of the polyhalite of the Ishimbai region, it must be borne in mind that the Kungurian saline rocks are here characterized by an absence of terrigenous débris. At the same time there is clear evidence of some discharge from streams entering from the Urals in the tongues of clastic material in the Kungurian sequence in the so-called Allakaev facies (see above). The Allakaev facies is traced in the area of the eastern reef mass not far from Yar-bishkadak (Strakhov 1947), where a stream of freshened water, carrying considerable quantities of $CaCO_3$ and clay, moved northward from that area to Yar-bishkadak, during the entire "salt-producing" period.

"The coexistence of rock salt and polyhalite-bearing rocks in the same horizons of a saline sequence and the fact that they more or less grade into

each other demonstrate the simultaneous precipitation of halite and polyhalite. The close paragenetic association of polyhalite with anhydrite,
magnesite, and clay in the peripheral zones of a salt deposit . . . attests to the
formation of polyhalite during constant introduction of fresh water carrying
calcium and clastic material into the basin.* The absence of any potassium
minerals other than polyhalite in the saline rocks of the Yar-bishkadak
deposit is no reason for supposing the polyhalite to be the result of diagenetic
alteration of unknown potassium minerals. The bedded structure of the
polyhalite and the alternation of polyhalite with rock salt are features that
indicate a syngenetic origin. Diagenetic processes here were not so intense

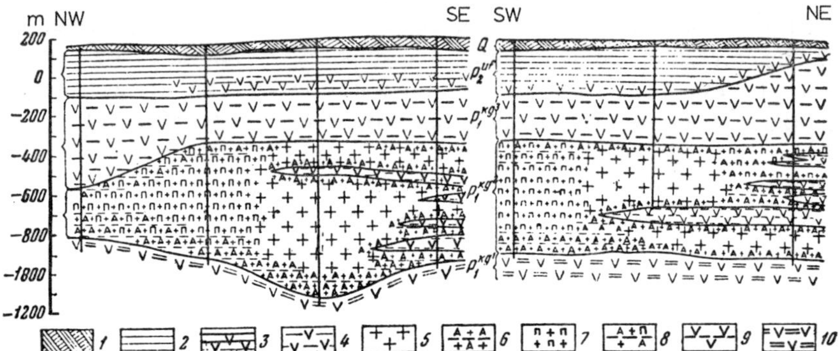

Fig. 222. The Yar-bishkadak polyhalite deposit (from Gerasimova). 1. Quaternary deposits
(Q); 2. variegated sequence of the Ufimian series (P_2^{uf}); 3. the same variegated sequence,
containing sulfates; 4. upper Kungurian member (P_1^{kg})—anhydrite with interlayers of marl
and clay; 5 – 9. middle Kungurian member (P_1^{kg2}): 5. rock salt; 6. anhydrite-halite, mag-
nesite-anhydrite-halite, and clay-magnesite-anhydrite-halite rocks; 7. polyhalite-halite and
magnesite-polyhalite-halite rocks; 8. anhydrite-polyhalite-halite and magnesite-anhydrite-
polyhalite-halite rocks; 9. anhydrite and clay-anhydrite rocks; 10. lower, Kungurian member
(P_1^{kg1})—anhydrite, clay-anhydrite, and dolomite-anhydrite rocks interbedded with clay,
marl, and dolomite.

that they would completely destroy all traces of the primary sedimentational
structure of the rocks" (Gerasimova 1956, pp. 359-360).

That polyhalite can form by this process was demonstrated by Valyashko
and Nechaeva (1952). These authors synthesized solutions saturated with
periodic salts† at 25°C, corresponding in composition to the different stages of
concentration of sea water from a point near the beginning of epsomite
crystallization to that of incipient crystallization of carnallite. In one series of
experiments a saturated solution of $Ca(HCO_3)_2$ was added to the solution, in
another a saturated solution of $CaSO_4$. The solutions were then evaporated to
their initial volumes. The procedure was repeated many times when, after
separation it was found that in some cases the solid phase was gypsum, in

* A similar conclusion was reached earlier by A. A. Ivanov for the Ciscarpathian
deposit.
† Periodic solid phases in the salt industry are minerals that separate out during one
season and are dissolved in another (such as mirabilite and halite) when suitable conditions
for accumulation of a salt bed have not yet been reached in the basin.

others polyhalite. Judging from the experiments, polyhalite crystallizes from the concentrated products of sea water only in the range near the incipient separation of sylvite. Directly altered solutions (i.e. solutions poor in SO_4^{2-}) do not yield polyhalite in the sylvite range of crystallization; they yield gypsum.

It is thus reliably established that *at high stages of salinity, when potassium salts are near saturation, modification of brine by Ca salts, introduced with inflowing fresh water, leads to the formation of polyhalite.* The ability of polyhalite to crystallize during interaction of slightly mineralized water with solutions saturated in potassium and magnesium salts also explains its common development during weathering of potassium deposits. As Yarzhemskii (1949) has shown, polyhalite replaces almost all the minerals of the potassium series: sylvite, kainite, langbeinite, and others.

10. The Law Governing Rock Distribution within Marine Saline Formations

In studying humid-climate lithogenesis (Vols. 1 and 2), it was shown that the position of a rock within a formation is determined by the facies type of rock and the facies type of formation. The distribution of rocks with marine saline formations essentially follows the same law, though with some modification. It is a fact that *in basins of arid zones the facies type of sediment is uniquely connected with the composition.* For example, dolomite as an independent deposit may form only where the water has comparatively low salinity, i.e. about the margins of the saline basin; when the water is strongly mineralized, gypsum forms. This leads to the conclusion that *the position of a rock within a saline formation is determined by the composition and by the facies type of that formation.* Therefore, it is futile to seek potassium salts in lagoonal formations; they form only in gulfs, marginal seas, and intracontinental saline basin. Potassium salts will always be found to occupy a restricted part of the rock salt field. Beds of potassium salts will always occur in the upper parts of a saline horizon of any member. In symmetrical formations, the potassium salts are in the central, thickest part; in simple asymmetrical formations they lie at the headward part of the basins, and in complexly asymmetrical formations, they are scattered, but always correspond to the thickest evaporite succession. The same may be said of any rock occurring in a saline formation.

This clear connection between rock composition and rock position within a saline formation indicates that such a formation is a combination of closely and genetically interrelated products.

C. MINOR CONSTITUENTS IN MARINE SALINE FORMATIONS

In continental saline formations the occasional occurrence of minor constituents—Br, B, and others—is well established. In marine saline formations, by contrast, F, B, Br, Sr, and some other elements are always present, and they follow a systematic distribution pattern with increasing salinity. In addition, Fe, Mn, P. V, Cr, Ni, Co, Cu, and some others are also present, and their distribution and mode of occurrence are also systematic. Their distribution and origin in marine saline formations will therefore be considered.

1. THE DISTRIBUTION AND DEPOSITION OF STRONTIUM

The strontium content of streams (according to V. I. Vernadskii) is about $1·3 \times 10^{-5}\%$, but in sea water about $10^{-3}\%$, i.e. almost 70 times as great. Despite this, even in the form of $SrCO_3$, strontium falls far short of saturating sea water, and marine sediments contain almost no deposits of it. The minor strontium content normally observed in calcareous sediments and rocks is due to biogenic extraction of the element along with $CaCO_3$ during shell formation.

In saline basins of marine origin the strontium content in the brine increases with salinity, finally reaching saturation, at which point strontium is precipitated, forming an independent solid phase—celestite $(SrSO_4)$—that becomes mixed with the principal salt phases. Occurrences of celestite in saline rocks have been known, the Cambrian deposits of the Siberian platform, the Devonian Kungurian, Kazanian, and other rocks of the Russian platform, the Zechstein of Germany, and elsewhere. However, determination of the precise moment in the history of a saline basin when $SrSO_4$ begins to precipitate has become possible only in recent years.

The distribution of strontium in the Kungurian deposits of the Ishimbai region was studied by the present author in cooperation with I. D. Borneman-Starynkevich (1946), working on 75 varied petrographic types (Table 44). On

TABLE 44

Distribution of Strontium in Kungurian Rocks of the Ishimbai Region in %

Sample locality	Insoluble residue	Sulfates	Carbon-ates	Sr
I. Sandstones				
Borehole 3/25, depth 729 m . .	51·79	0·91	16·18	0·10
Zigan River, base of Kungurian .	—	1·22	29·30	0·12
The same	—	0·49	34·86	0·03
			Average	0·08
II. Clays and marls				
Zigan River, base of Kungurian .	—	1·27	73·52	0·37
Borehole 3/25, depth 1252 m . .	42·14	0·39	39·95	0·76
Borehole 3/25, depth 1318 m . .	45·25	24·33	17·58	0·39
Borehole 3/25, depth 1323 m . .	55·04	8·23	19·86	0·25
Borehole 3/25, depth 1326 m . .	54·87	11·26	20·74	0·41
Borehole 3/25, depth 1329 m . .	60·02	5·97	18·51	0·64
Borehole 3/25, depth 1335 m . .	47·05	20·00	20·67	0·59
Borehole 3/25, depth 1403 m . .	55·87	0·57	32·53	None
Borehole 3/25, depth 1408 m . .	52·50	0·59	22·94	0·22
Borehole 3/25, depth 1420 m . .	12·89	66·20	15·99	None
			Average	0·36

TABLE 44 (continued)

Sample locality	Insoluble residue	Sulfates	Carbonates	Sr
III. Calcareous dolomites				
Borehole 5/11, depth 375·4–383·1 m .	6·46	10·15	—	0·20
Borehole 5/11, depth 90·5–398·2 m .	2·96	6·29	—	0·94
Borehole 5/11, depth 409·6–411·9 m .	0·74	0·66	—	Trace
Borehole 5/11, depth 519·9–528·5 m .	0·30	1·53	—	None
Borehole 5/11, depth 519·9–528·5 m .	0·20	0·95	—	3·78
Borehole 5/11, depth 519·9–528·5 m .	1·18	2·21	—	2·22
Borehole 5/11, depth 519·9–528·5 m .	0·34	4·68	—	0·84
Borehole 5/11, depth 528·5–536·7 m .	0·46	—	—	1·91
Borehole 5/11, depth 567·5–578·1 m .	0·28	0·83	—	0·99
Borehole 5/11, depth 578·1–581·1 m .	0·30	1·84	—	0·10
Borehole 10/20, depth 1033–1043 m .	0·10	1·12	—	None
Borehole Ryazanovka 1, depth 540–546·2 m	—	—	—	1·50
Borehole Ryazanovka 1, depth 540–546·2 m	—	—	—	0·90
Borehole Ryazanovka 1, depth 546·2–553 m	—	—	—	7·80
Borehole Timashevka, top of Kungurian	—	—	—	Trace–0·50
The same	—	—	—	0·40
Zigan River, base of Kungurian .	—	—	—	0·60
The same	—	—	—	0·02
Borehole Timashevka, top of Kungurian	—	1·63	—	None
			Average	?
IV. Sulfate-bearing dolomites				
Borehole 42/2, depth 658·4–663·4 m .	—	17·86	—	3·42
Borehole 10/20, depth 971–978 m .	5·36	26·30	—	11·72
Borehole 10/20, depth 971–978 m .	—	—	—	12·46
Borehole 10/20, depth 978·9–982 m .	—	—	—	15·58
Borehole 10/20, depth 1001–1008·3 m .	—	—	—	12·51
Borehole 10/20, depth 1001–1008·3 m .	—	—	—	19·95
Borehole 10/20, depth 1033–1043 m .	—	—	—	0·20
			Average	10·64
V. Anhydrite rocks				
Borehole 5/11, depth 346·5–349·5 m .	1·10	—	1·33	0·10
Borehole 5/11, depth 464·5–472·9 m .	0·98	—	29·33	None
Borehole 3/25, depth 1420 m . .	12·89	—	15·98	None
Borehole 48/20, depth 590 m . .	2·68	—	6·82	12·64
Borehole 10/20, depth 738·5–748·4 m .	0·93	—	1·72	0·74
Borehole 10/20, depth 738·5–748·4 m .	Trace	—	None	0·21
Borehole 10/20, depth 756·9–778·2 m .	2·09	—	9·30	0·25
Borehole 10/20, depth 906–912 m .	0·10	—	1·54	0·46
Borehole 10/20, depth 932–942 m .	0·25	—	4·74	0·10
Borehole 42/2, depth 245·9–255 m .	—	—	8·24	0·29
Borehole 42/2, depth 410·2–416·5 m .	—	—	9·34	0·20
Borehole 1/23, depth 429·5–431·8 m .	4·15	—	11·83	None
Section along Seleuk River . .	—	—	—	0·20
Section along Seleuk River . .	—	—	—	None
			Average	0·20

2 K

average, the sandstones are characterized by a minimal strontium content; the clays and marls have four times as much, calcareous dolomites and dolomites still more, and the maximum is in the sulfate-bearing dolomites, which form a transitional link between dolomitic and anhydritic sedimentation. In pure anhydrite rocks, the Sr content again falls significantly, to 0·20% on average; and in rock salt and potassium-salt rocks only negligible traces are found. This distribution is clear, however, only if averaged values are used. Individual analyses show sharply divergent values within each type of analyzed rock, resulting from the active redistribution of celestite during diagenesis. If we ignore these secondary factors, however, it is clear that *precipitation of strontium in celestite is characteristic of the middle stages of salinity, and reaches a maximum in the early phases of the sulfate stage*, when the formation of dolomite gives way to gypsum deposition. On both sides of this maximum the Sr concentration decreases. Towards the lower concentration, where marls, clays, and sandstones are deposited, the decrease is less sharp than on the side of higher mineralization, where rock salt and potassium salts are precipitated. In other words, *the distribution curve of strontium in the sedimentary series of a saline basin is asymmetrical, the gentler slope occurring at the initial stages of salinity, the steeper slope at high salinities.*

The distribution of strontium in saline rocks of other strata of the region supported these conclusions, and also brought to light some new details. Vinogradov and Borovik-Romanova (1945) studied the distribution of strontium (in celestite) in four boreholes in the Kungurian gypsum-dolomite deposits, the Artinskian dolomite-limestones, and part of the Upper Carboniferous carbonate sequence. The average $SrSO_4$ content in limestones (15 determinations) was 0·20%, in dolomites (36 determinations) 1·23%, and in anhydrite and gypsum rocks (17 determinations) 0·70%. In the limestones the range of $SrSO_4$ content is generally 0·06 to 0·40%, rarely less; one sample had as much as 2·78%. In the dolomites the range is about the same, but values of 2·8–5·0% were frequent, with maxima of 14·25%; i.e. local concentrations of strontium occurred, as in the Kungurian rocks of the Ishimbai region. In the sulfate rocks, the $SrSO_4$ content is relatively uniform, of the order of 0·30–0·40%, only occasionally reaching 1·00–2·8%. It thus seems clear that maximum strontium concentration occurs at the end of dolomite precipitation and at the beginning of the sulfate stage.

A well-defined *horizon of strontium enrichment*, therefore, lies in all four boreholes at the boundary between Artinskian and Kungurian deposits, i.e. at the stage when dolomite formation gave way to sulfate precipitation. The highest concentration of $SrSO_4$ is here found most frequently in the dolomites, but in places it lies in the overlying sulfate rocks; this is perfectly compatible with the conditions of celestite precipitation.

Yarzhemskii (1938), who studied the Cambrian evaporites of the Angara region, has described an accumulation of strontium in deposits that formed during the progressive reduction of salinity in the sequence above the saline beds. The deposition of celestite during dilution of a brine that is already poor in strontium has clearly been possible only by inflow of *marine* water into the basin, rather than stream water, because strontium occurs in higher concentrations in sea water than in stream water (see above).

In summary we should recognize that the deposition of strontium in sediments has taken place at the boundary between the carbonate and sulfate stages, sometimes shifted slightly in one direction or the other, depending upon local factors. Generally speaking, therefore, two intervals of strontium deposition may occur in saline sedimention: *infrasaline*, corresponding to the change from carbonate (dolomite) deposition to sulfate; and *suprasaline*, at the change from anhydrite to carbonate (dolomite) deposition. The first horizon forms with increasing salinity of the marine basin; the second occurs during dilution of the basin with seawater and not fresh water.

2. THE DISTRIBUTION OF FLUORINE, BROMINE, AND IODINE

The distribution of the accessory halogens F, Br, and I is substantially different and much more complex. In this group fluorine forms a single mineral, fluorite, characterized by a relatively low solubility. In sea water fluorine is unsaturated and consequently is not precipitated, but when marine basins become highly saline, the solution rapidly becomes saturated and CaF_2 is precipitated.

Quantitative determinations of F in sediments of saline basins are few. Almost all have been made by Correns (1956) on German material (Table 45)

TABLE 45

Fluorine Content in Deposits of Saline Basins, in %
(from Correns 1956)

Rock and sample locality	F
Zechstein dolomite; Hartzfeld, Harz	0·011
Zechstein dolomite, containing fluorite; Römerstein	0·025
Main anhydrite; Stassfurt	0·080
Upper anhydrite; Reiersgausen	0·081
Basal anhydrite; Reiersgausen	0·013
Anhydrite (insoluble residue of rock salt); Reiersgausen	0·089
Gypsum; Indersachswerfen	0·087
Gypsum breccia; Hundelshausen	0·013
Rock salt; Stassfurt	0·0002
Rock salt; Reiersgausen	0·0057
Rock salt; Reiersgausen	0·0007
Rock salt; Reiersgausen	0·0005
Rock salt; Reiersgausen	0·0006
Polyhalite; Stassfurt	0·0020
Older potassium bed; Reiersgausen	0·0300

and by V. V. Danilova on Soviet material. Unfortunately, the nature of fluorine distribution proves to be irregular. From Correns' analyses the maximal deposition of F in the form CaF_2 takes place at the sulfate stage; the preceding dolomite stage has only a low F content, and in the subsequent chloride stage it is negligible; and where it does occur in rock salt or potassium salts, it is most intimately associated with interbedded anhydrite.

The low content of fluorine in dolomites indicates that at the dolomite stage the solution has not yet become saturated with CaF_2, whereas the negligible amounts of fluorine in rock salt and potassium salts reflect the very rapid sedimentation rate at these stages, causing a marked dilution of CaF_2 by the chloride component.

TABLE 46

Fluorine Content in Deposits of Saline Basins, in % (from Danilova 1949)

Rock	Age	Borehole No.	Depth, m	F
Dolomite	P_2	4	93–101	0·03
Dolomite	P_2	4	39–40	0·02
Dolomite	P_2	4	132·25–134	0·02
Dolomite	C_3^3	9	180–184	0·02
Dolomite	C_3^3	4	164–167·8	0·03
Dolomite	C_3^3	4	211·3–212	0·025
Dolomite	C_2^3	4	193·12–200·41	0·027
Dolomite	P_2	4	104·7–106·35	0·03
Dolomite	P_2	4	140·9–142	0·02
Dolomite	C_3^3	4	245·1–246	0·03
			Average	0·025
Anhydrite	P_1	2	114·7–121·06	0·015
Anhydrite	P_1	4	106·35–115·25	0·015
Anhydrite	P_1	3	116·75–119·25	0·015
Anhydrite	P_1	4	48·52–52·52	0·012
Anhydrite	P_1	4	30·3–36·03	0·013
Anhydrite	P_1	4	37·5–38·00	0·020
Anhydrite	P_1	4	101·7–114·2	0·012
			Average	0·0142
Gypsum	P_1	4	96·95–98·95	0·015
Gypsum	P_1	4	20·30–28·25	0·012
Gypsum	P_1	2	101·4–103·8	0·013
Gypsum	P_1	2	114·39–117·64	0·015
Gypsum	P_1	9	120·7–128·64	0·010
			Average	0·0125

The distribution of fluorine (Danilova 1949) is substantially different (Table 46). According to Danilova the sulfate stage is characterized by a low fluorine content as compared with the dolomite stage. Furthermore, the fluorine content is essentially the same in dolomites as in limestones: 0·025% as against 0·0227%. In other words, the separation of fluorine exhibits no general maximum as marine water becomes more saline. This circumstance scarcely reflects the actual history of fluorine and would appear to be due to inaccurate analyses.

Under these circumstances, increased reliance must be placed on the microscopic evidence. These observations are not entirely compatible with the distributional picture appearing from the analyses of Correns. Judging from thin sections, fluorite is most widespread in dolomites and sulfate-bearing dolomites, is much less abundant in anhydrite rocks, and is very rare in rock salt and potassium salts. V. P. Baturin recorded fluorite in the insoluble residue of rock salt in the Ural-Emba region but only in negligible proportions. Abundant fluorite is known in the dolomites of the Kashirian horizon along the western margin of the Moscow syneclise, where it occurs in segregations that have all the features of primary deposits. From this and other records of fluorite in dolomite it is reasonable to conclude that *on the salinity scale the maximum of fluorite deposition occurs at the very end of the dolomite stage and the beginning of the sulfate stage.*

In contrast to fluorine, bromine forms no independent minerals at any salinity stage; it separates in the solid phase only as isomorphous admixtures in chloride minerals: halite, sylvite, carnallite, kainite, and bischofite. The intensity of bromine deposition in solid phases and in brines follows certain laws:—

(1) Each given chloride is characterized by a distributional coefficient for bromine, indicating the relationship between the solid phase and the brine:

$$J = C_{sd}^{Br} / C_{i.r.}^{Br}$$

where C_{sd}^{Br} is the concentration by weight of bromine in solid crystals, and $C_{i.r.}^{Br}$ is the concentration of bromine by weight in the dry (salt) residue of the liquid phase. For halite $J = 0.037$, for sylvite 0.20, for carnallite 0.32, and for bischofite 0.42; J is less than 1 for all chlorides that crystallize from sea water. It follows, therefore, that despite the extraction of bromine by chlorides that crystallize from the brine, the concentration of bromine in the solution increases continuously as the salinity increases. This important conclusion is supported by the data in Table 47.

TABLE 47

Bromine Content in Sea Water at different Stages of Concentration
(from Valyashko and Mandrykina)

Sp. gr.	°Bé	Cl⁻, %	Br⁻, %	$\dfrac{Br^- \times 10^3}{Cl^-}$	Solid rock
1·015	2·2	0·951	0·002	2·11	—
1·223	26·3	14·36	0·044	3·07	Halite
1·307	34·0	10·07	0·223	22·22	Halite+epsomite
1·320	35·1	10·67	0·317	29·80	Halite+epsomite+carnallite

The bromine content in a solution is at first a straight-line function of the concentration of the brine. Later, characteristic breaks appear on the curve (Fig. 223), the first when NaCl crystallizes. The second marks the beginning of precipitation of sylvite (or epsomite, which is precipitated at practically the same stage), whereas the third marks the precipitation of carnallite. The last break occurs when the solution reaches the eutonic point, i.e. at the beginning of bischofite precipitation.

"The intensity of bromine deposition during evaporation of sea water steadily increases, the curve of each new solid phase becoming progressively steeper. This is because the segregation of each new solid phase does not stop precipitation of the preceding phases; the phases combine in precipitation, and in this way more rapidly deplete the liquid phase" (Valyashko 1956, p. 37).

The "distribution coefficients" of bromine also indicate that, even under optimum conditions of chloride crystallization from evaporating water and even when this process reaches its fullest development, only a small part of the bromine is preserved in the sediment. Most of it disappears without trace with the residual brine.

(2) It is of utmost importance to note that the bromine content at the nodal points of the curve, obtained by evaporating the brines of various salt basins, does not remain constant, but fluctuates appreciably (Table 48).

TABLE 48

Bromine Content at Nodal Points, in wt. %

Basin	Beginning of halite precipitation		Beginning of sylvite precipitation		Beginning of carnallite precipitation		Eutonic	
	Sp. gr. of brine	Br	Sp. gr. of brine	Br	Sp. gr. of brine	Br	Sp. gr. of brine	Br
Lake No. 8 .	1·224	0·062	1·308	0·222	1·319	0·259	1·361	0·431
Lake No. 5 .	1·220	0·051	1·308	0·236	1·325	0·342	0·359	0·520
Sea of Azov .	1·221	0·044	1·308	0·223	1·321	0·357	1·357	0·608
Gulf . .	1·220	0·033	—	—	—	—	—	—

The greatest differences in bromine content are observed at the eutonic points; the smallest are found at the point of initial crystallization of carnallite. The bromine content is very nearly the same in all samples at the crystallization point of sylvite (epsomite). Differences are somewhat greater at the crystallization point of halite. These differences in bromine concentrations at the points of discontinuity are due to differences in composition of the initial brines and to different amounts of salts that have precipitated at particular intervals of concentration. This experimentally established fact is of great significance for understanding the variations in bromine content of saline rocks of apparently the same type, as observed in salt deposits. It can hardly be doubted that, because of local peculiarities of drainage areas, the initial brines in ancient saline basins had different bromine content, even as modern basins vary in this respect. Variations in bromine content in the past may have been even greater than those proved in modern analogues.

(3) If two chlorides, with which bromine forms isomorphous mixtures, crystallize simultaneously from a solution, the distribution of bromine in each chloride is independent of the other.

(4) The bromine content in chlorides is determined not only by its intrinsic

TABLE 49

The Normal Bromine Content and Value of the Ratio (Br⁻ × 10³)/Cl⁻ in Halite, Sylvite, and Carnallite precipitated from Sea Water (from Valyashko 1956a)

Liquid phase $\dfrac{Br^- \times 10^3}{Cl^-}$	Halite		Sylvite		Carnallite	
	Br^-, wt. %	$\dfrac{Br^- \times 10^3}{Cl^-}$	Br^-, wt. %	$\dfrac{Br^- \times 10^3}{Cl^-}$	Br^-, wt. %	$\dfrac{Br^- \times 10^3}{Cl^-}$
3·24	0·0068	0·11	—	—	—	—
5·34	0·011	0·18	—	—	—	—
9·00	0·017	0·288	—	—	—	—
11·70	0·022	0·36	—	—	—	—
13·70	0·025	0·41	—	—	—	—
14·30	0·026	0·43	—	—	—	—
14·80	0·027	0·44	0·137	2·9	—	—
16·50	0·033	0·54	0·176	3·7	—	—
17·50	0·037	0·61	0·190	4·0	—	—
17·80	0·037	0·60	0·199	4·2	0·318	8·3
20·01	0·045	0·74	(0·242)	(5·1)	0·387	10·1
21·40	0·049	0·81	(0·266)	(5·6)	0·425	11·3
22·40	0·052	0·86	—	—	0·523	13·7
26·70	0·063	1·03	—	—	0·514	14·1

distribution coefficient but also by the quantity of bromine in the solution from which crystallization takes place. This means that the percentage content of bromine in halite will increase continuously from the beginning of halite crystallization to the eutonic point (Table 49 and Fig. 223).

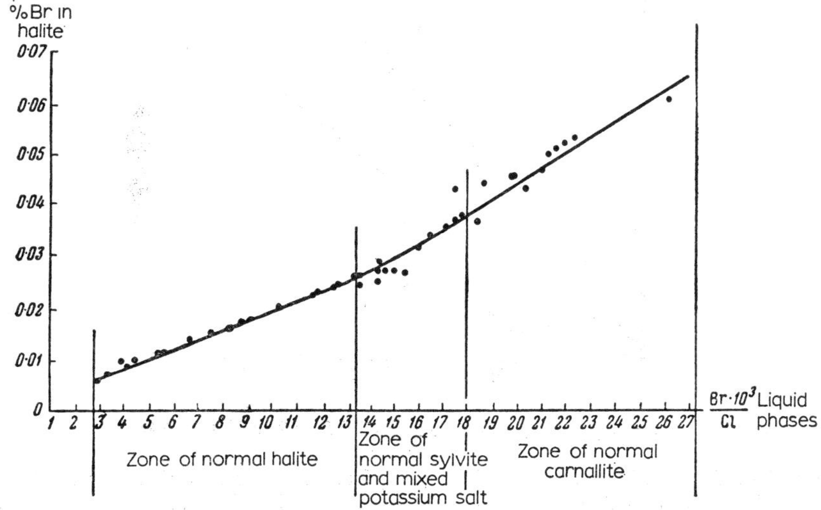

FIG. 223. Dependence of bromine content in solid halite on the bromine-chlorine coefficient of the liquid phase during progressive concentration of sea water (from Valyashko).

In the first crystals of halite, the bromine content is only about 0·005–0·006%, whereas halite that crystallizes from eutonic brine contains approximately ten times this amount. The increase in bromine content in sylvite from the beginning to the end of crystallization is only twofold, and in carnallite is about 1·6 times.

These relations lead to the conclusion that, by determining bromine in halite from different deposits or from different beds of a single deposit, the salinity conditions of the basin in which the halite crystallized may be determined (Fig. 223). This method will be admissible, however, only for precipitated halite that has not later recrystallized, because during recrystallization, according to the distribution coefficient, halite loses bromine through the brine, which may be to some extent removed.

(5) The actual bromine content in rock salt and potassium-salt rocks varies appreciably. In some deposits the amount of bromine corresponds to the

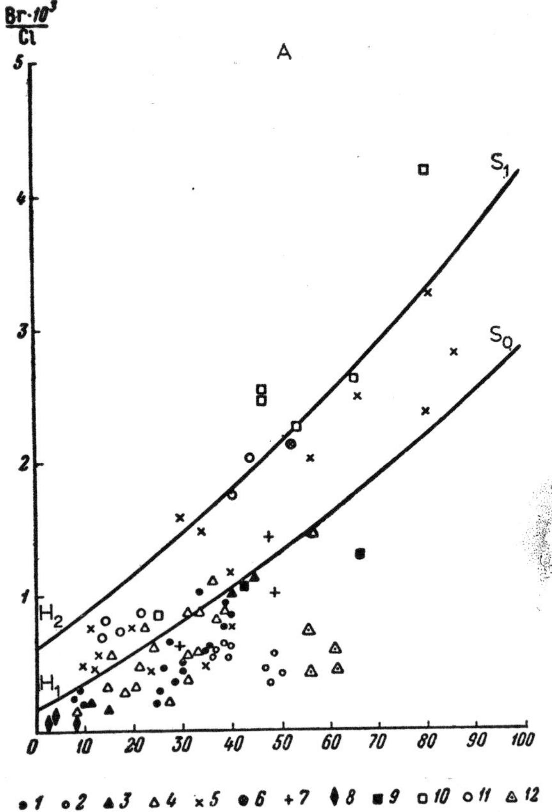

Fig. 224. The chlorine-bromine coefficients in potassium salts of various deposits (from Valyashko). *A.* In sylvite rocks: 1. Solikamsk, red sylvite, 2. Solikamsk, variegated sylvite; 3. Inder, sylvite; 4. Inder, variegated sylvite; 5. Kalush, 6. Stebnik, 7. Gaurdak; 8. Ozinki; 9. Stassfurt; 10. Werra; 11. Utah; 12. New Mexico. *B* (opposite page). In carnallite rocks; 1. Solikamsk; 2. Inder; 3. Kalush; 4. Ozinki; 5. Stassfurt; 6. Werra; 7. Utah.

concentration as determined by the mineral composition of the rock, i.e. according to the ratios of halite, sylvite, and carnallite. In other deposits (the larger number) the actual content is less than that theoretically expected. Rarely is the opposite true.

These relationships are graphically represented in Figs. 224*A* and *B* (Valyashko 1956) for sylvite and carnallite rocks. The abscissa axis represents the relations of chloride minerals in these rocks; the ordinate represents the value of the bromine-chlorine coefficient. The H_1—S_0 curve defines the values of the chlorine-bromine coefficient at the beginning of sylvite crystallization (and correspondingly of carnallite on diagram *B*). The H_2—S_1 curve indicates the same values at the terminal stage of crystallization of this mineral. Thus, the field H_1—S_0—H_2—S_1 forms the field of normal bromine content during

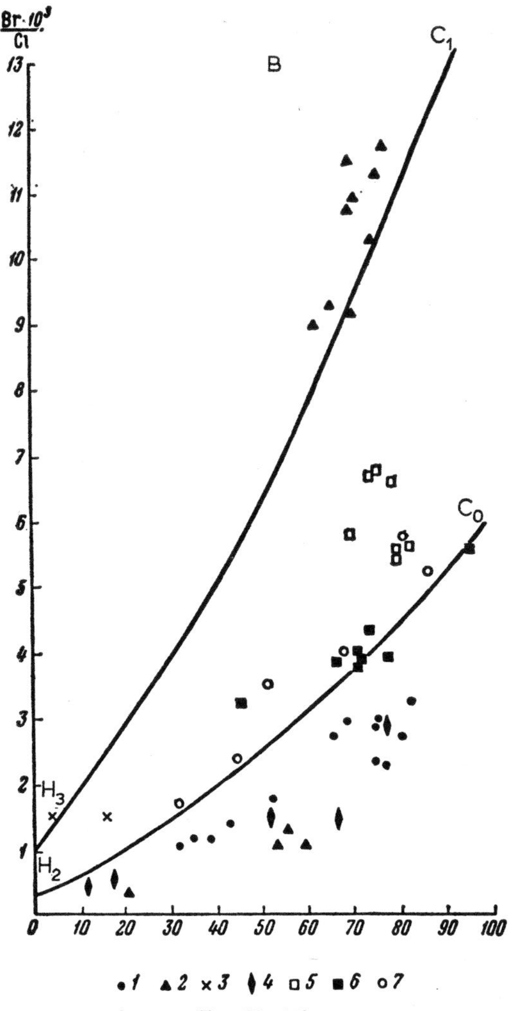

Fig. 224 (*B*)

precipitation of potassium salts. Points corresponding to observed values of the bromine-chlorine coefficient in different deposits are then plotted. The spottiness of the picture and the considerable deviations from the computed norm are clear.

The causes of the deviations are several. One is the varying content of bromine in the initial water supplying the saline sedimentation, as already noted. Another cause, according to Valyashko, may be the recrystallization of potassium salts during effluent movement of the parent brine. This process must be inevitably accompanied by decline in Br content (according to distribution laws) in the newly forming solid chloride phases. The relative importance of these two factors is not yet known.

A distinctive feature of iodine is that in saline deposits it forms neither independent solid phases nor isomorphous replacements in chlorides. This follows from the great difference between the ionic radii of chlorine (1·81 Å) and iodine (2·2 Å). Sea water iodine is therefore preserved in the residual solution, and is found in saline deposits only to the extent that brine is preserved in pores and intercrystalline spaces. This explains the minute quantities of iodine in all types of saline deposits: Correns (1956) shows from $0·0155$ to $0·1365 \times 10^{-4}\%$ iodine in anhydrite rocks, as analyzed by Roeber. Fellenberg found $0·3 \times 10^{-4}\%$ iodine in anhydrite of Bex. According to Roeber the iodine content in rock salt is $0·0965 \times 10^{-4}\%$, in sylvite rock $0·36 \times 10^{-4}\%$, and in carnallite rock $0·29 \times 10^{-4}\%$. No detectable pattern of iodine distribution among the different types of saline rocks has been noted, but this is to be expected in view of the above mechanism by which iodine is emplaced.

In summary, the following scheme for the distribution of halogens may be noted. (1) Fluorine forms an independent solid phase in saline deposits. It has a low solubility and therefore accumulates in sediments at relatively low salinities, reaching its maximum in dolomites, in deposits forming at the end of the carbonate stage. Later its concentration in the sediments decreases. (2) Bromine is concentrated in the solid phase only as isomorphous admixtures in chlorides. It therefore appears in saline deposits only at high salinities ($S_0 > 25\%$), simultaneously with NaCl precipitation, and is then progressively concentrated in the salts as the brine approaches the eutonic point. Only a small part of the bromine present in the brine goes into the solid phase; most of it remains unprecipitated. (3) Iodine forms neither independent mineral solid phases nor isomorphous admixtures with chlorides. It is therefore not extracted from the brine at any stage of crystallization, and it appears in saline rocks only when the brine is preserved in the accumulating saline sediments.

3. THE DISTRIBUTION OF BORON

Three features distinguish boron from other elements and give a specific character to its geochemistry.

In sea-water boron is present in smaller quantities than bromine ($4·02 \times$

$10^{-4}\%$), but it may nevertheless occur in independent, even large, accumulations in the sediments of saline lagoons. Furthermore, boron forms a large number of mineral species with Na, K, Mg, Ca, and Sr, so that its mineralogical range is greater than that of any other component of saline deposits in its saline occurrences. It must be remembered, however, that the great majority of boron minerals from saline deposits are of secondary origin and are essentially stages of weathering of the few primary mineral species in the zone of weathering. Precise separation of these two genetic groups of minerals has not yet been made, and this deficiency has prevented any clear synthesis of boron geochemistry in the saline process.

TABLE 50

The Boron Content in Sedimentary Rocks of the Moscow Region
(from Golovko 1959)

Rock	Boron content, in %											
	None	Traces	0·001	0·0012	0·0013	0·002	0·003	0·004	0·005	0·006	0·007	0·012
Devonian Rocks						Samples						
Dolomite	2	2	2	—	—	3	2	1	2	1	—	1
Carbonate-sulfate rock	—	—	1	—	—	—	—	—	—	—	—	—
Anhydrite	18	5	1	—	—	—	—	—	—	—	—	—
Gypsum	6	5	—	—	—	—	—	—	—	—	—	—
Dolomitic marl	—	—	—	—	—	1	—	—	—	—	—	—
Chert	—	—	—	—	—	—	—	1	—	—	—	—
Lower and Middle Carboniferous rocks												
Limestone	—	4	—	—	—	—	—	—	—	—	—	—
Primary dolomite	—	—	—	—	2	1	—	—	—	—	—	—
Secondary dolomite	—	2	—	—	—	—	—	—	—	—	—	—
Marl	—	—	—	—	—	—	—	1	—	—	—	—

Golovko (1960) has made spectroscopic determinations of boron in rocks of the beginning and intermediate salinity stages (Table 50). According to him: "It is absent for the most part, or is found only as traces, in anhydrite and gypsum rocks, only in individual cases being present in amounts up to 0·001%. In calcareous dolomites and dolomites the picture is reversed: in individual samples boron is absent or is present only in traces, but most samples contain 0·001–0·012%." At the beginning and middle stages of salinity, boron thus shows a preference for the dolomite stage, sharply declining in the sulfate stage.

This behavior of boron, according to Golovko (1959), follows from some of its chemical peculiarities. "Despite the fact that $(BO_3)^{3-}$ is not a completely typical complex anion, it may apparently exist in aqueous basins. The radius of the complex anion $(BO_3)^{3-}$ is more than four times that of the complex anion $(CO_3)^{2-}$, but it has a larger charge, which compensates for the disparity of their radii. At the same time a reverse relationship has been observed between the radii of $(BO_3)^{3-}$ and $(SO_4)^{2-}$, leading to more complex examples of compensation, requiring supplementary participation of other anions. Isomorphous admixture of complex anions in sulfates is therefore curtailed sharply, and the boron content in sulfates drops."

The occurrence of boron in dolomites rather than in limestones is caused by the fact that "Mg, having a small ionic radius and a large energy coefficient, is clearly able to effect the greatest isomorphous replacement of $(CO_2)^{2-}$ by $(BO_3)^{3-}$ in magnesian carbonates without disturbing the energy equilibrium of the crystal lattice. Magnesian carbonates are followed by iron carbonates, then by Mn, and finally by Ca, Cr, and Ba. These latter have the smallest ionic radii and the lowest energy coefficients, and they will clearly allow the least isomorphism of $(BO_3)^{4-}$ in the crystal lattices of carbonates of these elements." Whether this is the reason that some accumulations of boron are confined to dolomitic rocks is not yet clear, but the fact that boron does tend to concentrate in dolomites can hardly be doubted.

Another mechanism operates at the same time as isomorphous replacement: the sorption of boron by finely dispersed clay material. This latter process in boron accumulation, however, is of secondary importance, as is shown very clearly by Golovko's data: in carbonates with $0{\cdot}005$–$0{\cdot}006\%$ boron, the total insoluble residue amounts to $1{\cdot}3$–$1{\cdot}4\%$, but in sulfate rocks that contain twice as much insoluble residue boron is entirely absent or is present in amounts below $0{\cdot}001\%$.

At higher salinities, corresponding to the sulfate and almost all the chloride stage, including the sylvite and carnallite intervals, the progressive accumulation of boron in the brine is evident. The increase of boron in the sediments takes place in negligible amounts, the basic mechanism here being the sorption of borates by finely dispersed clay particles.

Nevertheless, in the history of some saline marine basins, a time arrived when the brine began mass precipitation of boron, and large primary accumulations of lenses and beds formed. It is significant that all known primary occurrences of borate accumulations of this type were localized in the carnallite and bischofite zones. They clearly correspond to salinities near the eutonic point. The fact that some borate deposits have formed somewhat below the eutonic point, whereas others have formed at this salinity, is apparently caused by variations in the intensity of supply of boron compounds to the basin. That borate sedimentation properly belongs to the final moments of a saline marine basin, to salinities near the eutonic point, has been confirmed by experiment.

These experiments have shown that the chemical precipitation of borates is possible only at salinities near the eutonic point, when magnesium is practically the only cation remaining. Since the attainment of eutonic solutions and the prolonged existence and evaporation of such solutions are possible only when a rare combination of physico-geographic conditions is satisfied, it follows that sedimentary deposits of borates are themselves rare.

When salt structures are exposed to erosion, boron-bearing potassium horizons are subjected to leaching and to removal of potassium salts and halite. A distinctive weathering crust of saline rocks develops in their place. The borates are subjected to numerous chemical and mineralogical alterations, giving rise to a series of more stable and less soluble compounds with Ca, Sr, and Na. These are concentrated in great measure in a saline cap, generally gypseous, giving a secondary deposit of boron.

4. The Distribution of Fe, Mn, P, and some Minor Elements

Iron, manganese, phosphorus, and a number of heavy minor elements (V, Cr, Co, Ni, and Cu) migrate chiefly in mechanical suspensions, with only minor amounts, if any, by solution. This distinctive manner of migration in streams predetermines the characteristic occurrences of these elements in saline deposits. It is clear that the main bulk of Fe, Mn, P, and the heavy minor elements must be concentrated with the terrigenous constituents of a rock, and that quantitatively they will vary with this clastic fraction. However, that part of the Fe, Mn, P, and other elements introduced into the basin in solution, may be captured within the lattice of the precipitating solid salt phases as has recently been demonstrated.

In our work on the Kungurian deposits of the Ishimbai region (Strakhov, Zalmanzon and Senderova 1944) the content of Fe, Mn, P, and a number of minor elements in a group of rocks was determined (Table 51). Sandstones and clays contain Fe, Mn, P, and the other elements in amounts comparable to those found in the same types of rocks in humid regions. In dolomites having only small insoluble fractions, the content of these elements falls sharply, commonly to values that cannot be analytically determined, and V and Ni disappear entirely. The same is true of anhydrites, in which the clastic admixture is as small as, even lower than, that in dolomites. The connection between the clastic fraction of a rock and its content of Fe, Mn, P, and heavy minor elements is undoubtedly close in this region.

Similar results have been obtained by Zelenov (1956) for the Lower Cambrian rocks of the Siberian platform. In order of decreasing clastic content those examined were: dolomitic marls (average 31·55% insoluble residue), dolomites (3·65% i.r.), dolomites with anhydrite admixtures (2·76% i.r.), anhydrite-dolomites (1·24% i.r.), and anhydrite rocks (0·66% i.r.). It is clear that the Fe, Mn, F, Ti, Cr, Ni, Co, Cu, V, Mo, Sr, and Ba content of these rocks, as determined spectrographically, varies directly with the insoluble content (Fig. 225). The same type of distribution of Fe, Mn, P, and the minor elements of the iron family was established by Golovko (1960) in the Devonian rocks of the Moscow region.

It is therefore undoubted that the contents of the elements here considered depend on the insoluble fraction of rocks that form at the beginning and middle stages of salinity.

TABLE 51

Content of Fe, Mn, P, Cr, Ni, V, and Cu in Kungurian Deposits of the Bashkirian-Ural Region (in %)

Rocks and sample localities		Fe	Mn	P	Cr	Ni	V	Cu
Sandstones								
Borehole	Depth							
3/25	729 m	—	—	0·070	0·006	0·006	0·008	None
3/25	1107 m	1·15	0·043	—	0·007	0·002	0·007	0·002
3/25	1403 m	3·18	0·720	—	0·008	0·040	0·010	0·040
Clays								
Borehole	Depth							
42/2	455–457 m	3·51	0·082	0·073	0·014	—	None	—
1/23	664–667 m	4·39	0·041	0·174	—	—	—	0·002
3/25	1408 m	3·73	0·413	0·080	0·008	0·013	0·012	0·002
3/25	1421 m	0·90	0·101	0·040	None	None	0·003	0·001
3/25	1323 m	3·68	0·660	0·080	0·006	0·006	0·010	0·001
3/25	1318 m	3·18	0·062	—	0·003	0·030	0·010	0·001
3/25	1326 m	3·07	0·071	0·023	0·004	0·004	0·010	0·001
3/25	1323 m	4·17	0·053	0·022	0·003	—	0·012	0·001
3/25	1395 m	4·17	0·061	—	—	—	—	0·003
3/25	1252 m	2·96	0·102	0·061	0·006	0·013	0·012	—
3/25	1325 m	4·06	0·062	0·082	0·006	None	0·010	0·002
3/25	1324 m	—	—	0·012	0·001	—	0·005	None
3/25	1336 m	2·85	0·032	0·051	0·004	None	0·010	0·002
3/25	727 m	3·40	0·141	0·174	0·021	—	None	0·001
Dolomites and calcareous dolomites								
Borehole	Depth							
5/11	403–411 m	0·020	None	—	—	None	None	0·005
5/11	375–383 m	0·083	None	0·035	None	None	None	0·010
5/11	390–398 m	0·042	0·062	0·026	None	None	None	0·002
5/11	578–581 m	0·063	None	—	None	None	None	0·005
10/20	738–748 m	0·440	None	0·018	0·0007	None	None	—
10/20	989–1001 m	0·007	None	0·065	None	None	None	0·001
10/20	1033–1043 m	0·440	0·380	0·065	0·0003	—	None	0·001
42/2	658–663 m	0·720	0·300	0·109	None	—	None	—
Anhydrite rocks								
Borehole	Depth							
5/11	346–349 m	0·350	0·015	0·035	0·0030	None	None	None
5/11	414–472 m	0·040	None	0·033	0·0007	None	None	0·002
10/20	738–748 m	0·001	None	0·018	None	None	None	None
10/20	906–912 m	0·003	None	0·015	None	None	None	0·001
10/20	932–942 m	0·021	None	0·018	None	None	None	None
42/2	245–255 m	0·098	None	0·018	None	None	None	0·009
42/2	410–416 m	0·880	0·020	0·087	0·0030	None	None	—
42/2	410–416 m	0·820	0·068	0·033	0·0030	None	None	0·013
42/2	432–440 m	0·105	0·008	0·003	0·0007	None	None	0·001
1/23	429–431 m	0·035	0·004	0·035	None	None	None	0·001

No systematic work on rock salt and potassium salts has yet been undertaken, though a similar relationship is likely. Support for this view may be found in the distribution of vanadium. Morachevskii and Tikhomirova (1940) determined V_2O_5 in the clay and the insoluble residues of salts (Table 52), when it was shown that vanadium occurs in approximately equal proportions in both groups of material. This again indicates that the clastic fraction is the source of the vanadium and not the salt phase itself.

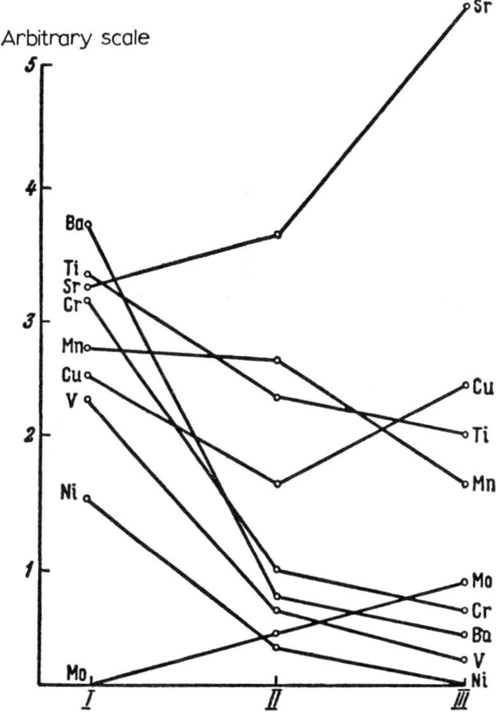

Fig. 225. Geochemical profile of elements in dolomite-sulfate rocks (from Zelenov). *I.* Argillaceous dolomitic rocks (15 samples); *II.* dolomite-sulfate rocks (17 samples); *III.* sulfate rocks (5 samples).

There must, nevertheless, be a very small amount of Fe, Mn, P, and minor elements in the structural lattice of the salts themselves. Their presence has been demonstrated by Morachevskii (1940) and Galakhovskaya (1953). In both investigations, aqueous solutions were obtained from large samples of salt, filtered, and evaporated to dryness. Fe, Mn, Cu, and other elements were determined chemically by Morachevskii and spectrographically by Galakhovskaya (Table 53).

The content of these elements in the structural lattices is negligible: at most only a few milligrams, or even parts of a milligram, per kilogram of rock, or $n \times 10^{-7}\%$. In the insoluble residues of the salts these minor elements are of the order of $n \times 10^{-3}\%$, i.e. approximately four times as great. These values may serve as a standard for the relative proportions of Fe, Mn, P, and the minor elements associated with the clastic fraction of the rock as compared with the same elements in the crystal lattices of the salts.

TABLE 52

Distribution of V_2O_5 in Saline Deposits

Rock	V_2O_5 content, %	Rock	V_2O_5 content, %
Clays		Sample 9 . . .	0·0060
In carnallite rock . .	0·0200	Sample 10 . . .	0·0150
In carnallite rock . .	0·0150		
In carnallite rock . .	0·0090	*Insoluble residue of*	
In sylvite rock . . .	0·0150	*potassium salts*	
In sylvite rock . . .	0·0025	Bed Kr III . . .	0·0100
In lower rock salt . .	0·0020	Bed Kr III . . .	0·0080
		Bed Kr II . . .	0·0240
Clays from sylvite		Bed Kr I . . .	0·0080
zone		Bed A'	Trace
Sample 1	0·0005	Bed A	Trace
Sample 2	0·0055	Bed B, upper layer 1 .	0·0090
Sample 3	0·0160	Bed C, sylvite rock, layer 2	0·0120
Sample 4	0·0090	Bed C, sylvite rock, layer 4	0·0120
Sample 5	0·0025	Bed C, sylvite rock, layer 6	0·0300
Sample 6	0·0070	Bed C, carnallite rock 1 .	Trace
Sample 7	0·0050	Bed C, carnallite rock 2 .	0·0200
Sample 8	0·0120	Bed C, carnallite rock 6 .	0·0120

5. THE DISTRIBUTION AND ACCUMULATION OF MINOR CONSTITUENTS IN SALINE ROCKS

From the above data a scheme of distribution and accumulation of minor constituents during saline deposition may be determined.

The accessory constituents of saline deposits may be divided into two groups. The first, the clastophile, group includes Fe, Mn, P, V, Cr, Ni, Co, Cu, Pb, Be, and some others, generally the same elements as are found in the sediments of humid climates. These enter saline basins as terrigenous material transported by either eolian or fluvial processes. They have been introduced in mechanical suspension, with only a small part in solution. In keeping with this type of migration, *the assemblage of clastophile elements and their mutual relationships are constant at all salinity stages in the basin; they do not depend on salt paragenesis, though they are subordinate to it.* The amount of Fe, Mn, P, and other elements varies with the terrigenous content. *Elements of the clastophile group occur in only very small amounts in the lattices of saline minerals.*

The second, the halophile, group includes F, Br, I, Sr, and B, introduced into the saline basin from the sea, where they had been previously concentrated

TABLE 53

Contents of Clastophile Components in Water-soluble Part of Solid Phases,
in mg/kg

Sample	Fe	Mn	Ni	Cu	Zn	Pb
Upper Kama salts						
Carnallite C	9·91	0·70	0·10	0·52	4·70	Undet.
Carnallite C	1·92	1·04	0·02	0·37	0·75	Undet.
Carnallite C	0·55	0·13	None	0·34	Undet.	Undet.
Carnallite C	0·27	0·01	None	0·24	0·02	Undet.
Carnallite C	1·42	0·07	0·02	0·43	0·22	Undet.
Sylvite C	1·54	0·01	None	0·20	0·07	Undet.
Sylvite C	1·92	0·01	Trace	0·32	1·36	Undet.
Sylvite C	1·91	0·97	0·03	0·36	0·24	Undet.
Sylvite C	1·97	None	0·04	0·36	0·06	Undet.
Sylvite B	4·63	0·92	0·01	0·43	0·42	Undet.
Sylvite A	None	None	None	0·46	None	Undet.
Sylvite Kr I	None	None	None	0·40	None	Undet.
Sylvite Kr II	None	None	None	0·27	None	Undet.
Sylvite Kr III	None	None	None	0·27	None	Undet.
Salts of various ages						
Freshly precipitated halite, Lake Sasyk-Sivash	None	3·00	None	0·05	Undet.	None
Older and buried halite, Lake Inder	4·20	0·02	None	0·05	Undet.	Trace
Older salt, Lake Inder	3·50	0·10	None	0·15	Undet.	None
Rock salt, Berezniki	None	0·08	None	0·15	Undet.	Trace
Sylvite rock, Kalush	0·76	0·31	None	0·06	Undet.	Trace
Sylvite rock, Kalush	1·95	0·25	None	0·25	Undet.	None
Variegated sylvite rock, Solikamsk	2·65	0·12	None	2·70	Undet.	Trace
Red sylvite rock, Solikamsk	1·50	0·05	None	0·09	Undet.	Trace

Note: Data on Upper Kama salts from Morachevskii and Tikhomirova (1940); other data from Galakhovskaya (1953).

to some extent. At the present day, sea water, representing the source material for the saline process, contains much more Sr, B, F, and Br than stream water does:

	Content in streams (%)	Content in sea water (%)
Sr	$1·3 \times 10^{-5}$	$1·0 \times 10^{-3}$
B	?	$4·5 \times 10^{-4}$
Br	?	$6·5 \times 10^{-3}$
F	$6·0 \times 10^{-5} - 2·0 \times 10^{-6}$	$1·0 \times 10^{-4}$

This relationship was probably similar in early Mesozoic time. The water was further concentrated in the marine basin, and ultimately all the compounds of Sr, B, F, and Br reached saturation and began to be precipitated, being concentrated in the early rock types of the saline sequence. Fluorine in fluorite accumulated in some of the first of these, at the end of the dolomite stage and during the transition to the anhydrite stage, locally forming notable deposits, as in the Kashirian dolomites of the Moscow region. At almost the same time, but at somewhat higher salinities, strontium began to appear in celestite, commonly forming well-defined horizons of sufficient concentration

2 L

to be of commercial importance. Two maxima have been noted for boron deposition: the first, only weakly expressed, coincides with the formation of dolomite and is related to the isomorphous replacement of $(CO_3)^{2-}$ by $(BO_3)^{-}$ in the dolomite lattice, the presence of magnesium being favorable for this type of replacement (Golovko 1960). The second maximum, very well defined, is found at salinites near the eutonic point. In the interval between these two points, chemical precipitation of boron does not occur and the element is taken into the sediment only by sorption on clay minerals. Bromine forms concentrations in the sediments later than the other elements, at the beginning of the chloride stage, forming isomorphous substitutions in halite, sylvite, carnallite, and bischofite. Because the distribution coefficient is always less than unity, only an insignificant amount of this element accumulates in the solid salts and most remains unprecipitated in the brine. Iodine does not participate in sedimentation as such. It accumulates in the brine and is found in saline deposits only to the extent that the deposits are impregnated with brine.

Although in present-day saline deposition the appearance and accumulation of these micro-elements in lakes reflect the composition of the rocks in the drainage areas, it is not known if this applies to marine saline formations in the stratigraphic column, because of lack of data. It is nevertheless highly probable that this is so. It is no accident, therefore, that the occasional high copper values in Permian sulfates of the Russian platform correspond to the erosion of a cupriferous zone in the Urals, with the formation of cupriferous sandstones in the depositional area. The distribution of the halophile micro-elements is more complex. Their introduction from the ocean would seem to assure their uniform distribution in the sediments, but modification of the marine salts by the introduction of sedimentary material from the drainage areas is always possible.

When, because of the specific composition of rocks of the drainage area or the manifestations of thermal springs, the introduction of F, Br, B, Sr, and other related elements from the shore has been greatly increased, not only would the concentrations of these elements in the saline deposits be increased, but also the stage of their accumulation would be shifted to somewhat lower salinities than normal for such deposition. This may control the variable Br and F values in dolomites and anhydrite-dolomitic rocks. The Sr content in chloride salts may correspond to the computed norm, but it may also vary. The accumulation of boron, even near the eutonic point, is at times very strong; but it may be absent. If this explanation is valid, the geochemistry of the minor halophile elements must supply a new and significant aspect to the close connection between a basin and its surrounding drainage area, even if the basin was marine.

6. ORGANIC MATERIAL IN SALINE DEPOSITS AND ITS EFFECT ON THE OCCURRENCE OF SOME ELEMENTS IN SALTS

Although in present-day saline basins organisms become fewer in species as the water becomes progressively more saline, they persist quantitatively for a long time. For this reason the content of C_{org} dissolved in the water and the mass of "living material" may continue to be large up to very high salinities. It is very likely that the same relationship held also in ancient basins, and the C_{org} content in different petrographic types of sediments in saline basins is therefore of interest.

Unfortunately, only a few determinations of C_{org} have been made on saline deposits. These include analyses from a wide spectrum of Kungurian sediments in the Ishimbai region (Strakhov and Zalmanzon 1944) and on Lower Cambrian lagoonal deposits of the Siberian platform (Zelenov 1956). Table 54 shows that the C_{org} content in the Kungurian sediments, from sandstone and clay to marl and limestone, is the same as for similar rocks of humid climates, even in such detail as the markedly poorer C_{org} content in red beds than in grey beds. All these deposits formed in parts of the basins where salinity was lowest. For saline deposits themselves the picture is entirely different: anhydrite rocks are essentially free from C_{org}. No determinations on organic carbon were made on rock salt or potassium salts, but even where these deposits are not red, their transparency and the composition of insoluble residues, chiefly of anhydrite and carbonate rocks, leave no doubt that the relation of C_{org} is also the same for these saline rocks.

Thus, despite the fact that *saline deposits have formed in basins known to have been colonized by well-developed organic communities, organic material in the deposits is essentially absent.* It seems likely that the cause of this may lie in the great rapidity of the saline process, especially in its later stages when the ever-increasing annual deposition of mineral salts has relatively lessened the percentage of C_{org}. However, even this minute amount of organic material had its effect on those components in the deposit that were susceptible to reduction, particularly Fe, Mn, and some others. This is confirmed by the forms in which iron occurs in the deposits of saline basins.

Hydrotroilite and pyrite have been recorded in modern saline basins at all stages of salinity up to the highest (26–28%). In ancient saline rocks and interbedded clays, pistomesite has been found along with pyrite, i.e. iron carbonate enriched in $MgCO_3$. Furthermore, Zelenov (1956) has shown that, at least at the gypsum-depositing stage, the authigenic forms of iron are clearly dependent on the amount of available organic material. When there is very little organic material, Fe^{2+} is absent and Fe^{3+}_{HCl} predominates over Fe^{2+}_{HCl}. As the amount of C_{org} increases, the mass of Fe^{2+}_{HCl} increases at the expense of ferric iron, and pyrite appears (Fig. 226). This indicates that diagenetic transformations of ferruginous compounds in the sediments of saline basins take place through a very wide range of mineralization, as in normal marine basins. A specific feature of the salt basin is lack of sensitivity. Irregularities occur only near the eutonic point, when the potassium salts sylvite and carnallite are precipitated. At this time the high salinity of the brine reduces the organic content. The C_{org} consequently proves to be increasingly inadequate for reducing Fe^{3+} to Fe^{2+} and the main line of iron diagenesis is

TABLE 54

Distribution of C_{org} in Kungurian Saline Rocks of the Ishimbai Region

Rock and sample locality	Content, %		
	Insoluble mineral residue	Sulfates	C_{org}
Sandstones			
At contact with Artinskian deposits along the Zigan River at Armet-Rakhimovo . .	62·52	1·22	0·01
The same, above lower gypsum layer . .	26·48	1·22	0·11
The same, above second gypsum layer . .	17·86	Trace	0·22
The same, at top of section . . .	58·30	0·49	0·26
At Timashevka, top of Kungurian . .	35·90	Trace	0·12
Clays			
Sandy clay	38·76	37·00	1·60
Borehole 3/25, cherry red clay . . .	38·5	26·43	None
Borehole 3/25, cherry red clay . . .	44·38	26·61	None
Borehole 3/25, cherry red clay . . .	63·07	2·30	0·10
Marls			
Borehole 6/11, grey marl, highly calcareous .	28·17	6·16	0·17
Borehole 1/24, black marl, highly calcareous.	31·31	1·90	1·23
Borehole 1/23, black	48·30	2·69	1·09
Borehole 48/20, highly gypseous marl . .	24·65	24·90	1·17
Limestone			
Timashevka, top of Kungurian . . .	3·26	Trace	0·05
Timashevka, top of Kungurian . . .	18·20	Trace	0·80
Calcareous dolomites and dolomites			
Borehole 5/21	0·36	1·02	0·05
Borehole 3/4	4·33	2·81	0·09
Borehole 3/4	3·90	2·77	0·02
Borehole 10/20	10·26	1·12	0·65
Borehole 10/20	4·08	10·20	0·30
Anhydrite-clastic rocks			
Borehole 48/2, argillaceous anhydrite . .	17·42	—	0·60
Borehole 48/2, sandy anhydrite . .	22·73	—	0·19
Anhydrite rocks			
Borehole 48/20	0·36	—	None
Borehole 48/20	0·16	—	None
Borehole 3/4	1·31	—	None
Borehole 3/4	0·22	—	None
Borehole 10/20	0·39	—	0·03
Borehole 10/20	0·10	—	0·01
Borehole 3/4	5·24	—	0·02
Borehole 48/20	13·29	—	None
Borehole 10/20	0·95	—	0·01

changed. If the parent brine is acid (from hydrolysis of $MgCl_2$), the suspended iron goes into solution, forming $FeCl_3$:

$$Fe_2O_3 = 6 \; HCl \rightarrow 2 \; FeCl_3 + 3 \; H_2O.$$

During the subsequent hydrolysis of $FeCl_3$ the iron is again precipitated in the form of fine needles of hematite, included in the sylvite and carnallite and making these minerals red (E. E. Razumovskaya). This diagenetic sequence, first determined by Morachevskii (1939), is a specific saline feature, which develops only at very high stages of salinity and is associated with the progressive suppression of life in the bottom brine and with a consequent ever-increasing deficit of C_{org} in the sediment. In different basins this saline type of iron transformation apparently begins at somewhat earlier stages of salinity.

One interesting detail was established by Dellwig in the salts of the Michigan basin. In the central part of the formation, seasonal anhydrite-dolomite

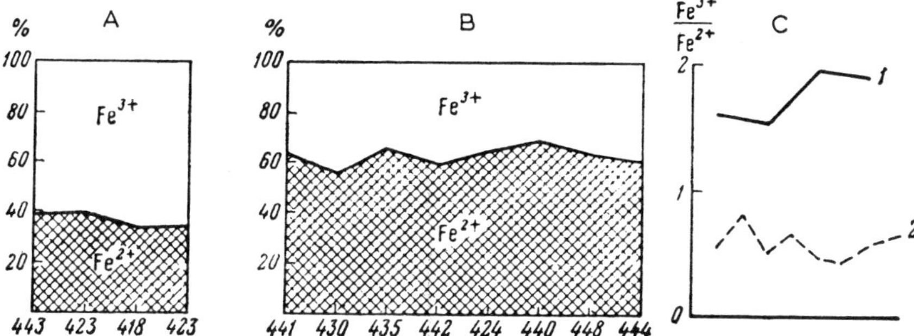

Fig. 226. *A*. Red rocks; *B*. green and grey rocks; *C*. Argillaceous dolomites of different colors: 1. red; 2. greenish-grey and grey.

laminae 0·5–1 mm thick are rich in organic material; the only authigenic form of iron now found in these beds is pyrite. By contrast, the halite beds (5–10 cm thick) separating these anhydrite-dolomites contain no detectable organic material, and the authigenic iron is here hematite. Pyrite crystallites are exceptional and, as Dellwig properly states, they probably formed by the penetration of small amounts of sulfate from the sulfate horizons into the interbedded halite. This clearly shows that the processes of mineral diagenesis may differ in adjacent deposits of substantially different material. In other words, it shows how completely diagenetic changes are predetermined by the initial composition of the sediment.

D. AN OUTLINE OF MARINE HALOGENESIS

It is now possible to formulate the basic patterns of composition and origin of marine saline rocks, and to outline a sequence of marine halogenesis. A convenient form is the compositional cyclogram (Fig. 227). To make the description complete, the cyclogram includes deposits for the initial stages of salinity. The constructional principles of the cyclogram are the same as was used for describing the cyclograms for humid-climate rocks and will not be repeated here.

1. A Cyclogram for Marine Saline Rocks

In the cyclogram as a whole, the following systematic relationships may be recognized.

(1) *The rocks of saline basins form a series, arranged according to the geochemical mobility of the constituents.* At the left-hand side of the cyclogram,

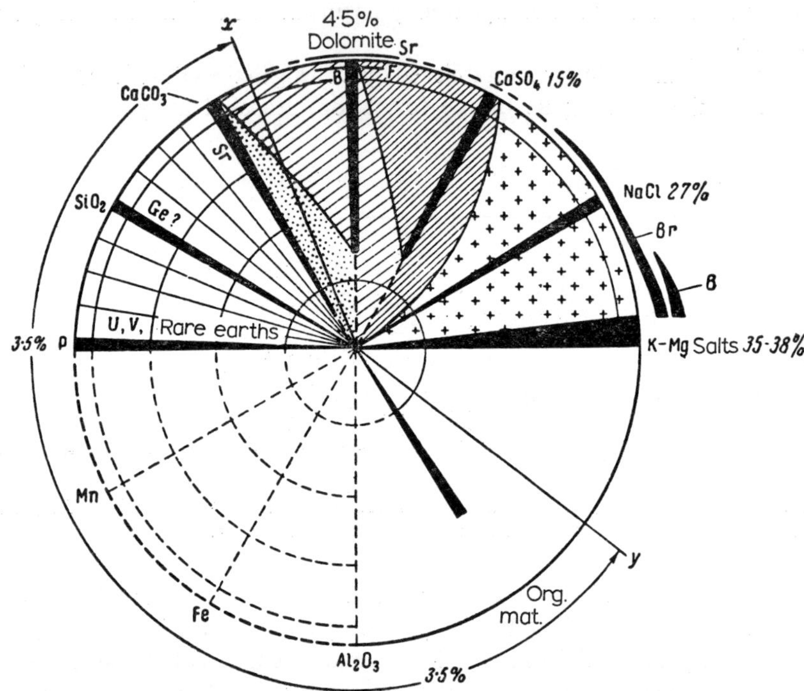

Fig. 227. Cyclogram of arid-climate chemically precipitated rocks.

the deposits are made of the least soluble components, those with the lowest geochemical mobility (P, SiO$_2$); at the right are the most soluble components, those with the highest geochemical mobility (K and Mg sulfates and chlorides). These relations are underlined more clearly if we turn to salinity values of the water at the time when particular components precipitated. The appropriate data are shown round the circumference of the cyclogram. The figure is divided into two parts. The first includes bituminous shales, accumulations of phosphorus, authigenic silica and limestones (pure or somewhat dolomitized, patchy-metasomatic). All these rocks form at salinities of 3–4%, i.e. at values characteristic of basins in humid regions. Rocks in the second half of the cyclogram form as basins become increasingly salty, from 4% to the eutonic point, i.e. up to 38–40%. These rocks are normal sedimentational, bedded dolomite, gypsum (anhydrite) rock, rock salt, and potassium salts. Dolomite begins to form at salinities near normal (4–4·5%) and continues in a pure form to salinities of about 15%, when gypsum begins to precipitate. Rock salt begins to form at a salinity of about 27%, potassium salts at 30–32%.

Accumulations of minor elements are formed along with those of the macrocomponents. For example, fluorine (as fluorite) and strontium (as celestite) accumulate in dolomites and sulfate-bearing dolomites. Beginning with the precipitation of halite and continuing to the eutonic stage, *bromine* accumulates in the solid phase in ever-increasing degree, forming isomorphous substitutions in NaCl, KCl, carnallite, and other chloride salts of K and Mg. Boron is concentrated in dolomites in very small quantities, but solid phases apparently begin to form only at very high salinities, near the eutonic point, making large primary concentrations in the carnallite and bischofite zones of potassium deposits.

(2) In the sequential change in rock composition from the left side of the diagram to the right, the characteristic changes in total constitution of the solid phases are shown along with the transition from less mobile to more mobile component. *The precipitation of simple anhydrous salts with progressive salinity is replaced in increasingly greater degree by the formation of double, perhaps triple, salts in which crystal-hydrate water becomes a significant factor.* In this process the crystallization capacity of a substance generally increases, and because of this the saline rocks on the right side of the cyclogram tend to be much coarser-grained than rocks on the left side. Whereas phosphorites and siliceous rocks are commonly aphanitic, limestones and dolomites are normally microgranular or fine-grained, anhydrite rocks are fine, medium, and coarse-grained, and beginning with rock salt the rocks are not only coarse-grained but giant-grained, with crystals millimeters or even centimeters in diameter.

(3) All gradations exist between pure accumulations of individual components, corresponding to the ends of the radii on the cyclogram; but the frequency with which these intermediate "mixed" and pure rocks occur is variable. *Over most of the cyclogram, beginning with deposits of phosphorus and extending to halite, the most widespread rocks in nature are those consisting predominantly of one chemical constituent, and containing only insignificant amounts of another.* Two-component chemically precipitated rocks, in which the two components occur in approximately equal amounts are rare. In rock salt and the potassium-magnesium rocks, on the other hand, the relations are radically different; mixed, multicomponent rocks dominate. This is because the volumes of precipitating material during crystallization of halite remain large, approximately equal to the volumes of potassium-magnesium salts.

The large left part of the cyclogram is thus characterized by rocks that show a tendency toward constant differentiation of the macrocomponents; the right side of the diagram represents rocks showing a marked reduction of this tendency.

(4) *The relations among the chemical components of rocks on different parts of the diagram are substantially different.*

On the left part of the diagram, corresponding to weak mineralization in the water (up to 3·5%), i.e. up to and including the limestone stage, the components accumulate in complete independence of each other. For example, siliceous rocks may contain appreciable amounts, even high amounts, of P; but as a rule they contain only the average content, or even less. In exactly the same way, limestones may or may not contain large amounts of SiO_2 and P.

On the right side of the diagram, beginning with dolomite, the relationship of the chemical macrocomponents is different. In dolomites some $CaCO_3$ (up to 5%) is always present. Anhydrite rocks always contain admixtures of dolomite and $CaCO_3$. Rock salt has impurities of anhydrite and carbonates, and potassium salts have large admixtures of halite, less of anhydrite and carbonates. The same may be said of accumulations of the microcomponents as soon as they begin to form solid phases. In other words, *in any saline rock, inherited solid phases, reflecting the chemical sedimentation of the preceding stages of salinity, are always present along with the chemical ("new") phase specific for the particular stage.* This is of fundamental importance because it represents basic differences between the saline basins and the rocks of humid regions, even among rocks deposited in arid regions if the salinities of the basins have been low (below 3·5%). *In humid-climate rocks and in arid-climate rocks of low-salinity basins, there are no features of rock composition inherited from deposits preceding them in the mobility series. In saline rocks, beginning with dolomite, inherited features are invariably present.* It is clear that compositional relationships among saline rocks are incomparably closer than those observed among humid-climate rocks. *This close compositional relationship among saline rocks is the simple consequence of the stage-like process of saline accumulation,* and also of the fact that no component on precipitation disappears entirely from the solution; a component continues to be precipitated to the highest stages of salinity.

(5) *When inherited components accumulate, they generally preserve the mineral forms in which they were originally precipitated during the basin's history; but they do sometimes change, giving new phases.* Minerals that have preserved their phases are fluorite, celestite, halite, kieserite, carnallite, and others. Minerals that change are the Ca and Mg carbonates. Dolomite, as a basic carbonate, is generally preserved in anhydrite rocks, but in rock salt and potassium salts it is replaced by a mixture of calcite and magnesite. Gypsum, being precipitated at the stage of halite deposition, near the eutonic stage, is replaced by anhydrite. These changes in inherited solid phases do not, however, alter the essential phenomena of inheritance.

(6) The introduction of terrigenous material into a modern basin and the mixing of this material with precipitating solid phases lead to the appearance

of terrigenous-saline rocks in saline marine formations. Distinctive reactions generally arise getween the chemical and the clastic components.

The introduction of clastic particles by fresh stream water causes shifts in the physicochemical equilibria of the saline basin. These shifts commonly retard chemical sedimentation in the zone of abundant precipitation of the terrigenous component. On the other hand, clay material in a saline basin is itself altered, becoming enriched in alkalies and MgO and changing its mineralogical character.

The freshening effect of stream water, carrying in terrigenous material, is reflected on the cyclogram in the following ways: (1) In the field between the calcite and dolomite boundaries the precipitation of dolomite is shifted to high salinities toward the center of the cyclogram, and precipitation of dolomite ceases altogether near the center; (2) in the field between dolomite and gypsum a similar shift is experienced by gypsum; (3) in the field between gypsum and halite the incipient precipitation of halite is similarly shifted.

(7) In parallel with the change in rock composition, the relationships between saline deposits and the bottom water alter as the marine basin becomes progressively more saline. At the beginning and intermediate stages of salinity, the solid phases precipitate from the bottom water, and the volume of precipitating solid phases is only a negligible part of this total mass. At high salinities the thickness of the bottom brine decreases sharply, and the volumes of precipitating solid phases begin to approach the volumes of parent brine. This leads to the result that the brine itself is gradually concentrated in the spaces between crystals. It then becomes intercrystalline brine. Further sedimentation, leading to the accumulation of potassium minerals, is related to movement of the intercrystalline brine toward persistent depressed zones in the field of rock salt. The terminal stages of the marine saline process are thus developed essentially in brine lakes in a field of rock salt, which have no direct connection with the sea. To put it differently, the saline process in marine basins terminates in an environment specifically lacustrine (Fig. 228).

Thus, in their *compositional and genetic patterns the rocks of marine saline basins are intimately associated, forming a single physicochemical whole. The marked increase in physicochemical interrelations among the formation-forming rocks and microcomponents is a specific characteristic of saline formations as compared with humid-climate formations.* In the latter such interrelations have merely a facies character, but in saline formations they represent not only facies phenomena but also physicochemical and compositional characteristics. Saline formations are thus classic examples of formations as paragenetic associations of rocks, considering paragenesis in the genetic sense.

2. THE PHYSICOCHEMICAL MECHANISM BY WHICH SALINE ROCKS FORM

In describing the paleogeography of ancient saline basins, it was noted that the types of such basins have been extremely varied, at times markedly different from any modern basins in size, outline, or depth. This has given ancient marine halogenesis very characteristic features and has made it difficult to apply views concerning the morphology of modern saline basins to ancient basins.

The matter is completely different when we turn to the saline process of ancient saline marine basins. Here, on the contrary, are examples of striking similarity, if not essentially identical.

The saline process in modern basins proceeds through annual hydrochemical cycles, the formation of salt beds becoming possible only after the

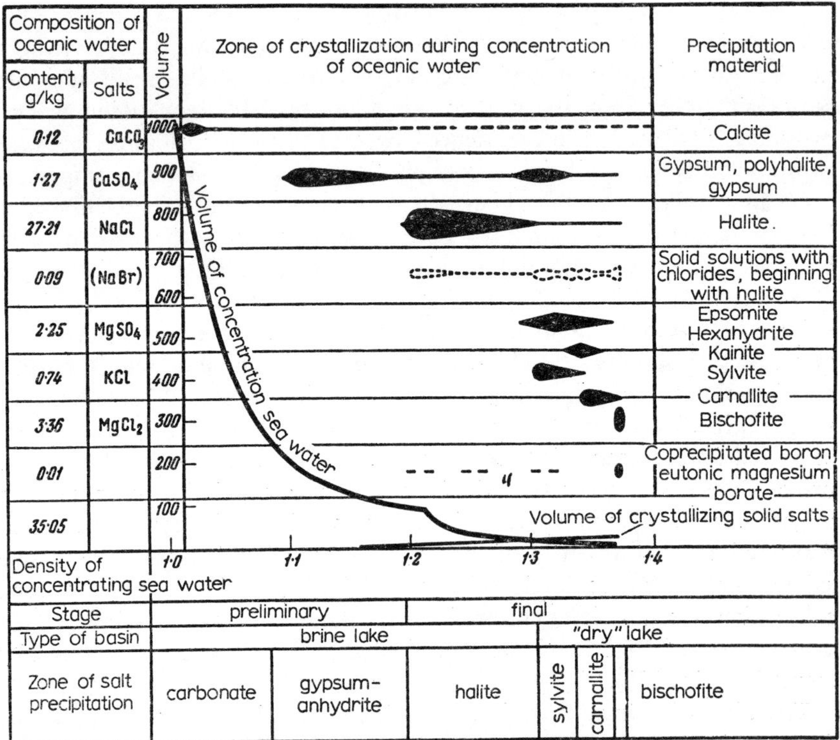

FIG. 228. Change in volume of oceanic water and precipitating solid salts during natural concentration of the water (from Valyashko).

bottom water and interstitial brine have reached saturation in a given solid phase. The constant presence of well-defined seasonal bedding in anhydrite rocks, rock salt, and potassium salts indicates beyond doubt that the same mechanism of annual hydrochemical cycles was operative when ancient saline formations were deposited. There is no doubt that salt beds formed in the past only after the bottom and intercrystalline brine became saturated in a given solid phase. This condition is required by the physicochemical nature of the process by which highly soluble salt minerals are formed.

The progressive salinity of modern basins is accompanied by continuous alteration of the brine through clay particles and dissolved $CaCO_3$ introduced into the basin. The existence of the sulfate and chloride groups of potassium rocks in ancient saline formations and the presence of coarse local accumulations of magnesite and polyhalite leave no doubt that ancient saline basins were characterized by this process no less than modern basins.

Halogenesis in modern saline basins is solar; i.e. it follows the metastable path. There is every reason to believe that the saline process followed the same course in the huge ancient saline basins. Early diagenesis of sediments, along with sedimentational processes, is a very important factor in producing the final mineral composition of modern saline deposits. This involves the conversion of metastable phases to stable by reactions of cation exchange, by dehydration of solid crystal-hydrate phases, and by replacement of stenostable minerals by eurystable. Redistribution of components, general compaction of sediments and lithification accompany this process. Similar processes, as indicated above, also took place in the accumulation of saline deposits in ancient basins. Early diagenetic processes converted metastable solid phases to stable phases along the solar path of sedimentation. Recrystallization is demonstrated by the loss of primary pinnate structure by many minerals. Redistribution of substances during diagenesis is very clearly proved by the highly variable content of many components, especially microcomponents, in saline rocks. Fluorite and celestite in dolomites and anhydrite rocks, for example, show extremely irregular distribution: along with zones practically free from these minerals are other zones where masses of them have accumulated, forming irregular patches, concretionary lenses, and nodules. The quantitative variations in strontium content in dolomites is striking, and may change a hundredfold within short distances.

Concretions of anhydrite and other minerals have been identified in rock salt. According to Yarzhemskii (1957, pp. 94-95): "In the 'spring' clay layers in rock salt, nodules of anhydrite are frequent, from sizes scarcely perceptible to the naked eye to larger spherical and other forms, perhaps 3 cm across, consisting of fine-grained anhydrite. . . . Recently numerous lenticular and nodular masses of carnallite have been found in rock salt in cores from the Inder saline sequence. . . . On the surfaces of many cores are thin (one millimeter thick) layers of anhydrite in the rock salt which abut against carnallite nodules or irregular inclusions. The boundaries between the carnallite layers of nodular or irregular inclusions and the enclosing rock salt are not always clear; they tend to be sinuous or labyrinthine."

For the sake of completeness, we may add that nodules of isotropic boracite (chlormagnesium borate), mined as a boron ore, have long been known from the carnallite beds of the German Zechstein.

The indicated nodules of micro- and macrocomponents in saline rocks are characterized by several general features. They are generally small, and consequently only a small amount of material is concentrated in them. As a rule they occur along the bedding planes of the rock. The thin layers of the enclosing beds bend around the nodules in some places, at other places pass through them, becoming thicker, and in still other places they disappear at the nodules, clearly being entirely replaced by them. The nodules are massive, having neither concentric nor radial-radiating structure. The primary texture is microgranular, the grains being generally smaller than 0·01 mm. Subsequent processes of recrystallization lead to the appearance of varigrained texture, in which remains of the primary microgranular texture may still be found.

There is no doubt that all these nodular accumulations of the microelements in salts result from diagenetic redistribution of material that was previously diffusely disseminated. At least some local concentrations of these elements became possible, apparently, because of secondary (late) permeation of the highly concentrated parent brine through the sediment, bringing new components into the sediment.

Modern halogenesis, as we known, is characterized by its great rapidity, the rate increasing progressively as the basin becomes more salty. Exactly the same occurred in the saline process of ancient marine saline basins. The thickness of the annual rhythms shows that even the absolute rates of the modern and ancient processes at like stages of salinity are identical.

As we see, all the critical features of the physicochemical mechanism of the present-day saline process are found without difficulty in the ancient process, despite the fact that there is commonly a fundamental difference in geomorphic aspect between modern and ancient saline marine basins.

This fact is fundamental in assessing the possible use of data on modern lithogenesis for understanding the ancient saline process. It has been shown that *the geomorphology of ancient saline basins may be reconstructed only in part from the geomorphology of modern representatives, but the physicochemical mechanism observed in modern halogenesis can be more directly applied to interpret the ancient process.* The essential difference lies in the geomorphology of a basin which itself is a product chiefly of the tectonic activity of the region, and this changes with time. The physicochemical mechanism of saline accumulation is due to the nature of precipitating solid phases, and this predetermines the mechanism throughout the entire span of geologic history.

CHAPTER 6

THE DISTRIBUTION OF SALINE FORMATIONS IN ARID ZONES
DEVELOPMENT OF HALOGENESIS THROUGHOUT EARTH HISTORY

In analyzing the distributional patterns of saline formations in arid regions, two questions, which arose also in studying humid-climate deposits, are of primary importance. (1) Is there any systematic relationship between the location of the formations and variations in the arid climate? (2) Is there any systematic relationship in the distribution of saline formations and the structural elements of the crust?

1. CLIMATIC LOCALIZATION OF SALINE FORMATIONS IN ARID REGIONS

In order to determine the distributional patterns of saline formations in arid regions of different ages, distributional maps of gypsum, rock salt, and potassium-salt deposits have been prepared for the largest and best authenticated of these regions. The Devonian, especially the Middle Devonian, was one of the most significant periods of saline deposition in both the northern and southern arid zones (Fig. 229). Gypsum deposits may be found in the northern zone along almost any traverse across it as is well illustrated by North America, at the western end of the zone, and Siberia, at its eastern end. Rock salt deposits in Siberia accumulated in approximately the middle of the belt, but the rock salt deposits of North America clearly formed at the southern margin. The potassium-bearing beds in the Williston basin of North America were deposited in approximately the middle of the arid belt. In the southern arid zone, marine rock salt lies along its middle part, but rock salt of the Dnieper-Donets basin was clearly deposited at the southern margin. Furthermore, the potassium salts of the Pripyat basin (D_3^2) also lie at the southern boundary of the Devonian arid belt. Thus, *of six large deposits of highly soluble salts, three formed in approximately the central part of either the north or south arid zone, and the other three at the margins of these zones.* In addition, *not only gypsum, but rock salt and potassium salts also, formed in Devonian time in "latitudinal" segments across these belts.* What is the significance of such distribution of saline deposits?

Ivanov (1953) suggested that gypsum, rock salt, and potassium salts require different degrees of aridity for their deposition: gypsum requires slight

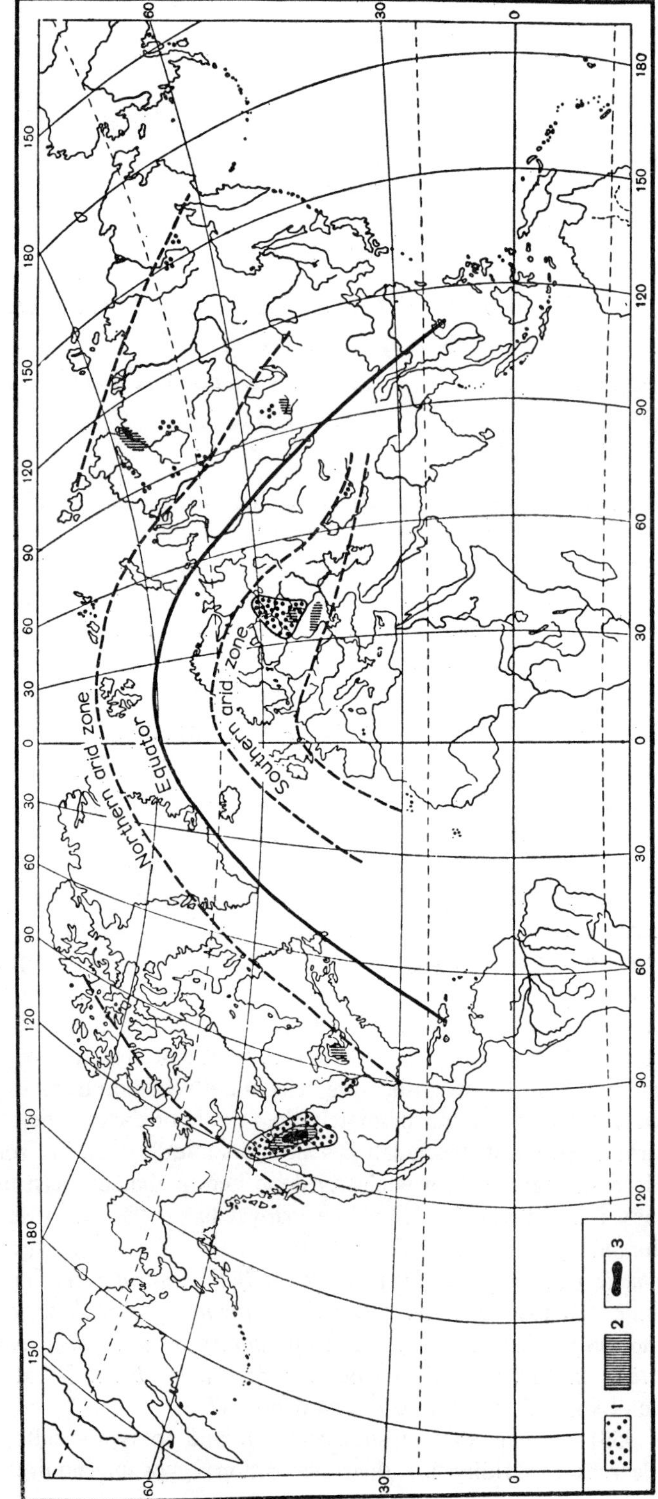

FIG. 229. Distribution of saline deposits in arid zones of Middle Devonian time. 1. Gypsum; 2. rock; 3. potassium salts.

aridity, which increases for rock salt to a maximum aridity for potassium salts. From this point of view, it is easy to understand the formation of rock salt and potassium salts in the middle segments of arid belts, because here, especially if the arid zones lie in a platform region of plains, arid conditions are most sharply defined. Other locations, nevertheless, remain inexplicable, particularly the occurrence of rock salt and potassium salts at the margins of arid belts. Following Ivanov, it becomes necessary to assume that markedly arid conditions must have been established for some reason at separate localities along the margins of the arid belts. This kind of climatic anomaly could be caused only by relief. Uplifts surrounding saline basins that lie along the margins of arid belts isolate the basins from regional atmospheric circulation, and create local, markedly arid conditions at the very edge of the arid belt, where a much less arid climate would normally prevail. This is seen in the Tuva and Minusinsk basins, which in Devonian times lay in the humid tropical zone (Vol. 1). Nevertheless, rock salt accumulated in the Tuva basin, and gypsum in the Minusinsk. The cause of these anomalies lies in the fact that both these basins were intermontane in Devonian time, isolated by mountain chains from general atmospheric circulation of the Devonian tropical zone. Clearly it would have been even easier, under such conditions of isolation, for marked increase in aridity to develop along the margins of ancient arid belts.

If this view is valid, the rock salt and potassium deposits that have accumulated along the margins of arid belts should bear indications of highly dissected drainage areas nearby, surrounding the particular locality. The Tuva rock salt and the Minusinsk gypsum, having accumulated in intermontane basins, are related to terrigenous sequences. The potassium deposits of the Pripyat basin contain a large amount of terrigenous material, but much less than is found in the Tuva or Minusinsk basins. The Pripyat basin can scarcely therefore be referred to the class of intermontane basins. This also applies to the Salina formation of the Michigan basin. The regional geologic data concerning the territory surrounding the Pripyat basin in Middle Devonian time (D_2^2) and that surrounding the Michigan basin in Givetian time indicate that though these regions were by no means mountainous, they were hilly, as are many platform regions.

These facts indicate that the marginal parts of arid belts, during times of rock salt and potassium salt deposition, could scarcely have been extremely arid, since the necessary geomorphic conditions were lacking. *Both rock salt and potassium deposits were formed in an arid climate, but the range of climatic conditions in Devonian time was wide, embracing both extremely arid conditions at the center of the zone and moderately arid conditions at the margins.* In these examples, the critical factor that determined the location of deposits of highly soluble salts within arid regions was not a local climatic modification, but rather the influence of tectonic movement, which caused prolonged and rapid subsidence of local depressions, necessary for the completion of the saline process, within the arid belts. *It is this tectonic aspect of halogenesis that has* *controlled the accumulation of rock salt and potassium salts, at times in the central part of the arid belt, at other times along the margins.**

By contrast, the Lower Carboniferous was very poor in saline deposition, and its arid belts have been reconstructed only with difficulty and in few areas. The largest of the Lower Carboniferous saline deposits was in North America (Fig. 230). Here gypsum was dominant among the saline components, having been deposited throughout the arid belt. Rock salt accumulated only in West Virginia, New Brunswick, and Nova Scotia, being clearly restricted to the southern margin of the arid belt. There is no evidence of intermontane basins, or of the mountain chains necessary to separate the basins. Low-lying plains with low subdued hills dominated the landscape; the folded and strongly

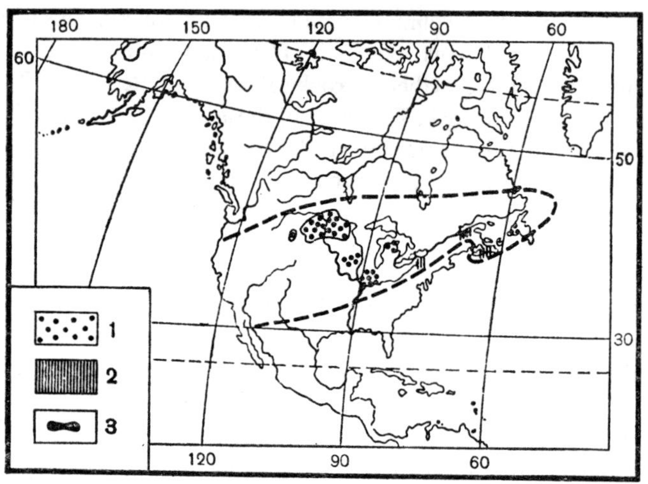

FIG. 230. Distribution of saline deposits within Lower Carboniferous arid zones. 1. Gypsum; 2. rock salt; 3. potassium salts.

uplifted Appalachian chains were not uplifted until Middle and Late Carboniferous time; and when they did form, it was coal, and not salt, that accumulated on the land to the west of the mountains; i.e. the climate was moist. These comparisons lead to the conclusion that saline deposition in the North American arid belt during Early Carboniferous time was controlled by tectonic factors rather than by a local modification of the climate. *Rock salt accumulated where local depressions subsided rapidly and for long periods of time, permitting the precipitation of highly soluble salts on top of previously deposited gypsum.*

The distribution of saline deposition at the end of the Early Permian and

* To avoid misunderstanding, let me state that Ivanov also recognized the role of tectonic movements in the location of saline deposits, although, along with this factor, he pointed out the necessity of extremely arid conditions for accumulation of potassium deposits. While I agree completely with Ivanov's point of view regarding the importance of tectonism (as do many other salt specialists, among them M. P. Fiveg), I criticize merely the idea that extreme aridity has been essential for the complete development of the process.

during Late Permian was highly characteristic (Fig. 231). Three foci of saline accumulation are clearly seen. One lies in North America, where rock salt accumulated both in the northern part of the arid belt (Delaware basin) and along its southeastern margins. Potassium salts formed only in the inner parts of the arid belt. In Europe the distribution of rock salt and potassium salts was substantially different. These salts accumulated strictly along the edge of the belt. In the West German basin they formed right at the southern margin of the arid belt. Saline deposition during Kungurian time took place immediately next to the easternmost and southeasternmost boundaries of the belt. These peripheral zones, with their intense saline deposition, stand in contrast to the barren central parts of the arid belt, where the climate is known to have been markedly more arid. Since folded Hercynian mountain structures existed near all the regions of Permian halogenesis, it is natural to inquire whether they played a decisive role in developing extremely arid conditions, with which the accumulation of Permian saline formations might have been associated. If it is assumed that the winds in the lower latitudes of the Permian northern hemisphere were like the present trade winds, blowing in general from northeast to southwest, the mountains may very well have had a marked effect by producing extremely arid conditions. Some facts, however, indicate that though the Hercynian chains actually caused local conditions of extreme aridity, these conditions were not the principal cause of the saline process. For example, the mountain chains of the Ural-Siberian Hercynian structures existed in approximately the same form in the Permian as in Middle and Late Carboniferous time, but no rock salt or potassium salts accumulated on the eastern part of the Russian platform then, and gypsum accumulated only in very small quantities. The same may be said for the North American platform. Consequently, if the Hercynian chains framing these platforms gave rise to extreme aridity on the eastern part of the Russian platform and on the southwestern part of the North American platform, this must have been a small factor in causing accumulation of saline deposits. Saline deposition began only when zones next to the Hercynian chains were cut off from easy access to the ocean because of tectonic movements, and when such movements also succeeded in producing local depressions that subsided rapidly and persisted for a long time. Even if extremely arid conditions existed on the Russian and North American platforms on the edge of the Permian arid belt because of geomorphic relations, it was not this factor but rather the specific tectonic activity at the end of the Early Permian and the beginning of the Late Permian and the corresponding landscape relations that led to thick saline accumulations. As for the German Zechstein saline accumulation, the Hercynian chains of Europe had already been considerably lowered by the time these salts were deposited, as may be discovered from the petrographic character of the saline rocks. The region adjacent to the zone of salt deposition was a hilly plain. The arid conditions in this zone were probably normal for the marginal parts of the arid belt. Moreover, the saline process was exceptionally intense and complete.

2 M

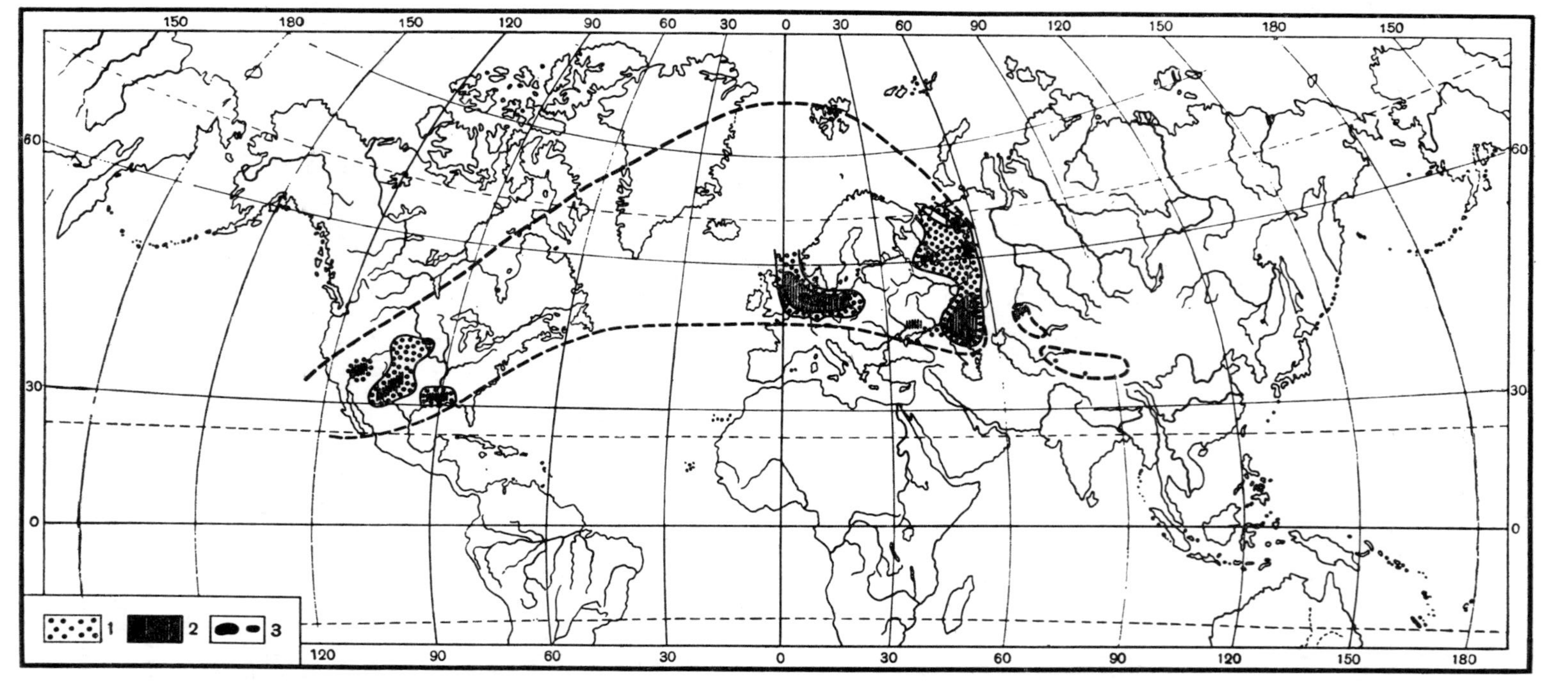

FIG. 231. Distribution of saline deposits within the Middle Permian arid belts: 1. Gypsum; 2. rock salt; 3. potassium salts.

Thus, *during Permian times also, during an epoch of tremendous accumulation of salt, tectonic conditions and not different modifications of the arid climate proved to exert the dominant effect on the localization of saline formations.* If extremely arid conditions prevailed in eastern Europe and southwestern North America because of the mountain chains, they could only have accentuated to some extent the effect of the specific tectonic conditions that made saline deposition possible.

During Triassic times, rock salt and potassium salts accumulated only in small amounts and only in Europe (Fig. 232). In Early Triassic time the deposits consisted of rock salt with minor amounts of polyhalite in northwestern and southwestern Germany, a thick salt sequence in the Pyrenees foredeep in France, and saline deposits in the southeastern foothills of the Pyrenees of Spain. During the Middle Triassic, a saline sequence accumulated in Western Germany, in a zone that extended to Switzerland. During the Late Triassic, restricted accumulation of saline rocks took place in Thuringia, England, the southern marginal depression of the Pyrenees, the Balearic Islands, western Sardinia, and the northern Alps. Some of these deposits were located in the inner parts of the arid belt (Mediterranean), where the climate might be said to have been extremely arid, but many formed along the northern margin of the belt, where it is by no means clear that the climate was arid. It is true that the Triassic period in Europe was characterized by active tectonic movements and by rugged relief, especially in the Early Triassic and the Keuper, as seen from the clastic character of the Lower and Upper Triassic deposits. It is possible that the rugged relief also brought about an increased aridity in the basins between uplifted segments. But if these basins had no communication with the open sea to supply them with salt water and if they were not characterized by prolonged and, at times, rapid subsidence, no conditions of extreme aridity could have given rise to saline formations. Thus, *in the Triassic also the chief factor bringing about the formation and localization of saline formations within the northern margin of the arid belt of Europe was tectonism and not extreme aridity.* If the latter existed generally in the depressions of Germany, it played only a secondary role.

Essentially the same evaluation holds for the distribution of Upper Jurassic saline formations (Fig. 233). Here the difference between abundant saline deposits in the relatively small arid zone of North America and the meager occurrence of such deposits in the huge Euro-Afro-Asian arid belt is striking. Within this latter belt, in the great central part, lying on the African platform and enjoying the greatest aridity, saline deposits are generally lacking because the dominant upward movements here permitted neither marine transgression nor the development of small depressions in a protractedly persistent basin, The saline process was displaced toward the northern half of the arid belt. where the relief was low, as may be determined from the nature of the rock salt and gypsum deposits, and local centers of extreme aridity could not therefore develop, but where segments persistently maintained connection with the sea and, at times, subsided rapidly.

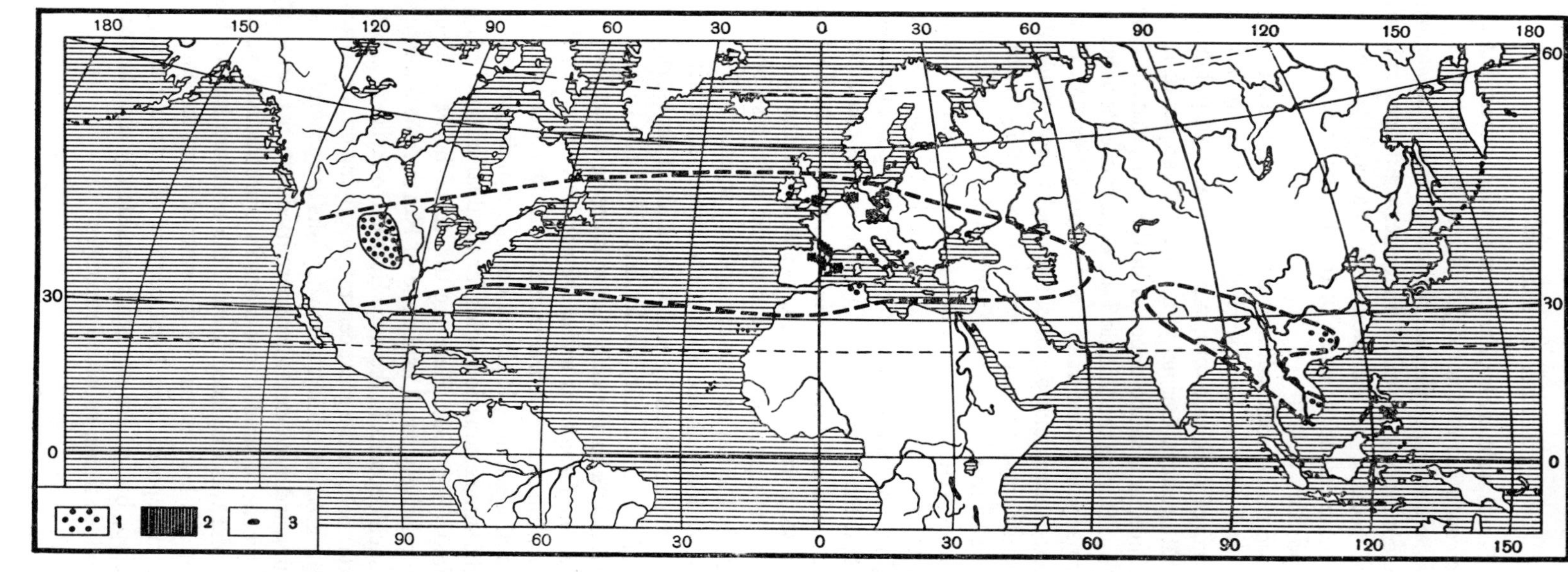

Fig. 232. The distribution of saline sediments within the Triassic arid belts. 1. Gypsum; 2. rock salt; 3. potassium salts.

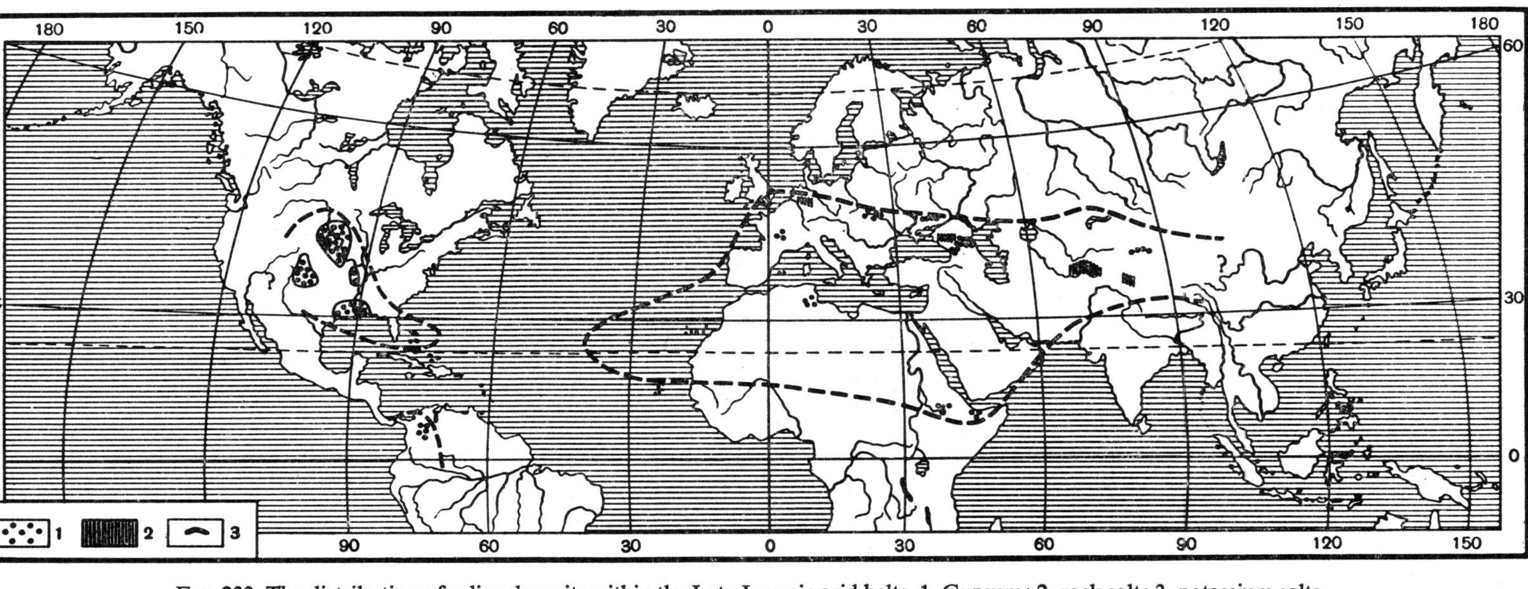

FIG. 233. The distribution of saline deposits within the Late Jurassic arid belts. 1. Gypsum; 2. rock salt; 3. potassium salts.

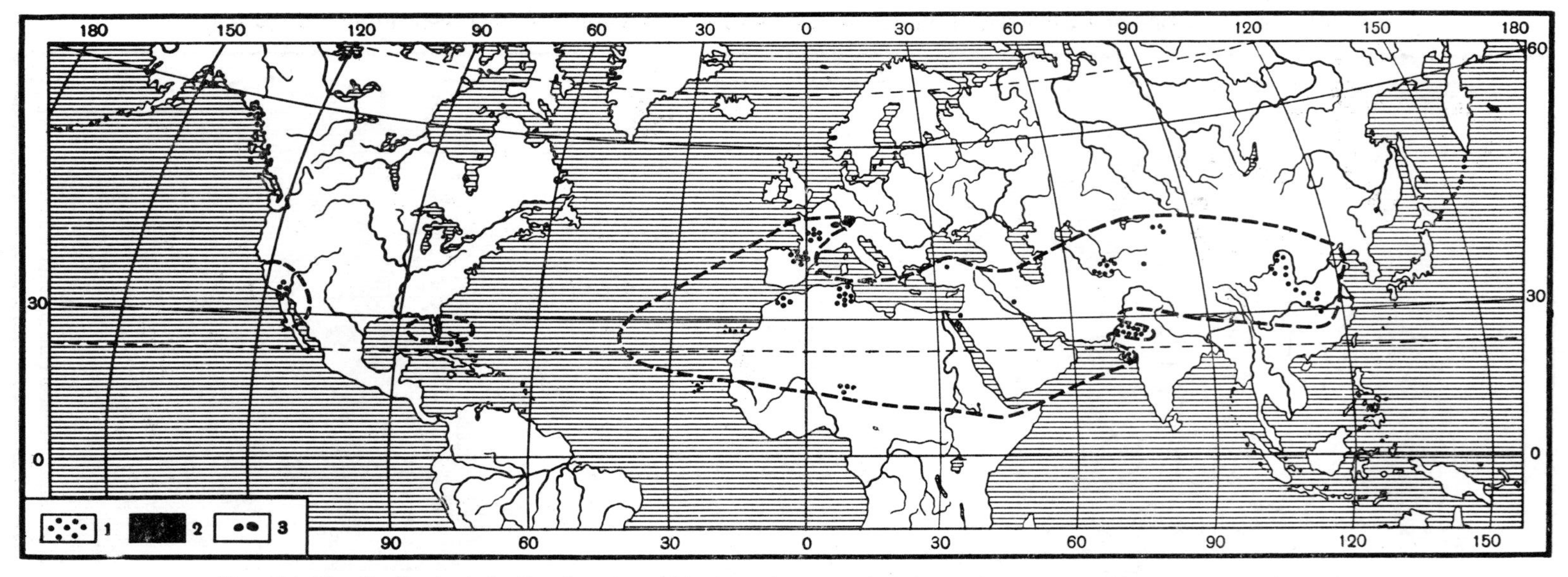

Fig. 234. The distribution of saline deposits within the Paleogene arid belts. 1. Gypsum; 2. rock salt; 3. potassium salts.

The Paleogene (Fig. 234) was characterized by a relatively large number of zones where the initial stage of the saline process was reached, represented now by gypsum deposits, and by a very small number of areas where the process reached the stage of halite or potassium-salt precipitation. Rock salt and potassium salts accumulated in the Rhine graben, where a well-defined intermontane depression undoubtedly existed. Potassium chlorides were also deposited in Spain, in the Eocene Ebro basin, again a well-defined geomorphic basin. Data indicating extreme aridity during the Oligocene in the Rhine graben are lacking. By contrast, floral and faunal evidence points to a semiarid climate (Wagner 1953). Here again, the most important factor controlling the distribution of saline formations was tectonism, not modifications of aridity within the arid belt.

The Neogene differed substantially from the Paleogene in the nature of its saline deposits (Fig. 235). The saline process reached the stage of halite precipitation, at times even potassium-salt precipitation (Ciscarpathia), at many localities in the Euro-Afro-Asian arid belt.

It is of interest to note that the zones where potassium salts accumulated, and where many rock salt deposits were localized, lay directly along the northern margin of the arid belt. The other deposits of rock salt formed in the middle of the arid belt or near it (Sicily, Calabria, Mesopotamia, Central Asian glauberite-halite formations). As a rule, all these accumulations formed in intermontane depressions of mountain foredeeps in regions of Alpine folding. Relict basins were generally preserved in these tectonic structures, with some, rather restricted, communication with the normal sea, and with depressions that warped downward for long periods and, at times, rather rapidly. These conditions permitted halogenesis occasionally to reach even the very highest stages of development. Whether these basins where the salts accumulated were in zones of extreme aridity is not known; in any case this factor had no marked effect, and it appears more likely that the effect was negative (Ciscarpathia, the Erivan and Nakhichevan basins).

Thus, on the basis of this survey of the distribution of saline formations in arid regions, the following facts are established.

(1) *Neither gypsum nor rock salt nor potassium salts exhibit any restriction to a particular, constantly preferred part of an arid belt. Each may be found in any part of the zone—marginal, intermediate, or central.*

(2) It has been reasonably established for at least some deposits, particularly the Upper Devonian of the Pripyat trough, the Lower Carboniferous of North America, the Upper Permian of Germany, and the Paleogene of the Rhine graben, that regions where the saline process was most completely developed were characterized, not by extreme aridity, but merely by normal aridity (or even semiaridity) proper to the margins of the arid belt. It might be that *extreme aridity was by no means necessary for the full development of the saline process.*

(3) Analyses of the conditions for saline accumulation invariably lead to the same conclusion—that a definite type of tectonic activity is essential,

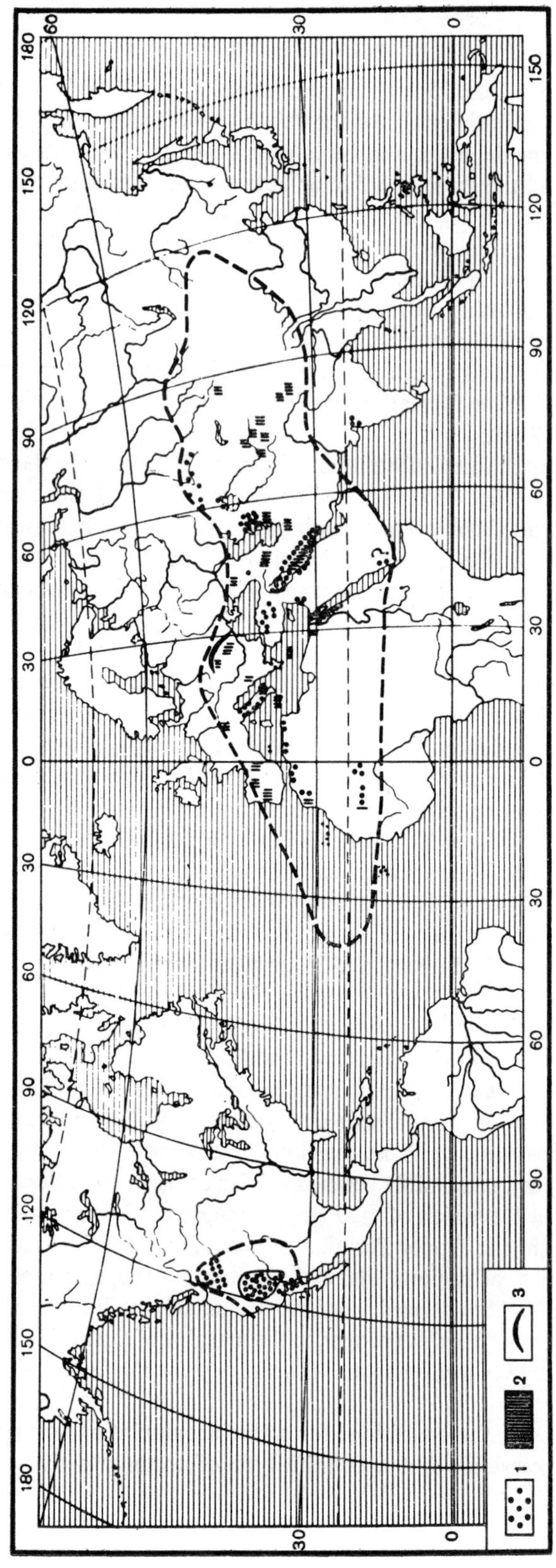

Fig. 235. The distribution of saline deposits within the Neogene arid belts. 1. Gypsum; 2. rock salt; 3. potassium salts.

bringing about restricted connection with the open sea for long periods of time and ensuring rapid, even if episodic, subsidence of the basin floor followed by the rapid accumulation of highly soluble salts.

The significance of these facts is clear. *The decisive factor in the distribution of saline formations in arid belts is not the distribution of different modifications of the arid climate, and certainly not the distribution of extreme aridity, but rather the tectonic activity and the topography of the different parts of the arid belt.* Arid climate with all its varieties is a necessary but by no means sufficient cause for the accumulation of saline deposits. Tectonic conditions therefore become the decisive factors in the distribution of arid-climate formations within arid belts, and are *a more potent factor than changes in aridity from one area of the arid belt to another.*

But, if this is so, a further question arises: on which of the principal structural units of the crust have tectonic conditions favorable to the development of saline process been most easily and most frequently realized?

2. The Tectonic Distribution of Saline Formations in Arid Regions

The answer to the above question may be found on maps showing the distribution of saline formations superimposed upon those of the principal structural elements of the crust for the Caledonian tectonic stage as a whole, for the Devonian, Carboniferous, and Permian, for the Mesozoic as a whole, and for the Cenozoic. These maps show a single clear pattern on the map of the Caledonian stage (Fig. 236), which is essentially repeated on the others.

The map for the Caledonian shows that *the great bulk of saline formations accumulated on platforms: the North American, Russian, and Siberian, and part of Gondwanaland. Only individual deposits, in the Anti-Atlas Mountains of North Africa and on the eastern shore of the Persian Gulf, are known to have formed in geosynclinal zones. Saline deposits are completely absent from most of the areas of geosynclinal zones.* Thus, *saline formations of the earliest saline-producing epoch for which data are reliable exhibit a remarkably well-defined preference for stable parts of the crust, avoiding mobile belts.*

Within the platforms, saline formations tend to occur in syneclises, and they are most fully developed in those that were characterized in the Caledonian stage by prolonged and, at times, rapid subsidence. Such were the Michigan basin, with salts 1000 m thick, on the North American platform and the Irkutsk amphitheater, in which the Usol'e saline formation accumulated to a thickness of about 1600 m, on the Siberian platform. Syneclises that subsided during saline accumulation characteristically gave rise to the most complete development of the saline process, receiving not only gypsum-anhydrite rocks but also rock salt. Some of the rock salt contains admixtures of potassium minerals, such as that in the lower macrocycle of the Usol'e formation. More weakly subsiding syneclises were characterized not only by thinner deposits of saline rocks but also by incompleteness of their development, the formations consisting only of gypsum and dolomite. Such formations are found in the Silurian of the Williston basin (8–10 m), Iowa (30 m), and West Virginia

(260 m). The accumulation of gypsum was even feebler during the Ordovician and Silurian on the Siberian platform, and it was entirely negligible on the Russian platform.

The tectonic distribution of Devonian saline formations, which accumulated under conditions of a more complex crustal structure, follows in essential outlines the Caledonian pattern, but also exhibits additional features. Devonian saline deposits accumulated either on areas of the Precambrian platform (the North American, Russian, or Siberian) or in regions of intense Caledonian folding where some segments of Caledonian geosynclines were compressed to form massifs of platform type (Fig. 237). The Caledonian structures of Europe, Central Kazakhstan, and the Kokchetav-Ulutau massif with its continuation in the Karatau Range and on to the southeast, the Caledonian chains of Sayan and the Kuznetsk Alatau, and the Caledonian structures of Taimyr and Severnaya Zemlya belong to this latter group. Another member is probably the Central Kolyma massif of northeastern Asia, where individual beds of gypsum occur. *The development of evaporites in regions of Caledonian structures of different types,* some within the still active geosynclinal core, others in zones attached to platforms, *is a new feature in the distribution of Devonian saline formations.* Intensely crumpled segments of a geosyncline, being converted to stable consolidated massifs, form essentially the same kind of tectonic framework for saline accumulation as those found on the ancient Precambrian platforms.

Quantitatively, however, basins on Precambrian platforms produced at first incomparably larger masses of saline rocks than those on post-Caledonian platform massifs. Furthermore, the saline process on Precambrian platforms went farther towards completion than those on epi-Caledonian massifs. Six large rock salt deposits accumulated on platforms, two of them (Williston and Pripyat basins) accompanied by vast accumulations of potassium salts. As a rule, only gypsum-anhydrite formations were deposited on Caledonian massifs; only in the Tuva basin did the saline process approach the stage of potassium sedimentation (in Eifelian time), and even there it did not reach it (rock salts of Tuz-tag).

Saline formations within Precambrian and epi-Caledonian platforms were concentrated in syneclises. The more intensely these basins subsided, the more completely the saline process developed. The potassium-bearing Prairie formation in the Williston basin has a thickness of about 200 m. The rock salt in the Michigan basin is about 400 m thick, the potassium-bearing Upper Devonian formation in the Pripyat basin 450 m thick, and the rock salt in Tuva 370 m thick. But the gypsum formation of Iowa is only about 20 m thick, and in southeastern Montana it is about the same. The gypsum-carbonate Morsovo group in the Moscow syneclise is 25–50 m thick; the Dankov-Lebedyan formation in the same locality reaches a maximum of 250 m, and the gypsum deposits in Central Asia, in the Minusinsk basin, and on the Siberian platform have thicknesses measured in tens, or only a few hundreds, of meters.

It should also be noted that in the Devonian saline deposition *is repeated in the same individual structure*. In the Williston basin, saline deposits were formed in the Silurian and the process was repeated twice in the Devonian (Middle and Late). In the Michigan basin, saline rocks were deposited at the end of Late Silurian and again in Middle Devonian times. Accumulation in the Moscow syneclise took place in Eifelian (or Early Givetian) time and in Dankov-Lebedyan time, the localization of both deposits being closely similar (see maps). In the Dnieper-Donets basin, saline rocks were deposited in Givetian, Frasnian, and Famennian times. This circumstance turned individual, especially large, syneclises into distinctive "nodes of halogenesis" within the huge arid belt. Outside these nodes, the saline process either did not develop or it advanced only slightly.

The tectonic distribution of Carboniferous saline deposits (Fig. 238) is fundamentally the same as that of the Devonian: all deposits were formed on ancient Precambrian platforms and in zones of Caledonian folding. Salt deposits are completely lacking in geosynclinal zones. Within the ancient and epi-Caledonian platforms, the Carboniferous saline deposits are mostly localized in syneclises as were the Devonian; and the Carboniferous "nodes of halogenesis" duplicate those of the Devonian to a considerable degree. At the same time, the completeness of the saline process was much less in Carboniferous time than in Devonian time. In most deposits the process stopped at the stage of sulfate deposition, and only in individual deposits, almost exclusively in North America, was rock salt precipitated. Potassium salts, such as accumulated in the Devonian, entirely failed to form in the Carboniferous. Correspondingly, the thickness of Carboniferous saline deposits is slight or insignificant. As before, the maximal thicknesses were attained by rock salt deposits. According to Krumbein (1951), the Lower Carboniferous formations of the U.S.A. have the following thicknesses: anhydrite rocks in the Williston basin 300 m, in southeastern Montana 30 m, in Iowa 10 m, and in Illinois 70 m; rock salt in the Michigan basin 120 m, in West Virginia 330 m. The Pennsylvanian rock salt in the Paradox basin is 1330 m thick. Gypsum and anhydrite rock in Colorado is 170 m thick, in Minnesota 17 m, and in the Michigan basin 17 m. On the Russian platform, Middle and Upper Carboniferous gypsum deposits range from a few meters to a few tens of meters in thickness. The same applies to the rocks of the central Kazakhstan Caledonian massif, but the lower figure is more common.

Three features distinguish saline deposition in Permian time (Fig. 239) from that in the Carboniferous: *the extraordinary intensity of the process*, the mass of saline rocks that accumulated in Permian time far exceeding what accumulated in any other period in geologic history; *the exceptional completeness of salt precipitation*, resulting not only in tremendous accumulations of gypsum and rock salt but also of potassium salts with horizons of eutonic sediments—bischofite and borates; and *the spatial compactness of the saline process*, the deposits being concentrated essentially in three regions—southern U.S.A., western Europe (Germany), and the eastern part of European U.S.S.R.

Beyond the limits of the indicated regions the accumulation of saline deposits was negligible (Dzhezkazgan, South America, Australia). This spatial restriction of salt accumulation led to a more precise definition of the "nodes of halogenesis" than observed in other periods of earth history. The tectonic character of the salt-accumulating zones in each of the three nodes varied. In North America the basin of deposition lay partly on the platform and partly on a structure that formed during Carboniferous folding. In Western Europe, the North German basin developed by subsidence of adjoining segments of Caledonian and Hercynian structures and of the contiguous margin of the Russian platform. In the European part of the U.S.S.R., Permian salt deposition took place at the extreme southeastern corner of the platform, in the Caspian basin, which was the most mobile segment of the platform, and in the Ural foredeep, which was also subsiding strongly.

The intensity and completeness of the Permian saline process were such that the thicknesses of the saline deposits were very great. Again the rock salt and potassium units are generally much thicker than the gypsum rocks. The thickness of the salt rocks that accumulated in the Zechstein basin of Germany ranges from 800 to 1000 m; the Kungurian salt beds of the Caspian basin are more than 2000 m thick; other values are approximately 1400 m in the Ural foredeep, 750–1000 m in the Upper Kama potassium deposit, and 600–700 m in Pechora. The thickness of the Kungurian salt deposits in the areas where gypsum-anhydrite rocks occur ranges from 100 to 150 m. In North America salt deposits in the Grand Canyon region are 300 m thick; they are 700 m thick in southeastern New Mexico, 500 m in Oklahoma, 270 m in west central Kansas, and about 500 m along the shores of the Gulf of Mexico. The Black Hill gypsum deposits are only 33 m thick. Gypsum beds are 170 m thick in the gypsum basin of Colorado, 33 m in New Mexico, 500–700 m in central Texas, 330 m in Oklahoma, and 17 m in Iowa.

During the Mesozoic (Fig. 240) the saline process was more widely developed: instead of three large and well-defined nodes of halogenesis, at least six appeared—in North America, where the node was perhaps most clearly defined, in South America, in Western Europe, in North Africa, along the margin of the Red Sea, and in a belt extending from the Caucasus to China. *Tectonically, most of the saline formations were formed on platforms and in zones of Hercynian folding, accumulating in syneclises. But some gypsum and salt formations accumulated in geosynclinal zones* (in the northern and middle parts of the Andean geosyncline, and on the western and eastern ends of the Mediterranean geosynclinal zone. The known thicknesses of the saline formations more or less conform to the pattern discussed above. Rock salt and potassium formations are generally thicker than gypsum deposits, though exceptions are known.

During the Cenozoic (Fig. 241) saline deposits were more abundant than in the Mesozoic, and their distribution was more restricted. They formed in four well-defined nodes of halogenesis—in North America, South America, Central Africa, and a long belt extending along the Mediterranean geosynclinal zone and adjacent segments of Precambrian and epi-Paleozoic plat-forms. Halogenesis also tended farther toward completion, leading to the formation of many rock- salt- and even potassium-bearing formations. Tectonically, these formations occur only in part and to a small degree on Precambrian and epi-Hercynian platforms. Most are found in zones of Alpine folding, being located in intermontane basins and foredeeps. Such deposits are found in the Cordillera of North America and the Andes of South America, in North Africa and Sicily, along both sides of the Apennines, in Calabria, in the Ciscarpathian depression and the Transcarpathian region, in the Erivan and Nakhichevan basins, in the Mesopotamian foredeep and elsewhere. *This notable shift of saline deposition from ancient and epi-Hercynian platforms to regions of Alpine folding is characteristic of the Cenozoic.* However, there is nothing essentially new in this. Saline deposition in intermontane depressions and marginal downwarps during folding has occurred before, as in the Permian salts of the Dzhezkazgan basin and the foredeep of the Urals, and the Devonian rocks of the Tuva basin. During the Cenozoic, consequently, the same trend was followed, but quantitatively was greatly intensified. The cause of this intensification probably lies in the different disposition of the arid belt relative to deforming geosynclinal zones. During the Paleozoic and Mesozoic, zones of active deformation lay almost exclusively in humid regions. Only small parts of developing fold systems extended into arid belts. During the Cenozoic almost the entire area of the Tethys trough was in an arid region, and a large part of the Andean cordilleran geosyncline was also within an arid belt. This explains the much wider development of saline deposits in Alpine intermontane basins and marginal depressions than in previous earth history.

It is therefore established that *saline formations accumulated mostly on stable segments of the crust: platforms and zones of Paleozoic folding. Much more rarely they were deposited in geosynclinal zones during the final stage and subsequent folding, and only on rare occasions did they accumulate in geo-synclines during normal geosynclinal development.*

The cause of this clearly defined and distinctive localization of saline accumulations lies in the topography and tectonic controls in geosynclinal zones and on platforms. The geosynclinal zones have been regions of well-defined and dominant marine conditions. The seas have been marked by islands, small and mountainous, separating broad and frequently deep basins. In such environment the only possible types of saline deposition have been of lagoonal and gulf origin. The high relief and the small areas of the islands have predetermined that sedimentary formations be rare and small. It must be kept in mind that the mobility of individual structural elements of the geo-synclinal zones, particularly the positive elements (since these gave rise to the islands) led to repeated erosion of the newly-formed inshore deposits, including saline formations of lagoons and gulfs. The formations that we have been able to observe undoubtedly represent only a small proportion of the number of saline formations that have actually formed. These factors are together responsible for the great rarity of geosynclinal saline deposits. The topographic conditions during the closing stages of geosynclinal history, when abundant intermontane basins and foredeeps have been present, undoubtedly favor saline deposition, especially in view of the fact that sedimentation takes place in a regressive environment. This is responsible for the reduced communication between depressions of the residual basin and the open sea, and it favors prolonged and rapid subsidence of the depression floors; i.e. it gives rise to the very conditions necessary for the successful and complete development of the saline process. Conditions were also more favorable for preservation of these saline deposits, since they have normally been much larger than geosynclinal deposits and have formed in the central parts of the basin rather than along the margins. For this reason, saline deposits of the closing geosynclinal stage are more widespread than those of the normal geosynclinal stage. A negative factor relative to saline formations of intermontane basins and foredeeps has been that their accumulation has taken place during strong upward movements of the surrounding region. This has led to the increased influx of clastic material into the basin and to strong contamination of the saline deposits, as in the Neogene formations of Ciscarpathian depression and the Transcarpathian region, in the Erivan and Nakhichevan basins, in Dzhezkazgan, in Tuva, and elsewhere.

The topographic-tectonic conditions characterizing the ancient Precambrian and epi-Paleozoic platform regions of arid belts were completely different. The platforms were typically marked by moderately large syneclises and intervening anticlises, modified by smaller uplifts. During marine transgression, large gulfs with restricted access to the open sea readily formed (the Michigan basin, the Middle Devonian Moscow syneclise, and others). When the long axis of the syneclise was parallel to the shore line, a broad marginal strip was separated from the main body of the sea by both exposed and underwater uplifts (Irkutsk amphitheater of Early Cambrian time and others). When the long axis of the syneclise was transverse to the shore line, long and often irregular intracontinental basins were formed, such as the Middle and Late Devonian Dnieper-Donets syneclise, the Kungurian sea of the Russian platform, and others. Once such basins were established, they commonly persisted for a long time during transgression and relative standstill of the sea. They formed very readily during regression. When connection with the open sea was restricted, the basins became highly saline, with subsequent deposition. Huge syneclises and their corresponding minor salt-producing depressions naturally gave rise to saline accumulations of tremendous mass and great areal extent. When low plains of the surrounding continental area prevailed during arid climatic conditions, contamination by clastic material was very small and saline deposits of great purity resulted. It may readily be seen that the preservation from subsequent erosion was incomparably more probable for platform saline formations than for geosynclinal ones, because of the difference in size.

These factors together led to the observed result that platform structures became the principal, and indeed the almost exclusive, locations of saline formations preserved in the stratigraphic record.

3. The Factors Controlling the Distribution of Saline Formations on the Earth's Surface

On the basis of the above discussion the mechanism that controls the distribution of saline deposits on the surface of the earth may be considered.

Zones of arid climate, forming belts or large areas on the earth's surface, are regions in which saline lithogenesis has been possible. This potential has been realized, however, only in those places within the arid belts where the necessary topographic-tectonic environments have also existed. These conditions have involved the formation of a marine basin on a subsiding floor, a restricted connection between the basin and the open sea, and prolonged and irregular subsidence of the floor, episodically becoming very rapid. These times of rapid subsidence have been marked by sequences of rock salt and potassium salts. Such conditions have been realized most frequently and most widely on stable platforms, in syneclises where the great bulk of modern saline deposits have been deposited. They have been realized much more rarely and over much smaller areas in geosynclinal zones during the closing stages of geosynclinal development, and still more rarely and in more restricted areas during normal geosynclinal development.

Since platforms are normally widely separated, saline formations within a single arid belt have naturally formed in groups or "nodes of halogenesis", with intervening areas free from such deposits. On the whole belts and zones of arid climate are termed "belts of halogenesis" in the separate geologic epochs and periods; the tectonic factor gives rise to "nodes of halogenesis". This mechanism corresponds exactly to that controlling the distribution of ore deposits in humid belts, as discussed in Vol. 2.

4. The Stratigraphic Distribution of Halogenic Rocks in Formations

The general development of halogenesis in the history of the earth may now be stratigraphically summarized (Fig. 242). From these data the following cardinal facts may be established.

(1) *A late appearance of saline rocks in the stratigraphic succession of the crust.* The oldest known saline rocks occur in the Salt Range of Punjab, at the base of the Lower Cambrian. There has been considerable controversy concerning their age—whether the deposits may not be Eocene with Cambrian rocks thrust over them—but the clear conformity of the overlying Cambrian upon the saline succession precludes this interpretation. Thus, saline deposits appear in the composite stratigraphic succession of the crust from the very end of the Precambrian or the beginning of the Cambrian, the latter usually being considered the more probable.

Attempts have been made repeatedly, and most recently by Rankama, to prove that saline deposits were actually formed in Precambrian time, at great depth. Inclusions of halite and veinlets of gypsum in rocks of the Grenville

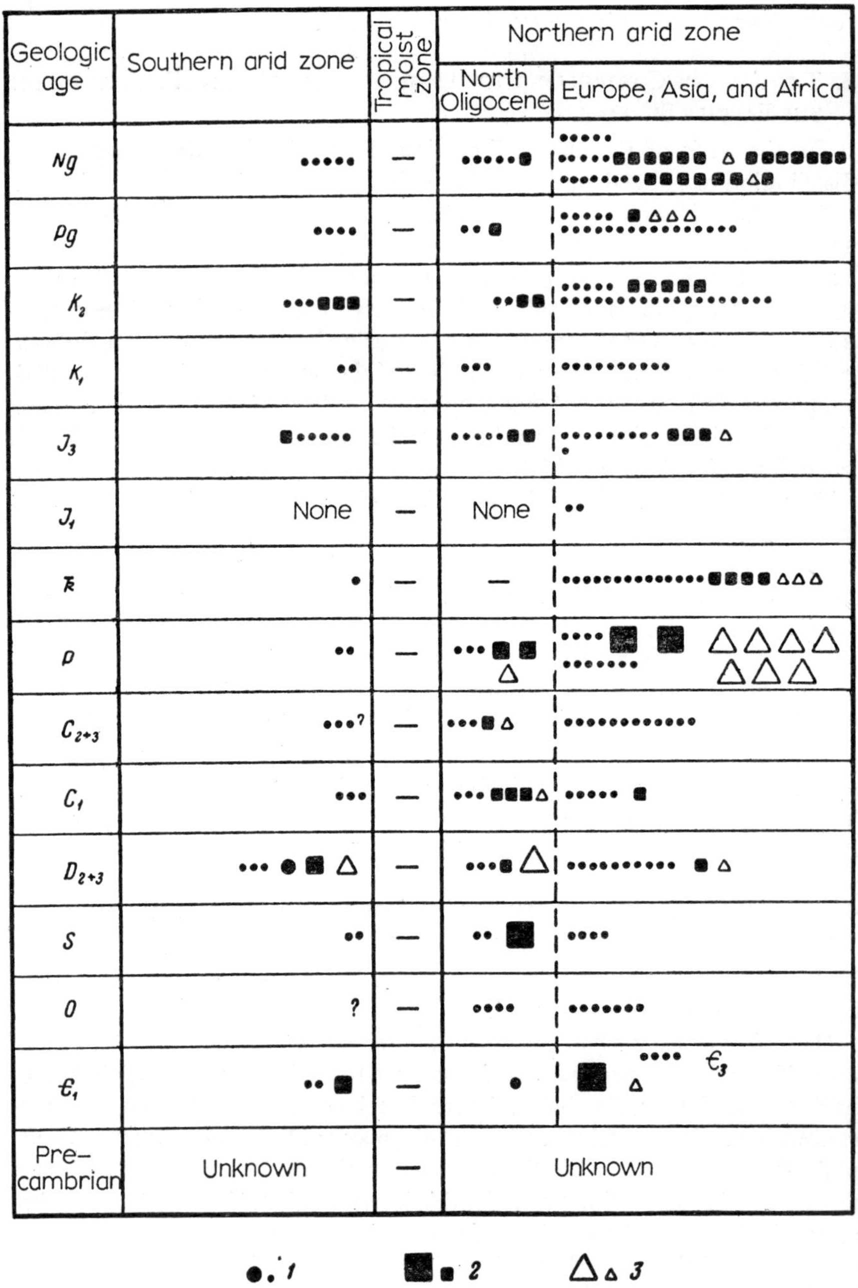

Geologic age	Southern arid zone	Tropical moist zone	Northern arid zone	
			North Oligocene	Europe, Asia, and Africa
Ng	•••••	—	••••• ■	••••• / ••••••■■■■■■ △ ■■■■■■■■ / ••••••••• •■■■■■■△■
Pg	••••	—	•• ■	•••••• ■ △△△ / •••••••••••••••••
K_2	•••■■■	—	••■■	•••••• ■■■■■ / ••••••••••••••••
K_1	••	—	•••	••••••••••
J_3	■••••••	—	••••••■■	•••••••••• ■■■ △
J_1	None	—	None	••
T	•	—	—	••••••••••••••■■■■ △△△
P	••	—	•••■■ ■ △	•••• ■ ■ △△△△ / ••••••• △△△
C_{2+3}	•••?	—	•••■ △	•••••••••••
C_1	•••	—	••• ■■■△	••••• ■
D_{2+3}	••• ● ■ △	—	••••■ △	••••••••••• ■ △
S	••	—	•• ■	••••
O	?	—	••••	•••••••
ε_1	•• ■	—	•	■ △ •••• ε_3
Pre-cambrian	Unknown	—	Unknown	

• •′ 1   ■ ▪ 2   △ ᴀ 3

FIG. 242. Stratigraphic distribution of saline deposits. 1. Gypsum; 2. rock salt; 3. potassium salts; size of symbols represents relative abundance of the corresponding types of saline rocks.

Series in the U.S.A. and saline springs in areas of the Precambrian near Abo in Finland, in southern Sweden, and in South America have been claimed to support this view. It has been assumed for this purpose that the veinlets and the ground water received their saline components by redistribution from ancient saline formations that have by now essentially disappeared. But, as Lotze (1957) has properly pointed out, this interpretation is by no means the only one possible. The salt might have been of secondary origin, from juvenile or vadose water; i.e. it might have been carried to depth and then deposited by surface waters. Clearly *there is no reliable evidence of saline deposits in Precambrian rocks. The saline process can be proved for only the last* 5×10^8 *years of earth history and has not been definitely detected during the preceding* 45×10^8 *years. It is possible merely to give a hypothetical representation of the character of the saline process during that great interval of time.* Some ideas concerning this will be discussed below.

(2) *From the Cambrian to the present halogenesis has operated essentially without break, because saline deposits are known from every period of the Paleozoic, Mesozoic, and Cenozoic, and from almost every epoch. Nevertheless, the intensity of the process clearly has not remained constant.*

During some epochs saline rocks either failed to accumulate or accumulated only in very small quantities; such were the Ordovician, Carboniferous, Lower Jurassic, and Lower Cretaceous. At other times the saline process was exceptionally active and vast amounts of different salts accumulated. One of these was the *Early Cambrian*, when the enormous Usol'e Series accumulated on the Siberian Platform, the rock salt deposits of the Salt Range in Pakistan formed, and huge masses of salts were precipitated along the Persian Gulf. Another was the middle Late Devonian, when the Michigan salt, the Williston potassium deposit in North America, the rock salt and potassium salts of the Pripyat basin, the rock salt of the Dnieper-Donets basin, two huge gypsum-anhydrite sequences in the Moscow syneclise (rock salt also in the Givetian), the rock salt of the Ha-tan basin, and the gypsum rocks of the Siberian platform accumulated. A third saline epoch included the end of the *Early Permian and the beginning of Late Permian.* During this interval there accumulated the impressive deposits of rock salt and potassium salts on the North American platform (the Delaware basin and elsewhere), in the West German basin, on the eastern part of the Russian platform, and in other places. During the Mesozoic and Cenozoic no epoch of accumulation on such a grand scale occurred, but indications of the process are not lacking. In particular, more intense accumulation of salt occurred in the Triassic, especially the Early Triassic, in the Late Jurassic, especially the Tithonian, in the Late Cretaceous, and in the Neogene. Thus, *the saline process, like ore emplacement in humid belts, has been characterized in the past by a well-defined periodic fluctuation of intensity, epochs of marked saline deposition alternated with phases of little accumulation.*

(3) Halogenesis in strongly saline epochs has been characterized by several distinctive features.

2 N

(*a*) Increase in salt accumulation during earlier epochs did not always take place in the same way. During the Mesozoic and Cenozoic the intensification took place by *increase in the number of basins of deposition*. The deposits themselves were restricted in area and volume, and in this respect were more or less the equivalents of some others. During the Paleozoic, by contrast, the intensification took place by the concentration of deposition in a limited number of basins, but each of colossal size. The first process of high saline deposition may be termed the *integration* (summation) process, the second the *pulsation* process. Intensification of ore deposition in humid zones took place essentially in the same way.

(*b*) Some investigators, particularly Lotze, have strongly emphasized the strict simultaneity of salt accumulation during saline epochs in widely

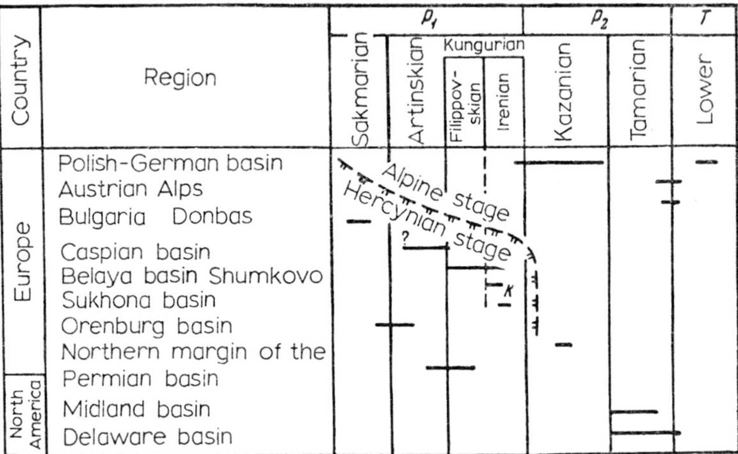

FIG. 243. Permian halogenesis in Europe and North America (from Fiveg).

separated regions—at different "nodes of halogenesis". M. P. Fiveg (1960) tested this statement for three different epochs and has shown it to be inaccurate. *There is no exact simultaneity of salt accumulation in scattered regions but only an approximation to it.* The varied time relationships are clearly shown by Fiveg (Figs. 243-245). When the rate of saline deposition has been high, the chronological relationships among the salt deposits of a single saline epoch have not only been perfectly natural, they may have been the only ones possible.

(*c*) *Epochs of accentuated saline deposition are for the most part characterized by marked completeness of the saline process* (Fig. 242). Not only have huge masses of rock salt accumulated at these times, but potassium salts have also formed. The precipitation of potassium salts in Siberia during the Early Cambrian, moreover, was very small, though individual boreholes, such as the Zhegalovo in the Usol'e sequence, have proved potassium-salt horizons.

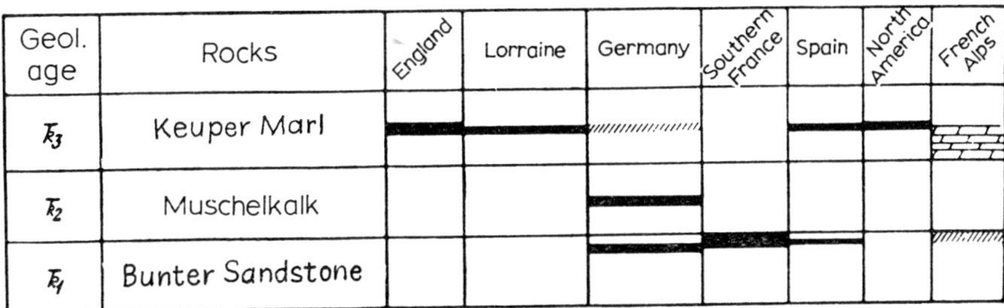

Geol. age	Rocks	England	Lorraine	Germany	Southern France	Spain	North America	French Alps
T_3	Keuper Marl							
T_2	Muschelkalk							
T_1	Bunter Sandstone							

Fig. 244. Saline rocks in the German Triassic succession (from Fiveg). 1. Evaporites with rock salt; 2. predominantly anhydrite; 3. predominantly dolomite.

(4) Many favor the view that saline epochs and marine halogenesis in general mark moments in geologic history of extensive marine regression. This is reflected in the diagram by Lotze (1957, pp. 189-190), who states: "A comparison of the course of salt accumulation with the course of epeirogenic and orogenic events demonstrates a clear interrelation. Maxima on the saline curve generally coincide with well-defined regression-epeirogenic epochs; i.e. *salt deposition increases either simultaneously with, or immediately after, the development of large-scale orogenic movements.* Relevant examples are the saline deposits at the end of the Silurian, at the end of the late Carboniferous, and in the Permian, late Triassic, late Jurassic, and Tertiary." To these may be added the early Cambrian salts and, more doubtfully, those of middle late Devonian age. It is necessary, nevertheless, to point out that this can be only a generalization. As already stated in describing the facies types of saline formations, *these deposits accumulate not only during regression but also during transgression; they are also found in stable basins, when water exchange with*

Regions	Paleogene		Eocene	
	Eocene	Oligocene	Miocene	Pliocene
Ebro basin	—		—	
Rhine graben		—		
Sicily			—	
Ukrainian Ciscarpathia			- - -	
Romanian Ciscarpathia			—	
Polish Ciscarpathia			—	
Transcarpathia			—	
Transcaucasia			—	
Tien Shan basin			—	

Fig. 245. Tertiary saline deposition (from Fiveg).

the open sea has been restricted by crustal movements. Thus, although saline epochs in their general form are more likely to occur at times of orogeny and regression, the complete picture of saline deposition is much more complex.

(5) *Despite the considerable extent of Phanerozoic time (about* 6×10^8 *years) no signs of irreversible evolution have been detected in the mineral composition of saline rocks* (Table 55).

Essentially the same minerals have occurred in potassium rocks throughout Phanerozoic time, though in different proportions and quantitatively showing no directional trend. At low salinities gypsum has been precipitated through-

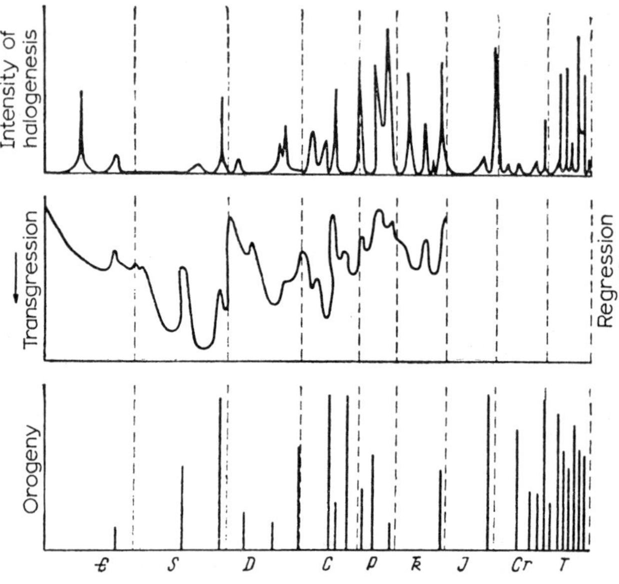

FIG. 246. Relationship between salt deposition and important tectonic phenomena (from Lotze 1927).

out this entire time, and at high salinities halite has formed. Thus the saline process has given rise to the same minerals, varying quantitatively but with no indication of irreversible evolution. *This must be emphasized, because in chemico-biogenic rocks of humid belts, in contrast, irreversible evolution is undoubted. Even in deposits of the initial stage of arid lithogenesis (e.g. in the development of limestone-dolomite rocks) this same process has been recognized.*

(6) A substantially different picture is presented by the stratigraphic distribution of different types of saline formations (Table 56). *Continental formations are relatively common from the Lower Carboniferous,* where they are represented by gypsum rocks (Shcherbina 1945, 1953). For a very long interval of time after this they are absent, to appear again in abundance in the Miocene of the Tien Shan and in the Pliocene of the Tsaidam basin. The sporadic occurrence and extreme rarity of observed continental saline formations are due to the fact that such deposition requires a very rare combination

TABLE 55

Potassium Minerals in Saline Deposits throughout Geologic Time
(from Valyashko 1956b)

Geologic period	Potassium and magnesium minerals	Deposit
Quaternary	(Kainite, carnallite) (Carnallite, kainite) (Carnallite) (Sylvite, carnallite) *Carnallite, sylvite*	Salt lakes of the Ural region Gulf of Kara-bogaz Inder Lake Spring deposits in Ethiopia Salt lakes in Nevada, U.S.A.; Tsaidam basin, China
Tertiary	(Kainite) (Polyhalite, sylvite) (Carnallite) Sylvite, langbeinite, kainite, kieserite, polyhalite, carnallite (Kainite) *Sylvite*, carnallite	Uzun-su, Turkmenia Azerbaijan, Iran Galasibetta, Sicily Eastern Ciscarpathia Alsace, France Ebro, Spain
Cretaceous	?	?
Jurassic	Sylvite, carnallite	Tadzhikistan
Triassic	Polyhalite, (langbeinite) (Sylvite) Polyhalite *Sylvite*, carnallite Carnallite, sylvite (?)	Salzburg, Tirol Lorraine Aquitanian basin, France ⎫ ⎬Northern Spain ⎭
Permian	*Sylvite, carnallite* Polyhalite, sylvite *Sylvite, kieserite, carnallite* *Polyhalite*, (kainite, bischofite), borates *Sylvite, kieserite, carnallite*, (kainite), langbeinite, (bischofite), borates Sylvite, carnallite *Sylvite*, carnallite, *polyhalite*, langbeinite	Upper Kama Bashkiria Caspian lowland ? ⎫ ⎬Zechstein salts of Germany ⎭ Yorkshire ⎫ ⎬Delaware basin, U.S.A. ⎭
Carboniferous	Sylvite Sylvite, carnallite	Nova Scotia Utah
Devonian	Sylvite, carnallite	⎫ Nordvik ⎪ Belorussia ⎨ Williston basin, U.S.A., ⎩ Canada
Silurian	(?)	?
Cambrian	(Sylvite, carnallite) Sylvite, kieserite, langbeinite	Usol'e series Salt Range, Pakistan
Precambrian	?	?

Note: Parentheses indicate accessory minerals; names without parentheses indicate rock forming minerals; italics represent commercial accumulations.

of conditions: isochoric behavior of evaporation combined with prolonged and, at times, rapid subsidence of the basin. Of course, the Lower Carboniferous gypsum rocks are not regarded as marking the beginning of continental saline accumulation. Such deposits might have formed earlier, at least as early as the Cambrian; but these older representatives either have not been preserved; or have not yet been recognized.

Lagoonal saline formations are known successively from rocks of Late Cambrian age (Upper Lena series of the Siberian platform), on through the Neogene, in Recent deposits, where they are widely developed. They are among the most persistent and representative facies types of formations. For a long time all marine saline deposits were incorrectly regarded as lagoonal. The oldest *formations of large gulfs* are found in the uppermost part of the Silurian section (Michigan basin). Younger representatives occur in Devonian strata (marine deposits of the Moscow syneclise, Givetian salts of the Dnieper-Donets basin, and potassium deposits of Byelorussia), the Upper Jurassic (rock salt beds of northwestern Germany), the Cretaceous (gypsum deposits of North Africa), and the Tertiary (Rhenish rock salt and potassium salts, the saline sequences of Ciscarpathia and Transcarpathia and the Erivan and Nakhichevan basins, and the middle Miocene salts of the trans-Caspian region). Saline formations of large gulfs are thus quite as widespread stratigraphically as lagoonal formations, but deposits are fewer.

Saline formations that formed in the marginal zones of huge epicontinental seas are still less abundant. The oldest known deposit of this type, and the largest representative, is the Lower Cambrian Usol'e sequence of the Siberian platform. After a considerable break, follow the Devonian Dankov-Lebedyan beds of the Moscow syneclise and the Middle Devonian deposits of the Williston basin in North America. The Triassic saline sequence of Germany also belongs to this class of deposits. The Jurassic potassium-bearing formation of Tadzhikistan may possibly belong to the same class, but reliable evidence of later representatives is lacking.

Saline formations of huge intracontinental saline seas were formed exclusively in the Permian; they are absent from older and younger stratigraphic horizons. Representatives of this type are the saline formations of the North German plain and the Kungurian and other horizons on the eastern part of the Russian platform.

A generalized view of the historical development of the saline process shows that three large-scale stages may be recognized. The first, Precambrian, can be represented only hypothetically, because no reliable evidence of saline deposits has been preserved. *Apparently the Precambrian was a time of weak saline development.* The second stage occupied the Paleozoic. It was characterized by an intensity and completeness of saline deposition that has never been surpassed. This is true particularly of the Permian, when eutonic precipitation took place. The greatest variety of facies types of saline formations also belongs to this stage. Nodes of salt accumulation are most clearly marked on the southern part of the North American platform, in the Hercynian structures of Europe, and on the Russian and Siberian platforms. Salt deposition took place in three magnificent surges: in the Cambrian, the Devonian, and the Permian. The saline process decreased sharply during the third stage, which included the Mesozoic and Cenozoic, and which was quantitatively much less than the preceding stage. Formations of intracontinental saline marine basins and of marginal parts of epicontinental seas died out one after the other, and only continental (lacustrine), lagoonal, and gulf formations remained. Since these are characteristically small and spatially widely separated, the nodes of salt accumulation lost their clear definition. The completeness of the process was clearly reduced, potassium salts being relatively rare, and eutonic sediments lacking entirely. All these quantitative and facies changes in the saline process have taken place with no substantial change in the composition of the solid salt phases.

The development of the saline process during Phanerozoic time has thus been characterized by a combination of constant mineral composition of the solid salt phases and a well-defined fluctuation in intensity and completeness of the process.

Furthermore, *the principal site of saline accumulation throughout earth history has been the northern arid zone, where the great bulk of gypsum rock, rock salt, and potassium salts accumulated. Saline deposits in the southern arid zone are few, especially rock salt and potassium salts, which are found only in the Upper Devonian deposit in the Pripyat basin.* We see a clearly defined asymmetry in the spatial localization of saline deposition on the surface of the earth, a new example to add to the many manifestations already known of asymmetry in the structure and composition of the lithosphere (Fig. 247).

5. THE MECHANISM CONTROLLING THE HISTORICAL DEVELOPMENT OF HALO-
    GENESIS

In any understanding of the development of the saline process through geological time it is first necessary to interpret the remarkable constancy of composition of the solid salt phases in saline rocks and the absence of any traces of irreversible evolution. There are apparently two causes. Saline rocks are known from only the past 6×10^8 years. No known saline beds have been preserved from the earlier development of the earth about $40–50 \times 10^8$ years. During this time the hydrosphere and atmosphere were developed, and processes developed whereby the saline composition of oceanic waters was evolved. *It is very likely that the salinity and ion content of sea water, as a result of this very prolonged development, was essentially the same by the end of Precambrian time as it is now.* To this we must add that the *metamorphization of sea water in basins that have grown highly saline took place in Cambrian time by the same processes as occurred subsequently. This is proved by the fact that potassium salts of both sulfate and chloride types were deposited in the Early Cambrian;* the former type is found in the saline formation of the Salt Range, the latter in the Early Cambrian Usol'e saline sequence of the Siberian platform. But if the composition of the initial sea water and the metamor-

TABLE 56

Stratigraphic Distribution of Saline Formations of Different Facies

Geologic age	Conti-nental	Lagoonal	Gulf type	Marginal marine zone	Intra-continental saline sea	Locality of formation
Ng	▲4 ▲1	●6 ●5 ●7	▲3 ▲2	None None	None None	1. Tian Shan halite-glauberite formation—salt-bearing strata; 2. Ciscarpathia and Transcarpathia; 3. Meso-potamia; 4. Tsaidam basin—gypsiferous strata; 5. Sicily; 6. Algeria; 7. Tunisia
Pg		●6 ●7 ●5 ●3 ●4	●2 ▲1	None	None	1. Ebro basin; 2. Rhine graben; 3. Florida; 4. India; 5. California; 6 and 7. Central Asia
K		●1 ●2 ●3 ●4	●5	None	None	1. Tuakyr (arch); 2. Kuba-dag; 3. Kuritang-tau; 4. Kuritang-tau (Okuzbulak series); 5. Tunisia
J		●1 ●2 ●3 ●4 ●5 ●6 ●7 ●9 ●10	▲11	▲8	None	1. Carpathians; 2. Moldavia; 3. Donbas; 4. Northern slope of Caucasus; 5. Southern slope of Caucasus; 6. Little Caucasus; 7. Kuba-dag; 8. Gaurdak and adjacent regions; 9. Balkhany; 10. Kopet-dag;

phization processes, as salinity increased in the saline basin, had acquired the same features at the beginning of the Cambrian as they possessed in later geologic history, it follows that evaporation must inevitably have given rise to the same solid mineral phases of salts, precipitated in the same solid sequence.

In the light of the maturity of the saline process as early as the Early Cambrian, it is pertinent to consider the causes of the late appearance of saline formations in the stratigraphic column. It is clear that these were by no means physicochemical—the physicochemical mechanism of salt deposition had long been operative—and they were not climatic—glacial, humid, and arid climatic zones had developed far back in Precambrian time. A possible cause may have been the tectonic conditions in combination with climatic factors.

During the early Precambrian, according to current opinion, the crust was dominantly in a geosynclinal stage, platform nuclei not yet having developed. Since geosynclinal conditions are by their very nature unsuited to development of the saline process, saline formations did not normally accumulate in early Precambrian time. At best they were rare and were probably lagoonal and very small. As the paleogeography changed markedly, as is customary in geosynclinal zones, such formations were destroyed. In later Precambrian time the vast Gondwana platform appeared, and toward the end of the Precambrian the smaller North American, Russian, Siberian, and Chinese platforms came into existence, apparently giving rise to the structural frameworks necessary for the development of saline formations. But, judging from what is known of the Proterozoic, particularly of South America and Africa, Gondwanaland lay chiefly in regions of glacial and temperate humid climates, thus inhibiting saline deposition. The platforms of North America, Russia, Siberia, and China were then in the lower latitudes, within the southern and northern arid zones. Saline deposition might well have occurred here, were it not that these platforms were tectonically stable, with only weakly developed syneclises and anticlises. Consequently prolonged and, occasionally rapid, subsidence of segments, so necessary for the accumulation of saline deposits, was lacking. In other words, no combination of a belt of arid climate and the necessary tectonic activity appeared on platforms in the northern hemisphere. As a result, these northern platforms also failed to develop sites of any notable saline deposition, and such small lagoonal formations as may have accumulated were quickly destroyed. Consequently the very long Precambrian time proved unfavorable for the deposition of saline rocks.

The sharp increase in saline deposition during the Paleozoic was due to the fact that the North American, Russian, Siberian, and Chinese platforms, continuing to lie in low latitudes with vast regions of arid climate, now developed large syneclises, which persisted for long periods of time. These at times subsided rapidly, some of them to form large saline gulfs, others to form saline zones marginal to saline seas because the subsiding segments lay parallel to the shore. Some syneclises combined into huge intracontinental seas that grew progressively saltier. Most Paleozoic salt was deposited in depressions of these types. Prolonged evaporation resulted in almost complete saline lithogenesis, up to the precipitation of potassium salts, sometimes to and beyond the eutonic stage of precipitation. The great distances between platforms and the huge size of the syneclises led to the development of well-defined nodes of saline accumulation. These are very clear in the saline rocks of the early Cambrian, Middle and Late Devonian, and Early and Late Permian.

During the Triassic, climatic zones that are now in high and middle latitudes were in low latitudes. The northern arid zone in the eastern hemisphere then lay chiefly in the region of the Mediterranean geosynclinal zone, including also a narrow border of epi-Paleozoic and Precambrian platform on the north, and a considerable part of the North African platform. The southern arid zone lay partly along the Andean geosyncline, partly on the fragments of Gondwanaland that here occupied but very limited areas. The coincidence of the arid belts over large areas with geosynclinal zones and small adjacent segments of Gondwanaland reduced the possibilities of saline accumulation. In North Africa, which fell within the arid belt, no long-persisting and rapidly subsiding syneclises, so typical of platforms of the present hemisphere, developed either in the Mesozoic or Cenozoic. Here again, tectonic conditions were unfavorable to saline deposition, and the intensity of Alpine halogenesis fell sharply. The tremendous "paleotypical" formations in marginal zones of open seas and in huge intracontinental saline seas no longer formed, leaving saline deposition only in lagoons and gulfs. These carry only small deposits of salts usually widely separated. Furthermore the saline process was less complete; potassium salts appeared in relatively fewer basins, and eutonic salts were not developed.

This interrelationship between climatic zonation and principal structural elements of the crust, varying with time, was responsible for the very small deposition of saline rocks in Precambrian time, for marked increase of the process during the Paleozoic, and for a decrease in the Mesozoic and Cenozoic. This further led to a marked asymmetry in the distribution of saline rocks in the sedimentary cover of the earth; such rocks were confined almost exclusively to what is now the northern hemisphere, and were lacking in the southern hemisphere.

The cause of periodic increase and decline of saline deposition during Phanerozoic time lies in the periodicity of epeirogeny and orogeny. *Saline epochs appeared during or immediately after active orogeny, usually during times of marine regression.* This relationship is to be expected, for it is during regression that relict basins with restricted marine connection—fundamental condition for saline deposition—most readily appear in syneclises and depressions in general. It is essential to repeat, however, that exceptions to this rule are known. Saline formations have accumulated during transgression and during relatively stable conditions, but, in general, they show a strong preference for periods of regression.

The absence of strict simultaneity in the formation of saline deposits during a single saline epoch follows from the fact that orogenic movements and marine regressions in different parts of the earth, especially widely separated regions, are not themselves simultaneous but only approximately so.

It should therefore be recognized that the development of the saline process has been controlled not by physicochemical factors or, particularly, by changes with time of the salt content of the initial oceanic waters (the composition of these waters has remained almost constant for the last 6×10^8 years) but solely by tectonic controls, and by changes in the associated climatic zones.

In saline evolution one may see very clearly how critical the tectonic factor in arid-climate lithogenesis is, how profound its effect on the variety and detail of lithogenesis, how it has become the essential control in the saline process.

6. Special Aspects of Saline Development as Compared with Lithogenesis in Humid Regions

In order to outline more distinctly the specific features of the saline process, it seems advisable to compare its history with the history of lithogenesis in humid zones and also with sedimentary deposition during the initial stage of arid-climate lithogenesis.

An essential feature of the composition of rocks from humid zones is the *irreversible evolution with time.* This is clearly demonstrated in the formation of iron ores, carbonate and siliceous rocks, and organic substances during diagenesis. It is less clearly seen in bauxites, manganese ores, and clay. Together with this irreversible evolution, however, *episodic fluctuations of the sedimentary process* are also prominent. These are seen primarily in the formation of a specific rock type, at one time in certain parts of a humid belt, at another time in other parts of the zone. This applies to coal, which at first accumulated in the tropics (Devonian-Carboniferous), later (Permian-Mesozoic) was concentrated almost exclusively in the north temperate zone and only weakly in the subtropics, and the Cenozoic was located in the tropics and the temperate belt. Similar shifts, in a somewhat different form, occurred in the history of the ore triad Al-Fe-Mn. Here, together with a spatial shift, there is also found a quantitative fluctuation in sedimentation, with temporary changes in the dominant rock type being deposited, similar fluctuations are also seen in the history of organic material (coal), phosphorus, carbonate rocks, and others. These fluctuations have caused the appearance of iron-ore epochs and iron-free epochs, manganese-ore epochs and barren epochs, coal epochs and coal-free epochs, and so forth. Ore epochs of closely related elements (such as Al-Fe-Mn) are intimately associated in time, and may at times coincide. Thus, *in the history of rocks in humid climates there is an irreversible evolution combined with purely quantitative fluctuations and with geographical migrations of the process from one part of the humid belt to another.*

In the historical development of rocks during the initial stage of arid lithogenesis, this same feature may also be detected. In particular, dolomite limestone, and siliceous rocks clearly demonstrate features of irreversible evolution, together with purely quantitative variations in the sedimentary process.

By contrast, in the history of saline deposition indications of irreversible evolution are absent. The mineral composition of saline rocks since the beginning of the Cambrian has been restricted to a definite assemblage of minerals, which has been broader at times and more restricted at others, but has not changed substantially with time. *The entire history of saline accumulation is, for practical purposes, expressed by the quantitative fluctuations in intensity and completeness of the saline process and by the geographical shifts in the nodes of salt accumulation within arid belts.* Thus, the history of the saline process is much simpler than that of humid-climate lithogenesis, even than the history of rocks that formed during the initial stage of arid lithogenesis. This is due to that fact that reliable information on saline rocks is available for only the later stage of earth development. The salt content of the oceans was essentially the same in Cambrian time as it is now, and it gave rise to the same processes of change with increasing salinity. Tectonic movements, changing in intensity and position, together with shifts of climatic zones have caused corresponding variations in the course of halogenesis, accentuating it to produce saline epochs, weakening it in the intervening epochs, causing the nodes of salt accumulation to migrate in different epochs to different geographic regions.

REFERENCES

RUSSIAN REFERENCES

ABRAMOVICH, E. L. 1957a. The deposition of lead in Middle and Upper Devonian deposits of the Tashkent region. *Dokl. Akad. Nauk. USSR*, **116** (No. 5).
1957b. Distributional features of iron, manganese, copper, and the minor elements in Middle and Upper Devonian sedimentary rocks of the Tashkent region. *Dokl. Akad. Nauk. USSR*, **116** (No. 3).
1959. The origin of bedded lead-zinc ores in the dolomites of Kalkan-Ata (Tashkent region). *Izv. Akad. Nauk. USSR, ser. geol.* (No. 11).

ALEKIN, O. A. (Editor). 1954. *Hydrochemical Data.* Izd. Akad. Nauk., USSR, Moscow (in Russian).

ALEKIN, O. A. 1954. A simplified method of computing activity coefficients during investigations of natural waters. *Gidrokhim. Mater.*, **22**.

ALEKIN, O. A. and MORICHEVA, N. P. 1957. Calcium-carbonate equilibrium in water of the Volga. *Gidrokhim. Mater.*, **26**.
1959. The stability of carbonate equilibrium as observed in the River Don. *Gidrokhim. Mater.*, **29**.

ALEKIN, O. A. and TARASOV, M. N. 1957. Hydrochemistry of Lake Balkhash. *Gidrokhim. Mater.*, **26**.

ALIEV, A. G. 1949. *Petrography of the Tertiary Rocks of Azerbaidzhan.* Baku, Izd. Akad. Nauk. Azerb. SSR (in Russian).

ANASTAT'EVA, O. M. 1958a. Upper Jurassic anhydrite and gypsum rocks on the southwestern border of the Russian platform. In *Questions on the Mineralogy of Sedimentary Formations.* Vol. 5. Izd. L'vov. Univ. L'vov. (in Russian).
1958b. Dolomites and the dolomitization of Upper Jurassic carbonate rocks. In *Questions on the Mineralogy of Sedimentary Formations.* Vol. 5. Izd. L'vov. Univ., L'vov. (in Russian).

ANDRIANOVSKAYA, K. N. 1956. The genetic sequence of microstructure development in anhydrite and gypsum rocks as observed in the Inder salt uplift. *Mater. vses. nauchno-issled. geol. Inst., Litologiya,* **1**.

APRODOVA, A. A. 1945. Thermonatrite in the Kama region. *Dokl. Akad. Nauk. SSSR,* **48**, No. 5.

ARKHANGEL'SKAYA, N. A. and GRIGOR'EV, V. N. 1960. Conditions for formation of salt-producing zones in marine basins as observed for the Lower Cambrian evaporite basin on the Siberian platform. *Izv. Akad. Nauk. SSSR, ser. geol.,* No. 4.

ARKHANGEL'SKAYA, N. A., GRIGOR'EV, V. N. and ZELENOV, K. K. 1960. Facies of the Lower Cambrian deposits on the southern and western borders of the Siberian platform. *Tr. Inst. geol. Akad. Nauk. SSSR,* **33**.

ARKHANGEL'SKII, A. D. and STRAKHOV, N. M. 1938. *Geologic Structure and Developmental History of the Black Sea.* Izd. Akad. Nauk. SSSR, Moscow and Leningrad (in Russian).

BAKHMAN, V. I. and PROKOF'EVA, E. F. 1953. Soluble gases in muds. *Gidrokhim. Mater.*, **21**.

BAKUN, N. N., VOLODIN, R. N. and KRENDELEV, P. F. 1958. Basic features of geologic structure in the Udokan deposit of cupriferous sandstones, and its future exploration. *Izv. vyssh. ucheb. Zaved., geol. i razved.* (No. 5).

BARENBOIM, M. I. 1955. Structural scheme of the sub-salt bed in the southeastern part of the Caspian basin. *Dokl. Akad. Nauk. SSSR,* **101** (No. 1).

BELYAKOVA, E. E. 1961. Patterns in the aqueous migration of copper, lead, and zinc, and their significance for prospecting. *Sov. Geol.* (No. 1).

BERDICHEVSKAYA, M. E. and LEITES, A. M. 1960. Some structural features of bedded copper deposits as observed in cupriferous sandstones in eastern Siberia. In *Questions on Sedimentology.* Moscow (in Russian).

BERGMAN, A. G. and LUZHNAYA, K. P. 1951. *The Physicochemical Basis for the Study and Use of Chloride-Sulfate Salt Deposits.* Izd. Akad. Nauk. SSSR, Moscow (in Russian).

BERGMAN, A. G., VALYASHKO, M. G. and FEIGEL'SON, I. B. 1953. Salt lakes in the northwestern Aral region, the Ustyurt Plateau, and the lower reaches of the Amu-Darya River. *Tr. Lab. Ozerov. Akad. Nauk. SSSR,* **2**.

BLINOV, L. K. 1946. Origin of the salt content of sea water. *Meteorologiya Gidrol.* (No. 4).

BLYUMBERG, YA. B., NIKOLAEV, V. I. and EGOROV, V. S. 1940. The hydrochemical regimen of the gulf of Kara-bogaz during 1936-38. *Tr. Kom. komil. Izuch. Kasp. Morya, Karabogazskii sektor.* (No. 11).

BOGDANOV, A. A. 1949. The chief tectonic features of the Eastern Carpathians. *Sov. Geol.,* Collection 40.

BOGDANOV, A. A. and SEROVA, M. YA. 1956. The stratigraphic position of saline series in the Miocene section of Ciscarpathia. *Uchen. Zap. mosk. Univ.* (No. 176).

BOGOMAZOV, G. P. 1957. Some questions on the geology and origin of lead deposits in the Dzhergalan region. *Tr. Inst. Geol. Akad. Nauk. Kirg. (Frunze) SSR* (No. 9).

BRODSKAYA, N. G. 1949. Carbonate formation in the Aral Sea. *Izv. Akad. Nauk. USSR, ser. geol.* (No. 6).
— 1952. Bottom deposits and the process of sedimentation in the Aral Sea. *Tr. Inst. geol. nauk., Akad. Nauk. SSSR,* 57, *geol. ser.* (No. 115).

BRUNS, E. P. 1956. Developmental history of the Pripyat basin during the Paleozoic. *Tr. vses. nauchno-issled. geol. Inst.* (No. 14).

BUINEVICH, D. V. *et al.* 1959. A physicochemical description of the present state of the Gulf of Kara-bogaz and its potential for industrial use. In *Problems of the Joint Use of the Mineral Wealth of the Gulf of Kara-bogaz.* Izd. Akad. Nauk. Turk. SSR, Ashkhabad (in Russian).

BUKSHTEIN, V. M., GARKAVI, M. YU. and GARKAVI, N. S. 1962. Metamorphization of brine in the Gulf of Karabogaz. *Zh. prikl. Khim. Leningr.,* 25 (No. 8).

BUNEEV, A. N. 1947. History of the waters of sedimentation. *Sov. Geol.,* 19.
— 1956. *Fundamentals of Hydrochemistry of the Mineral Waters in Sedimentary Deposits.* Medgiz., Moscow (in Russian).

BURKOVSKAYA, E. G. 1956. Potassium salts in the Aktyubinsk section of the Ural region. *Mater. Geol. polez. Iskop., Yuzhnogo Urala* (No. 1).

BUROVA, E. G. 1956. Rock-salt deposits in the southern part of the Moscow region. *Tr. vses. neftegaz. nauchno-issled. Inst.* (No. 9).

BURYKHINA, Z. E. 1957. Some remarks on the origin of ores in lead deposits of the Dzhergalan region. *Tr. Inst. geol. Akad. Nauk. Kirgiz. SSR, (Frunze)* (No. 9).

CHUDINOV, N. K. 1958. The biochemical characteristics of the conditions of precipitation, diagenesis, and secondary alteration of potassium rocks. *Tr. vses. nauchno-issled. Inst. Galurgii.*

CHUKHROV, F. V. 1955. *Colloids in the Earth's Crust.* Izd. Akad. Nauk. SSSR, Moscow (in Russian).

DANIL'CHENKO, P. T. 1932. Data on the hydrochemistry of lakes along the Black Sea coast; Donuzlav Lake. *Tr. krȳm. nauchno-issled. Inst.,* 3 (No. 2).

DANIL'CHENKO, P. T. and RAVICH, V. I. 1927. The alkali content of natural brines of marine origin. *Tr. krȳm. nauchno-issled. Inst.,* 1 (No. 2).

DANILOVA, V. V. 1949. The fluorine content in rocks. *Tr. Biogeokhim. Lab. Akad. Nauk. SSSR,* 9.

Data on Studies of Salt Deposits and Mineral Waters. 1955. Goskhimizd., Leningrad (*Tr. vses. nauchno-issled. Inst. Galurgii,* No. 30) (in Russian).

DEMOKIDOV, K. K. *et al.* 1958. Stratigraphy and facies of the Cambrian rocks on the Siberian platform. *Tr. arkt., Nauchno-issled. Inst.,* 80.

DOMAREV, V. S. 1948. The origin of deposits such as cupriferous sandstones. *Mater. vses. nauchno-issled. geol. Inst., polez. Iskop.,* sb. 4.
— 1958. The origin of cupriferous sandstones in Northern Rhodesia (according to foreign geologists). *Zap. vseross. miner. Obshch.,* 87 (No. 1).

DOMINIKOVSKII, V. N. and LIBROVICH, V. L. 1957. Types of shallow-water phosphorite-bearing deposits in the Middle Ordovician of the Irkutsk amphitheater. *Razv. Okhr. Nedr* (No. 8).

DRUZHININ, I. G. 1946. Sulfate Lake Ebeity. In *Sodium Sulphates in the USSR.* Izd. Akad. Nauk. SSSR, Moscow and Leningrad.

DUBININA, V. N. 1951a. Halite from the Upper Kama deposit. *Dokl. Akad. Nauk. SSSR,* 79 (No. 5).
— 1951b. The origin of sylvite. *Dokl. Akad. Nauk. SSSR,* 80 (No. 2).
— 1954. The mineralogy and petrography of the Upper Kama (potassium) deposit. *Tr. vses. nauchno-issled. Inst. Galurgii* (No. 29).

DZENS-LITOVSKII, A. I. 1936. Geologic age of bottom saline deposits in mineral lakes. *Priroda Mosk.* (No. 12).
— 1938. Mineral lakes of the USSR, their types and geographic distribution. *Priroda, Mosk.* (Nos. 11-12).
— 1944. The zone of mineral lakes in the USSR. *Izv. vses. geogr. Obshch.,* 76 (No. 4).

1951. Mineral lakes of the USSR. *Problemỹ fiz. Geogr.*, **17**.

1959. Geologic and geographical patterns of distribution of fresh, brackish, and salt lakes. In *Transactions of the All-Union Hydrogeological Congress*, Vol. 4. Gidrometeoizdat, Leningrad (in Russian).

DZENS-LITOVSKII, A. I., ELOVSKAYA, L. V. and GARKAVI, M. YU. 1956. Geology of the bottom deposits and the hydrochemical regimen of Kurguzul'skaya Bay and Satasskii Gulf of Kara-bogaz. *Tr. geol. Inst. Akad. Nauk. Turk. SSR*, **1**.

DZENS-LITOVSKII, A. I. *et al.* 1959. Bottom salt deposits and buried brines of the Gulf of Kara-bogaz as a new kind of sulfate raw material. *Tr. vses. nauchno-issled. Inst. Galurgii* (No. 35).

EGANOV, É. A. 1959. The Tsaidam basin. *Priroda, Mosk.* (No. 9).

ÉPSHTEIN, V. V. 1939. Metamorphization of continental mineral lakes. *Gidrokhim. Mater* (No. 9).

1940. Metamorphization of continental mineral lakes. *Gidrokhim. Mater* (No. 11).

EREMENKO, V. YA. 1955a. The solubility of calcium carbonate in natural waters at different partial pressures of CO_2 at 25°C. *Gidrokhim. Mater*, **25**.

1955b. The solubility of calcium carbonate in $MgSO_4$ solutions at a partial pressure of CO_2 equal to atmospheric. *Gidrokhim. Mater*, **25**.

1957. Values of the dissociation constant of carbonic acid. *Gidrokhim. Mater*, **27**.

EZROKHI, L. L 1953. Viscosity of solutions in the system $Na—K·(Cl—SO_4)·H_2O$. *Tr. vses. nauchno-issled. Inst. Galurgii* (No. 27).

FEIGEL'SON, I. B. 1939. The hydrochemistry of Lake Elton. In *Problems of Lake Elton*, Part 2. Leningrad (in Russian).

FEIGEL'SON, I. B., BEFORT, E. I. and LARINA, A. P. 1936. Composition of the waters in streams feeding Lake Elton and their freshening effect on the lake. *Tr. tsent. nauchnoissled. Sta. geokhim. solei* (No. 1).

FIVEG, M. P. 1948. The annual sedimentational cycle of rock salt in the Upper Kama deposit. *Dokl. Akad. Nauk. SSSR*, **61** (No. 6).

1952. Geologic features in the accumulation of saline sequences. *Tr. vses. nauchnoissled. Inst. Galurgii* (No. 23).

1954. Time required for accumulation of saline sequences. *Tr. vses. nauchno-issled. Inst. Galurgii* (No. 29).

1955. Conditions of formation of the Upper Kama saline series. *Tr. vses. nauchnoissled. Inst. Galurgii* (No. 30).

1956a. Geological environment during the sedimentational stage of saline accumulation. In *Questions on the Mineralogy of Sedimentary Formations*, Vols. 3 and 4. Izd. L'vov. Univ., L'vov. (in Russian).

1956b. Types of salt-producing basins. *Tr. vses. nauchno-issled. Inst. Galurgii* (No. 32).

1958. Patterns in the formation and distribution of potassium accumulations in saline formations. In *Systematic Distributional Patterns of Mineral Deposits*, Vol. 1. Izd. AN SSSR, Moscow (in Russian).

1959. Structures in the lower rock salt of the Upper Kama deposit. *Tr. vses. nauchnoissled. Inst. Galurgii* (No. 35).

1960a. *Geologic Conditions for the Formation of Saline Sequences and Potassium Beds* (in Russian).

1960b. The facies series of saline rocks and features of the spatial distribution of its members. In *Systematic Distributional Patterns of Mineral Deposits*, Vol. 3. Izd. Akad. Nauk SSSR, Moscow (in Russian).

FLATOV, K. V. 1956. *The Gravitational Hypothesis for Development of the Chemical Composition of Ground Water in Platform Depressions*. Izd. Akad. Nauk SSSR, Moscow.

FORSH, N. N. 1955. Permian deposits. The Ufimian and Kazanian stages. *Tr. vses. neft. nauchno-issled. geol.-razv. Inst.* (No. 92).

FROLOVA, E. K. 1955. Magnesite in the Lower Permian deposits of the Kuibyshev and Saratov districts of the trans-Volga region. *Izv. Akad. Nauk. SSSR, ser. geol.* (No. 5).

GABOVICH, R. D. 1949. Fluorine in potable waters of the Ukraine and its genetic significance. *Gidrokhim. Mater.*, **16**.

GALAKHOVSKAYA, T. V. 1953. A method of spectral determination of small quantities of Cu, Pb, Mn, Fe, and Ni in the soluble parts of galena, sylvite, and sylvinite. In *Investigations of the Physicochemical Properties of Salts and Salt Solutions*. Goskhimizdat, Moscow and Leningrad.

GALUSHKO, V. V. 1956. Sketch of the geologic history of the Ciscarpathian marginal depression. *Tr. vses. nauchno-issled. Inst. Galurgii* (No. 32).

GEDROITS, K. K. 1933. Studies on the Absorption Capacity of Soils (4th Ed.). Sel'khozgiz, Moscow (in Russian).

GEKKER, R. F. 1934. The Main Devonian Field (in Russian).
— 1948. Jurassic fauna and flora in the Karatau deposit. *Tr. paleontol. Inst. Akad. Nauk. SSSR*, **15** (No. 1).
GERASIMOV, I. P. 1935. Data on the geomorphology of the Kulunda Steppe. In *The Kulunda Expedition of the Academy of Sciences USSR*, Part 3. Moscow and Leningrad. (*Tr. Sov. Izuch. proizv. Sil Ser.*, No. 10).
GERASIMOVA, V. V. 1954. Boron occurrences in anhydrite-dolomite rocks of the Kungurian sequence in the Umentovskii uplift. *Tr. vses. nauchno-issled. Inst. Galurgii.* (No. 29).
— 1956. Geologic structure, lithology, and conditions of formation of the Yar-Bishkadak saline sequence. *Tr. vses. nauchno-issled. Inst. galurgii* (No. 32).
GIGNOUX, M. 1952. *Stratigraphic Geology.* Izd. inostr. lit., Moscow (trans. from the French).
GINZBURG, I. I. 1957. Development of Theoretical Grounds for Geochemical Methods of Prospecting for Ores of Nonferrous and Rare Metals. Gosgeoltekhizd., Moscow (in Russian).
GOLOVKO, V. A. 1960. Some data on the petrographic and geochemical characteristics of the Ozerki-Khovanshchina deposits in the Moscow region. *Dokl. Akad. Nauk. SSSR*, **130** (No. 4).
GORBANEV, A. I. and NIKOLINA, V. YA. 1954. Sodium Sulfate. Goskhimizdat, Moscow (in Russian).
GORBOV, A. F. 1950. The formation of thenardite in lakes of the Kulunda Steppe. *Dokl. Akad. Nauk. SSSR*, **74** (No. 5).
GORETSKII, YU. K. 1954. Conditions of formation and some distributional patterns of sedimentary and metasedimentary ore deposits. *Izv. Akad. Nauk. SSSR, ser. geol.* (No. 1).
GORKUN, O. P. 1959. The stratigraphic position and lithology of potassium lenses in the section of the Upper Vorotyshcha series in the Stebnik region. *Tr. vses. nauchno-issled. Inst. Galurgii* (No. 35).
GRIDNEV, N. I. 1955. Lithology of the Cenozoic Molasse in the Surkhan-Darya Depression. Izd. Uzb. SSR, Tashkent (in Russian).
IL'INSKII, V. P. 1927. The alkali content of salt solutions from Saki Lake. *Zh. khim. Prom.* (Nos. 7-8).
IL'INSKII, V. P., KLEBANOV, G. S. and BADER, F. F. 1932. Saline Lake Kuuli. *Tr. sol. Lab. vses. Inst. Galurgii* (No. 3).
IL'INSKII, V. P., KLEBANOV, G. S. and BLYUMBERG, YA. B. 1936. The hydrochemistry of the Gulf of Kara-bogaz. *Tr. sol. Lab. vses. Inst. Galurgii* (No. 5).
Investigation of the physicochemical properties of salts and salt solutions. 1953. *Tr. vses. nauchno-issled. Inst. Galurgii* (No. 27).
ISACHENKO, B. L. 1934. Biogenic processes in lakes of the Kulunda Steppe. *Tr. Sov. Izuch. proizv. Sil ser. Yakutsk.* (No. 8). Kulunda Expedition, Part 1, Leningrad.
IVANKOV, L. I. *et al.* 1957. Some aspects of the geology of the Dzhezkazgan deposit. *Tr. Inst. Geol. Akad. Nauk. Kirg. SSR*, Frunze (No. 9).
IVANOV, A. A. 1932. The origin of sylvinite in the Upper Kama deposit of potassium salts. *Zap. vseross. miner. Obshch.*, **61** (No. 2).
— 1934. Data on deposits underlying the saline sequence in the Kama region, and sediments parallel to them. *Problemy sov. Geol.* (No. 3).
— 1935a. Carnallite in the Upper Kama deposit. In *The Solikamsk Carnallite Rocks.* O.N.T.I., Moscow and Leningrad.
— 1935b. Principal features of the stratigraphy and tectonics of the sylvinite zone in the Upper Kama deposit in the zone of the first mine. *Tr. tsent. nauchno-issled. geologo-razv. Inst.* (No. 5).
— 1948. Occurrences of salt in the Transcarpathian region. *Mater. vses. nauchno-issled. geol.-razv. Inst. polezn. Iskop.*, sb. 4.
— 1950a. Intraformational tectonic breccia of some salt deposits. *Tr. vses. nauchno-issled. geol. Inst.* (No. 2).
— 1950b. Potassium salts in the Angara-Lena salt basin. *Zap. vseross. miner. Obshch.*, **79** (No. 4).
— 1953. *The Basic Geology and Method of Prospecting, Exploring, and Evaluating Salt Deposits.* Gosgeolizdat, Moscow (in Russian).
— 1959. Distribution and types of potassium-salt deposits. *Geol. rudnykh m-nii* (No. 4).
IVANOV, A. A. and LEVITSKII, YU. F. 1960. *Geology of Saline Deposits in the USSR.* Gosgeolizdat, Moscow (*Tr. vses. nauchno-issled. Inst.*, **35**).
IVANOV, A. A. and YARZHEMSKII, YA. YA. 1954. Occurrences of boron in the salt-bearing sequence of the Lena-Angara basin. *Tr. vses. nauchno-issled. Inst. Galurgii* (No. 29).

IVANOV, P. T. 1930. Petrographic and chemical investigations of the Saki medicinal mud by layers. *Izv. Inst. fiz.-khim. Analiza*, **4** (No. 2).

KASHIRSKII, P. *et al.* 1931. Mud and brine of eight lakes. *Gidrokhim. Mater.*, **7**.

KASHKAROV, O. D. 1956a. Surface brine of salt lakes and its alteration with time. *Tr. vses. nauchno-issled. Inst. Galurgii* (No. 32).
1956b. The precipitation of salts in salt lakes. *Tr. vses. nauchno-issled. Inst. Galurgii* (No. 32).

KASHKAROV, D. D., KARPYUK, I. D. and TYCHINO, YA. I. 1952. Teniz Lake. *Tr. vses. nauchno-issled. Inst. Galurgii* (No. 23).

KAZAKOV, A. V. and SOKOLOVA, E. I. 1950. Conditions of formation of fluorite in sedimentary rocks. *Tr. Inst. geol. nauk. Akad. Nauk. SSSR* (No. 114), *geol. ser.* (No. 40).

KAZAKOV, A. V., TIKHOMIROVA, M. M. and PLOTNIKOVA, V. I. 1957. The system of carbonate equilibria (dolomite, magnesite) and the paragenesis of siderite and phosphorites. *Tr. Inst. Geol. nauk. Akad. Nauk. SSSR*, No. 152, *geol. ser.* (No. 64).

KAZAKOV, A. V. *et al.* 1957. Mineralogical and physicochemical investigations of some sedimentary rocks and mineral deposits. *Tr. Inst. Geol. nauk. Akad. Nauk. SSSR*, No. 152, *geol. ser.* (No. 64).

KHOD'KOV, A. E. 1956. The origin of replaced zones in the Upper Kama deposit. *Tr. vses. nauchno-issled. Inst. Galurgii* (No. 32).

KHVOROVA, I. V. 1933. Developmental history of the Middle and Late Carboniferous sea in the western part of the Moscow syneclise. *Tr. paleont. Inst. Akad. Nauk. SSSR*, **43**, Part 2.
1956. Carboniferous dolomites and Lower Permian marine rocks on the western slope of the Southern Urals. *Tr. geol. Inst. Akad. Nauk. SSSR* (No. 4).
1960. Flysch and lower molasse formations of the Southern Urals. *Izv. vyssh. ucheb. Zaved., geol. razv.* (No. 2).

KIREEV, V. A. 1956. *Physical Chemistry.* Goskhimizdat, Moscow (in Russian).

KIRIKOV, V. P. 1959. Rhythmic structures in the salt sequence of the Pripyat basin. *Inf. Sb. vses. nauchno-issled. geol. Inst.*, **11**.

KIRSANOV, N. V. 1954. The origin of sulfide minerals in the Devonian rocks of Tataria. *Uchen. Zap. kazan. gos., Univ.*, **114**, Vol. 7.

KLEBANOV, G. S., KORF, D. M. and ELOVSKAYA, L. V. 1937. The Dzhaksy-Klych salt lake. *Sol. Labor. vses. Inst. Galurgii* (No. 12).

KNIPOVICH, N. M. 1938. *The Hydrogeology of Seas and Brackish Waters.* M.-L., Izd. Vses. nauch.-issled. inst. morsk. rybn. knozva, Moscow and Leningrad (in Russian).

KOLOTUKHINA, S. E. 1960. Lithology of ore-bearing layers of the Margalimsai deposit. *Izv. vyssh. ucheb. Zaved. geol. razved* (No. 8).

KONOVALOV, G. S., and OGURTSOVA, O. S. 1959. Fluorine in streams. *Gidrokhim. Mater.*, **29**.

KONSTANTINOV, M. M. 1951. The sedimentary origin of some lead and zinc deposits. *Razv. Nedr* (No. 5).
1952. The roles of diagenesis and metamorphism in forming sedimentary deposits of lead and zinc. *Razv. Nedr* (No. 5).
1954. Exogene lead and zinc sulfates. In *Questions on the Mineralogy of Sedimentary Formations*, Vol. 1. Izd. L'vov. Univ., L'vov. (in Russian).

KONSTANTINOV, M. M. and RAFAL'SKII, S. P. 1960. The solution of galena and the transfer of lead under near-surface conditions. *Geokhimiya* (No. 3).

KORENEVSKII, S. M. 1954. Structural features of syngenetic and diagenetic alterations of halite and sylvite deposits. *Zap. vses. miner. Obshch.*, **83** (No. 1).
1956a. Geologic environment of the sedimentational stage in the formation of saline rocks. In *Questions on the Mineralogy of Sedimentary Formations.* Vols. 3 and 4 Izd. L'vov. Univ., L'vov. (in Russian).
1956b. Geologic structures in fields of potassium accumulations in the Kalush-Golyn deposit. *Tr. vses. nauchno-issled. Inst. Galurgii* (No. 32).
1956c. Stratigraphy of the potassium-bearing series in the Kalush-Golyn region. *Tr. vses. nauchno-issled. Inst. Galurgii* (No. 32).
1956d. New Data on potassium occurrences in Ciscarpathia. In *Questions on the Geology of Agronomical Ores.* Izd. Akad. Nauk. SSSR, Moscow (in Russian).
1956e. Conditions of formation of the saline rocks and potassium deposits in the Stryi-Pystritsa interfluve of the Nadvornaya region. In *Questions on the Mineralogy of Sedimentary Formations*, Vols. 3 and 4. Izd. L'vov. Univ., L'vov. (in Russian).
1957. Some Questions on the Geology and Origin of Potassium Deposits in Ciscarpathia (in Russian). *Geological Collection L'vov. geol. Obshch.* (No. 4).
1959. Geologic conditions of formation of sodium sulfate deposits in Ciscarpathia. *Tr. vses. nauchno-issled. Inst. Galurgii* (No. 35).

KOROLYUK, I. K. 1959. Undulatory-bedded stromatolites (Stratifera) in the Cambrian rocks of southeastern Siberia. *Byull. mosk. Obshch. Ispȳt. Prir., otd. geol.*, **34** (No. 3).

KOSSOVSKAYA, A. G. 1954. Lithology, mineralogy, and conditions of formation of clays in the productive sequence of Azerbaidzhan. *Tr. Inst. Geol. nauk. Akad. Nauk. SSSR* (No. 153), *geol. ser.* (No. 64).

KOVDA, V. A. 1946-7. *The Origin and Behavior of Saline Soils*, Vols. 1-2. Izd. Akad. Nauk. SSSR, Moscow and Leningrad (in Russian).
1954. *Geochemistry of the Deserts of the USSR.* Report to the Fifth International Congress on Soil Science (in Russian). M., Izd. Akad. Nauk. SSSR, Moscow.

KRAUSKOPF, K. 1958. Sedimentary deposits of rare metals. In *Problems of Ore Deposits.* Izd. inostr. lit-ry, Moscow (translated from English).

KROTOV, B. P. 1925. Dolomites, their formation, stability conditions in the earth's crust, and alteration, in connection with a study on the upper horizons of the Kazan group in the vicinity of Kazan. *Tr. kazan. Obshch. estest. lispȳt.*, **50** (No. 6).

KUDENKO, A. A. 1955. Conditions of formation of sedimentary-metamorphic deposits of lead in central Kazakhstan. *Razv. Nedr* (No. 1).

KUDRIN, L. N. 1960. An Upper Tortonian bar between a salt-producing basin and the open sea within the southwestern border of the Russian platform. *Dokl. Akad. Nauk. SSSR*, **131** (No. 4).

KUNIN, V. N. 1959. *Local Water in Deserts and Problems of Using it.* Izd. Akad. Nauk. SSSR, Moscow (in Russian).

KURNAKOV, N. S. 1896. Metamorphization of brines in Crimean salt lakes. *Zap. S. Petersb. miner. Obshch.*, **34** (No. 2).

KURNAKOV, N. S. and ZHEMCHUZHNYI, S. F. 1917. Magnesium lakes of the Perekop group. Izv. Ross. AN, ser. 6, **11**, No. 2.

KURNAKOV, N. S. and NIKOLAEV, V. I. 1932. Evaporation of sea water and lake brines. In *Transactions of the All-Ukraine Salt Conference in Odessa, June 27-30, 1930.* Kharkov-Odessa.
1938. Solar evaporation of sea water and lake brines. *Izv. Sekt. fiz.-khim. Analiza Akad. Nauk. SSSR*, **10**.

KURNAKOV, N. S. *et al.* 1936. *Salt Lakes of the Crimea.* Izd. Akad. Nauk. SSSR, Leningrad and Moscow (in Russian).

KUZNETSOV, I. G. 1928. Deposits of bituminous limestones in Balkaria. *Izv. geol. Kom.*, **47** (No. 8).

KUZNETSOVA, A. I. and BERGMAN, A. G. 1958a. Salt deposits in southern Tadzhikistan. (Communication 5). On the classification of natural saline waters. *Tr. Akad. Nauk. Tadzh. SSR*, **84**.
1958b. Salt deposits in southern Tadzhikistan. (Communication 6). Salt deposits in the Kafirnigan River basin. *Tr. Akad. Nauk. Tadzh. SSR*, **84**.
1958c. Salt deposits in southern Tadzhikistan. (Communication 7). Salt deposits in the Vakhs-Yavan district and salt lakes of the Dzhilikul'-Lower Pandzha group. *Tr. Akad. Nauk. Tadzh. SSR*, **84**.

LEBCHENKO, V. M. 1950. Solubility of calcium sulfate. *Gidrokhim. Mater.*, **17**.

LEBEDEV, V. I. 1959. Reasons for the absence of potassium salts in saline sediments of continental lakes. *Tr. vses. nauchno-issled. Inst. Galurgii* (No. 35).

LEBENKO, A. I. 1956. New data on the age of saline deposits in Tuva (Devonian). *Dokl. Akad. Nauk. SSSR*, **109** (No. 5).

LEONT'EV, V. K. and LEONT'EV, O. K. 1956. Principal geomorphic features of the Sivash lagoon. *Vest. mosk. gos. Univ., ser. biol.*, pochv., *Geol. Geogr.* (No. 2).

LEONT'EV, O. K. and LEONT'EV, V. K. 1961. Origin and systematic development of lagoonal shores. *Tr. okeanogr. Kom.*, **11**.

LEPESHKOV, I. N. 1946. *Potassium Salts of the Volga-Emba and Carpathian Regions.* Izd. Akad. Nauk. SSSR, Moscow and Leningrad (in Russian).

LEPESHKOV, I. N. and BADALEVA, N. V. 1952. The sequence of crystallization among salts during evaporation of water from the Aral Sea. *Dokl. Akad. Nauk. SSSR*, **83** (No. 4).

LEPESHKOV, I. N. and FRADKINA, KH. B. 1959. Crystallization of salts during evaporation of brines from the Gulf of Kara-bogaz. In *Problems of Joint Use of the Mineral Wealth of the Gulf of Kara-bogaz.* Izd. Akad. Nauk. Turkm. SSR, Ashkhabad (in Russian).

LIBROVICH, V. L. 1957a. Lithology of the Ordovician Phosphorite-Bearing Deposits of the Irkutsk Amphitheater. Author's abstract of dissertation at the Competition for University Degrees as a candidate in the Geological and Mineralogical Sciences (in Russian).
1957b. Cyclicity in the Middle Ordovician phosphate-bearing deposits on the southern part of the Siberian platform. In *Questions on the Geomorphology and Geology of Bashkiria.* Collection 1. Izd. Bash. fil. Akad. Nauk. SSSR, Ufa (in Russian).

LITVINENKO, V. A. 1954. Petrography of the Yar-Bishkadak salt deposit. *Tr. vses. nauchno-issled. Inst. Galurgii* (No. 29).

LOBANOVA, V. V. 1949. Petrography of the potassium deposits in the eastern Carpathian region. *Dokl. Akad. Nauk. SSSR*, **66** (No. 6).
1956. Petrography of the potassium deposits in eastern Ciscarpathia. *Tr. vses. nauchno-issled. Inst. Galurgii* (No. 32).
1959. The role of pyroclastic material in forming the saline sequence on the uplift in western Azgir. *Dokl. Akad. Nauk. SSSR*, **125** (No. 5).

LUR'E, A. M. 1960. Geologic Structure and Distributional Pattern of Lead-Zinc Mineralization in the Region of the Sumsar Deposit (in Russian). Author's abstract of dissertation at the Competition for University Degrees as a candidate in the Geological and Mineralogical Sciences.

MAKAROV, S. Z. 1935. Data on the physicochemical study of salt lakes on the Kulunda Steppe. *Tr. soveta Izuch. proizv. Sil SSSR, Sib. ser.* (No. 9). Kulunda Expedition, Part 2, L.

MAKAROV, S. Z. and ENIKEEV, F. S. 1935. Dehydration of Glauber's salt by the brines of Lake Bol'shoe Mormyshanskoe. *Tr. soveta Izuch. proizv. Sil SSSR, Sib. ser.* (No. 9). Kulunda Expedition, Part 2, L.

MAKHLAEV, V. G. 1954. The rock-forming significance of stromatolites in the Dankov-Lebedyan beds. *Dokl. Akad. Nauk. SSSR*, **99** (No. 1).
1956. A brief description of the Dankov-Lebedyan beds. *Tr. vses. neft. nauchno-issled. geol.-razv. Inst.* (No. 7).
1957. Dedolomitized rocks in the Dankov-Lebedyan beds. *Dokl. Akad. Nauk. SSSR*, **117** (No. 2).
1958. Stromatolites as indicators of underwater discontinuities in the accumulation of sediments. *Nauch. Dokl. vyssh. Shk. geol.-geogr. nauk* (No. 3).
1959. Cyclic structure in the Dankov-Lebedyan beds in the central Devonian field. *Izv. vyssh. ucheb Zaved., geol. razv.* (No. 10).
1961. The Dankov-Lebedyan Beds in the Central Part of the Russian Platform. Author's abstract of dissertation at the Competition for University Degrees as a candidate for the doctorate in the Geological and Mineralogical Sciences, Moscow (in Russian).

MANUILOVA, N. S. 1954. Some questions on the origin of cupriferous sandstones in Dzhezkazgan. *Zap. vses. miner. Obshch.*, **83** (No. 4).

MASLOV, V. P. 1950. Geologic-lithologic investigation of the reef facies on the Ufa Plateau. *Tr. Inst. Geol. nauk. Akad. Nauk. SSSR* (No. 118), *geol. ser.* (No. 42).

MEIDZHIS, P. 1955. Distribution of arid and semiarid homoclimates on the earth. In *Hydrogeology and Hydrology of the Arid Zone of the Earth.* Izd. inostr. lit-ry, Moscow (Russian translation).

MIKHALEVSKII, A. A. 1932. Hydrological sketch of the Gulf of Kara-bogaz. *Zap. Gidrogr.* (No. 3).

MIROPOL'SKII, L. M. 1956. *Topogeochemical Investigation of the Permian Deposits in Tataria.* Izd. Akad. Nauk. SSSR, Moscow (in Russian).

MIROPOL'SKII, L. M. and GERASIMOVA, E. T. 1949. The distribution of fluorite in the Lower Artinskian deposits of Tataria and Chuvashia. *Dokl. Akad. Nauk. SSSR*, **66** (No. 2).

MORACHEVSKII, YU. V. 1930. Chemical composition of the Solikamsk salt deposit. *Izv. Inst. fiz.-khim.* Analiza, **4** (No. 2).
1940. Conditions of formation of sediments in the Solikamsk basin. *Byull. vses. nauchno-issled. Inst. Galurgii* (Nos. 6-7).

MORACHEVSKII, YU. V. et al. 1939. Geochemical sketches of the Upper Kama Salt Deposits. *Tr. vses. nauchno-issled. Inst. Galurgii* (No. 17).
1940. Composition of natural salts and lake brines. *Tr. tsent. nauchno-issled. sol. Lab.* (No. 3).

NIKIFOROVA, O. I. 1955. New data on the stratigraphy and paleogeography of the Ordovician and Silurian on the Siberian platform. *Mater. vses. nauchno-issled. geol. Inst.* (No. 7).

NIKOLAEV, A. V. 1934. General results of the Kulunda Expedition of 1931. *Tr. Sov. Izuch. proizv. Sil SSSR, Sibir. ser.*, No. 8. Kulunda Expedition, Part 1, L.
1936. Alluvial sulfate accumulation. *Tr. Kazakh. Bazy Akad. Nauk. SSSR* (No. 3).
1947. *Physicochemical Study of Natural Borates.* Izd. Akad. Nauk. SSSR, Moscow and Leningrad (in Russian).
1949. Potassium in natural waters and the mechanism of diffuse dissemination of elements. *Izv. Sekt. fiz.-khim. Analiza Akad. Nauk. SSSR*, **17**.

NIKOLAEV, V. I. 1935. Some considerations and data on the origin of the Solikamsk nonsulfate potassium deposit. *Izv. Sekt. fiz.-khim. Analiza Akad. Nauk. SSSR*, **7**.

NIKOLAEV, V. I. and BLYUMBERG, YA. B. 1940. Crystallization paths of salts from Kara-bogaz brines in connection with predicting the hydrochemical regimen of the Gulf of Kara-bogaz. *Tr. Kom. kompl. Izuch. Kasp. Mor. Karabogaz. Sektor* (No. 2).

NIKOLAEV, V. I. and KUZNETSOV, D. I. 1935a. The conditions of formation and the nature of blödite. *Izv. Inst. fiz.-khim. Analiza Akad. Nauk. SSSR*, 7.
1935b. *Salt Lakes on the Delta of the Volga River.* Izd. Akad. Nauk. SSSR, Moscow and Leningrad.

NIKOLAEV, V. I., BUYALOV, N. I. and LEPESHKOV, I. N. 1937. Origin of the Permian salt deposits. *Izv. Akad. Nauk. SSSR, ser. khim.* (No. 2).

NIZAMETDINKHODZHAEV, N. I. 1957. The sedimentary origin of lead mineralization in limestones of the Kalkan-Ata mountains. *Izv. Akad. Nauk. Uzb. SSR, ser. geol.* (No. 3).

NOINSKII, M. É. 1913. Samarskaya luka. Geologic investigation. *Tr. Obshch. Estest. imp. kazan. Univ.*, 45 (Nos. 4-6).

ODINTSOV, M. M. 1948. Geology of the Lower Paleozoic copper and lead ores of the Irkutsk amphitheater. *Zap. vses. miner. obshch*, 77 (No. 4).

OFFMAN, P. E. 1959. Tectonics and Volcanic Pipes in the Central Part of the Siberian Platform (in Russian). Izd. AN SSSR, Moscow. (*Tectonics of the USSR*, Vol. 4) (in Russian).

ORLOVA, E. V. 1951. *Phosphorite Basins of Foreign Countries.* Gosgeolizdat, Moscow. (*Min. resursy zarubezh. stran*, No. 19) (in Russian).

OSICHKINA, R. G. and BERGMAN, A. G. 1958a. Salt deposits of southern Tadzhikistan, (Communication 3). Salt deposits of the Kulyab group. *Tr. Akad. Nauk. Tadzh. SSR*, 84.
1958b. Salt deposits of southern Tadzhikistan, (Communication 4). Salt deposits of the region between the Tair-Su and Kizyl-Su Rivers and deposits in the region of the Vakhsh River. *Tr. Akad. Nauk. Tadzh. SSR*, 84.

OSIPOVA, A. I. 1956. Conditions of dolomite formation in the Fergana gulf of the Paleogene sea. *Tr. geol. Inst. Akad. Nauk. SSSR* (No. 4).

OVSYANNIKOVA, K. A. 1951. Oxidation-reduction potential and pH of brine and mud from Lake Saki. *Gidrokhim. Mater*, 19.

PEL'SH, A. D. 1937. Dynamics of the desulfatization process. *Tr. sol. Lab. vses. nauchno-issled. Inst. Galurgii* (No. 14).
1952a. Concentrated magnesian brines of basin operation at the Gulf of Kara-bogaz. *Tr. vses. nauchno-issled. Inst. Galurgii* (No. 24).
1952b. Brine composition and crystallization of blödite in the Gulf of Kara-bogaz in 1943-45. *Tr. vses. nauchno-issled. Inst. Galurgii* (No. 23).

PEL'SH, G. K. and VALYASHKO, M. G. 1953. Experimental study of metamorphization of saturated chloride solutions. *Tr. vses. nauchno-issled. Inst. Galurgii* (No. 27).

PEREL'MAN, A. I. 1951. Calcareous concretions of Karakumy and Kizylkumy. *Dokl. Akad. Nauk. SSSR*, 78 (No. 5).

PERVOL'F, YU. V. 1953. Muds and the conditions of their formation in Crimean salt lakes. *Tr. Lab. Ozerov. Akad. Nauk. SSSR*, 2.

PETROV, N. P. 1953. Lithology of the Upper Jurassic saline formation in southwestern Gissar. *Tr. Inst. Geol. Akad. Nauk. Uz. SSR* (No. 9).

PISARCHIK, YA. K. 1958a. Gypsum and anhydrite rocks. In *Comparative Handbook on the Petrography of Sedimentary Rocks*, Vol. 2. Gostoptekhizdat, Leningrad (in Russian).
1958b. Lithology of the Cambrian Rocks in the Irkutsk Amphitheater (according to Deep Drilling Data). *Tr. vses. nauchno-issled. geol. Inst.* (in Russian).

PISTRAK, R. M. and SYTOVA, V. A. 1951. Devonian and lower Paleozoic deposits in the western part of the Moscow syneclise. In *Geology of the Central Parts of the Russian Platform*. Gosgeolizdat, Moscow (in Russian).

POPOV, V. M. 1953. Intraformational conglomerates of Dzhezkazgan and the nature of mineralization in them. *Dokl. Akad. Nauk. SSSR*, 91 (No. 4).
1954. Diagenetic and epigenetic phenomena in cupriferous sandstones of the Donets basin. *Izv. Kirg. fil. Akad. Nauk. SSSR*, 1 (No. 11).
1955a. Some specific geologic features of the cupriferous sandstones in central Kazakhstan. *Tr. Inst. Geol. Akad. Nauk. Kirg. SSR* (No. 6).
1955b. Facies and paragenetic relations of cupriferous red-beds to gypsiferous and saliniferous deposits. *Tr. Inst. Geol. Akad. Nauk. Kirg. SSR* (No. 6).
1956. The origin of cupriferous sandstones in northern Kirgizia and central Kazakhstan. *Izv. Akad. Nauk. kirgiz. SSR* (No. 2).
1957. Periodic accumulation of copper in the geologic history of the earth. *Tr. Frunzen. politekhn. Inst.* (No. 1).

1959. Geologic distributional pattern of cupriferous sandstones in central Kazakhstan and the northern Tien Shan. In *Distributional Patterns of Mineral Deposits*, Vol. 2. Izd. Akad. Nauk. SSSR, Moscow (in Russian).

1960. Rhythmicity in the accumulation of cupriferous red beds. *Izv. Akad. Nauk. Kirg. SSR, ser. estestv. tekhn. nauk*, **2** (No. 1).

POSOKHOV, E. V. 1955. Salt Lakes of Kazakhstan. Izd. Akad. Nauk. SSSR, Moscow (in Russian).

PREOBRAZHENSKII, P. I. 1933. *The Solikamsk Potassium Deposit*. Goskhimtekhizdat, Leningrad (in Russian).

PREOBRAZHENSKII, P. I. and POLENOVA, T. B. 1939. Salt deposits in the Ishimbai petroleum region. *Byull. vses. nauchno-issled. Inst. Galurgii* (Nos. 4-5).

PUSTOVALOV, L. V. 1937. Ratofkite (Fluorite) in the Upper Volga Region (Lithogenetic Sketch of the Deposits). Izd. Akad. Nauk. SSSR, Moscow and Leningrad (in Russian).

RADISHCHEV, V. P. 1930. Data on the hydrochemistry of salt lakes in the Caspian basin. 1. The hydrochemistry of Lake Elton. 2. Lake Gorki. 3. A study of Inder Lake. *Rab. volzh. biol. Sta.*, **11** (No. 1).

RATEEV, M. A. 1956. Distribution of clay minerals in the Upper Givetian and Lower Shchigry deposits of the Russian platform. In *Questions on the Mineralogy of Sedimentary Formations*, Books 3-4. Izd. L'vov. Univ., L'vov. (in Russian).

1958. Spatial distribution of clay minerals in modern and ancient basins and their genetic relations to physico-geographic factors of sedimentation. In *Data on the Geology, Mineralogy and Use of Clays in the USSR*. Izd. Akad. Nauk. SSSR, Moscow (in Russian).

RATEEV, M. A. and OSIPOVA, A. I. 1958. Clay minerals in deposits of the Paleogene arid zone of Fergana. *Dokl. Akad. Nauk. SSSR*, **123** (No. 1).

RAVICH, M. G. 1927. Some data on the effect of the ground on composition of salt lake brines. *Tr. krym. nauchno-issled. Sta.* (No. 2).

RAVICH, M. I. 1936. The absorbtive capacity of lake muds. In *Salt Lakes of the Crimea*. Izd. Akad. Nauk. SSSR, Moscow and Leningrad (in Russian).

RAVICH, M. I. and KUZNETSOV, V. T. 1936. Muds of Crimean salt lakes. In *Salt Lakes of the Crimea*. Izd. Akad. Nauk. SSSR, Moscow and Leningrad (in Russian).

RAZUMOVA, V. N. 1960. Nature of red and green colors in red beds of Mesozoic and Cenozoic age in central and southern Kazakhstan. *Izv. Akad. Nauk. SSSR* (No. 5).

RAZUMOVSKAYA, E. É. 1927. Causes and nature of the red coloration in potassium compounds of the Solikamsk deposit. *Mater. obshch. prikl. Geol.* (No. 105).

1931. Description of the saline sequence of the Solikamsk deposit. *Tr. glav. geol.-razv. Uprav. V.S.N. Kh.* (No. 54).

1956. Distribution of saline formations in Siberia. *Mater. vses. nauchno-issled. geol.-razv. Inst.* (No. 8).

1958. Saline rocks. In *Comparative Handbook on the Petrography of Sedimentary Rocks*, Vol. 2. Gostoptekhizdat, Leningrad (in Russian).

REINEKE, V. I. 1932. A deposit of thenardite at Uzun-Su. *Tr. vses. geol.-razv. Ob"ed. NKTP* (No. 129).

RONOV, A. B. 1949. History of sedimentation and fluctuating movements in European USSR (from volumetric studies). *Tr. geofiz. Inst.* No. 3 (130).

1951. Chemical composition and conditions of formation of the Paleozoic carbonate sequences on the Russian platform. In *Types of Dolomitic Rocks and Their Origin*. Izd. Akad. Nauk. SSSR, Moscow (*Tr. geol. Inst. Akad. Nauk. SSSR*, No. 4) (in Russian).

RONOV, A. B. and KORZINA, G. A. 1960. Phosphorus in sedimentary rocks. *Geokhimia* (No. 8).

RONOV, A. B. and KHAIN, V. E. 1957. History of sedimentation during the middle and late Paleozoic in connection with the Hercynian stage of crustal development. *Sov. Geol.*, sb. 8.

ROZHKOVA, E. V. and SHCHERBAK, O. V. 1956. The sorption of lead on different rocks and the possibility of this process in the formation of lead deposits. *Izv. Akad. Nauk. SSSR, ser. geol.* (No. 2).

RYKOVSKOV, A. E. 1932. Problems of the lack of sulfate in the Solikamsk potassium deposit. *Tr. glav. geol-razv. Uprav. V.S.N.Kh.* (No. 43).

SAPOZHNIKOV, D. G. 1948. Cupriferous sandstones in the western part of central Kazakhstan. *Tr. Inst. Geol. nauk. Akad. Nauk. SSSR, ser. geol.* No. 93, (No. 28).

1951. Modern sediments and the geology of Lake Balkhash. *Tr. Inst. Geol. nauk. Akad. Nauk. SSSR, geol. ser.* No. 132, (No. 53).

1955. Stages in the formation of sedimentary ores. *Izv. Akad. Nauk. SSSR, ser. geol.* (No. 2).

SATPAEV, K. I. 1925. Some methodological questions on the theory of ore formation as observed in the origin of the so-called "cupriferous" sandstones of hydrothermal type. In *Questions on the Geology of Asia*, Vol. 2. Izd. Akad. Nauk. SSSR, Moscow (in Russian).

SATPAEV, T. A. 1954. Metamorphism of cupriferous sandstones. *Izv. Akad. Nauk. SSSR ser. geol.* (No. 2).

SAVICH, V. G. 1950. Physicochemical characteristics of basins and sediments on the Taman Peninsula. In *Modern Analogues of Oil-Bearing Facies*. Gostoptekhizdat, Moscow (in Russian).

SAVICH-ZABLOTSKII, K. N. and LANKIN, I. YU. 1949. The origin of cupriferous sandstones of the Donets basin. *Byull. mosk. Obshch. Ispȳt. Prir., otd. geol.*, **24** (No. 1).

SCHUCHERT, C. 1957. *Paleographic Atlas of North America.* Izd. inostr. lit-ry, Moscow (trans. from English).

SERGEEV, V. A. 1948. Geology of the Kara-bogaz basin. *Vest. leningr. gos. Univ.* (No. 6).

SHASHKIN, V. L. 1954. The source of material for the production of sedimentary deposits. *Tr. Inst. Geol. kirg. Fil. Akad. Nauk. SSSR* (No. 5).

SHAT-SKII, N. S. 1955. Phosphorite-bearing formations and the classification of phosphorite deposits. In *Conference on Sedimentary Rocks*, No. 2. Izd. Akad. Nauk. SSSR, Moscow (in Russian).

SHCHERBAK, O. V. 1957. The paths followed in the formation and accumulation of lead sulfide under natural conditions. *Geokhimiya* (No. 8).

SHCHERBINA, V. N. 1945. Genetic and age types of gypsum deposits in the Kazakh SSR. *Izv. kazakh. Fil. Akad. Nauk. SSSR, ser. geol.* (No. 4-5).
1949. The origin of mirabilite in the salt deposits of northern Kirgizia. *Dokl. Akad. Nauk. SSSR*, **67** (No. 2).
1952. Glauberite Deposits, Glauberitic Rocks, and Their Weathering Crusts. Izd. Kirg. fil. Akad. Nauk. SSSR, Frunze.
1953. The historical and geographic conditions for accumulation of gypsum deposits. *Tr. Inst. Geol. kirg. Fil. Akad. Nauk. SSSR* (No. 4).
1956a. *Mineralogical, Petrographic, and Genetic Features of the Tertiary Continental Saliniferous and Gypsiferous Deposits in the Intermontane Basins of the Tien Shan.* Izd. Akad. Nauk. Kirg. SSR, Frunze.
1956b. The Pripyat salt basin. *Izv. Akad. Nauk. BSSR, ser. fiz.-tekhn. nauk* (No. 2).
1959. Types of sylvinite rocks in the Pripyat saline basin. *Izv. Akad. Nauk. BSSR, ser. fiz.-tekhn. nauk* (No. 3).

SHCHUKAREV, S. A. and TOLMACHEVA, T. A. 1930. The colloidal-chemical theory of salt lakes. *Zh. russk. fis.-khim. Obshch.*, **62** (No. 4).

SHLESINGER, N. A. and SAMGINSKAYA, A. P. 1936. Oxidation and reduction processes in Elton mud. *Tr. tsent. nauchno-issled. Sta. geokhim. solei* (No. 1).

SHLESINGER, N. A. and FEIGEL'SON, M. B. 1936. The geochemistry of Lake Elton. *Tr. tsent. nauchno-issled. Sta. geokhim. solei* (No. 1).

SHNITNIKOV, A. V. 1949. Sketch of the cyclic fluctuations of lake level and humidity of a region in connection with solar activity. *Byull. Kom. Issled. Sol* (Nos. 3 and 4).
1950. Short-period fluctuations in levels of lakes on the steppes of Western Siberia and Northern Kazakhstan and the dependence of these fluctuations on climate. *Tr. Lab. Ozerov. Akad. Nauk. SSSR*, **1**.
1953. Short-period fluctuations in levels of lakes on the steppes of southeastern European USSR. *Tr. Lab. Ozerov. Akad. Nauk. SSSR* (No. 2).
1957. Variability of absolute humidity on the continents of the northern hemisphere. *Zap. geogr. Obshch.*, **16**.

SIDORENKO, A. V. 1949. Two types of eolian sands. *Dokl. Akad. Nauk. SSSR*, **69** (No. 3).
1950. Constructional and destructional deserts of Central Asia. *Dokl. Akad. Nauk. SSSR*, **70** (No. 5).
1953a. Silicification in deserts of Kara-Kum and Kizyl-Kum. *Izv. Akad. Nauk. SSSR, ser. geol.* (No. 3).
1953b. Continental deposits of eastern Karakumy and their origin. *Dokl. Akad. Nauk. SSSR*, **92** (No. 3).
1956a. Basic features of mineral growth in the desert (as observed in Kara-Kum). In *Questions on the Mineralogy of Sedimentary Formations*, Vols. 3-4. Izd. L'vov. Univ., L'vov.
1956b. Eolian differentiation of substances in the desert. *Izv. Akad. Nauk. SSSR, ser. geogr.* (No. 3).
1958. Calcareous accumulations (caliche) in the deserts of Mexico. *Izv. Akad. Nauk. SSSR, ser. geogr.* (No. 1).

SMEKHOV, E. M. 1953. The Sary-Su dome. *Dokl. Akad-Nauk. SSSR*, **92** (No. 6).

SMIRNOV, A. I. 1958. The origin of phosphorites. *Dokl. Akad. Nauk. SSSR*, **119** (No. 4).

SOLOV'EVA, E. F. 1953. Determining the vapor tension in aqueous solutions of the reciprocal system $Na_2SO_4+MgCl_2 \rightleftharpoons 2NaCl+MgSO_4$. *Tr. vses. nauchno-issled. Inst. Galurgii* (No. 27).

SPIRO, N. S. and BONCH-OSMOLOVSKAYA, K. S. 1956. Composition of the absorbent complex of clays for equilibrium with solutions of sea water. *Tr. nauchno-issled. Inst. Geol. Arkt.*, **86**.

STRAKHOV, N. M. 1934. A type of bituminous shale. *Problemy sov. Geol.*, **3** (No. 8).
1945. The paragenesis of carbonate minerals in deposits of saline lagoonal basins. In *Data on Lithology*. Izd. mosk. Obshch. Ispyt. Prir., Moscow (in Russian).
1947a. Carbonates in modern lagoons and their significance in the problem of dolomite formation. *Byull. mosk. Obshch. Ispyt. Prir., otd. geol.*, **22** (No. 4).
1947b. *Geologic Sketches of the Kungurian of the Ishimbai Oil Region.* Part 1: Stratigraphy and Tectonics, Moscow. (*Mater. Pozn. geol. Stroen. SSSR* (No. 5).)
1948. The true role of bacteria in the formation of carbonate rocks. *Izv. Akad. Nauk. SSSR, ser. geol.* (No. 3).
1949. Periodicity and irreversible evolution of sedimentation in earth history. *Izv. Akad. Nauk. SSSR, ser. geol.* (No. 6).
1951. Limestone-dolomite facies of modern and ancient basins. *Tr. Inst. Geol. nauk. Akad. Nauk. SSSR*, No. 124, *geol. ser.* (No. 45).
1956. Some errors of method in studying chemico-biological sedimentation and diagenesis. *Byull. mosk. Obshch. Ispyt. Prir. otd. geol.*, **31** (No. 2).
1958. Facts and hypotheses concerning the formation of dolomitic rocks. *Izv. Akad. Nauk. SSSR, ser. geol.* (No. 6).

STRAKHOV, N. M. and BORNEMAN-STARYNKEVICH, I. D. 1946. Strontium, boron, and bromine in rocks of the Lower Permian saline sequence of the Bashkirian Ural region. In *Questions on Mineralogy, Geochemistry, and Petrography*. M.-L., Izd. Akad. Nauk. SSSR, Moscow and Leningrad (in Russian).

STRAKHOV, N. M. and ZALMANZON, É. S. 1944. Contents and forms of organic substances in sediments of the Lower Permian saline lagoon of the Bashkirian Ural region. *Dokl. Akad. Nauk. SSSR*, **45** (No. 8).

STRAKHOV, N. M. and OSIPOV, S. S. 1935. Bituminous rocks along Yurezan River. *Byull. mosk. Obshch. Ispyt. Prir., otd. geol.*, **13** (No. 1).

STRAKHOV, N. M. and TSVETKOV, A. I. 1944. The distribution of magnesite in sedimentary rocks. *Zap. vseross miner. Obshch.*, **73** (No. 4).

STRAKHOV, N. M. and TSVETKOV, A. I. 1945. The paragenesis of carbonate minerals in deposits of saline lagoonal basins. In *Data on Lithology*. Izd. Mosk. Obsh. Ispyt. Prir., Moscow. (*Mater. Pozn. geol. Stroen. SSSR*, No. 3) (in Russian).

STRAKHOV, N. M., ZALMANZON, É. S. and GLAGOL'EVA, M. A. 1959. Geochemical sketches of humid-climate deposits of the Upper Paleozoic. (An experiment in facies-geochemical investigation.) *Tr. geol. Inst. Akad. Nauk. SSSR* (No. 23).

STRAKHOV, N. M., ZALMANZON, É. S. and SENDEROVA, R. E. 1944. Iron, manganese, phosphorus and some minor elements in rocks of the Lower Permian saline sequence of the Bashkirian Ural region. *Dokl. Akad. Nauk. SSSR*, **43** (No. 6).

TELENTYUK, E. S. 1952. Hydrochemistry of Lake Tanatar and the waters supplying it. *Tr. vses. nauchno-issled. Inst. Galurgii* (No. 24).

TEODOROVICH, G. M. 1950. *Lithology of the Paleozoic Carbonate Rocks in the Volga-Ural Region.* Izd. Akad. Nauk. SSSR, Moscow and Leningrad (in Russian).

TIKHVINSKII, I. N. 1955. Evolution of the saline conditions in the Assel-Sakmarian sea of southern Tataria and adjacent regions. *Dokl. Akad. Nauk. SSSR*, **126** (No. 5).
1959a. Stratigraphy and Conditions of Formation of the Lower Permian Deposits in the Region of Southern Tataria and Adjoining Regions. Author's abstract of dissertation at the Competition for University Degrees as a candidate in the Geological and Mineralogical Sciences. Kazan (in Russian).
1959b. Structure and conditions of formation of the Lower Kazanian deposits in the Ulyanovsk district of the Volga region. *Izv. kazan. Fil. Akad. Nauk. SSSR, ser. geol.* (No. 7).
1960. Possible accumulation of sulfate sediments in low bottom relief in epicontinental seas. *Dokl. Akad. Nauk. SSSR*, **130** (No. 6).

TOLSTIKHINA, M. M. 1952. Devonian Deposits on the Central Part of the Russian Platform and the Development of the Infra-Devonian Basement during the Paleozoic. Gosgeolizdat, Moscow (in Russian).

TUGARINOV, A. I. 1957. Isotopic composition of lead as a possible geochemical prospecting and evaluation indicator. In *Geochemical Prospecting for Ore Deposits in the USSR*. Gosgeoltekhizdat, Moscow (in Russian).

TYCHINO, YA. I. 1953. Some aspects of the thermal conditions in intercrystalline brine. *Tr. Lab. Ozerov. Akad. Nauk. SSSR* (No. 2).

1955. Evaporation for an ice and snow surface on the shore of Lake Inder (Kazakhstan). *Tr. vses. nauchno-issled. Inst. Galurgii* (No. 30).

TYCHINO, YA. I. and VALYASHKO, M. G. 1952. Evaporation from the surface of freshly precipitated halite at Lake Inder. *Tr. vses. nauchno-issled. Inst. Galurgii* (No. 24).

URAZOV, G. G. 1932. The order of deposition of salts in the Solikamsk potassium deposit. *Tr. glav. geol-razv. Uprav. V.S.N.Kh.* (No. 43).

VAKHRAMEEVA, V. A. 1954a. The origin of glauberite in the salt deposits of the Gulf of Kara-bogaz. *Dokl. Akad. Nauk. SSSR,* **99** (No. 2).

1954b. Origin of the variegated sylvite rocks in the Upper Kama deposit. *Tr. vses. nauchno-issled. Inst. Galurgii,* **29.**

1956a. Mineralogy and petrography of the salt deposits in the Gulf of Kara-bogaz. *Tr. vses. nauchno-issled. Inst. Galurgii,* **32.**

1956b. Stratigraphy and tectonics of the Upper Kama deposit. *Tr. vses. nauchno-issled. Inst. Galurgii,* **32.**

1959a. Comparative characteristics of rocks in sulfate deposits (Gulf of Kara-bogaz and the Tien Shan intermontane region). *Tr. vses. nauchno-issled. Inst. Galurgii* (No. 35).

1959b. Fracturing in the salt rocks of the carnallite zone in the Upper Kama deposit. *Tr. vses. nauchno-issled. Inst. Galurgii* (No. 35).

VALYASHKO, M. G. 1939. Toward an understanding of the basic physicochemical patterns in a region of salt lakes. *Dokl. Akad. Nauk. SSSR,* **23** (No. 7).

1947. Some anomalies in the distribution of salt lakes and salt deposits and their causes. *Dokl. Akad. Nauk. SSSR,* **58** (No. 8).

1949. Blödite in the fresh precipitates of salt lakes. *Priroda* (No. 1).

1950. Anomalous stratigraphy of modern salt deposits and its causes. *Priroda* (No. 3).

1951a. Volumetric relations between liquid and solid phases during evaporation of sea water as a factor controlling the formation of potassium-salt deposits. *Dokl. Akad. Nauk. SSSR,* **77** (No. 6).

1951b. Structural features in modern rock salt deposits. *Mineralog. Sb., L'vov, geol. obshch.,* **5.**

1952a. Halite, its principal varieties found in salt lakes, and its structural features. *Tr. vses. nauchno-issled. Inst. Galurgii,* **23.**

1952b. Geochemical patterns in modern accumulations of salt, and the formation of modern salt deposits in the USSR. *Tr. vses. nauchno-issled. Inst. Galurgii,* **23.**

1952c. Classification features of salt lakes. *Tr. vses. nauchno-issled. Inst. Galurgii,* **23.**

1952d. Definitions of basic concepts and choice of a method of investigation. *Tr. vses. nauchno-issled. Inst. Galurgii,* **23.**

1952e. A physicochemical investigation of the regimen in Lake Inder. *Tr. vses. nauchno-issled. Inst. Galurgii,* **23.**

1952f. Experimental investigation of metamorphization. 1. Definitions of basic concepts and choice of a method of investigation. *Tr. vses. nauchno-issled. Inst. Galurgii,* **23.**

1954. General systematic patterns in developing the chemical content of natural waters. *Gidrokhim. Mater.,* **21.**

1955. The principal chemical types of natural waters and the conditions leading to their formation. *Dokl. Akad. Nauk. SSSR,* **102** (No. 2).

1956a. The geochemistry of bromine during halogenesis and the use of bromine content as a genetic and prospecting criterion. *Geokhimiya,* **1** (No. 6).

1956b. Geochemistry of potassium-salt deposits. In *Questions on the Geology of Agronomic Ores.* Izd. Akad. Nauk. SSSR, Moscow (in Russian).

1956c. A method for determining the origin of potassium rocks by chemical composition, and the application of the method to Ciscarpathian deposits. In *Questions on the Mineralogy of Sedimentary Formations,* Vols. 3 and 4. Izd. L'vov. Univ., L'vov (in Russian).

1957. Geochemical Patterns in the Formation of Potassium-Salt Deposits. Author's abstract of dissertation for the Doctorate in Chemical Sciences. Leningrad (in Russian).

1961. *Geochemical Patterns in the Formation of Potassium-Salt Deposits.* Izd. Mosk. Univ., Moscow (in Russian).

VALYASHKO, M. G. and MANDRYKINA, T. V. 1952. Bromine in salt deposits as an indicator of origin and for prospecting. *Tr. vses. nauchno-issled. Inst. Galurgii* (No. 23).

VALYASHKO, M. G. and NECHAEVA, A. A. 1952. Experimental investigation of the conditions controlling the formation of polyhalite. *Min. sb. L'vov., geol. obshch.* (No. 6).

VALYASHKO, M. G. and PEL'SH, G. K. 1952. Metamorphization of saturated sulfate solutions of calcium bicarbonate. *Tr. vses. nauchno-issled. Inst. Galurgii* (No. 23).

VALYASHKO, M. G. and PETROVA, E. M. 1952. Metamorphization by calcium ions of carbonate saline waters. *Tr. vses. nauchno-issled. Inst. Galurgii* (No. 23).

VALYASHKO, M. G. and SOLOV'EV, E. F. 1950. The investigation of metastable equilibria in the system of $2Na^+$, $2K^+$, Mg^{2+}, SO_4^{2-}, and $2Cl^-/H_2O$. *Tr. vses. nauchno-issled. inst. galurgii* (No. 21).

—— 1953. Crystallization of sylvite during evaporation of sea water. *Tr. vses. nauchno-issled. inst. Galurgii* (No. 27).

VALYASHKO, M. G., NECHAEVA, A. A. and PEL'SH, G. K. 1953. Experimental investigation of metamorphization by calcium ions of dilute sulfate solutions. *Tr. vses. nauchno-issled. Inst. Galurgii* (No. 27).

VALYASHKO, M. G., NECHAEVA, A. A. and POLENOVA, T. B. 1952. Salt lakes in the Dzhambul region. *Tr. vses. nauchno-issled. Inst. Galurgii* (No. 24).

VALYASHKO, M. G. *et al.* 1952. Experimental investigation of solutions of natural polyhalite in water and in solutions of common salt. *Tr. vses. nauchno-issled. Inst. Galurgii* (No. 24).

VASIL'EV, G. A. 1953. Stream water as a source of salt accumulations in Elton Lake. *Tr. vses. nauchno-issled. Inst. Galurgii* (No. 28).

—— 1955. Quaternary deposits in Elton Lake and the history of their accumulation. *Tr. vses. nauchno-issled. Inst. Galurgii* (No. 30).

—— 1956. The hydrochemical characteristics of surface brines in Elton Lake. *Tr. vses. nauchno-issled. Inst. Galurgii* (No. 32).

VASIL'EV, YU. M. 1957. Facies peculiarities of Kungurian rocks in the northern Caspian region in connection with the nature of the southeastern framework of the Russian platform. *Dokl. Akad. Nauk. SSSR*, **112** (No. 1).

—— 1958. Inherited folding between the Southern Urals and Mangyshlak. *Dokl. Akad. Nauk. SSSR*, **119** (No. 4).

VASIL'EV, YU. M., ZVYAGEL'SKII A. A. and PODGORBUNSKII, S. L. 1958. The Chelkar salt massif in the northern Caspian region. *Dokl. Akad. Nauk. SSSR*, **121** (No. 6).

VERIGO, A. A. 1880. Nature of the salts in the brine of the Kuyal'nitskii and Khadzhibaevskii estuaries. *Gorn. Zh., Mosk.*, **3** (No. 9).

VINOGRADOV, A. P. and BOROVIK-ROMANOVA, T. F. 1945. The geochemistry of strontium. *Dokl. Akad. Nauk. SSSR*, **46** (No. 5).

VINOGRADOV, A. P. and RONOV, A. B. 1956. Composition of the sedimentary rocks on the Russian platform in connection with the history of tectonic movements of the platform. *Geokhimiya*, **6**.

VINOGRADOV, A. P., RONOV, A. B. and RATINSKII, V. M. 1932. Evolution of the chemical composition of the carbonate rocks. In *Conference on Sedimentary Rocks*. Vol. 1. Izd. Akad. Nauk. SSSR, Moscow.

VISHNYAKOV, S. G. 1956. Genetic types of dolomitic rocks on the northwestern border of the Russian platform. *Tr. geol. Inst. Akad. Nauk. SSSR* (No. 4).

VITAL', D. A. 1950. Modern carbonate concretions in salt lakes of the Kulunda Steppe and their origin. *Tr. Inst. geol. nauk. Akad. Nauk. SSSR* (No. 125), *geol. ser.* (No. 46).

VORONOVA, M. L. 1953. On the appearance of polyhalite and kainite in sulfate strata at Uzun-Su. *Dokl. Akad. Nauk. SSSR*, **99** (No. 3).

—— 1954. Some data on the petrography of the salt-bearing sequence in the Lower Cambrian of eastern Siberia. *Tr. vses. nauchno-issled. Inst. Galurgii* (No. 29).

YAGOVKIN, I. S. 1932. Cupriferous sandstones and shales (universal types). *Tr. vses. geol.-razv. Ob"ed.* (No. 185).

YAKOVLEVA, M. N. 1952. Experimental investigation on the accumulation of copper in sedimentary rocks. *Byull. mosk. Obshch. Ispȳt. Prir. otd. geol.*, **27** (No. 6).

YARZHEMSKAYA, E. A. 1954. Essential composition of saline pelites. *Tr. vses. nauchno-issled. Inst. Galurgii* (No. 29).

YARZHEMSKII, YA. YA. 1938. Lithology of the vicinity of Polovinka Station on the Eastern Siberian Railroad. *Tr. vost.-sib. geol-razv. Tresta* (No. 25).

—— 1949. The origin of polyhalite in deposits of potassium salts. *Dokl. Akad. Nauk. SSSR*, **66** (No. 6).

—— 1954a. The origin of polyhalite in potassium deposits. *Tr. vses. nauchno-issled. Inst. Galurgii* (No. 29).

—— 1954b. The mineral composition of the muds of Lake Inder. *Tr. vses. nauchno-issled. Inst. Galurgii* (No. 29).

—— 1959. Petrography and Origin of Borates of Inder. Author's abstract of dissertation at the Competition for a doctorate in the Geological and Mineralogical Sciences. Leningrad (in Russian).

YUAN CHIEN-CH'I. 1959. Types of salt lakes in the Tsaidam basin. *Acta geol. sin.*, **39** (No. 3).

ZAITSEV, N. S. 1946. Conditions of formation of some Paleozoic gypsum rocks in central Kazakhstan. *Izv. Akad. Nauk. SSSR, ser. geol.* (No. 4).

ZALMANZON, É. M. 1951. Toward an understanding of sedimentation in Lake Balkhash. *Byull. mosk. Obshch. Ispȳt. Prir., otd. geol.*, **26** (No. 4).

ZDANOVSKII, A. B. 1953. The hydrochemistry of Lake Ebeity. *Tr. vses. nauchno-issled. Inst. Galurgii* (No. 27).

ZDANOVSKII, A. B., LYAKHOVSKAYA, E. I. and SHLEIMOVICH, R. É. 1953. *Handbook of Experimental Data on Solubilities of Multicomponent Water-Salt Systems*, Vol. 1. Three-Component Systems. Goskhimizdat, Moscow and Leningrad (in Russian). 1954. *Handbook of Experimental Data on Solubilities of Multicomponent Water-Salt Systems*, Vol. 2. Four-Component and more Complex Systems. Goskhimizdat, Moscow and Leningrad (in Russian).

ZELENOV, K. K. 1955. Lower Cambrian marine bituminous rocks on the northern slope of the Aldan massif. *Tr. Inst. geol. nauk. AN SSSR*, ser. geol. (No. 155), No. 66. 1956. Lithology and geochemistry of Lower Cambrian lagoonal deposits on the northern slope of the Aldan massif. *Byull. mosk. Obsh. Ispȳt. prir. otd. geol.*, **31** (No. 5). 1957. Lithology of the Lower Cambrian deposits on the northern slope of the Aldan massif. *Tr. geol. Inst. Akad. Nauk. SSSR* (No. 8).

ZENIN, A. A. and KONOVALOV, G. S. 1953. The content of arsenic, boron, bromine, iodine, and fluorine in the Kuban River. *Gidrokhim. Mater.*, **20**.

ZENKOVICH, V. P. 1952. Evolution of lagoonal water. *Izv. vses. geogr. Obshch.*, **84** (No. 5). 1957. The origin of offshore bars and lagoonal shores. *Tr. Inst. Okeanol.*, **21**.

OTHER REFERENCES

BAILLIE, A. 1955. Devonian system Williston Basin. *Bull. Am. Ass. Petrol. Geol.* **39**, No. 5.

BARON, G. 1960. Sur la synthèse de la dolomite application au phénomène de dolomitisation. *Revue Inst. fr. Pétrole*, **15**, No. 1.

BARON, G. and FAVRE, J. 1958. État actuel des recherches en direction de la synthèse de la dolomite. *Revue Inst. fr. Pétrole*, **13**, No. 7-8.

BEYSCHLAG, F. 1922. Zur Frage der Entstehung des Kupferschiefers. *Z. dt. geol. Ges.*, **72**.

BRAITSCH, O. 1959. Über den Mineralbestand der wasserunloslochen Ruckstande von Salzen der Stassfurtserien im südlichen Leinetal. *Freiberger ForschHeft.*, Reihe A, No. 123.

BRIGGS, L. 1958. Evaporite Facies. *J. sedim. Petrol.*, No. 1.

BRUMMER, L. L. 1955. The geology of the Roan Antelope orebody. *Bull. Inst. Min. Metall.*, No. 580.

BUNDY, W. M. 1956. Petrology of gypsum-anhydrite deposits in Southwestern Indiana. *J. sedim. Petrol.*, No. 3.

CHILLINGAR, G. V. 1956. Relationship between Ca/Mg ratio and geological age. *Bull. Am. Ass. Petrol. Geol.*, No. 9.

CORRENS, K. W. 1956. The geochemistry of the halogens. In: *Physics and Chemistry of the Earth*. London.

DALY, R. A. 1907. The limeless ocean of Precambrian time. *Am. J. Sci.*, No. 133. 1909. First calcareous fossils and evolution of limestone. *Bull. geol. Soc. Am.*, 20.

DAVIS, G. R. 1954. The origin of the Roan Antelope copper deposits of Northern Rhodesia. *Econ. Geol.*, **49**, No. 6.

DELLWIG, L. F. 1955. Origin of the Salina Salt of Michigan. *J. sedim. Petrol.*, **25**, No. 2.

EISENHUT, K. H. and KAUTSCH, E. 1954. *Handbuch für den Kupferschieferbergbau*. Leipzig.

FINCH, I. W. 1928. Sedimentary metalliferous deposits of the red beds. *Trans. Am. Inst. Min. metall. Engrs.*, **76**. 1935. Sedimentary copper deposits in the Western states. In: *Copper resources of the world*, Vol. I. Washington.

FISCHER, R. P. 1937. Sedimentary deposits of copper, vanadium—uranium and silver in southwestern United States. *Econ. Geol.*, **32**.

GALE, H. 1914. Late development of magnesite deposits in California and Nevada. *Bull. U.S. geol. Surv.*, No. 540.

GARLICK, W. G. 1953. Reflections on prospecting and ore genesis in Northern Rhodesia. *Bull. Inst. Min. metall. Engrs*, No. 564.

GIGNOUX, M. 1952. Géographie stratigraphique (trans. into Russian).

GILLITZER, G. 1935. Durch welche Bedingungen oder Einflusse sind Metallanreicherungen im mitteldeutschen Kupferschiefer gebildet worden? *Metall Erz*, **32**.

GRABAU, A. W. 1920. *Principles of salt deposition.*

HEIDE, F. and THIELE, A. 1958. Zur Geochemie des Bors. *Chemie Erde*, **29**, Pt. 4.

HERRMAN, A. G. 1958. Geochemische Untersuchungen an Kalisalzlagerstätten im Südharz. *Freiberger ForschHeft.*, No. 43.
— 1959. Die Bedeutung der Spuren-Elementenanalyse für salzlagerstätten-kundliche Unteruchungen. *Freiberger ForschHeft.*, Reihe A., No. 123.

HOFFMAN, W. 1924. *Erzführung und Erzverteilung des Mansfelder Kupferschiefer und hieraus sich ergebenden mineralbildenden und umbildenden Vorgänge im Kupferschiefer.* Halle (Saale).

KAUFMANN, D. W. and SLAWSON, C. B. 1950. Ripple-marks in rock salt of the Salina formation. *J. Geol.* No. 1.

KAUTSCH, E. 1942. Untersuchungsergebnisse über die Metallverteilung im Kupferschiefer. *Arch. Lagerstätt Forsch.*, **74**.

KELVEY, V. E., SWANSON, R. W. and SHELDON, R. P. 1952. The Permian phosphorite deposits of western United States. In: *C. r. XIX séanc. Congr. géol. int.*, Algiers.

KLÄHN, H. 1928. Die Genese lakuster Dolomite und Kieselanscheidungen und ihre Übertragung auf die Entstehung mariner Dolomite und Kieselanscheidungen. *Neues Jb. Miner. Geol. Paläont. Beil Bd.*, **61**.

KRUMBEIN, W. C. 1951. Occurrence and lithological association of evaporites of United States. *J. sedim. Petrol.*, No. 1.

KÜHN, R. 1955. Tiefenberechung des Zechsteimeeres nach dem Bromgehalt der Salze. *Z. dt. Geol. Ges.*, **105**, Th. 4.

LONGWELLS, C. 1928. Geology of the Muddy Mountains, Nevada. *Bull. U.S. geol. Surv.*, No. 798.

LOTZE, F. 1938 (2nd edn. 1957). Steinsalz und Kalisalze. *Geologie.* Berlin.

MARR, UL. 1959. Die Bildung des Stassfurt-Flozes unter Berücksichtigung geochemischer Untersuchungen (III). *Freiberger ForschHeft.*, Reihe A., No. 123.

MORRIS, R. C. and DICKEY, P. A. 1957. Modern evaporite deposition in Peru. *Bull. Am. Ass. Petrol. Geol.*, **41**, No. 11.

MENNIG, J. I. and VATAN, A. 1959. Repartition des dolomites dans les dinantiens des Ardennes. *Revue Inst. fr. Pétrole*, **14**, No. 4-5.

POMPECKI, I. W. 1921. Kupferschiefer und Kupferschiefermeer. *Z. dt. geol. Ges.*, **73**. Relations entre mode de gisements et propriétés physicochimiques des dolomies. 1959. *Revue Inst. fr. Pétrole*, **14** (No. 4-5), pp. 475-518.

RICHTER, G. 1941. Geologische Gesetzmassigkeiten in der Metallführung des Kupferschiefers. *Arch.* Lagerstatt Forsch., **73**.

RICHTER-BERNBURG, G. 1953a. Stratigraphische Gliederung des deutschen Zechsteins. *Z. dt. geol. Ges.*, **105**, Pt. 4.
— 1953b. Über salinare Sedimentation. *Z. dt. geol. Ges.*, **105**, pt. 4.

SCHNEIDERHÖHN, H. 1921. Chalkografische Untersuchung des Mansfelden Kupferchiefers. *Neues Jb. Miner. Geol. Paläont. Beil Bd.*, **47**.

SCHÜLLER, A. 1957. The Copper shales of Mansfeld, Germany. *Acta geol. sin.*, **57**, No. 4.

SCHÜLLER, A. 1958a. Die Metallisation in Kupferschiefer und Dolomit des unteren Zechsteins in den Bohrungen Spremberg 13E/57 und 3/54. *Geologie*, **7**, pt. 3-6.

SCHÜLLER, A. 1958b. Die Metallisation und Genese des Kupferschiefers von Mansfeld. *Abh. dt. Akad. Wiss. Berl., Chemie*, **6**.

SCRUTON, P. C. 1953. Deposition of evaporites. *Bull. Am. Ass. Petrol. Geol.*, **37**, No. 11.

SINGEWALD, I. T. 1935. Corocoro copper district Bolivia. In: *Copper Resources of the World. I.* Washington.

SLOSS, L. L. 1953. Significance of evaporites. *J. sedim. Petrol.*, **23**, No. 2.

STEDTMAN, E. 1911. Evolution of limestone and dolomite. *J. Geol.*

TRASK, P. 1937. *Relation of Salinity to the Calcium Carbonate Content of Marine Sediments.* Washington.

WAGNER, W. 1953. Die tertiären Salzlagerstätten im Oberrheintal-Graben. *Z. dt. geol. Gese.*, **105**, pt. 4.

WEBB, T. B. 1951. Geological history of plains of Western Canada. *Bull. Am. Ass. Petrol. Geol.*, **35**, No. 11.

WHITE, C. H. 1942. Notes on the origin of the Mansfeld copper deposits. *Econ. Geol.*, **37**.